Inspired by Finance

Yuri Kabanov • Marek Rutkowski •
Thaleia Zariphopoulou
Editors

Inspired by Finance

The Musiela Festschrift

Springer

Editors
Yuri Kabanov
Laboratoire de mathématiques
Université de Franche-Comté
Besançon, France

International Laboratory of Quantitative Finance
Higher School of Economics
Moscow, Russia

Marek Rutkowski
School of Mathematics & Statistics
University of Sydney
Sydney, New South Wales, Australia

Thaleia Zariphopoulou
Depts. of Mathematics and IROM
McCombs School of Business
The University of Texas at Austin
Austin, USA

ISBN 978-3-319-02068-6 ISBN 978-3-319-02069-3 (eBook)
DOI 10.1007/978-3-319-02069-3
Springer Cham Heidelberg New York Dordrecht London

Library of Congress Control Number: 2013952730

Mathematics Subject Classification: 91GXX, 91G10, 91G20, 91G30, 91G40, 91G80

Printed on acid-free paper

Springer is part of Springer Science+Business Media (www.springer.com)

Introduction

The present volume contains 25 papers, contributed by 47 authors, and dealing with hot topics of modern mathematical finance. They cover a broad spectrum of areas, including: pricing and hedging of derivative securities, modeling of term structure of interest rates, optimal stopping problems and pricing of contingent claims of American style, performance criteria and portfolio optimization problems, counterparty credit risk and valuation of defaultable securities.

In the paper "Forward Start Foreign Exchange Options under Heston's Volatility and the CIR Interest Rates", Rehez Ahlip and Marek Rutkowski examine the valuation of forward start foreign exchange options in the Heston stochastic volatility model for the exchange rate combined with the Cox–Ingersoll–Ross dynamics for the domestic and foreign interest rates. They derive semi-analytical formulae for such contracts.

In "Real Options with Competition and Incomplete Markets", Alain Bensoussan and Sing Ru (Celine) Hoe consider a Stackelberg leader-follower game for exploiting an irreversible investment opportunity with payoffs of a continuous stochastic income stream for a fixed cost.

In the article "Dynamic Hedging of Counterparty Exposure", Tomasz Bielecki and Stéphane Crépey study mathematical aspects of dynamic hedging of Credit Valuation Adjustment in a portfolio of OTC financial derivatives. Their analysis justifies rigorously some market practice, thus making precise the proper definition of the Expected Positive Exposure (EPE) and the way the EPE should be used in the hedging strategy.

Luciano Campi in "A Note on Market Completeness with American Put Options" shows that any contingent claim on a possibly incomplete two-asset market, satisfying some natural hypotheses, can be approximated by investing dynamically in the underlying stock and statically in all American put options of every strike price k and with the same maturity T.

The paper "An f-Divergence Approach for Optimal Portfolios in Exponential Lévy Models" by Susanne Cawston and Ludmila Vostrikova develops a unified approach to derivation of explicit formulae for utility maximizing strategies in exponential Lévy models. This approach is related to f-divergence minimal martingale measures and is based on a new concept of preservation of the Lévy property by f-divergence minimal martingale measures. For a certain class of f-divergences functions, they give conditions for the existence of corresponding maximizing strategies as well as explicit formulae.

Bénamar Chouaf and Serguei Pergamenchtchikov consider, in their paper "Optimal Investment with Bounded VaR for Power Utility Functions", the classical Merton problem with a constraint involving Value-at-Risk. They obtain explicit expressions for the Bellman function and the optimal control.

In "Three Essays on Exponential Hedging with Variable Exit Times", Tahir Choulli, Junfeng Ma and Marie-Amélie Morlais address three main problems related to exponential hedging with variable exit times. The first problem is to explicitly parameterize the exponential forward performances and describing the optimal solution for the corresponding utility maximization problem. The second problem deals with the horizon-unbiased exponential hedging. The authors are interested in describing the dynamic payoffs for which there exists an admissible strategy that minimizes the risk—in the exponential utility framework—whenever the investor exits the market at stopping times. Furthermore, they explicitly describe the optimal strategy when it exists. The third contribution deals with the optimal selling problem, where the investor is simultaneously looking for the optimal portfolio and the optimal time to liquidate the assets.

In the paper "Mean Square Error and Limit Theorem for the Modified Leland Hedging Strategy with a Constant Transaction Costs Coefficient", Sébastien Darses and Emmanuel Denis obtain delicate results on the rate of convergence for the approximate hedging strategy. This strategy was recently suggested by the second author and it turns out that it performs well—in contrast to the Leland strategy—without rescaling.

In his paper "Yield Curve Smoothing and Residual Variance of Fixed Income Positions", Raphaël Douady treats the yield curve as an object lying in an infinite-dimensional Hilbert space, the evolution of which is driven by a cylindrical Brownian motion. He proves that the principal component analysis (PCA) can be applied and he provides the best approximation of the yield curve evolution by the Gaussian Heath–Jarrow–Morton model with a predetermined number of factors.

In the paper "Maximally Acceptable Portfolios", Ernst Eberlein and Dilip Madan consider an optimization problem, in a non-Gaussian setting, which performance criterion is the Cherny–Madan index of accessibility. Using back-testing on real data, they show that the corresponding optimal portfolios outperform those based on the maximal Sharpe ratio.

The paper "Conditional Default Probability and Density", co-authored by Nicole El Karoui, Monique Jeanblanc, Ying Jiao, and Benhaz Zargari, is dedicated to the study of some interesting mathematically and practically important questions arising in the theory of defaultable securities.

In "Some Extensions of Norros Lemma in Models with Several Defaults", Pavel Gapeev extends the result mentioned in the title to the case of credit risk models in which the reference filtration is not trivial. He shows that if the reference filtration satisfies the so-called immersion property with respect to every filtration which is progressively enlarged by any particular default time, then the terminal values of the compensators of the associated default processes are independent of the observations. The author also provides links between various kinds of immersion properties and (conditional) independence of the terminal values of the compensators (with respect to the reference filtration).

Pavel Gapeev and Neofytos Rodosthenous in their paper "On the Pricing of Perpetual American Compound Options" present, in the framework of the Black–Scholes model, explicit pricing formulae for financial contracts which give their holders the right to buy or sell some other options at certain times in the future. The rational pricing problems for such contracts are embedded into two-step optimal stopping problems for the underlying asset price processes. Their method consists of decomposing these two-step problems into ordinary one-step ones and, in turn, solve them sequentially.

Emmanuel Gobet and Ali Suleiman in "New Approximations in Local Volatility Models" propose new approximation formulae for the price of call options, more precise and numerically efficient than the existing ones. They extend previous results where stochastic expansions were combined with the Malliavin calculus to obtain approximations based on the local volatility at-the-money and they derive alternative expansions involving the local volatility at strike.

The paper "Low-Dimensional Partial Integro-Differential Equations for High-Dimensional Asian Options" by Peter Hepperger deals with problems of pricing Asian options with their payoffs depending on large numbers of securities (for example, an option on a stock basket index) whose prices are modeled by jump-diffusion processes.

Constantinos Kardaras contributes the work titled "A Time Before Which Insiders Would not Undertake Risk". The numéraire portfolio is the unique strictly positive wealth process that, when used as a benchmark to denominate all other wealth, makes all wealth processes local martingales. If the minimum of the numéraire portfolio is known then risk-averse insider traders would refrain from investing in the risky assets before that time. This and other results of the paper shed light on the importance of the numéraire portfolio as an indicator of an overall market performance.

The authors of "Sensitivity with Respect to the Yield Curve: Duration in a Stochastic Setting", Paul Kettler, Frank Proske, and Mark Rubtsov, study an extension of the concept of bond duration to stochastic setting. They define stochastic duration as a Malliavin derivative in the direction of a stochastic yield surface modeled by the Musiela equation. Using this concept, they propose a mathematical framework for the construction of immunization strategies (or delta hedges) of portfolios of interest rate securities with respect to the evolution of the whole yield surface.

In the paper "On the First Passage Time Under Regime-Switching with Jumps", Masaaki Kijima and Chi Chung Siu present the analytical solution for the Laplace transform of the joint distribution of the first passage time and undershoot/overshoot value under a regime-switching jump-diffusion model. Their methodology can be applied to a variety of stopping time problems under a regime-switching model with jump risks.

The article "Strong Consistency of the Bayesian Estimator for the Ornstein-Uhlenbeck Process" by Arturo Kohatsu-Higa, Nicolas Vayatis, and Kazuhiro Yasuda deals with a theoretical basis of a computational intensive parameter estimation method for Markov models. This method can be considered as an approximate Bayesian estimator method or a filtering problem approximated using particle methods.

The question how to retrieve the probability distributions of the underlying asset from the corresponding derivatives quotes is the main subject of the paper "Multi-asset Derivatives and Joint Distributions of Asset Prices" by Ilya Molchanov and Michael Schmutz. Their work is related to a geometric interpretation of multi-asset derivatives as support functions of convex sets. Various symmetry properties for basket, maximum and exchange options are discussed alongside with their geometric interpretations.

The paper "A Class of Homothetic Forward Investment Performance Processes with Non-zero Volatility" by Sergey Nadtochiy and Thaleia Zariphopoulou is a contribution to the new and promising theory of forward investment. This approach allows for dynamic update of the investor's investment criterion and offers an alternative to the classical maximal expected utility objective, which is defined only at a single instant. The underlying object is a stochastic process, the so-called forward investment performance process, which is defined for all times.

Alexander Novikov, Timothy Ling, and Nino Kordzakhia contributed to the volume by the paper "Pricing of Volume-Weighted Average Options: Analytical Approximations and Numerical Results". The volume weighted average price (VWAP), over rolling number of days in the averaging period, is used as a benchmark price by market participants and can be regarded as an estimate for the price that a passive trader will pay to purchase securities in a market. The VWAP is commonly used in brokerage houses as a quantitative trading tool and also appears in Australian taxation law to specify the price of share-buybacks of publicly-listed companies. The volume process is modeled via a shifted squared Ornstein-Uhlenbeck process and a geometric Brownian motion is used to model the asset price. The authors derive analytical formulae for moments of VWAP and use the moment matching approach to approximate a distribution of VWAP. Numerical results for moments of VWAP and call option prices are verified by Monte Carlo simulations.

In the paper "Solution of Optimal Stopping Problem Based on a Modification of Payoff Function", Ernst Presman compares the idea of the Sonin algorithm of space reduction and sequential modification of the Markov chain with the one of the algorithm of modification of the payoff function without modification of the chain. He provides some examples showing that the second approach can be extended to the continuous time models and that, in turn, it leads to a better understanding of solutions of optimal stopping problems.

The aim of the paper "A Stieltjes Approach to Static Hedges" by Michael Schmutz and Thomas Zürcher is to extend the Carr–Madan approach to hedging fairly general path-independent contingent claims by static positions in standard traded assets like bonds, forwards, and plain vanilla call and put options.

The paper "Optimal Stopping of Seasonal Observations and Projection of a Markov Chain" by Isaac Sonin is dedicated to an application of the state elimination algorithm, which was proposed by the author in his earlier work, and a study of the relationship of the fundamental matrices of the initial chain and its modification in the reduced state space.

Besançon, France
Sydney, Australia
Oxford, UK

Yuri Kabanov
Marek Rutkowski
Thaleia Zariphopoulou

Inspired by Finance

Marek Musiela graduated with M.Sc. degree in Mathematics from the University of Wrocław in 1973 and was awarded the Ph.D. degree from the Polish Academy of Sciences in 1976. During the first period of his academic career, his research interests focussed on statistics of stochastic processes and functionals of diffusion processes ([1, 2]). After a period of employment 1976–1980 at the Polish Academy of Sciences, he moved to France where he spent five years at the Institute National Polytechnique de Grenoble. During this period, he was awarded the degree of Docteur d'Etat in 1984. During his stay in France and afterwards, he very actively collaborated with Alain Le Breton with whom he has published several papers on estimation problems for diffusion processes and general semimartingales ([3, 4]).

In 1985 he took the position at the University of New South Wales, where he stayed till 2000. Encouraged by Alan Brace, he started research on the theory of term structure of interest rates, as well as practical implementations of various Gaussian Heath-Jarrow-Morton type models. In the first stage, his academic contributions were concerned with development and deepening of the HJM methodology ([5, 6]). In particular, he proposed and developed a novel way of analyzing an HJM-type model that hinges on introducing infinite-dimensional processes representing the yield curve and the study of the so-called Musiela's SPDE governing the dynamics of the yield curve. This highly innovative approach underpinned further studies of consistency problems for HJM models for the next decade.

The next exciting step in Marek's research was the development of original approaches to arbitrage-free modeling of market rates. His research in this area originally started in collaboration with Dieter Sondermann from the University of Bonn and was subsequently continued by the group concentrated around Marek at UNSW in Sydney. Their joint efforts and parallel studies by a group of researchers lead by Sondermann at the University of Bonn resulted in what is now well-known as the LIBOR Market Model. The ground-breaking papers ([7, 8, 9]), which were completed in 1995 and published in 1997, completely revised the traditional paradigm of term structure modeling with continuous compounding. Before 1995, virtually all continuous-time term structure models used in the valuation of derivatives were invariably based on either the concept of the short-term rate or the instantaneous forward rate. The influence of the new paradigm on further research was immense; it suffices to mention that each of these works was since then cited in hundreds of papers by other researchers. In retrospective, one can make an opinion that this was the last major development in the field of term structure modeling.

After a highly successful academic career at universities in France and Australia, Marek made in 2000 a bold decision to leave the academia and start a new exciting period in his life as the head quant with BNP Paribas in London. After several years of experience in consulting for investment banks in Australia and Europe, he was very well prepared to the new challenge of leading the Fixed Income Research and Support Team.

Around this time, Marek began a collaboration with Thaleia Zariphopoulou on indifference valuation in incomplete markets and forward investment performance criteria. This was also the time that he had started being interested in utility-based pricing in incomplete markets ([12, 13]). Subsequently Marek and Thaleia focussed the evolution of risk preferences and their connection with numeraire and risk premia. The goal was to understand the structure of indifference prices and what they tell us about pricing and optimal investment choice. This in turn generated many questions on the interface of derivative valuation and portfolio management and, gradually, led them to the development of the concept of forward investment performance measurement ([16, 17]). At the same time, Marek studied with Pierre-Louis Lions the fundamental properties of stochastic volatility models ([14, 15]).

All his colleagues were always struck by his constant drive for a better understanding and his uncanny ability to raise interesting and pertinent mathematical issues. They were very impressed and stimulated by Marek's inquisitive mind. He questioned almost everything in the classical setting and challenged many ideas and standardized formulations. We look forward to getting inspired by him for many more years to come.

References

1. Musiela, M.: Divergence, convergence and moments of some integral functionals of diffusions. Z. Wahrscheinlichkeitstheorie Verw. Geb. **70**, 49–65 (1985)
2. Musiela, M.: On Kac functionals of one-dimensional diffusions. Stoch. Process. Appl. **22**, 79–88 (1986)
3. Musiela, M., Le Breton, A.: Strong consistency of least squares estimates in linear regression models driven by semimartingales. J. Multivar. Anal. **23**, 77–92 (1987)
4. Musiela, M., Le Breton, A.: Laws of large numbers for semimartingales with applications to stochastic regression. Probab. Theory Relat. Fields **81**, 275–290 (1989)
5. Musiela, M.: A multifactor Gauss-Markov implementation of Heath, Jarrow and Morton. Math. Finance **4**(3), 259–283 (1994)
6. Brace, A., Musiela, M.: Swap derivatives in a Gaussian HJM framework. In: Dempster, M.A.H., Pliska, S.R. (eds.) Mathematics of Derivative Securities. Cambridge University Press (1996)
7. Brace, A., Gątarek, D., Musiela, M.: The market model of interest rate dynamics. Math. Finance **7**, 127–154 (1997)
8. Miltersen, K., Sandmann, K., Sondermann, D.: Closed form solutions for term structure derivatives with log-normal interest rates. J. Finance **52**, 409–430 (1997)

9. Musiela, M., Rutkowski, M.: Continuous-time term structure models: Forward measure approach. Finance Stoch. **1**, 261–291 (1997)
10. Musiela, M., Rutkowski, M.: Martingale Methods in Financial Modeling. Springer, Berlin, New York, First edition, 1997; Second edition, 2005.
11. Goldys, B., Musiela, M., Sondermann, D.: Lognormality of rates and term structure models. Stoch. Anal. Appl. **18**(3), 375–396 (2000)
12. Musiela, M., Zariphopoulou, T.: An example of indifference prices under exponential preferences. Finance Stoch. **8**, 229–239 (2004)
13. Musiela, M., Zariphopoulou, T.: A valuation algorithm for indifference prices in incomplete markets. Finance Stoch. **8**, 399–414 (2004)
14. Musiela, M., Lions, P.L.: Some properties of diffusion processes with singular coefficients. Commun. Appl. Anal. **1**, 109–125 (2006)
15. Musiela, M., Lions, P.L.: Correlations and bounds for stochastic volatility models. Ann. IHP, Analyse Nonlinéaire **24**(1), 1–16 (2007)
16. Musiela, M., Zariphopoulou, T.: Portfolio choice under dynamic investment performance criteria. Quant. Finance **9**(2), 161–170 (2009)
17. Musiela, M., Zariphopoulou, T.: Portfolio choice under space-time monotone performance criteria. SIAM J. Finance Math. **1**, 326–365 (2010).

Contents

Forward Start Foreign Exchange Options Under Heston's Volatility and the CIR Interest Rates

Rehez Ahlip and Marek Rutkowski

Abstract We examine the valuation of forward start foreign exchange options in the Heston (Rev. Financ. Stud. 6:327–343, 1993) stochastic volatility model for the exchange rate combined with the CIR (see Cox et al. in Econometrica 53:385–408, 1985) dynamics for the domestic and foreign interest rates. The instantaneous volatility is correlated with the dynamics of the exchange rate, whereas the domestic and foreign short-term rates are assumed to be independent of the dynamics of the exchange rate volatility. The main results are derived using the probabilistic approach combined with the Fourier inversion technique developed in Carr and Madan (J. Comput. Finance 2:61–73, 1999). They furnish two alternative semi-analytical formulae for the price of the forward start foreign exchange European call option. As was argued in Ahlip and Rutkowski (Quant. Finance 13:955–966, 2013), the setup examined here is the only analytically tractable version of the foreign exchange market model that combines the Heston stochastic volatility model for the exchange rate with the CIR dynamics for interest rates.

Keywords Option pricing · Heston stochastic volatility model · Forward start options · Interest rates

Mathematics Subject Classification (2010) 91G20 · 91G30

1 Introduction

Forward start options are financial derivatives belonging to the class of path-dependent contingent claims, in the sense that their pay-off depends not only on

R. Ahlip
School of Computing and Mathematics, University of Western Sydney, Penrith South, NSW 1797, Australia
e-mail: R.Ahlip@uws.edu.au

M. Rutkowski (✉)
School of Mathematics and Statistics, University of Sydney, Sydney, NSW 2006, Australia
e-mail: m.rutkowski@usyd.edu.au

Y. Kabanov et al. (eds.), *Inspired by Finance*, DOI 10.1007/978-3-319-02069-3_1,

the final value of the underlying asset, but also on the asset price at an intermediate time between the initiation date of a contract and its expiry date. Typically, a forward start contract gives the holder the right to enter into a call (or put) option with a strike level that will be a fixed percentage of the underlying asset price at a future date, termed the *strike determination date*.

Forward start options can be seen as building blocks to so-called *cliquets* or *ratchets*. Cliquet options are equivalent to a series of forward start at-the-money options with a single premium determined upfront. These are often sold by investment banks to institutional investors who seek to benefit from market oscillations in the price of the underlying during the lifetime of the contract. Cliquets are usually tailored to provide protection against downside risk, while retaining significant upside potential; see, for instance, Lipton [12] or Windcliff et al. [19]. However, in principle, it is also possible to design cliquet options to profit from bear markets.

In the financial literature, the most widely popular model for stochastic volatility is Heston's [9] model. Valuation of forward start equity options under a stochastic volatility model was addressed by several authors. Kruse and Nögel [11] derived closed-form solutions for the forward start call option in Heston's stochastic volatility model by integrating the call pricing formula with respect to the conditional density of the variance value at strike determination date. A numerical evaluation of their expression is rather complicated, however, since in order to obtain the desired distribution function, it introduces another level of integration to already complex integrals in Heston's formula. Independently, Lucic [13] established an exact pricing formula for forward start options in Heston's stochastic volatility model by representing the distribution functions in the form of a single integral. Amerio [2] provided a general framework for pricing forward start derivatives using Monte Carlo simulations and demonstrated the sensitivity with respect to future volatility. All of the above mentioned results have been obtained assuming a constant interest rate and for the case of equity call options.

More recently, Van Haastrecht et al. [17] extended the stochastic volatility model of Schöbel and Zhu [15] to equity/currency derivatives by including stochastic interest rates and assuming all driving model factors to be instantaneously correlated. It is notable that their model is based on Gaussian processes and thus it enjoys analytical tractability, even in the most general case of a full correlation structure. By contrast, when the squared volatility is driven by Heston's model and the interest rate is driven either by the Vasicek's [18] process or by the CIR process introduced by Cox et al. [4], a full correlation structure leads to intractability of equity options even under a partial correlation of the driving factors. This feature has been documented, among others, by Van Haastrecht and Pelsser [16] and Grzelak and Oosterlee [6] who examined, in particular, the Heston/Vasicek and Heston/CIR hybrid models (see also Grzelak and Oosterlee [7] and Grzelak et al. [8], where the Schöbel-Zhu/Hull-White and Heston/Hull-White models for foreign-exchange and equity derivatives are studied).

The goal of this work is to derive semi-analytical solutions for the price of the forward start foreign exchange option in a model in which the instantaneous volatility of the exchange rate is specified by Heston's model, whereas the short-term interest rate processes for the domestic and foreign economies are assumed to follow mutually independent CIR processes. It is worth noting that we extend here the pricing formula for the plain-vanilla foreign exchange option that was established in a recent paper by Ahlip and Rutkowski [1].

The paper is organized as follows. In Sect. 2, we set the foreign exchange model considered in this paper (see also Ahlip and Rutkowski [1]). The forward start option pricing problem is introduced in Sect. 3. In Sect. 4, we recall valuation formulae for zero-coupon bonds in the CIR short-term rate model. In Sect. 5, we introduce auxiliary probability measures and we examine the dynamics of relevant processes under these measures. Section 6 furnishes some preliminary results that are subsequently used in Sect. 7 to derive the main results, Theorems 1 and 2, that provide two alternative pricing formulae for the forward start foreign exchange call option. The paper concludes by deriving the put-call parity relationship for forward start foreign exchange options within the postulated setup.

2 Foreign Exchange Model

Let $(\Omega, \mathscr{F}, \mathbb{P})$ be an underlying probability space. We postulate that the dynamics of the exchange rate $Q = (Q_t)_{t\in[0,T]}$, its instantaneous squared volatility $v = (v_t)_{t\in[0,T]}$, the domestic short-term interest rates $r = (r_t)_{t\in[0,T]}$, and the foreign short-term interest rate $\widehat{r} = (\widehat{r}_t)_{t\in[0,T]}$ are governed by the stochastic differential equations

$$\begin{cases} dQ_t = \big(r_t - \widehat{r}_t\big) Q_t\, dt + Q_t\sqrt{v_t}\, dW_t^Q, \\ dv_t = \big(\theta - \kappa v_t\big)\, dt + \sigma_v \sqrt{v_t}\, dW_t^v, \\ dr_t = \big(a_d - b_d r_t\big)\, dt + \sigma_d \sqrt{r_t}\, dW_t^d, \\ d\widehat{r}_t = \big(a_f - b_f \widehat{r}_t\big)\, dt + \sigma_f \sqrt{\widehat{r}_t}\, dW_t^f. \end{cases} \tag{1}$$

We work throughout under the following standing assumptions:

(A.1) $W^Q = (W_t^Q)_{t\in[0,T]}$ and $W^v = (W_t^v)_{t\in[0,T]}$ are correlated Brownian motions with a constant correlation coefficient, so that the quadratic covariation of W^Q and W^v satisfies $d[W^Q, W^v]_t = \rho\, dt$ for some constant $\rho \in [-1, 1]$,

(A.2) $W^d = (W_t^d)_{t\in[0,T]}$ and $W^f = (W_t^f)_{t\in[0,T]}$ are independent Brownian motions and they are also independent of the Brownian motions W^Q and W^v (hence, in particular, the processes v, r and $\widehat{r}$ are independent),

(A.3) the model's parameters satisfy the stability conditions (see, e.g., Wong and Heyde [20])

$$\frac{2\theta}{\sigma_v^2} > 1, \quad \frac{2a_d}{\sigma_d^2} > 1, \quad \frac{2a_f}{\sigma_f^2} > 1.$$

It is worth stressing again that we postulate here that the squared volatility process v, the domestic short-term interest rate, denoted as r, and its foreign counterpart, denoted as $\widehat{r}$, are independent CIR processes. As argued in Ahlip and Rutkowski [1], this assumption is indeed crucial and thus it cannot be relaxed.

In our computations, we will usually adopt the domestic perspective, which will be sometimes represented by the subscript d. Similarly, we will use the subscript f when referring to a foreign denominated variable.

3 Forward Start Foreign Exchange Options

The *forward start foreign exchange option* is a contract in which the holder receives (at no additional cost) at the strike determination time $T_0 < T$ an option with expiry date T and some $\mathscr{F}_{T_0}$-measurable strike $\widehat{K}$. Typically, we have that $\widehat{K} = kQ_{T_0}$ for some positive constant k. For any strike $\widehat{K}$, the terminal payoff at expiry of the forward start foreign exchange call option is given by the following expression

$$C_T(T, \widehat{K}) = (Q_T - \widehat{K})^+ = Q_T \mathbb{1}_D - \widehat{K} \mathbb{1}_D$$

where we denote $D = \{Q_T > \widehat{K}\}$.

We denote by $\mathbb{F}$ the filtration generated by the Brownian motions W^Q, W^v, W^d, W^f and we write $\mathbb{E}_t^{\mathbb{P}}(\cdot)$ and $\mathbb{P}_t(\cdot)$ to denote the conditional expectation and the conditional probability under $\mathbb{P}$ with respect to the σ-field $\mathscr{F}_t$, respectively.

Let the process B represent the domestic savings account, that is, $dB_t = r_t B_t \, dt$ with $B_0 = 1$. The underlying probability measure $\mathbb{P}$ is interpreted as the *domestic martingale measure*. Hence the price of the option at time t equals, for all $t \in [0, T]$,

$$C_t(T, \widehat{K}) = B_t \, \mathbb{E}_t^{\mathbb{P}}\big[B_T^{-1} C_T(T, \widehat{K})\big] = B_t \, \mathbb{E}_t^{\mathbb{P}}\big[B_T^{-1}(Q_T - \widehat{K})^+\big]$$

or, equivalently,

$$C_t(T, \widehat{K}) = B_t \, \mathbb{E}_t^{\mathbb{P}}(B_T^{-1} Q_T \mathbb{1}_D) - B_t \, \mathbb{E}_t^{\mathbb{P}}(B_T^{-1} \widehat{K} \mathbb{1}_D).$$

Formula above is valid for any strike $\widehat{K}$. However, in what follows it will be always assumed that $\widehat{K} = kQ_{T_0}$. Since the process Q is governed under $\mathbb{P}$ by (1), the random variable Q_t satisfies, for all $t \in [0, T]$,

$$Q_t = Q_0 \exp\bigg(\int_0^t \sqrt{v_u}\, dW_u^Q + \int_0^t \big(r_u - \widehat{r}_u - (1/2)v_u\big)\, du\bigg). \tag{2}$$

4 Bond Pricing and Forward Exchange Rate

We make the standard assumption that the zero-coupon bond prices discounted by the domestic spot rate are martingales under $\mathbb{P}$, that is, the bond price equals

$B_d(t,T) = B_t\,\mathbb{E}_t^{\mathbb{P}}(B_T^{-1})$ for all $t \in [0,T]$. An analogous formula holds for the price process $B_f(t,T)$ of the foreign discount bond under the foreign spot martingale measure (see, e.g., Chap. 14 in Musiela and Rutkowski [14]).

We recall the well-known pricing result for zero-coupon bonds (see, e.g., Cox et al. [4] or Chap. 10 in Musiela and Rutkowski [14]). It is worth stressing that we use here, in particular, the postulated independence of Brownian motions W^Q and W^f driving the foreign interest rate $\widehat{r}$ and the exchange rate Q. Under Assumption (A.2), the dynamics of the foreign bond price $B_f(t,T)$ under the domestic spot martingale measure $\mathbb{P}$ can thus be obtained from formula (14.3) in Musiela and Rutkowski [14].

Proposition 1 *The prices at date t of a domestic and foreign discount bonds maturing at time $T \geq t$ in the CIR model are given by*

$$B_d(t,T) = \exp\big(m_d(t,T) - n_d(t,T)r_t\big),$$
$$B_f(t,T) = \exp\big(m_f(t,T) - n_f(t,T)\widehat{r}_t\big),$$

where for $i \in \{d, f\}$

$$m_i(t,T) = \frac{2a_i}{\sigma_i^2}\log\left[\frac{\gamma_i\, e^{\frac{1}{2}b_i(T-t)}}{\gamma_i\cosh(\gamma_i(T-t)) + \frac{1}{2}b_i\sinh(\gamma_i(T-t))}\right],$$
$$n_i(t,T) = \frac{\sinh(\gamma_i(T-t))}{\gamma_i\cosh(\gamma_i(T-t)) + \frac{1}{2}b_i\sinh(\gamma_i(T-t))},$$

and

$$\gamma_i = \frac{1}{2}\sqrt{b_i^2 + 2\sigma_i^2}.$$

The dynamics of the domestic and foreign bond prices under the domestic spot martingale measure $\mathbb{P}$ are given by

$$dB_d(t,T) = B_d(t,T)\big(r_t\,dt - \sigma_d n_d(t,T)\sqrt{r_t}\,dW_t^d\big),$$
$$dB_f(t,T) = B_f(t,T)\big(\widehat{r}_t\,dt - \sigma_f n_f(t,T)\sqrt{\widehat{r}_t}\,dW_t^f\big).$$

The following result is also well known (see, for instance, Sect. 14.1.1 in Musiela and Rutkowski [14]).

Lemma 1 *The forward exchange rate $F(t,T)$ at time t for settlement date T equals, for all $t \in [0,T]$,*

$$F(t,T) = \frac{B_f(t,T)}{B_d(t,T)}\,Q_t. \tag{3}$$

5 Auxiliary Probability Measures

Since manifestly $Q_T = F(T,T)$, the option's payoff at its expiration can also be expressed as follows

$$C_T(T,\widehat{K}) = F(T,T)\mathbb{1}_{\{F(T,T)>\widehat{K}\}} - \widehat{K}\mathbb{1}_{\{F(T,T)>\widehat{K}\}}.$$

Hence the option's price admits the following representation, for all $t \in [0,T]$,

$$\begin{aligned} C_t(T,\widehat{K}) &= \mathbb{E}_t^{\mathbb{P}}\left\{\exp\left(-\int_t^T r_u\,du\right)F(T,T)\mathbb{1}_{\{F(T,T)>\widehat{K}\}}\right\} \\ &\quad - \mathbb{E}_t^{\mathbb{P}}\left\{\widehat{K}\exp\left(-\int_t^T r_u\,du\right)\mathbb{1}_{\{F(T,T)>\widehat{K}\}}\right\}. \end{aligned}$$

When pursuing the probabilistic approach to the valuation of foreign exchange options, we are going to employ several auxiliary probability measures equivalent to the domestic spot martingale measure $\mathbb{P}$. Let us first recall the classical concept of the *domestic forward martingale measure* $\mathbb{P}_T$.

Definition 1 The *domestic forward martingale measure* $\mathbb{P}_T$ is the probability measure equivalent to $\mathbb{P}$ on $(\Omega, \mathscr{F}_T)$ with the Radon-Nikodým derivative process $\eta = (\eta_t)_{t\in[0,T]}$ given by

$$\eta_t = \frac{d\mathbb{P}_T}{d\mathbb{P}}\bigg|_{\mathscr{F}_t} = \exp\left(-\int_0^t \sigma_d n_d(u,T)\sqrt{r_u}\,dW_u^d - \frac{1}{2}\int_0^t \sigma_d^2 n_d^2(u,T) r_u\,du\right).$$

Under our assumptions, the process η can be checked to be a (true) martingale; one can use to this end the arguments given in the appendix in Kruse and Nögel [11]. Hence it follows from the Girsanov theorem that the process $W^T = (W_t^T)_{t\in[0,T]}$, which is given by the equality

$$W_t^T = W_t^d + \int_0^t \sigma_d n_d(u,T)\sqrt{r_u}\,du,$$

is the standard Brownian motion under the domestic forward martingale measure $\mathbb{P}_T$. It is also clear that the dynamics of r under $\mathbb{P}_T$ are

$$dr_t = \left(a_d - \widehat{b}_d(t) r_t\right)dt + \sigma_d\sqrt{r_t}\,dW_t^T \tag{4}$$

where the function $\widehat{b}_d : [0,T] \to \mathbb{R}$ equals $\widehat{b}_d(t) = b_d + \sigma_d^2 n_d(t,T)$. The following result is borrowed from Ahlip and Rutkowski [1].

Lemma 2 *Under Assumptions* (A.1)–(A.3), *the dynamics of the forward exchange rate* $F(t,T)$ *under the domestic forward martingale measure* $\mathbb{P}_T$ *are given by the stochastic differential equation*

$$dF(t,T) = F(t,T)\left(\sqrt{v_t}\,dW_t^Q + \sigma_d n_d(t,T)\sqrt{r_t}\,dW_t^T - \sigma_f n_f(t,T)\sqrt{\widehat{r}_t}\,dW_t^f\right)$$

or, equivalently,

$$F(T,T) = F(t,T)\exp\left(\int_t^T \widetilde{\sigma}_F(u,T)\cdot d\widetilde{W}_u^T - \frac{1}{2}\int_t^T \|\widetilde{\sigma}_F(u,T)\|^2\,du\right)$$

where the dot $\cdot$ *represents the inner product in* $\mathbb{R}^3$, *by* $(\widetilde{\sigma}_F(t,T))_{t\in[0,T]}$ *we denote the* $\mathbb{R}^3$*-valued process given by*

$$\widetilde{\sigma}_F(t,T) = \left[\sqrt{v_t},\, \sigma_d n_d(t,T)\sqrt{r_t},\, -\sigma_f n_f(t,T)\sqrt{\widehat{r}_t}\,\right]$$

and $\widetilde{W}^T = (\widetilde{W}_t^T)_{t\in[0,T]}$ *stands for the three-dimensional standard Brownian motion under* $\mathbb{P}_T$ *that is given by* $\widetilde{W}^T = [W^Q,\, W^T,\, W^f]^*$.

Using the classical change of a numéraire technique, one can check that under the probability measure $\mathbb{P}_T$ the time t price of the forward start foreign exchange call option equals, for all $t \in [T_0, T]$,

$$C_t(T,\widehat{K}) = B_d(t,T)\,\mathbb{E}_t^{\mathbb{P}_T}\big(F(T,T)\mathbb{1}_{\{F(T,T)>\widehat{K}\}}\big) - \widehat{K} B_d(t,T)\,\mathbb{E}_t^{\mathbb{P}_T}\big(\mathbb{1}_{\{F(T,T)>\widehat{K}\}}\big).$$

After the strike determination date the forward start foreign exchange call option becomes a plain-vanilla foreign exchange call option and thus it can be dealt with as in Ahlip and Rutkowski [1]. To compute the first term in the right-hand side in the formula above, we introduce an auxiliary probability measure $\widehat{\mathbb{P}}_T$.

Definition 2 The probability measure $\widehat{\mathbb{P}}_T$, equivalent to $\mathbb{P}_T$ on $(\Omega, \mathscr{F}_T)$, is defined by the Radon-Nikodým derivative process $\widehat{\eta} = (\widehat{\eta}_t)_{t\in[0,T]}$ where

$$\widehat{\eta}_t = \frac{d\widehat{\mathbb{P}}_T}{d\mathbb{P}_T}\Big|_{\mathscr{F}_t} = \exp\left(\int_0^t \widetilde{\sigma}_F(u,T)\cdot d\widetilde{W}_u^T - \frac{1}{2}\int_0^t \|\widetilde{\sigma}_F(u,T)\|^2\,du\right).$$

As a first step towards general valuation results presented in Sect. 7, we will now derive some preliminary results related to the pricing of the forward start foreign exchange call option prior to the strike determination date. In what follows, we present two alternative pricing methods. We will argue that each of them has some advantages, but also certain drawbacks.

5.1 Bond Price Numéraire

We define the process $\xi = (\xi_t)_{t\in[0,T]}$ by setting $\xi_t = \xi_{T_0}$ for all $t \in [T_0, T]$ and

$$\xi_t = \frac{Q_t B_f(t,T_0)}{Q_0 B_f(0,T_0) B_t}, \quad \forall t \in [0, T_0]. \tag{5}$$

In view of the postulated independence of processes $\widehat{r}$ and r, the foreign bond price $B_f(t, T_0)$ satisfies under the domestic martingale measure $\mathbb{P}$ are (see Proposition 1)

$$dB_f(t, T_0) = \widehat{r}_t B_f(t, T_0)\, dt - B_f(t, T_0)\sigma_f n_f(t, T)\sqrt{\widehat{r}_t}\, dW_t^f.$$

By combining this formula with the dynamics of the exchange rate Q, we obtain the following result.

Lemma 3 *The process $(\xi_t)_{t\in[0,T]}$ is a positive martingale under $\mathbb{P}$ stopped at T_0. Specifically,*

$$\xi_t = \exp\left(\int_0^{t\wedge T_0} \sqrt{v_u}\, dW_u^Q - \frac{1}{2}\int_0^{t\wedge T_0} v_u\, du\right)$$
$$\times \exp\left(-\int_0^{t\wedge T_0} \sigma_f n_f(u, T)\sqrt{\widehat{r}_u}\, dW_u^f - \frac{1}{2}\int_0^{t\wedge T_0} \sigma_f^2 n_f^2(u, T)\widehat{r}_u\, du\right).$$

Due to Lemma 3, we are in the position to define the probability measure $\mathbb{P}_N$, equivalent to $\mathbb{P}$ on $(\Omega, \mathscr{F}_T)$, by postulating that the Radon-Nikodým density process of $\mathbb{P}_N$ with respect to $\mathbb{P}$ equals ξ.

Definition 3 The probability measure $\mathbb{P}_N$ is equivalent to $\mathbb{P}$ on $(\Omega, \mathscr{F}_T)$ with the Radon-Nikodým density process with respect to $\mathbb{P}$ given by the formula

$$\xi_t = \frac{d\mathbb{P}_N}{d\mathbb{P}}\bigg|_{\mathscr{F}_t} = \exp\left(\int_0^{t\wedge T_0} \sqrt{v_u}\, dW_u^Q - \frac{1}{2}\int_0^{t\wedge T_0} v_u\, du\right)$$
$$\times \exp\left(-\int_0^{t\wedge T_0} \sigma_f n_f(u, T)\sqrt{\widehat{r}_u}\, dW_u^f - \frac{1}{2}\int_0^{t\wedge T_0} \sigma_f^2 n_f^2(u, T)\widehat{r}_u\, du\right).$$

Note that the process $\widehat{W}^Q = (\widehat{W}_t^Q)_{t\in[0,T]}$ that is given by

$$\widehat{W}_t^Q = W_t^Q - \int_0^{t\wedge T_0} \sqrt{v_u}\, du$$

is the standard Brownian motion under the auxiliary probability measure $\mathbb{P}_N$. The following useful result is an immediate consequence of the Girsanov theorem and Assumptions (A.1)–(A.3).

Lemma 4 *The processes $\widehat{W}^v$, $\widehat{W}^f$ and $\widehat{W}^d$ that are given by the equalities, for all $t \in [0, T]$,*

$$\widehat{W}_t^v = W_t^v - \rho\int_0^{t\wedge T_0} \sqrt{v_u}\, du,$$

$$\widehat{W}_t^f = W_t^f + \int_0^{t\wedge T_0} \sigma_f n_f(u, T_0)\sqrt{\widehat{r}_u}\, du,$$

$$\widehat{W}_t^d = W_t^d,$$

are independent standard Brownian motions under $\mathbb{P}_N$. *The processes* v, r *and* $\widehat{r}$, *with dynamics under* $\mathbb{P}$ *given by* (1), *are governed under* $\mathbb{P}_N$ *by the following stochastic differential equations, for all* $t \in [0, T_0]$,

$$\begin{aligned} dv_t &= \big(\theta - \widehat{\kappa} v_t\big)\, dt + \sigma_v \sqrt{v_t}\, d\widehat{W}_t^v, \\ dr_t &= \big(a_d - b_d r_t\big)\, dt + \sigma_d \sqrt{r_t}\, d\widehat{W}_t^d, \\ d\widehat{r}_t &= \big(a_f - \widehat{b}_f(t)\widehat{r}_t\big)\, dt + \sigma_f \sqrt{\widehat{r}_t}\, d\widehat{W}_t^f, \end{aligned} \tag{6}$$

where we denote $\widehat{\kappa} = \kappa - \rho\sigma_v$ *and we set* $\widehat{b}_f(t) = b_f + \sigma_f^2 n_f(t, T_0)$ *for all* $t \in [0, T_0]$.

Our next goal is to show that by changing the probability from $\mathbb{P}$ to $\mathbb{P}_N$ we can essentially simplify the pricing formula for the forward start foreign exchange option. Let the auxiliary process $(\widehat{Q}_t)_{t \in [T_0, T]}$ be given by

$$\widehat{Q}_t = \frac{Q_t}{Q_{T_0}} = \exp\bigg(\int_{T_0}^t \sqrt{v_u}\, dW_u^Q + \int_{T_0}^t \big(r_u - \widehat{r}_u - (1/2)v_u\big)\, du \bigg).$$

Equivalently, the process $(\widehat{Q}_t)_{t \in [T_0, T]}$ is the unique solution to the stochastic differential equation

$$d\widehat{Q}_t = \big(r_t - \widehat{r}_t\big)\widehat{Q}_t\, dt + \widehat{Q}_t \sqrt{v_t}\, dW_t^Q \tag{7}$$

with the initial condition $\widehat{Q}_{T_0} = 1$. The following lemma underpins the computation of the price of the forward start foreign exchange call option in Theorem 1.

Lemma 5 *The price of the forward start foreign exchange call option equals, for all* $t \in [0, T_0]$,

$$C_t(T, \widehat{K}) = Q_t B_f(t, T_0)\, \mathbb{E}_t^{\mathbb{P}_N} \Big[B_{T_0}\, \mathbb{E}_{T_0}^{\mathbb{P}} \big(B_T^{-1} (\widehat{Q}_T - k)^+ \big) \Big].$$

Consequently,

$$C_t(T, \widehat{K}) = Q_t B_f(t, T_0)\, \mathbb{E}_t^{\mathbb{P}_N} \big(\widehat{C}_{T_0}(T, k) \big) \tag{8}$$

where we denote

$$\widehat{C}_{T_0}(T, k) = B_{T_0}\, \mathbb{E}_{T_0}^{\mathbb{P}} \big(B_T^{-1} (\widehat{Q}_T - k)^+ \big). \tag{9}$$

Proof Recall that $\widehat{K} = kQ_{T_0}$. Using the Bayes formula and recalling that $\xi_t = \xi_{T_0}$ for $t \in [T_0, T]$, we obtain, for all $t \in [0, T_0]$,

$$C_t(T, \widehat{K}) = B_t\, \mathbb{E}_t^{\mathbb{P}} \Big[B_T^{-1} (Q_T - \widehat{K})^+ \Big]$$

$$
\begin{aligned}
&= \xi_t B_t \,\mathbb{E}_t^{\mathbb{P}_N}\Big[\,\xi_T^{-1} B_T^{-1}(Q_T - \widehat{K})^+\Big] \\
&= \frac{Q_t B_f(t,T_0)}{Q_0 B_f(0,T_0)}\,\mathbb{E}_t^{\mathbb{P}_N}\Big[\,\xi_{T_0}^{-1} B_T^{-1}(Q_T - \widehat{K})^+\Big] \\
&= Q_t B_f(t,T_0)\,\mathbb{E}_t^{\mathbb{P}_N}\Big[\,Q_{T_0}^{-1} B_{T_0} B_T^{-1}(Q_T - kQ_{T_0})^+\Big] \\
&= Q_t B_f(t,T_0)\,\mathbb{E}_t^{\mathbb{P}_N}\Big[\,B_{T_0} B_T^{-1}(\widehat{Q}_T - k)^+\Big] \\
&= Q_t B_f(t,T_0)\,\mathbb{E}_t^{\mathbb{P}_N}\Big[\,B_{T_0}\,\mathbb{E}_{T_0}^{\mathbb{P}_N}\big(B_T^{-1}(\widehat{Q}_T - k)^+\big)\Big].
\end{aligned}
$$

In view of the definition of the probability measure $\mathbb{P}_N$ and Lemma 4, we have that

$$
B_{T_0}\,\mathbb{E}_{T_0}^{\mathbb{P}_N}\big(B_T^{-1}(\widehat{Q}_T - k)^+\big) = B_{T_0}\,\mathbb{E}_{T_0}^{\mathbb{P}}\big(B_T^{-1}(\widehat{Q}_T - k)^+\big) = \widehat{C}_{T_0}(T,k)
$$

and thus formula (8) is established. □

5.2 Savings Account Numéraire

Let the process B^f represent the foreign savings account, so that $dB_t^f = \widehat{r}_t B_t^f\,dt$ with $B_0^f = 1$. We define the process $\widetilde{\xi} = (\widetilde{\xi}_t)_{t\in[0,T]}$ by setting $\widetilde{\xi}_t = \widetilde{\xi}_{T_0}$ for $t \in [T_0, T]$ and

$$
\widetilde{\xi}_t = \frac{Q_t B_t^f}{Q_0 B_t}, \quad \forall\, t \in [0, T_0]. \tag{10}
$$

By combining formula (10) with the dynamics of the exchange rate Q under $\mathbb{P}$, we obtain, for all $t \in [0, T_0]$,

$$
d\widetilde{\xi}_t = \widetilde{\xi}_t \sqrt{v_t}\, dW_t^Q
$$

and thus we arrive at the following explicit representation for the process $\widetilde{\xi}$

$$
\widetilde{\xi}_t = \exp\bigg(\int_0^{t\wedge T_0} \sqrt{v_u}\, dW_u^Q - \frac{1}{2}\int_0^{t\wedge T_0} v_u\, du\bigg).
$$

The process $\widetilde{\xi}$ is a positive martingale under $\mathbb{P}$ stopped at time T_0, and thus it can be used to define an equivalent probability measure, denoted as $\widetilde{\mathbb{P}}_N$.

Definition 4 The probability measure $\widetilde{\mathbb{P}}_N$ is equivalent to $\mathbb{P}$ on $(\Omega, \mathscr{F}_T)$ with the Radon-Nikodým density process with respect to $\mathbb{P}$ given by the formula

$$
\widetilde{\xi}_t = \frac{d\widetilde{\mathbb{P}}_N}{d\mathbb{P}}\bigg|_{\mathscr{F}_t} = \exp\bigg(\int_0^{T_0} \sqrt{v_u}\, dW_u^Q - \frac{1}{2}\int_0^{T_0} v_u\, du\bigg).
$$

It is clear that the process $\widehat{W}^Q = (\widehat{W}^Q_t)_{t\in[0,T]}$ given by the equality

$$\widehat{W}^Q_t = W^Q_t - \int_0^{t\wedge T_0} \sqrt{v_u}\, du$$

is the standard Brownian motion under $\widetilde{\mathbb{P}}_N$. In view of Assumptions (A.1)–(A.3), the following counterpart of Lemma 4 is rather obvious.

Lemma 6 *The processes $\widehat{W}^v$, $\widetilde{W}^f$ and $\widetilde{W}^d$ that are given by the equalities, for all $t \in [0, T]$,*

$$\begin{aligned}\widehat{W}^v_t &= W^v_t - \rho \int_0^{t\wedge T_0} \sqrt{v_u}\, du,\\ \widetilde{W}^f_t &= W^f_t,\\ \widetilde{W}^d_t &= W^d_t,\end{aligned}$$

are independent standard Brownian motions under $\widetilde{\mathbb{P}}_N$. The processes v, r and $\widehat{r}$, with dynamics given by (1), are governed under $\widetilde{\mathbb{P}}_N$ by the following stochastic differential equations, for all $t \in [0, T_0]$,

$$\begin{aligned}dv_t &= \big(\theta - \widehat{\kappa} v_t\big)\, dt + \sigma_v \sqrt{v_t}\, d\widehat{W}^v_t,\\ dr_t &= \big(a_d - b_d r_t\big)\, dt + \sigma_d \sqrt{r_t}\, d\widetilde{W}^d_t,\\ d\widehat{r}_t &= \big(a_f - b_f \widehat{r}_t\big)\, dt + \sigma_f \sqrt{\widehat{r}_t}\, d\widetilde{W}^f_t,\end{aligned} \tag{11}$$

where $\widehat{\kappa} = \kappa - \rho\sigma_v$.

The following result will be used in the proof of Theorem 2.

Lemma 7 *The price of the forward start foreign exchange call option at time t equals, for all $t \in [0, T_0]$,*

$$C_t(T, \widehat{K}) = Q_t B^f_t\, \mathbb{E}^{\widetilde{\mathbb{P}}_N}_t \Big[(B^f_{T_0})^{-1} B_{T_0}\, \mathbb{E}^{\mathbb{P}}_{T_0} \big(B_T^{-1} (\widehat{Q}_T - \widehat{K})^+ \big) \Big].$$

Consequently, we have that

$$C_t(T, \widehat{K}) = Q_t B^f_t\, \mathbb{E}^{\widetilde{\mathbb{P}}_N}_t \big[(B^f_{T_0})^{-1} \widehat{C}_{T_0}(T, k) \big] \tag{12}$$

where we denote

$$\widehat{C}_{T_0}(T, k) = B_{T_0}\, \mathbb{E}^{\mathbb{P}}_{T_0} \big(B_T^{-1} (\widehat{Q}_T - k)^+ \big).$$

Proof Recall that $\widehat{K} = k Q_{T_0}$. Using the abstract Bayes formula, we obtain, for all $t \in [0, T_0]$,

$$C_t(T, \widehat{K}) = B_t\, \mathbb{E}^{\mathbb{P}}_t \Big[B_T^{-1} (Q_T - \widehat{K})^+ \Big]$$

$$
\begin{aligned}
&= \widetilde{\xi}_t B_t \, \mathbb{E}_t^{\widetilde{\mathbb{P}}_N}\Big[\widetilde{\xi}_T^{-1} B_T^{-1}(Q_T - \widehat{K})^+\Big] \\
&= Q_0^{-1} Q_t B_t^f \, \mathbb{E}_t^{\widetilde{\mathbb{P}}_N}\Big[\widetilde{\xi}_{T_0}^{-1} B_T^{-1}(Q_T - \widehat{K})^+\Big] \\
&= Q_t B_t^f \, \mathbb{E}_t^{\widetilde{\mathbb{P}}_N}\Big[(Q_{T_0} B_{T_0}^f)^{-1} B_{T_0} B_T^{-1}(Q_T - kQ_{T_0})^+\Big]
\end{aligned}
$$

so that

$$
C_t(T, \widehat{K}) = Q_t B_t^f \, \mathbb{E}_t^{\widetilde{\mathbb{P}}_N}\Big[(B_{T_0}^f)^{-1} B_{T_0} \, \mathbb{E}_{T_0}^{\widetilde{\mathbb{P}}_N}\big(B_T^{-1}(\widehat{Q}_T - k)^+\big)\Big].
$$

The definition of the probability measure $\widetilde{\mathbb{P}}_N$ and Lemma 6 yield

$$
B_{T_0} \, \mathbb{E}_{T_0}^{\widetilde{\mathbb{P}}_N}\big(B_T^{-1}(\widehat{Q}_T - k)^+\big) = B_{T_0} \, \mathbb{E}_{T_0}^{\mathbb{P}}\big(B_T^{-1}(\widehat{Q}_T - k)^+\big) = \widehat{C}_{T_0}(T, k).
$$

This completes the proof of the lemma. □

6 Preliminary Results

We will need the following auxiliary lemma borrowed from Ahlip and Rutkowski [1] (see also Duffie et al. [5]). Note that the dynamics of the exchange rate process Q are not relevant for this result. Let us set $\widehat{\tau} = T - t$. For any complex numbers μ, λ, $\widetilde{\mu}$, $\widetilde{\lambda}$, $\widehat{\mu}$ and $\widehat{\lambda}$, we denote by $F(\widehat{\tau}, v_t, r_t, \widehat{r}_t)$ the conditional expectation

$$
\mathbb{E}_t^{\mathbb{P}}\left\{\exp\left(-\lambda v_T - \mu \int_t^T v_u \, du - \widetilde{\lambda} r_T - \widetilde{\mu} \int_t^T r_u \, du - \widehat{\lambda}\widehat{r}_T - \widehat{\mu} \int_t^T \widehat{r}_u \, du\right)\right\}.
$$

Lemma 8 *Let the dynamics of processes v, r and $\widehat{r}$ under the probability measure $\mathbb{P}$ be given by stochastic differential equations* (1) *with independent standard Brownian motions W^v, W^d and W^f. Then*

$$
\begin{aligned}
F(\widehat{\tau}, v_t, r_t, \widehat{r}_t) = \exp\big[&-G_1(\widehat{\tau}, \lambda, \mu)v_t - G_2(\widehat{\tau}, \widetilde{\lambda}, \widetilde{\mu})r_t - G_3(\widehat{\tau}, \widehat{\lambda}, \widehat{\mu})\widehat{r}_t \\
&- \theta H_1(\widehat{\tau}, \lambda, \mu) - a_d H_2(\widehat{\tau}, \widetilde{\lambda}, \widetilde{\mu}) - a_f H_3(\widehat{\tau}, \widehat{\lambda}, \widehat{\mu})\big]
\end{aligned}
$$

where

$$
G_1(\widehat{\tau}, \lambda, \mu) = \frac{\lambda[(\gamma + \kappa) + e^{\gamma\widehat{\tau}}(\gamma - \kappa)] - 2\mu(1 - e^{\gamma\widehat{\tau}})}{\sigma_v^2 \lambda \big(e^{\gamma\widehat{\tau}} - 1\big) + \gamma - \kappa + e^{\gamma\widehat{\tau}}(\gamma + \kappa)},
$$

$$
H_1(\widehat{\tau}, \lambda, \mu) = -\frac{2}{\sigma_v^2} \ln\left(\frac{2\gamma e^{\frac{(\gamma+\kappa)\widehat{\tau}}{2}}}{\sigma_v^2 \lambda \big(e^{\gamma\widehat{\tau}} - 1\big) + \gamma - \kappa + e^{\gamma\widehat{\tau}}(\gamma + \kappa)}\right),
$$

$$
G_2(\widehat{\tau}, \widetilde{\lambda}, \widetilde{\mu}) = \frac{\widetilde{\lambda}[(\widetilde{\gamma} + b_d) + e^{\widetilde{\gamma}\widehat{\tau}}(\widetilde{\gamma} - b_d)] - 2\widetilde{\mu}(1 - e^{\widetilde{\gamma}\widehat{\tau}})}{\sigma_d^2 \widetilde{\lambda} \big(e^{\widetilde{\gamma}\widehat{\tau}} - 1\big) + \widetilde{\gamma} - b_d + e^{\widetilde{\gamma}\widehat{\tau}}(\widetilde{\gamma} + b_d)},
$$

$$H_2(\widehat{\tau},\widetilde{\lambda},\widetilde{\mu}) = -\frac{2}{\sigma_d^2}\ln\left(\frac{2\widetilde{\gamma}e^{\frac{(\widetilde{\gamma}+b_d)\widehat{\tau}}{2}}}{\sigma_d^2\widetilde{\lambda}\left(e^{\widetilde{\gamma}\widehat{\tau}}-1\right)+\widetilde{\gamma}-b_d+e^{\widetilde{\gamma}\widehat{\tau}}\left(\widetilde{\gamma}+b_d\right)}\right),$$

$$G_3(\widehat{\tau},\widehat{\lambda},\widehat{\mu}) = \frac{\widehat{\lambda}[(\widehat{\gamma}+b_f)+e^{\widehat{\gamma}\widehat{\tau}}(\widehat{\gamma}-b_f)]-2\widehat{\mu}(1-e^{\widehat{\gamma}\widehat{\tau}})}{\sigma_f^2\widehat{\lambda}\left(e^{\widehat{\gamma}\widehat{\tau}}-1\right)+\widehat{\gamma}-b_f+e^{\widehat{\gamma}\widehat{\tau}}\left(\widehat{\gamma}+b_f\right)},$$

$$H_3(\widehat{\tau},\widehat{\lambda},\widehat{\mu}) = -\frac{2}{\sigma_f^2}\ln\left(\frac{2\widehat{\gamma}e^{\frac{(\widehat{\gamma}+b_f)\widehat{\tau}}{2}}}{\sigma_f^2\widehat{\lambda}\left(e^{\widehat{\gamma}\widehat{\tau}}-1\right)+\widehat{\gamma}-b_f+e^{\widehat{\gamma}\widehat{\tau}}\left(\widehat{\gamma}+b_f\right)}\right),$$

where we denote $\gamma=\sqrt{\kappa^2+2\sigma_v^2\mu}$, $\widetilde{\gamma}=\sqrt{b_d^2+2\sigma_d^2\widetilde{\mu}}$ *and* $\widehat{\gamma}=\sqrt{b_f^2+2\sigma_f^2\widehat{\mu}}$.

Note that Lemma 8 yields, in particular, alternative (but equivalent to formulae of Proposition 1) representations for the bond prices $B_d(t,T)$ and $B_f(t,T)$, specifically,

$$B_d(t,T) = \exp\left(-a_d H_2(\widehat{\tau},0,1) - G_2(\widehat{\tau},0,1)r_t\right), \tag{13}$$

$$B_f(t,T) = \exp\left(-a_f H_3(\widehat{\tau},0,1) - G_3(\widehat{\tau},0,1)\widehat{r}_t\right). \tag{14}$$

Recall that the dynamics of the auxiliary process $(\widehat{Q}_t)_{t\in[T_0,T]}$ under $\mathbb{P}$ are given by equation (7). Hence the next result is a straightforward consequence of Theorem 4.1 in Ahlip and Rutkowski [1]. For the sake of conciseness, we write here $\tau_0=T-T_0$. Recall also that the bond prices $B_d(T_0,T)$ and $B_f(T_0,T)$ are given in Proposition 1.

Proposition 2 *Assume that the foreign exchange model is given by stochastic differential equations* (1) *under Assumptions* (A.1)–(A.3). *Then the conditional expectation* $\widehat{C}_{T_0}(T,k)$ *defined by* (9) *is given by the following expression*

$$\widehat{C}_{T_0}(T,k) = B_f(T_0,T)P_1\big(T_0,v_{T_0},r_{T_0},\widehat{r}_{T_0},k\big) - kB_d(T_0,T)P_2\big(T_0,v_{T_0},r_{T_0},\widehat{r}_{T_0},k\big).$$

The functions P_1 *and* P_2 *are given by*

$$P_j\big(T_0,v_{T_0},r_{T_0},\widehat{r}_{T_0},k\big) = \frac{1}{2}+\frac{1}{\pi}\int_0^\infty \mathrm{Re}\left(f_j(\phi)\,\frac{\exp(-i\phi\ln k)}{i\phi}\right)d\phi$$

where the $\mathscr{F}_{T_0}$*-conditional characteristic functions*

$$f_j(\phi) = f_j(\phi,T_0,v_{T_0},r_{T_0},\widehat{r}_{T_0}), \quad j=1,2,$$

of the random variable $\ln\widehat{Q}_T$ *under the probability measures* $\widehat{\mathbb{P}}_T$ *(see Definition* 2) *and* $\mathbb{P}_T$ *(see Definition* 1), *respectively, satisfy*

$$\ln(f_1(\phi)) = i\phi\big(m_f(T_0,T)-m_d(T_0,T)\big) - (1+i\phi)\frac{\rho}{\sigma_v}\left(v_{T_0}+\theta\tau_0\right)$$

$$
\begin{aligned}
&- i\phi \int_{T_0}^{T} a_d n_d(u,T)\,du + n_f(T_0,T)\widehat{r}_{T_0} \\
&+ (1+i\phi) \int_{T_0}^{T} a_f n_f(u,T)\,du \\
&- G_1(\tau_0,s_1,s_2)v_{T_0} - G_2(\tau_0,s_3,s_4)r_{T_0} - G_3(\tau_0,s_5,s_6)\widehat{r}_{T_0} \\
&- \theta H_1(\tau_0,s_1,s_2) - a_d H_2(\tau_0,s_3,s_4) - a_f H_3(\tau_0,s_5,s_6) \qquad (15)
\end{aligned}
$$

and

$$
\begin{aligned}
\ln(f_2(\phi)) = {}& i\phi\big(m_f(T_0,T) - m_d(T_0,T)\big) - \frac{i\phi\rho}{\sigma_v}\big(v_{T_0} + \theta\tau_0\big) + (1-i\phi) \\
&\times \int_{T_0}^{T} a_d n_d(u,T)\,du + n_d(T_0,T)r_{T_0} + i\phi \int_{T_0}^{T} a_f n_f(u,T)\,du \\
&- G_1(\tau_0,q_1,q_2)v_{T_0} - G_2(\tau_0,q_3,q_4)r_{T_0} - G_3(\tau_0,q_5,q_6)\widehat{r}_{T_0} \\
&- \theta H_1(\tau_0,q_1,q_2) - a_d H_2(\tau_0,q_3,q_4) - a_f H_3(\tau_0,q_5,q_6) \qquad (16)
\end{aligned}
$$

where the functions $G_1, G_2, G_3, H_1, H_2, H_3$ are defined in Lemma 8. *The constants $s_1, s_2, s_3, s_4, s_5, s_6$ are given by*

$$
\begin{aligned}
s_1 &= -\frac{(1+i\phi)\rho}{\sigma_v}, \\
s_2 &= -\frac{(1+i\phi)^2(1-\rho^2)}{2} - \frac{(1+i\phi)\rho\kappa}{\sigma_v} + \frac{1+i\phi}{2}, \qquad (17) \\
s_3 &= 0, \quad s_4 = -i\phi, \quad s_5 = 0, \quad s_6 = 1 + i\phi,
\end{aligned}
$$

and the constants $q_1, q_2, q_3, q_4, q_5, q_6$ equal

$$
\begin{aligned}
q_1 &= -\frac{i\phi\rho}{\sigma_v}, \\
q_2 &= -\frac{i\phi\rho\kappa}{\sigma_v} - \frac{(i\phi)^2(1-\rho^2)}{2} + \frac{i\phi}{2}, \qquad (18) \\
q_3 &= 0, \quad q_4 = 1 - i\phi, \quad q_5 = 0, \quad q_6 = i\phi.
\end{aligned}
$$

7 Valuation of Forward Start Foreign Exchange Options

In this section, we establish the main results of this work, Theorems 1 and 2. Before stating these results, we need to introduce some notation. For the sake of brevity, in what follows we write $\tau = T_0 - t$ and $\tau_0 = T - T_0$. Recall also that we denote

$\widehat{\kappa} = \kappa - \rho\sigma_v$. Assume that the functions $\widehat{H}_1, \widehat{G}_1, \widehat{H}_2, \widehat{G}_2$ solve the following ODEs

$$\begin{aligned}
\frac{\partial \widehat{G}_1(\tau, \lambda)}{\partial \tau} &= -\frac{1}{2}\sigma_v^2 \widehat{G}_1^2(\tau, \lambda) - \widehat{\kappa}\widehat{G}_1(\tau, \lambda), \\
\frac{\partial \widehat{H}_1(\tau, \lambda)}{\partial \tau} &= \widehat{G}_1(\tau, \lambda), \\
\frac{\partial \widehat{G}_2(\tau, \widetilde{\lambda})}{\partial \tau} &= -\frac{1}{2}\sigma_d^2 \widehat{G}_2^2(\tau, \widetilde{\lambda}) - b_d \widehat{G}_2(\tau, \widetilde{\lambda}), \\
\frac{\partial \widehat{H}_2(\tau, \widetilde{\lambda})}{\partial \tau} &= \widehat{G}_2(\tau, \widetilde{\lambda}),
\end{aligned}$$

with initial conditions: $\widehat{G}_1(0, \lambda) = \lambda$, $\widehat{G}_2(0, \widetilde{\lambda}) = \widetilde{\lambda}$ and $\widehat{H}_1(0, \lambda) = \widehat{H}_2(0, \widetilde{\lambda}) = 0$.

From the proof of Lemma 8, which is given in Ahlip and Rutkowski [1], it is easy to deduce that the functions $\widehat{H}_1$, $\widehat{G}_1$, $\widehat{H}_2$, $\widehat{G}_2$, $\widehat{H}_3$ are given by Lemma 8 with $\mu = \widetilde{\mu} = \widehat{\mu} = 0$ and κ replaced by $\widehat{\kappa} = \kappa - \rho\sigma_v$. More explicitly,

$$\begin{aligned}
\widehat{G}_1(\tau, \lambda) &= \frac{2\lambda\widehat{\kappa}}{\sigma_v^2 \lambda\left(e^{\widehat{\kappa}\tau} - 1\right) + 2\widehat{\kappa}e^{\widehat{\kappa}\tau}}, \\
\widehat{G}_2(\tau, \widetilde{\lambda}) &= \frac{2\widetilde{\lambda} b_d}{\sigma_d^2 \widetilde{\lambda}\left(e^{b_d\tau} - 1\right) + 2b_d e^{b_d\tau}}, \\
\widehat{H}_1(\tau, \lambda) &= -\frac{2}{\sigma_v^2} \ln\left(\frac{2\widehat{\kappa}e^{2\widehat{\kappa}\tau}}{\sigma_v^2 \lambda\left(e^{\widehat{\kappa}\tau} - 1\right) + 2\widehat{\kappa}e^{\widehat{\kappa}\tau}}\right), \\
\widehat{H}_2(\tau, \widetilde{\lambda}) &= -\frac{2}{\sigma_d^2} \ln\left(\frac{2b_d e^{b_d\tau}}{\sigma_d^2 \widetilde{\lambda}\left(e^{b_d\tau} - 1\right) + 2b_d e^{b_d\tau}}\right).
\end{aligned} \tag{19}$$

7.1 *Options Pricing Formula in the Bond Numéraire*

We are in the position to prove the first main result of this work. According to the method developed in Sect. 5.1, the price of this option prior to the strike determination date T_0 can be expressed in terms of the foreign zero-coupon bond $B_f(t, T_0)$ and the exchange rate Q_t, as well as a certain conditional expectation (see formula (8)) that we will now evaluate in Heston's stochastic volatility model for the exchange rate combined with independent CIR models for domestic and foreign interest rates.

Unfortunately, since the foreign interest rate $\widehat{r}$ is a non-homogeneous process under $\mathbb{P}_N$, the quasi-analytical representation obtained in Theorem 1 still involves conditional expectations of the form $\mathbb{E}_t^{\mathbb{P}_N}[\exp(-s\widehat{r}_{T_0})]$. The presence of these terms will be avoided if we apply instead the alternative approach presented in the next subsection. In order to implement the pricing formula of Theorem 1, one needs to compute the conditional expectation $\mathbb{E}_t^{\mathbb{P}_N}[\exp(-s\widehat{r}_{T_0})]$ for a real number s, where the process $\widehat{r}$ is governed by the non-homogeneous stochastic differential equation

$$d\widehat{r}_t = \big(a_f - \widehat{b}_f(t)\widehat{r}_t\big)\,dt + \sigma_f\sqrt{\widehat{r}_t}\,d\widehat{W}_t^f \tag{20}$$

with $\widehat{b}_f(t) = b_f + \sigma_f^2 n_f(t, T_0)$ (see Lemma 4). For this purpose, one can use the property that, since

$$d(l(t)\widehat{r}_t) = a_f l(t)\,dt + \sigma_f l(t)\sqrt{\widehat{r}_t}\,d\widehat{W}_t^f,$$

the process $\widehat{r}$ has the same probability distribution as the process $\widehat{\rho}$ given by (see, for instance, Jeanblanc et al. [10])

$$\widehat{\rho}_t = \frac{1}{l(t)}\,\rho\bigg(\frac{\sigma_f^2}{4}\int_0^t l(u)\,du\bigg) \tag{21}$$

where $l(t) = \exp(\int_0^t \widehat{b}_f(u)\,du)$ and $\rho = (\rho(t))_{t\in\mathbb{R}_+}$ is the squared Bessel process with dimension $4a_f/\sigma_f^2$ started at $\widehat{r}_0$. From representation (21), it follows that the transition probability function of the Markov process $\widehat{r}$ under $\mathbb{P}_N$ is known explicitly.

The pricing formula of Theorem 1 is an extension of the pricing formula for the plain-vanilla foreign exchange option established in Ahlip and Rutkowski [1]. Hence it suffices to focus here on the valuation of the forward start foreign exchange option prior to the strike determination date T_0.

Theorem 1 *Consider the forward start foreign exchange call option with maturity T, strike determination date T_0 and strike $\widehat{K} = kQ_{T_0}$ where k is a positive constant. Assume that the foreign exchange model is given by stochastic differential equations* (1) *under Assumptions* (A.1)–(A.3). *Then the option's price equals, for all $t \in [0, T_0]$,*

$$C_t(T,\widehat{K}) = Q_t B_f(t,T_0)\Big(\widehat{P}_1\big(t,v_t,r_t,\widehat{r}_t,k\big) - k\widehat{P}_2\big(t,v_t,r_t,\widehat{r}_t,k\big)\Big). \tag{22}$$

The function $\widehat{P}_1$ equals, for all $t \in [0, T_0]$,

$$\widehat{P}_1(t,v_t,r_t,\widehat{r}_t,k) = \frac{1}{2}\widehat{V}_1(t,\widehat{r}_t) + \frac{1}{\pi}\int_0^\infty \mathrm{Re}\bigg(\widehat{f}_1(\phi)\,\frac{\exp(-i\phi\ln k)}{i\phi}\bigg)\,d\phi$$

where the function $\widehat{f}_1(\phi) = \widehat{f}_1(\phi,t,v_t,r_t,\widehat{r}_t)$ equals

$$\widehat{f}_1(\phi) = \widehat{c}_1\,\exp\big[-\widehat{G}_1(\tau,\widehat{s}_2)v_t - \widehat{G}_2(\tau,\widehat{s}_4)r_t - \theta\widehat{H}_1(\tau,\widehat{s}_2) - a_d\widehat{H}_2(\tau,\widehat{s}_4)\big]$$
$$\times\,\mathbb{E}_t^{\mathbb{P}_N}\big[\exp(-\widehat{s}_6\widehat{r}_{T_0})\big]$$

where in turn

$$\ln(\widehat{c}_1) = (1+i\phi)m_f(T_0,T) - i\phi m_d(T_0,T) - (1+i\phi)\tau_0\frac{\rho\theta}{\sigma_v}$$

$$- i\phi \int_{T_0}^{T} a_d n_d(u,T)\,du + (1+i\phi)\int_{T_0}^{T} a_f n_f(u,T)\,du$$
$$- \theta H_1(\tau_0, s_1, s_2) - a_d H_2(\tau_0, s_3, s_4) - a_f H_3(\tau_0, s_5, s_6).$$

The functions H_1, H_2, H_3 *are given by Lemma* 8, *the functions* $\widehat{G}_1, \widehat{G}_2, \widehat{H}_1, \widehat{H}_2$ *are given by* (19), *the constants* $s_1, s_2, s_3, s_4, s_5, s_6$ *are given by* (17), *and*

$$\widehat{s}_2 = \frac{(1+i\phi)\rho}{\sigma_v} + G_1(\tau_0, s_1, s_2),\ \widehat{s}_4 = G_2(\tau_0, s_3, s_4),\ \widehat{s}_6 = G_3(\tau_0, s_5, s_6),$$

with the functions G_1, G_2, G_3 *given by Lemma* 8. *Moreover, the function* $\widehat{V}_1(t, \widehat{r}_t)$ *is given by the formula*

$$\widehat{V}_1(t, \widehat{r}_t) = \mathbb{E}_t^{\mathbb{P}_N}\big[\exp(-G_3(\tau_0, 0, 1)\widehat{r}_{T_0} - a_f H_3(\tau_0, 0, 1))\big]. \tag{23}$$

The function $\widehat{P}_2$ *equals, for all* $t \in [0, T_0]$,

$$\widehat{P}_2(t, v_t, r_t, \widehat{r}_t, k) = \frac{1}{2}\widehat{V}_2(t, r_t) + \frac{1}{\pi}\int_0^\infty \mathrm{Re}\left(\widehat{f}_2(\phi)\frac{\exp(-i\phi \ln k)}{i\phi}\right) d\phi$$

where the function $\widehat{f}_2(\phi) = \widehat{f}_2(\phi, t, v_t, r_t, \widehat{r}_t)$ *is given by the expression*

$$\widehat{f}_2(\phi) = \widehat{c}_2 \exp\big[-\widehat{G}_1(\tau, \widehat{q}_2)v_t - \widehat{G}_2(\tau, \widehat{q}_4)r_t - \theta \widehat{H}_1(\tau, \widehat{q}_2) - a_d \widehat{H}_2(\tau, \widehat{q}_4)\big]$$
$$\times \mathbb{E}_t^{\mathbb{P}_N}\big[\exp(-\widehat{q}_6 \widehat{r}_{T_0})\big]$$

where in turn

$$\ln(\widehat{c}_2) = i\phi m_f(T_0, T) + (1-i\phi)m_d(T_0, T) - i\phi\tau_0\frac{\rho\theta}{\sigma_v}$$
$$+ (1-i\phi)\int_{T_0}^{T} a_d n_d(u,T)\,du + i\phi\int_{T_0}^{T} a_f n_f(u,T)\,du$$
$$- \theta H_1(\tau_0, q_1, q_2) - a_d H_2(\tau_0, q_3, q_4) - a_f H_3(\tau_0, q_5, q_6).$$

The constants $q_1, q_2, q_3, q_4, q_5, q_6$ *are given by* (18) *and*

$$\widehat{q}_2 = i\phi\frac{\rho}{\sigma_v} + G_1(\tau_0, q_1, q_2), \quad \widehat{q}_4 = G_2(\tau_0, q_3, q_4), \quad \widehat{q}_6 = G_3(\tau_0, q_5, q_6).$$

Finally, the function $\widehat{V}_2(t, r_t)$ *is given by*

$$\ln(\widehat{V}_2(t, r_t)) = -a_d H_2(\tau_0, 0, 1) - \widehat{G}_2(\tau, G_2(\tau_0, 0, 1))r_t - a_d \widehat{H}_2(\tau, G_2(\tau_0, 0, 1)).$$

Proof Let us fix $t \in [0, T_0]$. By combining Proposition 2 with Lemma 5, we obtain

$$C_t(T, \widehat{K}) = Q_t B_f(t, T_0)\,\mathbb{E}_t^{\mathbb{P}_N}\big(\widehat{C}_{T_0}(T, k)\big) = Q_t B_f(t, T_0)\big(\widehat{J}_t^1 - k\widehat{J}_t^2\big)$$

where we denote

$$\widehat{J}_t^1 = \mathbb{E}_t^{\mathbb{P}_N}\Big(B_f(T_0,T)P_1\big(T_0,v_{T_0},r_{T_0},\widehat{r}_{T_0},k\big)\Big)$$

and

$$\widehat{J}_t^2 = \mathbb{E}_t^{\mathbb{P}_N}\Big(B_d(T_0,T)P_2\big(T_0,v_{T_0},r_{T_0},\widehat{r}_{T_0},k\big)\Big).$$

We will first compute the conditional expectation $\widehat{J}_t^1$, that is,

$$\widehat{J}_t^1 = \widehat{P}_1(t,v_t,r_t,\widehat{r}_t) := \mathbb{E}_t^{\mathbb{P}_N}\Big(B_f(T_0,T)P_1(T_0,v_{T_0},r_{T_0},\widehat{r}_{T_0},k)\Big). \quad (24)$$

From Proposition 2, we know that

$$P_1(T_0,v_{T_0},r_{T_0},\widehat{r}_{T_0},k) = \frac{1}{2} + \frac{1}{\pi}\int_0^\infty \mathrm{Re}\bigg(f_1(\phi)\,\frac{\exp(-i\phi\ln k)}{i\phi}\bigg)d\phi \quad (25)$$

where the function $f_1(\phi)$ is given in the statement of Proposition 2. In view of (24) and (25), we obtain

$$\widehat{J}_t^1 = \frac{1}{2}\,\widehat{V}_1(t,\widehat{r}_t) + \frac{1}{\pi}\int_0^\infty \mathrm{Re}\bigg(\widehat{f}_1(\phi)\,\frac{\exp(-i\phi\ln k)}{i\phi}\bigg)d\phi$$

where $\widehat{V}_1(t,\widehat{r}_t)$ and $\widehat{f}_1(\phi)$ denote the following conditional expectations:

$$\widehat{V}_1(t,\widehat{r}_t) := \mathbb{E}_t^{\mathbb{P}_N}[B_f(T_0,T)] = \mathbb{E}_t^{\mathbb{P}_N}\big[\exp(-G_3(\tau_0,0,1)\widehat{r}_{T_0} - a_f H_3(\tau_0,0,1))\big]$$

and

$$\widehat{f}_1(\phi) := \mathbb{E}_t^{\mathbb{P}_N}[B_f(T_0,T)f_1(\phi)] = \mathbb{E}_t^{\mathbb{P}_N}[g_1(\phi)].$$

The function $g_1(\phi)$ is in turn given by the formula

$$g_1(\phi) := B_f(T_0,T)f_1(\phi) = \widehat{c}_1\exp\big(-\widehat{s}_2 v_{T_0} - \widehat{s}_4 r_{T_0} - \widehat{s}_6\widehat{r}_{T_0}\big) \quad (26)$$

with the constants $\widehat{c}_1,\widehat{s}_2,\widehat{s}_4,\widehat{s}_6$ given in the statement of the theorem. It is worth stressing that the second equality in (26) is an immediate consequence of (14) and (15). Recall that the dynamics of the process $\widehat{r}$ under $\mathbb{P}_N$ are given by equation (20). In particular, the drift term in these dynamics is time-dependent, specifically, $\widehat{b}_f(t) = b_f + \sigma_f^2 n_f(t,T_0)$. Hence a straightforward application of Lemma 8 for an explicit computation of $\widehat{V}^1(t,\widehat{r}_t)$ is not possible, although some approximations based on formulae of Lemma 8 are readily available. Alternatively, one can use the transition probability density function of $\widehat{r}$ under $\mathbb{P}_N$. To compute the conditional expectation

$$\widehat{f}_1(\phi) = \widehat{c}_1\,\mathbb{E}_t^{\mathbb{P}_N}\big[\exp(-\widehat{s}_2 v_{T_0} - \widehat{s}_4 r_{T_0} - \widehat{s}_6\widehat{r}_{T_0})\big],$$

we apply Lemma 8 and we use the dynamics of v, r, and $\widehat{r}$ under $\mathbb{P}_N$, as given in Lemma 4. We recall that the standard Brownian motions $\widehat{W}^v$, $\widehat{W}^d$ and $\widehat{W}^f$ are independent under $\mathbb{P}_N$, and thus

$$\mathbb{E}_t^{\mathbb{P}_N}\big[\exp(-\widehat{s}_2 v_{T_0} - \widehat{s}_4 r_{T_0} - \widehat{s}_6 \widehat{r}_{T_0})\big] = \mathbb{E}_t^{\mathbb{P}_N}\big[\exp(-\widehat{s}_2 v_{T_0})\big]\mathbb{E}_t^{\mathbb{P}_N}\big[\exp(-\widehat{s}_4 r_{T_0})\big] \times \mathbb{E}_t^{\mathbb{P}_N}\big[\exp(-\widehat{s}_6 \widehat{r}_{T_0})\big].$$

By an application of Lemma 8, we obtain the stated formula for $\widehat{f}_1(\phi)$ and thus also the required expression for $\widehat{J}_t^1$. To complete the proof, it remains to evaluate the conditional expectation

$$\widehat{J}_t^2 = \widehat{P}_2(t, v_t, r_t, \widehat{r}_t, k) := \mathbb{E}_t^{\mathbb{P}_N}\Big(B_d(T_0, T) P_2(T_0, v_{T_0}, r_{T_0}, \widehat{r}_{T_0}, k)\Big) \tag{27}$$

where

$$P_2(T_0, v_{T_0}, r_{T_0}, \widehat{r}_{T_0}, k) = \frac{1}{2} + \frac{1}{\pi}\int_0^\infty \mathrm{Re}\left(f_2(\phi)\, \frac{\exp(-i\phi \ln k)}{i\phi}\right) d\phi \tag{28}$$

and the function $f_2(\phi)$ is given in Proposition 2. Using (27) and (28), we obtain the following equality

$$\widehat{J}_t^2 = \frac{1}{2}\,\widehat{V}_2(t, \widehat{r}_t) + \frac{1}{\pi}\int_0^\infty \mathrm{Re}\left(\widehat{f}_2(\phi)\, \frac{\exp(-i\phi \ln k)}{i\phi}\right) d\phi$$

where $\widehat{V}_2(t, r_t)$ and $\widehat{f}_2(\phi)$ stand for the following conditional expectations:

$$\widehat{V}_2(t, r_t) := \mathbb{E}_t^{\mathbb{P}_N}[B_d(T_0, T)] = \mathbb{E}_t^{\mathbb{P}_N}\big[\exp(-G_2(\tau_0, 0, 1) r_{T_0} - a_d H_2(\tau_0, 0, 1))\big]$$

and

$$\widehat{f}_2(\phi) := \mathbb{E}_t^{\mathbb{P}_N}[B_d(T_0, T) f_2(\phi)] = \mathbb{E}_t^{\mathbb{P}_N}[g_2(\phi)].$$

The function $g_2(\phi)$ is given by the formula

$$g_2(\phi) := B_d(T_0, T) f_2(\phi) = \widehat{c}_2 \exp\big(-\widehat{q}_2 v_{T_0} - \widehat{q}_4 r_{T_0} - \widehat{q}_6 \widehat{r}_{T_0}\big) \tag{29}$$

with the constants $\widehat{c}_2, \widehat{q}_2, \widehat{q}_4, \widehat{q}_6$ given in the statement of the theorem. We remark that the second equality in (29) is obtained by invoking (13) and (16).

Using Lemma 4 and applying Lemma 8 under $\mathbb{P}_N$, we obtain the desired expressions for the function $\widehat{V}_2$, as reported in the statement of the theorem. To compute $\widehat{f}_2(\phi)$, we argue as in the case of $\widehat{f}_1(\phi)$. To be more specific, we note that

$$\mathbb{E}_t^{\mathbb{P}_N}\big[\exp(-\widehat{q}_2 v_{T_0} - \widehat{q}_4 r_{T_0} - \widehat{q}_6 \widehat{r}_{T_0})\big] = \mathbb{E}_t^{\mathbb{P}_N}\big[\exp(-\widehat{q}_2 v_{T_0})\big]\mathbb{E}_t^{\mathbb{P}_N}\big[\exp(-\widehat{q}_4 r_{T_0})\big] \times \mathbb{E}_t^{\mathbb{P}_N}\big[\exp(-\widehat{q}_6 \widehat{r}_{T_0})\big]$$

and we compute the first two conditional expectations in the right-hand side using Lemma 8. □

7.2 Options Pricing Formula in the Savings Account Numéraire

In this section, the price of the option is expressed in terms of savings accounts and the exchange rate. Although the option pricing formula is simpler and more explicit, in our opinion, the drawback of this approach is that it refers to a quantity that is not directly observed in the market, namely, the foreign market savings account.

Theorem 2 *Consider the forward start foreign exchange call option with maturity T, strike determination date T_0 and strike $\widehat{K} = kQ_{T_0}$. Let the foreign exchange model be given by stochastic differential equations* (1) *and Assumptions* (A.1)–(A.3). *Then the option's price equals, for all $t \in [0, T_0]$,*

$$C_t(T, \widehat{K}) = Q_t B_t^f \Big(\widetilde{P}_1\big(t, v_t, r_t, \widehat{r}_t, k\big) - k\widetilde{P}_2\big(t, v_t, r_t, \widehat{r}_t, k\big)\Big). \tag{30}$$

The function $\widetilde{P}_1$ equals, for all $t \in [0, T_0]$,

$$\widetilde{P}_1(t, v_t, r_t, \widehat{r}_t, k) = \frac{1}{2}\widetilde{V}_1(t, \widehat{r}_t) + \frac{1}{\pi}\int_0^\infty \mathrm{Re}\left(\widetilde{f}_1(\phi)\, \frac{\exp(-i\phi \ln k)}{i\phi}\right) d\phi$$

where the function $\widetilde{f}_1(\phi) = \widetilde{f}_1(\phi, t, v_t, r_t, \widehat{r}_t)$ satisfies

$$\begin{aligned}\ln(\widetilde{f}_1(\phi)) = \ln(\widehat{c}_1) - \widehat{G}_1(\tau, \widehat{s}_2)v_t - \widehat{G}_2(\tau, \widehat{s}_4)r_t - G_3(\tau, \widehat{s}_6, 1)\widehat{r}_t \\ - \theta \widehat{H}_1(\tau, \widehat{s}_2) - a_d \widehat{H}_2(\tau, \widehat{s}_4) - a_f H_3(\tau, \widehat{s}_6, 1)\end{aligned}$$

where in turn

$$\begin{aligned}\ln(\widehat{c}_1) &= (1 + i\phi)m_f(T_0, T) - i\phi m_d(T_0, T) - (1 + i\phi)\tau_0 \frac{\rho\theta}{\sigma_v} \\ &\quad - i\phi \int_{T_0}^T a_d n_d(u, T)\, du + (1 + i\phi)\int_{T_0}^T a_f n_f(u, T)\, du \\ &\quad - \theta H_1(\tau_0, s_1, s_2) - a_d H_2(\tau_0, s_3, s_4) - a_f H_3(\tau_0, s_5, s_6).\end{aligned}$$

The functions H_1, H_2, H_3 are given by Lemma 8, *the functions $\widehat{G}_1, \widehat{G}_2, \widehat{H}_1, \widehat{H}_2$ are given by* (19), *the constants $s_1, s_2, s_3, s_4, s_5, s_6$ are given by* (17), *and*

$$\widehat{s}_2 = \frac{(1 + i\phi)\rho}{\sigma_v} + G_1(\tau_0, s_1, s_2),\ \widehat{s}_4 = G_2(\tau_0, s_3, s_4),\ \widehat{s}_6 = G_3(\tau_0, s_5, s_6),$$

with the functions G_1, G_2, G_3 given by Lemma 8 *and the function $\widetilde{V}_1(t, \widehat{r}_t)$ equals*

$$\begin{aligned}\ln(\widetilde{V}_1(t, \widehat{r}_t)) = -a_f H_3(\tau_0, 0, 1) - G_3(\tau, G_3(\tau_0, 0, 1), 1)\widehat{r}_t \\ - a_f H_3(\tau, G_3(\tau_0, 0, 1), 1).\end{aligned}$$

The function $\widetilde{P}_2$ *equals, for all* $t \in [0, T_0]$,

$$\widetilde{P}_2(t, v_t, r_t, \widehat{r}_t, k) = \frac{1}{2}\,\widetilde{V}_2(t, r_t, \widehat{r}_t) + \frac{1}{\pi}\int_0^\infty \mathrm{Re}\left(\widetilde{f}_2(\phi)\,\frac{\exp(-i\phi\ln k)}{i\phi}\right)d\phi$$

where the function $\widetilde{f}_2(\phi) = \widetilde{f}_2(\phi, t, v_t, r_t, \widehat{r}_t)$ *satisfies*

$$\begin{aligned}\ln(\widetilde{f}_2(\phi)) &= \ln(\widehat{c}_2) - \widehat{G}_1(\tau, \widehat{q}_2)v_t - \widehat{G}_2(\tau, \widehat{q}_4)r_t - G_3(\tau, \widehat{q}_6, 1)\widehat{r}_t\\ &\quad - \theta\widehat{H}_1(\tau, \widehat{q}_2) - a_d\widehat{H}_2(\tau, \widehat{q}_4) - a_f H_3(\tau, \widehat{q}_6, 1)\end{aligned}$$

where in turn

$$\begin{aligned}\ln(\widehat{c}_2) &= i\phi m_f(T_0, T) + (1 - i\phi)m_d(T_0, T) - i\phi\tau_0\frac{\rho\theta}{\sigma_v}\\ &\quad + (1 - i\phi)\int_{T_0}^T a_d n_d(u, T)\,du + i\phi\int_{T_0}^T a_f n_f(u, T)\,du\\ &\quad - \theta H_1(\tau_0, q_1, q_2) - a_d H_2(\tau_0, q_3, q_4) - a_f H_3(\tau_0, q_5, q_6).\end{aligned}$$

The constants $q_1, q_2, q_3, q_4, q_5, q_6$ *are given by* (18) *and*

$$\widehat{q}_2 = i\phi\frac{\rho}{\sigma_v} + G_1(\tau_0, q_1, q_2),\ \widehat{q}_4 = G_2(\tau_0, q_3, q_4),\ \widehat{q}_6 = G_3(\tau_0, q_5, q_6).$$

Finally, the function $\widetilde{V}_2(t, r_t, \widehat{r}_t)$ *is given by*

$$\begin{aligned}\ln(\widetilde{V}_2(t, r_t, \widehat{r}_t)) &= -a_d H_2(\tau_0, 0, 1) - \widehat{G}_2(\tau, G_2(\tau_0, 0, 1))r_t\\ &\quad - a_d\widehat{H}_2(\tau, G_2(\tau_0, 0, 1)) - G_3(\tau, 0, 1)\widehat{r}_t - a_f H_3(\tau, 0, 1).\end{aligned}$$

Proof Let us fix $t \in [0, T_0]$. By combining Proposition 2 with Lemma 7, we obtain

$$C_t(T, \widehat{K}) = Q_t\,\mathbb{E}_t^{\mathbb{P}_N}\big(\beta_f(t, T_0)\widehat{C}_{T_0}(T, k)\big) = Q_t\big(\widetilde{J}_t^1 - k\widetilde{J}_t^2\big)$$

where we denote $\beta_f(t, T_0) = B_t^f\,(B_{T_0}^f)^{-1}$ and we set:

$$\widetilde{J}_t^1 = \mathbb{E}_t^{\widetilde{\mathbb{P}}_N}\Big(\beta_f(t, T_0)B_f(T_0, T)P_1\big(T_0, v_{T_0}, r_{T_0}, \widehat{r}_{T_0}, k\big)\Big)$$

and

$$\widetilde{J}_t^2 = \mathbb{E}_t^{\widetilde{\mathbb{P}}_N}\Big(\beta_f(t, T_0)B_d(T_0, T)P_2\big(T_0, v_{T_0}, r_{T_0}, \widehat{r}_{T_0}, k\big)\Big).$$

We will first compute the conditional expectation $\widetilde{J}_t^1$, that is,

$$\widetilde{J}_t^1 = \widetilde{P}_1(t, v_t, r_t, \widehat{r}_t) := \mathbb{E}_t^{\mathbb{P}_N}\Big(\beta_f(t, T_0)B_f(T_0, T)P_1(T_0, v_{T_0}, r_{T_0}, \widehat{r}_{T_0}, k)\Big). \quad (31)$$

From Proposition 2, we know that P_1 equals

$$P_1(T_0, v_{T_0}, r_{T_0}, \widehat{r}_{T_0}, k) = \frac{1}{2} + \frac{1}{\pi} \int_0^\infty \mathrm{Re}\Big(f_1(\phi) \frac{\exp(-i\phi \ln k)}{i\phi} \Big) d\phi \tag{32}$$

where the function $f_1(\phi)$ is given in the statement of Proposition 2.

Using equalities (31) and (32), we obtain

$$\widetilde{J}_t^1 = \frac{1}{2} \widetilde{V}_1(t, \widehat{r}_t) + \frac{1}{\pi} \int_0^\infty \mathrm{Re}\Big(\widetilde{f}_1(\phi) \frac{\exp(-i\phi \ln k)}{i\phi} \Big) d\phi$$

where $\widetilde{V}_1(t, \widehat{r}_t)$ and $\widetilde{f}_1(\phi)$ stand for the following conditional expectations:

$$\begin{aligned} \widetilde{V}_1(t, \widehat{r}_t) &:= \mathbb{E}_t^{\widetilde{\mathbb{P}}_N}[\beta_f(t, T_0) B_f(T_0, T)] \\ &= \mathbb{E}_t^{\widetilde{\mathbb{P}}_N}\Big[\exp\Big(-\int_t^{T_0} \widehat{r}_u \, du - G_3(\tau_0, 0, 1)\widehat{r}_{T_0} - a_f H_3(\tau_0, 0, 1)\Big)\Big] \end{aligned}$$

and

$$\widetilde{f}_1(\phi) := \mathbb{E}_t^{\widetilde{\mathbb{P}}_N}[\beta_f(t, T_0) B_f(T_0, T) f_1(\phi)] = \mathbb{E}_t^{\widetilde{\mathbb{P}}_N}[\widetilde{g}_1(\phi)].$$

The function $\widetilde{g}_1(\phi)$ is in turn given by the formula

$$\begin{aligned} \widetilde{g}_1(\phi) &:= \beta_f(t, T_0) B_f(T_0, T) f_1(\phi) \\ &= \widehat{c}_1 \exp\Big(-\int_t^{T_0} \widehat{r}_u \, du - \widehat{s}_2 v_{T_0} - \widehat{s}_4 r_{T_0} - \widehat{s}_6 \widehat{r}_{T_0}\Big) \end{aligned} \tag{33}$$

with the constants $\widehat{c}_1, \widehat{s}_2, \widehat{s}_4, \widehat{s}_6$ given in the statement of the theorem. It is worth stressing that the second equality in (26) is an immediate consequence of formulae (14) and (15).

The dynamics of $\widehat{r}_t$ under $\widetilde{\mathbb{P}}_N$ are given by equation (11). Hence a straightforward application of Lemma 8 yields the stated formula for $\widetilde{V}_1(t, \widehat{r}_t)$. Similarly, to compute the conditional expectation

$$\widetilde{f}_1(\phi) = \widehat{c}_1 \, \mathbb{E}_t^{\widetilde{\mathbb{P}}_N}\Big[\exp\Big(-\int_t^{T_0} \widehat{r}_u \, du - \widehat{s}_2 v_{T_0} - \widehat{s}_4 r_{T_0} - \widehat{s}_6 \widehat{r}_{T_0}\Big)\Big],$$

we apply Lemma 8 and we use the dynamics of processes v, r, and $\widehat{r}$ under $\widetilde{\mathbb{P}}_N$, as given by Lemma 6. In this manner, we obtain the stated formula for $\widetilde{f}_1(\phi)$ and thus also the required expression for the term $\widetilde{J}_t^1$.

To complete the proof, it remains to evaluate the conditional expectation

$$\widetilde{J}_t^2 = \widetilde{P}_2\big(t, v_t, r_t, \widehat{r}_t, k\big) := \mathbb{E}_t^{\widetilde{\mathbb{P}}_N}\Big(B_d(T_0, T) P_2\big(T_0, v_{T_0}, r_{T_0}, \widehat{r}_{T_0}, k\big)\Big) \tag{34}$$

where

$$P_2(T_0, v_{T_0}, r_{T_0}, \widehat{r}_{T_0}, k) = \frac{1}{2} + \frac{1}{\pi} \int_0^\infty \mathrm{Re}\Big(f_2(\phi) \frac{\exp(-i\phi \ln k)}{i\phi} \Big) d\phi \tag{35}$$

and the function $f_2(\phi)$ is given in Proposition 2. In view of (34) and (35), we obtain the following equality

$$\widetilde{J}_t^2 = \frac{1}{2}\,\widetilde{V}_2(t,r_t,\widehat{r}_t) + \frac{1}{\pi}\int_0^\infty \mathrm{Re}\bigg(\widetilde{f}_2(\phi)\,\frac{\exp(-i\phi\ln k)}{i\phi}\bigg)d\phi$$

where $\widetilde{V}_2(t,r_t,\widehat{r}_t)$ and $\widetilde{f}_2(\phi)$ stand for the following conditional expectations:

$$\begin{aligned}\widetilde{V}_2(t,r_t,\widehat{r}_t) &:= \mathbb{E}_t^{\widetilde{\mathbb{P}}_N}[\beta_f(t,T_0)B_d(T_0,T)]\\ &= \mathbb{E}_t^{\widetilde{\mathbb{P}}_N}\Big[\exp\Big(-\int_t^{T_0}\widehat{r}_u\,du - G_2(\tau_0,0,1)r_{T_0} - a_d H_2(\tau_0,0,1)\Big)\Big]\end{aligned}$$

and

$$\widetilde{f}_2(\phi) := \mathbb{E}_t^{\widetilde{\mathbb{P}}_N}[B_d(T_0,T)f_2(\phi)] = \mathbb{E}_t^{\widetilde{\mathbb{P}}_N}[\widetilde{g}_2(\phi)].$$

The function $\widetilde{g}_2(\phi)$ is in turn given by the formula

$$\widetilde{g}_2(\phi) := B_d(T_0,T)f_2(\phi) = \widehat{c}_2\exp\Big(-\int_t^{T_0}\widehat{r}_u\,du - \widehat{q}_2 v_{T_0} - \widehat{q}_4 r_{T_0} - \widehat{q}_6\widehat{r}_{T_0}\Big) \quad (36)$$

with the constants $\widehat{c}_2, \widehat{q}_2, \widehat{q}_4, \widehat{q}_6$ given in the statement of the theorem. We remark that the second equality in (36) is obtained by invoking (13) and (16). Using Lemma 6 and applying Lemma 8 under $\widetilde{\mathbb{P}}_N$, we obtain the desired expressions for the functions $\widetilde{f}_2$ and $\widetilde{V}_2$, as given in the statement of the theorem. □

8 Put-Call Parity for Forward Start Foreign Exchange Options

Our final goal is to establish the put-call parity relationship and thus to obtain a convenient representation for the price of the forward start foreign exchange put. Recall that the strike determination date T_0 satisfies $T_0 < T$ and the strike $\widehat{K}$ is given as $\widehat{K} = kQ_{T_0}$ for a positive constant k. The payoff at expiry date T of the forward start foreign exchange put option equals $P_T(T,\widehat{K}) = (\widehat{K} - Q_T)^+$. Therefore, at expiry date T we obtain

$$C_T(T,\widehat{K}) - P_T(T,\widehat{K}) = Q_T - \widehat{K} = Q_T - kQ_{T_0}. \quad (37)$$

Formula (37) is the starting point in derivation of the relationship between prices of call and put options at any date $t \in [0,T]$. The following result furnishes the put-call parity relationships for the market model (1) under Assumptions (A.1)–(A.3).

Proposition 3 (i) *For $t \in [T_0,T]$, that is, after the strike determination date, the put-call parity relationship is given by the following equality*

$$C_t(T,\widehat{K}) - P_t(T,\widehat{K}) = B_f(t,T)Q_t - kB_d(t,T)Q_{T_0}.$$

(ii) *For $t \in [0, T_0]$, that is, prior to the strike determination date, the put-call parity relationship becomes*

$$C_t(T, \widehat{K}) - P_t(T, \widehat{K}) = B_f(t, T)Q_t - kB_d(t, T)F(t, T_0)J_t \tag{38}$$

where the forward exchange rate $F(t, T_0)$ is given by formula (3).

The term J_t equals

$$J_t = \mathbb{E}_t^{\widetilde{\mathbb{P}}_{T_0}}\left(\exp\left(\int_t^{T_0} \widetilde{\mu}(u) r_u \, du\right)\right) \tag{39}$$

where the function $\widetilde{\mu} : [t, T_0] \to \mathbb{R}$ is given by the equality

$$\widetilde{\mu}(u) = \sigma_d^2 n_d(u, T_0)(n_d(u, T) - n_d(u, T_0)).$$

Finally, the dynamics of the process r under the probability measure $\widetilde{\mathbb{P}}_{T_0}$ are

$$dr_t = \left(a_d - \widetilde{b}_d(t) r_t\right) dt + \sigma_d \sqrt{r_t}\, d\widetilde{W}_t^{T_0}$$

where the function $\widetilde{b}_d : [t, T_0] \to \mathbb{R}$ is given by the equality

$$\widetilde{b}_d(t) = b_d + \sigma_d^2 n_d(t, T_0) + \sigma_d^2 n_d(t, T)$$

and the process $\widetilde{W}^{T_0}$ is the standard Brownian motion under $\widetilde{P}_{T_0}$.

Proof We start by noting that (37) yields, for all $t \in [0, T_0]$,

$$C_t(T, \widehat{K}) - P_t(T, \widehat{K}) = B_d(t, T)\,\mathbb{E}_t^{\mathbb{P}_T}(Q_T) - kB_d(t, T)\,\mathbb{E}_t^{\mathbb{P}_T}(Q_{T_0}).$$

Part (i). Let us first consider the case $t \in [T_0, T]$. To derive the relationship between prices of call and put options, it suffices to recall that the forward exchange rate $F(t, T)$ is a martingale under the domestic forward martingale measure $\mathbb{P}_T$ (cf. Lemma 2). Since the random variable Q_{T_0} is $\mathscr{F}_t$-measurable, it is easy to see that the put-call parity relationship takes the usual form (see, for instance, formula (4.20) in Musiela and Rutkowski [14])

$$C_t(T, \widehat{K}) - P_t(T, \widehat{K}) = B_f(t, T)Q_t - kB_d(t, T)Q_{T_0}.$$

Part (ii). We now focus on more challenging case where $t \in [0, T_0]$. Then

$$C_t(T, \widehat{K}) - P_t(T, \widehat{K}) = B_f(t, T)Q_t - kB_d(t, T)\,\mathbb{E}_t^{\mathbb{P}_T}(Q_{T_0})$$

and the standard change of measure arguments yield

$$B_d(t, T)\,\mathbb{E}_t^{\mathbb{P}_T}(Q_{T_0}) = B_d(t, T_0)\,\mathbb{E}_t^{\mathbb{P}_{T_0}}\left(B_d(T_0, T)Q_{T_0}\right).$$

To compute the conditional expectation in the right-hand side of the formula above, we note that $B_d(T_0, T)Q_{T_0} = Z_{T_0}$, where the process $(Z_t)_{t\in[0,T_0]}$ is given by the formula

$$Z_t = \frac{B_d(t,T)}{B_d(t,T_0)} F(t,T_0) = F_d(t,T,T_0)\, F(t,T_0).$$

Note that the quantity $F_d(t,T,T_0) = \frac{B_d(t,T)}{B_d(t,T_0)}$ represents the forward price at time t of the T-maturity domestic bond for settlement at time T_0.

From Lemma 2, it follows that the forward exchange rate $F(t,T_0)$ satisfies, under the domestic forward martingale measure $\mathbb{P}_{T_0}$,

$$F(T_0,T_0) = F(t,T_0)\exp\left(\int_t^{T_0} \widetilde{\sigma}_F(u,T_0)\cdot d\widetilde{W}_u^{T_0} - \frac{1}{2}\int_t^{T_0} \|\widetilde{\sigma}_F(u,T_0)\|^2\, du\right)$$

where the process $(\widetilde{\sigma}_F(t,T_0))_{t\in[0,T_0]}$ is given by

$$\widetilde{\sigma}_F(t,T_0) = \left[\sqrt{v_t},\ \sigma_d n_d(t,T_0)\sqrt{r_t},\ -\sigma_f n_f(t,T_0)\sqrt{\widehat{r}_t}\,\right]$$

and $\widetilde{W}^{T_0} = (\widetilde{W}_t^{T_0})_{t\in[0,T_0]}$ is the three-dimensional standard Brownian motion under $\mathbb{P}_{T_0}$, which is represented as $\widetilde{W}^{T_0} = [W^Q,\ W^{T_0},\ W^f]$. It is also well known that the forward price of the T-maturity domestic bond satisfies, under the domestic forward martingale measure $\mathbb{P}_{T_0}$,

$$\begin{aligned} &F_d(T_0,T,T_0)\\ &\quad = F_d(t,T,T_0)\exp\left(\int_t^{T_0} \psi(u,T,T_0)\, dW_u^{T_0} - \frac{1}{2}\int_t^{T_0} \psi^2(u,T,T_0)\, du\right)\end{aligned}$$

where

$$\psi(u,T,T_0) = b(u,T) - b(u,T_0) = \big(\sigma_d n_d(u,T_0) - \sigma_d n_d(u,T)\big)\sqrt{r_u}$$

where in turn we denote, for any maturity U,

$$b(u,U) = -\sigma_d n_d(u,U)\sqrt{r_u}.$$

Using the independence of processes W^{T_0} and (W^Q, W^f) under $\mathbb{P}_{T_0}$, we thus obtain

$$\mathbb{E}_t^{\mathbb{P}_{T_0}}\big(B(T_0,T)Q_{T_0}\big) = F_d(t,T,T_0)F(t,T_0)\,\mathbb{E}_t^{\mathbb{P}_{T_0}}\big(\zeta(t,T_0)\big)$$

where $\zeta(t,T_0)$ is given by the following expression

$$\begin{aligned}\ln(\zeta(t,T_0)) &= \int_t^{T_0} b(u,T)\, dW_u^{T_0} - \frac{1}{2}\int_t^{T_0} b^2(u,T)\, du\\ &\quad + \int_t^{T_0} b(u,T_0)(b(u,T) - b(u,T_0))\, du.\end{aligned}$$

To complete the derivation of relationship (38), it remains to compute the conditional expectation $J_t = \mathbb{E}_t^{\mathbb{P}_{T_0}}(\zeta(t, T_0))$. The dynamics of the process $(r_t)_{t\in[0,T_0]}$ under $\mathbb{P}_{T_0}$ are (see formula (4))

$$dr_t = \big(a_d - \widehat{b}_d(t) r_t\big)\, dt + \sigma_d \sqrt{r_t}\, dW_t^{T_0}$$

where the continuous function $\widehat{b}_d : [0, T_0] \to \mathbb{R}$ is given by the formula

$$\widehat{b}_d(t) = b_d + \sigma_d^2 n_d(t, T_0).$$

Hence, using the Girsanov theorem, we obtain

$$\begin{aligned} J_t &= \mathbb{E}_t^{\widetilde{\mathbb{P}}_{T_0}}\bigg(\exp\bigg(\int_t^{T_0} b(u, T_0)(b(u, T) - b(u, T_0))\, du\bigg)\bigg) \\ &= \mathbb{E}_t^{\widetilde{\mathbb{P}}_{T_0}}\bigg(\exp\bigg(\int_t^{T_0} \sigma_d^2 n_d(u, T_0)(n_d(u, T) - n_d(u, T_0)) r_u\, du\bigg)\bigg) \end{aligned}$$

where the dynamics of the process $(r_t)_{t\in[0,T_0]}$ under $\widetilde{\mathbb{P}}_{T_0}$ are given by the following expression

$$dr_t = \big(a_d - \widetilde{b}_d(t) r_t\big)\, dt + \sigma_d \sqrt{r_t}\, d\widetilde{W}_t^{T_0} \tag{40}$$

where in turn the continuous function $\widetilde{b}_d : [0, T_0] \to \mathbb{R}$ equals

$$\widetilde{b}_d(t) = b_d + \sigma_d^2 n_d(t, T_0) + \sigma_d^2 n_d(t, T)$$

and $\widetilde{W}^{T_0}$ is the standard Brownian motion under $\widetilde{\mathbb{P}}_{T_0}$. To complete the proof, it suffices to observe that

$$J_t = \mathbb{E}_t^{\widetilde{\mathbb{P}}_{T_0}}\bigg(\exp\bigg(\int_t^{T_0} \widetilde{\mu}(u) r_u\, du\bigg)\bigg)$$

where $\widetilde{\mu} : [0, T_0] \to \mathbb{R}$ is a continuous function given by the expression

$$\widetilde{\mu}(t) = \sigma_d^2 n_d(t, T_0)(n_d(t, T) - n_d(t, T_0))$$

and the dynamics of r under $\widetilde{\mathbb{P}}_{T_0}$ are given by (40). □

Although we do not provide here any closed-form expression for the term J_t defined by formula (39), it is clear that this quantity can be easily approximated by combining Lemma 8 with suitable piecewise constant approximations of continuous functions $\widetilde{b}_d$ and $\widetilde{\mu}$. In conclusion, it is fair to say that the numerical implementations of pricing formulae for forward start foreign exchange options established in this work are yet to be examined, so that the practical importance of these formulae and comparative analysis with alternative numerical approaches proposed recently in the literature are left for a future research.

Acknowledgements The research of M. Rutkowski was supported under Australian Research Council's Discovery Projects funding scheme (project number DP0881460). The paper is in the final form and no similar paper has been or is being submitted elsewhere. The authors are grateful to anonymous referees for their detailed and insightful reports.

References

1. Ahlip, R., Rutkowski, M.: Pricing of foreign exchange options under the Heston stochastic volatility model and the CIR interest rates. Quant. Finance **13**, 955–966 (2013)
2. Amerio, E.: Forward start option pricing with stochastic volatility: a general framework. In: Locke, E. (ed.) Financial Engineering and Applications: Proceedings of the Fourth IASTED International Conference, pp. 44–53. Acta Press, Calgary (2007)
3. Carr, P., Madan, D.: Option valuation using the fast Fourier transform. J. Comput. Finance **2**, 61–73 (1999)
4. Cox, J.C., Ingersoll, J.E., Ross, S.A.: A theory of term structure of interest rates. Econometrica **53**, 385–408 (1985)
5. Duffie, D., Pan, J., Singleton, K.: Transform analysis and asset pricing for affine jump-diffusions. Econometrica **68**, 1343–1376 (2000)
6. Grzelak, L.A., Oosterlee, C.W.: On the Heston model with stochastic interest rates. SIAM J. Financ. Math. **2**, 255–286 (2011)
7. Grzelak, L.A., Oosterlee, C.W.: On cross-currency models with stochastic volatility and correlated interest rates. Appl. Math. Finance **19**, 1–35 (2012)
8. Grzelak, L.A., Oosterlee, C.W., Van Weeren, S.: Extension of stochastic volatility equity models with the Hull-White interest rate process. Quant. Finance **12**, 89–105 (2012)
9. Heston, S.L.: A closed-form solution for options with stochastic volatility with applications to bond and currency options. Rev. Financ. Stud. **6**, 327–343 (1993)
10. Jeanblanc, M., Yor, M., Chesney, M.: Mathematical Methods for Financial Markets. Springer, Berlin (2009)
11. Kruse, S., Nögel, U.: On the pricing of forward starting options in Heston's model on stochastic volatility. Finance Stoch. **9**, 233–250 (2005)
12. Lipton, A.: Mathematical Methods for Foreign Exchange Options: A Financial Engineer's Approach, pp. 608–611. World Scientific, New Jersey (2001)
13. Lucic, V.: Forward start options in stochastic volatility models. In: Wilmott, P. (ed.) The Best of Wilmott 1, pp. 413–420. John Wiley, Chichester (2004)
14. Musiela, M., Rutkowski, M.: Martingale Methods in Financial Modelling, 2nd edn. Springer, Berlin (2005)
15. Schöbel, R., Zhu, J.: Stochastic volatility with an Ornstein-Uhlenbeck process: an extension. Eur. Finance Rev. **3**, 23–46 (1999)
16. Van Haastrecht, A., Pelsser, A.: Generic pricing of foreign exchange, inflation and stock options under stochastic interest rates and stochastic volatility. Quant. Finance **11**, 665–691 (2011)
17. Van Haastrecht, A., Lord, R., Pelsser, A., Schrager, D.: Pricing long-maturity equity and FX derivatives with stochastic interest rates and stochastic volatility. Insur. Math. Econ. **45**, 436–448 (2009)
18. Vasicek, O.: An equilibrium characterisation of the term structure. J. Financ. Econ. **5**, 177–188 (1977)
19. Windcliff, H.A., Forsyth, P.A., Vetzal, K.R.: Numerical methods and volatility models for valuing cliquet options. Appl. Math. Finance **13**, 353–386 (2006)
20. Wong, B., Heyde, C.C.: On the martingale property of stochastic exponentials. J. Appl. Probab. **41**, 654–664 (2004)

Real Options with Competition and Incomplete Markets

Alain Bensoussan and SingRu (Celine) Hoe

Abstract Ever since the first attempts to model capital investment decisions as options, financial economists have sought more accurate, more realistic real options models. Strategic interactions and market incompleteness are significant challenges that may render existing classical models inadequate to the task of managing the firm's capital investments. The purpose of this paper is to address these challenges. The issue of incompleteness comes in for the valuation of payoffs due to absence of a unique martingale measure. One approach is to valuate assets by considering a rational utility-maximizing consumer/investor's joint decisions with respect to portfolio investment strategy and consumption rule. In our situation, we add the stopping time as an additional decision. We employ variational inequalities (V.I.s) to solve the optimal stopping problems corresponding to times to invest. The regularity of the obstacle (payoffs received at the decision time) is a major element for defining the optimal strategy. Due to the lack of smoothness of the obstacle raised by the game problem, the optimal strategy is a two-interval solution, characterized by three thresholds.

Keywords Stackelberg leader-follower game · Utility maximization · Bellman equation · Optimal stopping

Mathematics Subject Classification (2010) 91G80 · 91A30 · 91A15

A. Bensoussan (✉)
International Center for Decision and Risk Analysis, School of Management, University of Texas at Dallas, 800 West Cambell Rd, SM30, Richardson, TX 75080-3021, USA
e-mail: axb046100@utdallas.edu

A. Bensoussan
City University of Hong Kong, Hong Kong, China

S.(C.) Hoe
Texas A&M University-Commerce, Commerce, TX 75429, USA
e-mail: hoceline02@yahoo.com

Y. Kabanov et al. (eds.), *Inspired by Finance*, DOI 10.1007/978-3-319-02069-3_2,

1 Investment Game Problems and General Model Assumptions

We consider a Stackelberg leader-follower game for exploiting an irreversible investment opportunity with payoffs of a continuous stochastic income stream $Y(t)$ for a fixed cost K. We limit the flexibility in the investment decisions to the times when to invest. The roles of leader and follower are predetermined by regulations. Each firm chooses its individual stopping time to invest over an infinite horizon with the constraint that the follower be forbidden to undertake the investment until the leader has already done so. By investing K, the leader receives $\delta_1 Y(t)$ per unit time till the follower's entry. Once both have entered, each gets a continuous cashflow stream $\delta_2 Y(t)$ per unit time, with $\delta_2 < \delta_1$.

Consider a probability space $(\Omega, \mathscr{F}, \mathrm{Q})$ with $W^*(t) = (W(t), W^0(t))^T$ a standard Wiener process. The asset S representing the market and the cashflow process Y evolve as follows:

$$dS(t) = rS(t)dt + \sigma S(t)\left(\lambda dt + dW(t)\right), \tag{1}$$

$$dY(t) = Y(t)\left(\alpha dt + \varsigma\left(\rho dW(t) + \sqrt{1-\rho^2}dW^0(t)\right)\right), \tag{2}$$

where $W(t)$ and $W^0(t)$ are independent Wiener processes, $\rho^2 < 1$ is the correlation coefficient between market uncertainty and the cashflow process uncertainty, and r (risk-free rate), $\sigma, \lambda, \alpha, \varsigma$ are all constants. The market is incomplete since the market asset S can span only the portion of the stochastic cashflow risk driven by the Wiener process $W(t)$, leaving the remaining risk driven by $W^0(t)$ unhedgeable. There is no unique martingale measure, so the risk-neutral pricing is no longer appropriate, and an alternative must be developed in this framework.

We adopt utility-based pricing in which a risk averse investor/firm maximizes the expected utility of consumption. We assume that the investor's risk preferences are characterized by a constant absolute risk aversion utility function

$$U(C) = -\frac{1}{\gamma}e^{-\gamma C} \tag{3}$$

where the argument C is the investor's consumption, and γ is his/her risk aversion parameter, $\gamma > 0$.

Remark 1 We allow for negative consumption. For $C \in \mathbf{R}$, U increases from $-\infty$ to 0. As $C \to -\infty$, it leads to huge negative values. We interpret this effect as a penalty to the utility maximization investor. We could of course impose the constraint of non-negative consumption. However, imposing non-negativity on the consumption would rule out the analytical solutions for further developments, a property we would like to retain for the full analysis. Therefore, we choose to accept for negative consumption which could lead huge negative utility values (big penalties for our utility maximization investor) instead of imposing the non-negativity constraint on the consumption. We also note that the negative consumption occurs when x becomes very negative and we cannot avoid this situation since $x \in \mathbf{R}$.

Each firm maximizes its expected discounted utility from consumption over an infinite horizon, subject to choice over investment timing, consumption, hedge position in the market asset, and allocation in the riskless bond. Thus, each firm considers undertaking the investment as an additional decision besides portfolio investment and consumption decisions. The decision remains a stopping time, for which the right approach is that of variational inequality (V.I.) [1, 5]. Our duopoly game requires us to solve two V.I.s corresponding to the leader's and the follower's optimal stopping respectively. As such, we will need two obstacles corresponding to each V.I. We obtain the obstacles from solving continuous control problems, i.e., portfolio investment and consumption decisions, and we call this as solutions to postinvestment utility maximization. Employing the obstacles obtained, we then form V.I.s to solve the optimal stopping problems, and we call this as solutions to preinvestment utility maximization.

One point to note is that we need to consider an auxiliary problem of which the cashflow process (2) hits zero; the problem will then be reduced to classical investment-consumption portfolio decisions. We next summarize the general notations used in the paper to facilitate reading:

- τ for the follower's stopping time and θ for the leader's stopping time;
- $F^1(x, y)$ for the follower's obstacle, i.e., solution to follower's postinvestment utility maximization, and $F(x, y)$ for the follower's solution to the V.I., i.e., solution to the follower's preinvestment utility maximization;
- $L^1(x, y)$ for the leader's obstacle, i.e., solution to leader's postinvestment utility maximization, and $L(x, y)$ for the leader's solution to the V.I., i.e., solution to the leader's preinvestment utility maximization;
- $F(x)$ for the solution to the classical investment-consumption utility maximization, i.e., no augmented stochastic income stream $Y(t)$.

We detail follower's problem and solution in Sect. 2 and the leader's in Sect. 3. We conclude in Sect. 4. We omit most of the proofs except the main result.

2 Follower's Problem and Solution

We start with the follower's investment problem. Given the initial wealth, x, the follower optimizes his portfolio by dynamically choosing allocations in the market asset S, the riskless bond, and the consumption rate, C. The follower's wealth, X, evolves as follows:

$$\begin{cases} dX(t) = \pi(t)X(t)\sigma(\lambda dt + dW(t)) + rX(t)dt - C(t)dt, \quad t < \tau, \\ X(\tau) = X(\tau - 0) - K, \\ dX(t) = \pi(t)X(t)\sigma(\lambda dt + dW(t)) + rX(t)dt - C(t)dt + \delta_2 Y(t)dt, \quad t > \tau, \\ dY(t) = Y(t)\left(\alpha dt + \varsigma\left(\rho dW(t) + \sqrt{1-\rho^2}dW^0(t)\right)\right), \\ X(0) = x, \quad Y(0) = y, \end{cases} \tag{4}$$

where $\pi(t)$ is the proportion of wealth invested in asset S, $C(t)$ is the consumption rate, and τ is the stopping time to undertake the investment, chosen optimally by the follower. The wealth process is discontinuous at τ. From (4), we observe that the wealth process has two possible evolution regimes. To facilitate further exposition, we introduce the processes X^0 and X^1 (regime 0 and regime 1, respectively):

$$dX^0(t) = \pi(t)X^0(t)\sigma(\lambda dt + dW(t)) + rX^0(t)dt - C(t)dt, \tag{5}$$

$$dX^1(t) = \pi(t)X^1(t)\sigma(\lambda dt + dW(t)) + rX^1(t)dt - C(t)dt + \delta_2 Y(t)dt. \tag{6}$$

The follower's problem is to maximize his expected discounted utility from consumption by choosing stopping time τ, consumption rate C, and investment strategy π. We have to solve the problem in two steps, beginning with the utility maximization after τ (postinvestment utility maximization) and then solving the complete utility maximization prior to τ (preinvestment utility maximization). The rationale behind this two-step procedure is because we need a clearly defined obstacle function when solving the stopping time problem.

2.1 Postinvestment Utility Maximization

After τ, the follower solves his utility maximization as a control problem of portfolio selections and consumption rules augmented by a stochastic cashflow stream $\delta_2 Y(t)$ per unit time.

To facilitate representation, for $\mathscr{F}_t$-adapted processes $\pi(t)$, $C(t)$, we introduce the local integrability conditions

$$I^i = \begin{cases} \mathrm{E}\int_0^T \left(\pi(t)X^i(t)\right)^2 dt < \infty, & \forall\, T, \\ \mathrm{E}\int_0^T (C(t))^2\, dt < \infty, & \forall\, T, \end{cases} \tag{7}$$

and define

$$\tau_N^i = \inf\{t:\ X^i(t) < -N\}, \qquad i = 0, 1. \tag{8}$$

The follower reveals his preference through his expected discounted utility of consumption, and so, to the pair $(C(\cdot), \pi(\cdot))$, we introduce the objective function

$$J\big(C(\cdot)\big) = \mathrm{E}\int_0^\infty e^{-\mu t} U\big(C(t)\big)dt, \tag{9}$$

where μ, a constant, is the discount rate. This function is well-defined, but it may take the value $-\infty$. Since the follower can manage his investment-consumption portfolio, we consider the following control problem:

$$F^1(x,y) = \sup_{\{\pi(\cdot), C(\cdot)\} \in \mathscr{U}^1_{x,y}} J\big(C(\cdot)\big), \tag{10}$$

where

$$\mathscr{U}^1_{x,y} = \Big\{(\pi, C):\ I^1;\ \tau^1_N \uparrow \infty \text{ as } N \uparrow \infty;\ e^{-\mu T}\mathrm{E}e^{-r\gamma(X^1(T)+f(Y(T)))} \to 0,$$
$$\text{as } T \uparrow \infty\Big\},$$

and $f(y)$ is a positive function of linear growth with $f(0) = 0$ which will be made precise later (cf. (17), (16)).

We associate the value function $F^1(x, y)$ with the Bellman equation:

$$-\mu F^1 \frac{\partial F^1}{\partial x}(rx + \delta y) + \frac{\partial F^1}{\partial y}\alpha y + \frac{1}{2}\frac{\partial^2 F^1}{\partial y^2} y^2 \varsigma^2 + \sup_C\left\{U(C) - C\frac{\partial F^1}{\partial x}\right\}$$
$$+ \sup_\pi\left\{\pi x \sigma\left(\lambda \frac{\partial F^1}{\partial x} + y\varsigma\rho \frac{\partial^2 F^1}{\partial xy}\right) + \frac{1}{2}\frac{\partial^2 F^1}{\partial x^2}\pi^2 x^2 \sigma^2\right\} = 0. \tag{11}$$

The domain is $x \in \mathbf{R},\ y > 0$.

We note that if $y = 0$, then $Y(t) = 0$ for all t. The problem reduces to the classical investment-consumption problem with the solution given by:

$$F(x) = -\frac{1}{r\gamma}\exp\left\{-r\gamma x + 1 - \frac{\mu + \frac{\lambda^2}{2}}{r}\right\}. \tag{12}$$

We thus have:

$$F^1(x, 0) = F(x)\,. \tag{13}$$

We look for a solution of (10) in the form

$$F^1(x, y) = -\frac{1}{r\gamma}\exp\left\{-r\gamma(x + f(y)) + 1 - \frac{\mu + \frac{\lambda^2}{2}}{r}\right\} \tag{14}$$

in which, by (13),

$$f(0) = 0\,. \tag{15}$$

By (14), defining the optimal feedback,

$$\widehat{C}(x, y) = -\frac{1}{\gamma}\ln\frac{\partial F^1}{\partial x}$$

and

$$\widehat{\pi}(x, y) = -\frac{\lambda\frac{\partial F^1}{\partial x} + y\varsigma\rho\frac{\partial^2 F^1}{\partial x\partial y}}{\frac{\partial^2 F^1}{\partial x^2}\sigma x},$$

we reduce the Bellman equation (11) to:

$$\frac{1}{2}y^2\varsigma^2 f'' + (\alpha - \lambda\varsigma\rho) y f' - \frac{1}{2} r\gamma y^2 \varsigma^2 (1 - \rho^2) f'^2 - rf + \delta_2 y = 0\,. \tag{16}$$

Proposition 1 *The value function,*

$$f(y) = \inf_{\{v(\cdot)\in\mathscr{U}_y\}} \mathrm{E}\int_0^\infty e^{-rt}\left(\delta_2 Y_y(t) + \frac{1}{2}v^2(t)\right)dt \tag{17}$$

with

$$\begin{cases} dY_y(t) = Y_y(t)\big[\alpha - \lambda\varsigma\rho + \varsigma\sqrt{r\gamma(1-\rho^2)}v(t)\big]dt + Y_y(t)\varsigma dW(t), \;\; Y_y(0) = y\,, \\ \mathscr{U}_y = \{v(\cdot):\; E\int_0^\infty e^{-rt}v^2(t)dt < \infty,\; e^{-rT}EY_y(T) \to 0 \text{ as } T \to \infty\}\,, \end{cases} \tag{18}$$

is the unique function in $C^2(0,\infty)$ *solving* (16), (15) *on the interval* $[0, \varepsilon y + M_\varepsilon]$,[1] *and such that* $f(y) \uparrow \infty$ *as* $y \uparrow \infty$.

Proposition 2 *The function* $f'(y)$ *is bounded.*

We now state the result that the value function given by (10) is indeed of the form (14).

Theorem 1 *The function* $F^1(x, y)$ *given by* (14) *coincides with the value function given by* (10).

2.2 Preinvestment Utility Maximization

We now turn to the problem of optimal stopping with the obstacle defined by $F^1(x, y)$, the solution to the postinvestment utility maximization. Before the stopping time τ, the wealth process is governed by (5) and the cashflow process evolves as (2). Set $\theta^0 = \inf\{t:\; Y(t) = 0\}$. At time $\tau \wedge \theta^0$ the follower stops. If $\theta^0 \le \tau$, the investment never takes place and the follower receives $F(X^0(\theta^0))$, where $F(x)$ is given by (12). If $\tau < \theta^0$, the follower receives $F^1(X^0(\tau) - K, Y(\tau))$ at the stopping time τ, where $F^1(x, y)$ is given by (14). Therefore, the objective function is:

$$\begin{aligned} &J_{x,y}\big(C(\cdot), \pi(\cdot), \tau\big) \\ &\quad = \mathrm{E}\bigg[\int_0^{\tau\wedge\theta^0} e^{-\mu t}U\big(C(t)\big)dt + F^1\big(X^0(\tau) - K, Y(\tau)\big)e^{-\mu\tau}\mathbf{1}_{\tau<\theta^0} \\ &\qquad + F\big(X^0(\theta^0)\big)e^{-\mu\theta^0}\mathbf{1}_{\theta^0\le\tau}\bigg], \end{aligned} \tag{19}$$

---

[1] M_ε is defined as: $M_\varepsilon = \dfrac{\left(\frac{\delta_2}{\varepsilon} - r + \alpha - \lambda\varsigma\rho\right)^2}{2\varsigma^2 r^2\gamma(1-\rho^2)}$. If $r + \lambda\varsigma\rho - \alpha > 0$, we can take $M_\varepsilon = 0$, hence $\varepsilon = \frac{\delta_2}{r-\alpha+\lambda\varsigma\rho}$. Note that ε can be arbitrarily small.

and we define the associated value function:

$$F(x,y) = \sup_{\{\pi(\cdot),C(\cdot),\tau\}\in\mathscr{U}^0_{xy}} J_{x,y}\big(C(\cdot),\pi(\cdot),\tau\big), \tag{20}$$

where

$$\mathscr{U}^0_{x,y} = \Big\{(C,\pi,\tau):\ I^0;\ \tau\wedge\theta^0 < \infty \text{ a.s.};\ \tau^* = \lim\uparrow\tau^0_N \geq \tau\wedge\theta^0 \text{ a.s.}\Big\}.$$

As a consequence of Dynamic Programming, assuming sufficient smoothness of the function $F(x,y)$, we may write the strong formulation of V.I. that $F(x,y)$ must satisfy as follows:

$$\begin{cases} -\mu F + \frac{\partial F}{\partial x} rx + \frac{\partial F}{\partial y}\alpha y + \frac{1}{2}\frac{\partial^2 F}{\partial y^2}\varsigma^2 y^2 + \sup_C\left(U(C) - C\frac{\partial F}{\partial x}\right) \\ \quad + \sup_\pi\left[\pi x\sigma\left(\lambda\frac{\partial F}{\partial x} + y\varsigma\rho\frac{\partial^2 F}{\partial x\partial y}\right) + \frac{1}{2}\pi^2x^2\sigma^2\frac{\partial^2 F}{\partial x^2}\right] \leq 0, \\ F(x,y) \geq F^1(x-K,y), \\ \big(F(x,y) - F^1(x-K,y)\big)\Big[-\mu F + \frac{\partial F}{\partial x} rx + \frac{\partial F}{\partial y}\alpha y + \frac{1}{2}\frac{\partial^2 F}{\partial y^2}\varsigma^2 y^2 \\ \quad + \sup_C\left(U(C) - C\frac{\partial F}{\partial x}\right) + \sup_\pi\left[\pi x\sigma\left(\lambda\frac{\partial F}{\partial x} + y\varsigma\rho\frac{\partial^2 F}{\partial x\partial y}\right) + \frac{1}{2}\pi^2x^2\sigma^2\frac{\partial^2 F}{\partial x^2}\right]\Big] \\ \quad = 0. \end{cases} \tag{21}$$

We have the boundary condition:

$$F(x,0) = F(x). \tag{22}$$

We look for a solution of the form:

$$F(x,y) = -\frac{1}{r\gamma}\exp\left[-r\gamma\big(x+g(y)\big) + 1 - \frac{\mu + \frac{\lambda^2}{2}}{r}\right]. \tag{23}$$

Using (23) and (14) and defining the optimal feedback

$$\widehat{C}(x,y) = -\frac{1}{\gamma}\ln\frac{\partial F}{\partial x}$$

and

$$\widehat{\pi}(x,y) = -\frac{\lambda\frac{\partial F}{\partial x} + y\varsigma\rho\frac{\partial^2 F}{\partial x\partial y}}{\frac{\partial^2 F}{\partial x^2}\sigma x},$$

we transform V.I. (21) to the form:

$$\begin{cases} \frac{1}{2}y^2\varsigma^2 g'' + g' y(\alpha - \lambda\varsigma\rho) - \frac{1}{2}y^2\varsigma^2 r\gamma(1-\rho^2)g'^2 - rg \leq 0, \\ g(y) \geq f(y) - K, \\ \big(g(y) - f(y) + K\big)\left[\frac{1}{2}y^2\varsigma^2 g'' + g' y(\alpha - \lambda\varsigma\rho) - \frac{1}{2}y^2\varsigma^2 r\gamma(1-\rho^2)g'^2 - rg\right] \\ = 0, \\ g(0) = 0. \end{cases} \tag{24}$$

This V.I. cannot be interpreted as a control problem because the non-linear operator is connected to a minimization problem, while the inequalities are connected to a maximization problem. So, $g(y)$ is more appropriately the value function of a differential game rather than of a control problem. Define

$$u(y) = g(y) - f(y) + K.$$

Then (24) becomes (using the equation of $f(y)$ (cf.(16)):

$$\begin{cases} -\frac{1}{2}y^2\varsigma^2u'' - yu'\big(\alpha - \lambda\varsigma\rho - yf'\varsigma^2 r\gamma(1-\rho^2)\big) + \frac{1}{2}y^2\varsigma^2 r\gamma(1-\rho^2)u'^2 + ru \\ \quad \geq -\delta_2 y + rK, \\ u \geq 0, \\ u\big[-\frac{1}{2}y^2\varsigma^2u'' - yu'\big(\alpha - \lambda\varsigma\rho - yf'\varsigma^2 r\gamma(1-\rho^2)\big) + \frac{1}{2}y^2\varsigma^2 r\gamma(1-\rho^2)u'^2 \\ \quad + ru + \delta_2 y - rK\big] = 0, \\ u(0) = K. \end{cases} \tag{25}$$

We study (25) by the threshold approach. Let $\hat{y}$ be fixed, to be determined below. We consider the Dirichlet problem

$$\begin{cases} -\frac{1}{2}y^2\varsigma^2u'' - yu'\big(\alpha - \lambda\varsigma\rho - yf'\varsigma^2 r\gamma(1-\rho^2)\big) + \frac{1}{2}y^2\varsigma^2 r\gamma(1-\rho^2)u'^2 + ru \\ \quad = -\delta_2 y + rK, \quad 0 < y < \hat{y}, \\ u(0) = K, \quad u(\hat{y}) = 0. \end{cases} \tag{26}$$

For $\hat{y}$ fixed, this problem is a classical Bellman equation. Similar to Proposition 1, equation (26) is a Bellman equation of the following control problem with the controlled diffusion:

$$\begin{cases} dY_y(t) = Y_y(t)\big[\alpha - \lambda\varsigma\rho - Y_y(t)f'(Y_y(t))\varsigma^2 r\gamma(1-\rho^2) \\ \qquad\qquad + \varsigma\sqrt{r\gamma(1-\rho^2)}v(t)\big]dt + Y_y(t)\varsigma dW(t), \\ Y_y(0) = y, \quad 0 < y < \hat{y}, \end{cases} \tag{27}$$

and the value function

$$u(y) = \inf_{v(\cdot)} E\left[\int_0^{\theta_y(v(\cdot))} e^{-rt}\left[-\delta_2 Y_y(t) + rK + \frac{1}{2}v^2(t)\right]dt \right.$$
$$\left. + e^{-r\theta_y(v(\cdot))}K\mathbf{1}_{Y_y(\theta_y(v(\cdot)))=0}\right], \tag{28}$$

where $\theta_y(v(\cdot)) = \inf\{t : \ Y_y(t) \text{ is outside } (0, \hat{y})\}$ and it is finite (a.s.). Obviously, $u(y) > 0$ if $\hat{y} < \frac{Kr}{\delta_2}$, and we also have $u(y) \leq K$.[2]

[2]For $v(t)$, we can take the same class as in problem (17)–(18).

Theorem 2 *There exists a unique value $\hat{y}$ such that*

$$u'(\hat{y}) = 0, \quad \hat{y} \geq \frac{Kr}{\delta_2}. \tag{29}$$

The value function $u(y)$ (cf. (28)) extended by zero beyond $\hat{y}$ is the unique solution of V.I. (25). It is C^1 and piecewise C^2.

Referring back to (24), from Theorem 2, we have obtained that there exists a unique solution of (24) such that $g(y) \in C^1$ and piecewise C^2. There exists a unique $\hat{y}$ such that

$$\begin{cases} -\frac{1}{2}y^2\varsigma^2 g'' - g'y(\alpha - \lambda\varsigma\rho) + \frac{1}{2}y^2\varsigma^2 r\gamma(1-\rho^2)g'^2 + rg = 0, & y < \hat{y}, \\ g(y) = f(y) - K, \quad y \geq \hat{y}, \\ g'(\hat{y}) = f'(\hat{y}), \\ g(0) = 0. \end{cases} \tag{30}$$

Note that $g(y) \geq 0$ since $u(y) \geq -f(y) + K$. We generate the main result that the value function given by (20) is indeed of the form (23).

Theorem 3 *The function $F(x, y)$ defined by* (23) *coincides with the value function given by* (20).

2.3 Follower's Optimal Stopping Rule

We next define the optimal stopping rule as:

$$\hat{\tau}(y) = \inf\{t : Y_y(t) \geq \hat{y}\}, \tag{31}$$

where $Y_y(t)$ is the process defined in (2) and $\hat{y}$ is the unique value defined by the V.I. (29) (the smooth matching point). We must note that the follower's stopping time $\hat{\tau}(y)$ is the follower's optimal entry if he can enter in the market at time zero. Since the follower can enter only after the leader (who starts at time θ), for finite θ, the follower will enter at time:[3]

$$\hat{\tau}_\theta = \theta + \hat{\tau}\big(Y_y(\theta)\big). \tag{33}$$

[3]For any test function $\Psi(x, s)$, we have the formula:

$$\mathrm{E}\big[\Psi(Y_y(\hat{\tau}_\theta), \hat{\tau}_\theta)|\mathscr{F}_\theta\big] = \Psi(Y_y(\theta), \theta)\mathbf{1}_{Y_y(\theta)\geq\hat{y}} + \mathbf{1}_{Y_y(\theta)<\hat{y}}\mathrm{E}\big[\Psi(\hat{y}, t + \hat{\tau}(y))\big]\,|_{y=Y_y(\theta), t=\theta}\,. \tag{32}$$

3 Leader's Problem and Solution

After solving the follower's optimal policy, we are now ready to solve the leader's problem, which is complicated by the fact that he must share the market (project value) upon the follower's optimal entry at $\hat{\tau}_\theta$. Thus, by investing K, the leader expects to receive a continuous cash flow $\delta_1 Y(t)$ per unit time prior to the follower's entry, and $\delta_2 Y(t)$ per unit time afterwards. The leader's wealth evolves according to the following system of stochastic equations:

$$\begin{cases} dX(t) = \pi(t)X(t)\sigma\big(\lambda dt + dW(t)\big) + rX(t)dt - C(t)dt, \quad t < \theta, \\ X(\theta) = X(\theta - 0) - K, \\ dX(t) = \pi(t)X(t)\sigma\big(\lambda dt + dW(t)\big) + rX(t)dt + \delta_1 Y(t)dt - C(t)dt, \\ \quad \theta < t < \hat{\tau}_\theta, \\ dX(t) = \pi(t)X(t)\sigma\big(\lambda dt + dW(t)\big) + rX(t)dt + \delta_2 Y(t)dt - C(t)dt, \quad t > \hat{\tau}_\theta, \\ dY(t) = Y(t)\big(\alpha dt + \varsigma\big(\rho dW(t) + \sqrt{1-\rho^2}dW^0(t)\big)\big), \\ X(0) = x, \quad Y(0) = y, \quad \text{with } x \in \mathbf{R},\ y \geq 0, \end{cases} \tag{34}$$

where θ and $\hat{\tau}_\theta$ are stopping times chosen optimally by the leader and the follower, respectively. The leader's problem is to maximize his expected discounted utility from consumption by choosing stopping time θ, consumption rate C, and investment strategy π. As in the follower's case, we have to solve leader's complete utility maximization problem in two steps.

3.1 Postinvestment Utility Maximization

Suppose that $\theta = 0$, the leader's wealth is x, and the cash flow $y > 0$; then the leader's wealth becomes immediately $x - K$ since he must pay the fixed cost of entry, K. The leader must share the market upon follower's entry at $\hat{\tau}(y)$. Thus, for a generic initial wealth x, the leader's wealth evolves as follows:

$$\begin{cases} dX^{L1}(t) = \pi(t)X^{L1}(t)\sigma\big(\lambda dt + dW(t)\big) + \big(rX^{L1}(t) + \delta_1 Y(t) - C(t)\big)\,dt, \\ \quad t < \hat{\tau}(y), \\ X^{L1}(0) = x, \\ dX^2(t) = \pi(t)X^2(t)\sigma\big(\lambda dt + dW(t)\big) + rX^2(t)dt + \delta_2 Y(t)dt - C(t)dt, \\ \quad t > \hat{\tau}(y), \\ X^2\big(\hat{\tau}(y)\big) = X^{L1}\big(\hat{\tau}(y)\big). \end{cases} \tag{35}$$

If $\theta = 0$ and $y \geq \hat{y}$, the follower enters immediately, and the leader's problem is identical to the follower's, i.e., (10). So, we consider the function

$$L^2(x,y) = -\frac{1}{r\gamma} e^{-r\gamma\left(x+f(y)\right)+1-\frac{\mu+\frac{\lambda^2}{2}}{r}}, \tag{36}$$

where f is the solution of (16), (15) on the interval $[0, \varepsilon y + M_\varepsilon]$.[4]

If $\theta = 0$ and $y < \hat{y}$, the leader's problem is described as follows. The wealth process is described by X^{L1} in (35) and the cash flow process follows (2). Recall that $\theta^0 = \inf\{t : \ Y_y(t) = 0\}$ and $\hat{\tau}(y) = \inf\{t : \ Y_y(t) \geq \hat{y}\}$.[5] If $\theta^0 < \hat{\tau}(y)$, the follower never invests, and the leader's value function at time θ^0 corresponds to $F(X^{L1}(\theta^0))$ (cf. (12)). If $\hat{\tau}(y) < \theta^0$, the leader's value function corresponds to $L^2(X^{L1}(\hat{\tau}(y)), Y(\hat{\tau}(y)))$, at the follower's entry time, $\hat{\tau}(y)$. Thus, to a pair of $(C(\cdot), \pi(\cdot))$, we associate the objective function

$$\begin{aligned} J\big(C(\cdot),\pi(\cdot)\big) = E\bigg[\int_0^{\hat{\tau}(y)\wedge\theta^0} & e^{-\mu t}U(C(t))dt + F\big(X^{L1}(\theta^0)\big)e^{-\mu\theta^0}\mathbf{1}_{\theta^0\leq\hat{\tau}(y)} \\ & + L^2\big(X^{L1}\big(\hat{\tau}(y)\big), Y\big(\hat{\tau}(y)\big)\big)e^{-\mu\hat{\tau}(y)}\mathbf{1}_{\hat{\tau}(y)<\theta^0}\bigg], \end{aligned} \tag{37}$$

and we consider the value function:

$$L^1(x,y) = \sup_{\{\pi(\cdot),C(\cdot)\}\in\mathscr{U}^1_{x,y}} J\big(C(\cdot),\pi(\cdot)\big), \tag{38}$$

where

$$\mathscr{U}^1_{x,y} = \{(\pi, C) : \ I^1; \ \tau^* = \lim \uparrow \tau^1_N \geq \hat{\tau}(y) \wedge \theta^0 \text{ a.s.}\}$$

with I^1 and τ^1_N defined in (7) and (8) by replacing X^1 with X^{L1} respectively. We associate the value function with the Bellman equation:

$$\begin{cases} -\mu L^1 + \frac{\partial L^1}{\partial x}(rx + \delta_1 y) + \frac{\partial L^1}{\partial y}\alpha y + \frac{1}{2}\frac{\partial^2 L^1}{\partial y^2}\varsigma^2 y^2 + \sup_C \left(U(C) - C\frac{\partial L^1}{\partial x}\right) \\ \quad + \sup_\pi \left[\pi x\sigma\left(\lambda\frac{\partial L^1}{\partial x} + \varsigma\rho y\frac{\partial^2 L^1}{\partial x\partial y}\right) + \frac{1}{2}\pi^2 x^2\sigma^2\frac{\partial^2 L^1}{\partial x^2}\right] = 0\,, \\ L^1(x,\hat{y}) = L^2(x,\hat{y})\,. \end{cases} \tag{39}$$

We study the Bellman equation (39) for $y \in\,]0, \hat{y}[$ and we define:

$$L^1(x,y) = L^2(x,y), \quad \text{if} \quad y > \hat{y},$$

where $L^2(x,y)$ is defined in (36). The extension is continuous but not C^1. Also, we note that for $y = 0$, then $Y(t) = 0$ for all t, the problem then reduces to the

[4]See footnote 1 for the definition of M_ε.

[5]Here $Y_y(t)$ is the process defined in (2).

classical investment-consumption portfolio optimization problem; thus we have the boundary condition:

$$L^1(x,0) = F(x). \tag{40}$$

We look for a solution of the form

$$L^1(x,y) = -\frac{1}{r\gamma} e^{-r\gamma\left(x+q(y)\right)+1-\frac{\mu+\frac{\lambda^2}{2}}{r}} \tag{41}$$

with q solving the problem

$$\begin{cases} \frac{1}{2}y^2\varsigma^2 q'' + (\alpha - \lambda\varsigma\rho)yq' - \frac{1}{2}r^2\gamma^2 y^2\varsigma^2(1-\rho^2)q'^2 - rq + \delta_1 y = 0, \\ \quad 0 < y < \hat{y}, \\ q(0) = 0, \quad q(\hat{y}) = f(\hat{y}). \end{cases} \tag{42}$$

where $f(y)$ is the solution of (16), (15) on the interval $[0, \varepsilon y + M_\varepsilon]$. We extend $q(y)$ by $f(y)$ for $y > \hat{y}$.

The function $L^1(x, y)$ is continuous but not C^1. The study of (42) is similar to (16), but it is simpler because it is defined on a bounded interval. Similar to the study of (16), we can show that $q(y)$ may be interpreted as a function of a control problem, and there exists a unique solution which is $C^2(0, \hat{y})$. For $\delta_1 > \delta_2$, we have:

$$q(y) \geq f(y)\,. \tag{43}$$

Theorem 4 *The function $L^1(x, y)$ defined by* (41) *coincides with the value function given in* (38).

3.1.1 The Leader's Pre-investment Utility Maximization

We now turn to the leader's optimal stopping problem (i.e., choice of θ) with obstacle defined by $L^1(x, y)$, the solution to the postinvestment utility maximization. Before the stopping time θ, the leader's wealth and the cashflow process evolve as (5) and (2) respectively.

At time $\theta \wedge \theta^0$, the leader stops. If $\theta^0 \leq \theta$, the leader never takes the investment and receives $F(X^0(\theta^0))$ (cf. (12)). If $\theta < \theta^0$, the leader receives

$$L^1(X^0(\theta) - K, Y(\theta))$$

(cf. (41)) at θ. Therefore, the objective function is

$$J_{x,y}\big(C(\cdot), \pi(\cdot), \theta\big) = \mathrm{E}\bigg[\int_0^{\theta\wedge\theta^0} U\big(C(t)\big)e^{-\mu t}dt + L^1\big(X^0(\theta) - K, Y(\theta)\big)e^{-\mu\theta}\mathbf{1}_{\theta<\theta^0} + F\big(X^0(\theta^0)\big)e^{-\mu\theta^0}\mathbf{1}_{\theta^0\leq\theta}\bigg]. \tag{44}$$

We define the value function

$$L(x,y) = \sup_{\{\pi(\cdot),C(\cdot),\theta\}\in\mathscr{U}^0_{x,y}} J_{x,y}\big(C(\cdot),\pi(\cdot),\theta\big), \tag{45}$$

where

$$\mathscr{U}^0_{x,y} = \{(C,\pi,\theta) : \ I^0;\ \theta\wedge\theta^0 < \infty \text{ a.s.};\ \tau^* = \lim\uparrow\tau^0_N \geq \theta\wedge\theta^0 \text{ a.s.}\}$$

with I^0 and τ^0_N defined in (7) and (8) respectively. As a consequence of Dynamic Programming, assuming sufficient smoothness of L, we can associate the strong formulation of V.I. to the value function $L(x,y)$ as:

$$\begin{cases} -\mu L + rx\frac{\partial L}{\partial x} + \alpha y\frac{\partial L}{\partial y} + \frac{1}{2}\varsigma^2y^2\frac{\partial^2 L}{\partial y^2} + \sup_c\big(U(C) - C\frac{\partial L}{\partial x}\big) \\ \quad + \sup_\pi\big[\pi\sigma x\big(\lambda\frac{\partial L}{\partial x} + \varsigma\rho y\frac{\partial^2 L}{\partial x\partial y}\big) + \frac{1}{2}\pi^2x^2\sigma^2\frac{\partial^2 L}{\partial x^2}\big] \leq 0, \\ L(x,y) \geq L^1(x-K,y), \\ \big(L(x,y) - L^1(x-K,y)\big)\big[-\mu L + rx\frac{\partial L}{\partial x} + \alpha y\frac{\partial L}{\partial y} + \frac{1}{2}\varsigma^2y^2\frac{\partial^2 L}{\partial y^2} \\ \quad + \sup_C\big(U(C) - C\frac{\partial L}{\partial x}\big) + \sup_\pi\big[\pi\sigma x\big(\lambda\frac{\partial L}{\partial x} + \varsigma\rho y\frac{\partial^2 L}{\partial x\partial y}\big) + \frac{1}{2}\pi^2x^2\sigma^2\frac{\partial^2 L}{\partial x^2}\big]\big] = 0. \end{cases} \tag{46}$$

We have the boundary condition:

$$L(x,0) = F(x). \tag{47}$$

We look for a solution of the form

$$L(x,y) = -\frac{1}{r\gamma}e^{-r\gamma\left(x+h(y)\right)+1-\frac{\mu+\frac{\lambda^2}{2}}{r}}, \tag{48}$$

with $h(y)$ satisfying the following V.I.:

$$\begin{cases} \frac{1}{2}y^2\varsigma^2h'' + h'y(\alpha - \lambda\varsigma\rho) - \frac{1}{2}y^2\varsigma^2r\gamma(1-\rho^2)h'^2 - rh \leq 0, \\ h(y) \geq q(y) - K, \\ \big(h(y) - q(y) + K\big)\big[\frac{1}{2}y^2\varsigma^2h'' + h'y(\alpha - \lambda\varsigma\rho) - \frac{1}{2}y^2\varsigma^2r\gamma(1-\rho^2)h'^2 - rh\big] \\ \quad = 0, \\ h(0) = 0. \end{cases} \tag{49}$$

We encounter a new difficulty that does not occur in the follower's problem. We observe that the leader's obstacle $q(y) - K$ is C^0 but not C^1. We cannot as in (25) consider $u(y) = h(y) - q(y) + K$ since $q(y)$ is not sufficiently smooth. We will consider nonetheless the function

$$u(y) = h(y) - f(y) + K$$

which satisfies the following problem:

$$\begin{cases} -\frac{1}{2}y^2\varsigma^2 u'' - y\big(\alpha - \lambda\varsigma\rho - y\varsigma^2 r\gamma(1-\rho^2)f'\big)u' + \frac{1}{2}\varsigma^2 r\gamma y^2(1-\rho^2)u'^2 + ru \\ \quad \geq -\delta_2 y + rK, \\ u \geq m, \\ (u-m)\big[-\frac{1}{2}y^2\varsigma^2 u'' - y\big(\alpha - \lambda\varsigma\rho - y\varsigma^2 r\gamma(1-\rho^2)f'\big)u' \\ \quad + \frac{1}{2}\varsigma^2 r\gamma y^2(1-\rho^2)u'^2 + ru + \delta_2 y - rK\big] = 0, \\ u(0) = K. \end{cases} \tag{50}$$

In (50), the function $m = q(y) - f(y)$ is the solution of the problem

$$\begin{cases} -\frac{1}{2}y^2\varsigma^2 m'' - y\big(\alpha - \lambda\varsigma\rho - y\varsigma^2 r\gamma(1-\rho^2)f'\big)m' + \frac{1}{2}\varsigma^2 r\gamma y^2(1-\rho^2)m'^2 + ry \\ \quad = (\delta_1 - \delta_2)y, \quad 0 < y < \hat{y}, \\ m(0) = m(\hat{y}) = 0, \end{cases} \tag{51}$$

and $m(y)$ is extended by 0 for $y > \hat{y}$. The function m is continuous but its derivative is discontinuous at $\hat{y}$. The difficulty is that one cannot interpret $u(y)$ as the value function of a control problem. Instead, it is, more appropriately, the value function of a stochastic differential game.

Theorem 5 *We assume* $\frac{r-\alpha+\lambda\varsigma\rho}{\gamma\varsigma^2(1-\rho^2)} > \delta_1\hat{y}$. *There exists a unique* $u(y) \in C^1(0,\infty)$, *piecewise* C^2, *solving* (50). *This function vanishes for y sufficiently large. Moreover, it is the value function given by*

$$u(y) = \inf_{v(\cdot)}\sup_{\theta} J_y\big(v(\cdot),\theta\big) \tag{52}$$

with the controlled diffusion and objective function given by

$$\begin{cases} dY_y(t) = Y_y(t)\big(\alpha - \lambda\varsigma\rho - \varsigma^2 r\gamma(1-\rho^2)Y_y(t)f'(Y_y(t)) \\ \qquad\qquad + v(t)\varsigma\sqrt{r\gamma(1-\rho^2)}\big)dt + \varsigma Y_y(t)dW(t), \\ Y_y(0) = y, \\ J_y\big(v(\cdot),\theta\big) = \mathrm{E}\big[\int_0^{\theta^0\wedge\theta}\big(-\delta_2 Y_y(t) + rK + \frac{1}{2}v^2(t)\big)e^{-rt}dt + Ke^{-r\theta^0}\mathbf{1}_{\theta^0<\theta} \\ \qquad\qquad + m\big(Y_y(\theta)\big)e^{-r\theta}\mathbf{1}_{\theta<\theta^0}\big], \end{cases} \tag{53}$$

where $\theta^0 = \inf\{t : Y_y(t) = 0\}$, *and m is the solution of* (51) *extended by zero for* $y > \hat{y}$.[6]

We next state that the solution of (50) is characterized by two intervals.

[6]For $v(t)$, we can take the same class as in problem (17)–(18).

Theorem 6 *The solution* $u(y)$ *of* (50) *is of the form*

$$-\frac{1}{2}y^2\varsigma^2u'' - y\big(\alpha - \lambda\varsigma\rho - y\varsigma^2 r\gamma(1-\rho^2)f'\big)u' + \frac{1}{2}\varsigma^2 r\gamma y^2(1-\rho^2)u'^2 + ru$$
$$= -\delta_2 y + rK, \quad 0 < y < y_1 \quad \textit{and} \quad y_2 < y < y_3, \tag{54}$$

with the value matching and smooth pasting conditions:

$$\begin{cases} u(y_1) = m(y_1), & u'(y_1) = m'(y_1), \\ u(y_2) = m(y_2), & u'(y_2) = m'(y_2), \\ u(y_3) = 0, & u'(y_3) = 0, \end{cases} \tag{55}$$

where $m(y) = g(y) - f(y)$, *the solution of* (51) *and extended by* 0 *for* $y > \hat{y}$. *There exists a unique triple* $y_1,\ y_2,\ y_3$ *with* $0 < y_1 < y_2 < \hat{y} < y_3$ *such that* (54), (55) *hold.*

Proof We know that u, the solution to (50), vanishes for $\bar{y} > \hat{y}$, $\bar{y}$ sufficiently large. Since $u(0) > m(0)$ and $u(\bar{y}) = m(\bar{y}) = 0$, there exists a first point $y_1 < \bar{y}$ such that $u(y_1) = m(y_1)$. We must have $y_1 < \hat{y}$. Otherwise, $y_1 = \bar{y}$ and u coincides with the solution of (25), i.e., the same system (50) with $m = 0$. But then $\bar{y} = \hat{y}$, hence $y_1 = \hat{y}$. In this case, $\tilde{u} = u - m$ satisfies the equation

$$-\frac{1}{2}y^2\varsigma^2\tilde{u}'' - y\big(\alpha - \lambda\varsigma\rho - y(f' + m')\varsigma^2 r\gamma(1-\rho^2)\big)\tilde{u}'$$
$$+ \frac{1}{2}y^2\varsigma^2 r\gamma(1-\rho^2)(\tilde{u}')^2 + r\tilde{u} = -\delta_1 y + rK \tag{56}$$

with the boundary conditions

$$\tilde{u}(0) = K, \qquad \tilde{u}(\hat{y}) = 0,$$

and since $u'(\hat{y}) = 0$, $\tilde{u}'(\hat{y} - 0) = -m'(\hat{y} - 0)$ which implies $\tilde{u}'(\hat{y} - 0) > 0$. It follows that $\tilde{u}(y) < 0$ for y close to $\hat{y}$, which is impossible since it must be positive. Therefore, $y_1 < \hat{y}$. We claim also that $\delta_1 y_1 \geq rK$. Indeed, set $\tilde{u}(y) = u(y) - m(y)$, then it satisfies (56) with the boundary conditions

$$\tilde{u}(0) = K, \qquad \tilde{u}(y_1) = 0, \qquad \tilde{u}'(y_1) = 0. \tag{57}$$

The matching of the derivatives comes from the fact that $\tilde{u}(y)$ is C^1 and $\tilde{u}(y) > 0$, $\tilde{u}(y_1) = 0$. So y_1 is a local minimum, hence $\tilde{u}'(y_1) = 0$.

Suppose $\delta_1 y_1 < rK$, then using (56), we see that $\tilde{u}''(y_1 - 0) < 0$; hence, $\tilde{u}(y) < 0$ for $y < y_1$, close to y_1. This is impossible.

Since $u(\hat{y}) > m(\hat{y}) = 0$, there exists an interval in which $\hat{y}$ is contained and such that the equation holds on this interval. One of the extremities of this interval is $y_3 = \bar{y}$. Call y_2 the other extremity, such that $u(y_2) = m(y_2)$. Therefore, $y_1 \leq y_2 < \hat{y}$. Necessarily, $y_2 > y_1$. Otherwise, u will be the solution of the equation on $(0, y_3)$, which is the case studied at the beginning of the proof, which is impossible. But then we have $u(y_2) = m(y_2),\ u'(y_2) = m'(y_2)$.

On the other hand, on the interval (y_1, y_2), m satisfies (51) and the right-hand side $(\delta_1 - \delta_2)y > -\delta_2 y + rK$, since $\delta_1 y > rK$, by virtue of $y > y_1$ and $\delta_1 y_1 > rK$.

Thus, m satisfies all conditions on $(y_1,\ y_2)$. Therefore, $u = m$ on $(y_1,\ y_2)$. By the uniqueness of u (Theorem 5), the triple $y_1,\ y_2,\ y_3$ is necessarily unique. □

We note the property that $u(y) \geq -f(y) + K$, which implies $h(y) \geq 0$. It remains to show that $L(x, y)$ defined by (48) is the value function (45).

Theorem 7 *The function* $L(x, y)$ *defined by* (48) *coincides with the value function* (44).

3.2 *Leader's Optimal Stopping Rule*

The optimal stopping rule for the leader is defined as:

$$\hat{\theta}(y) = \begin{cases} \inf\{t : Y_y(t) \geq y_1\}, & \text{if } 0 < y < y_1, \\ 0, & \text{if } y_1 \leq y \leq y_2, \\ \inf\{t : Y_y(t) \leq y_2 \text{ or } Y_y(t) \geq y_3\}, & \text{if } y_2 < y < y_3, \\ 0, & \text{if } y \geq y_3, \end{cases} \tag{58}$$

where $Y_y(t)$ is the process defined in (2).

4 Conclusion

We study a problem similar to the one presented in Bensoussan et al. [2]. Although we consider the investment payoffs governed by a geometric Brownian motion dynamics like the lump-sum payoff case in Bensoussan et al. [2], we do not encounter additional regularity issues encountered in the lump-sum payoff case, which results from indifference consideration for overcoming the comparison of gains and losses at different times in the incomplete markets. On the contrary, we are able to characterize a two-interval solution for the leader's optimal investment rule as the arithmetic Brownian motion cashflow payoff case presented in Bensoussan et al. [2]. The choice of a geometric Brownian motion cashflow process is motivated by the specification of an uncertain payoff arising from a stochastic demand process for the project's output, common in the financial economics literature (see, for example, Dixit and Pindyck [3] and Grenadier [4]). We note that to study cashflow process in terms of a geometric Brownian motion process rather than an arithmetic Brownian motion process invokes additional nontrivial mathematical consideration. Comparing with the arithmetic Brownian motion cashflow payoff case, the current study requires additional absorbing barrier consideration as well as an additional intermediate study of non-linear 2nd order differential equation, which turns out to be a solution to a minimization problem.

The economic interpretation of the leader's two-interval solution for the Stackelberg game is interesting. Below the lower threshold, neither player will invest because the output value is too low. Above the upper threshold, both players invest as soon as possible because output value is very high. Around the middle threshold, output value is attractive to the follower, who invests as soon as possible. As a result the leader will have little or no time to exploit their monopoly position in the output market. Since output value is below the upper threshold, the leader prefers to invest at a lower threshold value, thus decreasing the follower's interest. This allows the leader to maintain a monopoly position in the output market for a longer time. This result, understandable but not necessarily intuitive, can be revealed only through the mathematics of the V.I.

Acknowledgement The first author acknowledges support of National Science Foundation DMS-1303775 and of the Research Grants Council of HKSAR (CityU 500113).

References

1. Bensoussan, A.: Applications of Variational Inequalities in Stochastic Control. Elsevier/North-Holland, Amsterdam (1978)
2. Bensoussan, A., Diltz, J.D., Hoe, S.: Real options games in complete and incomplete markets with several decision makers. SIAM J. Financ. Math. **1**(1), 666–728 (2010)
3. Dixit, A., Pindyck, R.S.: Investment Under Uncertainty. Princeton University Press, Princeton (1994)
4. Grenadier, S.: The strategic exercise of options: development cascades and overbuilding in real estate markets. J. Finance **51**(5), 1653–1679 (1996)
5. Kinderlenher, D., Stampacchia, G.: An Introduction to Variational Inequalities and Their Applications. Academic Press, San Diego (1980)

Dynamic Hedging of Counterparty Exposure

Tomasz R. Bielecki and Stéphane Crépey

Abstract We study mathematical aspects of dynamic hedging of Credit Valuation Adjustment (CVA) in a portfolio of OTC financial derivatives. Since the sub-prime crisis, the counterparty risk and the wrong way risk are crucial issues in connection with valuation and risk management of credit derivatives. In this work we first derive a general model-free equation for the dynamics of the CVA of a portfolio of OTC derivatives. We then particularize these dynamics to the counterparty risk of a portfolio of credit derivatives including, for instance, CDSs and/or CDOs, possibly netted and collateralized, considered in the so-called Markovian copula model. Wrong way risk is represented in the model by the possibility of simultaneous defaults. We establish a rigorous connection between the CVA, which represents the price of the counterparty risk, and a suitable notion of Expected Positive Exposure (EPE). Specifically, the EPE emerges as the key ingredient of the min-variance hedging ratio of the CVA by a CDS on the counterparty. Related notions of EPE have actually long been used in an ad-hoc way by practitioners for hedging their CVA. Our analysis thus justifies rigorously this market practice, making also precise the proper definition of the EPE which should be used in this regard, and the way in which the EPE should be used in the hedging strategy.

Keywords Counterparty risk · Credit risk · Credit Valuation Adjustment · Expected Positive Exposure · Collateralization · Markov copula · Joint defaults · Hedging

Mathematics Subject Classification (2010) 91G80

T.R. Bielecki (✉)
Department of Applied Mathematics, Illinois Institute of Technology, Chicago, IL 60616, USA
e-mail: bielecki@iit.edu

S. Crépey
Equipe Analyse et Probabilité, Université d'Évry Val d'Essonne, 91025 Évry Cedex, France
e-mail: stephane.crepey@univ-evry.fr

Y. Kabanov et al. (eds.), *Inspired by Finance*, DOI 10.1007/978-3-319-02069-3_3,

1 Introduction

Counterparty risk is the most primitive risk in any financial contract involving cash-flows/liabilities distributed over time. This is the risk that the future contractual obligations will not be fulfilled by at least one of the two parties to such a financial contract.

There has been a lot of research activity in the recent years devoted to valuation of counter-party risk (we refer to [1] for a comprehensive survey of literature). In contrast, almost no attention has been payed to quantitative studies of the problem of dynamic hedging of this form of risk. There is some discussion devoted to dynamic hedging of counterparty exposure in Cesari et al. [10] and in Gregory [15].

In this paper we build upon the model developed in [1] for the purpose of valuation of CVA, and we present formal mathematical results that provide an analytical basis for the quantitative methodology of dynamic hedging of counterparty risk.

In Sects. 2 and 3 we recall and give new ramifications to the general CVA results of [1], integrating to the set-up important practical notions related to the modeling of the collateral. This is a key counterparty risk modeling issue since, for instance, AIG's bailout was largely triggered by its inability to face increasing margin calls on its sell-protection CDS positions (on the distressed Lehman in particular). In Sect. 4 we present a variant of the common shocks portfolio credit risk model of [2], more specifically tailored to the application of valuation and hedging of the counterparty risk on a portfolio of credit derivatives. We proceed, in Sect. 5, with a mathematical study of dynamic hedging of counterparty risk on a portfolio of credit derivatives, in the common shocks model of Sect. 4. In particular, we provide a formula for the risk-neutral min-variance delta of the portfolio CVA with respect to a counterparty clean CDS on the counterparty which is used to hedge the counterparty's jump-to-default exposure component of the CVA. Notably, we establish the connection between this delta, and a suitable notion of Expected Positive Exposure (EPE), providing ground to the market intuition of using EPE to hedge CVA. We make precise the proper definition of the EPE which should be used in this regard, and the way in which EPE should be used in the hedging strategy. Implementation issues and numerics will be considered in a follow-up paper.

1.1 General Set-up

We consider two parties of a financial contract. We call them *the investor* and *the counterparty*. We denote by τ_{-1} and τ_0 the default times of the investor and of the counterparty, respectively. In [1] (see also [12]) we studied the problem of valuation of the unilateral counterparty risk (as seen from the perspective of the investor, i.e. $\tau_{-1} = \infty$ and $\tau_0 < \infty$), as well as valuation of the bilateral counterparty risk (i.e. $\tau_{-1} < \infty$ and $\tau_0 < \infty$). In particular, we formulated various ways to represent and to compute the counterparty value adjustment (CVA).

Here we focus on the problem of dynamic hedging of the counterparty risk. CVA can be thought of as the price of an exotic derivative, sometimes referred to as the contingent credit default swap (CCDS, see e.g. [10], [15]). In this paper, by hedging of the counterparty risk, we shall mean dynamic hedging of CVA (or, dynamic hedging of the corresponding CCDS).

We start by recalling from [1] a general representation formula for bilateral counterparty risk valuation adjustment, for a fully netted and collateralized portfolio of contracts between the investor and his/her counterparty. This result can be considered as general since, for any partition of a portfolio into netted sub-portfolios, the results of this section may be applied separately to every sub-portfolio. The exposure at the portfolio level is then simply derived as the sum of the exposures of the sub-portfolios. Moreover, this holds for a general portfolio, not necessarily made of credit derivatives.

It needs to be emphasized that we do not exclude simultaneous defaults of the investor and his/her counterparty, since in Sects. 4–5, we shall actually use simultaneous defaults, in the manner of [1], to implement defaults dependence and wrong way risk. We do assume however that the default times cannot occur at fixed times, which is for instance satisfied in all the intensity models of credit risk.

For $i = -1$ or 0, representing the two counterparties, let H^i stand for the default indicator processes of τ_i, so $H^i_t = \mathbb{1}_{\tau_i \le t}$. By default time, we mean the effective default time in the sense of the time at which promised dividends and margin calls, cease to be paid by the distressed party. We also denote $\tau = \tau_{-1} \wedge \tau_0$, with related default indicator process denoted by H. In the case where unilateral counterparty risk is considered, one simply sets $\tau_{-1} = \infty$, so in this case $\tau = \tau_0$. We fix the portfolio time horizon $T \in \mathbb{R}_+$, and we fix an underlying risk-neutral pricing model $(\Omega, \mathbb{F}, \mathbb{P})$ such that τ_{-1} and τ_0 are $\mathbb{F}$-stopping times. All processes are $\mathbb{F}$-adapted.[1]

We assume that all the random times are $[0, T] \cup \{\infty\}$-valued. We denote by $\mathbb{E}_\theta$ the conditional expectation under $\mathbb{P}$ given A_θ, for any $\mathbb{F}$-stopping time θ. All the cash flows and prices (mark-to-market values of cash flows) are considered from the perspective of the investor. In accordance with the usual convention regarding *ex-dividend* valuation, $\int_a^b$ is to be understood as $\int_{(a,b]}$, so in particular $\int_a^b = 0$ whenever $a \ge b$.

In the rest of the paper, β will denote a finite variation and continuous *risk-free discount factor* process.

2 Cashflows

We let D and $\mathscr{D}$ represent, respectively, the counterparty clean and the counterparty risky cumulative dividend processes of the portfolio over the time horizon $[0, T]$, assumed to be of finite variation. For future convenience, we extend these processes to the interval $[0, \infty]$ by constancy, that is setting them equal to D_T and $\mathscr{D}_{T+\delta}$ on the intervals $(T, \infty]$ and $(T + \delta, \infty]$, respectively.

[1] See Remark 3.2 for filtration issues.

By *counterparty clean cumulative dividend process* we mean the cumulative dividend process that does not account for the counterparty risk, whereas by *counterparty risky cumulative dividend process* we mean the cumulative dividend process that does account for the counterparty risk.

We shall consider collateralized portfolios. In this regard we shall consider a *cumulative margin process* and we shall assume that no lump margin cash-flow can be asked for at time τ. Accordingly, given a finite variation cumulative margin process γ, we define the *cumulative discounted margin process* by

$$\beta\Gamma = \int_{[0,\cdot)} \beta_t(1 - H_t)d\gamma_t \,. \tag{1}$$

So, in particular, $\Gamma_0 = \gamma_0 - \gamma_{0-}$, and one has for $\tau < \infty$,

$$\beta_\tau \Gamma_\tau = \int_{[0,\tau)} \beta_t d\gamma_t.$$

In our notation the collateral process Γ is the algebraic amount given to the investor -1 by the counterparty 0 at time τ. Thus, a positive Γ_t means cash and/or collateral assets already transferred to the account of the investor but still owned by the counterparty.[2] These funds will actually become property of the investor in case of default of the counterparty at time τ. It is worth stressing that, according to industry standards, in the case of default of the investor at time τ, these funds will also become property of the investor, unless a special segregation procedure is in force (see Sect. 2.1). Symmetric remarks apply to negative Γ_t (swap the roles of the counterparty and investor in the above description).

Three reference collateralization schemes are the naked scheme $\Gamma = 0$, and the so-called perfect scheme and ISDA scheme to be defined in Sect. 3.2.

We assume for notational simplicity that γ and Γ are killed at T (so $\Gamma_t = \gamma_t = 0$ for $t \geq T$) and we define an $\mathscr{F}_\tau$-measurable random variable χ as

$$\chi = P_{(\tau)} + \Delta D_\tau - \Gamma_\tau, \tag{2}$$

in which, for $\tau < \infty$, $\Delta D_\tau = D_\tau - D_{\tau-}$ denotes the jump of D at τ, and where the so called *legal value* $P_{(\tau)}$ is a A_τ-measurable random variable representing the 'fair value', in a sense to be agreed upon between the two parties at the contract's inception, of the portfolio at time τ.

From the point of view of financial interpretation, χ represents the (*algebraic*) *debt* of the counterparty to the investor at the first time of default τ of either party, accounting for the legal value of the portfolio at that time, plus any bullet dividend which should be paid at time τ by the counterparty to the investor, less the margin amount Γ_τ which is already in the hands of the counterparty (cf. the term Γ_τ in the first line of Eq. (3) below).

---

[2]Consequently, any cash flows, such as dividends paid by the collateral assets, are thus channeled back to the counterparty.

Let D^* denote the dividend process corresponding to the cash flows of D 'stopped at $\tau-$', that is

$$D^* = (1 - H)D + HD_{\tau-}.$$

We model the counterparty risky portfolio cumulative dividend process as

$$\begin{aligned}\mathscr{D} = D^* + \mathbb{1}_{\tau<T}\big(\Gamma_\tau H + (R_0\chi^+ - \chi^-)[H, H^0] \\ - (R_{-1}\chi^- - \chi^+)[H, H^{-1}] - \chi[[H, H^0], H^{-1}]\big),\end{aligned} \quad (3)$$

where in the *close out cash-flow* corresponding to the second line of (3), the [0, 1]-valued A_{τ_0}- and $A_{\tau_{-1}}$-measurable random variables R_0 and R_{-1}, respectively denote the recovery rates of the investor and of its counterparty upon default, and $[\cdot, \cdot]$ is the covariation process, which in the present case of the default indicator processes, reduces to the indicator process of the simultaneous defaults.

So, if the investor defaults first at time $\tau_{-1} < \tau_0$, then, at time $\tau = \tau_{-1}$, the close out cash-flow takes place in the amount of $\Gamma_\tau - (R_{-1}\chi^- - \chi^+)$; if the investor's counterparty defaults first at time $\tau_0 < \tau_{-1}$, then, at time $\tau = \tau_0$, the close out cash-flow takes place in the amount of $\Gamma_\tau + R_0\chi^+ - \chi^-$; if the investor and the counterparty default simultaneously at time $\tau_0 = \tau_{-1}T$, then, at time $\tau = \tau_0 = \tau_{-1}$ the close out cash-flow takes place in the amount of $\Gamma_\tau + R_0\chi^+ - R_{-1}\chi^-$.

2.1 Re-hypothecation Risk and Segregation

Re-hypothecation refers to the possibility for the investor (the symmetric issue arises relatively to her counterparty) to use as collateral, assets that were already posted to her as collateral, in the context of another transaction with a third party. In this case, setting-up the collateral is at no cost for the investor, and can even be beneficial in certain cases. This explains the popularity of re-hypothecation among market participants. But, on the other hand, re-hypothecation raises a new counterparty risk, namely, the risk of not getting back one's collateral at time where this should be the case (one's position having appreciated), because the collateral that one has posted to a counterparty is not by this counterparty anymore, but stuck to some third party to which it has been re-hypothecated. This practically means that the counterparty defaults at this time, and that the colletaral is then lost, up to a fractional recovery.

However, in practice, the collateral is typically kept in a segregated, third-party account, and under certain collateral conventions, clauses are that, should the counterparty default first at time $\tau = \tau_0 < \tau_{-1}$ and the investor be in-the money at that time because of the collateralization scheme, so $\chi = P_{(\tau)} + \Delta D_\tau - \Gamma_\tau > 0$ but $\chi^0 := P_{(\tau)} + \Delta D_\tau < 0$, then the investor will be fully compensated on the segregated collateral and will incur no loss at default in this case (see Durand and Rutkowski [14]).

This means in this case that the collateral posted in excess by the investor will be returned to her, and that the close out cashflow will be $P_{(\tau)} + \Delta D_\tau$, instead of $\Gamma_\tau H + (R_0\chi^+ - \chi^-) < P_{(\tau)} + \Delta D_\tau$ (assuming a 'nominal' recovery rate $R_0 < 1$).

Note that this can be accounted for in the above formalism, by working with an 'effective' (as opposed to nominal) recovery rate R_0 of the counterparty, equal to one on the event that $P_{(\tau)} + \Delta D_\tau$ is negative.

Segregation in this sense thus eliminates the investor's re-hypothecation risk.

Likewise, the symmetric case regarding the counterparty can be accounted for by letting an effective recovery rate R_{-1} be equal to one on the event that $P_{(\tau)} + \Delta D_\tau$ is positive, to the effect of eliminating the counterparty's re-hypothecation risk.

2.2 Cure Period

In practice there is a time lag $\delta > 0$, called the cure period, and typically taken to be $\delta =$ two weeks, between the default time τ and the close out cash flow, which thus occurs at time $t + \delta$. The exact interpretation of the cure period depends on the CSA (Credit Support Annex) which is in force regarding the particular portfolio at hand.

More generally, one calls *margin period of risk*, the time lag between the last margin call preceding τ, and the time $\tau + \delta$ of the close-out cash flow. The cure period thus constitutes the second part of the margin period of risk, the first part of the margin period of risk consisting of the time lag between the default time τ and the last margin call preceding it. These two components of the margin period of risk play rather distinct roles in the modeling. The role of the first component will be analyzed in Sect. 3.2.

Let $\hat{H}_t = \mathbb{1}_{t \geq \tau+\delta}$, and let similar notations $\hat{H}^0$ and $\hat{H}^{-1}$ hold for H^0 and H^{-1}.

In a first interpretation, the cure period accounts for the time that is needed to liquidate collateral assets in case of the default of one of the two parties, so

$$\begin{aligned}\mathscr{D} &= D^* + \mathbb{1}_{\tau<T}\big(\Gamma_\tau \hat{H} + \big(R_0\chi^+ - \chi^-\big)[\hat{H}, \hat{H}^0] \\ &\quad - \big(R_{-1}\chi^- - \chi^+\big)[\hat{H}, \hat{H}^{-1}] - \chi[[\hat{H}, \hat{H}^0], \hat{H}^{-1}]\big) \\ &=: D^* + \Upsilon. \end{aligned} \tag{4}$$

For example, if the investor defaults first at time $\tau = \tau_{-1} < \tau_0 \wedge T$, then, at time $\tau + \delta$ the close out cash-flow takes place in the amount of $\Gamma_\tau - (R_{-1}\chi^- - \chi^+)$.

In a second interpretation, the cure period represents a time period δ between the 'effective' default time τ in the sense of the time at which promised dividends and margin calls actually cease to be paid by the distressed party, and the 'legal' default time $\tau + \delta$ of the close-out cashflow (whereas the effective and the legal default time are both equal to τ in the first interpretation). The counterparty risky cash-flows are thus still given by (4), but for χ in (4) now given, instead of (2), by

$$\beta_{\tau+\delta}\chi = \beta_{\tau+\delta} P_{(\tau+\delta)} + \int_{[\tau,\tau+\delta]} \beta_t dD_t - \beta_{\tau+\delta}\Gamma_\tau, \tag{5}$$

for an $\mathcal{A}_{\tau+\delta}$-measurable legal value $P_{(\tau+\delta)}$. Also, the recoveries R_0 and R_{-1} are now given as $\mathcal{A}_{\tau_0+\delta}$- and $\mathcal{A}_{\tau_{-1}+\delta}$-measurable random variables (instead of $\mathcal{A}_{\tau_0}$- and $\mathcal{A}_{\tau_{-1}}$-measurable previously).

For example, if the investor stops payments at time $\tau = \tau_{-1} < \tau_0$, then, at time $\tau + \delta$ the close out cash-flow takes place in the amount of $\Gamma_\tau - (R_{-1}\chi^- - \chi^+)$, for χ therein given by (5).

Of course, in the case $\delta = 0$, both interpretations reduce to the above no-cure-period case.

3 Pricing

The definitions below are consistent with the standard theory of arbitrage (cf. [13]).

Definition 3.1 (i) The *counterparty clean price process, or counterparty clean mark-to-market process*, of the portfolio, is given by $P_t = \mathbb{E}_t[p^t]$, where the random variable $\beta_t p^t$ represents the *cumulative discounted cash flows of the portfolio* on the time interval $(t, T]$, not accounting for counterparty risk. So, for $t \in [0, T]$,

$$\beta_t p^t = \int_t^T \beta_s dD_s. \tag{6}$$

The *cumulative counterparty clean value process* of the portfolio is given by

$$\widehat{P}_t = P_t + p_t, \tag{7}$$

where p_t represents the discounted cumulative dividend process up to time t, so

$$\beta_t p_t = \int_0^t \beta_s dD_s. \tag{8}$$

(ii) The *counterparty risky mark-to-market* process of the portfolio is given by $\Pi_t = \mathbb{E}_t[\pi^t]$, where the random variable π^t represents the *cumulative discounted cash flows of the portfolio adjusted for the counterparty risk* on the time interval $(t, T]$. So, for $t \in [0, T]$,

$$\beta_t \pi^t = \int_t^T \beta_s d\mathscr{D}_s. \tag{9}$$

The *cumulative counterparty risky price process* of the portfolio is given by

$$\widehat{\Pi}_t = \Pi_t + \pi_t, \tag{10}$$

where

$$\beta_t \pi_t = \int_0^t \beta_s d\mathscr{D}_s. \tag{11}$$

Recall $\tau = \tau_{-1} \wedge \tau_0$, $H_t = \mathbb{1}_{\tau \le t}$. In the counterparty risky case there are no cash flows after $\tau \wedge T$, so the (A_T-measurable) random variable π^t is in fact $A_{\tau \wedge T}$-measurable, and one has that $\pi^t = \Pi_t = 0$ for $t \ge \tau \wedge T$.

Remark 3.2 In principle, when dealing with CVA, one should consider not one but two filtered pricing models relatively to a given risk-neutral measure $\mathbb{P}$: $(\Omega, \mathbb{F}, \mathbb{P})$ and $(\Omega, \widetilde{\mathbb{F}}, \mathbb{P})$. Here the filtration $\widetilde{\mathbb{F}} = (\widetilde{A}_t)_{t \in [0,T]}$ would represent the counterparty risk free filtration, not carrying any direct information about the default times τ_{-1} and τ_0, nor about any factors that might be specific to evolution of credit standards ('ratings') of the counterparties. This is the proper filtration that would normally be used for pricing the counterparty risk free contracts,[3] which serve as a reference so to assess the counterparty riskiness of actual contracts being priced and hedged. Mathematically speaking, we have that $\sigma(\tau_{-1} \wedge t) \nsubseteq \widetilde{A}_t$ and $\sigma(\tau_0 \wedge t) \nsubseteq \widetilde{\mathscr{F}}_t$. The filtration $\mathbb{F} = (A_t)_{t \in [0,T]}$ would represent the counterparty risky filtration, and it is a filtration such that $\widetilde{A}_t \vee \sigma(\tau_{-1} \wedge t) \vee \sigma(\tau_0 \wedge t) \subseteq A_t$.

The discount factor β, the counterparty clean cumulative dividend process D and the cumulative margin process γ, would thus be assumed to be $\widetilde{\mathbb{F}}$ adapted, and the *counterparty clean price process (or counterparty clean mark-to-market process)* of the portfolio would be given by $P_t = \widetilde{\mathbb{E}}_t[p^t]$, where we denote by $\widetilde{\mathbb{E}}_t$ the conditional expectation under $\mathbb{P}$ given $\widetilde{A}_t$.

The fact that we only work with one, counterparty risky, filtration $\mathbb{F}$ in this paper, as, incidentally, is the case with all the counterparty risk literature that we know of, simply means that we work under the implicit assumption that, for any $t \le T$, the time-t price of any $\widetilde{A}_T$ measurable, integrable cumulative cash flow, can be computed by evaluating appropriate conditional expectation (under $\mathbb{P}$) either conditioned on $\widetilde{A}_t$ or conditioned on A_t. See discussion in [9].

It thus holds for instance that $P_t = \widetilde{\mathbb{E}}_t[p^t] = \mathbb{E}_t[p^t]$. This property implies, in particular, that process $\beta\widehat{P}$ is an $\widetilde{\mathbb{F}}$- as well as an $\mathbb{F}$-martingale, whereas the process $\beta\widehat{\Pi}$ is only an $\mathbb{F}$-martingale.

Models of this type are for example models where so called immersion property is satisfied between filtrations $\widetilde{\mathbb{F}}$ and $\mathbb{F}$ (see [7] for a general reference). Another example is provided by *Markov copula models* [5, 6] such as the one to be considered in Sect. 4.

3.1 CVA

We introduce now the (cumulative) CVA process on the time interval $[0, \tau \wedge T]$ (we do not define the CVA beyond $\tau \wedge T$ since it is not needed there).

---

[3] We emphesize again that the clean price process P above is the process of the 'clean contract,' that is the contract in which any counterparty risk is disregarded.

Definition 3.3 The CVA process Θ is given as, for $t \in [0, \tau \wedge T]$,

$$\Theta_t = \hat{P}_t - \hat{\Pi}_t \, .$$

Lemma 3.1 *The martingale $\beta\Theta$ can be represented as, for every $t \in [0, \tau \wedge T]$:*

$$\beta_t \Theta_t = \mathbb{E}_t\big[\beta_{\tau+\delta} \mathbb{1}_{\tau<T} \xi\big], \tag{12}$$

where:

(i) *In the case $\delta = 0$ (no cure period),*

$$\xi = P_\tau - P_{(\tau)} + (1 - R_0)\mathbb{1}_{\tau=\tau_0}\chi^+ - (1 - R_{-1})\mathbb{1}_{\tau=\tau_{-1}}\chi^-;$$

(ii) *In the first interpretation of a cure period δ,*

$$\xi = \beta^{-1}_{\tau+\delta}\beta_\tau \left(P_\tau + \Delta D_\tau - B(\tau, \tau+\delta)\left(\Gamma_\tau + \mathbb{1}_{\tau=\tau_0}\big(R_0\chi^+ - \chi^-\big) - \mathbb{1}_{\tau=\tau_{-1}}\big(R_{-1}\chi^- - \chi^+\big) - \mathbb{1}_{\tau_0=\tau_{-1}}\chi\right)\right),$$

where $B(s,t)$ is the time-s price of zero coupon bond expiring at time t;

(iii) *In the second interpretation of a cure period δ,*

$$\xi = P_{\tau+\delta} - P_{(\tau+\delta)} + (1 - R_0)\mathbb{1}_{\tau=\tau_0}\chi^+ - (1 - R_{-1})\mathbb{1}_{\tau=\tau_{-1}}\chi^-$$

with χ as of (5) *therein.*

Proof (i) See [1].

(ii) First observe that for $t \in [0, \tau \wedge T]$ we have (recalling that $dD_t = 0$ for $t > T$, so that $P_t = 0$ and $\Delta D_t = 0$ for $t > T$)

$$\mathbb{E}_\tau \int_t^\infty \beta_s\big(dD_s - dD^*_s\big) = \mathbb{E}_\tau \int_t^T \beta_s\big(dD_s - dD^*_s\big) = \mathbb{E}_\tau \int_{[\tau,T]} \beta_s dD_s = \beta_\tau\left(P_\tau + \Delta D_\tau\right).$$

Consequently, in the first interpretation of a cure period δ, one has by Definition 3.1 and in view of (4), for $t \in [0, \tau \wedge T]$,

$$\begin{aligned}
\beta_t\Theta_t &= \beta_t(\widehat{P_t} - \widehat{\Pi}_t) \\
&= \mathbb{E}_t\left(\mathbb{E}_\tau\big(dD_s - dD^*_s\big) - \int_t^\infty \beta_s \int_t^\infty \beta_s d\Upsilon_s\right) \\
&= \mathbb{E}_t\left(\mathbb{E}_\tau \int_t^T \beta_s\big(dD_s - dD^*_s\big) - \int_t^\infty \beta_s d\Upsilon_s\right) \\
&= \mathbb{E}_t\Big\{\mathbb{1}_{\tau<T}\beta_\tau(P_\tau + \Delta D_\tau) - \mathbb{1}_{\tau<T}\beta_{\tau+\delta}\big(\Gamma_\tau + \mathbb{1}_{\tau=\tau_0}\big(R_0\chi^+ - \chi^-\big) \\
&\quad - \mathbb{1}_{\tau=\tau_{-1}}\big(R_{-1}\chi^- - \chi^+\big) - \mathbb{1}_{\tau_0=\tau_{-1}}\chi\big)\Big\}
\end{aligned}$$

$$= \mathbb{E}_t\{\mathbb{1}_{\tau<T}\beta_\tau(P_\tau + \Delta D_\tau)\} - \mathbb{E}_t\{\mathbb{1}_{\tau<T}\beta_\tau\big(\Gamma_\tau + \mathbb{1}_{\tau=\tau_0}(R_0\chi^+ - \chi^-)$$
$$- \mathbb{1}_{\tau=\tau_{-1}}(R_{-1}\chi^- - \chi^+) - \mathbb{1}_{\tau_0=\tau_{-1}}\chi\big)\mathbb{E}_\tau\beta_\tau^{-1}\beta_{\tau+\delta}\}$$
$$= \mathbb{E}_t\{\mathbb{1}_{\tau<T}\beta_\tau(P_\tau + \Delta D_\tau)\}$$
$$- \mathbb{E}_t\{\mathbb{1}_{\tau<T}\beta_\tau B(\tau, \tau+\delta)\big(\Gamma_\tau + \mathbb{1}_{\tau=\tau_0}(R_0\chi^+ - \chi^-)$$
$$- \mathbb{1}_{\tau=\tau_{-1}}(R_{-1}\chi^- - \chi^+) - \mathbb{1}_{\tau_0=\tau_{-1}}\chi\big)\}.$$

(iii) In the second interpretation of a cure period δ, the result follows by a straightforward adaptation of the no-cure-period computations of [1]. □

For simplicity we assume henceforth that $\delta = 0$. We also assume that the legal value of the portfolio is given by its counterparty clean value, so $P_{(\tau)} = P_\tau$. This simplifying assumption is common in the counterparty risk literature. Note however that in practice, the quantity $P_{(\tau)}$ should account not only for the clean mark-to-market value of the contract, but also for replacement costs as well as for the systemic risk (via modified funding rates).

Consequently the random variables χ and ξ are the values at time τ of the progressively measurable processes (χ_t) and (ξ_t) defined by, for $t \in [0, T]$,

$$\begin{aligned} \chi_t &= P_t + \Delta D_t - \Gamma_t, \\ \xi_t &= (1 - R_0)\mathbb{1}_{t\ge\tau_0}\chi_t^+ - (1 - R_{-1})\mathbb{1}_{t\ge\tau_{-1}}\chi_t^-. \end{aligned} \tag{13}$$

In the theoretical part of the paper we assume henceforth nil interest rates so that the discount factor β is one. Time-deterministic interest-rates will be used in the numerical part, the extension of all results to constant or time-deterministic interest rates being straightforward (but more cumbersome notationally, especially regarding hedging).

3.1.1 CVA Dynamics

The next step consists in deriving dynamics of the CVA Θ, which, under the current zero interest rates environment, is a martingale over $[0, \tau \wedge T]$.

Lemma 3.2 *For any $t \in [0, \tau \wedge T]$, we have*

$$\begin{aligned} d\Theta_t &= (1 - H_{t-})(d\hat{P}_t - d\hat{\Pi}_t) \\ &= (1 - H_t)(dP_t - d\Pi_t) + (\Delta\hat{P}_\tau - \Delta\hat{\Pi}_\tau)dH_t \\ &= (1 - H_t)(dP_t - d\Pi_t) + (\xi_\tau - \Theta_{\tau-})dH_t \\ &= (1 - H_t)(dP_t - d\Pi_t) + (\xi_t - \Theta_{t-})dH_t. \end{aligned} \tag{14}$$

Proof The first line holds by definition of Θ and by application of Itô's formula. The second one follows from the fact that $p^0 - p^t = \pi^0 - \pi^t$ for any $t < \tau$. The remaining three equalities follow easily. □

Equation (14) is the key to hedging of counterparty risk. The dynamics of Θ splits into the 'pre-counter-party-default' part $(1 - H_t^0)(dP_t - d\Pi_t)$, and the 'at-counter-party-default' part $(\xi_t - \Theta_{t-})dH_t^0$.

3.2 Collateral Modeling

Three reference collateralization schemes are the *naked scheme* $\Gamma = 0$, the *perfect scheme* $\Gamma = P_{\cdot -}$, and the *ISDA scheme* to be studied now. According to ISDA document [16], page 57, the paradigm for the level of collateral amount is the following:

> "Collateral value = (i) the [Collateral Taker]'s Exposure plus (ii) the aggregate of all Independent Amounts applicable to the [Collateral Provider], if any, minus (iii) the aggregate of all Independent Amounts applicable to the Collateral Taker, if any, minus (iv) the [Collateral Provider]'s Threshold"

The 'exposure' in the above terminology refers to the counterparty risk free mark-to-market value of the reference portfolio. Here, we propose an algorithm that is meant to generate the collateral process, which, right after every margin call time, conforms to the above paradigm. That is to say, since there are no 'Independent Amounts' as of items (ii) and (iii) in our set-up, 'Collateral value = Mark-to-Market minus Threshold', where Threshold refers to bounds which are set on the admissible values of χ (so the parties need to adjust the collateral Γ in case χ leaves these bounds).

Towards this end, we denote by $t_0 = 0 < t_1 < \cdots < t_n < T$ the margin call dates. Thus, we assume, as it is done in practice that margin calls are executed according to a discrete tenor of dates. Note that the time interval $\tau - t_\iota$ between the effective default time τ and the last margin call date t_ι preceding it, constitutes the first part of the margin period of risk, the second part consisting of the cure period δ already dealt with in Sect. 2.2.

In order to construct the collateral process we need to introduce the following quantities,

- the nominal threshold for the counterparty: $\bar{\chi}^0 \geq 0$,
- the nominal threshold for the investor: $\bar{\chi}^{-1} \leq 0$,
- the minimum transfer amount for the counterparty: $\varepsilon^0 \geq 0$,
- the minimum transfer amount for the investor: $\varepsilon^{-1} \leq 0$,
- the effective threshold for the counterparty: $\hat{\chi}^0 = \bar{\chi}^0 + \varepsilon^0$,
- the effective threshold for the investor: $\hat{\chi}^{-1} = \bar{\chi}^{-1} + \varepsilon^{-1}$.

In the ISDA collateralization scheme we construct the left continuous, piecewise-constant collateral process Γ by setting $\Gamma_0 = 0$ and by postulating that at every $t_i < \tau$,

$$
\begin{aligned}
\Delta\Gamma_{t_i} := \Gamma_{t_i+} - \Gamma_{t_i} &= \mathbb{1}_{\chi_{t_i} > \hat{\chi}^0}(\chi_{t_i} - \bar{\chi}^0)^+ - \mathbb{1}_{\chi_{t_i} < \hat{\chi}^{-1}}(\chi_{t_i} - \bar{\chi}^{-1})^- - \Delta D_{t_i} \\
&= \mathbb{1}_{\chi_{t_i} > \hat{\chi}^0}(\chi_{t_i} - \bar{\chi}^0) + \mathbb{1}_{\chi_{t_i} < \hat{\chi}^{-1}}(\chi_{t_i} - \bar{\chi}^{-1}) - \Delta D_{t_i}. \qquad (15)
\end{aligned}
$$

Then we let Γ be constant on every interval (t_i, t_{i+1}).

Note that the amount of collateral transferred at the call times according to the above collateralization scheme, that is the quantity $\Delta\Gamma_{t_i}$, satisfies natural properties. For instance, assuming no bullet dividend paid at t_i (so $\Delta D_{t_i} = 0$):

- $\Delta\Gamma_{t_i} > 0$ if the collateralized exposure χ_{t_i} exceeds the counterparty's threshold $\hat{\chi}^0$; this means that at time t_i the investor makes a margin call, and the counterparty delivers $\Delta\Gamma_{t_i}$ worth of (cash) collateral; intuitively, the counterparty thus brings χ_t down to $\bar{\chi}^0$ at t_i+ if it exceeded $\hat{\chi}^0$ at t_i,
- $\Delta\Gamma_{t_i} < 0$ if the collateralized exposure χ_{t_i} is less than the investor's threshold $\hat{\chi}^1$; this means that at time t_i the counterparty makes a margin call, and the investor delivers $-\Delta\Gamma_{t_i}$ worth of (cash) collateral; intuitively, the investor brings it up to $\bar{\chi}^{-1}$ at t_i+ if it was lower than $\hat{\chi}^{-1}$ at t_i,
- $\Delta\Gamma_{t_i} = 0$ if the collateralized exposure χ_{t_i} is within the bounds $[\hat{\chi}^{-1}, \hat{\chi}^0]$; this means that at time t_i no margin call is made and no collateral is transferred by any of the two parties; collateralized exposure remains unadjusted.

More generally, identity (16) in the following result indeed shows that, right after the margin call times, the ISDA collateralization scheme conforms to the requirements of ISDA.

Proposition 3.3 *One has, at every t_i,*

$$\Gamma_{t_i+} = P_{t_i} - \left(\mathbb{1}_{\chi_{t_i} > \hat{\chi}^0}\bar{\chi}^0 + \mathbb{1}_{\chi_{t_i} < \hat{\chi}^{-1}}\bar{\chi}^{-1} + \mathbb{1}_{\hat{\chi}^{-1} \le \chi_{t_i} \le \hat{\chi}^0}\chi_{t_i}\right), \tag{16}$$

or, equivalently,

$$\chi_{t_i+} = \left(\mathbb{1}_{\chi_{t_i} > \hat{\chi}^0}\bar{\chi}^0 + \mathbb{1}_{\chi_{t_i} < \hat{\chi}^{-1}}\bar{\chi}^{-1} + \mathbb{1}_{\hat{\chi}^{-1} \le \chi_{t_i} \le \hat{\chi}^0}\chi_{t_i}\right). \tag{17}$$

In particular,

$$\Gamma_\tau \in [P_{t_\iota} - \hat{\chi}^0, P_{t_\iota} - \hat{\chi}^{-1}], \tag{18}$$

where t_ι denotes the greatest t_i less or equal to τ.

Proof Recall (13): $\chi_t = P_t + \Delta D_t - \Gamma_t$. From (15), one thus has at every t_i,

$$\Gamma_{t_i+} = \Gamma_{t_i} - \left(\mathbb{1}_{\chi_{t_i} > \hat{\chi}^0}\bar{\chi}^0 + \mathbb{1}_{\chi_{t_i} < \hat{\chi}^{-1}}\bar{\chi}^{-1} + \mathbb{1}_{\hat{\chi}^{-1} \le \chi_{t_i} \le \hat{\chi}^0}\chi_{t_i}\right) + \chi_{t_i} - \Delta D_{t_i},$$

which is (16). Now, P does not jump at fixed times, and one has by càdlàg regularity of D that $D = D_{\cdot+} = (D_{\cdot-})_{\cdot+}$, so $(\Delta D)_{\cdot+} = D_{\cdot+} - (D_{\cdot-})_{\cdot+} = 0$. Thus

$$\chi_{t_i+} = \chi_{t_i} - \Delta D_{t_i} - (\Gamma_{t_i+} - \Gamma_{t_i}) = P_{t_i} - \Gamma_{t_i+},$$

hence (16) is equivalent to (17). Finally (18) is an immediate consequence of (16), which implies in particular

$$\Gamma_\tau = P_{t_\iota} - \left(\mathbb{1}_{\chi_{t_\iota} > \hat{\chi}^0}\bar{\chi}^0 + \mathbb{1}_{\chi_{t_\iota} < \hat{\chi}^{-1}}\bar{\chi}^{-1} + \mathbb{1}_{\hat{\chi}^{-1} \le \chi_{t_\iota} \le \hat{\chi}^0}\chi_{t_\iota}\right) \in [P_{t_\iota} - \hat{\chi}^0, P_{t_\iota} - \hat{\chi}^{-1}].$$

□

Note that our construction above is implicitly cash based. Translations to cash from a portfolio of assets needs to be done via haircuts. That is, if the collateral transferred at time t_i is posted in some asset different from cash, then the total value of that asset that needs to be posted is $(1+h_{t_i})\Delta\Gamma_{t_i}$, where h_{t_i} is the appropriate haircut to be applied at time t_i. In case of a portfolio of assets, one distributes $\Delta\Gamma_{t_i}$ among the assets and applies appropriate haircut to each portion.

4 Common Shock Model of Counterparty Credit Risk

4.1 Unilateral Counterparty Credit Risk

We shall more specifically focus henceforth on the issue of counterparty credit risk. We consider a bank that holds a portfolio of credit contracts referencing various credit names. This portfolio is subject to a counterparty credit risk with regard to a single counterparty. A bank typically disregards its own counterparty risk when assessing the counterparty risk of a portfolio with another party. Thus, we are led to considering unilateral counterparty risk from the perspective of the bank.

Towards this end, we postulate that the contracts comprising the portfolio between the investor (the bank) and its counterparty, reference defaultable credit names. We denote by τ_i, for $i \in \mathbb{N}_n^* = \{1, \dots, n\}$, the default times of n credit names underlying the portfolio's contracts. For $i \in \mathbb{N}_n = \{0, 1, \dots, n\}$, we let H^i stand for the default indicator process of τ_i, so $H_t^i = \mathbb{1}_{\tau_i \le t}$, and we denote $\mathbf{H} = (H^i)_{i \in \mathbb{N}_n}$.

More precisely, our final aim is to study the hedge of the unilateral counterparty risk exposure of a portfolio credit derivative by means of a counterparty clean CDS contract referencing the counterparty. We assume that the CDS contracts which are used therein for hedging are entered into with counterparties that are remote from default. So, there is no counterparty risk associated with the hedging instruments.

Let, for $t \in [0, T]$,

$$D_t = \int_{[0,t]} \varphi(\mathbf{H}_s)ds + \phi(\mathbf{H}_t), \qquad D_t^i = \int_{[0,t]} \varphi_i(H_s^i)ds + \phi_i(H_t^i) \tag{19}$$

represent the cumulative cash flow processes of a portfolio credit derivative on all names, and of a single-name credit derivative on name $i \in \mathbb{N}_n$. Here the idea is that φ and φ_i correspond to the fees (also called premium) leg of a swapped credit derivative, with continuous-time premium payments for notational simplicity, whereas ϕ and ϕ_i correspond to the default leg.

A practically important class of portfolio credit derivatives consists of the *portfolio loss derivatives* with the cash flows in (19) depending only on $\mathbf{H}_t = (H_t^i)_{i \in \mathbb{N}_n}$ through the number of defaults $N_t = \sum_{i \in \mathbb{N}_n} H_t^i$ in the portfolio, or $N_t^* = \sum_{i \in \mathbb{N}_n^*} H_t^i$

in a more standard situation of a counterparty not belonging to the pool of credit names underlying the contracted derivative, so

$$\varphi(\mathbf{k}) = \hat{\varphi}(|\mathbf{k}|), \quad \phi(\mathbf{k}) = \hat{\phi}(|\mathbf{k}|) \text{ or } \varphi(\mathbf{k}) = \hat{\varphi}(|\mathbf{k}|^*), \quad \phi(\mathbf{k}) = \hat{\phi}(|\mathbf{k}|^*), \tag{20}$$

where we let $|\mathbf{k}| = \sum_{i \in \mathbb{N}_n} k_i$, $|\mathbf{k}|^* = \sum_{i \in \mathbb{N}_n^*} k_i$, for every $\mathbf{k} = (k_i)_{i \in \mathbb{N}_n} \in \{0, 1\}^n$. For instance, one has in the case of a payer CDO tranche on names 1 to n, with contractual spread Σ and normalized attachment/detachment point L/U:

$$\varphi(\mathbf{k}) = -\Sigma\big(U - L - \phi(\mathbf{k})\big), \quad \phi(\mathbf{k}) = \left((1 - \mathscr{R})\frac{|\mathbf{k}|^*}{n} - L\right)^+ \wedge (U - L) \tag{21}$$

where a constant and homogeneous recovery $\mathscr{R}$ on the underlying CDSs is assumed.

As for single-name credit derivatives, one has in the case of a payer CDS with contractual spread S on name $i \in \mathbb{N}_n$:

$$\varphi_i(k_i) = -S(1 - k_i), \;\; \phi_i(k_i) = (1 - R_i)k_i. \tag{22}$$

Note that in the unilateral counterparty risk case we have $\tau = \tau_0$ and thus $H = H^0$. For simplicity we assume a constant recovery rate in case counterparty defaults; specifically we set $R_0 = R$. By application of (13), one thus has,

$$\Theta_t = \mathbb{E}_t\big[\mathbb{1}_{\tau < T}\xi\big] \tag{23}$$

with

$$\xi = (1 - R)\chi^+, \;\; \chi = P_\tau + \Delta\phi(\mathbf{H}_\tau) - \Gamma_\tau \tag{24}$$

in which $P_t = \mathbb{E}_t(D_T - D_t)$ is the counterparty clean price process of the portfolio credit derivative, and Γ represents the collateral process.

4.2 *Model of Default Times*

We now propose a Markovian model of counterparty credit risk, that will be able to put the above general results to work. This model is a variant of the common shocks portfolio credit risk model of [2], more specifically tailored to the application of valuation and hedging of the counterparty risk on a portfolio of credit derivatives.

In order to describe the defaults we define a certain number m (typically small: a few units) of groups $I_l \subseteq \mathbb{N}_n$, of obligors who are likely to default simultaneously, for $l \in \mathbb{N}_m$. More precisely, the idea is that at every time t, there will be a positive probability that the survivors of the group of obligors I_l (obligors of group I_l still alive at time t) default simultaneously. Let $\mathscr{I} = \{I_0, \dots, I_m\}$, $\mathscr{Y} = \{\{0\}, \dots, \{n\}, I_0, \dots, I_m\}$. Let *group intensity processes* X^Y be given in the form of extended CIR processes as, for every $Y \in \mathscr{Y}$,

$$dX_t^Y = a(b_Y(t) - X_t^Y)dt + c\sqrt{X_t^Y}dW_t^Y, \tag{25}$$

where the Brownian motions $W^{\{i\}}$ s for $0 \le i \le n$ are correlated at the level ρ, and the Brownian motions W^Is for $I \in \mathscr{I}$ are independent between themselves and from everything else. Given $\mathbf{X} = (\mathbf{X}^Y)_{Y \in \mathscr{Y}}$s, we would like a model in which the predictable intensity of a jump of $\mathbf{H} = (H^i)_{i \in \mathbb{N}_n}$ from $\mathbf{H}_{t-} = \mathbf{k}$ to $\mathbf{H}_t = \mathbf{l}$, with $\mathrm{supp}(\mathbf{k}) \subsetneq \mathrm{supp}(\mathbf{l})$ in $\{0, 1\}^{n+1}$, is given by

$$\sum_{\{Y \in \mathscr{Y};\, \mathbf{k}^Y = \mathbf{l}\}} X_t^Y, \tag{26}$$

where $\mathbf{k}^Y$ denotes the vector obtained from $\mathbf{k} = (k_i)_{i \in \mathbb{N}_n}$ by replacing the components k_i, $i \in Y$, by numbers one. The intensity of a jump of $\mathbf{H}$ from $\mathbf{k}$ to $\mathbf{l}$ at time t is thus equal to the sum of the intensities of the groups $Y \in \mathscr{Y}$ such that, if the default of the survivors in group Y occurred at time t, the state of $\mathbf{H}$ would move from $\mathbf{k}$ to $\mathbf{l}$.

To achieve this, we classically construct $\mathbf{H}$ by an $\mathbf{X}$-related change of probability measure, starting from a continuous-time Markov chain with intensity one (see [2, 11]). As a result (see [2, 11]), the pair-process $(\mathbf{X}, \mathbf{H})$ is a Markov process with respect to the filtration $\mathscr{F}$ generated by the Brownian Motion $\mathbf{W}$ and the random measure counting the jumps of $\mathbf{H}$, with infinitesimal generator $\mathscr{A}$ of $(\mathbf{X}, \mathbf{H})$ given as, for $u = u(t, \mathbf{x}, \mathbf{k})$ with $t \in \mathbb{R}_+$, $\mathbf{x} = (\lambda_Y)_{Y \in \mathscr{Y}}$ and $\mathbf{k} = (k_i)_{i \in \mathbb{N}_n}$:

$$\begin{aligned}\mathscr{A}_t u(t, \mathbf{x}, \mathbf{k}) = &\sum_{Y \in \mathscr{Y}} \Big(a(b_Y(t) - \lambda_Y)\partial_{\lambda_Y} u(t, \mathbf{x}, \mathbf{k}) + \frac{1}{2} c^2\, \lambda_Y \partial^2_{\lambda_Y^2} u(t, \mathbf{x}, \mathbf{k}) \Big) \\ &+ \sum_{0 \le i < j \le n} \rho_{i,j}(t) c^2 \sqrt{\lambda_{\{i\}}\lambda_{\{j\}}}\, \partial^2_{\lambda_{\{i\}}, \lambda_{\{j\}}} u(t, \mathbf{x}, \mathbf{k}) \\ &+ \sum_{Y \in \mathscr{Y}} \lambda_Y \delta u^Y(t, \mathbf{x}, \mathbf{k}),\end{aligned} \tag{27}$$

for non-negative constants a, c and non-negative functions $b_Y(t)$s, $[-1, 1]$-valued correlation functions $\rho_{i,j}(t)$, and where we denote, for $Y \in \mathscr{Y}$,

$$\delta u^Y(t, \mathbf{x}, \mathbf{k}) = u(t, \mathbf{x}, \mathbf{k}^Y) - u(t, \mathbf{x}, \mathbf{k}).$$

One also has the following expression for the predictable intensity ℓ_t^Z of the indicator process H_t^Z of the event of a joint default of names in the set Z and only in Z, for every subset Z of $\mathbb{N}_n$ (see [2]):

$$\ell_t^Z = \ell_Z(t, \mathbf{X}_t, \mathbf{H}_{t-}) = \sum_{Y \in \mathscr{Y};\, Y_t = Z} X_t^Y, \tag{28}$$

where Y_t stands for the set of survivors of set Y right before time t, for every $Y \in \mathscr{Y}$. So $Y_t = Y \cap \mathrm{supp}^c(\mathbf{H}_{t-})$, where $\mathrm{supp}^c(\mathbf{k}) = \{i \in \mathbb{N}_n;\, k_i = 0\}$, for $\mathbf{k} = (k_i) \in \{0, 1\}^{n+1}$. One denotes by M^Z the corresponding *compensated set-event martingale*, so for $t \in [0, T]$,

$$dM_t^Z = dH_t^Z - \ell_Z(t, \mathbf{X}_t, \mathbf{H}_t)dt. \tag{29}$$

We refer the reader to [4] for a two-obligors preliminary version of this model dedicated to valuation and hedging of counterparty risk on a CDS. The numerical results of [4] illustrate that using such 'fully stochastic' specifications of the intensities potentially leads to a better behaved CVA than the intensities specification of [2], in which the X^Is are deterministic functions of time.

4.2.1 Markov Copula Properties

Note that the SDEs for factors X^Y have the same coefficients except for the $b_Y(t)$, to the effect that

$$X^i := \sum_{\mathscr{Y}\ni Y\ni i} X^Y = X^{\{i\}} + \sum_{\mathscr{I}\ni I\ni i} X^I ,$$

for $i \in \mathbb{N}_n$, is again an extended CIR process, with parameters a, c and

$$b_i(t) := \sum_{\mathscr{Y}\ni Y\ni i} b_Y(t) = b_{\{i\}}(t) + \sum_{\mathscr{I}\ni I\ni i} b_I(t),$$

driven by the Brownian motion W^i such that

$$\sqrt{X_t^i}dW_t^i = \sum_{Y\ni i}\sqrt{X_t^Y}dW_t^Y, \qquad dW_t^i = \sum_{Y\ni i}\frac{\sqrt{X_t^Y}}{\sqrt{\sum_{Y\ni i} X_t^Y}}dW_t^Y. \tag{30}$$

One can then check, as is done in [2], that the so-called *Markov copula property* holds (see [6]), in the sense that for every $i \in \mathbb{N}_n$, (X^i, H^i) is an $\mathscr{F}$ – Markov process admitting the following generator, for $u_i = u_i(t, \lambda_i, k_i)$ with $(\lambda_i, k_i) \in \mathbb{R} \times \{0, 1\}$:

$$\begin{aligned}&\mathscr{A}_t^i u_i(t,\lambda_i,k_i)\\&\quad = a(b_i(t)-\lambda_i)\partial_{\lambda_i} u_i(t,\lambda_i,k_i) + \frac{1}{2}c^2\,\lambda_i\partial^2_{\lambda_i^2} u_i(t,\lambda_i,k_i)\\&\qquad + \lambda_i\big(u_i(t,\lambda_i,1) - u_i(t,\lambda_i,k_i)\big).\end{aligned} \tag{31}$$

Also, the $\mathscr{F}$-intensity process of H^i is given by $(1 - H_t^i)X_t^i$. In other words, the process M^i defined by,

$$M_t^i = H_t^i - \int_0^t (1 - H_s^i)X_s^i ds\,, \tag{32}$$

is an $\mathscr{F}$-martingale. Finally, the conditional survival probability function of name $i \in \mathbb{N}_n$ is given by, for every $t_i > t$,

$$\mathbb{P}(\tau_i > t_i \,|\, \mathscr{F}_t) = \mathbb{E}\left\{\exp\left(-\int_t^{t_i} X_s^i ds\right) \Big|\, X_t^i\right\} \tag{33}$$

so that, in particular,

$$\mathbb{E}\exp\left\{-\int_0^t X_s^{\{i\}}ds\right\} = \exp\left\{-\left(\Gamma_i(t) - \sum_{i\in I}\int_0^t X_s^I ds\right)\right\}, \tag{34}$$

where $\Gamma_i = -\ln \mathbb{P}(\tau_i > t)$ is the *hazard function* of name i.

4.3 Credit Derivatives Prices and Price Dynamics in the Common Shocks Model

The following pricing results also follow by straightforward adaptation of the proofs of the analogous results in [2].

Let $\mathscr{Z}_t$ stands for the set of all non-empty sets of survivors of sets Y in $\mathscr{Y}$ right before time t. We denote $\nabla u(t, \mathbf{x}, \mathbf{k}) = (\partial_{\lambda_Y} u(t, \mathbf{x}, \mathbf{k}))_{Y\in\mathscr{Y}}$, and by $\sigma(t, \mathbf{x})$ the diagonal matrix with diagonal $(c\sqrt{x_Y})_{Y\in\mathscr{Y}}$.

Proposition 4.1 (i) *The price process P and the cumulative value $\hat{P}$ of the portfolio credit derivative are such that, for $t \in [0, T]$,*

$$\begin{aligned} P_t &= u(t, \mathbf{X}_t, \mathbf{H}_t), \\ d\hat{P}_t &= \nabla u(t, \mathbf{X}_t, \mathbf{H}_t)\sigma(t, \mathbf{X}_t)d\mathbf{W}_t + \sum_{Z\in\mathscr{Z}_t} \delta u^Z(t, \mathbf{X}_t, \mathbf{H}_{t-})dM_t^Z, \end{aligned} \tag{35}$$

where the pricing function $u(t, \mathbf{x}, \mathbf{k})$ is given by

$$u(t, \mathbf{X}_t, \mathbf{H}_t) = \mathbb{E}\left[\int_t^T \varphi(\mathbf{H}_s)ds + \phi(\mathbf{H}_T) - \phi(\mathbf{H}_t)|\mathscr{F}_t\right]; \tag{36}$$

(ii) *The price Q^i and the cumulative value $\hat{Q}^i$ of the single-name credit derivative on name i are such that, for $t \in [0, T]$,*

$$\begin{aligned} Q_t^i &= (1 - H_t^i)v_i(t, X_t^i), \\ d\hat{Q}_t^i &= (1 - H_t^i)\partial_{\lambda_i} v_i(t, X_t^i)\sigma_i(t, X_t^i)dW_t^i \\ &\quad + \sum_{Z\in\mathscr{Z}_t} \mathbf{1}_{i\in Z}\big(\phi_i - v_i(t, X_t^i)\big)dM_t^Z, \end{aligned} \tag{37}$$

for a pre-default pricing function $v_i(t, \lambda_i)$ such that

$$v_i(t, x_i) = \mathbb{E}\left(\int_t^T e^{-\int_t^s X_\zeta^i d\zeta}\left(\varphi_i + \phi_i X_s^i\right) \,\Big|\, X_t^i = x_i\right). \tag{38}$$

Now, an important practical point is that in the affine factor specification of this paper, all the expectations and conditional expectations that arise in the single-name formulas (33), (34) and (38), can be computed explicitly (see [4] for details).

Moreover, a common shocks interpretation of the model analogous to the one developed in [2], also allows one to compute the conditional expectations in 36 in a fast and exact way, for all the *portfolio loss derivatives* as of (20).

To get the formulas for the conditional expectations in (36), simply 'add' expectations in front of the terms $\mathbb{E}\exp(\pm\int_0^t X_s^I ds)$ in the corresponding formulas in [2] (in which the X^Is are deterministic).

The min-variance hedging formula of [2] also still holds true, using the pricing functions of Proposition 4.1 therein.

5 Hedging Counterparty Credit Risk in the Common Shocks Model

Our final aim is a study of the hedging problem of the unilateral CVA on a portfolio credit derivative, in the common shocks model of the previous section.

5.1 Min-Variance Hedging by a Rolling CDS on the Counterparty

We first study the problem of hedging the CVA by a single counterparty clean CDS on the counterparty. Note however that a fixed CDS (of a given contractual spread in particular) cannot be traded dynamically in the market. Indeed, only freshly emitted CDSs can be entered into, at no cost and at the related fair market spread, at any given time. To address this issue we shall thus actually use a *rolling CDS* as our hedging instrument. The practical concept of a rolling CDS, introduced in [8] and already used for hedging purposes in [3], is essentially a self-financing trading strategy in market CDSs. So, much like with futures contracts, the value of a rolling CDS is null at any point in time, yet due to the trading gains of the strategy the related cumulative value process is not zero.

We now derive the dynamics of the CVA and of the rolling CDS in the common shocks model.

Note that $\mathbf{H}_{t-}^Z$ below stands for the vector obtained from $\mathbf{H}_{t-}$ by replacing its components with indices in Z by units (cf. the generic notation $\mathbf{k}^Y$ introduced with equation (26)), whereas M_t^Z is the compensated set-event martingale of (29).

Proposition 5.1

(i) *One has, for* $t \in [0, \tau \wedge T]$,

$$\Theta_t = \mathbb{E}_t\left[\mathbb{1}_{\tau<T}\xi\right] \textit{ with } \xi = (1-R)\chi^+, \quad \chi = u(\tau, \mathbf{X}_\tau, \mathbf{H}_\tau) + \Delta\phi(\mathbf{H}_\tau) - \Gamma_\tau,$$
$$d\Theta_t = (1-R)\Big(\mu_t d\mathbf{W}_t + \sum_{Z\in\mathscr{Z}_t} \nu_t^Z dM_t^Z\Big) \tag{39}$$

for suitable integrands μ *and* ν^Y *with predictable* $\nu^Y s$. *Moreover, for* $t \in [0, \tau \wedge T]$,

$$\sum_{Z \in \mathscr{Z}_t;\, 0 \in Z} \nu_t^Z dM_t^Z = \sum_{Z \in \mathscr{Z}_t;\, 0 \in Z} \left(\chi_t^{Z,+} - (1-R)^{-1}\Theta_{t-}\right) dM_t^Z, \tag{40}$$

where, for every $Z \subseteq \mathbb{N}_n$, *the symbol* $\chi_t^{Z,+}$ *stands for the positive part of* χ_t^Z *defined as, for* $t \in [0, T]$,

$$\chi_t^Z = u(t, \mathbf{X}_t, \mathbf{H}_{t-}^Z) + \phi(t, \mathbf{H}_{t-}^Z) - \phi(t, \mathbf{H}_{t-}) - \Gamma_t. \tag{41}$$

(ii) *The value* Q *and the cumulative value* $\hat{Q}$ *of the rolling CDS on the counterparty are such that, for* $t \in [0, \tau \wedge T]$,

$$\begin{aligned} Q_t &= 0, \\ d\hat{Q}_t &= (1-R)\left(\partial_{x_0} v(t, X_t^0) c\sqrt{X_t^0} dW_t^0 + \sum_{Z \in \mathscr{Z}_t;\, 0 \in Z} dM_t^Z\right), \end{aligned} \tag{42}$$

where $\partial_{x_0} v(t, X_t^0)$ *is an abbreviation for* $\partial_{x_0} p(t, X_t^0) - (1-R)^{-1} S(t, X_t^0) \partial_{x_0} \times f(t, X_t^0)$, *and where* p *and* f *denote the pre-default pricing functions of the unit protection and fees legs of the CDS initiated at time* t, *so*

$$f(t, X_t^0) = \mathbb{E}\left(\int_t^T e^{-\int_t^s X_\zeta^0 d\zeta} ds \,|\, X_t^0\right),$$

$$p(t, X_t^0) = \mathbb{E}\left(\int_t^T e^{-\int_t^s X_\zeta^0 d\zeta} X_s^0 ds \,|\, X_t^0\right),$$

and $S = p/f$ *is the corresponding CDS fair spread function.*

Proof

(i) Formula (39) is the predictable representation of the martingale Θ in our model. Note that this martingale representation indeed holds in virtue of our model construction by change of measure, starting from a measure under which $\mathbf{H}$ is a time-continuous Markov chain with intensity one: see, for instance, Proposition 24 in Crépey [11] or Proposition 7.6 in the online preprint version, for analogous results with detailed proofs. Moreover, one has that $\Delta\Theta_\tau = \xi - \Theta_{\tau-}$ on $\{\tau \le T\}$. Recalling $P_t = u(t, \mathbf{X}_t, \mathbf{H}_t)$ and $\Delta\Gamma_\tau = 0$, also observe that $\chi = \chi_\tau^Z$ on the intersection $\{\Delta H_\tau^Z = 1\} \cap \{\tau \le T\}$, namely, for $\tau \le T$ coinciding with the default time of the names in Z (including the counterparty) and only with them. The lhs and the rhs local martingales in (40) thus differ by an integral with respect to time, so that their difference is, in fact, constant.

(ii) In view of the Markov copula property of our model, this can be shown as in the proof of Lemma 2.2 of [3] (see also Proposition 4.1(ii) for comparison with the case of a standard, non-rolling CDS on the counterparty). □

Now, let ζ be an $\mathbb{R}$-valued process, representing the number of units held in the rolling CDS which is used along with the constant asset in a self-financing hedging strategy for the counterparty risk of the portfolio credit derivative. Given (42) and (39), the tracking error (e_t) of the hedged portfolio satisfies $e_0 = 0$ and, for t in $[0, \tau \wedge T]$,

$$(1-R)^{-1} de_t = (1-R)^{-1}(d\Theta_t - \zeta_t d\hat{Q}_t) \tag{43}$$

$$= \mu_t d\mathbf{W}_t - \zeta_t \partial_{x_0} v(t, X_t^0) c\sqrt{X_t^0} dW_t^0$$
$$+ \sum_{Z \in \mathscr{Z}_t;\, 0 \in Z} (\nu_t^Z - \zeta_t) dM_t^Z + \sum_{Z \in \mathscr{Z}_t;\, 0 \notin Z} \nu_t^Z dM_t^Z, \tag{44}$$

where the Brownian terms and the jump terms can be interpreted as the market and/or spread risk component and the jump-to-default risk component of the hedging error, the last sum representing the counterparty jump-to-default risk component of the hedging error.

Theorem 5.2 *The strategy which minimizes the risk-neutral variance of the jump-to-default risk component of the hedging error, or, equivalently, which minimizes the risk-neutral variance of the counterparty jump-to-default risk component of the hedging error, is given by, for $t \le \tau \wedge T$ (and $\zeta^{jd} = 0$ on $(\tau \wedge T, T]$)*

$$\zeta_t^{jd} = \sum_{Y \in \mathscr{Y};\, 0 \in Y_t} w_t^Y \left(\chi_t^{Y,+} - (1-R)^{-1}\Theta_{t-}\right) = \theta_t - (1-R)^{-1}\Theta_{t-}, \tag{45}$$

where $\theta_t = \sum_{Y \in \mathscr{Y};\, 0 \in Y_t} w_t^Y \chi_t^{Y,+}$, for weights w_t^Y defined as $\frac{X_t^Y}{\sum_{Z \in \mathscr{Y};\, 0 \in Z_t} X_t^Z}$, for every $Y \in \mathscr{Y}$ with $0 \in Y_t$. In particular, on $\{\tau < T\}$,

$$\zeta_\tau^{jd} = \theta - (1-R)^{-1}\Theta_{\tau-}, \tag{46}$$

for the so called Expected Positive Exposure $\theta = \theta_\tau = \mathbb{E}(\chi^+ \,|\, \mathscr{F}_{\tau-})$.

Proof The strategy minimizing the risk-neutral variance of the counterparty jump-to-default risk component of the hedging error is given by, for $t \le \tau$, $\zeta_t^{jd} = \frac{d\langle M, \hat{Q}\rangle_t}{d\langle \hat{Q}\rangle_t}$, with

$$M = \int_0^{\cdot} \sum_{Z \in \mathscr{Z}_t;\, 0 \in Z} \nu_t^Z dM_t^Z = \sum_{Z \in \mathscr{Z}_t;\, 0 \in Z} \int_0^{\cdot} \left(\chi_t^{Z,+} - (1-R)^{-1}\Theta_{t-}\right) dM_t^Z, \tag{47}$$

by (40). So, in view of the dynamics of $\hat{Q}$ in (42) (note that all the jump martingales integrands are predictable in (42) and (47)) and of the expression (28) of the intensities ℓ^Zs of the M^Zs,

$$\zeta_t^{jd} = \frac{\sum_{Z \in \mathscr{Z}_t;\, 0 \in Z} \ell_t^Z (\chi_t^{Z,+} - (1-R)^{-1}\Theta_{t-})}{\sum_{Z \in \mathscr{Z}_t;\, 0 \in Z} \ell_t^Z}$$

$$= \frac{\sum_{Y \in \mathscr{Y};\, 0 \in Y_t} X_t^Y (\chi_t^{Y_t,+} - (1-R)^{-1}\Theta_{t-})}{\sum_{Y \in \mathscr{Y};\, 0 \in Y_t} X_t^Y},$$

from which (45) follows by noting that one has $\chi^{Y_t} = \chi^Y$ for every $Y \in \mathscr{Y}$. Moreover, the M^Z are pure jump processes which do not jump simultaneously, so $\hat{Q}$ is orthogonal to $\tilde{M} = \int_0^{\cdot} \sum_{Z \in \mathscr{Z}_t;\, 0 \in Z} \nu_t^Z dM_t^Z$. One thus also has that

$$\zeta_t^{jd} = \frac{d\langle M + \tilde{M}, \hat{Q}\rangle_t}{d\langle \hat{Q}\rangle_t},$$

hence ζ^{jd} also minimizes the risk-neutral variance of the overall jump-to-default risk component of the hedging error.

Finally, to deduce (46) from (45), one only needs to show is that

$$\sum_{Y \in \mathscr{Y};\, 0 \in Y_\tau} w_\tau^Y \chi_\tau^{Y,+} = \theta = \mathbb{E}(\chi^+ \,|\, \mathscr{F}_{\tau-}). \tag{48}$$

Now, by the classical expression for the conditional jump law of a finitely-valued pure jump process mitigated by a diffusion, the law of $\mathbf{H}_\tau$ conditional on $\mathscr{F}_{\tau-}$ is supported by $\{\mathbf{H}_{\tau-}^Z;\ Z \in \mathscr{Z}_\tau, 0 \in Z\}$, and it is given by, for every such Z (cf. (26))

$$\mathbb{P}\Big(\mathbf{H}_\tau = \mathbf{H}_{\tau-}^Z \,|\, \mathscr{F}_{\tau-}\Big) = \sum_{Y \in \mathscr{Y};\, Y_\tau = Z} w_\tau^Y .$$

So,

$$\theta = \sum_{Z \in \mathscr{Z}_\tau;\, 0 \in Z} \Big(\sum_{Y \in \mathscr{Y};\, Y_\tau = Z} w_\tau^Y \Big) (u(\tau, \mathbf{X}_\tau, \mathbf{H}_{\tau-}^Z) + \phi(\tau, \mathbf{H}_{\tau-}^Z) - \phi(\tau, \mathbf{H}_{\tau-}) - \Gamma_\tau)^+, \tag{49}$$

and (48) follows from the fact that $\mathbf{H}_{\tau-}^Z = \mathbf{H}_{\tau-}^Y$, for every $Y \in \mathscr{Y}$ with $Y_\tau = Z$. □

One thus retrieves in (46) the definition of the hedging ratio which is often advocated by CVA desks for hedging the counterparty jump-to-default component of the counterparty risk. In fact, this hedging ratio is commonly referred to as the Expected Positive Exposure, loosely defined as $\theta^\sharp = \mathbb{E}(\chi_t^+ \,|\, \tau)$. But, the above min-variance hedging analysis reveals that, from a dynamic hedging point of view, this hedging ratio should really be defined as $\theta - (1-R)^{-1}\Theta_{\tau-}$, where the second term accounts for the value of the portfolio CVA right before the default time τ of the counterparty, and where the Expected Positive Exposure corresponding to the first term should really defined as θ, rather than by its 'proxy' $\theta^\sharp$.

Note that in the course of the derivation of this result, Proposition 5.2 exploits two model-dependent features of our set-up:

- First, the fact that the cumulative value process of the rolling CDS only jumps at the default time τ of the counterparty, as opposed to jumps at other defaults too in a 'general' model of credit risk.

- Second, our assumption of a constant recovery of the counterparty (and of other obligors too, but this is irrelevant here).

Also observe that without hedging, so for $\zeta = 0$, one would have

$$\Delta e_\tau = \Delta e^0_\tau = \xi - \Theta_{\tau-}$$

on $\{\tau < T\}$. With the ζ^{jd} strategy, one has $\Delta e_\tau = \Delta e^{jd}_\tau = (1-R)\theta - \Theta_{\tau-}$ on $\{\tau < T\}$. So, basically, the ζ^{jd} hedging strategy changes the counterparty jump-to-default exposure from ξ to $(1-R)\theta = \mathbb{E}(\xi_\tau \mid \mathscr{F}_{\tau-})$, the 'best guess' of ξ available right before τ. Note however that this strategy, which is 'optimal' as far as the counterparty jump-to-default component (or altogether jump-to-default component) of the counterparty risk is concerned, is disregarding the market risk component of the hedging error. In fact, this strategy typically creates some additional market risk.

Given (30), one can also rewrite in (42), (44):

$$\partial_{x_0} v(t, X^0_t) c \sqrt{X^0_t} dW^0_t = \partial_{x_0} v(t, X^0_t) \sum_{Y \ni 0} c \sqrt{X^Y_t} dW^Y_t = \partial_{x_0} v(t, X^0_t) \Sigma_t d\mathbf{W}_t, \quad (50)$$

where Σ_t is the row-vector indexed by $\mathscr{Y}$ such that $\Sigma^Y_t = c \mathbb{1}_{Y \ni 0} \sqrt{X^Y_t}$, $Y \in \mathscr{Y}$. It is then rather straightforward to write the formula for the strategy which minimizes the risk-neutral variance of the hedging error altogether (see [2]). However; in practice it will typically be difficult to compute all the terms that appear in this formula (unless we are in the pure jump case with no factors of a time-deterministic intensities model, for which $\zeta^{va} = \zeta^{jd}$).

Finally, recall that Proposition 5.2 deals with the issue of hedging unilateral counterparty risk. The issue of hedging bilateral counterparty risk seems more involved, since in this case instruments sensitive to the default times of the investor and the counterparty should clearly be used (ideally, an instrument sensitive to their first default time, like a first-to-default swap on both names).

5.1.1 Case of One CDS

Let us consider the special case of a CDS on name one chosen as a special case of the above portfolio credit derivative, without collateralization. So $\Gamma = 0$, and

$$\varphi(\mathbf{k}) = f_1(k_1) = -S_1(1-k_1), \qquad \phi(\mathbf{k}) = g_1(k_1) = (1-R_1)k_1. \quad (51)$$

In virtue of the Markov copula properties of the model one may and do forget about names 2 to n and take $\mathscr{Y} = \{\{0\}, \{1\}, \{0, 1\}\}$, without loss of generality. Then (see Proposition 4.1(ii)),

$$u(t, \mathbf{X}_t, \mathbf{H}_t) = (1 - H^1_t) v_1(t, X^1_t)$$

with

$$v_1(t, X_t^1) = \mathbb{E}\left(\int_t^T e^{-\int_t^u X_v^1 dv}\left((1-R_1)X_u^1 - S_1\right) du \mid X_t^1\right). \quad (52)$$

Thus by (24), $\xi = (1-R)\chi^+$, in which χ here assumes the following form:

$$\chi = \mathbb{1}_{\tau_1 < \tau} v_1(\tau, X_\tau^1) + \mathbb{1}_{\tau_1 = \tau}(1 - R_1). \quad (53)$$

Moreover, one has by application of formula (41):

$$\begin{aligned}
\chi_t^{\{0\}} &= u(t, \mathbf{X}_t, \mathbf{H}_{t-}^{\{0\}}) + \phi(t, \mathbf{H}_{t-}^{\{0\}}) - \phi(t, \mathbf{H}_{t-}) = u(t, \mathbf{X}_t, \mathbf{H}_{t-}^{\{0\}}) = \mathbb{1}_{t \le \tau_1} v_1^+(t, X_t^1), \\
\chi_t^{\{0,1\}} &= u(t, \mathbf{X}_t, \mathbf{H}_{t-}^{\{0,1\}}) + \phi(t, \mathbf{H}_{t-}^{\{0,1\}}) - \phi(t, \mathbf{H}_{t-}) = 0 + (1-R_1) - (1-R_1)H_{t-}^1 \\
&= \mathbb{1}_{t \le \tau_1}(1 - R_1).
\end{aligned}$$

Therefore, (45) yields, for $t \le \tau \wedge \tau_1 \wedge T$ (and $\zeta^{jd} = 0$ on $(\tau \wedge \tau_1 \wedge T, \tau \wedge T]$),

$$\zeta_t^{jd} = \theta_t - (1-R)^{-1}\Theta_{t-},$$

where

$$\theta_t = \sum_{Y \in \mathscr{Y};\, 0 \in Y_t} w_t^Y \chi_t^{Y,+} = w_t^{\{0\}} v_1^+(t, X_t^1) + w_t^{\{0,1\}}(1 - R_1)$$

in which

$$w_t^{\{0\}} = \frac{X_t^{\{0\}}}{X_t^{\{0\}} + X_t^{\{0,1\}}}, \qquad w_t^{\{0,1\}} = \frac{X_t^{\{0,1\}}}{X_t^{\{0\}} + X_t^{\{0,1\}}}.$$

See also [4] for the entire specification of the dynamics of the CVA process on this example, in a related model with X^Is given as deterministic functions of time.

5.2 Multi-instruments Hedge

We now consider the situation where additional instruments can be used for hedging the 'market' or 'spread risk' component (diffusive part) of the counterparty risk exposure. More specifically, we suppose that there exists an $\mathbb{R}^m$-valued martingale price process $\mathbf{Q} = (Q^j)_{1 \le j \le m}$ of hedging instruments with $\mathbb{Q}$-dynamics

$$d\mathbf{Q}_t = \varsigma_t \, d\mathbf{W}_t, \quad (54)$$

for a left-invertible diffusion matrix-process ς_t, with left-inverse denoted by ς_t^{-1}. Let η be an $\mathbb{R}^{1 \otimes m}$-valued process, representing the number of units held in every of the Q^j's, which are used along with the rolling CDS on the counterparty and the constant asset for hedging the counterparty risk exposure. The tracking error (e_t) of the hedged portfolio now satisfies $e_0 = 0$ and, for $t \in [0, \tau \wedge T]$,

$$de_t = d\Theta_t - \zeta_t d\hat{Q}_t - \eta_t d\mathbf{Q}_t. \tag{55}$$

Proposition 5.3 *The strategy which minimizes the risk-neutral variance of the hedging error is given by ζ^{jd} as of* (45), *and, for $t \in [0, \tau \wedge T]$ (recall* (50)),

$$\eta_t^{va} = \left(\mu_t - \zeta_t^{jd} \partial_{x_0} v(t, X_t^0) \Sigma_t\right) \varsigma_t^{-1}. \tag{56}$$

The residual hedging error satisfies $e_0^{va} = 0$ and, for $t \in [0, T]$,

$$(1 - R)^{-1} de_t^{va} = (\chi^+ - \theta) \sum_{Z \in \mathscr{Z}_t;\, 0 \in Z} \nu_t^Z dM_t^Z + \sum_{Z \in \mathscr{Z}_t;\, 0 \notin Z} \nu_t^Z dM_t^Z. \tag{57}$$

Proof One has

$$\min_{\zeta,\eta} \mathbb{V}\mathrm{ar}\, e(\zeta, \eta) = \min_{\eta} \left(\min_{\zeta} \mathbb{V}\mathrm{ar}\, e(\zeta, \eta) \right),$$

where, given any η, the solution of the inner minimization problem is given by ζ^{jd}, independently of η. So

$$\min_{\zeta,\eta} \mathbb{V}\mathrm{ar}\, e(\zeta, \eta) = \min_{\eta} \mathbb{V}\mathrm{ar}\, e(\zeta^{jd}, \eta),$$

where the minimum in the right-hand-side is obviously achieved by η^{va}, the residual hedging error being then given by (57). □

Acknowledgements The authors warmly thank Giovanni Cesari for stimulating discussions throughout the preparation of this work. We also thank an anonymous referee for a very careful reading of the manuscript and for valuable comments and remarks.

The research of T.R. Bielecki was supported by NSF Grant DMS-0603789 and NSF Grant DMS-0908099.

The research of S. Crépey benefited from the support of the 'Chaire Risque de crédit', Fédération Bancaire Française, and of the DGE.

References

1. Assefa, S., Bielecki, T.R., Crepey, S., Jeanblanc, M.: CVA computation for counterparty risk assessment in credit portfolios. In: Bielecki, T.R., Brigo, D., Patras, F. (eds.) Credit Risk Frontiers: Subprime Crisis, Pricing and Hedging, CVA, MBS, Ratings and Liquidity. Wiley, New York (2011)
2. Bielecki, T.R., Cousin, A., Crépey, S., Herbertsson, A.: Dynamic hedging of portfolio credit risk in a Markov copula model. JOTA **157**(3) (2013)
3. Bielecki, T.R., Crépey, S., Jeanblanc, M., Rutkowski, M.: Convertible bonds in a defaultable diffusion model. In: Kohatsu-Higa, A., Privault, N., Sheu, S.J. (eds.) Stochastic Analysis with Financial Applications. Birkhäuser, Basel (2010)

4. Bielecki, T., Crépey, S., Jeanblanc, M., Zargari, B.: Valuation and hedging of CDS counterparty exposure in a Markov copula model. IJTAF **15**(1) (2012)
5. Bielecki, T.R., Vidozzi, A., Vidozzi, L.: A Markov copulae approach to pricing and hedging of credit index derivatives and ratings triggered step-up bonds. J. Credit Risk **4**, 1 (2008)
6. Bielecki, T.R., Vidozzi, A., Vidozzi, L., Jakubowski, J.: Study of dependence for some stochastic processes. Stoch. Anal. Appl. **26**(4), 903–924 (2008)
7. Bielecki, T.R., Rutkowski, M.: Credit Risk: Modeling, Valuation and Hedging. Springer, Berlin (2002)
8. Bielecki, T.R., Jeanblanc, M., Rutkowski, M.: Pricing and trading credit default swaps. Ann. Appl. Probab. **18**(6), 2495–2529 (2008)
9. Blanchet-Scalliet, C., Jeanblanc, M.: Hazard rate for credit risk and hedging defaultable contingent claims. Finance Stoch. **8**(1), 145–159 (2008)
10. Cesari, G., Aquilina, J., Charpillon, N.: Modelling, Pricing, and Hedging Counterparty Credit Exposure. Springer Finance (2010)
11. Crépey, S.: About the pricing equations in finance. Forthcoming in Paris-Princeton Lectures in Mathematical Finance. Lecture Notes in Mathematics. Springer (2010) (Preprint version available online at http://www.maths.univ-evry.fr/crepey)
12. Crépey, S., Jeanblanc, M., Zargari, B.: Counterparty risk on a CDS in a Markov chain copula model with joint defaults. In: Kijima, M., Hara, C., Muromachi, Y., Tanaka, K. (eds.) Recent Advances in Financial Engineering 2009. World Scientific, Singapore (2010). Available on http://grozny.maths.univ-evry.fr/pages_perso/crepey/
13. Delbaen, F., Schachermayer, W.: The Mathematics of Arbitrage. Springer Finance (2006)
14. Durand, C., Rutkowski, M.: Credit value adjustment for bilateral counterparty risk of collateralized contracts under systemic risk. Working paper
15. Gregory, J.: Counterparty Credit Risk: The New Challenge for Global Financial Markets. Wiley, New York (2009)
16. ISDA Collateral Steering Committee: Market Review of OTC Derivative Bilateral Collateralization Practices, ISDA Swaps and Derivatives Association, March 2010

A Note on Market Completeness with American Put Options

Luciano Campi

Abstract We consider a non necessarily complete financial market with one bond and one risky asset, whose price process is modeled by a suitably integrable, strictly positive, càdlàg process S on $[0, T]$. Every option price is defined as the conditional expectation under a given equivalent (true) martingale measure $\mathbb{P}$, the same for all options. We show that every positive contingent claim on S can be approximately replicated in L^2-sense by investing dynamically in the underlying and statically in all American put options (of every strike price k and with the same maturity T). We also provide a counterexample to static hedging with European call options of all strike prices and all maturities $t \leq T$.

Keywords Market completeness · American put · Tanaka formula · European call · Marginals

Mathematics Subject Classification (2010) 91B28 · 60G40 · 60G44 · 60G48

1 Introduction

The aim of this paper is to investigate the issue of hedging in a market where agents are allowed to invest continuously in time in a risky asset and statically in American put options of all strike prices and with a fixed maturity T. Such additional investment opportunities are of particular interest in incomplete markets, in which case the payoffs cannot, in general, be replicated by a trading strategy in the underlying.

Since in real financial markets many types of options are becoming more and more liquid, it is very natural to reformulate hedging and optimal investment problems incorporating those larger trading opportunities. Indeed, in the recent years many papers treated problems like absence of arbitrage, hedging, optimal portfolio choice in a financial market where investors are allowed to trade in the underlying assets as well as to assume static positions in some class of derivatives. Here, we

---

L. Campi (✉)
Département de Mathématiques, Institut Galilée, Université Paris 13, 99, avenue Jean-Baptiste Clément, 93430 Villetaneuse, France
e-mail: campi@math.univ-paris13.fr

Y. Kabanov et al. (eds.), *Inspired by Finance*, DOI 10.1007/978-3-319-02069-3_4,

recall only few of them: Campi [1] for no-arbitrage and completeness issues, the papers by Ilhan et al. [11, 12] and Carr et al. [5] for optimal investment problems, and the more recent papers by Schweizer and Wissel [19, 20] and by Jacod and Protter [13] where an HJM approach for European call options is developed.

Our paper is much closer in spirit to that part of literature initiated by Ross [18], where it has been shown that in a single period model with a finitely many states and n stocks, simple options (i.e. written on only one stock) can be replicated by trading in European call (or put) options on that stock. Ross [18] established also the existence of a portfolio in all n stocks such that call (or put) options written on that portfolio span the set of all options, simple as well as complex, i.e. written on all stocks at the time. This result has been generalized in an infinite state space setting by Green and Jarrow [10] (see also Nachman [15]). For a similar result in a dynamic discrete-time setting with a general state space, we quote only the paper by Rogge [17], where the market completeness is characterized in terms of a so-called "call-completeness": there exists a strike price $k > 0$ such that if a claim is attainable then so is a call written on it with the same strike.

A more explicit result has been established in Carr and Madan [4], where it is shown that a European option with pay-off $f(S_T)$ (f sufficiently regular) can be hedged by a unique initial position of $f(S_0) - f'(S_0)S_0$ unit discount bonds, $f'(S_0)$ stock shares, and $f''(k)dk$ out-of-the-money call and put options of all strikes. Even more importantly, this result is model-free, so that it holds also in a continuous-time framework. For static hedging of exotic options in a diffusion setting, we refer to Carr et al. [3]. Moreover, in a discontinuous setting, Corcuera et al. [7] proved that a Lévy market can be completed by trading statically in power-jump assets, which are somewhat related to contracts on realized variance. In [6], a link between power-jump assets and European call options is established, allowing Corcuera and Guerra to prove that a Lévy market can be completed by trading in a continuum of European call options along strikes. Finally, in the paper [8], Davis and Obloj provide necessary and sufficient conditions for a market model to be complete by trading in a given finite set of European-type derivatives driven by finitely many factors modeled as diffusion processes.

To sum up, all these papers deal with the problem of finding a good class of derivatives capable to span all contingent claims written on the same underlying. By "good class of derivatives" we mean a class of options sufficiently liquid in real financial markets in order to reproduce with a certain precision the hedging strategies suggested in these papers.

The present paper identifies American put options of all strike prices and with the same maturity as a good class of derivatives allowing an investor to replicate approximately in L^2-sense *every positive contingent claim* written on the same underlying in a very general frictionless financial market with one risk-free asset and one stock. Our main result, Theorem 1, can be viewed as a generalization to the continuous-time setting of previously quoted Ross's result in [18]—notice that in Ross's single period setting there is no need to distinguish between European and path-dependent options like American puts. Moreover, our result is model-free. Indeed, we will make only the very mild assumption that the price process is a square-integrable and possibly discontinuous martingale with no jump at maturity. The price we have to pay is that the agent has to trade in infinitely many securities, in contrast with, e.g., Davis and Obloj [8] where, since agents can trade only in finitely many assets, the authors need to assume more regularity as, e.g., analiticity of the coefficients together with a non-singularity condition, to get the market completeness.

The present paper is structured as follows: In Sect. 2, we give a short description of the model. Section 3 contains the main result on hedging with American put options. Finally, in Sect. 4, we exhibit a two-period market model where static investments in all European call (equivalently, put) options of all strike prices and all maturities are not sufficient for replicating all contingent claims, so justifying the use of their American counterparts.

2 The Model

We consider a financial market composed by a riskless asset and a stock. More precisely, let $(\Omega, \mathscr{F}, \mathbb{P})$ be a probability space equipped with a filtration $(\mathscr{F}_t)_{t\in[0,T]}$, where $T > 0$ is a finite horizon, $\mathscr{F} = \mathscr{F}_T$ and $\mathscr{F}_0$ is trivial. We assume that $(\mathscr{F}_t)_{t\in[0,T]}$ is the natural filtration—right-continuous and $\mathbb{P}$-saturated—generated by a strictly positive, càdlàg process $S = (S_t)_{t\in[0,T]}$ modelling the price of the risky asset. We assume without loss of generality that the price at time t of the riskless asset is $S_t^0 = e^{rt}$ where $r > 0$ is the spot interest rate (see Remark 3 for a straightforward generalization). In the sequel $\mathbb{E}[\cdot]$ will denote expectation with respect to $\mathbb{P}$ and, for any process X, we set $\tilde{X} := X/S^0$ and $\Delta X_t = X_t - X_{t-}$ for any $t \in]0, T]$. Finally, $\mathscr{H}^2(\mathbb{P})$ will denote the space of all martingales bounded in $L^2(\mathbb{P}) := L^2(\mathbb{P}, \mathscr{F}_T)$.

Our first assumption concerns the behavior of the risky asset process S.

Assumption 1 *The process $\tilde{S} = S/S^0$ belongs to $\mathscr{H}^2(\mathbb{P})$ and does not jump at T, i.e. $\Delta S_T = 0$ a.s.*

Since by assumption $(\mathscr{F}_t)_{t\in[0,T]}$ is the natural filtration of S, the equality $\Delta S_T = 0$ a.s. is equivalent to $\mathscr{F}_T = \mathscr{F}_{T-}$, i.e. the underlying filtration is left-continuous at T. We notice that this property is satisfied, e.g., by all diffusion models and also by all exponential Lévy models. Indeed, it is well-known that the natural filtration of any Lévy process is quasi-left continuous, i.e. it does not jump at predictable stopping times (see, e.g., Protter [16], p. 150, Exercises 8 and 9, and p. 191 for details).

In this setting, a contingent claim pay-off on S with maturity T is naturally modeled by a random variable $f \in L^2(\mathbb{P})$. This model is not necessarily complete, so there may exist infinitely many equivalent (local) martingale measures different from $\mathbb{P}$ and so, equivalently, not every contingent claim can be hedged by trading only in the underlying. Allowing an investor to trade also in some class of options can enlarge considerably his hedging opportunities.

The following assumption can be viewed as a kind of no-arbitrage consistency among prices of different options written on the same underlying S.

Assumption 2 *In this financial market every option price comes from the same equivalent martingale measure* $\mathbb{P}$.

Observe that Assumption 2 has a natural financial motivation. Indeed, let us forget about integrability issues for a while and consider two contingent claims with pay-offs f and f' and corresponding risk-neutral prices f_t and f_t' at time t. Assume that they come from two equivalent martingale measures $\mathbb{Q}$ and $\mathbb{Q}'$ respectively, i.e. $f_t = \mathbb{E}_{\mathbb{Q}}[f|\mathscr{F}_t]$ and $f_t' = \mathbb{E}_{\mathbb{Q}'}[f'|\mathscr{F}_t]$. If we want the enlarged market (S, f, f') to be arbitrage-free, there must be an equivalent martingale measure $\mathbb{Q}''$ for (S, f, f'), so that one has

$$f_t = \mathbb{E}_{\mathbb{Q}}[f|\mathscr{F}_t] = \mathbb{E}_{\mathbb{Q}''}[f|\mathscr{F}_t]$$

and the same for f'. This means that the risk-neutral prices of f and f' come from the same measure $\mathbb{Q}''$.

3 Hedging with American Put Options

In this section, we will prove that additional static investments in American put options of every strike price and with the same maturity T allow an investor to hedge any contingent claim written on the underlying S. We make the following model assumption:

Assumption 3 *American put options with any maturity T and any strike price $k > 0$ are available for trading and all are issued at $t = 0$.*

Let us denote $\mathscr{T}$ the set of all $[0, T]$-valued stopping times and $\widehat{\tau}_k$ the smallest optimal exercise time corresponding to a given strike price k. Such stopping times do exist (see Remark 1 for details), so that we have

$$P_0(k, \widehat{\tau}_k) := \sup_{\tau \in \mathscr{T}} \mathbb{E}[e^{-r\tau}(k - S_\tau)^+] = \mathbb{E}[e^{-r\widehat{\tau}_k}(k - S_{\widehat{\tau}_k})^+], \quad (1)$$

which is the price that an agent must pay for buying at time 0 an American put option with strike k and maturity T. More generally, we set

$$\tilde{P}_t(k, \tau) := \mathbb{E}[e^{-r\tau}(k - S_\tau)^+ \mid \mathscr{F}_t],$$

for *any* stopping time $\tau \in \mathscr{T}$. Notice that under our assumptions each price process $\tilde{P}(k, \tau)$ is a martingale in $\mathscr{H}^2(\mathbb{P})$.

Remark 1 Observe that $\widehat{\tau}_k$ exists for all strike prices $k > 0$. Indeed, the underlying discounted price process $\tilde{S}$ is a strictly positive martingale in $\mathscr{H}^2(\mathbb{P})$ on a finite time interval $[0, T]$, so that it is of class (D) and, obviously, constant in expectations. Thus, Theorem 2.43 in El Karoui's St. Flour Lecture Notes [9] can be applied, providing the existence of $\widehat{\tau}_k$ for all $k > 0$.

Now, we consider the set $\mathscr{R}_a$ of all discounted contingent claims which can be approximately replicated by investing dynamically in the underlying S and statically in finitely many American put options as follows: the American puts that the agent buys at time $t = 0$ can be exercised at any stopping time $\tau \in \mathscr{T}$, while the American puts that he sells will be exercised at their corresponding optimal times by their buyers.

Mathematically speaking, $\mathscr{R}_a$ is the set of all $\mathscr{F}_T$-measurable random variables of the form

$$x + \int_0^T \theta_t d\tilde{S}_t + \sum_{i=1}^n \alpha_i(\tilde{P}_T(k_i, \tau_i) - P_0(k_i, \widehat{\tau}_{k_i}))$$
$$- \sum_{j=1}^m \beta_{n+j}(\tilde{P}_T(k_{n+j}, \widehat{\tau}_{k_{n+j}}) - P_0(k_{n+j}, \widehat{\tau}_{k_{n+j}})) \tag{2}$$

where

- $x \in \mathbb{R}$ is the initial endowment of the agent,
- θ is a real-valued S-integrable predictable process such that $\int \theta d\tilde{S}$ is a martingale in $\mathscr{H}^2(\mathbb{P})$ modelling the dynamic investment strategy in the stock S,
- $n \geq 0$ is the number of American puts that the agent buys at time 0, while $m \geq 0$ is the number of American puts sold at time 0,
- each weight $\alpha_i \geq 0$ is a nonnegative real number representing the number of American puts with strike k_i, $i = 1, \ldots, n$, bought by the agent at time 0 paying the price $P_0(k_i, \widehat{\tau}_{k_i})$, while $\beta_{n+j} \geq 0$, $j = 1, \ldots, m$, represents the number of American puts with strike k_{n+j} sold by the agent at time 0 receiving the price $P_0(k_{n+j}, \widehat{\tau}_{k_{n+j}})$.

We adopt the convention that any summation over an empty set of indexes is equal to zero. Notice that $\mathscr{R}_a$ is a convex cone. We will denote $\mathscr{R}_a^*$ its (positive) dual, i.e.

$$\mathscr{R}_a^* := \{f \in L^2(\mathbb{P}) : \mathbb{E}[fg] \geq 0, \ \forall g \in \mathscr{R}_a\},$$

and by $\mathscr{R}_a^{**}$ its bidual, i.e. $\mathscr{R}_a^* := (\mathscr{R}_a^*)^*$. We recall that the bidual C^{**} of any convex cone C in a vector topological space coincides with the closure of C, and that for any pair of convex cones C and C' such that $C \subset C'$ one has $C^* \supset (C')^*$ (see, e.g, [21], Chap. 1).

Remark 2 Notice that in the first summation appearing in (2), denoting the final gain coming from a long position taken at time $t = 0$ in American puts with strikes k_i, the puts are not necessarily exercised at their optimal exercise times $\widehat{\tau}_{k_i}$. The agent, willing to hedge against the risk of a given final pay-off f, can in principle exercise his puts at any stopping time between today and the maturity T. The question if hedging purposes may lead an agent to exercise such options at sub-obtimal times remains open.

Our main result is that $\mathscr{R}_a \cap L_+^2(\mathbb{P})$ is dense in $L_+^2(\mathbb{P})$ which denotes the set of all positive random variables in $L^2(\mathbb{P})$. In financial terms, it means that every positive contingent claim can be (approximately) replicated by a mixed investment as in (2): dynamic in the underlying and static in American put options.

Theorem 1 *Under our assumptions, the closure of* $\mathscr{R}_a$ *contains* $L^2_+(\mathbb{P})$.

Proof As the closure of a convex cone C in a topological vector space equals its bidual C^{**}, all we need is to prove is that the bidual $\mathscr{R}_a^{**}$ contains $L^2_+(\mathbb{P})$. In order to do this, it suffices to show that $L^2_+(\mathbb{P})$ contains the dual cone $\mathscr{R}_a^*$, that is any random variable $N_T \in L^2(\mathbb{P})$ which is positive over $\mathscr{R}_a$, i.e. $\mathbb{E}[N_T f] \geq 0$ for all $f \in \mathscr{R}_a$, is positive itself a.s. Denote N the martingale in $\mathscr{H}^2(\mathbb{P})$ associated to N_T, i.e. N is the càdlàg version of the martingale $\mathbb{E}[N_T|\mathscr{F}_t]$, $t \in [0,T]$.

Since $\mathscr{R}_a$ contains $\mathbb{R}$, we have $\mathbb{E}[N_T] = N_0 = 0$. From the fact that $\mathbb{E}[N_T f] \geq 0$ for any f equal to the static hedging part appearing in (2), we can deduce that

$$\alpha_1 \mathbb{E}[N_T \tilde{P}_T(k_1, \tau_1)] - \alpha_2 \mathbb{E}[N_T \tilde{P}_T(k_2, \widehat{\tau}_{k_2})] \geq 0$$

for all nonnegative real numbers α_1, α_2, all strikes k_1, k_2, and all stopping times $\tau_1 \in \mathscr{T}$. Taking $\alpha_2 = 0$, we get that

$$\begin{aligned} 0 &\leq \mathbb{E}[N_T \tilde{P}_T(k_1, \tau_1)] = \mathbb{E}[N_T \tilde{P}_{\tau_1}(k_1, \tau_1)] = \mathbb{E}[N_{\tau_1} e^{-r\tau_1}(k_1 - S_{\tau_1})^+] \\ &= \mathbb{E}[\tilde{N}_{\tau_1}(k_1 - S_{\tau_1})^+], \end{aligned}$$

for every strike price k_1 and every stopping time $\tau_1 \in \mathscr{T}$. Doob's optional sampling theorem implies that the process $\tilde{N}(k - S)^+$ is a $\mathbb{P}$-submartingale for every $k > 0$. The integration by parts formula gives

$$\begin{aligned} \tilde{N}_t(k - S_t)^+ &= \tilde{N}_t(S_t - k)^- \\ &= \int_0^t (S_{u-} - k)^- d\tilde{N}_u + \int_0^t \tilde{N}_{u-} d(S - k)^-_u + [\tilde{N}, (S - k)^-]_t \end{aligned}$$

for all $t \in [0,T]$. By Tanaka's formula for discontinuous semimartingales (see, e.g., Protter [16], Theorem 68, p. 216) and since $dS_u = rS_u du + e^{ru} d\tilde{S}_u$, we have that

$$\begin{aligned} \int_0^t \tilde{N}_{u-} d(S - k)^-_u = &- \int_0^t \tilde{N}_{u-} \mathbf{1}_{\{S_{u-} \leq k\}} (rS_u du + e^{ru} d\tilde{S}_u) \qquad (3) \\ &+ \sum_{0<u\leq t} \tilde{N}_{u-} \left[\mathbf{1}_{\{S_{u-} > k\}} (S_u - k)^- + \mathbf{1}_{\{S_{u-} \leq k\}} (S_u - k)^+\right] \\ &+ \frac{1}{2} \int_0^t \tilde{N}_{u-} dL^k_u(S), \qquad (4) \end{aligned}$$

for every instant $t \in [0,T]$ and strike price $k > 0$.

The fact that N_T belongs to the dual of $\mathscr{R}_a$ implies also that N_T is weakly orthogonal to $\tilde{S}$ and so strongly orthogonal to $\tilde{S}$ as well,[1] i.e. $N\tilde{S} = \tilde{N}S$ belongs to $\mathscr{H}^1(\mathbb{P})$,

[1] We recall that, in this $\mathscr{H}^2$ setting, weak orthogonality between two martingales M and N in $\mathscr{H}^2(\mathbb{P})$ is equivalent to strong orthogonality, i.e. $MN \in \mathscr{H}^1(\mathbb{P})$. See, e.g., Lemma 2, Sect. IV, in Protter's book [16].

the space of all martingales bounded in $L^1(\mathbb{P})$ (use, e.g., Lemma 2, Sect. IV, in [16]). Moreover, being $d[\tilde{N},(S-k)^-]_t = e^{-rt}d[N,(S-k)^-]_t$ and $L^k(S)$ a continuous increasing process, one has

$$[\tilde{N},(S-k)^-]_t = \sum_{0<u\le t} \Delta\tilde{N}_u \Delta(S-k)_u^-,$$

which is a pure jump process. Notice that $\tilde{N}(k-S)^+$ is a $\mathbb{P}$-submartingale with the Doob–Meyer decomposition $\tilde{N}(k-S)^+ = M + B$, where M is a local martingale and B is a predictable increasing process. The local martingale part M includes certainly the following term

$$\int_0^t (S_{u-}-k)^- d\tilde{N}_u - \int_0^t \tilde{N}_{u-}\mathbf{1}_{\{S_{u-}\le k\}} e^{ru} d\tilde{S}_u$$

and, since any càdlàg local martingale with finite variation must be purely discontinuous (see, e.g., [14, Lemma 4.14b)]), M cannot contain the finite variation terms with continuous paths appearing in (3) and (4). Thus, those terms have to belong to the increasing part B, which could in principle contains some additional terms of pure jump type. An important consequence of it is that the process

$$\frac{1}{2}\int_0^t \tilde{N}_{u-} dL_u^k(S) - \int_0^t rk\tilde{N}_u \mathbf{1}_{\{S_u\le k\}} du, \quad t\in[0,T],$$

must be increasing, i.e.

$$\int_s^t rk\tilde{N}_u \mathbf{1}_{\{S_u\le k\}} du \le \frac{1}{2}\int_s^t \tilde{N}_{u-} dL_u^k(S), \tag{5}$$

for all $s,t\in[0,T]$ with $s\le t$ and all $k>0$. Using a standard monotone class argument, the inequality in (5) can be generalized as follows

$$\int_A rk\tilde{N}_u \mathbf{1}_{\{S_u\le k\}} du \le \frac{1}{2}\int_A \tilde{N}_{u-} dL_u^k(S), \tag{6}$$

for all Borel set A and all $k>0$. Observe that in the two integrals in (6) the same function $u\to\tilde{N}_{u-}$ is integrated with respect to two dt-a.e. mutually singular[2] measures $(1/2)dL_u^k(S)$ and $rk\mathbf{1}_{\{S_u\le k\}}du$, which implies that such an inequality is verified only if $N\le 0\; d\mathbb{P}\otimes dt$-a.e. on the set $\{(\omega,t): S_t(\omega)\le k\}$, for all $k>0$. As a consequence, one has $N\le 0\; d\mathbb{P}\otimes dt$-a.e. on $\Omega\times[0,T]$ and, since N is càdlàg, one has also that $N_{T-}\le 0$ a.s. Finally, notice that Assumption 1 consequence $\mathscr{F}_T = \mathscr{F}_{T-}$ implies that no martingale can jump at T (see Protter [16], p. 191, for details), so that one has also that $N_T\le 0$ a.s. To end the proof, it suffices to recall that N is a martingale with $N_0=0$, so that $\mathbb{E}[N_T]=0$. Hence, $N_T=0$ a.s. □

---

[2]Indeed, the support of $dL_u^k(S)$ is $\{u: S_u = S_{u-} = k\}$ (see, e.g., Protter's book [16], Theorem 69, p. 217) while that of $\mathbf{1}_{\{S_u\le k\}}du$ is $\{u: S_u\le k\}$. Thus, their intersection is contained in $\{u: S_u = k\}$ which is at most countable and so it has zero Lebesgue measure.

Remark 3 A careful inspection of our proof reveals that Theorem 1 holds true even if the spot interest rate r is not necessarily constant but a positive and bounded deterministic function of time. More precisely, if $S_t^0 = \exp(\int_0^t r(u)du)$ where $r(u)$ is a measurable positive function defined on $[0, T]$ and such that S_T^0 is bounded from above by some constant.

Remark 4 Note that if one considers contingent claims depending on some randomness source different from S, then the previous completeness result breaks down. Indeed, take a model whose price processes are identically equal to one, i.e. $S^0 \equiv S \equiv 1$, so that the natural filtration of S is trivial and the collection of all American put option pay-offs $\{(k-1)^+ : k > 0\}$ coincides with $[0, \infty)$. Thus, $\mathscr{R}_a = \mathbb{R}$. Then, consider a sufficiently large filtration $(\mathscr{F}_t)_{t\in[0,T]}$ such that $\mathscr{F}_T$ contains at least one non-degenerate square-integrable positive random variable f. It is now clear that, even if we allow the strategies to be $\mathbb{F}$-adapted, in such a market it is not possible to hedge f as in Theorem 1.

4 A Counterexample to Hedging with European Call Options

In this section, we describe a financial market model in discrete time with a finite horizon $T \in \mathbb{N}$ and a finite probability space, where it is not possible to hedge all contingent claims by trading dynamically in a given underlying and statically in all European call options of every strike price $k > 0$ and every maturity before T.

Let $(\Omega, \mathscr{F}, \mathbb{P})$ be a finite probability space supporting a martingale $S = (S_t)_{t=0}^T$ modelling the price evolution of a stock. This space is assumed to be equipped with the filtration $(\mathscr{F}_t)_{t=0}^T$ naturally generated by S and for which $\mathscr{F} = \mathscr{F}_T$. As usual, we denote S^0 the price process for a riskless asset. Assume $S^0 \equiv 1$. For a given discrete-time process X, we set $\Delta X_t := X_t - X_{t-1}, t = 1, \dots, T$.

Consider the linear space $\mathscr{R}_e$ spanned by all random variables f of the form

$$f = x + \sum_{t=1}^{T} \theta_t \Delta S_t + \sum_{i=1}^{n} \alpha_i (C_T(T_i, k_i) - C_0(T_i, k_i)), \tag{7}$$

where $x \in \mathbb{R}$ is an initial endowment, θ is any predictable process modelling the dynamic strategy in S, $\alpha = (\alpha_1, \dots, \alpha_n) \in \mathbb{R}^n$ is any static strategy in n European call options with maturities $T_i \leq T$ and strike prices k_i for $1 \leq i \leq n$, whose no-arbitrage prices are denoted by $C_t(T_i, k_i) := \mathbb{E}[(S_{T_i} - k_i)^+ | \mathscr{F}_t]$, and $n \geq 1$ is an arbitrary positive integer.

Our aim is to construct a process S such that $\mathscr{R}_e$ is not dense in the set of all positive $\mathscr{F}_T$-measurable random variables L_+^0, equipped with the usual scalar product $(f, g) = \mathbb{E}[fg]$. To do so, we use the following consequence of Theorem 3 in Campi [1]. We provide its short proof for reader's convenience.

Lemma 1 *Assume that $\mathscr{R}_e \cap L_+^0$ is dense in L_+^0. Then the set of all $\mathbb{P}$-equivalent martingale measures $\mathbb{Q}$ under which S has the same marginals as under $\mathbb{P}$ reduces to a singleton.*

Proof Let $\mathbb{Q}$ be an equivalent martingale measure under which S has the same marginals as under $\mathbb{P}$. In this case, for all postive random variables f as in (7) we have $\mathbb{E}[f] = x = \mathbb{E}_{\mathbb{Q}}[f]$ and, since the family of those random variables is assumed to be dense in L^0_+, we can conclude that $\mathbb{Q} = \mathbb{P}$ on $\mathscr{F}_T$. □

In the light of this result, it suffices now to find a process S admitting two *different* equivalent martingale measures $\mathbb{P}$ and $\mathbb{Q}$ under which S has the *same marginals*. Here it is: $T = 2$, $S_0 = 3/2$, the marginals at time $t = 1$ are given by $\mathbb{P}[S_1 = 1] = \mathbb{Q}[S_1 = 1] = 1/2$ and $\mathbb{P}[S_1 = 2] = \mathbb{Q}[S_1 = 2] = 1/2$, and S_2 takes the values $0, 1, 2, 3$ each one with probability $1/4$ under both $\mathbb{P}$ and $\mathbb{Q}$. To complete the description of $\mathbb{P}$ and $\mathbb{Q}$, we only need to assign the transition probabilities between $t = 1$ and $t = 2$. This can be done in many ways to get $\mathbb{P}, \mathbb{Q} \in \mathscr{M}$ and nonetheless keep them different. For instance, set $p_{ij} := \mathbb{P}[S_2 = j|S_1 = i]$ and $q_{ij} := \mathbb{Q}[S_2 = j|S_1 = i]$ for $i, j \in \{0, 1, 2, 3\}$, and consider $p_{23} = p_{10} = 0.4$, $p_{22} = p_{11} = 0.3$, $p_{21} = p_{12} = 0.2$ and $p_{20} = p_{13} = 0.1$ for the measure $\mathbb{P}$, and $q_{23} = q_{10} = 0.39$, $q_{22} = q_{11} = 0.33$, $q_{21} = q_{12} = 0.17$ and $q_{20} = q_{13} = 0.11$ for the measure $\mathbb{Q}$. It can be easily verified that this example is exactly what we were looking for.

Remark 5 As it is formulated, the model of this example does not satisfy the assumption $S_T = S_{T-}$. Indeed, $S_{T-} = S_{2-} = S_1 \neq S_2$. Nonetheless, it can be easily embedded in the framework of the previous section, where the price process S is assumed to be left-continuous at T by simply adding the date $T + 1$ and setting $S_{T+1} = S_T$.

Remark 6 By the call-put parity, trading in all European call options as in (7) is equivalent to trading in S and in all European put options. As a consequence, our example also shows that, in general, it is not possible to replicate each square-integrable positivee contingent claim by trading dynamically in the underlying and statically in all European put options.

Acknowledgements I wish to thank Sara Biagini, José M. Corcuera, Jerôme Renault and two anonymous referees for many valuable remarks. I also thank the "Chair Les Particuliers Face aux Risques", Fondation du Risque (Groupama-ENSAE-Dauphine), and the GIP-ANR "Croyances" project. The usual disclaimer applies.

References

1. Campi, L.: Arbitrage and completeness in financial markets with given N-dimensional distributions. Decis. Econ. Finance **27**(1), 57–80 (2004)
2. Carmona, R., Nadtochiy, S.: Local volatility dynamic models. Finance Stoch. **13**(1), 1–48 (2009)
3. Carr, P., Ellis, K., Gupta, V.: Static hedging of exotic options. J. Finance **53**(3), 1165–1190 (1998)
4. Carr, P., Madan, D.B.: Optimal positioning in derivative securities. Quant. Finance **1**, 19–37 (2001)

5. Carr, P., Jin, X., Madan, D.B.: Optimal investment in derivative securities. Finance Stoch. **5**(1), 33–59 (2001)
6. Corcuera, J.M., Guerra, J.: Dynamic complex hedging in additive markets. Preprint, IMUB, Universitat de Barcelona (2007)
7. Corcuera, J.M., Nualart, D., Schoutens, W.: Completion of a Lévy market by power-jump assets. Finance Stoch. **9**(1), 109–127 (2005)
8. Davis, M., Obloj, J.: Market completion using options. In: Stettner, L. (ed.) Advances in Mathematics of Finance. Banach Center Publications, vol. 43, pp. 49–60. Polish Academy of Sciences, Warsaw (2008)
9. El Karoui, N.: Les aspects probabilistes du contrôle stochastique. (French) [The probabilistic aspects of stochastic control] Ninth Saint Flour Probability Summer School 1979, pp. 73–238. Lecture Notes in Math., vol. 876, Springer, Berlin, (1981)
10. Green, R.C., Jarrow, R.A.: Spanning and completeness in markets with contingent claims. J. Econ. Theory **41**(1), 202–210 (1987)
11. Ilhan, A., Sircar, R.: Optimal static-dynamic hedges for barrier options. Math. Finance **16**(2), 359–385 (2006)
12. Ilhan, A., Jonsson, M., Sircar, R.: Optimal investment with derivative securities. Finance Stoch. **9**(4), 585–595 (2005)
13. Jacod, J., Protter, Ph.: Risk neutral compatibility with option prices. Finance Stoch. **14**(2), 285–315 (2010)
14. Jacod, J., Shiryaev, A.N.: Limit Theorems for Stochastic Processes. Springer, Berlin (2003)
15. Nachman, D.: Spanning and Completeness with Options. The Review of Financial Studies. Fall (1988)
16. Protter, Ph.: Stochastic Integration and Differential Equations, 2nd edn. Stochastic Modelling and Applied Probability, vol. 21, Springer, Berlin (2005). Version 2.1. Corrected third printing
17. Rogge, L.: Call completeness implies completeness in the n-period model of a financial market. Finance Stoch. **10**(2), 298–301 (2006)
18. Ross, S.: Options and efficiency. Q. J. Econ. **90** (1976)
19. Schweizer, M., Wissel, J.: Term structures of implied volatilities: absence of arbitrage and existence results. Math. Finance **18**(1), 77–114 (2008)
20. Schweizer, M., Wissel, J.: Arbitrage-free market models for option prices: the multi-strike case. Finance Stoch. **12**(4), 469–505 (2008)
21. Zălinescu, C.: Convex Analysis in General Vector Spaces. World Scientific, New Jersey (2002)

An f-Divergence Approach for Optimal Portfolios in Exponential Lévy Models

S. Cawston and L. Vostrikova

Abstract We present a unified approach to get explicit formulas for utility maximizing strategies in exponential Lévy models. This approach is related to f-divergence minimal martingale measures and based on a new concept of preservation of the Lévy property by f-divergence minimal martingale measures. For common f-divergences, i.e. functions which such that $f''(x) = ax^{\gamma}$, $a > 0$, $\gamma \in \mathbb{R}$, we give the conditions for the existence of corresponding u_f- maximizing strategies, as well as explicit formulas.

Keywords f-Divergence · Exponential Lévy models · Optimal portfolio

Mathematics Subject Classification (2010) 91B20 · 60G07 · 60G51

1 Introduction

Exponential Lévy models have been widely used since the 1990's to represent asset prices. In the case of continuous trajectories, this leads to the classical Black–Scholes model, but the class of Lévy models also contains a number of popular jump models including Generalized Hyperbolic models ([5]) and Variance-Gamma models [1]. The use of such processes allows for an excellent fit both for daily log-returns ([6]) and intra-day data ([6]). The class is also flexible enough to allow for processes with either finite or infinite variation and finite or infinite activity. However, contrary to the Black–Scholes case, Lévy models generally lead to incomplete financial markets: contingent claims cannot all be replicated by admissible strategies. Therefore, it is important to determine strategies which are, in a certain sense

S. Cawston
LAREMA, Département de Mathématiques, Université d'Angers, 2, Bd Lavoisier, 49045 Angers Cedex 01, France
e-mail: suzanne.cawston@univ-angers.fr

L. Vostrikova (✉)
Université d'Angers, rue de Rennes, 40, 49035 Angers, France
e-mail: vostrik@univ-angers.fr

Y. Kabanov et al. (eds.), *Inspired by Finance*, DOI 10.1007/978-3-319-02069-3_5,

optimal. Various criteria are used, some of which are linked to risk minimization (see [8, 20, 21]) and others consisting in maximizing certain utility functions (see [10, 13]). It has been shown (see [10, 15]) that such questions are strongly linked via the Fenchel–Legendre transform to dual optimization problems on the set of equivalent martingale measures, i.e. the measures which are equivalent to the initial physical measure and under which the stock price is a martingale. More precisely, we recall that the convex conjugate of a concave function u is defined by

$$f(y) = \sup_{x \in \mathbb{R}}\{u(x) - xy\} = u(I(y)) - yI(y)$$

where $I = (u')^{-1}$. In particular, we have the following correspondences:

$$\begin{cases} \text{if } u(x) = \ln(x), \text{ then } f(x) = -\ln(x) - 1, \\ \text{if } u(x) = \frac{x^p}{p},\ p < 1, \text{ then } f(x) = -\frac{p-1}{p} x^{\frac{p}{p-1}}, \\ \text{if } u(x) = 1 - e^{-x}, \text{ then } f(x) = 1 - x + x\ln(x). \end{cases}$$

Given a convex function f, the problem of minimizing the f-divergence $E[f(\frac{dQ_T}{dP_T})]$ of the restrictions of the measures P and Q on the time interval $[0, T]$ over the set of equivalent martingale measures has been well studied for a number of functions in [3, 4, 7, 9, 14, 17] and [12]. For properties of f-divergence see also [16]. It has been noted in [10] that if a solution Q^* to such a problem exists, there exists a predictable process $\hat{\phi}$ such that

$$-f'\left(\frac{dQ_T^*}{dP_T}\right) = x + \int_0^T \hat{\phi}_s dS_s,$$

where the process S which represents the risky asset, is a semimartingale and x is a constant. Moreover, under some assumptions, $\hat{\phi}$ will then define a u-optimal strategy. However, it is in general far from easy to obtain an explicit expression for $\hat{\phi}$, although results exist for a certain number of special cases. These special cases concern what we will call common f-divergences, i.e. functions f such that $f''(x) = ax^{\gamma}$ where $a > 0$.

Our aim here is to obtain, for a certain class of utility functions, an explicit expression for $\hat{\phi}$ both when the Gaussian part of the Lévy process is non-zero, i.e. $c \neq 0$, and when $c = 0$. We consider a class of f-divergences whose f-divergence minimal martingale measure Q^* preserves the Lévy property of the initial Lévy process. It is known that common f-divergences preserves Lévy property for all Levy processes and that the class of Levy preserving f-divergences for fixed Levy process is larger, in general, then common f-divergences as it was shown in [2]. In addition, this new approach permit us to suggest a unified way for finding $\hat{\phi}$. In particular, we deduce from this result a unified formula for $\hat{\phi}$ for all common f-divergences.

Let us denote by $Z_T = \frac{dQ_T^*}{dP_T}$ the Radon–Nikodym derivative of Q_T^* with respect to P_T and let (β, Y) be the Girsanov parameters for the change of measure from P_T to Q_T^* (cf. [11], p. 159). We exclude from our consideration a trivial case $P = Q^*$

in which $\hat{\phi} = 0$. We consider utility functions u such that their convex conjugate f_u used as an f-divergence gives us a Lévy property preserving f-divergence minimal equivalent martingale measure Q^*. Then, under some integrability conditions, we prove that if the Gaussian part of the initial Lévy process is not zero, then the asymptotically optimal strategy $\hat{\phi}$ is given by:

$$\hat{\phi}_s^{(i)} = -\frac{\lambda \beta^{(i)} Z_{s-}}{S_{s-}^{(i)}} E_{Q^*}[f''(\lambda x Z_{T-s}) Z_{T-s}]_{|x=Z_{s-}}$$

where $\lambda > 0$ is the unique solution to the equation $E_{Q^*}[-f'(\lambda Z_T^*)] = x$ and x is the initial capital. If the Gaussian part of the initial Lévy process is zero, the support of the Lévy measure is of non-empty interior, it contains zero and Y is not identically 1, then

$$\hat{\phi}^{(i)} = -\frac{\lambda \gamma^{(i)} Z_{s-}}{S_{s-}^{(i)}} E_{Q^*}[f''(\lambda x Z_{T-s}) Z_{T-s}]_{|x=Z_{s-}}$$

where $\gamma^{(i)}$ are constants related with the second Girsanov parameter and given by (12) (cf. Theorem 2).

In the particular case of common utility functions (corresponding to common f-divergences) we give conditions that ensure existence of the optimal strategy and we obtain also its expression. For example, for $c \neq 0$,

$$\hat{\phi}_s^{(i)} = \frac{\alpha_{\gamma+1}(x)\, \beta^{(i)}}{E_{Q^*}[Z_s^{\gamma+1}]} \frac{Z_{s-}^{\gamma+1}}{S_{s-}^{(i)}}$$

where $\alpha_{\gamma+1}(x)$ is given by (19) (cf. Proposition 1).

The paper is organized in the following way. In Sect. 2 we recall known facts about utility maximization. In Sect. 3 we prove (cf. Theorem 1) a decomposition needed to find optimal strategies, then in Sect. 4 we give a general result about optimal strategies (cf. Theorem 2). Finally, in Proposition 1 we obtain the results concerning common f-divergences.

2 Utility Maximization in Exponential Lévy Models

We start by describing our model in more detail. We assume that the financial market consists of a non-risky asset B whose value at time t is

$$B_t = B_0 e^{rt},$$

where $r \geq 0$ is the interest rate which we assume to be constant, and d risky assets whose prices are described by a d-dimensional stochastic process $S = (S_t)_{t \geq 0}$ with

$$S_t = (e^{X_t^{(1)}}, \ldots, e^{X_t^{(d)}})$$

where $X = (X_t^{(1)}, \dots, X_t^{(d)})_{t\geq 0}$ is a d-dimensional Lévy process defined on a filtered probability space $(\Omega, \mathscr{F}, \mathbb{F}, P)$ with the natural filtration $\mathbb{F} = (\mathscr{F}_t)_{t\geq 0}$ satisfying usual properties. We recall that Lévy processes form the class of càdlàg processes with stationary and independent increments and such that the law of X_t is given by the Lévy–Khintchine formula: for all $t \geq 0$, for all $u \in \mathbb{R}$

$$E[e^{i\langle u, X_t\rangle}] = e^{t\psi(u)}$$

with

$$\psi(u) = i\langle u, b\rangle - \frac{1}{2}{}^{\top}ucu + \int_{\mathbb{R}^d} [e^{i\langle u,y\rangle} - 1 - i\langle u, h(y)\rangle]\nu(dy)$$

where $b \in \mathbb{R}^d$ is a drift, c is a positive $d \times d$ symmetric matrix, ν is a positive measure on $\mathbb{R}^d \setminus \{0\}$ which satisfies

$$\int_{\mathbb{R}^d} 1 \wedge |y|^2 \nu(dy) < \infty$$

and $h(\cdot)$ is a truncation function. The triplet (b, c, ν) entirely determines the law of the Lévy process X, and is called the characteristic triplet of X. For more details see [18]. We also recall that if $S = e^X$, there exists a Lévy process $\hat{X}$ such that $S = \mathscr{E}(\hat{X})$, where $\mathscr{E}$ denotes the Doleans-Dade exponential. For more details see [11].

An investor will share out his capital among the different assets according to a strategy which is represented by a process $\Phi = (\eta, \phi)$, where η represents the quantity invested in the non-risky asset B, and $\phi = (\phi^{(1)}, \dots, \phi^{(d)})$ is the quantity invested in the risky assets. From now on, we will denote by

$$(\phi \cdot S)_t = \int_0^t \eta_s dB_s + \sum_{i=1}^d \int_0^t \phi_s^{(i)} dS_s^{(i)}$$

the variation of capital due to the investment in the risky assets. We now define more precisely our set of admissible strategies. We recall that an admissible strategy is a predictable process $\Phi = (\eta, \phi)$ taking values in $\mathbb{R}^{d+1}$, such that η is B-integrable, ϕ is S-integrable and for which there exists $a \in \mathbb{R}^+$ such that for all $t \geq 0$,

$$(\phi \cdot S)_t \geq -a.$$

We denote by $\mathscr{A}$ the set of all admissible strategies.

We are interested in strategies which are optimal in the sense of utility maximization. We recall that a utility function is a function $u :]\underline{x}, \infty[\to \mathbb{R}$, which is $\mathscr{C}^1$, strictly increasing, strictly concave and such that

$$\lim_{x\to\infty} u'(x) = 0, \qquad \lim_{x\to\underline{x}} u'(x) = \infty$$

where $\underline{x} = \inf\{x : x \in \operatorname{dom} u\}$. In particular, the most common utility functions are $u(x) = \ln(x)$, $u(x) = \frac{x^p}{p}$, $p < 1$, or $u(x) = 1 - e^{-x}$. We now recall the definition of u-optimal and u-asymptotically optimal strategies. This last notion was first introduced in [13]. It will allow us to consider in a unified way all utilities including those with $\underline{x} = \infty$.

We say that a strategy $\hat{\phi} \in \mathscr{A}$ is u-optimal on $[0, T]$ if

$$E[u(x + (\hat{\phi} \cdot S)_T)] = \sup_{\phi \in \mathscr{A}} E[u(x + (\phi \cdot S)_T)].$$

A sequence of admissible strategies $(\hat{\phi}^{(n)})_{n \geq 1}$ is asymptotically u-optimal on $[0, T]$ if

$$\lim_{n \to \infty} E[u(x + (\hat{\phi}^{(n)} \cdot S)_T)] = \sup_{\phi \in \mathscr{A}} E[u(x + (\phi \cdot S)_T)].$$

To avoid unnecessary complications and without loss of generality, we suppose up to now that the interest rate $r = 0$.

3 A Decomposition for Lévy Preserving Equivalent Martingale Measures

In this section, we consider a fixed strictly convex function f, $f \in C^3(\mathbb{R}^{+,*})$, and a Lévy preserving equivalent martingale measure Q whose density is given by the process $Z = (Z_t)_{t \geq 0}$. We recall that Q preserves the Lévy property if X remains a Lévy process under Q. We also recall that we characterize the change of measure from P into Q by the Girsanov parameters (β, Y). Then the fact that Q preserves the Lévy property can be seen as a change of measure such that the first Girsanov parameter β is a constant and the second parameter Y depends only on jump-sizes. As a consequence, the density of a Lévy preserving measure can be represented in the form $Z = \mathscr{E}(N)$, where

$$N_t = \beta X_t^{(c)} + \int_0^t \int_{\mathbb{R}^{d*}} (Y(x) - 1)(\mu^X - \nu^{X,P})(ds, dx).$$

In addition, if Q is a martingale measure then β and Y satisfy

$$b + \frac{1}{2} diag(c) + c\beta + \int_{\mathbb{R}^d} [(e^x - 1)Y(x) - h(x)]\nu(dx) = 0.$$

The last relation ensures that the drift of S under the measure Q is zero.

Our main aim in this section is to show that under certain integrability conditions, the decomposition given in Theorem 1 holds. We introduce càdlàg versions of the processes $(\xi_t(x))_{t \geq 0}$ and $(H_t(x, y))_{t \geq 0}$ where for $t \leq T$

$$\xi_t(x) = E_Q[f''(xZ_{T-t})Z_{T-t}] \tag{1}$$

and

$$H_t(x, y) = E_Q[f'(xZ_{T-t}Y(y)) - f'(xZ_{T-t})]. \tag{2}$$

Theorem 1 *Let f be a strictly convex function belonging to $\mathscr{C}^3(\mathbb{R}^{+,*})$. Let Z be the density of a Lévy preserving equivalent martingale measure Q. Assume that Q is such that: for all $\lambda > 0$ and all compact set $K \subseteq \mathbb{R}^+$*

$$E_P|f(\lambda Z_T)| < \infty, \quad E_Q|f'(\lambda Z_T)| < \infty, \quad \sup_{t\leq T}\sup_{\lambda\in K} E_Q[f''(\lambda Z_t)Z_t] < \infty. \tag{3}$$

Then, for all $\lambda > 0$ we have Q-a.s., for all $t \leq T$,

$$E_Q[f'(\lambda Z_T)|\mathscr{F}_t] = E_Q[f'(\lambda Z_T)] + \sum_{i=1}^{d} \lambda\beta^{(i)} \int_0^t \xi_s(\lambda Z_{s-})\, Z_{s-}\, dX_s^{(c),Q,i}$$
$$+ \int_0^t \int_{\mathbb{R}^d} H_s(\lambda Z_{s-}, y)\,(\mu^X - \nu^{X,Q})(ds, dy) \tag{4}$$

where $\beta = {}^\top(\beta_1, \ldots, \beta_d)$ is a first Girsanov parameter and $\nu^{X,Q}$ is the dual predictable projection or the compensator of the jump measure μ^X with respect to $(\mathbb{F}, Q)$.

This result is based on an application of the Ito formula, but it will require some technical lemmas.

We recall that as Q preserves the Lévy property, for all $t \leq T$, Z_t and $\frac{Z_T}{Z_t}$ are independent under P and that $\mathscr{L}(\frac{Z_T}{Z_t} \mid P) = \mathscr{L}(Z_{T-t} \mid P)$. Therefore

$$E_Q[f'(\lambda Z_T)|\mathscr{F}_t] = \rho(t, Z_t)$$

where $\rho(t, x) = E_Q[f'(\lambda x Z_{T-t})]$. Our integrability conditions do not allow us to apply the Ito formula directly to the function $\rho(t, Z_t)$. Therefore, we start by considering a sequence of bounded approximations of f', and will then obtain (4) by studying the convergence of analogous decompositions for the approximations of f'.

Lemma 1 *Let f be a strictly convex function belonging to $\mathscr{C}^3(\mathbb{R}^{+,*})$. There exists a sequence of bounded increasing functions $(\phi_n)_{n\geq 1}$, which are of class $\mathscr{C}^2$ on $\mathbb{R}^{+,*}$, such that for all $n \geq 1$, ϕ_n coincides with f' on the compact set $[\frac{1}{n}, n]$ and such that for n large enough and for all $x, y > 0$ the following inequalities hold:*

$$|\phi_n(x)| \leq 4|f'(x)| + \alpha, \quad |\phi_n'(x)| \leq 3f''(x), \qquad |\phi_n(x) - \phi_n(y)| \leq 5|f'(x) - f'(y)|, \tag{5}$$

where α is a real positive constant.

Proof We set, for $n \geq 1$,

$$A_n(x) = f'\left(\frac{1}{n}\right) - \int_{x\vee\frac{1}{2n}}^{\frac{1}{n}} f''(y)(2ny-1)^2(5-4ny)dy,$$

$$B_n(x) = f'(n) + \int_n^{x\wedge(n+1)} f''(y)(n+1-y)^2(1+2y-2n)dy,$$

and, finally,

$$\phi_n(x) = \begin{cases} A_n(x) & \text{if } 0 \le x < \frac{1}{n}, \\ f'(x) & \text{if } \frac{1}{n} \le x \le n, \\ B_n(x) & \text{if } x > n. \end{cases}$$

Here A_n and B_n are defined so that ϕ_n is of class $\mathscr{C}^2$ on $\mathbb{R}^{+,*}$. For the inequalities we use the fact that f' is increasing function and the estimations:

$$0 \le (2nx-1)^2(5-4nx) \le 1 \quad \text{for } x \in \left[\frac{1}{2n}, \frac{1}{n}\right]$$

and

$$0 \le (n+1-x)^2(1+2x-2n) \le 3 \quad \text{for } x \in [n, n+1].$$ □

We now introduce, for each $n \ge 1$, the function

$$\rho_n(t,x) = E_Q[\phi_n(\lambda x Z_{T-t})]$$

and we obtain the following version to Theorem 1, replacing f' with ϕ_n. For that we put

$$\xi_t^{(n)}(x) = E_Q[\phi_n'(xZ_{T-t})Z_{T-t}] \tag{6}$$

and

$$H_t^{(n)}(x,y) = E_Q[\phi_n(xZ_{T-t}Y(y)) - \phi_n(xZ_{T-t})]. \tag{7}$$

Lemma 2 *We have Q-a.s., for all $t \le T$,*

$$\rho_n(t,Z_t) = E_Q[\phi_n(\lambda Z_T)] + \sum_{i=1}^d \lambda\,\beta^{(i)} \int_0^t \xi_s^{(n)}(\lambda Z_{s-})\,Z_{s-}\,dX_s^{(c),Q,i}$$
$$+ \int_0^t\int_{\mathbb{R}^d} H_s^{(n)}(\lambda Z_{s-},y)\,(\mu^X - \nu^{X,Q})(ds,dy) \tag{8}$$

where $\beta = {}^\top(\beta_1,\ldots,\beta_d)$ is the first Girsanov parameter and $\nu^{X,Q}$ is the dual predictable projection or the compensator of the jump measure μ^X with respect to $(\mathbb{F}, Q)$.

Proof In order to apply the Ito formula to ρ_n, we need to show that ρ_n is twice continuously differentiable with respect to x and once with respect to t and that the corresponding derivatives are bounded for all $t \in [0, T]$ and $x \geq \epsilon$, $\epsilon > 0$. First of all, we note from the definition of ϕ_n that for all $x \geq \epsilon > 0$

$$\left|\frac{\partial}{\partial x}\phi_n(\lambda x Z_{T-t})\right| = |\lambda Z_{T-t}\phi_n'(\lambda x Z_{T-t})| \leq \frac{(n+1)}{\epsilon}\sup_{z>0}|\phi_n'(z)| < \infty.$$

Therefore, ρ_n is differentiable with respect to x and we have

$$\frac{\partial}{\partial x}\rho_n(t,x) = \lambda E_Q[\phi_n'(\lambda x Z_{T-t})\, Z_{T-t}].$$

Moreover, the function $(x,t) \mapsto \lambda\phi_n'(\lambda x Z_{T-t})Z_{T-t}$ is continuous P-a.s. and bounded. This implies that $\frac{\partial}{\partial x}\rho_n$ is continuous and bounded for $t \in [0, T]$ and $x \geq \epsilon > 0$.

In the same way, for all $x \geq \epsilon > 0$

$$\left|\frac{\partial^2}{\partial x^2}\phi_n(\lambda x Z_{T-t})\right| = \lambda^2 Z_{T-t}^2\phi_n''(\lambda x Z_{T-t}) \leq \frac{(n+1)^2}{\epsilon^2}\sup_{z>0}\phi_n''(z) < \infty.$$

Therefore, ρ_n is twice continuously differentiable in x and

$$\frac{\partial^2}{\partial x^2}\rho_n(t,x) = \lambda^2 E_Q[\phi_n''(\lambda x Z_{T-t})Z_{T-t}^2].$$

We can verify easily that it is again continuous and bounded function. In order to obtain differentiability with respect to t, we need to apply the Ito formula to ϕ_n:

$$\begin{aligned}\phi_n(\lambda x Z_t) &= \phi_n(\lambda x) + \sum_{i=1}^{d}\int_0^t \lambda x\phi_n'(\lambda x Z_{s-})\beta^{(i)}\, Z_{s-} dX_s^{(c),Q,i}\\ &+ \int_0^t\int_{\mathbb{R}^d}[\phi_n(\lambda x Z_{s-}Y(y)) - \phi_n(\lambda x Z_{s-})]\,(\mu^X - \nu^{X,Q})(ds,dy)\\ &+ \int_0^t \psi_n(\lambda x, Z_{s-})ds\end{aligned}$$

where

$$\begin{aligned}\psi_n(\lambda x, Z_{s-}) = {}^{\top}\beta c\beta&\left[\lambda x Z_{s-}\phi_n'(\lambda x Z_{s-}) + \frac{1}{2}x^2\lambda^2 Z_{s-}^2\phi_n''(\lambda x Z_{s-})\right]\\ &+ \int_{\mathbb{R}^d}[(\phi_n(\lambda x Z_{s-}Y(y)) - \phi_n(\lambda x Z_{s-}))\,Y(y)\\ &- \lambda x\phi_n'(\lambda x Z_{s-})Z_{s-}(Y(y)-1)]\nu(dy).\end{aligned}$$

Therefore, for fixed $t > 0$

$$E_Q[\phi_n(\lambda x Z_{T-t})] = \int_0^{T-t} E_Q[\psi_n(\lambda x, Z_{s-})]ds$$

so that ρ_n is differentiable with respect to t and

$$\frac{\partial}{\partial t}\rho_n(t,x) = -E_Q[\psi_n(\lambda x, Z_{s-})]_{|s=T-t}.$$

We can also easily check that this is again a continuous and bounded function. For this we use the fact that ϕ_n, ϕ_n', and ϕ_n'' are bounded functions and also that the Hellinger process of Q_T and P_T of the order $1/2$ is finite.

We can finally apply the Ito formula to ρ_n. For that we use the stopping times

$$s_m = \inf\{t \geq 0 : \ Z_t \leq m^{-1}\},$$

$m \geq 1$ and $\inf\{\emptyset\} = \infty$. Then, from the Markov property of Lévy process we have:

$$\rho_n(t \wedge s_m, Z_{t\wedge s_m}) = E_Q(\phi_n(\lambda Z_T) \mid \mathscr{F}_{t\wedge s_m}).$$

Note that $(E_Q(\phi_n(\lambda Z_T) \mid \mathscr{F}_{t\wedge s_m})_{t\geq 0}$ is a Q-martingale, uniformly integrable with respect to m. From the Ito formula we have:

$$\begin{aligned}\rho_n(t \wedge s_m, Z_{t\wedge s_m}) &= E_Q(\phi_n(\lambda Z_T)) + \int_0^{t\wedge s_m} \frac{\partial \rho_n}{\partial s}(s, Z_{s-})ds \\ &\quad + \int_0^{t\wedge s_m} \frac{\partial \rho_n}{\partial x}(s, Z_{s-})dZ_s + \frac{1}{2}\int_0^{t\wedge s_m} \frac{\partial^2 \rho_n}{\partial x^2}(s, Z_{s-})d\langle Z^c\rangle_s \\ &\quad + \sum_{0\leq s\leq t\wedge s_m} \rho_n(s, Z_s) - \rho_n(s, Z_{s-}) - \frac{\partial \rho_n}{\partial x}(s, Z_{s-})\Delta Z_s\end{aligned}$$

where $\Delta Z_s = Z_s - Z_{s-}$. After standard simplifications we get that

$$\rho_n(t \wedge s_m, Z_{t\wedge s_m}) = A_{t\wedge s_m} + M_{t\wedge s_m}$$

where $(A_{t\wedge s_m})_{t\leq T}$ is a predictable process

$$\begin{aligned}A_{t\wedge s_m} &= \int_0^{t\wedge s_m} \frac{\partial \rho_n}{\partial s}(s, Z_{s-})ds + \frac{1}{2}\int_0^{t\wedge s_m} \frac{\partial^2 \rho_n}{\partial x^2}(s, Z_{s-})d\langle Z^c\rangle_s \\ &\quad + \int_0^{t\wedge s_m}\int_{\mathbb{R}} [\rho_n(s, Z_{s-}+x) - \rho_n(s, Z_{s-}) - \frac{\partial \rho_n}{\partial x}(s, Z_{s-})x]\nu^{Z,Q}(ds,dx)\end{aligned}$$

and $(M_{t\wedge s_m})_{t\leq T}$ is a Q-martingale,

$$\begin{aligned}M_{t\wedge s_m} &= E_Q(\phi_n(\lambda Z_T)) + \int_0^{t\wedge s_m} \frac{\partial \rho_n}{\partial x}(s, Z_{s-})dZ_s^c \\ &\quad + \int_0^{t\wedge s_m}\int_{\mathbb{R}} [\rho_n(s, Z_{s-}+x) - \rho_n(s, Z_{s-})](\mu^Z(ds,dx) - \nu^{Z,Q}(ds,dx)).\end{aligned}$$

Then, we pass to the limit as $m \to \infty$. Note that the sequence $(s_m)_{m\geq 1}$ tends to infinity as $m \to \infty$. From [19], Corollary 2.4, p. 59, we obtain that

$$\lim_{m\to\infty} E_Q(\phi_n(Z_T)\,|\,\mathscr{F}_{t\wedge s_m}) = E_Q(\phi_n(Z_T)\,|\,\mathscr{F}_t)$$

and by the definition of local martingales we get:

$$\lim_{m\to\infty}\int_0^{t\wedge s_m}\frac{\partial\rho_n}{\partial x}(s,Z_{s-})dZ_s^c = \int_0^t \frac{\partial\rho_n}{\partial x}(s,Z_{s-})dZ_s^c = \int_0^t \lambda\xi_s^{(n)}(Z_{s-})dZ_s^c$$

and

$$\lim_{m\to\infty}\int_0^{t\wedge s_m}\int_{\mathbb{R}}[\rho_n(s,Z_{s-}+x)-\rho_n(s,Z_{s-})](\mu^Z(ds,dx)-\nu^{Z,Q}(ds,dx))$$
$$=\int_0^t\int_{\mathbb{R}}[\rho_n(s,Z_{s-}+x)-\rho_n(s,Z_{s-})](\mu^Z(ds,dx)-\nu^{Z,Q}(ds,dx)).$$

Now, in each stochastic integral we pass from the integration with respect to the process Z to the one with respect to the process X. For that we observe that

$$dZ_s^c = \sum_{i=1}^d \beta^{(i)} Z_{s-}dX_s^{c,Q,i}, \qquad \Delta Z_s = Z_{s-}Y(\Delta X_s).$$

Lemma 2 is proved. □

We now turn to the proof of Theorem 1. In order to obtain the decomposition for f', we prove convergence in probability of the processes in (8).

Proof of Theorem 1 For $n \geq 1$ and a fixed $\lambda > 0$, we introduce the stopping times

$$\tau_n = \inf\{t\geq 0:\ \lambda Z_t \geq n \text{ or } \lambda Z_t \leq n^{-1}\} \tag{9}$$

and note that $\tau_n \to \infty$ (P-a.s.) as $n\to\infty$. We note also that

$$|E_Q[f'(\lambda Z_T)|\mathscr{F}_t] - \rho_n(t,Z_t)| \leq E_Q[|f'(\lambda Z_T)-\phi_n(\lambda Z_T)||\mathscr{F}_t]$$

As f' and ϕ_n coincide on the interval $[n^{-1},n]$, it follows from Lemma 2 that

$$|E_Q[f'(\lambda Z_T)|\mathscr{F}_t] - \rho_n(t,Z_t)| \leq E_Q[|f'(\lambda Z_T)-\phi_n(\lambda Z_T)|\mathbf{1}_{\{\tau_n\leq T\}}|\mathscr{F}_t]$$
$$\leq E_Q[(5|f'(\lambda Z_T)|+\alpha)\mathbf{1}_{\{\tau_n\leq T\}}|\mathscr{F}_t].$$

Now, for every $\epsilon > 0$, by the Doob inequality and the Lebesgue dominated convergence theorem we get:

$$\lim_{n\to\infty} Q\Big(\sup_{t\leq T} E_Q[(5|f'(\lambda Z_T)|+\alpha)\mathbf{1}_{\{\tau_n\leq T\}}|\mathscr{F}_t] > \epsilon\Big)$$

$$\leq \lim_{n\to\infty} \frac{1}{\epsilon} E_Q[(5|f'(\lambda Z_T)| + \alpha)\mathbf{1}_{\{\tau_n \leq T\}}] = 0.$$

Therefore, we have

$$\lim_{n\to\infty} Q\Big(\sup_{t\leq T} |E_Q[f'(\lambda Z_T) - \rho_n(t, \lambda Z_t)|\mathscr{F}_t]| > \epsilon\Big) = 0.$$

We now turn to the convergence of the three elements on the rhs of (8). We have that $\lim_{n\to+\infty} \phi_n(\lambda Z_T) = f'(\lambda Z_T)$ almost surely, and $|\phi_n(\lambda Z_T)| \leq 4|f'(\lambda Z_T)| + \alpha$ for all $n \geq 1$. Therefore, it follows from the dominated convergence theorem that

$$\lim_{n\to\infty} E_Q[\phi_n(\lambda Z_T)] = E_Q[f'(\lambda Z_T)].$$

We now prove the convergence of the continuous martingale parts of (8). It follows from Lemma 1 that

$$\begin{aligned} Z_t|\xi_t^{(n)}(\lambda Z_t) - \xi_t(\lambda Z_t)| &\leq E_Q[Z_T|\phi_n'(\lambda Z_T) - f''(\lambda Z_T)|\mathscr{F}_t] \\ &\leq 4E_Q[Z_T|f''(\lambda Z_T)|\mathbf{1}_{\{\tau_n \leq T\}}|\mathscr{F}_t]. \end{aligned}$$

Hence, we have as before for $\epsilon > 0$

$$\lim_{n\to\infty} Q\Big(\sup_{t\leq T} Z_t|\xi_t^{(n)}(\lambda Z_t) - \xi_t(\lambda Z_t)| > \epsilon\Big) \leq \lim_{n\to\infty} \frac{4}{\epsilon} E_Q[Z_T f''(\lambda Z_T)\mathbf{1}_{\{\tau_n \leq T\}}] = 0.$$

Therefore, it follows from the Lebesgue dominated convergence theorem for stochastic integrals (see [11], Theorem I.4.31, p. 46) that for all $\epsilon > 0$ and $1 \leq i \leq d$

$$\lim_{n\to\infty} Q\Big(\sup_{t\leq T} \Big| \int_0^t Z_{s-}(\xi_s^{(n)}(\lambda Z_{s-}) - \xi_s(\lambda Z_{s-}))dX_s^{(c),Q,i} \Big| > \epsilon\Big) = 0.$$

It remains to show the convergence of the discontinuous martingales to zero as $n \to \infty$. We start from the identity

$$\int_0^t \int_{\mathbb{R}^d} [H_s^{(n)}(\lambda Z_{s-}, y) - H_s(\lambda Z_{s-}, y)](\mu^X - \nu^{X,Q})(ds, dy) = M_t^{(n)} + N_t^{(n)}$$

with

$$\begin{aligned} M_t^{(n)} &= \int_0^t \int_A [H_s^{(n)}(\lambda Z_{s-}, y) - H_s(\lambda Z_{s-}, y)](\mu^X - \nu^{X,Q})(ds, dy), \\ N_t^{(n)} &= \int_0^t \int_{A^c} [H_s^{(n)}(\lambda Z_{s-}, y) - H_s(\lambda Z_{s-}, y)](\mu^X - \nu^{X,Q})(ds, dy), \end{aligned}$$

where $A = \{y : |Y(y) - 1| < 1/4\}$.

For $p \geq 1$, we consider the sequence of stopping times τ_p defined by (9) with n replaced by a real positive p. We also introduce the processes

$$M^{(n,p)} = (M_t^{(n,p)})_{t\geq 0}, \ N^{(n,p)} = (N_t^{(n,p)})_{t\geq 0}$$

with $M_t^{(n,p)} = M_{t\wedge\tau_p}^{(n)}$, $N_t^{(n,p)} = N_{t\wedge\tau_p}^{(n)}$. Note that for $p \geq 1$ and $\epsilon > 0$

$$Q\Big(\sup_{t\leq T} |M_t^{(n)} + N_t^{(n)}| > \epsilon\Big) \leq Q(\tau_p < T) + Q\Big(\sup_{t\leq T} |M_t^{(n,p)}| > \frac{\epsilon}{2}\Big) + Q\Big(\sup_{t\leq T} |N_t^{(n,p)}| > \frac{\epsilon}{2}\Big).$$

Furthermore, we obtain from the Doob martingale inequalities that

$$Q\Big(\sup_{t\leq T} |M_t^{(n,p)}| > \frac{\epsilon}{2}\Big) \leq \frac{4}{\epsilon^2} E_Q[(M_T^{(n,p)})^2] \tag{10}$$

and

$$Q\Big(\sup_{t\leq T} |N_t^{(n,p)}| > \frac{\epsilon}{2}\Big) \leq \frac{2}{\epsilon} E_Q|N_T^{(n,p)}|. \tag{11}$$

Since $\tau_p \to \infty$ as $p \to \infty$, it is sufficient to show that $E_Q[M^{(n,p)}]^2$ and $E_Q|N^{(n,p)}|$ converge to 0 as $n \to \infty$. To do so we estimate $E_Q[(M_T^{(n,p)})^2]$ and prove that

$$E_Q[(M_T^{(n,p)})^2] \leq C\Big(\int_0^T \sup_{v\in K} E_Q^2[Z_s f''(vZ_s)\mathbf{1}_{\{\tau_{q_n}\leq s\}}]ds\Big) \times \Big(\int_A (\sqrt{Y(y)} - 1)^2 \nu(dy)\Big)$$

where C is a positive constant, K is a compact subset of $\mathbb{R}^{+,*}$, and $q_n = n/(4p)$.

Note that on the stochastic interval $[\![0, T \wedge \tau_p)]\!]$ we have $1/p \leq \lambda Z_{s-} \leq p$ and, hence,

$$\begin{aligned} E_Q[(M_T^{(n,p)})^2] &= E_Q \int_0^{T\wedge\tau_p} \int_A |H_s^{(n)}(\lambda Z_{s-}, y) - H_s(\lambda Z_{s-}, y)|^2 Y(y)\nu(dy)ds \\ &\leq \int_0^T \int_A \sup_{1/p\leq x\leq p} |H_{T-s}^{(n)}(x, y) - H_{T-s}(x, y)|^2 Y(y)\nu(dy)ds. \end{aligned}$$

To estimate the difference $|H_{T-s}^{(n)}(x, y) - H_{T-s}(x, y)|$ we observe that

$$\begin{aligned} &H_{T-s}^{(n)}(x, y) - H_{T-s}(x, y) \\ &\quad = E_Q[\phi_n(xZ_sY(y)) - \phi_n(xZ_s) - f'(xZ_sY(y)) + f'(xZ_s)]. \end{aligned}$$

From Lemma 1 we deduce that if $xZ_sY(y) \in [1/n, n]$ and $xZ_s \in [1/n, n]$ then the expression on the rhs of the previous equality is zero. But if $y \in A$ we have that $3/4 \leq Y(y) \leq 5/4$ and, hence,

$$
\begin{aligned}
&|H^{(n)}_{T-s}(x,y) - H_{T-s}(x,y)| \\
&\quad \le E_Q[\mathbf{1}_{\{\tau_{qn}\le s\}}|\phi_n(xZ_sY(y)) - \phi_n(xZ_s) - f'(xZ_sY(y)) + f'(xZ_s)|].
\end{aligned}
$$

Again from the inequalities of Lemma 1 we get that

$$
|H^{(n)}_{T-s}(x,y) - H_{T-s}(x,y)| \le 6E_Q[\mathbf{1}_{\{\tau_{qn}<s\}}|f'(xZ_sY(y)) - f'(xZ_s)|].
$$

Writing

$$
f'(xZ_sY(y)) - f'(xZ_s) = \int_1^{Y(y)} xZ_s f''(xZ_s\theta)d\theta
$$

we finally get

$$
|H^{(n)}_{T-s}(x,y) - H_{T-s}(x,y)| \le 6x\,|Y(y)-1| \sup_{3/4\le u\le 5/4} E_Q[\mathbf{1}_{\{\tau_{qn}<s\}}\, Z_s\, f''(xuZ_s)]
$$

and this gives us the estimate of $E_Q[(M_T^{(n,p)})^2]$ above.

We know that $P_T \sim Q_T$ and this means that the corresponding Hellinger process of order $1/2$ is finite:

$$
h_T\left(P, Q, \frac{1}{2}\right) = \frac{T}{2}{}^{\top}\beta c\beta + \frac{T}{8}\int_{\mathbb{R}}(\sqrt{Y(y)}-1)^2\nu(dy) < \infty.
$$

Then

$$
\int_A (\sqrt{Y(y)}-1)^2\nu(dy) < \infty.
$$

From the Lebesgue dominated convergence theorem and (3) we get:

$$
\int_0^T \sup_{v\in K} E_Q^2[Z_sf''(vZ_s)\mathbf{1}_{\{\tau_{qn}\le s\}}]ds \to 0
$$

as $n\to\infty$ and this information together with the estimate of $E_Q[(M_T^{(n,p)})^2]$ proves the convergence of $E_Q[(M_T^{(n,p)})^2]$ to zero as $n\to\infty$.

Now we check that $E_Q|N_T^{(n,p)}| \to 0$ as $n\to\infty$. For this we prove that

$$
E_Q|N_T^{(n,p)}| \le 2T E_Q[\mathbf{1}_{\{\tau_n<T\}}(5|f'(\lambda Z_T)| + \alpha)] \int_{A^c} Y(y)d\nu.
$$

We start by noticing that

$$
\begin{aligned}
E_Q|N_T^{(n,p)}| &\le 2E_Q[\int_0^{T\wedge\tau_p}\int_{A^c}|H_s^{(n)}(\lambda Z_{s-},y) - H_s(\lambda Z_{s-},y)|\,Y(y)\nu(dy)ds] \\
&\le 2\int_0^T\int_{A^c} E_Q|H_s^{(n)}(\lambda Z_{s-},y) - H_s(\lambda Z_{s-},y)|\,Y(y)\nu(dy)ds.
\end{aligned}
$$

To evaluate the rhs of the previous inequality we observe that

$$|H_s^{(n)}(\lambda x, y) - H_s(\lambda x, y)| \le E_Q|\phi_n(\lambda x Z_{T-s} Y(y)) - f'(\lambda x Z_{T-s} Y(y))| \\ + E_Q|\phi_n(\lambda x Z_{T-s}) - f'(\lambda x Z_{T-s})|.$$

We remark that in law with respect to Q

$$|\phi_n(\lambda x Z_{T-s} Y(y)) - f'(\lambda x Z_{T-s} Y(y))| = E_Q[|\phi_n(\lambda Z_T) - f'(\lambda Z_T)| \,|\, Z_s = x\, Y(y)]$$

and

$$|\phi_n(\lambda x Z_{T-s}) - f'(\lambda x Z_{T-s})| = E_Q[|\phi_n(\lambda Z_T) - f'(\lambda Z_T)| \,|\, Z_s = x].$$

Then

$$|H_s^{(n)}(\lambda x, y) - H_s(\lambda x, y)| \le 2E_Q|\phi_n(\lambda Z_T) - f'(\lambda Z_T)|.$$

From Lemma 1 we get that

$$E_Q|\phi_n(\lambda x Z_T) - f'(\lambda x Z_T)| \le E_Q[\mathbf{1}_{\{\tau_n \le T\}}|\phi_n(\lambda Z_T) - f'(\lambda Z_T)|] \\ \le E_Q[\mathbf{1}_{\{\tau_n \le T\}}(5|f'(\lambda Z_T)| + \alpha)]$$

and this proves the estimate for $E_Q|N_T^{(n,p)}|$.

Then, the Lebesgue dominated convergence theorem applied to the rhs of the previous inequality shows that the latter tends to zero as $n \to \infty$. On the other hand, from the fact that the Hellinger process is finite and also from the inequality $(\sqrt{Y(y)} - 1)^2 \ge Y(y)/25$ satisfied on A^c we get that

$$\int_{A^c} Y(y) d\nu < \infty.$$

This result with the previous convergence proves that $E_Q|N_T^{(n,p)}| \to 0$ as $n \to \infty$. Theorem 4 is proved. □

4 Utility Maximizing Strategies

We combine the decomposition of the previous section with Theorem 3.1 of [10] in order to get an explicit expression of the optimal strategy.

Theorem 2 *Let u be a $\mathscr{C}^3(]\underline{x}, \infty[)$ utility function and let f be its convex conjugate. Assume that there exists an f-minimal martingale measure Q^* which preserves the Lévy property and such that the integrability conditions* (3) *are satisfied. Then for any fixed initial capital $x > \underline{x}$ there exists an asymptotically u-optimal strategy $\hat{\phi}$. In addition, $\hat{\phi}$ defines a u-optimal strategy as soon as $\underline{x} > -\infty$.*

Furthermore, if $c \neq 0$, we have

$$\hat{\phi}_s^{(i)} = -\frac{\lambda \beta^{(i)} Z_{s-}}{S_{s-}^{(i)}} \xi_s(\lambda Z_{s-})$$

where $\beta =^\top (\beta^{(1)}, \dots, \beta^{(d)})$ is the first Girsanov parameter, the process $\xi_s(\cdot)$ is defined by (1), *and λ is the unique solution to the equation $E_{Q^*}(-f'(\lambda Z_T)) = x$. If $c = 0$, the interior of* supp ν *contains zero, and Y is not identically* 1, *then $f''(x) = ax^\gamma$ with $a > 0$ and $\gamma \in \mathbb{R}$, and*

$$\hat{\phi}_s^{(i)} = -\frac{\lambda \gamma^{(i)} Z_{s-}}{S_{s-}^{(i)}} \xi_s(\lambda Z_{s-})$$

where again λ is a unique solution to the equation $E_{Q^*}(-f'(\lambda Z_T)) = x$ and the constants $\gamma^{(i)}$ are related with the second Girsanov parameter Y by the formula:

$$\gamma^{(i)} = \exp(-y_{0,i}) Y(y_0)^\gamma \frac{\partial}{\partial y_i} Y(y_0) \tag{12}$$

where y_0 is an arbitrary point in supp ν.

Proof of Theorem 2 The first part of the theorem is a minor adaptation of a result from [13]. Because of some changes due to the use of asymptotically optimal strategies, for the reader's ease we recall the proof.

As the function f' is strictly increasing and continuous, due to (3) so is the function $\lambda \mapsto E_{Q^*}[f'(\lambda Z_T)]$. Furthermore, since $f' = -(u')^{-1}$, we have

$$\lim_{\lambda \to 0} E_{Q^*}[f'(\lambda Z_T)] = -\infty, \qquad \lim_{\lambda \to \infty} E_{Q^*}[f'(\lambda Z_T)] = -\underline{x}.$$

Hence, for all $x > \underline{x}$, there exists a unique $\lambda > 0$ such that $E_{Q^*}[f'(\lambda Z_T)] = -x$. As Q^* is minimal for the function $x \mapsto f(\lambda x)$, it follows from Theorem 3.1 of [10] that there exists a predictable process $\hat{\phi}$ such that

$$-f'(\lambda Z_T) = x + (\hat{\phi} \cdot S)_T \tag{13}$$

and, furthermore, $\hat{\phi} \cdot S$ defines a Q^*-martingale. By definition of the convex conjugate, we have that

$$u(x + (\hat{\phi} \cdot S)_T) = f(\lambda Z_T) - \lambda Z_T f'(\lambda Z_T)$$

and, hence,

$$E_P[|u(x + (\hat{\phi} \cdot S)_T|] \leq E_P|f(\lambda Z_T)| + \lambda E_P[Z_T|f'(\lambda Z_T)|] < \infty.$$

If ϕ denotes any admissible strategy, we have, by definition of f, that

$$u(x + (\phi \cdot S)_T) \leq (x + (\phi \cdot S)_T)\lambda Z_T + f(\lambda Z_T)$$

$$\leq (x + (\phi \cdot S)_T)\lambda Z_T + u(x + (\hat{\phi} \cdot S)_T) + \lambda Z_T f'(\lambda Z_T).$$

Taking the expectation, we obtain that

$$E_P[u(x + (\phi \cdot S)_T)] \leq E_P[u(x + (\hat{\phi} \cdot S)_T] + \lambda E_{Q^*}[(\phi \cdot S)_T].$$

Now, under Q^*, the process $(\phi \cdot S)$ is a local martingale which is bounded from below, hence it is a supermartingale, so that $E_{Q^*}[(\phi \cdot S)_T] \leq 0$. Therefore,

$$E_P[u(x + (\phi \cdot S)_T)] \leq E_P[u(x + (\hat{\phi} \cdot S)_T)].$$

Furthermore, if $\underline{x} > -\infty$, we have the bound $(\hat{\phi} \cdot S)_T \geq \underline{x} - x$, so that $\hat{\phi}$ defines an admissible strategy, and hence is a u-optimal strategy.

When $\underline{x} = -\infty$, we can construct using the definition of $\mathscr{A}$ a sequence of admissible strategies $\hat{\phi}^n$ such that $(\hat{\phi}^n \cdot S)_t \geq -n$ for all $t \leq T$ and such that

$$\lim_{n\to\infty} E[u(x + (\hat{\phi}^{(n)} \cdot S)_T)] = \sup_{\phi \in \mathscr{A}} E[u(x + (\phi \cdot S)_T)].$$

Finally, $\hat{\phi}$ is asymptotically u-optimal.

We now want to obtain a more explicit expression for $\hat{\phi}$. First of all, we note that the relation (13) may be rewritten as

$$-E_{Q^*}[f'(\lambda Z_T)|\mathscr{F}_t] = x + \sum_{i=1}^{d} \int_0^t \hat{\phi}_s^{(i)} S_{s-}^{(i)} dX_s^{(c)}$$
$$+ \int_0^t \int_{\mathbb{R}^{d*}} \hat{\phi}_s^{(i)} S_{s-}^{(i)} (e^{y_i} - 1)(\mu^X - \nu^{X,Q^*})(ds, dy).$$

We can then identify this decomposition with that obtained in Theorem 1. If $c \neq 0$, we identify the continuous components and obtain that Q^*-a.s, for all $t \leq T$,

$$\sum_{i=1}^{d} \lambda \beta^{(i)} \int_0^t \xi_s(\lambda Z_{s-}) Z_{s-} \, dX_s^{(c),i} = -\sum_{i=1}^{d} \int_0^t \hat{\phi}_s^{(i)} S_{s-}^{(i)} dX_s^{(c),i}.$$

Taking quadratic variation of the difference of the right- and left-hand sides in the previous equality, we obtain that Q^*-a.s. for all $s \leq T$

$${}^{\top}[\lambda\beta\xi_s(\lambda Z_{s-})Z_{s-} + \phi_s\, S_{s-}]\, c\, [\lambda\beta\xi_s(\lambda Z_{s-})Z_{s-} + \phi_s\, S_{s-}] = 0$$

where by convention $\phi_s\, S_{s-} = (\phi_s^{(i)}\, S_{s-}^{(i)})_{1 \leq i \leq d}$. Therefore, as c is a symmetric positive matrix, we have

$$\hat{\phi}_s S_{s-} = -\lambda\beta\xi_s(\lambda Z_{s-})Z_{s-} + V_s$$

where V_s belongs to the kernel of c. We may now write

$$-E_{Q^*}[f'(\lambda Z_T)|\mathscr{F}_t] = x - \sum_{i=1}^{d} \lambda\beta^{(i)} \int_0^t \xi_s(\lambda Z_{s-}) \frac{dS_s^{(i)}}{S_{s-}^{(i)}} + \sum_{i=1}^{d} \int_0^t V_s^{(i)} dX_t^{(c),i}.$$

As for all $s \geq 0$, $cV_s = 0$, we must have $\langle \sum_{i=1}^{d} \int_0^{\cdot} V_s^{(i)} dX^{(c),i} \rangle_s = 0$, and so Q^*-a.s.

$$-E_{Q^*}[f'(\lambda Z_T)|\mathscr{F}_t] = x - \sum_{i=1}^{d} \lambda\beta^{(i)} \int_0^t \xi_s(\lambda Z_{s-}) \frac{dS_s^{(i)}}{S_{s-}^{(i)}}.$$

It then follows from the first part of the proof that the process $\hat{\phi}$ defined in (13) defines an (asymptotically-) optimal strategy.

If we now assume that $c = 0$, we identify the discontinuous components and obtain that Q^*-a.s., for all $s \leq T$ and for almost all $y \in \operatorname{supp} \nu$,

$$\sum_{i=1}^{d} \hat{\phi}_s^{(i)} S_{s-}^{(i)} (e^{y_i} - 1) = -H_s(\lambda Z_{s-}, y). \tag{14}$$

In addition, since the interior of $\operatorname{supp} \nu$ contains zero and Y is not identically 1, we obtain from Theorem 3 of [2] that for $y \in \operatorname{supp} \nu$

$$f'(xY(y)) - f'(x) = \Phi(x) \sum_{i=1}^{d} \alpha^{(i)} (e^{y_i} - 1)$$

where

$$\Phi(x) = xf''(xY(y_0)), \ \ \alpha^{(i)} = \exp(-y_{0,i}) \frac{\partial}{\partial y_i} Y(y_0)$$

with any y_0 in the interior of $\operatorname{supp} \nu$. Again from Theorem 5 of [2], $f''(x) = ax^{\gamma}$. This implies, after taking the derivative of (14) with respect to y_i, the formula for optimal strategy. □

We finally give a unified expression of optimal strategies for all utility functions associated with common f-divergence functions.

Proposition 1 *Let X be a Lévy process with characteristics (b, c, ν) and let f be a function such that $f''(x) = ax^{\gamma}$ where $a > 0$, $\gamma \in \mathbb{R}$. Let u_f be its concave conjugate. Assume that there exist $\alpha, \beta \in \mathbb{R}^d$ and a Borel function $Y : \mathbb{R}^d \setminus \{0\} \to \mathbb{R}^+$ such that*

$$Y(y) = (f')^{-1}\left(f'(1) + \sum_{i=1}^{d} \alpha^{(i)} (e^{y_i} - 1) \right) \tag{15}$$

and the following properties hold:

$$Y(y) > 0 \quad \nu\text{-a.e.}, \tag{16}$$

$$\sum_{i=1}^{d} \int_{|y|\geq 1} (e^{y_i} - 1)Y(y)\nu(dy) < \infty, \tag{17}$$

$$b + \frac{1}{2}\operatorname{diag} c + c\beta + \int_{\mathbb{R}^d} ((e^y - 1)Y(y) - h(y))\nu(dy) = 0. \tag{18}$$

Then if $c \neq 0$, *there exists an asymptotically optimal strategy* $\hat{\phi}$ *whose coordinates are given by the formula*

$$\hat{\phi}_s^{(i)} = \alpha_{\gamma+1}(x) \frac{\beta^{(i)}}{E_{Q^*}[Z_s^{\gamma+1}]} \frac{Z_{s-}^{\gamma+1}}{S_{s-}^{(i)}},$$

where Z *is the density process of the change of measure from* P *into the* f*-minimal equivalent martingale measure* Q^* *and*

$$\alpha_{\gamma+1}(x) = -(\gamma + 1)(x + f'(1)) + a. \tag{19}$$

If $c = 0$, *the interior of* $\operatorname{supp} \nu$ *contains zero, and* Y *is not equal to* 1, *then*

$$\hat{\phi}_s^{(i)} = \alpha_{\gamma+1}(x) \frac{\gamma^{(i)}}{E_{Q^*}[Z_s^{\gamma+1}]} \frac{Z_{s-}^{\gamma+1}}{S_{s-}^{(i)}},$$

where the constants γ^i *are given by* (12). *In addition,* $\hat{\phi}$ *is optimal as soon as* $\gamma \neq -1$.

Proof We know from [2] that under the assumptions (16), (17), and (18), the Lévy model has an f-minimal martingale measure which preserves the Lévy property and whose Girsanov parameters are (β, Y) if $c \neq 0$, and $(0, Y)$ if $c = 0$. Let $\lambda > 0$ be such that $E_{Q^*}[f'(\lambda Z_T)] = -x$. It is easy to see that if $c \neq 0$, the decomposition of Theorem 1 can be written as

$$-E_{Q^*}[f'(\lambda Z_T)|\mathscr{F}_t] = x - a\lambda^{\gamma+1} \sum_{i=1}^{d} \beta^{(i)} \int_0^T Z_{s-}^{\gamma+1}\, E_{Q^*}[Z_{T-s}^{\gamma+1}] \frac{dS_s^{(i)}}{S_{s-}^{(i)}}.$$

As Q^* preserves the Lévy property, we have $E_{Q^*}[Z_{T-s}^{\gamma+1}]\, E_{Q^*}[Z_s^{\gamma+1}] = E_{Q^*}[Z_T^{\gamma+1}]$, so calculating λ we obtain that

$$-f'(\lambda Z_T) = x + \alpha_{\gamma+1}(x) \sum_{i=1}^{d} \beta^{(i)} \int_0^T \frac{Z_{s-}^{\gamma+1}}{E_{Q^*}[Z_s^{\gamma+1}]} \frac{dS_s^{(i)}}{S_{s-}^{(i)}}.$$

The analogous procedure can be also applied for the case $c = 0$. It then follows from the proof of Theorem 2 that $\hat{\phi}$ defines an asymptotically optimal strategy. □

Acknowledgements This work was supported in part by ECOS project M07M01 and ANR-09-BLAN-0084-01 of Auto-similarity of Department of Mathematics of Angers University.

References

1. Carr, P., Geman, H., Madan, D., Yor, M.: The fine structure of asset returns: an empirical investigation. J. Bus. **2**, 61–73 (2002)
2. Cawston, S., Vostrikova, L.: Lévy preservation and associated properties for f-minimal equivalent martingale measures. In: Shiryaev, A., Presman, E., Yor, M. (eds.) Prokhorov and Contemporary Probability Theory. Springer, Berlin (2012)
3. Choulli, T., Stricker, C.: Minimal entropy-Hellinger martingale measure in incomplete markets. Math. Finance **15**, 465–490 (2005)
4. Choulli, T., Stricker, C., Li, J.: Minimal Hellinger martingale measures of order q. Finance Stoch. **11**, 399–427 (2007)
5. Eberlein, E.: Application of generalizes hyperbolic Lévy motions to finance. In: Lévy Processes: Theory and Applications. Birkhauser, Basel (2001)
6. Eberlein, E., Keller, U.: Hyperbolic distributions in finance. Bernoulli **1**, 281–299 (1995)
7. Essche, F., Schweizer, M.: Minimal entropy preserves the Lévy property: how and why. Stoch. Process. Appl. **115**, 299–327 (2005)
8. Föllmer, H., Schweizer, M.: Hedging of contingent claims under incomplete information. In: Davis, M.H., Eliott, R.J. (eds.) Applied Stochastic Analysis. Stochastic Monographs, vol. 5, pp. 389–414. Gordon and Breach, London (1991)
9. Fujiwara, T., Miyahara, Y.: The minimal entropy martingale measures for geometric Lévy processes. Finance Stoch. **7**, 509–531 (2003)
10. Goll, T., Rüschendorf, L.: Minimax and minimal distance martingale measures and their relationship to portfolio optimisation. Finance Stoch. **5**, 557–581 (2001)
11. Jacod, J., Shyriaev, A.: Limit Theorems for Stochastic Processes. Springer, Berlin (1987)
12. Jeanblanc, M., Klöppel, S., Miyahara, Y.: Minimal f^q-martingale measures for exponential Levy processes. Ann. Appl. Probab. **17**, 1615–1638 (2007)
13. Kallsen, J.: Optimal portfolios for exponential Lévy process. Math. Methods Oper. Res. **51**, 357–374 (2000)
14. Klöppel, S.: Dynamic valuation in incomplete markets. Diss. ETH **16**, 666 (2006)
15. Kramkov, D., Schahermayer, V.: The asymptotic elasticity of utility functions and optimal investment in incomplete markets. Ann. Appl. Probab. **9**, 904–950 (1999)
16. Liese, F., Vajda, I.: Convex Statistical Distances. Teubner, Leipzig (1987)
17. Miyahara, Y.: Minimal entropy martingale measures of jump type price processes in incomplete assets markets. Asian–Pac. Financ. Mark. **6**(2), 97–113 (1999)
18. Sato, K.: Lévy Processes and Infinitely Divisible Distributions. Cambridge University Press, Cambridge (1999)
19. Revuz, D., Yor, M.: Continuous Martingales and Brownian Motion. Springer, Berlin (1999)
20. Schweizer, M.: On minimal martingale measure and Föllmer–Schweizer decomposition. Stoch. Anal. Appl. **13**, 573–599 (1995)
21. Schweizer, M.: A guided tour through quadratic hedging approaches. In: Jouini, E., Cvitanic, J., Musiela, M. (eds.) Option Pricing, Interest Rates and Risk Management, pp. 538–574. Cambridge University Press, Cambridge (1999)

Optimal Investment with Bounded VaR for Power Utility Functions

Bénamar Chouaf and Serguei Pergamenchtchikov

Abstract We consider an optimal investment problem for Black–Scholes type financial market with bounded VaR measure on the whole investment interval $[0, T]$. The explicit form for the optimal strategies is found.

Keywords Portfolio optimization · Stochastic optimal control · Risk constraints · Value-at-Risk

Mathematics Subject Classification (2010) 91B28 · 93E20

1 Introduction

We consider an investment problem aiming at optimal terminal wealth at maturity T. The classical approach to this problem goes back to Merton [11] and involves utility functions, more precisely, the expected utility serves as the functional which has to be optimized.

We adapt this classical utility maximization approach to nowadays industry practice: investment firms customarily impose limits on the risk of trading portfolios.

B. Chouaf
Laboratoire de Mathématiques Appliquées, Université de Sidi Bel Abbes, Sidi Bel Abbês, Algeria
e-mail: bchouaf@univ-sba.dz

S. Pergamenchtchikov (✉)
Laboratoire de Mathématiques Raphaël Salem, UMR 6085 CNRS-Université de Rouen, Avenue de l'Université, BP.12, Technopôle du Madrillet, 76801 Saint Etienne du Rouvray, France
e-mail: Serge.Pergamenchtchikov@univ-rouen.fr

S. Pergamenchtchikov
Laboratory of Quantitative Finance, National Research University-Higher School of Economics, Moscow, Russia

S. Pergamenchtchikov
Department of Mathematics and Mechanics, National Research Tomsk State University, Tomsk, Russia

Y. Kabanov et al. (eds.), *Inspired by Finance*, DOI 10.1007/978-3-319-02069-3_6,

These limits are specified in terms of downside Value-at-Risk (VaR) risk measures (see, for example, [1]).

As Jorion [6], p. 379 points out, VaR creates a common denominator for the comparison of different risk activities. Traditionally, position limits of traders are set in terms of notional exposure, which may not be directly comparable across treasuries with different maturities. In contrast, VaR provides a common denominator to compare various asset classes and business units. The popularity of VaR as a risk measure has been endorsed by regulators, in particular, the Basel Committee on Banking Supervision, which resulted in mandatory regulations worldwide.

Our approach combines the classical utility maximization with risk limits in terms of VaR. This leads to control problems under restrictions on uniform versions of VaR, where the risk bound is supposed to be intact throughout the duration of the investment. To our knowledge such problems have only been considered in dynamic settings, which reduce intrinsically to static problems. Emmer, Klüppelberg and Korn [5] consider a dynamic market, but maximize only the expected wealth at maturity under a downside risk bound at maturity. Basak and Shapiro [2] solve the utility optimization problem for complete markets with bounded VaR at maturity. Gabih, Gretsch and Wunderlich [4] solve the utility optimization problem for constant coefficients markets with bounded Expected Shortfall (ES) risk measure at maturity. Klüppelberg and Pergamenchtchikov [8, 9] considered the optimization problems with bounded VaR and ES risk measure on the whole time interval in the class of the nonrandom financial strategies. Note that this approach does not work in the general case, i.e. for the *random financial* strategies. Therefore, the question about the existence of the optimal strategies for the optimization problems with bounded risk measure uniformly on the whole time interval $[0, T]$ is open. It should be noted that it is impossible to calculate the explicit form of the VaR and ES risk measures for the random financial strategies. This is the main difficulty in such problems. Indeed, it is not clear how one can "see" optimal solution for the constrained problems if we can't calculate the constraints. To overcome this problem Cuoco, He and Isaenko [3] propose to replace the VaR by some discrete approximation. In this paper we work with the "true" VaR values and we find an explicit form for the optimal strategies for the VaR constrained optimization problem.

Our paper is organized as follows. In Sect. 2 we formulate the Black–Scholes model for the price processes. In Sect. 3 all optimization problems and their solutions are given. All proofs are summarized in Sect. 4 with the technical lemma postponed to the Appendix.

2 The Model

We consider a Black–Scholes type financial market consisting of one *riskless bond* and several *risky stocks*. Their respective price processes $(S_0(t))_{t\geq 0}$ and $(S_i(t))_{t\geq 0}$ for $i = 1, \dots, d$ evolve according to the equations:

$$\begin{cases} \mathrm{d}S_0(t) = r_t\, S_0(t)\, \mathrm{d}t\,, & S_0(0) = 1\,, \\ \mathrm{d}S_i(t) = S_i(t)\, \mu_i(t)\, \mathrm{d}t + S_i(t) \sum_{j=1}^{d} \sigma_{ij}(t)\, \mathrm{d}W_j(t)\,, & S_i(0) = s_i > 0. \end{cases} \tag{1}$$

Here $W_t = (W_1(t), \ldots, W_d(t))'$ is a standard d-dimensional Brownian motion; $r_t \in \mathbb{R}$ is the *riskless interest rate*, $\mu_t = (\mu_1(t), \ldots, \mu_d(t))' \in \mathbb{R}^d$ is the vector of *stock-appreciation rates* and $\sigma_t = (\sigma_{ij}(t))_{1 \le i, j \le d}$ is the matrix of *stock-volatilities*. We assume that the coefficients r_t, μ_t, and σ_t are deterministic functions which are right-continuous with left limits (càdlàg). We also assume that the matrix σ_t is non-singular for Lebesgue almost all $t \ge 0$.

We denote by $\mathscr{F}_t = \sigma\{W_s\,, s \le t\}$, $t \ge 0$, the filtration generated by the Brownian motion (augmented by the null sets). Furthermore, $|\cdot|$ denotes the Euclidean norm for vectors and the corresponding matrix norm for matrices.

For $t \ge 0$ let $\phi_t \in \mathbb{R}$ denote the amount of investment into bond and let

$$\varphi_t = (\varphi_1(t), \ldots, \varphi_d(t))' \in \mathbb{R}^d$$

be the amount of investment into risky assets. We recall that a *trading strategy* is an $\mathbb{R}^{d+1}$-valued $(\mathscr{F}_t)_{t \ge 0}$-progressively measurable process $(\phi_t, \varphi_t)_{t \ge 0}$ and that

$$X_t = \phi_t\, S_0(t) + \sum_{j=1}^{d} \varphi_j(t)\, S_j(t)\,, \quad t \ge 0\,,$$

is called the *wealth process*.

The trading strategy $((\phi_t, \varphi_t))_{t \ge 0}$ is called *self-financing*, if the wealth process satisfies the following equation

$$X_t = x + \int_0^t \phi_u\, \mathrm{d}S_0(u) + \sum_{j=1}^{d} \int_0^t \varphi_j(u)\, \mathrm{d}S_j(u)\,, \quad t \ge 0\,, \tag{2}$$

where $x > 0$ is the initial endowment.

In this paper we work with relative quantities, i.e., we define for $j = 1, \ldots, d$

$$\pi_j(t) := \frac{\varphi_j(t)\, S_j(t)}{\phi_t\, S_0(t) + \sum_{i=1}^{d} \varphi_i(t)\, S_i(t)}\,, \quad t \ge 0\,.$$

Then $\pi_t = (\pi_1(t), \ldots, \pi_d(t))'$, $t \ge 0$, is called the *portfolio process* and we assume throughout the paper that it is $(\mathscr{F}_t)_{t \ge 0}$-progressively measurable. We assume that for the fixed investment horizon $T > 0$

$$\|\pi\|_T^2 := \int_0^T |\pi_t|^2 \mathrm{d}t < \infty \quad \text{a.s.}$$

We also define, with $\mathbf{1} = (1, \ldots, 1)' \in \mathbb{R}^d$, the quantities

$$y_t = \sigma_t' \pi_t \quad \text{and} \quad \theta_t = \sigma_t^{-1}(\mu_t - r_t\, \mathbf{1})\,, \quad t \ge 0\,, \tag{3}$$

where it suffices that these quantities are defined for Lebesgue almost all $t \geq 0$. Taking these definitions into account we rewrite Eq. (2) for X_t as

$$\mathrm{d}X_t \,=\, X_t\,(r_t \,+\, y_t'\,\theta_t)\,\mathrm{d}t \,+\, X_t\,y_t'\,\mathrm{d}W_t\,, \quad X_0 = x > 0. \tag{4}$$

This implies in particular that any optimal investment strategy is equal to

$$\pi_t^* = \sigma_t'^{-1} y_t^*,$$

where y_t^* is the optimal control process for Eq. (4). We also require for the investment horizon $T > 0$

$$\|\theta\|_T^2 = \int_0^T |\theta_t|^2\,\mathrm{d}t \,<\, \infty. \tag{5}$$

We assume that $(y_t)_{t\leq T}$ is any $(\mathscr{F}_t)_{t\leq T}$-adapted a.s. square integrable process, i.e.

$$\|y\|_T^2 = \int_0^T |y_t|^2\,\mathrm{d}t \,<\, \infty \quad \text{a.s.},$$

such that the stochastic equation (4) has a unique strong solution. We denote by $\mathscr{Y}$ the class of all such processes $y = (y_t)_{t\leq T}$. Note that for every $y \in \mathscr{Y}$, through Itô's formula, we represent Eq. (4) in the following form (to emphasize that the wealth process corresponds to some control process y we write X^y):

$$X_t^y \,=\, x\,e^{R_t + (y,\theta)_t}\,\mathscr{E}_t(y), \tag{6}$$

where $R_t = \int_0^t r_u \mathrm{d}u$, $(y,\theta)_t = \int_0^t y_u'\,\theta_u \mathrm{d}u$ and the process $(\mathscr{E}_t(y))_{t\leq T}$ is the stochastic exponent for y, i.e.

$$\mathscr{E}_t(y) = \exp\Big(\int_0^t y_u'\mathrm{d}W_u - \frac{1}{2}\int_0^t |y_u|^2\mathrm{d}u\Big).$$

Therefore, for every $y \in \mathscr{Y}$ the process $(X_t^y)_{t\geq 0}$ is a.s. positive and continuous.

For an initial endowment $x > 0$ and a control process $y = (y_t)_{t\geq 0}$ in $\mathscr{Y}$, we introduce the *cost function*

$$J(x, y) := \mathbf{E}_x\,\left(X_T^y\right)^\gamma, \tag{7}$$

where $\mathbf{E}_x$ is the expectation operator conditional on $X_0^y = x$.

For $\gamma \in (0, 1)$ the utility function $U(z) = z^\gamma$ is concave and is called the power (or HARA) utility function. We include the case of $\gamma = 1$, which corresponds to optimizing the expected terminal wealth. In combination with a downside risk bound this allows us in principle to disperse with the utility function, where in practice one has to choose the parameter γ.

3 Optimization Problems

3.1 The Unconstrained Problem

We consider two regimes with the cost functions (7) for $0 < \gamma < 1$ and for $\gamma = 1$.

$$\max_{y \in \mathcal{Y}} J(x, y) . \tag{8}$$

First we study Problem (8) for $\gamma \in (0, 1)$. The following result can be found in Example 6.7 on page 106 in Karatzas and Shreve [7]; it's proof there is based on the martingale method.

Theorem 1 *Consider Problem* (8) *for* $\gamma \in (0, 1)$. *The optimal value of* $J(x, y)$ *is given by*

$$J^*(x) = \max_{y \in \mathcal{Y}} J(x, y) = J(x, y^*) = x^\gamma \exp\Big(\gamma R_T + \frac{\gamma}{2(1-\gamma)} \|\theta\|_T^2\Big),$$

where the optimal control $y^* = (y_t^*)_{t \le T}$ *is of the form*

$$y_t^* = \frac{\theta_t}{1-\gamma} \quad \left(\pi_t^* = \frac{(\sigma_t \sigma_t')^{-1}(\mu_t - r_t \mathbf{1})}{1-\gamma}\right). \tag{9}$$

The optimal wealth process $(X_t^*)_{0 \le t \le T}$ *is given by*

$$\mathrm{d}X_t^* = X_t^* \left(r_t + \frac{|\theta_t|^2}{1-\gamma}\right) \mathrm{d}t + X_t^* \frac{\theta_t'}{1-\gamma} \mathrm{d}W_t, \quad X_0^* = x. \tag{10}$$

Let now $\gamma = 1$.

Theorem 2 [8] *Consider Problem* (8) *with* $\gamma = 1$. *Assume a riskless interest rate* $r_t \ge 0$ *for all* $t \in [0, T]$. *If* $\|\theta\|_T > 0$ *then*

$$\max_{y \in \mathcal{Y}} J(x, y) = \infty.$$

If $\|\theta\|_T = 0$, *then a solution exists and the optimal value of* $J(x, y)$ *is given by*

$$\max_{y \in \mathcal{Y}} J(x, y) = J(x, y^*) = x\, e^{R_T},$$

corresponding to an arbitrary deterministic square integrable function $(y_t^*)_{t \le T}$. *In this case the optimal wealth process* $(X_t^*)_{t \le T}$ *satisfies the following equation*:

$$\mathrm{d}X_t^* = X_t^* r_t \mathrm{d}t + X_t^* \, y_t^{*\prime} \mathrm{d}W_t, \quad X_0^* = x. \tag{11}$$

3.2 The Constrained Problem

As risk measures we use modifications of the Value-at-Risk as introduced in Emmer, Klüppelberg and Korn [5]. They can be summarized under the notion of Capital-at-Risk as they reflect the required capital reserve. To avoid non-relevant cases we consider only $0 < \alpha < 1/2$. We use here the definition as in [8, 9].

Definition 1 (Value-at-Risk (VaR)) Define for an initial endowment $x > 0$, a control process $y \in \mathscr{Y}$ and $0 < \alpha \le 1/2$ the *Value-at-Risk (VaR)* by

$$\mathrm{VaR}_t(x, y, \alpha) := x\, e^{R_t} - Q_t\,, \quad t \ge 0\,,$$

where $Q_t = Q_t(x, y, \alpha)$ is the $\mathscr{F}_t^y = \sigma\{y_s\,, s \le t\}$-measurable random variable such that

$$\alpha \quad \text{quantile of the ratio} \quad \widehat{X}_t^y = \frac{X_t^y}{Q_t} \quad \text{is equal to unit} \tag{12}$$

i.e.

$$\inf\{z \ge 0:\ \mathbf{P}(\widehat{X}_t^y \le z) \ge \alpha\} = 1.$$

Remark 1 Note that for the nonrandom financial strategies $(y_t)_{t \le T}$ the process Q_t is the usual α-quantile for the process X_t^y. To define the "random" quantile for the process X_t^y we consider the ratio process $\widehat{X}_t^y$ for which the α-quantile is equal to unit.

Corollary 1 *For every $y \in \mathscr{Y}$ with $\|y\|_t > 0$ the process Q_t defined in Definition* 1, *is given by*

$$Q_t = x \exp\left(R_t + (y, \theta)_t - \frac{1}{2}\|y\|_t^2 + \tau_t \|y\|_t \right), \quad t \ge 0,$$

where $\tau_t = \tau_t(\alpha, y)$ is the α-quantile of the normalized stochastic integral

$$\xi_t(y) = \frac{1}{\|y\|_t} \int_0^t y_u' \mathrm{d}W_u\,,$$

i.e.

$$\tau_t = \inf\{z \ge -\infty:\ \mathbf{P}\,(\xi_t(y) \le z) \ge \alpha\}\,. \tag{13}$$

It is clear that for any nonrandom function $(y_t)_{t \le T}$ the random variable

$$\xi_t \sim \mathscr{N}(0, 1),$$

i.e. in this case $\tau_t = -|z_\alpha|$, where z_α is the α-quantile of the standard normal distribution.

In fact, in this paper we work with a more strong constraint than VaR risk measure, we work with a upper bound for VaR risk measure, i.e. we consider

$$\mathrm{VaR}_t^*(x, y, \alpha) := x\, e^{R_t} - Q_t^*, \quad t \geq 0, \tag{14}$$

where

$$Q_t^* = x \exp\left(R_t + (y, \theta)_t - \frac{1}{2} \|y\|_t^2 + \tau_t^* \|y\|_t \right) \quad \text{with} \quad \tau_t^* = \min(z_\alpha, \tau_t).$$

Obviously,

$$\mathrm{VaR}_t(x, y, \alpha) \leq \mathrm{VaR}_t^*(x, y, \alpha),$$

i.e. the VaR^* constraint is more stable than VaR risk measure with respect to financial strategies.

We define the *level risk function* for some coefficient $\zeta \in (0, 1)$ as

$$\zeta_t(x) = \zeta\, x\, e^{R_t}, \quad t \in [0, T]. \tag{15}$$

The coefficient ζ introduces some risk aversion behavior into the model. In that sense it acts similarly as a utility function does. However, ζ has a clear interpretation, and every investor can choose and understand the influence of the risk bound ζ as a proportion of the riskless bond investment.

We consider only controls $y \in \mathcal{Y}$ for which the Value-at-Risk is a.s. bounded by this level function over the interval $[0, T]$. That is, we require

$$\sup_{t \leq T} \frac{\mathrm{VaR}_t^*(x, y, \alpha)}{\zeta_t(x)} \leq 1 \quad \text{a.s.} \tag{16}$$

The optimization problem is

$$\max_{y \in \mathcal{Y}} J(x, y) \quad \text{subject to} \quad \sup_{t \leq T} \frac{\mathrm{VaR}_t^*(x, y, \alpha)}{\zeta_t(x)} \leq 1 \quad \text{a.s.} \tag{17}$$

To describe the optimal strategies we need the following function:

$$g(a) := \sqrt{2a + \widetilde{z}_\alpha^2} - \widetilde{z}_\alpha \tag{18}$$

with

$$\widetilde{z}_\alpha = |z_\alpha| - \|\theta\|_T \quad \text{and} \quad 0 \leq a \leq a_{max} := -\ln(1 - \zeta).$$

Moreover, we set

$$a_0 = \frac{\|\theta\|_T^2}{2(1-\gamma)^2} + \widetilde{z}_\alpha \frac{\|\theta\|_T}{1-\gamma}. \tag{19}$$

Theorem 3 *Consider Problem* (17) *for* $\gamma \in (0, 1)$. *Assume that* $|z_\alpha| \geq 2\|\theta\|_T$. *Then the optimal value for the cost function is given by*

$$J(x, y^*) = x^\gamma \, e^{\gamma R_T + \gamma G(g^*)}, \tag{20}$$

where $G(g) = g\|\theta\|_T + (1-\gamma)g^2/2$, $g^* = g(a^*)$ *with*

$$a^* = \min(a_0, a_{max}), \tag{21}$$

and the optimal control y^* *is, for all* $t \leq T$, *of the form*

$$y_t^* = \frac{g^*}{\|\theta\|_T}\theta_t \, \mathbf{1}_{\{\|\theta\|_T > 0\}}. \tag{22}$$

Moreover, if $\|\theta\|_T > 0$ *then the optimal wealth process* $(X_t^*)_{t \leq T}$ *is given by*

$$\mathrm{d}X_t^* = X_t^* \left(r_t + \frac{g^* |\theta_t|^2}{\|\theta\|_T} \right) \mathrm{d}t + X_t^* \frac{g^*}{\|\theta\|_T} \theta_t' \mathrm{d}W_t, \qquad X_0^* = x; \tag{23}$$

if $\|\theta\|_T = 0$, *then* $X_t^* = x \, e^{R_t}$ *for* $t \leq T$.

Theorem 4 *Consider Problem* (17) *for* $\gamma = 1$. *Assume that* $|z_\alpha| \geq 2\|\theta\|_T$. *Then the optimal value for the cost function is given by*

$$J(x, y^*) = x \, e^{R_T + g(a_{max})\|\theta\|_T}, \tag{24}$$

and the optimal control y^* *is, for all* $t \leq T$, *of the form*

$$y_t^* = \frac{g(a_{max})}{\|\theta\|_T}\theta_t \, \mathbf{1}_{\{\|\theta\|_T > 0\}}. \tag{25}$$

Moreover, if $\|\theta\|_T > 0$ *then the optimal wealth process* $(X_t^*)_{t \leq T}$ *is given by*

$$\mathrm{d}X_t^* = X_t^* \left(r_t + \frac{g(a_{max}) |\theta_t|^2}{\|\theta\|_T} \right) \mathrm{d}t + X_t^* \frac{g(a_{max})}{\|\theta\|_T} \theta_t' \mathrm{d}W_t, \qquad X_0^* = x; \tag{26}$$

if $\|\theta\|_T = 0$, *then* $X_t^* = x \, e^{R_t}$ *for* $t \leq T$.

4 Proofs

4.1 Proof of Theorem 3

Let $\gamma \in (0, 1)$. By (6) we represent the γ power of the wealth process as

$$(X_T^y)^\gamma = x^\gamma \, e^{\gamma R_T + \gamma F_T(y)} \, \mathcal{E}_T(\gamma y),$$

where

$$F_T(y) = (\theta, y)_T - \frac{1-\gamma}{2}\|y\|_T^2 . \tag{27}$$

Moreover, we introduce the measure (generally, not a probability) by the following Radon–Nikodym density:

$$\frac{d\widetilde{\mathbf{P}}}{d\mathbf{P}} = \mathscr{E}_T(\gamma y).$$

By denoting $\widetilde{\mathbf{E}}$ the expectation with respect to this measure we get that

$$\mathbf{E}(X_T^y)^\gamma = x^\gamma \, e^{\gamma R_T} \, \widetilde{\mathbf{E}} e^{\gamma F_T(y)} . \tag{28}$$

If $\|\theta\|_T = 0$, then

$$\mathbf{E}(X_T^y)^\gamma = x^\gamma e^{\gamma R_T} \widetilde{\mathbf{E}} e^{-\frac{\gamma(1-\gamma)}{2}\|y\|_T^2} .$$

Taking into account that for any process y from $\mathscr{Y}$ (see, for example, p. 211 in [10])

$$\mathbf{E}\mathscr{E}_T(\gamma y) \le 1$$

we get for any $y \in \mathscr{Y}$

$$\mathbf{E}(X_T^y)^\gamma \le x^\gamma e^{\gamma R_T}$$

with the equality if and only if $y_t = 0$.

Therefore, in the sequel we assume that $\|\theta\|_T > 0$. Now we shall consider the almost sure optimization problem for the function $F_T(\cdot)$. First, we consider this constrained the last time moment $t = T$, i.e.

$$\sup_{y \in \mathscr{Y}} F_T(y) \quad \text{subject to} \quad \frac{\text{VaR}_T^*(x, y, \alpha)}{\zeta_T(x)} \le 1 \quad \text{a.s.} \tag{29}$$

This constraint is equivalent to

$$\frac{1}{2}\|y\|_T^2 - \tau_T^* \|y\|_T - (\theta, y)_T \le -\ln(1-\zeta) =: a_{max} .$$

By fixing the quantile as $\tau_T^* = -\beta$ for some $\beta \ge |z_\alpha|$ and denoting

$$K_T(y) = \frac{1}{2}\|y\|_T^2 + \beta\|y\|_T - (\theta, y)_T$$

we will consider more general problem than (29), i.e. we will find the optimal solution in the Hilbert space $L_2[0, T]$, that is

$$\sup_{y \in Ł_2[0,2]} F_T(y) \quad \text{subject to} \quad K_T(y) \le a_{max} .$$

To resolve this problem we have to resolve the following one:

$$\sup_{y\in L_2[0,T]} F_T(y) \quad \text{subject to} \quad K_T(y)=a \tag{30}$$

for some parameter $0 \le a \le a_{max}$. We use the Lagrange multipliers method, i.e. we pass to the Lagrange cost function $H_\lambda(y) = F_T(y) - \lambda K_T(y)$ and we have to resolve the optimization problem for this function:

$$\max_{y\in L_2[0,T]} H_\lambda(y)\,. \tag{31}$$

In this case

$$H_\lambda(y) = -\frac{\lambda+1-\gamma}{2}\|y\|_T^2 + (1+\lambda)(\theta, y)_T - \lambda\beta\|y\|_T,$$

where λ is Lagrange multiplier. It is clear that $\lambda > \gamma - 1$. Since the problem (31) has no finite solution for $\lambda \le \gamma - 1$, i.e.

$$\max_{y\in L_2[0,T]} H_\lambda(y) = \infty.$$

To this end we calculate the Gâteau derivative

$$\mathrm{d}_\lambda(y,h) = \lim_{\delta\to 0} \frac{H_\lambda(y+\delta h) - H_\lambda(y)}{\delta}\,.$$

It is easy to check directly that for any function y from $L_2[0,T]$ with $\|y\|_T > 0$

$$D_\lambda(y,h) = \int_0^T h_t'\left((1+\lambda)\theta_t - (1-\gamma+\lambda)y_t - \lambda\beta\overline{y}_t\right)\mathrm{d}t$$

with $\overline{y}_t = y_t/\|y\|_T$. Moreover, if $\|y\|_T = 0$, then

$$D_\lambda(y,h) = (1+\lambda)\int_0^T h_t'\theta_t\mathrm{d}t - \lambda\beta\|h\|_T\,.$$

It is clear that $D_\lambda(y,h) \ne 0$ for $h_t = -\mathrm{sign}(\lambda)\theta_t$. Therefore, to resolve the equation

$$D_\lambda(y,h) = 0 \tag{32}$$

for all $h \in L_2[0,T]$ we assume that $\|y\|_T > 0$. This implies that

$$(1+\lambda)\theta_t - (1-\gamma+\lambda)y_t - \lambda\beta\overline{y}_t = 0,$$

i.e.

$$y_t = \frac{(1+\lambda)\|y\|_T}{\lambda\beta + (1+\lambda-\gamma)\|y\|_T}\theta_t\,.$$

Therefore,

$$y_t^\lambda = \frac{\psi(\lambda)}{\|\theta\|_T}\theta_t \quad \text{with} \quad \psi(\lambda) = \frac{\|\theta\|_T + \lambda(\|\theta\|_T - \beta)}{1-\gamma+\lambda}. \tag{33}$$

The coefficient ψ must be positive, i.e.

$$\gamma - 1 < \lambda < \frac{\|\theta\|_T}{(\beta - \|\theta\|_T)_+}. \tag{34}$$

Now we have to verify that the solution of Eq. (32) gives the maximum solution for the problem (31). To this end for any function y from $L_2[0,T]$ with $\|y\|_T > 0$ we set

$$\Delta_\lambda(y,h) = H_\lambda(y+h) - H_\lambda(y) - D_\lambda(y,h).$$

Moreover, by putting

$$\delta(y,h) = \|y+h\|_T - \|y\|_T - (h,\overline{y})_T, \tag{35}$$

we obtain that

$$\Delta_\lambda(y,h) = -\frac{\lambda+1-\gamma}{2}\|h\|_T^2 - \lambda\beta\delta(y,h).$$

Now Lemma 1 implies that the function $\Delta(y,h) \le 0$ for all $h \in L_2[0,T]$. Therefore the solution of Eq. (32) gives the solution for the problem (31).

Now we chose the Lagrange multiplier λ to satisfy the condition in (30), i.e.

$$K_T(y^\lambda) = a,$$

i.e.

$$\psi^2(\lambda) + 2\psi(\lambda)(\beta - \|\theta\|_T) = 2a.$$

It follows that

$$\widetilde{\psi}(a) = \psi(\lambda(a)) = \sqrt{2a + (\beta - \|\theta\|_T)^2} - (\beta - \|\theta\|_T)$$

with

$$\lambda = \lambda(a) = \frac{\|\theta\|_T + (1-\gamma)(\beta - \|\theta\|_T)}{\sqrt{2a + (\beta - \|\theta\|_T)^2}} - 1 + \gamma.$$

One can check directly that the function $\lambda(a)$ satisfies the condition (34) for any $a > 0$. This means that the solution for the problem (30) is given by the function

$$\widetilde{y}_t^a = y_t^{\lambda(a)} = \frac{\widetilde{\psi}(a)}{\|\theta\|_T}\theta_t.$$

Now to chose the parameter $0 < a \le a_{max}$ in (30) we have to maximize the function (27), i.e.

$$\max_{0\le a\le a_{max}} F_T(\widetilde{y}^a)\,.$$

Note that

$$F_T(\widetilde{y}^a) = G(\widetilde{\psi}(a)) \quad \text{with} \quad G(\psi) = \psi\,\|\theta\|_T - (1-\gamma)\frac{\psi^2}{2}\,.$$

Moreover, note that for any $a > 0$ and $\beta \ge |z_\alpha|$

$$\widetilde{\psi}(a) \le g(a)\,,$$

where the function g is defined in (18). Therefore,

$$\max_{0\le a\le a_{max}} F_T(\widetilde{y}^a) \le \max_{0\le a\le a_{max}} G(g(a)) = G(g(a^*)),$$

where a^* is defined in (21). To obtain here the equality we take in (33) $\beta = |z_\alpha|$. Thus, the function (22) is the solution of the problem (29). Now to pass to the problem (17) we have to check the condition (16) for the function (22). To this end note that

$$\frac{1}{2}\|y^*\|_t^2 + |z_\alpha|\|y^*\|_t - (\theta, y^*)_t = \int_0^t \omega(s)\,\mathrm{d}s,$$

where

$$\omega_s = |\theta_s|^2\left(\frac{(g^*)^2}{2\|\theta\|_T^2} + \frac{g^*(|z_\alpha| - 2\|\theta\|_s)}{2\|\theta\|_T\|\theta\|_s}\right).$$

Taking into account here the condition $|z_\alpha| \ge \|\theta\|_T$ we obtain $\omega_t \ge 0$, i.e.

$$\begin{aligned}
&\frac{1}{2}\|y^*\|_t^2 + |z_\alpha|\|y^*\|_t - (\theta, y^*)_t \\
&\le \frac{1}{2}\|y^*\|_T^2 + |z_\alpha|\|y^*\|_T - (\theta, y^*)_T \\
&= a^* \le -\ln(1-\zeta).
\end{aligned}$$

This implies immediately that the function (22) is a solution of the problem (17). □

4.2 Proof of Theorem 4

Let now $\gamma = 1$. Note that in this case we can obtain the following upper bound:

$$\mathbf{E}\,X_T^y \le x e^{R_T}\,\mathbf{E}\,e^{\|\theta\|_T\|y\|_T}\,\mathscr{E}_T(y).$$

If $\|\theta\|_T = 0$, we obtain here the equality if and only if $y = 0$. Let now $\|\theta\|_T > 0$. Note that the condition

$$K_T(y) \le a_{max} \tag{36}$$

implies $\|y\|_T \le g(a_{max})$. Thus, for any function $(y_t)_{t\le T}$ satisfying this condition we have

$$\mathbf{E}\, X_T^y \le x e^{R_T + g(a_{max})\|\theta\|_T}.$$

Moreover, the function (25) transforms this inequality in the equality. By the same way as in the proof of Theorem 4 we check that the function (25) satisfies the condition (16). □

Acknowledgements This work was supported by the scientific cooperation CNRS/DPGRF, Project DZAC 19856, France-Algérie. The second author is partially supported by the RFBR-Grant 09-01-00172-a.

Appendix: Properties of the Function (35)

Lemma 1 *Assume that $y \in L_2[0,T]$ with $\|y\|_T > 0$. Then for every $h \in L_2[0,T]$ the function* (35) *is positive, i.e.* $\delta(y,h) \ge 0$.

Proof Obviously, if $h \equiv ay$ for some $a \in \mathbb{R}$, then $\delta(y,h) = (|1+a| - 1 - a)\|y\|_T \ge 0$. Let now the functions h and y be linearly independent. Then

$$\delta(y,h) = \frac{2(y\,,\,h)_T + \|h\|_T^2}{\|y+h\|_T + \|y\|_T} - (\overline{y},h)_T = \frac{\|h\|_T^2 - (\overline{y},h)_T((\overline{y},h)_T + \delta(y,h))}{\|y+h\|_T + \|y\|_T}.$$

It is clear that for all h

$$\|y+h\|_T + \|y\|_T + (\overline{y},h)_T \ge 0$$

with the equality if and only if $h \equiv ay$ for some $a \le -1$. Therefore, if the functions h and y are linearly independent, then

$$\delta(y,h) = \frac{\|h\|_T^2 - (\overline{y}\,,\,h)_T^2}{\|y+h\|_T + \|y\|_T + (\overline{y},h)_T} \ge 0.$$

□

References

1. Artzner, P., Delbaen, F., Eber, J.-M., Heath, D.: Coherent measures of risk. Math. Finance **9**, 203–228 (1999)
2. Basak, S., Shapiro, A.: Value at risk based risk management: optimal policies and asset prices. Rev. Financ. Stud. **14**(2), 371–405 (2001)

3. Cuoco, D., He, H., Isaenko, S.: Optimal dynamic trading strategies with risk limits. Working paper (2005)
4. Gabih, A., Grecksch, W., Wunderlich, R.: Dynamic portfolio optimization with bounded shortfall risks. Stoch. Anal. Appl. **23**, 579–594 (2005)
5. Emmer, S., Klüppelberg, C., Korn, R.: Optimal portfolios with bounded capital-at-risk. Math. Finance **11**, 365–384 (2001)
6. Jorion, P.: Value at Risk. McGraw-Hill, New York (2001)
7. Karatzas, I., Shreve, S.E.: Methods of Mathematical Finance. Springer, Berlin (2001)
8. Klüppelberg, C., Pergamenchtchikov, S.M.: Optimal consumption and investment with bounded downside risk for power utility functions. In: Delbaen, F., Rásonyi, M., Stricker, C. (eds.) Optimality and Risk—Modern Trends in Mathematical Finance, pp. 133–169. Springer, Heidelberg (2009)
9. Klüppelberg, C., Pergamenchtchikov, S.M.: Optimal consumption and investment with bounded downside risk measures for logarithmic utility functions. In: Albrecher, H., Runggaldier, W., Schachermayer, W. (eds.) Advanced Financial Modelling, pp. 245–273. Radon Ser. Comput. Appl. Math., vol. 8. Walter de Gruyter, Berlin (2009)
10. Liptser, R.S., Shirayev, A.N.: Statistics of Random Processes I. General Theory. Springer, New York (1977)
11. Merton, R.C.: Continuous Time Finance. Blackwell, Cambridge (1990)

Three Essays on Exponential Hedging with Variable Exit Times

Tahir Choulli, Junfeng Ma, and Marie-Amélie Morlais

Abstract This paper addresses three main problems that are intimately related to exponential hedging with variable exit times. The first problem consists of explicitly parameterizing the exponential forward performances and describing the optimal solution for the corresponding utility maximization problem. The second problem deals with the horizon-unbiased exponential hedging. Precisely, we are interested in describing the dynamic payoffs for which there exists an admissible strategy that minimizes the risk—in the exponential utility framework—whenever the investor exits the market at stopping times. Furthermore, we explicitly describe this optimal strategy when it exists. Our last contribution is concerned with the optimal sale problem, where the investor is looking simultaneously for the optimal portfolio and the optimal time to liquidate her assets.

Keywords Exponential hedging · Variable horizon · Utility maximization · Entropy-Hellinger process

Mathematics Subject Classification (2010) 91B28 · 93E20

1 Introduction

The impact of a variable horizon in financial markets has been drawing attention of economists since the early thirties of the twentieth century through the work of Fisher, [10]. Since then there has been an upsurge interest in this matter throughout the following decades, especially in the late sixties with the works of Yaari,

T. Choulli (✉) · J. Ma
Mathematical and Statistical Sciences Dept., University of Alberta, Edmonton, AB, T6G 261
Canada
e-mail: tchoulli@ualberta.ca

J. Ma
e-mail: jma@math.ualberta.ca

M.-A. Morlais
Département de Mathématiques, Université du Maine, 72085 Le Mans Cedex 9, France
e-mail: Marie-Amelie.Morlais@univ-lemans.fr

Y. Kabanov et al. (eds.), *Inspired by Finance*, DOI 10.1007/978-3-319-02069-3_7,

Hakansson, and others, see [11, 26], and the references therein. While in economics and empirical studies researchers have been actively discussing this issue of variable horizon, the mathematical structure/foundation that drives this impact of the horizon on market models was left open—up to our knowledge—and only recently the literature starts growing with the works of Choulli and Schweizer, [1], and Larsen and Hang [19]. Furthermore, during the recent decade, this horizon-dependence problem has been addressed in a different perspective which lead to the birth of forward utilities. These forward utilities were fathered and baptized (with their current name) by Musiela and Zariphopoulou in a series of papers starting with the multiperiod incomplete binomial model in [24]. Then, the concept was extended to diffusion models in [23]. For the economic motivations of the forward utilities, we refer the reader to the numerous papers of Musiela and Zariphopoulou on this topic. Around the birth time of these forward utilities, Choulli and Stricker introduced and constructed in [3] and [4] a class of optimal martingale measures that possess the feature of being robust with respect to the variation of the horizon. These martingale measures appeared to be the key in solving utility maximization when the optimal strategy needs to be robust with respect to (independent of) the horizon. Thus, these martingale measures constitute an efficient tool for providing examples of forward utilities. Intuitively, these authors (Choulli and Stricker) addressed a sort of a "dual" problem for the problem proposed by Musiela and Zariphopoulou through the forward utility concept. The concept of forward utility has been—since its birth—successfully developed, used in many aspects (see [24, 27] and the references therein), and very recently extended to a general context by Zitkovic in [28]. In this work, the author characterized the forward utilities through a dual problem for general utilities, while he gave explicit formula for exponential (or affine) utilities only when markets are driven by Brownian uncertainty. In the present work, we propose an explicit parameterization of exponential forward utilities (or affine forward utilities) in the semimartingale framework. This generalization of [28] is based on the entropy-Hellinger concept, which allows us to build-up directly our parameterization algorithm for these dynamic utilities.

A closely related problem is the optimal sale problem for real options with investments, where the agent has two optimal controls to determine (namely, the optimal time to sell the real asset and the optimal portfolio for her investment). Motivated by this problem, Henderson and Hobson proposed the horizon-unbiased utility concept in [13]. This concept—at least at the first glance—seems to be very close (or similar) to the forward utility, but in fact the two concepts appear to be different for some market models, as we will explain in Sect. 4. However, both concepts (forward utilities and horizon-unbiased utilities) are dealing with the issue of variable horizon, and certainly both are intimately related to the notion of minimal Hellinger martingale measures (when these utilities are of HARA type). This last statement was proved in the work of Choulli, Li and Stricker [5], which was developed during the same time as Henderson–Hobson's work. Herein, we view the Henderson-Hobson problem differently by interpreting the real asset as a dynamic payoff and, hence, calling the problem the horizon-unbiased hedging. To this end, we only focus on the exponential utility in analyzing this problem which represents the second essay herein. Precisely, we explicitly determine the optimal portfolio using again the entropy-Hellinger concept and we describe the payoffs for which the maximization problem admits a solution. As a consequence of our analysis, we easily explain how the two concepts of utilities mentioned above can differ.

The last essay addresses directly the optimal sale problem (or the investment timing). Here, again, we show that the entropy-Hellinger concept plays a crucial role. By characterizing the optimal value process intrinsically to the optimization problem via a dynamic programming equation, we describe explicitly the optimal investment timing and the optimal strategy via a pointwise equation that depends on the optimal value process.

This paper is organized as follows. In Sect. 2, we introduce the model, the notation, and the definitions that we will be using throughout. Then, in Sect. 3 we present the first essay, which is concerned with the exponential forward performances. The second essay, which deals with horizon-unbiased hedging for the exponential utility, will be detailed in Sect. 4. The last section concentrates on the optimal sale problem with investments. The paper contains two appendices. The first appendix contains all technical lemmas that we use throughout the main body of the paper, while the second appendix discusses the minimal entropy-Hellinger martingale densities under change of probability measures.

2 Mathematical Model and Preliminaries

The mathematical model starts with a given filtered probability space denoted by $(\Omega, \mathscr{F}, (\mathscr{F}_t)_{0\leq t\leq T}, P)$ where the filtration is complete and right-continuous, and T represents a fixed horizon for investments. In this setup, we consider a d-dimensional semimartingale $S=(S_t)$ which represents the discounted price processes of d risky assets.

Next, we recall the definition of the predictable characteristics of the semimartingale S (see Sect. II.2 of [15]). The random measure μ associated to its jumps is defined by

$$\mu(\mathrm{d}t, \mathrm{d}x) = \sum I_{\{\Delta S_s\neq 0\}}\delta_{(s,\Delta S_s)}(\mathrm{d}t, \mathrm{d}x),$$

with δ_a the Dirac measure at point a. The continuous local martingale part of S is denoted by S^c. This leads to the following decomposition, called the *canonical representation* (see Theorem 2.34, Sect. II.2 of [15]), namely,

$$S = S_0 + S^c + h(x)\star(\mu-\nu) + (x-h(x))\star\mu + B, \tag{1}$$

where the random measure ν is the compensator of the random measure μ, and $h(x)$ is the truncation function, usually, $h(x) = xI_{\{|x|\leq 1\}}$. For the matrix C with entries $C^{ij} := \langle S^{c,i}, S^{c,j}\rangle$, the triple (B, C, ν) is called *predictable characteristics* of S. Furthermore, we can find a version of the characteristics triple satisfying

$$B = b\cdot A, \quad C = c\cdot A \quad \text{and} \quad \nu(\omega, \mathrm{d}t, \mathrm{d}x) = \mathrm{d}A_t(\omega)F_t(\omega, \mathrm{d}x).$$

Here A is an increasing and predictable process, which is continuous if and only if S is quasi-left continuous, b and c are predictable processes, $F_t(\omega, \mathrm{d}x)$ is a predictable kernel, $b_t(\omega) \in \mathbb{R}^d$ and $c_t(\omega)$ is a symmetric $d \times d$-matrix, for all $(\omega, t) \in \Omega \times [0, T]$. In the sequel, we will often drop ω and t and write, for instance, $F(\mathrm{d}x)$ as a shorthand for $F_t(\omega, \mathrm{d}x)$. The characteristics, B, C, and ν, satisfy

$$F_t(\omega, \{0\}) = 0, \qquad \int (|x|^2 \wedge 1) F_t(\omega, \mathrm{d}x) \le 1,$$

$$\Delta B_t = \int h(x)\nu(\{t\}, \mathrm{d}x), \qquad c = 0 \quad \text{on } \{\Delta A \neq 0\}.$$

We set

$$\nu_t(\mathrm{d}x) := \nu(\{t\}, \mathrm{d}x), \qquad a_t := \nu_t(\mathbb{R}^d) = \Delta A_t F_t(\mathbb{R}^d) \le 1.$$

We denote by $\mathbb{P}_a$ (respectively, $\mathbb{P}_e$) the set of all probability measures that are absolutely continuous with respect to (respectively, equivalent to) P. The set of martingales under a probability Q is denoted by $\mathcal{M}(Q)$. Finally, $\mathcal{M}^e(S)$ is the set of probabilities $Q \in \mathbb{P}_e$ such that S is a Q-local martingale.

If $\mathcal{C}$ is a class of processes, we denote by $\mathcal{C}_0$ the set of processes X with $X_0 = 0$ and by $\mathcal{C}_{loc}$ the set of processes X such that there exists a sequence of stopping times, $(T_n)_{n\ge1}$, increasing stationarily to T (i.e., $P(T_n = T) \to 1$ as $n \to \infty$) and the stopped process X^{T_n} belongs to $\mathcal{C}$. We put $\mathcal{C}_{0,loc} = \mathcal{C}_0 \cap \mathcal{C}_{loc}$.

Definition 1 Let X be a RCLL (right-continuous with left limits) semimartingale, and Q be a probability measure.

(i) X is called a *σ-martingale* with respect to Q if there exists a bounded and positive predictable process ϕ such that $\phi \cdot X$ is a Q-local martingale. The set of all σ-martingales with respect to Q will be denoted hereafter by $\mathcal{M}_\sigma(Q)$.

(ii) X is called a *special semimartingale* if there exist a local martingale, $\overline{M}$, and a predictable process, $\overline{A}$, with finite variation such that $\overline{M}_0 = \overline{A}_0 = 0$ and

$$X = X_0 + \overline{M} + \overline{A}.$$

(iii) X is said to be *locally integrable* if there exists a sequence of stopping times, $(T_n)_{n\ge1}$, that increases stationarily to T such that

$$E\Big[\sup_{t\le T_n} |X_t|\Big] < \infty.$$

Because herein we focus on the exponential utility, we consider the σ-martingale measures with finite entropy. The set of these measures is given by

$$\mathcal{M}^e_f(S) = \left\{ Q \in \mathbb{P}_e :\ S \in \mathcal{M}_\sigma(Q), \text{ and } E\left[\frac{\mathrm{d}Q}{\mathrm{d}P} \log \frac{\mathrm{d}Q}{\mathrm{d}P}\right] < \infty \right\}.$$

Very frequently, throughout the paper, we will work with densities instead of probabilities. For this, we will use the following set of densities

$$\mathcal{Z}^e_{loc}(S) := \{Z \in \mathcal{M}_{loc}(P):\ Z > 0,\ Z\log Z \text{ is locally integrable},\ ZS \in \mathcal{M}_\sigma(P)\}. \tag{2}$$

As usual, $\mathcal{A}^+$ denotes the set of increasing, right-continuous, adapted and integrable processes.

On the set $\Omega \times [0, T]$, we define two σ-fields, denoted by $\mathcal{O}$ and $\mathcal{P}$, generated by the adapted and RCLL processes and the adapted and continuous processes, respectively. On the set $\Omega \times [0, T] \times \mathbb{R}^d$ we consider the σ-field $\widetilde{\mathcal{P}} = \mathcal{P} \otimes \mathcal{B}(\mathbb{R}^d)$ (resp. $\widetilde{\mathcal{O}} = \mathcal{O} \otimes \mathcal{B}(\mathbb{R}^d)$), where $\mathcal{B}(\mathbb{R}^d)$ is the Borel σ-field for $\mathbb{R}^d$.

For any $\widetilde{\mathcal{O}}$-measurable function g (hereafter denoted by $g \in \widetilde{\mathcal{O}}$), we define $M^P_\mu(g|\widetilde{\mathcal{P}})$ to be the unique $\widetilde{\mathcal{P}}$-measurable function, when it exists, such that for any bounded $W \in \widetilde{\mathcal{P}}$,

$$M^P_\mu(Wg) := E\Big(\int_0^T \int_{\mathbb{R}^d} W(s,x)g(s,x)\mu(\mathrm{d}s, \mathrm{d}x)\Big) = M^P_\mu\Big(W M^P_\mu(g \mid \widetilde{\mathcal{P}})\Big).$$

For the following representation theorem, we refer to [14] (Theorem 3.75, p. 103) and to [15] (Lemma 4.24, p. 185).

Theorem 1 *Let $N \in \mathcal{M}_{0,loc}$. Then there exist a predictable and S^c-integrable process ϕ, $N' \in \mathcal{M}_{0,loc}$ with $[N', S] = 0$ and functions $f \in \widetilde{\mathcal{P}}$ and $g \in \widetilde{\mathcal{O}}$ such that*

(i) $\big(\sum_{s=0}^t f(s, \Delta S_s)^2 I_{\{\Delta S_s \neq 0\}}\big)^{1/2} \in \mathcal{A}^+_{loc}$, $\big(\sum_{s=0}^t g(s, \Delta S_s)^2 I_{\{\Delta S_s \neq 0\}}\big)^{1/2} \in \mathcal{A}^+_{loc}$,
(ii) $M^P_\mu(g|\widetilde{\mathcal{P}}) = 0$,
(iii) *the process N is given by*

$$N = \phi \cdot S^c + W \star (\mu - \nu) + g \star \mu + N', \quad W = f + \frac{\widehat{f}}{1-a} I_{\{a<1\}},$$

where $\widehat{f}_t = \int f_t(x)\nu(\{t\}, \mathrm{d}x)$ and f has a version such that $\{a = 1\} \subseteq \{\widehat{f} = 0\}$.

Moreover

$$\Delta N_t = \Big(f_t(\Delta S_t) + g_t(\Delta S_t)\Big) I_{\{\Delta S_t \neq 0\}} - \frac{\widehat{f}_t}{1 - a_t} I_{\{\Delta S_t = 0\}} + \Delta N'_t.$$

In the remaining part of this section, we define the entropy-Hellinger concept that will play a crucial role in our analysis.

Definition 2 (i) Let $N \in \mathcal{M}_{0,\,loc}(P)$ such that $1 + \Delta N \geq 0$. If the non-decreasing adapted process

$$V^{(E)}_t(N) := \frac{1}{2}\langle N^c\rangle_t + \sum_{0<s\leq t} \big[(1 + \Delta N_s)\log(1 + \Delta N_s) - \Delta N_s\big]$$

is locally integrable (i.e. $V^E(N) \in \mathscr{A}^+_{loc}(P)$), then its compensator (with respect to the probability P) is called the *entropy-Hellinger process* of N, and is denoted by $h^E(N, P)$.

(ii) Let $Q \in \mathbb{P}_a$ with density $Z = \mathscr{E}(N)$. We define the *entropy-Hellinger process* of Q with respect to P by

$$h^E_t(Q, P) := h^E_t(Z, P) := h^E_t(N, P), \quad t \le T.$$

Next, we give the variation of this entropy-Hellinger concept towards the change of probability measures.

Definition 3 (i) Let Q be a probability measure and let Y be a Q-local martingale such that $1 + \Delta Y \ge 0$. If the RCLL nondecreasing process

$$V^E(Y) = \frac{1}{2}\langle Y^c\rangle + \sum \big[(1 + \Delta Y)\log(1 + \Delta Y) - \Delta Y\big]$$

is Q-locally integrable (i.e. $V^E(Y) \in \mathscr{A}^+_{loc}(Q)$), then its Q-compensator is called the *entropy-Hellinger process* of Y (or equivalently of $\mathscr{E}(Y)$) with respect to Q, and is denoted by $h^E(Y, Q)$ (respectively $h^E(\mathscr{E}(Y), Q)$).

(ii) Let $N \in \mathscr{M}_{0,loc}(P)$ such that $1 + \Delta N > 0$ and Y is a semimartingale such that $Y\mathscr{E}(N)$ is a P-local martingale and $1 + \Delta Y \ge 0$. Then, if the process

$$\frac{1}{2}\langle Y^c\rangle + \sum (1 + \Delta N)\Big[(1 + \Delta Y)\log(1 + \Delta Y) - \Delta Y + 1\Big]$$

is P-locally integrable, then its P-compensator is called the *entropy-Hellinger process* of $\mathscr{E}(Y)$ with respect to $\mathscr{E}(N)$, and is denoted by $h^E(\mathscr{E}(Y), \mathscr{E}(N))$.

Remark 1 Definition 2 was first given in [3], and to which we refer the reader for more details about the history of the entropy-Hellinger process of a probability measure (which is also called Leibler–Kullback process). Definition 3-(i) is a natural extension in probability as well as in mathematical finance areas, due to the popular and useful technique of change of probability measures. The last definition, Definition 3-(ii), which we will use throughout the paper, extends Definition 3-(i) to the case when the uniform integrability of the nonnegative local martingale $\mathscr{E}(N)$ may not hold. The relationship between the two definitions in Definition 3 is obvious. Indeed, let $(T_n)_{n\ge 1}$ be a sequence of stopping times that increases stationarily to T such that $\mathscr{E}(N)^{T_n}$ is a true martingale. Then, by putting $Q_n := \mathscr{E}_{T_n}(N) \cdot P$, we obtain

$$h^E_{t\wedge T_n}(\mathscr{E}(Y), \mathscr{E}(N)) = h^E_t(\mathscr{E}(Y^{T_n}), Q_n), \quad t \le T.$$

Definition 4 Let X and Y be two processes such that $X_0 = Y_0$. We write $X \preceq Y$ if $Y - X$ is a nondecreasing process.

Definition 5 (i) A *σ-martingale density* for S is any positive local martingale, Z, such that ZS is a σ-martingale. If furthermore, ZS is locally integrable, then Z is called a local martingale density for S.

(ii) We call the *minimal entropy-Hellinger σ-martingale density* (MEH σ-martingale density hereafter)—when it exists—the σ-martingale density, $\widetilde{Z}$, that minimizes $h^E(Z, P)$ over the set $\mathscr{Z}^e_{loc}(S)$ with respect to the order defined in Definition 4.

(iii) We call the *minimal entropy-Hellinger σ-martingale measure* (MEH σ-martingale measure hereafter), the σ-martingale measure, $\widetilde{Q}$, minimizing $h^E(Q, P)$ over the set $\mathscr{M}^e_f(S)$.

Lemma 1 *Let $(T_n)_{n\geq 0}$ ($T_0 = 0$) be a sequence of stopping times that increases stationarily to T. Suppose that for each n the process S^{T_n} admits the MEH σ-martingale density, denoted by $\widetilde{Z}^{(n)}$. Then S admits the MEH σ-martingale density, $\widetilde{Z}$, given by*

$$\widetilde{Z} := \mathscr{E}(\widetilde{N}) \quad and \quad \widetilde{N} := \sum_{n\geq 1} I_{]\!]T_{n-1},T_n]\!]} \frac{1}{\widetilde{Z}^{(n)}_-} \cdot \widetilde{Z}^{(n)}.$$

Proof It is clear that the process $\widetilde{N}$ is a local martingale (for more details about this, the reader is referred to [8], Chap. VI, pp. 96–97). Similar arguments as in [8] allow us to conclude that $\widetilde{Z}\log(\widetilde{Z})$ is locally integrable and $\widetilde{Z}(\phi \cdot S)$ is a local martingale. Here

$$\phi := \sum_n \frac{\phi^{(n)}}{1 + \|\phi^{(n)}\|_\infty} I_{]\!]T_{n-1},T_n]\!]},$$

where $\phi^{(n)}$ is such that $\widetilde{Z}^{(n)}(\phi^{(n)} \cdot S)^{T_n}$ is a local martingale. Thus, we deduce that $\widetilde{Z} \in \mathscr{Z}^e_{loc}(S)$. Then the proof of the lemma follows immediately from

$$\begin{aligned} h^E(Z, P) - h^E(\widetilde{Z}, P) &= \sum_{n=1} I_{]\!]T_{n-1},T_n]\!]} \cdot \left(h^E(Z, P) - h^E(\widetilde{Z}, P)\right) \\ &= \sum_{n=1} I_{]\!]T_{n-1},T_n]\!]} \cdot \left(h^E(Z^{T_n}, P) - h^E(\widetilde{Z}^{(n)}, P)\right) \succeq 0, \end{aligned}$$

for any $Z \in \mathscr{Z}^e_{loc}(S)$. □

We send the reader to [8, 14], and/or [15], for further details on probabilistic concepts, while for σ-martingale measures and arbitrage we suggest [7] and [16].

3 Complete Parameterization of Exponential Forward Performances

This section is devoted to the explicit parameterization of exponential forward utilities. We start by defining random utility fields, their associated set of admissible strategies, and the forward performances. Throughout this section, the main assumption on the process S is

$$\int_{\{|x|>1\}} |x| e^{\lambda^T x} F(dx) < \infty, \quad \text{for all } \lambda \in \mathbb{R}^d. \tag{3}$$

Definition 6 We call a *random utility field* a $\mathcal{B}([0,T]) \times \mathcal{B}(\mathbb{R}) \times \mathcal{F}$-measurable function, $U(t,x,\omega)$, such that, for any fixed x, the process $U(t,x,\omega)$ is a RCLL adapted process, and for any fixed (t,ω) the function $x \mapsto U(t,x,\omega)$ is strictly increasing and strictly concave.

Definition 7 For a random utility field, $U(t,x,\omega)$, any probability measure Q, any semimartingale X, and $x \in \mathbb{R}$ such that $U(t,x,\omega) < \infty$ we denote by

$$\mathcal{A}_{adm}(x,X,Q) := \left\{\theta \in L(X) : \sup_{\tau \in \mathcal{T}_T} E^Q\left[U\left(\tau, x + (\theta \cdot X)_\tau\right)^-\right] < \infty\right\}$$

the set of *admissible strategies* for the model (x,X,Q,U). Here $\mathcal{T}_T$ is the set of stopping times, τ, such that $\tau \le T$. When $X = S$ and $Q = P$, we simply write $\mathcal{A}_{adm}(x)$.

In the literature, there were a number of definitions proposed for forward utilities, however—here in the next definition—we will consider the original definition of forward utilities that was given by Musiela and Zariphopoulou in [23]. For the latest and general definition, we refer the reader to [28].

Definition 8 Consider a RCLL semimartingale, X, and a probability measure, Q. We call *forward dynamic utility* for (X,Q), a random utility field $U := (U(t,\omega,x))$, fulfilling the following self-generating property:

a) The function $U(0,x)$ is strictly increasing and concave.

b) There exists an admissible strategy θ^* (i.e. $\theta^* \in \mathcal{A}_{adm}(x,X,Q)$) such that

$$U\left(s, x + (\theta^* \cdot X)_s\right) = E^Q\left[U\left(t, x + (\theta^* \cdot X)_t\right) | \mathcal{F}_s\right], \quad \forall T \ge t \ge s \ge 0.$$

c) For any admissible strategy θ, for any $T \ge t \ge s$, we have:

$$U\left(s, x + (\theta \cdot X)_s\right) \ge E^Q\left[U\left(t, x + (\theta \cdot X)_t\right) | \mathcal{F}_s\right].$$

When $X = S$ and $Q = P$, we simply call U a forward dynamic utility.

Definition 9 Let X be a RCLL semimartingale and Q be a probability measure. Then, we call an *exponential forward utility* for (X,Q), any forward dynamic utility for (X,Q), $U(t,x,\omega)$, given by

$$U(t,x,\omega) = -\exp\left(-\frac{x - B_t(\omega)}{N_t(\omega)}\right),$$

where N is a positive process and B is a process.

In the forthcoming analysis, both the stopping rule and the change of probability measures play crucial roles. Thus, it is worth to state the following easy and useful lemma.

Lemma 2 *Let $U = U(t, \omega, x)$ be a forward dynamic utility for the process (S, P). Then the following assertions hold.*

(i) *For any stopping time τ, the process*

$$\overline{U}(t, \omega, x) := U(t \wedge \tau(\omega), \omega, x)$$

is a forward dynamic utility for (S^τ, P).

(ii) *Consider a probability measure Q that is absolutely continuous with respect to P with the density process denoted by Z. Then the random utility field*

$$U^Q(t, \omega, x) := U(t, \omega, x) Z_t(\omega)$$

is a forward dynamic utility for (S, P) if and only if U is forward dynamic utility for (S, Q).

Proof The proof of this lemma is straightforward. □

Now we present the main results of this section. To this end, we first assume that the process $N = 1$ and B is predictable with finite variation. While this may look restrictive, this assumption leads to some kind of "uniqueness" of the forward utility.

Theorem 2 *Suppose that S satisfies (3) and $B = (B_t)_{t \leq T}$ is a RCLL predictable process with finite variation. Then the following assertions are equivalent:*

(i) *The random utility field, $U(t, \omega, x) := -\exp(-x + B_t(\omega))$, is a forward performance.*

(ii) *The minimal entropy-Hellinger σ-martingale measure, $\widetilde{Q}$, exists and*

$$B = B_0 + h^E(\widetilde{Q}, P). \tag{4}$$

Proof (1) In this part, we will prove (ii) $\Longrightarrow$ (i). Suppose that (ii) holds. Then, due to Theorem 9 (or see Theorem 4.6 in [3]), the MEH σ-martingale measure $\widetilde{Q}$ has the density process $\widetilde{Z}$ with

$$\log \widetilde{Z} = \widetilde{\theta} \cdot S + h^E(\widetilde{Z}, P).$$

It is easy to check that $-\widetilde{\theta}$ is admissible and $U(t, -(\widetilde{\theta} \cdot S)_t) = -e^{B_0} \widetilde{Z}_t$ is a true martingale. Because of Lemma 5 in the appendix, it is also clear that for any admissible strategy $\theta \in \mathscr{A}_{adm}(x)$, the process

$$U(t, (\theta \cdot S)_t) = -e^{B_0} \widetilde{Z}_t \exp\left[-((\theta + \widetilde{\theta}) \cdot S)_t\right]$$

is a supermartingale. Hence assertion (i) follows immediately.

(2) Herein, we prove (i) $\Longrightarrow$ (ii) in several steps. To this end, we assume that there exists $\widehat{\theta} \in \mathscr{A}_{adm}(x)$ such that $-\exp(-(\widehat{\theta} \cdot S)_t + B_t)$ is a true martingale and for any $\theta \in \mathscr{A}_{adm}(x)$, the process $-\exp(-(\theta \cdot S)_t + B_t)$ is a supermartingale.

(a) We first show that the optimal strategy $\widehat{\theta}$ satisfies the pointwise equation that characterizes the MEH σ-martingale density when it exists. By Ito's formula, for any $\theta \in L(S)$,

$$-\exp\bigl(-(\theta \cdot S)_t + B_t\bigr) = -\mathrm{e}^{-(\theta \cdot S)_t}\mathrm{e}^{B_t} = -\mathrm{e}^{B_0}\mathscr{E}_t(X^\theta)\mathscr{E}_t(X^B),$$

$$X^\theta := -\theta \cdot S + \frac{1}{2}\theta^T c\theta \cdot A + (\mathrm{e}^{-\theta^T x} - 1 + \theta^T x) \star \mu,$$

$$X^B := B - B_0 + \sum(\mathrm{e}^{\Delta B} - 1 - \Delta B).$$

Therefore, for any admissible strategy θ, the process $-\exp(-(\theta \cdot S)_t + B_t)$ is a local supermartingale (respectively, a local martingale) if and only if the process $\mathrm{e}^{\Delta B} \cdot X^\theta + X^B$ is a local submartingale (respectively, a local martingale). This fact is equivalent to the statements (a.1) and (a.2) given by:

(a.1) the process $|\mathrm{e}^{-\theta^T x} - 1 + \theta^T h(x)| \star \mu$ is locally integrable, and

(a.2) the process $\mathrm{e}^{-\Delta B} \cdot X^B - K(\theta) \cdot A$ is nondecreasing (respectively is null), where

$$K(\theta) := \theta^T b - \frac{1}{2}\theta^T c\theta + \int \Bigl(-\mathrm{e}^{-\theta^T x} + 1 - \theta^T h(x)\Bigr) F(\mathrm{d}x), \quad \theta \in \mathbb{R}^d.$$

As a result, the optimal admissible strategy for the forward utility, $\widehat{\theta}$, maximizes the functional K over the set of admissible strategies, and

$$K(\widehat{\theta}) \cdot A = \mathrm{e}^{-\Delta B} \cdot X^B = \mathrm{e}^{-\Delta B} \cdot B + \sum(1 - \mathrm{e}^{-\Delta B} - \Delta B \mathrm{e}^{-\Delta B}). \tag{5}$$

Then, using the optimality of $\widehat{\theta}$ together with Lemma 9 (note that $K(\theta) = -\overline{K}(-\theta)$) we conclude that $-\widehat{\theta}$ is a root of the equation

$$b + c\theta + \int [\mathrm{e}^{\theta^T x} x - h(x)] F(\mathrm{d}x) = 0. \tag{6}$$

Furthermore, by combining (5) and (6), we deduce that the policy $\widetilde{\theta} := -\widehat{\theta}$ satisfies

$$\left[\frac{1}{2}\widetilde{\theta}^T c\widetilde{\theta} + \int (\widetilde{\theta}^T x \mathrm{e}^{\widetilde{\theta}^T x} - \mathrm{e}^{\widetilde{\theta}^T x} + 1) F(\mathrm{d}x)\right] \cdot A$$
$$= \mathrm{e}^{-\Delta B} \cdot B + \sum(1 - \mathrm{e}^{-\Delta B} - \Delta B \mathrm{e}^{-\Delta B}). \tag{7}$$

(b) In this step, we construct a positive local martingale, $\widetilde{Z}$, candidate to the MEH σ-martingale density. Since $\widetilde{\theta} := -\widehat{\theta} \in L(S^c)$, the process $\widetilde{\theta} \cdot S^c$ is a well-defined continuous local martingale that constitute the continuous part of $\widetilde{N} := (1/\widetilde{Z}_-) \cdot \widetilde{Z}$.

To define the purely discontinuous ingredient of $\widetilde{N}$, we consider the $\widetilde{\mathscr{P}}$-measurable function

$$W_t(x) := (\widetilde{\gamma}_t)^{-1}\left(\mathrm{e}^{\widetilde{\theta}_t^T x} - 1\right), \qquad \widetilde{\gamma}_t := 1 - a_t + \int \mathrm{e}^{\widetilde{\theta}_t^T x}\nu(\{t\}, \mathrm{d}x) \tag{8}$$

and we will prove that W is $(\mu - \nu)$-integrable (see [14] or [15] for the definition of this integrability), i.e. that $\left(\sum (W_t(\Delta S) I_{\{\Delta S_t \neq 0\}} - \widehat{W}_t)^2\right)^{1/2} \in \mathscr{A}^+_{loc}$. This will be carried out in several steps, see (b.1)–(b.5).

(b.1) Since $\widehat{\theta} \cdot S$ is a RCLL semimartingale, the process $I_{\{|\widetilde{\theta}^T \Delta S| \leq \alpha\}} \cdot [\widetilde{\theta} \cdot S, \widetilde{\theta} \cdot S]$ is locally bounded and, hence, locally integrable. Then, due to the inequalities

$$\begin{aligned}\sum (\mathrm{e}^{\widetilde{\theta}^T \Delta S} - 1)^2 I_{\{|\widetilde{\theta}^T \Delta S| \leq \alpha\}} &\preceq \mathrm{e}^{2\alpha} \sum (\widetilde{\theta}^T \Delta S)^2 I_{\{|\widetilde{\theta}^T \Delta S| \leq \alpha\}} \\ &\preceq \mathrm{e}^{2\alpha} I_{\{|\widetilde{\theta}^T \Delta S| \leq \alpha\}} \cdot [\widetilde{\theta} \cdot S, \widetilde{\theta} \cdot S],\end{aligned}$$

we deduce that $\sum (\mathrm{e}^{\widetilde{\theta}^T \Delta S} - 1)^2 I_{\{|\widetilde{\theta}^T \Delta S| \leq \alpha\}}$ is locally integrable.

(b.2) From (7) we deduce that the process $(\widetilde{\theta}^T x \mathrm{e}^{\widetilde{\theta}^T x} - \mathrm{e}^{\widetilde{\theta}^T x} + 1) \star \mu$ is locally integrable. This and the relation

$$\sum |\mathrm{e}^{\widetilde{\theta}^T \Delta S} - 1| I_{\{|\widetilde{\theta}^T \Delta S| > \alpha\}} \preceq \sum \left[\frac{\mathrm{e}^\alpha - 1}{\alpha} + \frac{\widetilde{\theta}^T \Delta S \mathrm{e}^{\widetilde{\theta}^T \Delta S} - \mathrm{e}^{\widetilde{\theta}^T \Delta S} + 1}{\alpha}\right] I_{\{|\widetilde{\theta}^T \Delta S| > \alpha\}}$$

imply the local integrability of $\sum |\mathrm{e}^{\widetilde{\theta}^T \Delta S} - 1| I_{\{|\widetilde{\theta}^T \Delta S| > \alpha\}}$.

(b.3) Using the relation

$$\left(\sum (\mathrm{e}^{\widetilde{\theta}^T \Delta S} - 1)^2 I_{\{|\widetilde{\theta}^T \Delta S| > \alpha\}}\right)^{1/2} \preceq \sum |\mathrm{e}^{\widetilde{\theta}^T \Delta S} - 1| I_{|\{\widetilde{\theta}^T \Delta S| > \alpha\}}$$

and parts (b.1)–(b.2), we obtain the local integrability of $\left(\sum (\mathrm{e}^{\widetilde{\theta}^T \Delta S} - 1)^2\right)^{1/2}$.

(b.4) Defining $\Gamma := \{x \in \mathbb{R}^d : |\widetilde{\theta}^T x| \leq \alpha\}$ and using the notation of (8), we derive that

$$\begin{aligned}\frac{1}{2}\sum (\widehat{W}_t)^2 &\preceq \sum \left(\frac{1}{\widetilde{\gamma}_t}\int_\Gamma (\mathrm{e}^{\widetilde{\theta}_t^T x} - 1)\nu_t(\mathrm{d}x)\right)^2 \\ &\quad + \sum \left((\widetilde{\gamma}_t)^{-1}\int_{\mathbb{R}^d \setminus \Gamma} (\mathrm{e}^{\widetilde{\theta}_t^T x} - 1)\nu_t(\mathrm{d}x)\right)^2 \\ &\preceq \sum (\widetilde{\gamma}_t)^{-2}\int_\Gamma (\mathrm{e}^{\widetilde{\theta}_t^T x} - 1)^2 \nu_t(\mathrm{d}x) \\ &\quad + \left(\sum (\widetilde{\gamma}_t)^{-1}\int_{\mathbb{R}^d \setminus \Gamma} |\mathrm{e}^{\widetilde{\theta}_t^T x} - 1|\nu_t(\mathrm{d}x)\right)^2 \\ &= (\widetilde{\gamma})^{-2}\left(\mathrm{e}^{\widetilde{\theta}^T x} - 1\right)^2 I_{\{|\widetilde{\theta}^T x| \leq \alpha\}} \star \nu + \left((\widetilde{\gamma})^{-1}|\mathrm{e}^{\widetilde{\theta}^T x} - 1| I_{\{|\widetilde{\theta}^T x| > \alpha\}} \star \nu\right)^2.\end{aligned}$$

Due to (b.1)–(b.2), the predictable nondecreasing processes

$$(\mathrm{e}^{\widetilde{\theta}^T x}-1)^2 I_{\{|\widetilde{\theta}^T x|\le\alpha\}}\star\nu,\qquad |\mathrm{e}^{\widetilde{\theta}^T x}-1|I_{\{|\widetilde{\theta}^T x|>\alpha\}}\star\nu$$

have finite variation and, thus, are locally bounded. This follows from the fact that these processes are the compensators of the two processes discussed in (b.1) and (b.2) respectively. Using similar arguments as in Lemma 2.1 of [2], we deduce the local boundedness of $\widetilde{\gamma}^{-1}$. Hence, the process $\sum(\widehat{W}_t)^2$ is locally bounded.

(b.5) Using once more the local boundedness of $\widetilde{\gamma}^{-1}$, parts (b.1)–(b.4), and

$$\begin{aligned}
&\Bigl(\sum(W_t(\Delta S)I_{\{\Delta S_t\neq 0\}}-\widehat{W}_t)^2\Bigr)^{1/2}\\
&\quad\preceq\Bigl(2\sum(W_t(\Delta S))^2 I_{\{\Delta S_t\neq 0\}}\Bigr)^{1/2}+\Bigl(2\sum(\widehat{W}_t)^2\Bigr)^{1/2}\\
&\quad=\bigl[2(\widetilde{\gamma})^{-2}\bigl(\mathrm{e}^{\widetilde{\theta}^T x}-1\bigr)^2\star\mu\bigr]^{1/2}+\Bigl(2\sum(\widehat{W}_t)^2\Bigr)^{1/2}
\end{aligned}$$

we deduce the local integrability of $(\sum(W_t(\Delta S)I_{\{\Delta S_t\neq 0\}}-\widehat{W}_t)^2)^{1/2}$. This ends the proof of the $(\mu-\nu)$-integrability of W. We conclude that $W\star(\mu-\nu)$ is a local martingale and the process $\widetilde{Z}:=\mathscr{E}(\widetilde{N})$ such that

$$\begin{aligned}
\widetilde{N}&:=\widetilde{\theta}\cdot S^c+W\star(\mu-\nu),\qquad \widetilde{\theta}=-\widehat{\theta},\\
W_t(x)&:=\frac{\mathrm{e}^{\widetilde{\theta}_t^T x}-1}{1-a_t+\int \mathrm{e}^{\widetilde{\theta}_t^T y}\nu(\{t\},\mathrm{d}y)}
\end{aligned}$$

is well-defined and is a σ-martingale density for S, due to (6).

(c) In this step, we prove (4). Considering (7) and (47), on $\{\Delta A=0\}$, we derive

$$\begin{aligned}
I_{\{\Delta A=0\}}\cdot B&=\bigl(\mathrm{e}^{-\Delta B}I_{\{\Delta A=0\}}\bigr)\cdot X^B\\
&=I_{\{\Delta A=0\}}\Bigl[\frac{1}{2}\widetilde{\theta}^T c\widetilde{\theta}+\int\Bigl(\widetilde{\theta}^T x\mathrm{e}^{\widetilde{\theta}^T x}-\mathrm{e}^{\widetilde{\theta}^T x}+1\Bigr)F(\mathrm{d}x)\Bigr]\cdot A\\
&=I_{\{\Delta A=0\}}\cdot h^E(\widetilde{Z},P).
\end{aligned}\tag{9}$$

Again, equality (7) together with (6) and (8), imply that

$$1-\mathrm{e}^{-\Delta B_t}=a_t-\int \mathrm{e}^{\widetilde{\theta}_t^T x}\nu(\{t\},\mathrm{d}x)=1-\widetilde{\gamma}_t$$

or, equivalently, that $\Delta B=-\log\widetilde{\gamma}$. By combining this with (48) (here $\lambda=\widetilde{\theta}$ and thus $\gamma=\widetilde{\gamma}$), we obtain that

$$\Delta B=\Delta h^E(\widetilde{Z},P).\tag{10}$$

Therefore, (4) follows immediately from (9) and (10).

(d) It remains to prove the optimality of $\widetilde{Z}$ and, consequently, to conclude the whole part (2). Thanks to Proposition 3.2 in [4] (see also Proposition 4.2 in [3] for the case of quasi-left continuity), it is enough to consider a positive σ-martingale density $Z = \mathscr{E}(N)$ of the form

$$N = \beta \cdot S^c + Y \star (\mu - \nu), \qquad Y_t(x) = k_t(x) + \frac{\widehat{k}_t}{1 - a_t} I_{\{a_t < 1\}},$$

$$\widehat{k}_t := \int k_t(x) \nu(\{t\}, \mathrm{d}x),$$

where $\beta \in L(S)$ and $\left(\sum k_t (\Delta S_t)^2 I_{\{\Delta S_t \neq 0\}}\right)^{1/2} \in \mathscr{A}^+_{loc}$. Then, due to the convexity of $z^T cz$ and $\phi(z) := (1+z)\log(1+z) - z$, we obtain on $\{\Delta A = 0\}$ that

$$\begin{aligned}
&\frac{\mathrm{d}h^E(Z, P)}{\mathrm{d}A} - \frac{\mathrm{d}h^E(\widetilde{Z}, P)}{\mathrm{d}A} \\
&\quad = \int \left[\phi(k(x)) - \phi\left(\mathrm{e}^{\widetilde{\theta}^T x} - 1\right)\right] F(\mathrm{d}x) + \frac{1}{2}(\beta^T c \beta - \widetilde{\theta}^T c \widetilde{\theta}) \\
&\quad \geq \widetilde{\theta}^T c(\beta - \widetilde{\theta}) + \int \widetilde{\theta}^T x \left(k(x) + 1 - \mathrm{e}^{\widetilde{\theta}^T x}\right) F(\mathrm{d}x) = 0.
\end{aligned} \tag{11}$$

Equality (11) is derived from a combination of (6) for $\widetilde{Z}$ and a similar equation for Z, that is,

$$0 = b + c\beta + \int [x(k(x) + 1) - h(x)] F(\mathrm{d}x)$$

since Z is a σ-martingale density for S. On the other hand, due to (46), we get

$$\begin{aligned}
\Delta h^E_t(Z, P) - \Delta h^E_t(\widetilde{Z}, P) &= \int \left[\phi(k_t(x)) - \phi\left(\frac{\mathrm{e}^{\widetilde{\theta}_t^T x}}{\widetilde{\gamma}} - 1\right)\right] \nu(\{t\}, \mathrm{d}x) \\
&\quad + (1 - a_t)\left[\phi\left(-\frac{\widehat{k}_t}{1 - a_t}\right) - \phi\left(\widetilde{\gamma}_t^{-1} - 1\right)\right] \\
&\geq \int \left[k_t(x) + 1 - \frac{\mathrm{e}^{\widetilde{\theta}_t^T x}}{\widetilde{\gamma}_t}\right]\left(\widetilde{\theta}_t^T x + \log\frac{1}{\widetilde{\gamma}_t}\right) \nu(\{t\}, \mathrm{d}x) \\
&\quad + (1 - a_t)\left(1 - \frac{\widehat{k}_t}{1 - a_t} - \frac{1}{\widetilde{\gamma}_t}\right) \log\frac{1}{\widetilde{\gamma}_t} \\
&= \int \left[(k_t(x) + 1) - (\widetilde{\gamma}_t)^{-1} \mathrm{e}^{\widetilde{\theta}_t^T x}\right] \widetilde{\theta}_t^T x \nu(\{t\}, \mathrm{d}x) = 0.
\end{aligned} \tag{12}$$

Equality (12) follows from the fact that Z and $\widetilde{Z}$ are σ-martingale densities for S. Thus, by combining (11) and (12), we deduce that $\widetilde{Z}$ is the MEH σ-martingale density for S. Furthermore, due to Theorem 9, (4) and $\widetilde{\theta} = -\widehat{\theta}$ (see (a), (b.1)–(b.5), and (c)), we get that

$$\widetilde{Z} = \mathrm{e}^{-B_0} \exp\left[B - (\widehat{\theta} \cdot S)\right].$$

Hence, it is a true martingale and this implies the existence of the MEH σ-martingale measure, $\widetilde{Q}$. This proves (ii) and completes the proof of the theorem. □

Remark 2

1. It is clear that the proof of the part (ii) $\Longrightarrow$ (i) of Theorem 2 follows easily from [3] and [4]. In fact, it was clearly stated in those papers that this kind of robustness with respect to the horizon is one of the important features of the minimal entropy-Hellinger σ-martingale measure that other σ-martingale measures lack to possess; see, also, [5] for a more explicit relationship between this horizon-robustness for σ-martingale measures and utility maximization for all HARA utilities.
2. The most original part of Theorem 2 lies in proving that the only forward utility of this kind (i.e. when B is predictable with finite variation) is the one given through the MEH σ-martingale measure and this σ-martingale measure in fact exists. Furthermore, this part of the theorem also gives necessary and sufficient conditions for the existence of MEH σ-martingale measure via the utility maximization problem with weaker conditions on S.

Theorem 2 looks restrictive due to the assumption on B, while—as we will illustrate in the proof of the next theorem—it is crucial and constitutes an important step for proving our general result. This result requires some preparations.

Definition 10 A RCLL semimartingale B is said to be *exponentially special*, if $\exp(B)$ is a special semimartingale, i.e.

$$\exp(B) = \exp(B_0) + M^{(B)} + A^{(B)},$$

where the process $M^{(B)}$ is a local martingale, the process $A^{(B)}$ is predictable with finite variation, $M_0^{(B)} = A_0^{(B)} = 0$.

Lemma 3 *Let B be a RCLL semimartingale. Then the following statements hold.*

(i) *If B is exponentially special, then there exist a unique positive local martingale, $Z^{(B)}$, and predictable process, B', with finite variation such that*

$$\mathrm{e}^B = \mathrm{e}^{B_0 + B'} Z^{(B)}, \quad B'_0 = 0, \quad Z_0^{(B)} = 1. \tag{13}$$

(ii) *Suppose that pB is exponentially special, for some $p \in (1, \infty)$, $Z^{(B)}$ is a true martingale, and* (3) *holds. Then,*

$$\int_{\{|x|>1\}} |x| \mathrm{e}^{\lambda^T x} F^Q(\mathrm{d}x) < \infty, \quad \textit{for all } \lambda \in \mathbb{R}^d, \tag{14}$$

where F^Q is the kernel measure for the jumps sizes of S under $Q := Z_T^{(B)} \cdot P$.

Proof Since e^B is a special semimartingale, then $\mathrm{e}^{-B_-} \cdot \mathrm{e}^B$ is also a special semimartingale and there exist a unique local martingale, $N^{(B)}$, and a predictable process, $C^{(B)}$, with finite variation such that

$$\mathrm{e}^{-B_-} \cdot \mathrm{e}^B = N^{(B)} + C^{(B)} \quad \text{and} \quad C_0^{(B)} = N_0^{(B)} = 0.$$

The above equation implies that

$$\mathrm{e}^B = \mathrm{e}^{B_0} \mathscr{E}\big(N^{(B)} + C^{(B)}\big), \qquad 1 + \Delta C^{(B)} > 0 \quad \text{and} \quad 1 + \frac{\Delta N^{(B)}}{1 + \Delta C^{(B)}} > 0.$$

As a result, the process $\frac{1}{1+\Delta C^{(B)}} \cdot N^{(B)}$ is a local martingale, $\mathscr{E}(\frac{1}{1+\Delta C^{(B)}} \cdot N^{(B)}) > 0$, and $\mathscr{E}(C^{(B)})$ is a positive predictable process with finite variation. Then, due to Yor's formula ($\mathscr{E}(X)\mathscr{E}(Y) = \mathscr{E}(X + Y + [X, Y])$ for semimartingales X, Y), we write

$$\mathrm{e}^B = \mathrm{e}^{B_0} \mathscr{E}\left(\frac{1}{1 + \Delta C^{(B)}} \cdot N^{(B)}\right) \mathscr{E}\big(C^{(B)}\big).$$

Now (i) follows directly by putting $Z^{(B)} := \mathscr{E}(\frac{1}{1+\Delta C^{(B)}} \cdot N^{(B)})$ and $B' := \log \mathscr{E}(C^{(B)})$.

Next, we will prove the assertion (ii). To this end, we suppose that pB is exponentially special. Thus, B is exponentially special and, hence, (i) holds. On the other hand, it is clear that $(Z^{(B)})^p$ is locally integrable (i.e. a special semimartingale), and $F^Q(\mathrm{d}x) = (1 + f(x))F(\mathrm{d}x)$, if $(\beta, f, g, \overline{M})$ are Jacod's components for $M^{(B)} := \frac{1}{Z_-^{(B)}} \cdot Z^{(B)}$. Using Lemma 10, we deduce that

$$\int_{\{|x|>1\}} |x| \mathrm{e}^{\lambda^T x} F^Q(\mathrm{d}x) = \int_{\{|x|>1\}} |x| \mathrm{e}^{\lambda^T x} (1 + f(x)) F(\mathrm{d}x)$$
$$\leq \left(\int I_{\{|x|>1\}} |x|^q \mathrm{e}^{q\lambda^T x} F(\mathrm{d}x)\right)^{\frac{1}{q}} \left(\int I_{\{|x|>1\}} (1 + f(x))^p F(\mathrm{d}x)\right)^{\frac{1}{p}} < \infty.$$

This proves the assertion (ii), and the proof of the lemma is complete. □

Now, we will state our main and general result of this section.

Theorem 3 *Suppose that S satisfies* (3) *and consider a RCLL semimartingale, B, such that pB is exponentially special for some $p \in (1, \infty)$. Then*:

(1) *The following assertions are equivalent*:

(i) *The random utility field, $U(t, \omega, x) = -\exp(-x + B_t(\omega))$, is a forward utility with optimal strategy $\widehat{\theta}$.*

(ii) *There exists a unique positive local martingale $Z^{(B)}$ satisfying*:

(a) *The MEH σ-martingale density with respect to $Z^{(B)}$ exists. It is denoted by $\widetilde{Z}^{(B)}$ and satisfies*

$$B - B_0 = \log Z^{(B)} + h^E(\widetilde{Z}^{(B)}, Z^{(B)}). \tag{15}$$

(b) *The process $\widehat{Z}^{(B)} := \widetilde{Z}^{(B)} Z^{(B)}$ is a true martingale, $\widehat{Q}^{(B)} := \widehat{Z}^{(B)}_T \cdot P$ is a σ-martingale measure, and $\widehat{Z}^{(B)} \log \widehat{Z}^{(B)}$ is locally integrable (i.e. a special semimartingale).*

(c) *We have:*

$$\log(\widetilde{Z}^{(B)}) = \widetilde{\theta}^{(B)} \cdot S + h^E(\widetilde{Z}^{(B)}, Z^{(B)}) \quad \textit{and} \quad \widetilde{\theta}^{(B)} = -\widehat{\theta}.$$

(2) *If the assertion* (i) *holds and furthermore, B is such that*

$$\sup_{\tau \in \mathcal{T}_T} E\left[e^{pB_\tau} \right] < \infty \quad \textit{for some } p \in (1, \infty), \tag{16}$$

then $Z^{(B)}$ is a true martingale. Moreover, the probability $\widehat{Q}^{(B)}$ has a finite P-entropy, i.e. $\widehat{Q}^{(B)} \in \mathcal{M}^e_f(S)$.

Proof The proof of this theorem will be given in three parts. Part I will prove (i) $\Longrightarrow$ (ii), Part II will prove the reverse, while Part III will prove the assertion (2). Notice that under the assumptions of this theorem, the assertions of Lemma 3 hold.

(I) Suppose that assertion (i) holds and consider a sequence of stopping times, $(T_n)_{n \geq 1}$, that increases stationarily to T such that $(Z^{(B)})^{T_n}$ is a true martingale and $B'_{t \wedge T_n}$ is bounded. Then, by putting $Q_n := Z^{(B)}_{T_n} \cdot P$, and using Lemma 2, we deduce that the process $U_n(t, \omega, x) := -\exp(-x + B'_{t \wedge T_n})$ is a forward dynamic utility for (S^{T_n}, Q_n). Therefore, assertion (ii) of Lemma 3 (precisely, condition (14)) guarantee a direct application of Theorem 2 to the model (S^{T_n}, Q_n, U_n). This implies the existence of the MEH σ-martingale measure with respect to Q_n, denoted by $\widetilde{Q}^n$, whose density $\widetilde{Z}^{(B,n)}$ satisfies

$$B'_{t \wedge T_n} = h^E_t\big(\widetilde{Z}^{(B,n)}, Q_n\big).$$

Using Lemma 1, we conclude that the MEH σ-martingale density with respect to $Z^{(B)}$, denoted by $\widetilde{Z}^{(B)}$, exists and satisfies

$$B'_t = h^E_t\big(\widetilde{Z}^{(B)}, Z^{(B)}\big).$$

Then, plugging this equation into (13), the assertion (ii)-(a) follows immediately.

From Theorem 11, we have

$$\log \widetilde{Z}^{(B)} = \widetilde{\theta}^{(B)} \cdot S + h^E\big(\widetilde{Z}^{(B)}, Z^{(B)}\big),$$

where the process $\widetilde{\theta}^{(B)}$ is explicitly described and coincides with $-\widehat{\theta}$; this follows by applying Theorem 2 to the model (S^{T_n}, Q_n, U_n). This proves (ii)-(c).

To prove assertion (ii)-(b), it is easy to note that—due to the definition of $\widetilde{Z}^{(B)}$—$\widehat{Z}^{(B)}$ is a σ-martingale density for S, and $Z^{(B)}\widetilde{Z}^{(B)}\log\widetilde{Z}^{(B)}$ is locally integrable.

Now, we will prove that $\widehat{Z}^{(B)}\log\widehat{Z}^{(B)}$ is locally integrable. Consider a sequence of stopping times, $(T_n)_{n\geq 1}$, that increases stationarily to T such that $E[(Z^{(B)}_{T_n})^p]<\infty$ (this is possible since pB is exponentially special) and $h^E_{t\wedge T_n}(\widetilde{Z}^{(B)},Z^{(B)})$ is bounded. Then, by putting $\varepsilon=p-1$ and $Q_n:=Z^{(B)}_{T_n}\cdot P$, and using Young's inequality (i.e. $xy\leq y\log(y)-y+e^x$), we derive that

$$
\begin{aligned}
E\left[Z^{(B)}_{T_n}\widetilde{Z}^{(B)}_{T_n}\log(Z^{(B)}_{T_n})\right] &= E^{Q_n}\left(\frac{1}{\varepsilon}\widetilde{Z}^{(B)}_{T_n}\log[(Z^{(B)}_{T_n})^{\varepsilon}]\right)\\
&\leq E^{Q_n}\left[\frac{\widetilde{Z}^{(B)}_{T_n}}{\varepsilon}\log\left(\frac{\widetilde{Z}^{(B)}_{T_n}}{\varepsilon}\right)-\frac{\widetilde{Z}^{B}_{T_n}}{\varepsilon}\right]+E^{Q_n}\left[(Z^{(B)}_{T_n})^{\varepsilon}\right]\\
&\leq E^{Q_n}\left[\frac{\widetilde{Z}^{(B)}_{T_n}}{\varepsilon}\log\left(\frac{\widetilde{Z}^{B}_{T_n}}{\varepsilon}\right)\right]-\frac{1}{\varepsilon}+e^{-pB_0}\sup_{\tau\in\mathscr{T}_T}E\left[e^{pB_\tau}\right].
\end{aligned}
$$

This proves that $Z^{(B)}\widetilde{Z}^{(B)}\log Z^{(B)}$ is locally integrable. Thus, by putting

$$
\widehat{Z}^{(B)}\log\widehat{Z}^{(B)}=Z^{(B)}\widetilde{Z}^{(B)}\log\widetilde{Z}^{(B)}+Z^{(B)}\widetilde{Z}^{(B)}\log Z^{(B)}, \tag{17}
$$

we deduce that $\widehat{Z}^{(B)}\log\widehat{Z}^{(B)}$ is locally integrable.

(II) Suppose that the assertion (ii) holds. Then, assertions (ii)-(b) and (ii)-(c) imply that $-\widetilde{\theta}^{(B)}$ is an admissible strategy, the process

$$
U\big(t,-(\widetilde{\theta}^{(B)}\cdot S)_t\big)=-\exp\big((\widetilde{\theta}^{(B)}\cdot S)_t+B_t\big)=-e^{B_0}\widehat{Z}^{(B)}_t
$$

is a true martingale, and $\widehat{Q}:=\widehat{Z}^{(B)}_T\cdot P$ is a σ-martingale measure for S. Then, for any admissible strategy θ, we have

$$
\sup_{\tau\in\mathscr{T}_T}E^{\widehat{Q}}\exp\big[-(\theta+\widetilde{\theta}^{(B)})\cdot S_\tau\big]=e^{-B_0}\sup_{\tau\in\mathscr{T}_T}E\exp\big[B_\tau-(\theta\cdot S)_\tau\big]<\infty.
$$

Thus, thanks to Lemma 5, we deduce that the process

$$
-\exp\big(-\theta\cdot S+B\big)=-e^{B_0}\widehat{Z}^{(B)}\exp\big(-(\widetilde{\theta}+\theta)\cdot S\big),
$$

is a supermartingale. This proves assertion (i).

(III) Thanks to (15), (17), and assertion (ii)-(c), we obtain

$$
\begin{aligned}
E\widehat{Z}^{(B)}_{\tau\wedge T_n}\log\widehat{Z}^{(B)}_{\tau\wedge T_n} &= E^{Q_n}\widetilde{Z}^{(B)}_{\tau\wedge T_n}\log\widetilde{Z}^{(B)}_{\tau\wedge T_n}\\
&\quad+E^{Q_n}\Big[\widetilde{Z}^{(B)}_{\tau\wedge T_n}\Big(B_{T_n\wedge\tau}-B_0-\log\widetilde{Z}^{(B)}_{\tau\wedge T_n}\Big)\Big]\\
&=E\left[\widehat{Z}^{(B)}_{\tau\wedge T_n}B_{\tau\wedge T_n}\right]-B_0.
\end{aligned}
$$

Hence, using again Young's inequality, we obtain

$$E\left(\widehat{Z}^{(B)}_{\tau\wedge T_n}\log \widehat{Z}^{(B)}_{\tau\wedge T_n}\right) \leq \frac{p}{p-1}\sup_{\sigma\in\mathcal{T}_T} E\left(e^{pB_\sigma}\right) - \frac{p}{p-1}B_0.$$

Then, using Fatou's lemma, the above inequality leads to $\widehat{Q}^{(B)} \in \mathcal{M}^e_f(S)$ and, hence, assertion (2) of the theorem follows. This ends the proof of the theorem. □

Theorem 4 *Let B be a RCLL semimartingale and $N := \mathcal{E}(\pi\cdot S)$ a numéraire. Then, there is equivalence between:*

(i) *The random utility field $U(t,\omega,x) = -\exp\left(-\frac{x+B_t(\omega)}{N_t(\omega)}\right)$ is a forward utility for the assets S.*

(ii) *The random utility field $\overline{U}(t,\omega,x) := -\exp\left(-x + \frac{B_t(\omega)}{N_t(\omega)}\right)$ is a forward utility for the assets*

$$\overline{S} := S - \frac{1}{1+\pi^T\Delta S}\cdot[S,\pi\cdot S].$$

Proof Due to Yor's formula, we deduce that

$$\frac{1}{N} = \mathcal{E}\left(-\pi\cdot S + \frac{1}{1+\pi^T\Delta S}\cdot[\pi\cdot S,\pi\cdot S]\right).$$

On the other hand, Ito's formula yields

$$\mathrm{d}\left(\frac{\theta\cdot S}{N}\right) = \phi(\theta)\mathrm{d}\overline{S},$$

where $\phi(\theta)$ is given by

$$\phi(\theta) := \frac{\theta - (\theta\cdot S)_-\pi}{N_-}, \quad \text{for any } \theta\in L(S).$$

As a result, we get

$$U\left(t, x + (\theta\cdot S)_t\right) = \overline{U}\left(t, x + (\phi(\theta)\cdot\overline{S})_t\right), \quad \text{for any } \theta\in L(S).$$

Therefore, for any process θ, $\theta\in\mathcal{A}_{adm}(x,S,U)$ if and only if $\phi(\theta)\in\mathcal{A}_{adm}(x,\overline{S},\overline{U})$. The proof of the theorem follows easily. □

Remark 3

1. Theorem 4 yields our complete and explicit parametrization for the exponential forward utilities. In fact, using a nice result of [28] that states that if $-\exp(-\frac{x+B}{N})$ is a forward utility, then N is a numéraire and B is a semimartingale. This gives us the first parametrization through the description of N. Then, by using Theorem 4, we transfer the self-generating property to the model $\overline{S}$ and the payoff $\overline{B} = \frac{B}{N}$ instead, and Theorem 3 completes the explicit parametrization

of the utility by describing the structure of $\overline{B}$. Thus, the parameters of a forward utility are $(\pi, N^{(B)}) \in L(S) \times \mathcal{M}_{loc}(P)$ or, equivalently, $(\pi, \beta, f, g, \overline{N}^{(B)})$.

2. The semimartingale property for B becomes obvious from the definition of forward dynamic utility if the set of admissible strategies $\mathcal{A}_{adm}(x)$ contains the null strategy for some $x \in \mathbb{R}$. This situation is realizable when more integrability conditions are imposed on the payoff B such as boundedness for instance.

The following remark discusses the originality of this section, and compares its results (mainly Theorems 2–3) with the most recent literature on the exponential forward dynamic utilities.

Remark 4 This remark, as suggested by an anonymous referee, discusses the originality of the results of this section and compares them with those obtained by Zitkovic in [28] (especially Theorem 4.4 of that paper). To this end, we focus on the case of $N = 1$, for simplicity. The result of Zitkovic in [28] characterizes the exponential forward utility relying on the relative conditional entropy concept. Precisely, for any $Q \in \mathbb{P}_a$ with density process Z^Q, and any $0 \leq t \leq T < \infty$,

$$H(Q, t, T) := E\left(\frac{Z_T^Q}{Z_t^Q} \log \frac{Z_T^Q}{Z_t^Q} \bigg| \mathcal{F}_t \right),$$

denotes the relative conditional entropy of Q with respect to P. Using this concept, Zitkovic derived the following characterization

$$B_t = - \operatorname*{ess\,inf}_{Q \in \mathcal{M}_f^a(S)} H(Q, t, T), \tag{18}$$

for the case of $N = 1$, for any $0 \leq t \leq T < \infty$. Here, $\mathcal{M}_f^a(S)$ denotes the set of $Q \in \mathbb{P}_a$ with finite entropy (i.e. $H(Q, 0, T) < \infty$ for any T) such that S is a Q-local martingale.

It is very clear—up to our knowledge—that for any T, the essential infimum in the rhs term of (18) is attainable under Zitkovic's assumptions (i.e. S locally bounded and $\mathcal{M}_f^a(S) \neq \emptyset$) by the minimal entropy martingale measure for the model S^T. It is, also, very clear that there is no a single result in the literature that describes *explicitly* this optimal martingale measure for the general semimartingale S. Thus, in our view, (18) is a characterization that is not applicable (at least, we do not see how to apply it) and it is not explicit for general case of locally bounded semimartingale S. Thus, this result does not parameterize the exponential forward utility, whereas our results presented in this section give a clear and explicit parameterization.

Furthermore—as was pointed out to us by an anonymous referee—our assumptions on the model S are much more general than those of Zitkovic. Indeed, in [28], the author assumed that S is locally bounded and $\mathcal{M}_f^a(S) \neq \emptyset$ (that is, assumption (19) below), while we obtain our parameterization under the assumption (3), which is essentially weaker.

In our view, the most practical result of [28]—besides the section that deals with the easiest case of Ito processes—is Proposition 4.7, where the author proved that the process N (denoted by γ in his paper) should be a numéraire. In other words, there exists $\pi \in L(S)$ such that $N = N_0\mathscr{E}(\pi \cdot S)$. Herein, we use this nice result to complete our full parameterization.

4 Horizon-Unbiased Exponential Hedging

Throughout this section, we assume that

$$S \text{ is locally bounded and } \mathscr{M}_f^e(S) \neq \emptyset. \tag{19}$$

For a process B and a stopping time τ, we denote by $Q^{(\tau,B)}$ the minimal entropy martingale measure for S^τ with respect to $P^{(\tau,B)}$, where $P^{(\tau,B)} := \frac{e^{B_\tau}}{Ee^{B_\tau}} \cdot P$. The set of admissible strategies that we consider in this section is given by

$$\Theta(S,B) := \left\{\theta \in L(S) : (\theta \cdot S)^\tau \in \mathscr{M}(Q^{(\tau,B)}) \quad \text{for all} \quad \tau \in \mathscr{T}_T\right\},$$

where $\mathscr{T}_T$ denotes the set of all stopping times bounded from above by T. This definition of strategies extends slightly the definition given by [6] to the case of a dynamic payoff B. For other sets of strategies, we refer the reader to this seminal paper.

Following the arguments from the previous section, we start addressing the horizon-unbiased hedging problem for the case when the payoff process is predictable with finite variation.

Theorem 5 *Suppose that* (19) *is satisfied and let B be a bounded predictable process with finite variation. Then the following assertions* (i) *and* (ii) *are equivalent*:

(i) *There exists $\widehat{\theta} \in \Theta(S,B)$ such that for any stopping time τ*

$$\min_{\theta \in \Theta(S,B)} E\big[\exp\big(B_\tau - (\theta \cdot S)_\tau\big)\big] = E\big[\exp\big(B_\tau - (\widehat{\theta} \cdot S)_\tau\big)\big]. \tag{20}$$

(ii) *For any $\theta \in \Theta(S,B)$*

$$I_{\{(\theta\cdot S)_- \neq 0\}} \cdot B = I_{\{(\theta\cdot S)_- \neq 0\}} \cdot h^E(\widetilde{Z}, P), \tag{21}$$

where $\widetilde{Z}$ is minimal entropy-Hellinger local martingale density described by (50), *i.e. $\widetilde{Z} = \exp(\widetilde{\theta} \cdot S + h^E(\widetilde{Z}, P))$.*

Furthermore, the optimal strategy $\widehat{\theta}$ coincides with $-\widetilde{\theta}$ obtained explicitly from $\widetilde{Z}$, i.e. $-\widehat{\theta}$ is a pointwise root of

$$0 = \begin{cases} b + c\theta + \int x(e^{\theta^T x} - 1)F(dx), & \text{on } \{\Delta A = 0\}, \\ \int x e^{\theta^T x} F(dx), & \text{on } \{\Delta A \neq 0\}. \end{cases} \tag{22}$$

Proof The most difficult part in proving this theorem is to prove that the optimal strategy $\widehat{\theta}$ in (20) can be derived from (22). We start with this statement.

Suppose that the assertion (i) holds. Note—see Lemma 6 for details—that (20) is equivalent to the fact that for any stopping time $T \geq \tau \geq \sigma$ and any $\theta \in \Theta(S, B)$, we have

$$E\left(\exp\left[B_\tau - \int_\sigma^\tau \theta_u \mathrm{d}S_u\right] \Big| \mathscr{F}_\sigma\right) \geq E\left(\exp\left[B_\tau - \int_\sigma^\tau \widehat{\theta}_u \mathrm{d}S_u\right] \Big| \mathscr{F}_\sigma\right), \quad P\text{-a.s.}$$

This, in turn, is equivalent to the fact that for any nonnegative left-continuous and bounded process H, and any finite and increasing sequence of stopping times $(\tau_i)_{i \leq n+1}$, we have

$$\sum_{i=0}^n H_{\tau_i} E\left(\exp\left[B_{\tau_{i+1}} - \int_{\tau_i}^{\tau_{i+1}} \theta_u \mathrm{d}S_u\right] \Big| \mathscr{F}_{\tau_i}\right)$$
$$\geq \sum_{i=0}^n H_{\tau_i} E\left(\exp\left[B_{\tau_{i+1}} - \int_{\tau_i}^{\tau_{i+1}} \widehat{\theta}_u \mathrm{d}S_u\right] \Big| \mathscr{F}_{\tau_i}\right). \tag{23}$$

Put $X_t^\theta := \exp[B_t - (\theta \cdot S)_t]$, and for any $\theta \in \Theta(S, B)$ consider a stationarily increasing sequence of stopping times $(T_n)_{n\geq 1}$ such that $X^{\widehat{\theta}}_{t\wedge T_n}$ and $X^{\theta}_{T_n \wedge t}$ are both special semimartingales with integrable martingale and predictable parts, and their left limit processes are bounded from below by $1/n$. Then, (23) is equivalent to the fact that for any nonnegative left-continuous and bounded adapted process H

$$E\left[\int_0^{T_n} H_u \mathrm{d}A_u^\theta\right] = E\left[\int_0^{T_n} \frac{H_u}{X_{u-}^\theta} \mathrm{d}X_u^\theta\right] \geq E\left[\int_0^{T_n} \frac{H_u}{X_{u-}^{\widehat{\theta}}} \mathrm{d}X_u^{\widehat{\theta}}\right] = E\left[\int_0^{T_n} H_u \mathrm{d}A_u^{\widehat{\theta}}\right], \tag{24}$$

where A^θ is a predictable process with finite variation given by

$$A^\theta := B - \theta^T b \cdot A + \frac{1}{2}\theta^T c\theta \cdot A + \sum (\mathrm{e}^{\Delta B} - 1 - \Delta B)(1 - a)$$
$$+ \left(\mathrm{e}^{\Delta B - \theta^T x} - 1 - \Delta B + \theta^T x\right) \star \nu.$$

Since the process H is arbitrary, we deduce that (24) is equivalent to the property:

$$A^{\widehat{\theta}} \preceq A^\theta, \quad \text{for any } \theta \in \Theta(S, B).$$

Or, equivalently, for any $\theta \in \Theta(S, B)$, we have $\overline{f}(\theta) \geq \overline{f}(\widehat{\theta})$, where

$$\overline{f}(\lambda) := -\lambda^T b + \frac{1}{2}\lambda^T c\lambda + \int \left(\mathrm{e}^{\Delta B - \lambda^T x} - 1 - \Delta B + \lambda^T x\right) F(\mathrm{d}x).$$

We easily deduce that, on the set $\{\Delta A = 0\}$, the function $\overline{f}(\theta)$ coincides with $\overline{K}(-\theta)$ of Lemma 9 (note that, in the present situation, the truncation function can be taken to be $h(x)=x$, due to the local boundedness of S). Hence we deduce that $-\widehat{\theta}$ is a root of the first equation in (22). On the set $\{\Delta A\neq 0\}$, we obtain that

$$\overline{f}_t(\theta_t)\Delta A_t = \mathrm{e}^{\Delta B_t}\int \mathrm{e}^{-\theta_t^T x}\nu(\{t\},\mathrm{d}x)-(1+\Delta B_t)a_t.$$

Thus, in this case, $\overline{f}(\theta)$ is a linear transformation of $\overline{K}(-\theta)$, and hence $-\widehat{\theta}$ is a root of the second equation in (22). This proves the last statement of the theorem.

Next, we prove the equivalence between assertions (i) and (ii). First, we assume that the assertion (i) holds. Put

$$\Theta_b := \{\theta\in L(S):\ (\theta\cdot S)_t(\omega)\quad \text{is uniformly bounded in } t \text{ and } \omega\}.$$

Then, for any $\theta\in\Theta_b$, both $(\theta\cdot S)^\tau$ and $(\widehat{\theta}\cdot S)^\tau$ are true $Q^{(\tau,B)}$-martingales, where $Q^{(\tau,B)}$ is given by

$$Q^{(\tau,B)} := \frac{\exp[B_\tau-(\widehat{\theta}\cdot S)_\tau]}{E\exp[B_\tau-(\widehat{\theta}\cdot S)_\tau]}P.$$

Therefore, for any $\theta\in\Theta_b$ and any stopping time τ, we obtain that

$$E\Big[(\theta\cdot S)_\tau\exp\big(B_\tau-(\widehat{\theta}\cdot S)_\tau\big)\Big]=E\Big[(\widehat{\theta}\cdot S)_\tau\exp\big(B_\tau-(\widehat{\theta}\cdot S)_\tau\big)\Big]=0.$$

Hence, the process $(\theta\cdot S)\exp(B-(\widehat{\theta}\cdot S))=\mathrm{e}^{B-h^E(\widetilde{Z},P)}\widetilde{Z}(\theta\cdot S)$ is a local martingale. A direct application of Ito's formula leads to (21), and the assertion (ii) follows.

Now, suppose that the assertion (ii) holds. Due to a direct application of Ito's formula, the assertion (ii) is equivalent to the statement that, for any $\theta\in\Theta(S,B)$, the process $Y^\theta := \exp\big[B+\widetilde{\theta}\cdot S\big](\theta\cdot S)$ is a local martingale. Indeed, this equivalence follows immediately from the fact that

$$\exp\big[B+\widetilde{\theta}\cdot S\big](\theta\cdot S)=\widetilde{Z}\exp\big[B-h^E(\widetilde{Z},P)\big](\theta\cdot S).$$

Let $\theta\in\Theta_b$ and let $(T_n)_{n\geq 1}$ be a sequence of stopping times increasing stationarily to T and such that $Y^{\widetilde{\theta}}_{t\wedge T_n}$ and $Y^{\theta}_{t\wedge T_n}$ are true martingales. Then, for any stopping time τ, we put $\tau_n := \tau\wedge T_n$ and obtain that

$$E\big[\mathrm{e}^{B_{\tau_n}-(\theta\cdot S)_{\tau_n}}\big]-E\big[\mathrm{e}^{B_{\tau_n}+(\widetilde{\theta}\cdot S)_{\tau_n}}\big]\geq E\Big\{\big((-\theta-\widetilde{\theta})\cdot S\big)_{\tau_n}\mathrm{e}^{B_{\tau_n}+(\widetilde{\theta}\cdot S)_{\tau_n}}\Big\}=0.$$

Due to Fatou's lemma and the boundedness of $\exp[B+(\theta\cdot S)]$ for any $\theta\in\Theta_b$, we get that

$$E\exp\big[B_\tau-(\theta\cdot S)_\tau\big]\geq E\exp\big[B_\tau+(\widetilde{\theta}\cdot S)_\tau\big].$$

The proof of the assertion (i) then follows from a direct application of the main result of [17] and by putting $\widehat{\theta}:=-\widetilde{\theta}$. □

Remark 5 Theorem 5 determines explicitly the optimal strategy in the horizon-unbiased exponential hedging when it exists. Furthermore, the theorem clearly illustrates the relationship between the horizon-unbiased hedging and a forward utility. In fact, we can easily conclude that, in general, the horizon-unbiased problem in (20) admits a solution while the corresponding random utility field, namely, $U(t,x) = -\exp(-x + B_t)$, may not be a forward utility. A simple example is when S is constant in a neighborhood of zero (i.e. $S_t = S_0$ for t close to zero), and B is neither an increasing nor a constant process. Furthermore, the equivalence between the existence of solution to the horizon-unbiased hedging problem and the property that $-\exp(-x + B)$ is a forward utility only holds only if there exists a strategy θ such that

$$\{(\omega, t) : (\theta \cdot S)_{t-}(\omega) \neq 0\} = \Omega \times [0, T].$$

In general, this equality does not hold. In fact, if S is constant in a neighborhood of zero (i.e. $S_t = S_0$ for t close to zero), then this equality is violated. Hence, in this case, the two concepts of forward utility and horizon-unbiased utility differ. Also, it is easy to see that the horizon-unbiased hedging problem admits a solution and its value function $v(\tau) := \min_{\theta\in\Theta(S,B)} E[\exp(B_\tau - (\theta \cdot S)_\tau)]$ is constant, i.e. $v(\tau) = v(T)$, if and only if $-\exp(B_t - x)$ is a forward dynamic utility.

Theorem 6 *Suppose that* (19) *holds and consider a semimartingale, B, satisfying* (16) *with the Doob-Meyer multiplicative decomposition given by*

$$\mathrm{e}^B = \mathrm{e}^{B_0} Z^{(B)} \mathrm{e}^{B'},$$

where $Z^{(B)}$ is a positive local martingale, B' is a predictable process with finite variation, $Z_0^{(B)} = 1$, $B'_0 = 0$. Then the following assertions are equivalent:

(i) *There exists $\widehat{\theta} \in \Theta(S, B)$ such that, for any stopping time τ,*

$$\min_{\theta\in\Theta(S,B)} E\Big[\exp(B_\tau - (\theta \cdot S)_\tau)\Big] = E\Big[\exp(B_\tau - (\widehat{\theta} \cdot S)_\tau)\Big]. \tag{25}$$

(ii) *The MEH local martingale density with respect to $Z^{(B)}$, denoted by $\widetilde{Z}^{(B)}$, satisfies*

$$I_{\{(\theta\cdot S)_- \neq 0\}} \cdot B' = I_{\{(\theta\cdot S)_- \neq 0\}} \cdot h^E\big(\widetilde{Z}^{(B)}, Z^{(B)}\big), \tag{26}$$

for any $\theta \in \Theta(S, B)$.

Furthermore, the optimal strategy in (25) *is given by*

$$\log \widetilde{Z}^{(B)} = -\widehat{\theta} \cdot S + h^E\big(\widetilde{Z}^{(B)}, Z^{(B)}\big).$$

Proof Consider a sequence of stopping times, $(T_n)_{n\geq 1}$ increasing stationarily to T and such that $(B')^{T_n}$ is bounded and $(Z^{(B)})^{T_n}$ is a true martingale. By putting $Q_n := Z_{T_n}^{(B)} \cdot P$, the assertion (i) implies that the horizon-unbiased hedging problem for $(S^{T_n}, (B')^{T_n}, Q_n)$ has a solution. Thus, a direct application of Theorem 5—to the model $(S^{T_n}, (B')^{T_n}, Q_n)$—implies that (26) holds for any $\theta \in \Theta(S, B)$, and the optimal strategy $\widehat{\theta}$ in (25) coincides with $-\widetilde{\theta}$, where $\widetilde{\theta}$ is the integrand that appears in the expression of $\widetilde{Z}^{(B)}$. This proves (ii). Next, assume that assertion (ii) holds, and notice that this assertion is equivalent to the statement that $\exp(B - \widehat{\theta} \cdot S)(\theta \cdot S)$ is a local martingale for any $\theta \in \Theta(S, B)$. Let $\theta \in \Theta_b$ and $(T_n)_{n \geq 1}$ be a stationarily increasing sequence of stopping times such that $((\widehat{\theta} - \theta) \cdot S)^{T_n} \exp\left(B^{T_n} - (\widehat{\theta} \cdot S)^{T_n}\right)$ is a true martingale.

Then, for any stopping time τ, we derive that

$$\begin{aligned} 0 &= E\Big[\exp\big(B_{\tau \wedge T_n} - (\widehat{\theta} \cdot S)_{\tau \wedge T_n}\big)\big((\widehat{\theta} - \theta) \cdot S\big)_{\tau \wedge T_n}\Big] \\ &\leq E \exp\big(B_{\tau \wedge T_n} - (\theta \cdot S)_{\tau \wedge T_n}\big) - E \exp\big(B_{\tau \wedge T_n} - (\widehat{\theta} \cdot S)_{\tau \wedge T_n}\big). \end{aligned}$$

Thus, due to Fatou's lemma, we get that

$$E \exp\big(B_\tau - (\widehat{\theta} \cdot S)_\tau\big) \leq \liminf E \exp\big(B_{\tau \wedge T_n} - (\theta \cdot S)_{\tau \wedge T_n}\big) = E \exp\big(B_\tau - (\theta \cdot S)_\tau\big).$$

The equality above follows because the set $\{\exp(B_\tau - (\theta \cdot S)_\tau), \quad \tau \in \mathcal{T}_T\}$ is uniformly integrable. Indeed, this fact follows from

$$\int_{\{B_\tau > c\}} \mathrm{e}^{B_\tau} \mathrm{d}P \leq \mathrm{e}^{-(p-1)c} E\big[\mathrm{e}^{pB_\tau}\big] \leq \mathrm{e}^{-(p-1)c} \sup_{\tau \in \mathcal{T}_T} E\big[\mathrm{e}^{pB_\tau}\big],$$

and for any $\theta \in \Theta_b$,

$$\mathrm{e}^{B_\tau - (\theta \cdot S)_\tau} \leq \mathrm{e}^{B_\tau} \exp\left[\Big\| \sup_{t \in [0,T]} |(\theta \cdot S)_t| \Big\|_\infty\right].$$

Hence, again due to [17], we obtain that

$$E \exp\big(B_\tau - (\widehat{\theta} \cdot S)_\tau\big) = \min_{\theta \in \Theta(S,B)} E \exp\big(B_\tau - (\theta \cdot S)_\tau\big) = \inf_{\theta \in \Theta_b} E \exp\big(B_\tau - (\theta \cdot S)_\tau\big).$$

This proves the assertion (i), and the proof of the theorem is complete. □

5 Optimal Portfolio and Investment Timing for Semimartingales

Throughout this section, we suppose that the following hold:

$$\mathcal{M}_f^e(S) \neq \emptyset \text{ and for any } \lambda \in \mathbb{R}^d, \quad \int_{\{|x|>1\}} |x| \mathrm{e}^{\lambda^T x} F(\mathrm{d}x) < \infty, \quad \mathrm{d}P \times \mathrm{d}A_t\text{-a.e.} \tag{27}$$

The payoff process B is a RCLL semimartingale for which the set of admissible strategies is

$$\Theta := \mathcal{A}_{adm}(0) := \Big\{\theta \in L(S) : \sup_{\tau \in \mathcal{T}_T} E \exp[B_\tau - (\theta \cdot S)_\tau] < \infty\Big\}.$$

This section is devoted to the following problem:

Problem 1 Find a pair $(\theta^*, \tau^*) \in \Theta \times \mathscr{T}_T$ such that

$$\min_{\theta\in\Theta,\tau\in\mathscr{T}_T} E\exp[B_\tau - (\theta\cdot S)_\tau] = E\exp[B_{\tau^*} - (\theta^*\cdot S)_{\tau^*}]. \tag{28}$$

Precisely, we will describe—as explicitly as possible—the optimal control solution to Problem 1. Our description is essentially based on the characterization of the optimal value process via a dynamic programming equation. This will constitute our first result in this section and is given by Theorem 7. The latter is based on the following

Lemma 4 *Suppose that the payoff process, B, is such that*

$$\sup_{\tau\in\mathscr{T}_T} E\big[\exp(B_\tau)\big] < \infty. \tag{29}$$

Then, for any $\theta \in \Theta$, the process

$$L_t(\theta) := V(t)\exp\left(-\int_0^t \theta_u \mathrm{d}S_u\right), \quad t \le T,$$

is a supermartingale. Here, V is the value process, given by the formula

$$V(t) := \operatorname*{ess\,sup}_{\theta\in\Theta,\ \tau\ge t} E\left(-\exp\left[B_\tau - \int_t^\tau \theta_u \mathrm{d}S_u\right]\Big|\mathscr{F}_t\right), \quad t \le T. \tag{30}$$

Proof For any $\theta \in \Theta$ and any stopping time $\tau \ge t$, we put

$$j_t(\theta,\tau) := E\big(-\mathrm{e}^{-\int_t^\tau \theta_u \mathrm{d}S_u + B_\tau}\big|\mathscr{F}_t\big)$$

and

$$J_t(\theta) := \operatorname*{ess\,sup}_{\tau\ge t} E\big(-\mathrm{e}^{-\int_t^\tau \theta_u \mathrm{d}S_u + B_\tau}\big|\mathscr{F}_t\big).$$

Notice that the process $J(\theta)\exp(-\theta\cdot S)$ is the Snell envelope of $-\exp(-\theta\cdot S + B)$ and, hence, it is a RCLL supermartingale (see [22], and [8]). Furthermore, we have for any $t \in [0,T]$, $\tau \ge t$, and any $\theta \in \Theta$,

$$V(t) \ge J_t(\theta) \ge j_t(\theta,\tau) \quad \text{and} \quad j_t(\theta,\tau) = j_t(\theta I_{]\!]t,T]\!]},\tau).$$

Consider $t \ge s \ge 0$, $\tau \ge t$, and $\theta,\ \overline{\theta} \in \Theta$ such that $\theta I_{[\![s,t]\!]} = \overline{\theta} I_{[\![s,t]\!]}$. Then, due to the above facts, we derive

$$\begin{aligned} V(s) \ge J_s(\overline{\theta}) \ge E\big(J_t(\overline{\theta})\mathrm{e}^{-\int_s^t \overline{\theta}_u \mathrm{d}S_u}\big|\mathscr{F}_s\big) &= E\big(J_t(\overline{\theta})\mathrm{e}^{-\int_s^t \theta_u \mathrm{d}S_u}\big|\mathscr{F}_s\big) \\ &\ge E\big(j_t(\overline{\theta},\tau)\mathrm{e}^{-\int_s^t \theta_u \mathrm{d}S_u}\big|\mathscr{F}_s\big). \end{aligned} \tag{31}$$

Note that for two pairs (θ_1, τ_1) and (θ_2, τ_2), there exists (θ_3, τ_3) such that

$$\max(j_t(\theta_1, \tau_1); j_t(\theta_2, \tau_2)) = j_t(\theta_3, \tau_3).$$

In fact, it is sufficient to consider

$$\theta_3 := \theta_1 I_{\{j_t(\theta_1,\tau_1)\geq j_t(\theta_2,\tau_2)\}\otimes]\!]t,T]\!]} + \theta_2 I_{\{j_t(\theta_1,\tau_1)< j_t(\theta_2,\tau_2)\}\otimes]\!]t,T]\!]},$$

which is predictable and belongs to Θ, and

$$\tau_3 = \begin{cases} \tau_1, & \text{on } \{j_t(\theta_1, \tau_1) \geq j_t(\theta_2, \tau_2)\}; \\ \tau_2, & \text{otherwise,} \end{cases}$$

which is a stopping time satisfying $\tau_3 \geq t$. Then—an application of Zorn's lemma leads to—for any t there exists a sequence of pairs (θ_n, τ_n) such that $j_t(\theta_n, \tau_n)$ increases to $V(t)$. By combining this fact with (31), we obtain that

$$V(s) \geq E\big(j_t(\theta_n, \tau_n)\mathrm{e}^{-\int_s^t \theta_u \mathrm{d}S_u} \big| \mathscr{F}_s\big).$$

Thus, due to the monotone convergence theorem, we deduce that

$$V(s) \geq E\big(V(t)\mathrm{e}^{-\int_s^t \theta_u \mathrm{d}S_u} \big| \mathscr{F}_s\big).$$

This ends the proof of the lemma. □

Theorem 7 *Suppose that* (29) *holds. Then*

(i) *V admits a RCLL modification and satisfies the following dynamic programming equation*:

$$V(t) = \max\left[-\mathrm{e}^{B_t};\ \operatorname*{ess\,sup}_{\theta\in\Theta,\tau>t} E\left(V(\tau)\exp\left[-\int_t^\tau \theta_u \mathrm{d}S_u\right]\middle|\mathscr{F}_t\right)\right]. \tag{32}$$

(ii) *If B is bounded from below and assumption* (27) *holds, then V is a RCLL negative semimartingale that has the following decomposition*

$$V(t) = V(0)\mathscr{E}_t(M^V)\mathrm{e}^{A_t^V} \tag{33}$$

where

$$M^V = \beta\cdot S^c + W\star(\mu-\nu) + g\star\mu + \overline{M^V}, \qquad W_t(x) := f_t(x) + \frac{\widehat{f}_t}{1-a_t} I_{\{a_t<1\}}. \tag{34}$$

Here, A^V is a predictable process with finite variation, M^V is a local martingale, and $(\beta, f, g, \overline{M^V})$ are its Jacod's components.

Proof (i) It is clear from Lemma 4, that the process $L(\theta)$ is a supermartingale for any $\theta \in \Theta$. Due to Theorem 2 in [8] (p. 73), we deduce that the process $L(\theta)$ admits right and left limits along the rationales and the process $L_{t+}(\theta)$ is a RCLL supermartingale with respect to the filtration $\mathscr{F}_{t+} = \mathscr{F}_t$. It follows that both processes $V(t+)$ and $V(t-)$ exist, and, moreover, that

$$V(t) \geq V(t+), \quad P\text{-a.s.} \tag{35}$$

On the other hand, since $L_{t+}(\theta) = V(t+)\exp[-(\theta \cdot S)_t]$ is a RCLL supermartingale and $V(t+) \geq -\mathrm{e}^{B_t}$, an application of the optional sampling theorem for supermartingales leads to the inequalities

$$\begin{aligned} V(t+) &\geq E\left(V(\tau+)\exp\left(-\int_t^\tau \theta_u \mathrm{d}S_u\right)\Big|\mathscr{F}_t\right) \\ &\geq E\left(-\exp\left[-\int_t^\tau \theta_u \mathrm{d}S_u + B_\tau\right]\Big|\mathscr{F}_t\right). \end{aligned}$$

Then, by taking the essential sup, we obtain that

$$V(t+) \geq V(t).$$

A combination of this with (35) implies the right-continuity of V. This proves that the process V admits a RCLL modification. We will consider this modification throughout the rest of this paper. As a result, the two processes $L(\theta)$ and $(\theta \cdot S)$ are RCLL semimartingales for any $\theta \in \Theta$. Therefore, an application of the optional sampling theorem implies

$$V(t) \geq \operatorname*{ess\,sup}_{\theta\in\Theta,\tau>t} E\left(V(\tau)\exp\left[-\int_t^\tau \theta_u \mathrm{d}S_u\right]\Big|\mathscr{F}_t\right).$$

Combining the above with the inequality $V(t) \geq -\mathrm{e}^{B_t}$, we conclude that

$$V(t) \geq \max\left[-\mathrm{e}^{B_t};\ \operatorname*{ess\,sup}_{\theta\in\Theta,\tau>t} E\left(V(\tau)\exp\left[-\int_t^\tau \theta_u d S_u\right]\Big|\mathscr{F}_t\right)\right].$$

To prove the reverse inequality, we write:

$$\begin{aligned} V(t) &= \operatorname*{ess\,sup}_{\theta\in\Theta,\tau\geq t} E\left(-\exp\left[B_\tau - \int_t^\tau \theta_u \mathrm{d}S_u\right]\Big|\mathscr{F}_t\right) \\ &= \max\left[-\mathrm{e}^{B_t};\ \operatorname*{ess\,sup}_{\theta\in\Theta,\tau>t} E\left(-\exp\left[B_\tau - \int_t^\tau \theta_u \mathrm{d}S_u\right]\Big|\mathscr{F}_t\right)\right] \\ &\leq \max\left[-\mathrm{e}^{B_t};\ \operatorname*{ess\,sup}_{\theta\in\Theta,\tau>t} E\left(V(\tau)\exp\left[-\int_t^\tau \theta_u \mathrm{d}S_u\right]\Big|\mathscr{F}_t\right)\right]. \end{aligned}$$

This ends the proof of (i).

(ii) Suppose that B is bounded from below by a constant $-C$. Then

$$
\begin{aligned}
-V(t) &= \operatorname*{ess\,inf}_{\theta\in\Theta,\tau\in\mathcal{T}_T} E\left(\exp\left(B_\tau - \int_t^\tau \theta_u \mathrm{d}S_u\right)\Big|\mathcal{F}_\tau\right)\\
&\geq \mathrm{e}^{-C} \operatorname*{ess\,inf}_{\theta\in\Theta,\tau\in\mathcal{T}_T} E\left(\exp\left(-\int_t^\tau \theta_u \mathrm{d}S_u\right)\Big|\mathcal{F}_\tau\right)\\
&= \mathrm{e}^{-C} \operatorname*{ess\,inf}_{\theta\in\Theta} E\left(\exp\left(-\int_t^T \theta_u \mathrm{d}S_u\right)\Big|\mathcal{F}_t\right)\\
&= \mathrm{e}^{-C} \exp\left(-\operatorname*{ess\,inf}_{Z\in\mathcal{Z}_f^e(S)} E\left(\frac{Z_T}{Z_t}\log\frac{Z_T}{Z_t}\Big|\mathcal{F}_t\right)\right).
\end{aligned}
$$

Here $\mathcal{Z}_f^e(S)$ denotes the set of martingale densities, Z, such that $Z\log Z$ is an integrable submartingale. Due to the assumption that $\mathcal{M}_f^e(S)$ is not empty or, equivalently, $\mathcal{Z}_f^e(S)\neq\emptyset$, we have:

$$
\operatorname*{ess\,inf}_{Z\in\mathcal{Z}_f^e(S)} E\left(\frac{Z_T}{Z_t}\log\frac{Z_T}{Z_t}\Big|\mathcal{F}_t\right) < \infty, \quad P\text{-a.s.}
$$

This together with the right continuity of V proves that the process V is a negative supermartingale (take $\theta=0$ in Lemma 4) or, equivalently, $\frac{V}{V(0)}$ is a positive exponential local submartingale. This leads to the existence of a local martingale M^V and a predictable process, A^V, with finite variation such that $V = V(0)\mathcal{E}(M^V)e^{A^V}$. These facts follow from the Doob–Meyer decomposition and the fact that $\frac{1}{V_-}\cdot V$ is a local submartingale. The decomposition for the local martingale M^V follows from Jacod's theorem; see Theorem 1. This completes the proof. □

Remark 6 Equation (32) describes the optimal cost process/optimal value process. This description resembles the dynamic maximum principle, which will lead, in the Markovian case, to a HJB equation. In a model driven by Brownian motions, this HJB equation can be solved explicitly, see [12]. The derivation of these HJB in a more general case than the Brownian one as well as their investigations, and their relationship to backward stochastic differential equations (BSDEs) are beyond the scope of this paper and are left to future research.

Once the process V is determined, the optimal investment timing and the optimal portfolio can be derived in the general semimartingale framework, as it will be illustrated in the following.

Theorem 8 *Consider the process V defined in* (30) *and its Jacod's components* $(\beta, f, g, \overline{M^V}, A^V)$ *given by* (33)–(34). *Suppose that Problem* 1 *admits a solution* (θ^*, τ^*), *and that the assumptions* (27) *and* (16) *are fulfilled. Then the following assertions hold.*

(i) *There exists a probability measure* $Q_V \sim P$ *such that the MEH* σ*-martingale measure with respect to* Q_V*—that we denote by* $\widetilde{Q}_V$ *and its density process by* $\widetilde{Z}_t^V := E(\frac{\mathrm{d}\widetilde{Q}_V}{\mathrm{d}Q_V} \mid \mathscr{F}_t)$*—exists and satisfies*

$$A^V_{t\wedge\tau^*} = h^E_{t\wedge\tau^*}(\widetilde{Q}_V, Q_V), \qquad \log \widetilde{Z}^V = \widetilde{\theta}^V \cdot S + h^E(\widetilde{Q}_V, Q_V). \tag{36}$$

(ii) *The optimal controls,* (θ^*, τ^*)*, solution to* (28)*, can be described as follows:*
(a) *The optimal investment* θ^* *coincides with* $-\widetilde{\theta}^V$ *on* $[\![0, \tau^*]\!]$*, i.e.* $-\theta^*$ *is a pointwise root to*

$$0 = \begin{cases} -b + c(\theta - \beta) + \int \left[h(x) - (f(x)+1)\mathrm{e}^{-\theta^T x} x\right] F(\mathrm{d}x), \\ \quad \text{on } \{\Delta A = 0\} \cap [\![0, \tau^*]\!] \\ \int (f(x)+1)\mathrm{e}^{-\theta^T x} x F(\mathrm{d}x), \quad \text{on } \{\Delta A \neq 0\} \cap [\![0, \tau^*]\!]. \end{cases}$$

(b) *The stopping time* τ^* *satisfies* $\tau^* \geq \widetilde{\tau}$ P*-a.s., where* $\widetilde{\tau}$ *is the smallest stopping times such that* $(\theta^* I_{[\![0,\widetilde{\tau}]\!]}, \widetilde{\tau})$ *is a solution to* (28)*, and is given by*

$$\widetilde{\tau} = \inf\{\leq t : \ V(t) = -\mathrm{e}^{B_t}, \text{ or } V(t-) = -\mathrm{e}^{B_{t-}} \} \wedge T, \tag{37}$$

i.e. $V(0) = \sup_{\theta\in\Theta} E[-\mathrm{e}^{-(\theta\cdot S)_{\widetilde{\tau}} + B_{\widetilde{\tau}}}] = E[-\mathrm{e}^{-(\theta^*\cdot S)_{\widetilde{\tau}} + B_{\widetilde{\tau}}}]$*. More generally, we have*:

$$V(t) = \operatorname*{ess\,sup}_{\theta\in\Theta} E\left(-\exp\left[-\int_t^{\tau_t} \theta_u \mathrm{d}S_u + B_{\tau_t}\right] \Big| \mathscr{F}_t\right),$$
$$\tau_t := \inf\{u \in [t, T[: \ V(u) = -\mathrm{e}^{B_u} \text{ or } V(u-) = -\mathrm{e}^{B_{u-}} \} \wedge T.$$

Proof Due to the main result of [2], we deduce that the set $\mathscr{Z}^e_{f,loc}(S, Q) \neq \emptyset$ for any $Q \sim P$ if and only if $\mathscr{Z}^e_{loc}(S) \neq \emptyset$. Here the set $\mathscr{Z}^e_{f,loc}(S, Q)$ denotes the set of positive Q-local martingale, Z^Q, (i.e. $Z^Q \in \mathscr{M}_{loc}(Q)$, $Z^Q > 0$) such that $Z^Q S$ is a σ-martingale under Q, (i.e. $Z^Q S \in \mathscr{M}_\sigma(Q)$) and $Z^Q \log Z^Q$ is Q-locally integrable.

It is obvious that $(-V(0))^{-1}V$ is a positive local submartingale and the inequality, $\frac{V}{V(0)} = \mathscr{E}(M^V)\mathrm{e}^{A^V} \leq \frac{\mathrm{e}^B}{-V(0)}$, holds. Thus, under assumption (16), we derive

$$\sup_{\tau\in\mathscr{T}_T} E\left[\mathscr{E}_\tau\left(M^V\right)^p\right] \leq (-V(0))^{-p} \sup_{\tau\in\mathscr{T}_T} E\left(\mathrm{e}^{pB_\tau}\right) < \infty,$$

and the uniform integrability of $\mathscr{E}(M^V)$ follows. Hence, $Q_V := \mathscr{E}_T(M^V)P \sim P$ is a probability measure. Furthermore, due to Lemma 10 and assumption (27), we get that

$$\int_{\{|x|>1\}} \mathrm{e}^{\lambda^T x} F^{Q_V}(\mathrm{d}x) < \infty,$$

where $F^{Q_V}(\mathrm{d}x)$ is the kernel corresponding to the jumps of S under the measure Q_V. Thus, under the assumption (27) and (16), we get that $\mathscr{Z}^e_{f,loc}(S, Q_V) \neq \emptyset$ holds and thus we can apply Theorem 3.3 of [4] for the model (S, Q_V). This proves the existence of the MEH σ-martingale density $\widetilde{Z}^V := \widetilde{Z}^{Q_V}$ with respect to Q_V, and, moreover, that $\log \widetilde{Z}^V = \widetilde{\theta}^V \cdot S + h^E(\widetilde{Z}^V, Q_V)$.

Since $L(\theta) = V\mathrm{e}^{-\theta\cdot S}$ is a supermartingale for any $\theta \in \Theta$, the process

$$\exp\left(A^V - h^E(\widetilde{Q}_V, Q_V) - (\theta + \widetilde{\theta}^V)\cdot S\right)$$

is a $\widetilde{Q}_V$-submartingale. As a result, the process

$$L(-\widetilde{\theta}^V) = V\mathrm{e}^{\widetilde{\theta}^V\cdot S} = V(0)\mathscr{E}(M^V)\mathrm{e}^{A^V+\widetilde{\theta}^V\cdot S} = V(0)\mathscr{E}(M^V)\widetilde{Z}^V\mathrm{e}^{A^V - h^E(\widetilde{Q}_V, Q_V)}$$

is a local supermartingale or, equivalently, the process $A^V - h^E(\widetilde{Z}^V, Q_V)$ is nondecreasing. Furthermore, a combination of the inequalities

$$E^{Q_V}\left(\mathrm{e}^{h^E_T(\widetilde{Z}^V, Q_V)}\right) \le E^{Q_V}\left(\mathrm{e}^{A^V_T}\right) = E\left(\frac{V(T)}{V(0)}\right) = E\left(\frac{-\mathrm{e}^{B_T}}{V(0)}\right) < \infty,$$

and Theorem III.1 of [20], implies that $\widetilde{Z}^V$ is a true Q_V-martingale. This proves the existence of the MEH σ-martingale measure for (S, Q_V), denoted by $\widetilde{Q}_V$. This proves the assertion (i) without the first equality of (36).

By combining the equality

$$V\exp(-(\theta^*\cdot S)) = V_0\mathscr{E}(M^V)\widetilde{Z}^V\exp\big(A^V - h^E(\widetilde{Q}_V, Q_V) - (\widetilde{\theta}^V + \theta^*)\cdot S\big),$$

the $\widetilde{Q}_V$-submartingale property of

$$A^V_{t\wedge\tau^*} - h^E_{t\wedge\tau^*}\big(\widetilde{Q}_V, Q_V\big) - (\widetilde{\theta} + \theta^*)\cdot S_{t\wedge\tau^*}, \tag{38}$$

and the strict convexity of e^z, we deduce that $V(t\wedge\tau^*)\exp(-(\theta^*\cdot S)_{t\wedge\tau^*})$ is a true martingale if and only if the process (38) is null, or, equivalently, that

$$A^V_{t\wedge\tau^*} = h^E_{t\wedge\tau^*}\big(\widetilde{Q}_V, Q_V\big), \qquad \widetilde{\theta} I_{[\![0,\tau^*]\!]} = -\theta^* I_{[\![0,\tau^*]\!]}.$$

This ends, simultaneously, the proof of the assertions (i) and (ii)-(a).

Next, we will prove assertion (ii)-(b). To this end, we consider the process $\widetilde{Y}$ and the stopping time $\widetilde{\tau}^*$ given, respectively, by

$$\widetilde{Y}(t) := \operatorname*{ess\,sup}_{\tau\ge t} E\left(-\exp\left[B_\tau + \int_t^\tau \widetilde{\theta}_u \mathrm{d}S_u\right] \Big| \mathscr{F}_t\right) \quad \text{and}$$

$$\widetilde{\tau}^* := \inf\left\{t \in [0, T[:\ \widetilde{Y}(t) = -\mathrm{e}^{B_t},\ \text{or}\ \widetilde{Y}(t-) = -\mathrm{e}^{B_{t-}}\right\} \wedge T.$$

Then, it is obvious that for any $t \in [0, T]$,

$$V(t) \ge \widetilde{Y}(t) \ge -\exp(B_t), \quad P\text{-a.s.} \tag{39}$$

Furthermore, since

$$V(0) = E\big[-V(\tau^*)\exp\big(-(\theta^* \cdot S)_{\tau^*}\big)\big] = E\big[-\exp\big(B_{\tau^*} - (\theta^* \cdot S)_{\tau^*}\big)\big]$$
$$\leq \sup_{\tau \in \mathcal{T}_T} E\big[-\exp\big(B_\tau - (\theta^* \cdot S)_\tau\big)\big]$$
$$=: \widetilde{Y}(0),$$

we derive that

$$V(0) = \widetilde{Y}(0) \quad \text{and} \quad \tau^* \geq \widetilde{\tau} \geq \widetilde{\tau}^* \quad P\text{-a.s.}$$

Combining these inequalities with the fact that $V(t \wedge \widetilde{\tau}^*)\exp(-(\theta^* \cdot S)_{t\wedge\widetilde{\tau}^*})$ and $\widetilde{Y}(t \wedge \widetilde{\tau}^*)\exp(-(\theta^* \cdot S)_{t\wedge\widetilde{\tau}^*})$ are martingales, we deduce that

$$EV(t \wedge \widetilde{\tau}^*)\exp(-(\theta^* \cdot S)_{t\wedge\widetilde{\tau}^*}) = E\widetilde{Y}(t \wedge \widetilde{\tau}^*)\exp(-(\theta^* \cdot S)_{t\wedge\widetilde{\tau}^*}).$$

This equality together with (39) prove that the processes $V(t \wedge \widetilde{\tau}^*)$ and $\widetilde{Y}(t \wedge \widetilde{\tau}^*)$ coincide. Thus, the stopping times $\widetilde{\tau}$ and $\widetilde{\tau}^*$ coincide also. Due to the result of [21] (see Théorème 4 therein), we deduce that the stopping time $\widetilde{\tau}$ is the smallest optimal stopping time, and the assertion (ii)-(b) follows. This ends the proof of the theorem. □

Remark 7

1. Our main results of this section (Theorems 7 and 8) contribute by giving the structure of the optimal value process V and the explicit description of τ^* and θ^* when they exist.
2. The financial problem that we consider in this section is the same as the one of [12]. Therefore, our two theorems generalize the results of that paper to the semimartingale framework. See also [9, 13] and the reference therein for the same financial problem with other utilities.
3. Concerning the mathematical formulation and/or technical aspects, Problem 1 is very close to the one considered in [18]. However, there are fundamental differences:

   a. Our running reward function (e^x) is multiplied to the terminal reward function ($g(S_t) = -e^{B_t}$), while in [18] they add up. Furthermore, the control θ appears in the expectation operation which is not the case in our situation.
   b. The terminal reward function, $g(x)$, is assumed to be bounded from below (positive), which does not correspond to our case ($g(S_t) = -e^{B_t} < 0$ might be unbounded from below). It is important to mention that this positivity assumption is crucial in the analysis of [18].
   c. Our framework is very general, dealing with semimartingales in which the predictable representation property may never hold. Furthermore, the additional feature of jumps in the model may add tremendous technical difficulties to the method used in [18].

As mentioned in the introduction, this optimal sale problem with investments (i.e. Problem 1) was the main motivation for the horizon-unbiased utility concept of Henderson-Hobson. Herein, Theorem 8—and mainly its proof—establishes the connection between the existence of solution to Problem 1 and the forward utility concept of Musiela–Zariphopoulou. This can be stated as follows.

Corollary 1 *Suppose that assumptions of Theorem* 8 *hold, and consider the following random utility field*

$$U(t,\omega,x) = V_t(\omega)\exp(-x).$$

Then there exists a stopping time τ *such that the random utility field* $U(t\wedge\tau(\omega),x)$ *is an exponential forward dynamic utility for the model* S^τ, *and* $\widetilde{\tau}$*—defined in* (37)*— is the smallest stopping satisfying this property.*

Proof The proof of the corollary follows directly from the proof of Theorem 8.

Acknowledgements This research was supported financially by the Natural Sciences and Engineering Research Council of Canada via Choulli's Grant G121210818.

The first and the second authors would like to thank anonymous referees for the careful reading and valuable inputs and suggestions. Both, the first and the second authors are grateful to Christoph Frei for his fruitful comments and advices. Any remaining error is of our responsibility.

Appendix 1: Some Auxiliary Lemmas

This section is contains six lemmas, which were used in previous sections. We note that some of these lemmas are interesting on their own right.

Lemma 5 *Let* Q *be a* σ*-martingale measure for* S, *and* $\theta \in L(S)$ *be such that*

$$\sup_{\tau\in\mathcal{T}_T} E^Q \exp\big[(\theta\cdot S)_\tau\big] < \infty. \tag{40}$$

Then the process $\theta\cdot S$ *is a* Q*-local martingale and the process* $\exp[\theta\cdot S]$ *is a positive* Q*-submartingale.*

Proof Since Q is a σ-martingale measure for S, there exists a positive, bounded and predictable process ϕ such that $\phi\cdot S$ is a Q-local martingale. As a result, $\theta\cdot S$ is σ-martingale under Q. On the other hand, it is clear that

$$X_t := \exp\left(\frac{1}{2}(\theta\cdot S)_t\right)$$

is a positive special semimartingale under Q with the Doob–Meyer decomposition

$$X = X_0 + \overline{N} + \overline{B}$$

where $\overline{N}$ is a Q-local martingale and $\overline{B}$ is a predictable process with finite variation such that $\overline{N}_0 = \overline{B}_0 = 0$. Let $(T_n)_{n\geq 1}$ be a sequence of stopping times that increases stationarily to T and

$$E^Q\left([\overline{N}, \overline{N}]_{T_n}^{1/2} + \mathrm{Var}_{T_n}\overline{B}\right) < \infty.$$

For any predictable process φ such that $|\varphi| \leq 1$, we have:

$$E^Q|\varphi \cdot X_{T_n}| \leq c E^Q\left([\overline{N}, \overline{N}]_{T_n}^{1/2} + \mathrm{Var}_{T_n}(\overline{B})\right), \tag{41}$$

where c is a constant that does not depend on φ.

Using Ito's formula, we obtain

$$X = 1 + \frac{1}{2} X_- \cdot (\theta \cdot S) + X_- \cdot V(\theta),$$

where $V(\theta)$ is a non-decreasing process given by

$$V(\theta) := \frac{1}{8}\theta^T c\theta \cdot A + \left(\exp\left(\frac{1}{2}\theta^T x\right) - 1 - \frac{1}{2}\theta^T x\right) \star \mu.$$

Since $\theta \cdot S$ is a σ-martingale under Q, then there exists a predictable process ϕ with values in the interval $(0, 1]$ such that $\phi\theta \cdot S$ is a Q-local martingale. Consider a sequence of stopping times, $(\sigma_n)_{n\geq 1}$ increasing stationarily to T such that $(\phi\theta \cdot S)^{\sigma_n}$ is a true Q-martingale. Then, for any $\varepsilon > 0$, the process

$$\left(\frac{\phi}{\phi + \varepsilon + \varepsilon X_-} X_- \theta \cdot S\right)^{\sigma_n}$$

is also a true Q-martingale. As a result,

$$\begin{aligned} E^Q \int_0^{\sigma_n \wedge T_n} X_{s-} \mathrm{d}V_s(\theta) &= \lim_{\varepsilon \downarrow 0} E^Q \int_0^{\sigma_n \wedge T_n} \frac{\phi_s}{\phi_s + \varepsilon + \varepsilon X_{s-}} X_{s-} \mathrm{d}V_s(\theta) \\ &= \lim_{\varepsilon \downarrow 0} E^Q \left(\frac{\phi}{\phi + \varepsilon + \varepsilon X_-} \cdot X_{\sigma_n \wedge T_n}\right) < \infty. \end{aligned}$$

The first equality follows from the monotone convergence theorem, while the finiteness of the last quantity is due to (41).

Hence, $V(\theta)$ is Q-locally integrable and thus $(\theta \cdot S)$ is Q-locally integrable. This proves that $(\theta \cdot S)$ is really a Q-local martingale. Furthermore, $\exp\left(\frac{1}{2}\theta \cdot S\right)$ is a positive Q-local submartingale. Then, the condition (40) and de la Vallée Poussin's argument imply that $\exp\left(\frac{1}{2}\theta \cdot S\right)$ is a positive Q-submartingale which is square integrable. Now the lemma follows from Jensen's inequality. □

For a random variable H, we denote by $\widetilde{Q}^{(H)}$ the minimal entropy martingale measure for S with respect to $P^{(H)} := \mathrm{e}^H (E(\mathrm{e}^H))^{-1} \cdot P$. Also, Θ_1 denotes the set of strategies considered in [6]:

$$\Theta_1 := \left\{\theta \in L(S) : (\theta \cdot S) \in \mathscr{M}(\widetilde{Q}^{(H)})\right\}.$$

Lemma 6 *Suppose that S is locally bounded and $\mathscr{M}_f^e(S) \neq \emptyset$. Let H be a random variable bounded from below with*

$$E\left[\mathrm{e}^{pH}\right] < \infty,$$

for some $p \in (1, \infty)$, and let $\widetilde{\theta} \in \Theta_1$. Then the assertions (i) *and* (ii) *are equivalent*:

(i)

$$1 - u_0 := \inf_{\theta \in \Theta_1} E \exp\bigl(H - (\theta \cdot S)_T\bigr) = E \exp\bigl(H - (\widetilde{\theta} \cdot S)_T\bigr).$$

(ii) *For any stopping time $\sigma \leq T$, we have*

$$\begin{aligned} 1 - u_\sigma &:= \operatorname*{ess\,inf}_{\theta \in \Theta_1} E\left(\exp\left[H - \int_\sigma^T \theta_u \mathrm{d}S_u\right] \Big| \mathscr{F}_\sigma\right) \\ &= E\left(\exp\left[H - \int_\sigma^T \widetilde{\theta}_u \mathrm{d}S_u\right] \Big| \mathscr{F}_\sigma\right). \end{aligned}$$

Proof Using the results in [6], we change the probability and work under Q instead of P, where

$$Q := \frac{\exp(H)}{E[\exp(H)]} P.$$

Suppose that the assertion (i) holds. Putting

$$J_t := \operatorname*{ess\,inf}_{Z \in \mathscr{Z}_f^e(S,Q)} E^Q\left(\frac{Z_T}{Z_t} \log \frac{Z_T}{Z_t} \Big| \mathscr{F}_t\right), \tag{42}$$

where

$$\mathscr{Z}_f^e(S, Q) := \left\{Z > 0 : Z \in \mathscr{M}_{loc}(Q),\ ZS \in \mathscr{M}_\sigma(Q),\ \text{and}\ E^Q[Z_T \log Z_T] < \infty\right\},$$

we obtain the existence of $\underline{\xi}$ that belongs to the set

$$\Xi := \left\{\xi > 0 : E(\xi) = 1,\ E(\xi\eta) = 0,\ \text{for any}\ \eta := (\theta \cdot S)_T,\ \theta \in \Theta_1\right\}$$

and satisfies

$$J_0 = \min_{\xi \in \Xi} E^Q(\xi \log \xi) = E^Q(\underline{\xi} \log \underline{\xi}).$$

Thus, Theorem 3.5 of [17] implies that

$$\underline{\xi} = \exp\bigl(-\log\bigl(E^Q \mathrm{e}^{-(\widetilde{\theta} \cdot S)_T}\bigr) - \widetilde{\theta} \cdot S_T\bigr) \quad \text{and} \quad u_0 = 1 - \mathrm{e}^{-J_0}. \tag{43}$$

It is clear that the set $\mathscr{Z}_f^e(S,Q)$ is stable under concatenation (for more detail about this see [17]), and due to Proposition 4.1 in [17] we conclude that the optimizer of J_t is given by $Z_t^* := E^Q(\underline{\xi}|\mathscr{F}_t) \in \mathscr{Z}_f^e(S,Q)$. Denoting $P^* := Z_T^* Q$ and using the first equation in (43), we derive that

$$J_t = E^Q\left(\frac{Z_T^*}{Z_t^*}\log\frac{Z_T^*}{Z_t^*}\Big|\mathscr{F}_t\right) = J_0 - \widetilde{\theta}\cdot S_t - \log Z_t^*.$$

Equivalently, we have

$$\frac{Z_T^*}{Z_\sigma^*} = \exp\left[-\int_\sigma^T \widetilde{\theta}_u \mathrm{d}S_u + J_\sigma\right]. \tag{44}$$

Due to Young's inequality ($xy \le \mathrm{e}^x + y\log y - y$), we obtain that

$$\left(-\int_\sigma^T \theta_u dS_u\right)\left(\frac{Z_T^*}{Z_\sigma^*}\mathrm{e}^{-J_\sigma}\right) \le \mathrm{e}^{-\int_\sigma^T \theta_u \mathrm{d}S_u} + \frac{Z_T^*}{Z_\sigma^*}\mathrm{e}^{-J_\sigma}\log\left(\frac{Z_T^*}{Z_\sigma^*}\mathrm{e}^{-J_\sigma}\right) - \frac{Z_T^*}{Z_\sigma^*}\mathrm{e}^{-J_\sigma}.$$

Therefore, by taking conditional expectation on both sides, and using the equalities

$$E^Q\left(\frac{Z_T^*}{Z_\sigma^*}\mathrm{e}^{-J_\sigma}\log\left(\frac{Z_T^*}{Z_\sigma^*}\mathrm{e}^{-J_\sigma}\right)\Big|\mathscr{F}_\sigma\right) = 0 = E^Q\left(\left(\int_\sigma^T \theta_u \mathrm{d}S_u\right)\frac{Z_T^*}{Z_\sigma^*}\mathrm{e}^{-J_\sigma}\Big|\mathscr{F}_\sigma\right),$$

and (44), we derive that

$$E^Q\left(1 - \mathrm{e}^{-\int_\sigma^T \theta_u \mathrm{d}S_u}\Big|\mathscr{F}_\sigma\right) \le 1 - \mathrm{e}^{-J_\sigma}.$$

Since there is equality for $\theta = \widetilde{\theta}$, due to

$$\frac{Z_T^*}{Z_\sigma^*}\mathrm{e}^{-J_\sigma} = \exp\left(-\int_\sigma^T \widetilde{\theta}_u \mathrm{d}S_u\right),$$

assertion (ii) follows. The converse is immediate by putting $\sigma = 0$. This ends the proof of the lemma. □

Lemma 7 *Let Z be a given positive local martingale such that $Z\log Z$ is locally integrable. There is a RCLL semimartingale X such that ZX is a local martingale and*

$$\log Z = X + h^E(Z,P).$$

Proof Since Z is a positive local martingale, there exists a local martingale N such that $N_0 = 0$ and $Z = \mathscr{E}(N)$. Due to Ito's formula, we deduce that

$$\log Z = \left(N - \frac{1}{1+\Delta N}\cdot[N,N]\right) + \frac{1}{2}\langle N^c\rangle + \sum\frac{(1+\Delta N)\log(1+\Delta N) - \Delta N}{1+\Delta N}.$$

Next, we note that $h^E(Z, P)$ is the compensator of the process $(1 + \Delta N) \cdot V$, where $Y := N - \frac{1}{1+\Delta N} \cdot [N, N]$ and $V := \log(Z) - Y$. Again Ito's formula implies that

$$ZY = (Y_- + 1)Z_- \cdot N \in \mathcal{M}_{0,loc}(P)$$

and

$$Z\overline{V} = \left(\overline{V}_- - \Delta h^E(Z, P)\right)Z_- \cdot N + Z_-(1+\Delta N)\cdot V - Z_- \cdot h^E(Z, P) \in \mathcal{M}_{0,loc}(P),$$

where $\overline{V} := V - h^E(Z, P)$. Thus, the conclusion follows immediately. □

Lemma 8 *Consider a positive σ-martingale density $Z = \mathcal{E}(N)$ with*

$$N = \lambda \cdot S^c + W \star (\mu - \nu), \;\; W_t(x) = \left(e^{\lambda_t^T x} - 1\right)\left(1 - a_t + \int e^{\lambda_t^T x}\nu(\{t\}, dx)\right)^{-1}. \quad (45)$$

Then

$$h^E(Z, P) = \frac{1}{2}\lambda^T c\lambda \cdot A + \phi(\gamma^{-1}e^{\lambda^T x} - 1) \star \nu + \sum (1 - a)\phi(\gamma^{-1} - 1) \quad (46)$$

$$= \frac{1}{2}\lambda^T c\lambda \cdot A + \frac{1}{\gamma}\phi\left(e^{\lambda^T x} - 1\right) \star \nu - \sum \frac{1}{\gamma}\left[\gamma \log(\gamma) - \gamma + 1\right], \quad (47)$$

$$\Delta h^E(Z, P) = -\log\gamma, \quad (48)$$

where $\gamma_t := 1 - a_t + \int e^{\lambda_t^T x}\nu(\{t\}, dx)$ and $\phi(z) := (1 + z)\log(1 + z) - z$.

Proof Notice that $h^E(Z, P)$ is the compensator of $V^E(N)$, where

$$V^E(N) = \frac{1}{2}\langle N^c\rangle + \sum \left[(1 + \Delta N)\log(1 + \Delta N) - \Delta N\right]. \quad (49)$$

From (45) we derive that

$$1 + \Delta N_t = \frac{e^{\lambda_t^T \Delta S_t}}{\gamma_t} I_{\{\Delta S_t \neq 0\}} + \frac{1}{\gamma_t} I_{\{\Delta S_t = 0\}}.$$

After simplification, this leads to the identity

$$\sum \left[(1 + \Delta N)\log(1 + \Delta N) - \Delta N\right]$$
$$= \phi\left(\gamma^{-1}e^{\lambda^T x} - 1\right) \star \mu + \sum \phi\left(\gamma^{-1} - 1\right) I_{\{\Delta S = 0\}}.$$

By plugging this representation into (49) and compensating, we obtain (46).

Inserting the expression

$$\phi\left(\gamma^{-1}e^{\lambda^T x} - 1\right) \star \nu = \gamma^{-1}\phi\left(e^{\lambda^T x} - 1\right) \star \nu + \sum \frac{(1 - a - \gamma)\log(\gamma) + (\gamma - 1)a}{\gamma}$$

into (46) and simplifying, we get (47).

Calculating the jumps in both sides of (47), we have

$$\Delta h^E(Z,P) = \frac{1}{\gamma}\left(a - \int e^{\lambda^T x} F(dx)\Delta A\right) - \frac{\gamma \log(\gamma) - \gamma + 1}{\gamma}$$
$$= \frac{-\gamma+1}{\gamma} - \frac{\gamma \log(\gamma) - \gamma + 1}{\gamma} = -\log(\gamma).$$

Note that the first equality follows because

$$\int x e^{\lambda^T x} F(dx)\Delta A = \int x e^{\lambda^T x} \nu(\{.\}, dx) = 0$$

in virtue of the fact that Z is a σ-martingale density for S. This ends the proof. □

Lemma 9 *Suppose that* (3) *holds. Then the function*

$$\overline{K}(\lambda) := b^T\lambda + \frac{1}{2}\lambda^T c\lambda + \int \left(e^{\lambda^T x} - 1 - \lambda^T h(x)\right) F(dx), \quad \lambda \in \mathbb{R}^d,$$

is convex, proper, closed, and continuously differentiable with

$$\nabla\overline{K}(\lambda) = b + c\lambda + \int \left(x e^{\lambda^T x} - h(x)\right) F(dx), \quad \lambda \in \mathbb{R}^d.$$

Proof The proof of this lemma is obvious. For the definitions of proper and closed convex functions, we refer the reader to [25]. □

Lemma 10 *The following assertions are equivalent*:

(i) *For any* $\lambda \in \mathbb{R}^d$,

$$\int_{\{|x|>1\}} e^{\lambda^T x} F(dx) < \infty.$$

(ii) *For any* $\lambda \in \mathbb{R}^d$,

$$\int_{\{|x|>1\}} |x| e^{\lambda^T x} F(dx) < \infty.$$

As a result, if (i) *holds, then for any* $\lambda \in \mathbb{R}^d$ *and* $q \in (0,\infty)$,

$$\int_{\{|x|>1\}} |x|^q e^{\lambda^T x} F(dx) < \infty.$$

Proof The implication (ii) $\Longrightarrow$ (i) is obvious. We focus on proving the reverse. Let e_i be the element of $\mathbb{R}^d$ that has the ith component equal to one and the other components null. Due to the equivalence of norms in $\mathbb{R}^d$, we may work with the norm $|x| = \sum_{i=1}^{d} |x_i|$. We get that

$$\int_{\{|x|>1\}} |x| e^{\lambda^T x} F(dx) = \sum_{i=1}^{d} \int_{\{|x|>1\}} \left((e_i^T x)^+ + (-e_i^T x)^+\right) e^{\lambda^T x} F(dx)$$
$$\leq \sum_{i=1}^{d} \int_{\{|x|>1\}} e^{(e_i+\lambda)^T x} F(dx) + \sum_{i=1}^{d} \int_{\{|x|>1\}} e^{(-e_i+\lambda)^T x} F(dx).$$

Due to (i) the last term in the rhs of the above string is finite for any $\lambda \in \mathbb{R}^d$. The proof of the remaining part of the lemma follows by the same arguments. □

Appendix 2: MEH σ-Martingale Density Under Change of Probability

In this section, we focus on describing the MEH σ-martingale density when we change probability. This case can be derived easily from a more general case where one works with respect to a positive local martingale density, Z, that may not be uniformly integrable. First, we generalize the characterization of the MEH σ-martingale density for the case when S may not be bounded nor quasi-left continuous. For the case of bounded and quasi-left continuous S, a more elaborate result is given in [3].

Theorem 9 *Suppose that $\mathcal{Z}_{loc}^{e}(S) \neq \emptyset$ and* (3) *holds. If $\widetilde{Z} \in \mathcal{Z}_{loc}^{e}(S)$ is the MEH σ-martingale density then there exists $\widetilde{H} \in L(S)$ such that*

$$\log(\widetilde{Z}) = \widetilde{H} \cdot S + h^E(\widetilde{Z}, P). \tag{50}$$

Furthermore, $\widetilde{H}$ can be described as root of the equation

$$0 = \begin{cases} b + c\theta + \int \left[e^{\theta^T x} x - h(x)\right] F(dx), & \text{on } \{\Delta A = 0\} \\ \int e^{\theta^T x} x F(dx), & \text{on } \{\Delta A \neq 0\}. \end{cases} \tag{51}$$

Proof Notice that the assumptions of Theorem 3.3 in [4] are fulfilled. Hence, a direct application of this theorem implies that $\widetilde{Z}$ is given by

$$\widetilde{Z} = \mathcal{E}(\widetilde{N}), \qquad \widetilde{N} := \widetilde{\beta} \cdot S^c + \widetilde{W} \star (\mu - \nu),$$
$$\widetilde{W}_t(x) := (\widetilde{\gamma}_t)^{-1}\left(e^{\widetilde{\beta}_t^T x} - 1\right), \qquad \widetilde{\gamma}_t := 1 - a_t + \int e^{\widetilde{\beta}_t^T x} \nu(\{t\}, dx),$$

where $\widetilde{\beta}$ is a root of (51). Therefore, in the remaining part of this proof we will focus on showing (50). Thus,

$$\begin{aligned}\log(\widetilde{Z}) &= \widetilde{N} - \frac{1}{2}\langle \widetilde{N}\rangle + \sum[\log(1+\Delta\widetilde{N}) - \Delta\widetilde{N}]\\ &= \widetilde{\beta}\cdot S^c + \widetilde{W}\star(\mu-\nu) - \frac{1}{2}\widetilde{\beta}^T c\widetilde{\beta}\cdot A + \sum\left[\log\frac{e^{\widetilde{\beta}^T x}}{\widetilde{\gamma}} - \frac{e^{\widetilde{\beta}^T x}}{\widetilde{\gamma}} + 1\right] I_{\{\Delta S\neq 0\}}\\ &\quad + \sum\left[\log\frac{1}{\widetilde{\gamma}} - \frac{1}{\widetilde{\gamma}} + 1\right] I_{\{\Delta S=0\}}\\ &= \widetilde{\beta}\cdot S^c + \widetilde{W}\star(\mu-\nu) - \frac{1}{2}\widetilde{\beta}^T c\widetilde{\beta}\cdot A + \sum\left[\frac{-\widetilde{\gamma}\log\widetilde{\gamma} + \widetilde{\gamma} - 1}{\widetilde{\gamma}}\right]\\ &\quad + \frac{\widetilde{\gamma}\widetilde{\beta}^T x - e^{\widetilde{\beta}^T x} + 1}{\widetilde{\gamma}}\star\mu.\end{aligned}$$

Note that

$$\begin{aligned}&\frac{1}{\widetilde{\gamma}}\big(\widetilde{\gamma}\widetilde{\beta}^T x - e^{\widetilde{\beta}^T x} + 1\big)\star\mu\\ &\quad = \widetilde{\beta}^T\big(x - h(x)\big)\star\mu + \frac{1}{\widetilde{\gamma}}\big(\widetilde{\gamma}\widetilde{\beta}^T h(x) - e^{\widetilde{\beta}^T x} + 1\big)\star(\mu-\nu)\\ &\quad\quad + \widetilde{\gamma}^{-1}\big(\widetilde{\gamma}\widetilde{\beta}^T h(x) - e^{\widetilde{\beta}^T x} + 1\big)\star\nu,\end{aligned}$$

since the function $\widetilde{\gamma}^{-1}(\widetilde{\gamma}\widetilde{\beta}^T h(x) - e^{\widetilde{\beta}^T x} + 1)$ is $(\mu-\nu)$-integrable which is due to the $(\mu-\nu)$–integrability of $\widetilde{\gamma}^{-1}(e^{\widetilde{\beta}^T x} - 1) = W(x)$ and the boundedness of $h(x)$. Therefore, we get that

$$\begin{aligned}\log(\widetilde{Z}) &= \widetilde{\beta}\cdot S^c + \widetilde{\beta}^T h(x)\star(\mu-\nu) + \widetilde{\beta}^T(x-h(x))\star\mu\\ &\quad + \widetilde{\gamma}^{-1}\big(\widetilde{\gamma}\widetilde{\beta}^T h(x) - e^{\widetilde{\beta}^T x} + 1\big)\star\nu - \frac{1}{2}\widetilde{\beta}^T c\widetilde{\beta}\cdot A\\ &\quad + \sum\widetilde{\gamma}^{-1}(-\widetilde{\gamma}\log(\widetilde{\gamma}) + \widetilde{\gamma} - 1).\end{aligned}$$

Equivalently, we deduce that

$$\log\widetilde{Z} = \widetilde{\beta}\cdot S + \frac{1}{2}\widetilde{\beta}^T c\widetilde{\beta}\cdot A + \frac{\widetilde{\beta}^T x e^{\widetilde{\beta}^T x} - e^{\widetilde{\beta}^T x} + 1}{\widetilde{\gamma}}\star\nu + \sum\widetilde{\gamma}^{-1}(-\widetilde{\gamma}\log(\widetilde{\gamma}) + \widetilde{\gamma} - 1),$$

since

$$\widetilde{\beta}\cdot S = \widetilde{\beta}\cdot S^c + \widetilde{\beta}^T b\cdot A + \widetilde{\beta}^T h(x)\star(\mu-\nu) + \widetilde{\beta}^T(x-h(x))\star\mu.$$

Therefore, by a direct application of Lemma 8 for $\lambda = \widetilde{\beta}$, (50) follows immediately from putting $\widetilde{H} = \widetilde{\beta}$. This ends the proof of the theorem. □

In what follows, we denote by $Z := \mathscr{E}(N)$ a positive local martingale with

$$N := \beta \cdot S^c + W \star (\mu - \nu) + g \star \mu + \overline{N}, \qquad W_t(x) := f_t(x) + \frac{\widehat{f}_t}{1 - a_t} I_{\{a_t < 1\}}, \quad (52)$$

where $\left(\beta, f, g, \overline{N}\right)$ are Jacod's components of N. Here, we define:

$$\mathscr{Z}^e_{loc}(S, Z) := \left\{\overline{Z} :\ \overline{Z} > 0,\ \overline{Z}Z \in \mathscr{Z}_{loc}(S)\right\},$$

where $\mathscr{Z}_{loc}(S)$ is given by (2).

Theorem 10 *Consider Z defined in* (52) *and suppose that*

$$\mathscr{Z}^e_{loc}(S, Z) \neq \emptyset, \qquad \int_{\{|x|>1\}} e^{\lambda^T x}(1 + f(x))F(dx) < \infty, \qquad \forall \lambda \in \mathbb{R}^d.$$

Then the minimization problem

$$\min_{\overline{Z} \in \mathscr{Z}^e_{loc}(S,Z)} h^E(\overline{Z}, Z), \qquad (53)$$

admits a solution $\widetilde{Z} = \mathscr{E}(\widetilde{N})$ given by

$$\widetilde{N} = \widetilde{\beta} \cdot S^{c,Z} + \widetilde{W} \star (\mu - \nu^Z), \qquad \widetilde{W}_t(x) = \frac{e^{\widetilde{\beta}_t^T x} - 1}{1 - a^Z_t + \int e^{\widetilde{\beta}_t^T y} \nu^Z(\{t\}, dy)},$$

where $\widetilde{\beta}$ is the root of the equation

$$0 = b^Z + c\lambda + \int (e^{\lambda^T x} x - h(x))F^Z(dx). \qquad (54)$$

Here $S^{c,Z}$, b^Z, a^Z, ν^Z and F^Z are given by

$$S^{c,Z} := S^c - c\beta \cdot A, \qquad b^Z := b + c\beta - \int f(x)h(x)F(dx),$$

$$a^Z_t := \nu^Z(\{t\}, \mathbb{R}^d \setminus \{0\})$$

and

$$\nu^Z(dt, dx) := F^Z_t(dx)dA_t, \qquad F^Z_t(dx) := (1 + f_t(x))F_t(dx).$$

Proof Consider a sequence of stopping times, $(T_n)_{n\geq 1}$, stationarily increasing to T and such that Z^{T_n} is a true martingale. For a fixed but arbitrary n we denote $Q := Z_{T_n} \cdot P$. Note that all the equations in the theorem are robust under stopping. Due to Lemma 1, it is sufficient to prove that the theorem holds on $[\![0, T_n]\!]$. Then, we obtain that $\nu^Q(dt, dx) = (1 + f_t(x))I_{\{t \leq T_n\}}\nu(dt, dx)$, $\mathscr{Z}^e_{loc}(S, Q) \neq \emptyset$, and $\int_{\{|x|>1\}} e^{\lambda^T x} F^Q(dx) < \infty$ for $\lambda \in \mathbb{R}^d$.

Therefore, the assumptions of Theorem 3.3 in [4] are fulfilled. By direct application of this theorem for S^{T_n} and under the measure $Q = Z_{T_n} \cdot P$, we deduce that the problem defined in (53) admits a solution $\widetilde{Z}^Q = \mathscr{E}(\widetilde{N}^Q)$, where $\widetilde{N}^Q$ is given, on $[\![0, T_n]\!]$, by

$$\widetilde{N}^Q = \widetilde{\beta} \cdot S^{c,Q} + \widetilde{W} \star (\mu - \nu^Q), \qquad \widetilde{W}_t(x) = \frac{e^{\widetilde{\beta}_t^T x} - 1}{1 - a_t^Q + \int e^{\widetilde{\beta}_t^T y} \nu^Q(\{t\}, dy)}.$$

Herein $S^{c,Q}$ is the continuous local martingale part of S under Q and ν^Q is the Q-compensator measure of μ, and $a_t^Q = \nu^Q(\{t\}, \mathbb{R}^d \setminus \{0\})$. Moreover, $\widetilde{\beta}$ is given by the equation

$$\begin{aligned} 0 &= b^Q + c\lambda + \int \left(e^{\lambda^T x} x - h(x)\right) F^Q(dx) \\ &= \left[b^Z + c\lambda + \int (e^{\lambda^T x} x - h(x)) F^Z(dx) \right] I_{[\![0,T_n]\!]}. \end{aligned} \tag{55}$$

It is then clear that $\widetilde{N}^Q$ coincides with $\widetilde{N}^{T_n}$ of the theorem and that the equation (55) is exactly the equation (54) on $[\![0, T_n]\!]$. This ends the proof of theorem. □

Theorem 11 *Let Z be a positive local martingale and let $\widetilde{Z} \in \mathscr{Z}_{loc}^e(S, Z)$. If the assumptions of Theorem* 10 *are fulfilled and $\widetilde{Z}$ is the MEH local martingale density with respect to Z, then*

$$\log \widetilde{Z} = \widetilde{\beta} \cdot S + h^E(\widetilde{Z}, Z)$$

and $\widetilde{\beta}$ is a root of (54).

Proof The proof of this theorem follows from the same arguments as in the proofs of Theorems 9 and 10. □

References

1. Choulli, T., Schweizer, M.: The mathematical structure of horizon-dependence in optimal portfolio choice. Preprint (2009)
2. Choulli, T., Schweizer, M.: Stability of sigma-martingale densities in LlogL under an equivalent change of measure. Preprint (2010)
3. Choulli, T., Stricker, C.: Minimal entropy—Hellinger martingale measure in incomplete markets. Math. Finance **15**(3), 465–490 (2005)
4. Choulli, T., Stricker, C.: More on minimal entropy—Hellinger martingale measures. Math. Finance **16**(1), 1–19 (2006)
5. Choulli, T., Stricker, C., Li, J.: Minimal Hellinger martingale measures of order q. Finance Stoch. **11**(3), 399–427 (2007)
6. Delbaen, F., Grandits, P., Rheinländer, T., Samperi, D., Schweizer, M., Stricker, C.: Exponential hedging and entropic penalties. Math. Finance **12**(2), 99–123 (2002)
7. Delbaen, F., Schachermayer, W.: The Mathematics of Arbitrage. Springer, Heidelberg (2006)

8. Dellacherie, C., Meyer, P.-A.: Théorie des Martingales. Hermann, Paris (1980). Chapter V to VIII
9. Evans, J., Henderson, V., Hobson, D.: Optimal timing for an indivisible asset sale. Math. Finance **18**, 545–567 (2008)
10. Fisher, I.: The impatience theory of interest. Am. Econ. Rev. **3**, 610–618 (1931)
11. Hakansson, N.H.: Optimal investment and consumption strategies under risk, an uncertain lifetime and insurance. Int. Econ. Rev. **10**, 443–466 (1969)
12. Henderson, V.: Valuing the option to invest in an incomplete market. Math. Finance Econ. **1**(2), 103–128 (2007)
13. Henderson, V., Hobson, D.: Horizon-unbiased utility functions. Stoch. Process. Appl. **117**(11), 1621–1641 (2006)
14. Jacod, J.: Calcul Stochastique et Problèmes de Martingales. Lecture Notes in Mathematics, vol. 714. Springer, Berlin (1979)
15. Jacod, J., Shiryaev, A.: Limit Theorems for Stochastic Processes, 2nd edn. Springer, Berlin (2002)
16. Kabanov, Y.: On the FTAP of Kreps–Delbaen–Schachermayer. Statistics and control of random processes. The Liptser Festschrift. In: Proceedings of Steklov Mathematical Institute Seminar, pp. 191–203. World Scientific, Singapore (1997)
17. Kabanov, Y., Stricker, C.: On the optimal portfolio for the exponential utility maximization: remarks to the six-author paper. Math. Finance Econ. **12**(2), 125–134 (2002)
18. Karatzas, I., Zamfirescu, M.: Martingale approach to stochastic control with discretionary stopping. Appl. Math. Optim. **53**, 163–184 (2006)
19. Larsen, K., Hang, Y.: Horizon dependence of utility optimizers in incomplete models. Preprint (2010)
20. Lépingle, D., Mémin, J.: Sur l'intégrabilité uniforme des martingales exponentielles. Z. Wahrscheinlichkeitstheor. Verw Geb. **42**, 175–203 (1978)
21. Maingueneau, M.A.: Temps d'arrêt optimaux et théorie générale. Sémin. Probab. **12**, 457–467 (1978)
22. Mertens, J.F.: Processus stochastiques généraux et surmartingales. Z. Wahrscheinlichkeitstheor. Verw Geb. **22**, 45–68 (1972)
23. Musiela, M., Zariphopoulou, T.: Portfolio choice under dynamic investment performance criteria. Quant. Finance **9**(2), 161–170 (2009)
24. Musiela, M., Zariphopoulou, T.: Backward and forward utilities and the associated indifference pricing systems: the case study of the binomial model. In: Carmona, R. (ed.) Indifference Pricing, pp. 3–43. Princeton University Press, Princeton (2009)
25. Rockafellar, R.T.: Convex Analysis. Princeton University Press, Princeton (1970)
26. Yaari, M.E.: Uncertain lifetime, life insurance and the theory of the consumer. Rev. Econ. Stud. **32**, 137–158 (1965)
27. Zariphopoulou, T., Zitkovic, G.: Maturity-independent risk measures. SIAM J. Financ. Math. **1**, 266–288 (2010)
28. Zitkovic, G.: A dual characterization of self-generation and log-affine forward performances. Ann. Appl. Probab. **19**(6), 2176–2270 (2009)

Mean Square Error and Limit Theorem for the Modified Leland Hedging Strategy with a Constant Transaction Costs Coefficient

Sébastien Darses and Emmanuel Lépinette

Abstract We study the modified Leland's strategy defined in Lépinette (Math. Finance 22(4):741–752, 2012) for hedging portfolios in the presence of a constant proportional transaction costs coefficient. We prove a limit theorem for the deviation between the real portfolio and the payoff. We identify the rate of convergence and the associated limit distribution. This rate can be improved using the modified strategy and non periodic revision dates.

Keywords Option pricing · Transaction costs · Leland strategy

Mathematics Subject Classification (2010) 91G20

1 Introduction

The present paper is concerned with the study of asymptotic hedging in the presence of transaction costs. The asymptotic replication of a given payoff is performed via a modified Leland's strategy recently introduced in [8].

Let us briefly recall the history and the main known results about Leland's strategy. In 1985 Leland suggested an approach to price contingent claims under proportional transaction costs. His main idea was to use the classical Black–Scholes formula with a suitably adjusted volatility for a periodically revised portfolio whose terminal value approximates the payoff. The intuition behind this practical method is to compensate for transaction cost by increasing the volatility in the following way:

$$\widehat{\sigma}_t^2 = \sigma^2 + \sigma \sqrt{n} k_n \sqrt{8/\pi} \sqrt{f'(t)}, \tag{1}$$

S. Darses
LATP, Université Aix-Marseille I, 13453 Marseille cedex 13, France
e-mail: darses@cmi.univ-mrs.fr

E. Lépinette (✉)
Ceremade, Université Paris Dauphine, Place du Maréchal De Lattre De Tassigny, 75775 Paris cedex 16, France
e-mail: emmanuel.lepinette@ceremade.dauphine.fr

Y. Kabanov et al. (eds.), *Inspired by Finance*, DOI 10.1007/978-3-319-02069-3_8,

where n is the number of the portfolio revision dates and $k_n = k_0 n^{-\alpha}$, $\alpha \in [0, \frac{1}{2}]$ is the transaction costs coefficient generally depending of n; f is an increasing and smooth function whose inverse $g := f^{-1}$ defines the revision dates $t_i^n := g(\frac{i}{n})$, $1 \leq i \leq n$.

The principal results on convergence for models with transaction costs can be described as follows. First consider the case of approximate hedging of the European call option using the strategy with periodic portfolio revisions (i.e. $g(t) = t$). We know the following results with $T = 1$:

- For $\alpha = \frac{1}{2}$, Lott gave the first rigorous result on the convergence of the approximating portfolio value V_1^n to the payoff $V_1 = (S_1 - K)_+$. The sequence $V_1^n - V_1$ tends to zero in probability [9], and a stronger result holds: $n\,\mathbf{E}\,(V_1^n - V_1)^2$ converges to a constant $A_1 > 0$ [4].
- For $\alpha \in (0, \frac{1}{2})$, the sequence $V_1^n - V_1$ tends to zero in probability (see [7]), and it is shown in [1] that $n^{p_\alpha}\mathbf{E}\,(V_1^n - V_1)^2 \to 0$ as $n \to \infty$, with $p_\alpha < \alpha$.
- For $\alpha = 0$, the terminal values of portfolios do not converge to the European call as shown by Kabanov and Safarian [7]. Namely, there is a negative $\sigma\{S_1\}$-measurable random variable ξ such that $V_1^n - V_1 \to \xi$ in probability. Pergamenshchikov [10] then analyzed the rate of convergence and proved a limit theorem: the sequence $n^{\frac{1}{4}}(V_1^n - V_1 - \xi)$ converges in law to a mixture of Gaussian distributions [10]. He noticed that one can increase the modified volatility to obtain the asymptotic replication. To do so, he utilizes the explicit form of the systematic hedging error for the European call option. For related results see also [5] and [11].

For models including uniform and non-uniform revision intervals one needs to impose certain conditions on the scale transform g. Generalizations of some of the above results to this more technical case as well as extensions to contingent claims of the form $h(S_1)$ can be found in [1, 4, 11]. In particular, $n^{1/2+\alpha}\mathbf{E}\,(V_1^n - V_1)^2$ converges to a constant in the case $\alpha > 0$. Moreover, for $\alpha = \frac{1}{2}$, the distributions of the process $Y_t^n := n^{\frac{1}{2}}(V_t^n - \widehat{V}_t)$ in the Skorohod space [1], $\mathscr{D}[0, 1]$, converges weakly to the distribution of a two-dimensional Markov diffusion process component (see [3]). Notice that the asymptotic replication does not hold for $\alpha = 0$ in this more general setting. For more details we refer to [1–3] and references therein.

We solve the case $\alpha = 0$ for a large class of payoffs and with specific revision dates (including uniform dates) by means of the modified strategy introduced in [8]. This one makes the portfolio's terminal value converge to the contingent claim as n tends to infinity, that is the approximation error vanishes. The analysis we performed here suggests that concentrating the revision dates near the maturity $T = 1$ accelerates the convergence rate.

---

[1]Note that Y_t^n corresponds to the deviation (up to a multiplicative constant) between the "real world" portfolio and the theoretical Leland's portfolio $\widehat{V}_t = \widehat{C}(t, S_t)$ where $\widehat{C}$ is the modified heat equation solution which terminal value is the payoff function. This approach was suggested by Leland.

The asymptotic behavior of the hedging error is a practical important issue. Since traders obviously prefer gains than losses, using the L^2-norm to measure hedging errors is strongly criticized. Of course, the limiting distribution of the hedging error is much more informative. Our present work also aims at tackling this issue: we prove that

$$n^{\frac{1}{4}+p}(V_1^n - h(S_1)) \xrightarrow[n\to\infty]{d} Z,$$

where the law of Z is explicitly identified, $EZ = 0$ and $p > 0$ depends on the chosen grid.

The paper is organized as follows. In Sect. 2, we introduce the basic notations, models and assumptions of our study. In particular, we recall the modified Leland strategy defined in [8]. In Sect. 3, we state our main result: a limit theorem for the renormalized asymptotic hedging error. In Sect. 4, we establish two lemmas concerning, on one hand, random variables constructed from the geometric Brownian motion, and on the other hand, some change of variables for the revision dates. These auxiliary results will be used repeatedly throughout the paper. In Sect. 5, we prove the main result. The Appendix recalls technical results we need in proofs.

2 Notations and Models

2.1 Black–Scholes Model and Hedging Strategy

We are given a filtered probability space $(\Omega, \mathscr{F}, (\mathscr{F}_t)_{t\in[0,1]}, \mathbb{P})$ on which a standard one-dimensional $(\mathscr{F}_t)$-adapted Brownian motion W is defined. As usual, we denote by $L^2(\Omega)$ the space of square integrable $\mathscr{F}_1$-measurable random variables endowed with its norm $\|X\|_2 := \sqrt{\mathbf{E}\, X^2}$.

We consider the classical Black–Scholes model composed of two assets without transaction costs, i.e. $k_0 = 0$ and $\widehat{\sigma} = \sigma$. The first one is riskless (bond) with the interest rate $r = 0$ and the second asset is $S = (S_t)$, $t \in [0, 1]$, a geometric Brownian motion that is $S_t = S_0 e^{\sigma W_t - \frac{1}{2}\sigma^2 t}$. It satisfies the SDE $dS_t = \sigma S_t dW_t$, with positive constants S_0, σ. It means that the risky asset is seen under the martingale measure.

The well-known Black and Scholes problem without transaction costs is to hedge a payoff $h(S_1)$, h being a continuous function of polynomial growth. The pricing function solves the terminal valued Cauchy problem

$$\begin{cases} C_t(t,x) + \frac{\sigma_t^2}{2}x^2 C_{xx}(t,x) = 0, & t \in [0,1],\ x > 0, \\ C(1,x) = h(x). \end{cases}$$

Its solution can be written as

$$C(t,x) = \int_{-\infty}^{\infty} h\Big(x e^{\rho_t y - \frac{\rho_t^2}{2}}\Big)\varphi(y)dy \tag{2}$$

where $\rho_t^2 = (1-t)\sigma^2$ and φ is the standard Gaussian density.

Without transaction costs ($\sigma = \widehat{\sigma}$) the self-financed portfolio process reads

$$V_t = C(0, S_0) + \int_0^t C_x(u, S_u) dS_u. \tag{3}$$

In the Itô formula for $C(t, S_t)$ the integral over dt vanishes and, therefore, we have that $V_t = C(t, S_t)$ for all $t \in [0, 1]$. In particular, $V_1 = h(S_1)$: at maturity the portfolio V replicates the terminal payoff of the option. Modeling assumptions of the above formulation include frictionless market and continuous trading for instance. However, an investor revises the portfolio at a finite set of dates

$$\mathcal{T}^n := \{t_i \in [0,1], i = 0, \cdots, n\}$$

and keeps $C_x(t_i, S_{t_i})$ units of the stock until the next revision date t_{i+1}. It is well known that this discretized model converges to the Black–Scholes one in the sense that the corresponding portfolio terminal value converges to the payoff as the number of revision dates tends to infinity.

2.2 Reminder About Leland's Strategy

We are now concerned with transaction costs. We directly work in a discrete-time setting. Leland suggested to replace σ in the Cauchy problem above by a suitably modified volatility $\widehat{\sigma}$. In the case where $\widehat{\sigma}$ does not depend on t, the solution $\widehat{C}$ satisfies $\widehat{C}(t,x) = C(t, x, \widehat{\sigma})$, i.e. practitioners do not need to rectify their algorithms to compute the strategy. Leland obtained an explicit expression of $\widehat{\sigma}$ by equalizing the transaction costs of the portfolio and the drift term generated by the additional term $\widehat{\sigma} - \sigma^2 > 0$ in the Ito expansion of the payoff $h(S_1) = \widehat{C}(1, S_1)$. In the general case, the pricing function can be written as

$$\widehat{C}(t,x) = \int_{-\infty}^{\infty} h\left(x e^{\rho_t^n y - (\rho_t^n)^2/2}\right) \varphi(y) dy \tag{4}$$

where

$$\begin{aligned} (\rho_t^n)^2 &:= \int_t^1 \widehat{\sigma}_s^2 ds, \\ \widehat{\sigma}_t^2 &:= \sigma^2 + \sigma \sqrt{n} k_n \sqrt{8/\pi} \sqrt{f'(t)}, \end{aligned} \tag{5}$$

φ is the Gaussian density and $g = f^{-1}$ is the revision date function.

2.3 A Possible Modification of Leland's Strategy

The practically interesting case $\alpha = 0$ (i.e., with constant k_0), where is a systematic error attracted a lot of attention. Limit theorems were obtained by Granditz and Schachinger [5] and Pergamenshchikov [10]. Zhao and Ziemba [12, 13] provides a numerical study of the limiting error for practical values of parameters. Sekine and Yano, [11] suggested some scheme to reduce it. In the paper [10] a modification of the Leland strategy was suggested for the call option eliminating the limiting error. Unfortunately, the approach is based on the explicit formulae and, seemingly, cannot be easily generalized for more general payoff functions. Our modification of the Leland strategy has the following features:

(1) we use the same enlarged volatility;

(2) the initial value of the portfolio V_0^n is exactly the same than for the initial Leland strategy (see [10] where the behavior of V_0^n is studied as $n \to \infty$ and a method is suggested to decrease the option price);

(3) the only difference is at the revision dates t_i; we apply not the modified "delta" of the Black–Scholes formula with enlarged volatility, but correct it on the basis of previous revisions, see the formula (7). This is a technical modification of Leland's strategy for which it is difficult to give an economical interpretation but which has an advantage to avoid the limiting error.

In the model with proportional transaction costs and a finite number of revision dates the current value of the portfolio process at time t is described as

$$V_t^n := V_0^n + \int_0^t D_u^n dS_u - \sum_{t_i < t} k_0 S_{t_i} |D_{i+1}^n - D_i^n| \tag{6}$$

where D^n is a piecewise-constant process with $D^n = D_i^n$ on the interval $(t_{i-1}, t_i]$, $t_i = t_i^n, i \le n$, are the revision dates, and D_i^n are $\mathscr{F}_{t_{i-1}}$-measurable random variables.

Recall that the transaction costs coefficient is a constant $k_0 > 0$ (that is $\alpha = 0$ in the Leland model) and the dates t_i are defined by a function g, namely $t_i = g(\frac{i}{n})$. Let us denote by f the inverse of g. Set for all $i_0 < n$

$$J_{i_0}^n(t) = \{i \gamma i_0,\ t_i \le t,\ t_i \in \mathscr{T}^n\}$$

and let us define the dates

$$t_n^-(t) = t_{(n-1) \wedge \max J_0^n(t)},$$

$$t_n^+(t) = t_{1+(n-1) \wedge \max J_0^n(t)}.$$

The "enlarged volatility", depending on n, is given by the formula (5). We modify the usual Leland strategy (see [8]) by considering the process D^n with

$$D_i^n := \widehat{C}_x(t_{i-1}, S_{t_i-1}) - \sum_{j=1}^{i-1} \int_{t_{j-1}}^{t_j} \widehat{C}_{xt}(u, S_{t_{j-1}}) du. \tag{7}$$

Moreover, let us define $K_t^n := \sum_{i \in J_1^n(t)} \Delta K_{t_i^n}^n$ where $\Delta K_{t_0^n}^n := 0$ and for $i \geq 1$,

$$\Delta K_{t_i^n}^n := -\int_{t_{i-1}}^{t_i} \widehat{C}_{xt}(u, S_u)du. \tag{8}$$

In the same way, we set $L_t^n := \sum_{i \in J_1^n(t)} \Delta L_{t_i^n}^n$, where $\Delta L_{t_0^n}^n := 0$ and for $i \geq 1$,

$$\Delta L_{t_i^n}^n := -\int_{t_{i-1}}^{t_i} \widehat{C}_{xt}(u, S_{t_{i-1}})du. \tag{9}$$

2.4 Assumptions and Notational Conventions

Throughout the paper, we adopt the following rules:

- we will often omit the indexes n and the variable t (especially in the Appendix) when there is no ambiguity;
- the constants C appearing in the various inequalities are independent of n and may vary from one line to another;
- we use the classical Landau notations O and o. These quantities will be always deterministic.

As shown in [2], recall that the solution $\widehat{C}$ of the Cauchy problem we consider is infinitely differentiable on $[0, T) \times (0, \infty)$. We use the abbreviations $\delta_t := \widehat{C}_x(t, S_t)$, $\gamma_t := \widehat{C}_{xx}(t, S_t)$. We denote by $(\delta_t^n)_t$ the process equal to $\delta_{t_i^n}$ on the interval $[t_i^n, t_{i+1}^n)$ and $(\gamma_t^n)_t$ is defined similarly. For an arbitrary process H, we set $\Delta H_{t_i} := H_{t_i} - H_{t_{i-1}}$.

We will work under the following assumptions:

(A1) The function g has the following form:

$$g(t) = 1 - (1-t)^{\mu}, \quad \mu \in \left[1, \frac{3+\sqrt{57}}{8}\right).$$

(A2) The function h is a convex and continuous function on $[0, \infty)$ which is twice differentiable except the points $K_1 < \cdots < K_{p_h}$ where h' and h'' admit right and left limits; $|h''(x)| \leq Mx^{-\beta}$ for $x \geq K_{p_h}$ where $\beta \geq 3/2$.

Assumption (A1) is not too restrictive. A trader can in particular choose $\mu = 1$ to balance its portfolio periodically. However, as we will see, it is more preferable to increase μ to obtain a better rate of convergence.

Note that $f(t) = 1 - (1-t)^{1/\mu}$, hence the derivative f' for $\mu > 1$ explodes at the maturity (see Fig. 1) date and so does the enlarged volatility. We define the increasing function

$$p := p(\mu) := \frac{\mu - 1}{4(1+\mu)}.$$

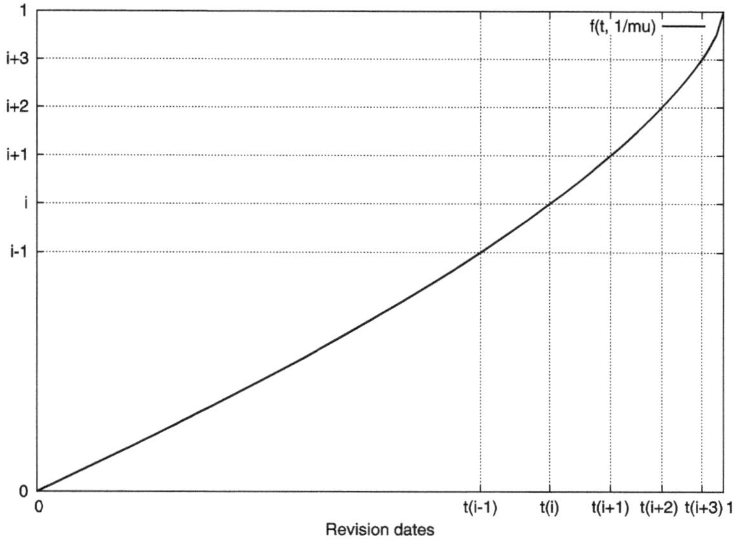

Fig. 1 Revision dates with $f(t) = 1 - (1-t)^{\frac{1}{\mu}}$, $\mu = 1.5$ and $n = 10$

Under Assumption (A1), we have $0 \leq p < 1/16$. In the sequel, will frequently appear the quantity

$$\mathcal{Q}(\mu) = \frac{\mu^{1/2-2p}(1+\mu)^{4p}}{2^{4p}(\sqrt{8/\pi})^{4p+1}}.$$

3 Main Result

In [8], it is proven that V_1^n converges in probability to $h(S_1)$. We recall this result:

Theorem 1 *Let* $k_0 > 0$. *Suppose that Assumption* (A2) *hold and* $g' > 0$, $g \in C^2[0,1]$. *Then* $P\text{-}\lim_n V_1^n = h(S_1)$.

Our main result here provides the rate of convergence for a specific family of revision dates functions including the uniform grid (i.e. $g(t) = t$) and identifies the associated limit distribution of the deviation:

Theorem 2 *Consider the portfolio* V^n *defined by* (6) *and* (7) *under Assumptions* (A1) *and* (A2). *The following convergence then holds*:

$$n^{\frac{1}{4}+p}(V_1^n - h(S_1)) \xrightarrow[n\to\infty]{d} Z, \tag{10}$$

where the law of Z is a mixed Gaussian distribution, i.e. $Z = \eta N$ where N is a standard normal independent of η and

$$\eta^2 := \mathscr{Q}(\mu)(k_0\sigma)^{1-4p} S_1^2 \times \int_0^\infty x^{4p} \left\{ \left(\int_x^\infty J(y, S_1)dy \right)^2 + \left(1 - \frac{2}{\pi}\right) \widetilde{J}(x, S_1)^2 \right\} dx,$$

and

$$J(x, S_1) := \frac{1}{2x} \int_{-\infty}^{\infty} h'(S_1 e^{\sqrt{x}y + x/2})(-y^2 - \sqrt{x}y + 1)\varphi(y)dy \tag{11}$$

$$\widetilde{J}(x, S_1) := \frac{1}{\sqrt{x}} \int_{-\infty}^{\infty} h'(S_1 e^{\sqrt{x}y + x/2}) y\varphi(y)dy. \tag{12}$$

Moreover $n^{\frac{1}{2}+2p}\mathbf{E}\,(V_1^n - h(S_1))^2 \xrightarrow[n\to\infty]{\mathbf{E}} \eta^2$.

Observe that $EZ = 0$. In the proof, Z will be identified by its characteristic function given by $\phi_Z(s) = \mathbf{E}\; e^{-\frac{s^2}{2}\eta^2}$. As we can see, concentrating the revision dates near the horizon date ($p > 0$) improves the convergence rate. Actually, we can observe that near $T = 1$, the derivative f' explodes if $p > 0$ and so increases the modified volatility, which confirms the main Leland idea; Artificially increase the volatility to compensate for transaction costs. The proof of the theorem above is given in Sect. 5; To do so, we decompose the difference $n^{\frac{1}{4}+p}(V_t^n - \widehat{C}(t, S_t))$ into a martingale which converges in L^2 and a residual term tending to 0 at $T = 1$. We conclude with $h(S_1) = \widehat{C}(1, S_1)$.

4 Auxiliary Results

4.1 Geometric Brownian Motion and Related Quantities

In the sequel, we shall use the decomposition given by Ito formula

$$\widehat{C}_x(t, S_t) = \widehat{C}_x(0, S_0) + \widehat{M}_t^n + \widehat{A}_t^n \tag{13}$$

where

$$\widehat{M}_t^n := \int_0^t \sigma_u S_u \widehat{C}_{xx}(u, S_u)dW_u,$$

$$\widehat{A}_t^n := \int_0^t \left[\widehat{C}_{xt}(u, S_u) + \frac{1}{2}\sigma_u^2 S_u^2 \widehat{C}_{xxx}(u, S_u) \right] du.$$

The process $\widehat{M}^n$ is a square integrable martingale on $[0, 1]$ by virtue of [8].

We set for $u < v$

$$\mathscr{E}_u^v = \frac{S_v}{S_u} - 1,$$

and

$$\left[\mathscr{E}_u^v\right]_c = \mathbf{E}\left(\left|\mathscr{E}_u^v\right|\right) - \left|\mathscr{E}_u^v\right|.$$

$$\{\mathscr{E}_u^v\}_s^2 := \left(\mathscr{E}_u^v\right)^2 \operatorname{sgn} \mathscr{E}_u^v.$$

In the sequel, we will use several times the following basic results.

Lemma 1 *For all i the following inequalities and expansions hold:*

$$\begin{aligned}
\mathbf{E}\left(\mathscr{E}_u^v\right)^{2m} &\le C_m(v-u)^m, \quad u \le v \\
\mathbf{E}\left(\mathscr{E}_{t_{i-1}}^{t_i}\right)^2 &= \sigma^2 \Delta t_i (1 + o(1)) \\
\mathbf{E}\left[\mathscr{E}_{t_{i-1}}^{t_i}\right]_c^2 &= \left(1 - \frac{2}{\pi}\right)\sigma^2 \Delta t_i (1 + o(1)) \\
\mathbf{E}\left[\mathscr{E}_{t_{i-1}}^{t_i}\right]_c^2 \operatorname{sgn} \mathscr{E}_u^v &= \left(1 - \frac{2}{\pi}\right)\sigma^2 (\Delta t_i)^{\frac{3}{2}} (1 + o(1)) \\
\mathbf{E}\,\{\mathscr{E}_{t_{i-1}}^{t_i}\}_s^2 &= k(\Delta t_i)^{3/2}\left(1 + o(n^{-1/4})\right).
\end{aligned} \tag{14}$$

Proof We refer to [1] or [3]. For the sake of completeness we recall the proof of the last one. Let us notice the equality in law

$$\{\mathscr{E}_{t_{i-1}}^{t_i}\}_s^2 \stackrel{d}{=} \left(\exp\left\{\sigma\sqrt{\Delta t_j}\xi - \sigma^2 \Delta t_j/2\right\} - 1\right)^2 \left(1_{\xi \ge \sigma\sqrt{\Delta t_j}/2} - 1_{\xi \le \sigma\sqrt{\Delta t_j}/2}\right),$$

where ξ is the standard Gaussian variable. Since ξ and $-\xi$ have the same law, this yields

$$\begin{aligned}
\mathbf{E}\,\{\mathscr{E}_{t_{i-1}}^{t_i}\}_s^2 = \mathbf{E}&\left[\left(e^{u\xi - u^2/2} - 1\right)^2 - \left(e^{-u\xi - u^2/2} - 1\right)^2\right] 1_{\xi \ge u/2} \\
&- \mathbf{E}\left(e^{-u\xi - u/2} - 1\right)^2 1_{|\xi| \le u/2},
\end{aligned}$$

where $u = \sigma\sqrt{\Delta t_j}$. Moreover, we have the inequality

$$\mathbf{E}\left(e^{-u\xi - u^2/2} - 1\right)^2 1_{|\xi| \le u/2} \le u^4.$$

From [4], we recall that

$$\mathbf{E}\left[\left(e^{u\xi - u^2/2} - 1\right)^2 - \left(e^{-u\xi - u^2/2} - 1\right)^2\right] 1_{\xi \ge u/2} = \frac{2}{\sqrt{2\pi}} u^3 + O(u^4).$$

This completes the proof. □

4.2 Basic Results Concerning the Revision Dates

The function ρ_t decreases from ρ_0 to 0. The following useful bounds are obvious:

$$\rho_t^2 \geq (\sigma^2 + cn^{\frac{1}{2}})(1-t), \tag{15}$$

$$\rho_t^2 \leq \sigma^2(1-t) + \sigma k_0 n^{\frac{1}{2}} \sqrt{8/\pi}(1-t)^{\frac{1}{2}}(1-f(t))^{\frac{1}{2}}. \tag{16}$$

Moreover, it is straightforward that

$$\rho_t^2 \geq cn^{\frac{1}{2}} \sqrt{f'(t)}(1-t), \tag{17}$$

provided that f' is not decreasing. Note that there is a constant C independent of n such that for all $i \leq n-1$, $\frac{1-t_{i-1}}{1-t_i} \leq C$. From this we obtain that

$$\frac{\Delta t_i}{1-t_i} \leq C. \tag{18}$$

We shall often use the inequality

$$\sum_{i=1}^{n-1} \frac{\Delta t_i}{1-t_i} \leq C \log n$$

where C is a constant independent of n.

Lemma 2 *Fix $x > 0$ and $t := t(n,x) \in [0,1)$ such that $x = \rho_t^2$. Set $x_{i-1} = \rho_{t_{i-1}}^2$ and $x_i = \rho_{t_i}^2$ where t_{i-1}, t_i are such that $t \in [t_{i-1}, t_i)$. Then, $x \in (x_i, x_{i-1}]$ with $|x_{i-1} - x_i| \leq cn^{-1/2}$, c is a constant. There exists a constant $C > 0$ such that*

$$\frac{\Delta t_i n^{1/2+2p}}{x_{i-1} - x_i} \leq C(x+1). \tag{19}$$

Moreover, for a given x, $(1-t) \leq cn^{-1/2}x \to 0$ as $n \to \infty$ and

$$\frac{\Delta t_i n^{1/2+2p}}{x_{i-1} - x_i} \underset{n\to\infty}{\longrightarrow} \frac{\mathcal{Q}(\mu)x^{4p}}{(\sigma k_0)^{4p+1}}. \tag{20}$$

Proof Let us write

$$\frac{\Delta t_i n^{1/2+2p}}{x_{i-1} - x_i} = \frac{n^{2p}}{\sigma^2 n^{-1/2} + \sigma k_0 \sqrt{8/\pi} \frac{1}{\Delta t_i} \int_{t_{i-1}}^{t_i} \sqrt{f'(u)} du}$$

$$\underset{n\to\infty}{\sim} \frac{n^{2p}}{\sigma k_0\sqrt{8/\pi}\sqrt{f'(\tilde{t}_i)}}$$

where $\tilde{t}_i \in [t_{i-1}, t_i]$. Moreover,

$$\begin{aligned} x = \rho_t^2 &= \sigma^2(1-t) + \sigma k_0\sqrt{8/\pi}n^{1/2}\int_t^1 \sqrt{f'(u)}du \\ &= \sigma^2(1-t) + \sigma k_0\sqrt{8/\pi}n^{1/2}\frac{2\mu^{1/2}}{1+\mu}(1-t)^{\frac{1+\mu}{2\mu}} \end{aligned}$$

and

$$1-t = \left(\frac{x-\sigma^2(1-t)}{\sigma k_0\sqrt{8/\pi}n^{1/2}}\frac{1+\mu}{2\mu^{1/2}}\right)^{\frac{2\mu}{1+\mu}}.$$

Note that $x \geq cn^{1/2}(1-t)$ so that $(1-t) \leq cn^{-1/2}x \to 0$. In a similar way, we have

$$x_{i-1} - x_i = \rho_{t_{i-1}}^2 - \rho_{t_i}^2 = \sigma^2\Delta t_i + cn^{1/2}\sqrt{f'(\check{t}_i)}\Delta t_i$$

where $\check{t}_i \in [t_{i-1}, t_i]$. We deduce that $x_{i-1} - x_i = \sigma^2\Delta t_i + cn^{1/2}\sqrt{f'(\check{t}_i)}g'(\theta_i)n^{-1}$ where $\theta_i \in [(i-1)/n, i/n]$. Moreover,

$$\sqrt{f'(\check{t}_i)}g'(\theta_i) = \frac{g'(\theta_i)}{\sqrt{g'(f(\check{t}_i))}}$$

is bounded since $f(\check{t}_i) \in [(i-1)/n, i/n]$, $i \leq n-1$. Hence, there is a constant c satisfying $x_{i-1} - x_i \leq cn^{-1/2}$. Since ρ is decreasing, $x \in [x_i, x_{i-1}]$.

Eventually, $\tilde{t}_i \in [t_{i-1}, t_i]$ is such that $\tilde{x}_i = \rho_{\tilde{t}_i}^2 \in [x_i, x_{i-1}]$ and $\tilde{x}_i \to x$. Similarly we have:

$$1-\tilde{t}_i = \left(\frac{\tilde{x}_i - \sigma^2(1-\tilde{t}_i)}{\sigma k_0\sqrt{8/\pi}n^{1/2}}\frac{1+\mu}{2\mu^{1/2}}\right)^{\frac{2\mu}{1+\mu}}$$

which yields

$$\sqrt{f'(\tilde{t}_i)} = \mu^{-1/2}\left(\frac{\tilde{x}_i - \sigma^2(1-\tilde{t}_i)}{\sigma k_0\sqrt{8/\pi}n^{1/2}}\frac{1+\mu}{2\mu^{1/2}}\right)^{\frac{1-\mu}{1+\mu}}$$

and

$$\begin{aligned} \frac{\Delta t_i n^{1/2+2p}}{x_{i-1}-x_i} &\underset{n\to\infty}{\sim} \frac{n^{2p}}{\sigma k_0\sqrt{8/\pi}\sqrt{f'(\tilde{t}_i)}} \\ &\underset{n\to\infty}{\sim} \frac{n^{2p}}{\sigma k_0\sqrt{8/\pi}}\mu^{1/2}\left(\frac{\sigma k_0\sqrt{8/\pi}n^{1/2}}{\tilde{x}_i - \sigma^2(1-\tilde{t}_i)}\frac{2\mu^{1/2}}{1+\mu}\right)^{\frac{1-\mu}{1+\mu}} \end{aligned}$$

$$\underset{n\to\infty}{\sim} \frac{1}{\sigma k_0\sqrt{8/\pi}}\mu^{1/2}\left(\frac{\sigma k_0\sqrt{8/\pi}}{\tilde{x}_i-\sigma^2(1-\tilde{t}_i)}\frac{2\mu^{1/2}}{1+\mu}\right)^{\frac{1-\mu}{1+\mu}}.$$

Since $\tilde{x}_i \to x$ and $\tilde{t}_i \to 0$, we get that

$$\frac{\Delta t_i n^{1/2+2p}}{x_{i-1}-x_i} \xrightarrow[n\to\infty]{} \frac{1}{\sigma k_0\sqrt{8/\pi}}\mu^{1/2}\left(\frac{\sigma k_0\sqrt{8/\pi}}{x}\frac{2\mu^{1/2}}{1+\mu}\right)^{\frac{1-\mu}{1+\mu}}.$$

Since $0 \le (\mu-1)/(1+\mu) < 1$, we also find a constant c such that

$$\frac{\Delta t_i n^{1/2+2p}}{x_{i-1}-x_i} \le c\left|\tilde{x}_i-\sigma^2(1-\tilde{t}_i)\right|^{\frac{\mu-1}{1+\mu}} \le c\,(x+1)$$

This completes the proof. □

We now give an important remark regarding a slight abuse of notation repeatedly used along the paper.

Remark 1 Throughout the sequel, we shall often use the change of variable $x=\rho_t^2$ with $dx=-\widehat{\sigma}_t^2 dt$. For ease of notation, we shall abuse of notation t instead of $t(n,x) := (\rho^2)^{-1}(x)$ when applying this change of variable in an integral.

Similarly, a direct computation yields the following lemma.

Lemma 3 *Set $y>0$ and $v := v(n,y)$ such that $y=\rho_v^2$. There exists a constant $C>0$ such that*

$$\frac{(1-v)n^{1/2+2p}}{y} \le C\,y.$$

Moreover, for a given y, $(1-v) \le cn^{-1/2}y \to 0$ as $n\to\infty$ and

$$\frac{(1-v)n^{1/2+2p}}{y} \xrightarrow[n\to\infty]{} \frac{\mu^{-1/2-2p}(1+\mu)^{4p+1}y^{4p}}{2^{4p}(\sigma k_0\sqrt{8/\pi})^{4p+1}}.$$

5 Proof of the Limit Theorem

The proof is divided into three parts. In Step 1 we split the hedging error into a martingale part M and a residual part ε. In Step 2 we show that the residual terms tend to zero in $L^2(\Omega)$ as n tends to infinity. The convergence rate $n^{\frac{1}{4}+p}$ is generated by the revision function f defining the modified volatility. We identify in Step 3 the asymptotic distribution of the martingale $n^{\frac{1}{4}+p}M^n$ and we complete the proof of the main result.

5.1 Step 1: Splitting of the Hedging Error

Comparing Expression (6) with the Ito expansion of $h(S_1) = \widehat{C}(1, S_1)$ yields the following decompositions. The hedging error reads

$$V_1^n - h(S_1) = M_1^n + \varepsilon_1^n, \tag{21}$$

where for all $n \in \mathbb{N}$, M^n is a martingale of terminal value

$$M_1^n := k_0 \sum_{i \le n-1} \gamma_{t_{i-1}} S_{t_{i-1}}^2 \left[\mathscr{E}_{t_{i-1}}^{t_i}\right]_c + \int_0^1 K_u^n dS_u. \tag{22}$$

The residual term can be splited as

$$\varepsilon_t^n = R_0^n(t) + R_1^n(t) + R_2^n(t) + R_3^n(t), \tag{23}$$

where

$$\begin{aligned}
R_0^n(t) &:= k_0 \sum_{i \in J_1^n(t)} \gamma_{t_{i-1}} S_{t_{i-1}}^2 \left(\sigma\sqrt{\frac{2}{\pi}}\sqrt{nf'(t_{i-1})}\Delta t_i - \mathbf{E}\left|\mathscr{E}_{t_{i-1}}^{t_i}\right|\right)\\
R_1^n(t) &:= \int_0^t (\delta_u^n - \delta_u) dS_u,\\
R_2^n(t) &:= k_0 \sum_{i \in J_1^n(t)} \left(|\Delta\,\delta_{t_i}^n + \Delta\,K_{t_i}^n| - |\Delta\,\delta_{t_i}^n + \Delta\,L_{t_i}^n|\right) S_{t_i},\\
R_3^n(t) &:= \int_0^t (L_u^n - K_u^n) dS_u.
\end{aligned} \tag{24}$$

5.2 Step 2: The Mean Square Residue Tends to 0 with Rate $n^{\frac{1}{2}+2p}$

The most technical part of this paper is the following. The deviation of the approximating portfolio from the payoff has been written in an integral form by virtue of the Ito formula. The "real world" portfolio may be interpreted as a discrete-time approximation of the theoretical portfolio $\widehat{C}(t, S_t)$ yielding the residual terms above. Consequently, the following analysis is mainly based on Taylor approximations involving the successive derivatives of $\widehat{C}$ and so heavily utilizes estimates of the appendix. Standard tools from stochastic calculus are also frequently used.

Theorem 3 *The following convergence holds*:

$$n^{\frac{1}{2}+2p}\mathbf{E}\,(\varepsilon_1^n)^2 \xrightarrow[n\to\infty]{} 0. \tag{25}$$

To prove this theorem, we show the convergence to zero of the terms R_j, $j \leq 3$.

Lemma 4 $n^{\frac{1}{2}+2p}\mathbf{E}\,(R_0^n)^2 \xrightarrow[n\to\infty]{} 0.$

Proof We have:

$$\mathbf{E}\left|\mathscr{E}_{t_{i-1}}^{t_i}\right| = 4\Phi\left(\frac{\sigma\sqrt{\Delta t_i}}{2}\right) - 2 = \sigma\sqrt{\frac{2}{\pi}}\sqrt{\Delta t_i} + (\Delta t_i)o(1),$$

$$\sigma\sqrt{\frac{2}{\pi}}n^{\frac{1}{2}}\sqrt{f'(t_{i-1})}\Delta t_i = \sigma\sqrt{\frac{2}{\pi}}\sqrt{\Delta t_i}\,\varepsilon_i,$$

where $\varepsilon_i = n^{\frac{1}{2}}\sqrt{\Delta t_i}\sqrt{f'(t_{i-1})}$ verifies $|\varepsilon_i - 1| \leq \frac{c\Delta t_i}{1-t_i}$ by virtue of Lemma 25. Hence, there is a constant $C > 0$ such that:

$$\sup_t |R_0^n(t)| \leq Ck_0 \sum_{i=1}^{n-1} \gamma_{t_{i-1}} S_{t_{i-1}}^2 \frac{(\Delta t_i)^{\frac{3}{2}}}{1-t_i}.$$

From Corollary 3 and inequalities (15–18), we obtain:

$$n^{\frac{1}{4}+p}\sqrt{\mathbf{E}\left(\sup_t |R_0^n(t)|\right)^2} \leq Cn^{\frac{1}{8}+p}\sum_{i=1}^{n-1}\frac{(\Delta t_i)^{\frac{3}{2}}}{(1-t_i)^{5/4}} \leq C\frac{n^{\frac{1}{8}+p}}{n^{\frac{1}{4}}}\log n \xrightarrow[n\to\infty]{} 0.$$

□

The Taylor formula leads to the following representation:

$$R_1^n = \sigma\left(R_{10}^n - R_{11}^n - R_{12}^n - R_{13}^n + 2R_{14}^n\right),$$

where

$$R_{10}^n(t) := \sum_{i\leq n}\gamma_{t_{i-1}} S_{t_{i-1}}^2 \int_{t_{i-1}\wedge t}^{t_i\wedge t} \mathscr{E}_{t_{i-1}}^u \frac{S_u}{S_{t_{i-1}}} dW_u,$$

$$R_{11}^n(t) := \sum_{i=1}^{n-1}\int_{t_{i-1}\wedge t}^{t_i\wedge t} \widehat{C}_{xt}(t_{i-1}, S_{t_{i-1}})(u - t_{i-1})S_u dW_u,$$

$$R_{12}^n(t) := \frac{1}{2}\sum_{i=1}^{n-1} S_{t_{i-1}}^3 \int_{t_{i-1}\wedge t}^{t_i\wedge t} \widehat{C}_{xxx}(\widetilde{t}_{i-1}, \widetilde{S}_{t_{i-1}})\left(\mathscr{E}_{t_{i-1}}^u\right)^2 \frac{S_u}{S_{t_{i-1}}} dW_u,$$

$$R_{13}^n(t) := \frac{1}{2}\sum_{i=1}^{n-1} S_{t_{i-1}} \int_{t_{i-1}\wedge t}^{t_i\wedge t} \widehat{C}_{xtt}(\widetilde{t}_{i-1}, \widetilde{S}_{t_{i-1}})(u - t_{i-1})^2 \frac{S_u}{S_{t_{i-1}}} dW_u,$$

$$R_{14}^n(t) := -\frac{1}{2}\sum_{i=1}^{n-1} S_{t_{i-1}}^2 \int_{t_{i-1}\wedge t}^{t_i\wedge t} \widehat{C}_{xxt}(\widetilde{t}_{i-1}, \widetilde{S}_{t_{i-1}})\mathscr{E}_{t_{i-1}}^u (u - t_{i-1}) \frac{S_u}{S_{t_{i-1}}} dW_u.$$

Lemma 5

$$n^{\frac{1}{2}+2p}\mathbf{E}\left(\sup_{t\in[0,1]} R^n_{10}(t)\right)^2 \xrightarrow[n\to\infty]{} 0. \tag{26}$$

Proof The Doob inequality yields $n^{\frac{1}{2}+2p}\mathbf{E}\ \sup_t(R^n_{10}(t))^2 \le 4n^{\frac{1}{2}+2p}\mathbf{E}\ (R^n_{10}(1))^2$ where the r.h.s. tends to 0 as shown below. Indeed, by the independence of the increments of the Wiener process, we write:

$$\mathbf{E}\ (R^n_{10}(1))^2 = \sigma^2 \sum_{i=1}^n \Lambda_{t_{i-1}} \int_{t_{i-1}}^{t_i} \mathbf{E}\ \left(\mathcal{E}^u_{t_{i-1}}\right)^2 \frac{S_u^2}{S^2_{t_{i-1}}} du$$

where $\Lambda_t := \mathbf{E}\ \widehat{C}^2_{xx}(t, S_t)S_t^4$. It is easy to check the following asymptotics:

$$\mathbf{E}\ \left(\mathcal{E}^u_{t_{i-1}}\right)^2 \frac{S_u^2}{S^2_{t_{i-1}}} = \sigma^2(u-t_{i-1}) + (u-t_{i-1})O(n^{-1}).$$

Therefore, $\mathbf{E}\ (R^n_{10}(1))^2 = \frac{\sigma^4}{2}\sum_{i\le n}\Lambda_{t_{i-1}}(\Delta t_i)^2(1+O(n^{-1}))$, where $\Delta t_i = g'(\theta_i)/n$ with $\theta_i \in [(i-1)/n, i/n]$. We then get that

$$n^{\frac{1}{2}+2p}\mathbf{E}\ (R^n_{10}(1))^2 = \frac{\sigma^4(1+O(n^{-1}))}{2} n^{-1} \sum_{i\le n} \Lambda_{t_{i-1}}(\Delta t_i n)\frac{\Delta t_i n^{\frac{1}{2}+2p}}{x_{i-1}-x_i}(x_{i-1}-x_i)$$

where $x_i = \rho^2_{t_i}$. So, we have:

$$n^{\frac{3}{2}+2p}\mathbf{E}\ (R^n_{10}(1))^2 = \frac{\sigma^4(1+O(n^{-1}))}{2} n^{-1} \int_0^{\rho_0^2} f_n(x)dx,$$

where

$$f_n(x) = \sum_{i=1}^n \Lambda_{t_{i-1}}(\Delta t_i n)\frac{\Delta t_i n^{\frac{1}{2}+2p}}{x_{i-1}-x_i} 1_{(x_i, x_{i-1}]}(x).$$

Let us remark the abuse of notations $\rho_0^2 = \rho_{0,n}^2$ and $t_i = t(n, x_i)$ as previously mentioned. First, let us show that f_n satisfies the dominated convergence bound condition. If $x \in (x_i, x_{i-1}]$ then from Corollary 3, we have

$$0 \le \Lambda_{t_{i-1}} \le \frac{C}{\sqrt{x_{i-1}}} e^{-x_{i-1}/4} \le \frac{C}{\sqrt{x}} e^{-x/4}.$$

Thus, from (19) we obtain that $f_n(x) \le \frac{C}{\sqrt{x}} e^{-x/4}(1+x)$.

Regarding the pointwise convergence of f_n, for a given $x \in (x_i, x_{i-1}]$, there exists $u \in [t_{i-1}, t_i)$ such that $x = \rho_u^2 \ge cn^{\frac{1}{2}}(1-u)$. It follows that not only $u \to 1$ but also $t_i, t_{i-1} \to 1$. Recall that $\Delta t_i = g'(\theta_i)n^{-1}$ where $\theta_i \in [(i-1)/n, i/n]$. Thus

$g(\theta_i) \to 1$ and $\theta_i \to 1$ since f is continuous. Therefore $\Delta t_i n \to g'(1) \in \{0, 1\}$. Moreover, note that

$$\Lambda_{t_{i-1}} = \frac{1}{x_{i-1}} \int_{-\infty}^{\infty} e^{2\sigma\sqrt{t_{i-1}}z - \sigma^2 t_{i-1}} \Upsilon_i(z)\varphi(z)dz,$$

where

$$\Upsilon_i(z) = \left(\int_{-\infty}^{\infty} h'\left(e^{\sigma\sqrt{t_{i-1}}z - \frac{\sigma^2 t_{i-1}}{2} + \sqrt{x_{i-1}}y + \frac{x_{i-1}}{2}} \right) y\varphi(y)dy \right)^2.$$

Applying the Lebesgue theorem, we deduce that $\Lambda_{t_{i-1}}$ converges to

$$\Lambda(x) := \frac{1}{x} \int_{-\infty}^{\infty} e^{2\sigma z - \sigma^2} \left(\int_{-\infty}^{\infty} h'\left(e^{\sigma z - \frac{\sigma^2}{2} + \sqrt{x}y + \frac{x}{2}} \right) y\varphi(y)dy \right)^2 \varphi(z)dz.$$

Finally, together with (20), $f_n \xrightarrow[n\to\infty]{} 0$ a.e. if $\mu > 1$ and $f_n \xrightarrow[n\to\infty]{} f_\infty$ a.e. where f_∞ is integrable if $\mu = 1$. We then apply the Lebesgue theorem to get the following limit:

$$\frac{\sigma^4(1 + O(n^{-1}))}{2} n^{-1} \int_0^{\rho_0^2} f_n(x)dx \xrightarrow[n\to\infty]{} 0. \qquad \square$$

Lemma 6 $n^{\frac{1}{2}+2p}\mathbf{E}\,(\sup_t R_{11}^n(t))^2 \xrightarrow[n\to\infty]{} 0.$

Proof Using the Doob inequality, we obtain that $\mathbf{E}\,(\sup_t R_{11}^n(t))^2 \le 4\mathbf{E}\,(R_{11}^n(1))^2$. By independence of the increments of the Wiener process, we deduce that

$$\begin{aligned} &n^{\frac{1}{2}+2p}\mathbf{E}\,(R_{11}^n(1))^2 \\ &\quad = n^{\frac{1}{2}+2p} \sum_{i=1}^{n-1} \mathbf{E}\,\widehat{C}_{xt}^2(t_{i-1}, S_{t_{i-1}})S_{t_{i-1}}^2 \int_{t_{i-1}}^{t_i} (u - t_{i-1})^2 \mathbf{E}\left(\frac{S_u}{S_{t_{i-1}}}\right)^2 du. \end{aligned}$$

It follows that

$$n^{\frac{1}{2}+2p}\mathbf{E}\,(R_{11}^n(1))^2 \le cn^{\frac{1}{2}+2p} \sum_{i=1}^{n-1} \mathbf{E}\,\widehat{C}_{xt}^2(t_{i-1}, S_{t_{i-1}})S_{t_{i-1}}^2 (\Delta t_i)^3 \le cn^{-\frac{1}{4}+2p}\log n,$$

since Corollary 5 gives

$$\mathbf{E}\,\widehat{C}_{xt}^2(t_{i-1}, S_{t_{i-1}})S_{t_{i-1}}^2 \le c\frac{n^{\frac{1}{4}} f'(t_{i-1})}{(1 - t_{i-1})^{\frac{3}{2}}},$$

where $nf'(t_{i-1})\Delta t_i$ is bounded. This completes the proof. $\square$

Lemma 7 $n^{\frac{1}{2}+2p}\mathbf{E}\,(\sup_t R_{12}^n(t))^2 \xrightarrow[n\to\infty]{} 0.$

Proof As previously, we have the Doob inequality $\mathbf{E}\,(\sup_t R^n_{12}(t))^2 \le 4\mathbf{E}\,(R^n_{12}(1))^2$ and the equality

$$4\mathbf{E}\,(R^n_{12}(1))^2 = \sum_{i=1}^{n-1}\int_{t_{i-1}}^{t_i} \mathbf{E}\left(\widehat{C}^2_{xxx}(\widetilde{t}_{i-1}, \widetilde{S}_{t_{i-1}}) S^6_{t_{i-1}} \left(1 - \frac{S_t}{S_{t_{i-1}}}\right)^4 \frac{S_t^2}{S^2_{t_{i-1}}}\right) dt.$$

From (64), there exists a constant C such that:

$$\mathbf{E}\,\widehat{C}^4_{xxx}(\widetilde{t}_{i-1}, \widetilde{S}_{t_{i-1}}) \le \frac{C}{\rho^8_{t_i}}.$$

Using the Cauchy–Schwarz inequality and (14) with $m = 8$, we deduce that

$$n^{\frac{1}{2}+2p}\mathbf{E}\,(R^n_{12}(1))^2 \le Cn^{\frac{1}{2}+2p}\sum_{i=1}^{n-1}\frac{(\Delta t_i)^3}{n(1-t_i)^2} \le C\frac{n^{2p}\log n}{n^{\frac{3}{2}}}$$

which proves the desired convergence to 0. □

Lemma 8 $n^{\frac{1}{2}+2p}\mathbf{E}\,(\sup_t R^n_{13}(t))^2 \xrightarrow[n\to\infty]{} 0.$

Proof We still consider the Doob inequality $\mathbf{E}\,(\sup_t R^n_{13}(t))^2 \le 4\mathbf{E}\,(R^n_{13}(1))^2$ and

$$4\mathbf{E}\,(R^n_{13}(1))^2 \le \sum_{i=1}^{n-1}\int_{t_{i-1}}^{t_i} \mathbf{E}\left(\widehat{C}^2_{xtt}(\widetilde{t}_{i-1}, \widetilde{S}_{t_{i-1}})(t - t_{i-1})^4 S_t^2\right) dt.$$

Moreover, using Lemma 28 and the Cauchy–Schwarz inequality, we deduce that

$$\mathbf{E}\left(\widehat{C}^2_{xtt}(\widetilde{t}_{i-1}, \widetilde{S}_{t_{i-1}}) S_t^2\right) \le \frac{c}{(1-t_i)^4}.$$

Then, we obtain

$$n^{\frac{1}{2}+2p}\mathbf{E}\,(R^n_{13}(1))^2 \le Cn^{\frac{1}{2}+2p}\sum_{i=1}^{n-1}\frac{(\Delta t_i)^5}{(1-t_i)^4} \le C\,n^{-\frac{1}{2}+2p}\log n.$$

The claim follows. □

Lemma 9 $n^{\frac{1}{2}+2p}\mathbf{E}\,(\sup_t R^n_{14}(t))^2 \xrightarrow[n\to\infty]{} 0.$

Proof We use the Doob inequality $\mathbf{E}\,(\sup_t R^n_{14}(t))^2 \le 4\mathbf{E}\,(R^n_{14}(1))^2$ and the equality

$$\begin{aligned}&4\mathbf{E}\,(R^n_{14}(1))^2\\ &= \sum_{i=1}^{n-1}\int_{t_{i-1}}^{t_i} \mathbf{E}\left(S^4_{t_{i-1}}\widehat{C}^2_{xxt}(\widetilde{t}_{i-1}, \widetilde{S}_{t_{i-1}})\left(1 - \frac{S_t}{S_{t_{i-1}}}\right)^2 (t - t_{i-1})^2 \frac{S_t^2}{S^2_{t_{i-1}}}\right) dt.\end{aligned}$$

From (65), we deduce that

$$\mathbf{E}\left(S_{t_{i-1}}^4 \widehat{C}_{xxt}^2(\widetilde{t}_{i-1}, \widetilde{S}_{t_{i-1}})\left(1-\frac{S_t}{S_{t_{i-1}}}\right)^2 \frac{S_t^2}{S_{t_{i-1}}^2}\right) \le c\frac{t-t_{i-1}}{(1-t_i)^3}.$$

Then,

$$n^{\frac{1}{2}+2p}\mathbf{E}\,(R_{14}^n(1))^2 \le cn^{\frac{1}{2}+2p}\sum_{i=1}^{n-1}\frac{(t_i-t_{i-1})^4}{(1-t_i)^3} \le c\,n^{-\frac{1}{2}+2p}\log n$$

and we conclude. □

Let us now study the residual term R_2^n. Again, the Taylor formula suggests to write that $R_2^n = R_{20}^n + \cdots + R_{24}^n$, where

$$\begin{aligned}
R_{20}^n(t) &:= \sigma k_0\sqrt{\frac{2}{\pi}}n^{\frac{1}{2}}\int_{t_n^-(t)}^t S_u^2\gamma_u\sqrt{f'(u)}du,\\
R_{21}^n(t) &:= \sigma k_0 n^{\frac{1}{2}}\sqrt{\frac{2}{\pi}}\sum_{i\in J_1^n(t)}\int_{t_{i-1}}^{t_i}\left(S_u^2\gamma_u\sqrt{f'(u)} - S_{t_{i-1}}^2\gamma_{t_{i-1}}\sqrt{f'(t_{i-1})}\right)du,\\
R_{22}^n(t) &:= k_n\sum_{i\in J_1^n(t)}\gamma_{t_{i-1}}|S_{t_i}-S_{t_{i-1}}|(S_{t_{i-1}}-S_{t_i}),\\
R_{23}^n(t) &:= k_0\sum_{i\in J_1^n(t)}\Theta_i(S_{t_i}-S_{t_{i-1}}),\\
R_{24}^n(t) &:= k_0\sum_{i\in J_1^n(t)}\Theta_i S_{t_{i-1}},\\
\Theta_i &:= \gamma_{t_{i-1}}|S_{t_i}-S_{t_{i-1}}| - |\widehat{C}_x(t_i,S_{t_i}) - \widehat{C}_x(t_{i-1},S_{t_{i-1}}) + \Delta K_{t_i}^n|.
\end{aligned} \tag{27}$$

Lemma 10 $n^{\frac{1}{2}+2p}\mathbf{E}\,(R_{20}^n(1))^2 \xrightarrow[n\to\infty]{} 0.$

Proof We have:

$$n^{\frac{1}{2}+2p}\mathbf{E}\,(R_{20}^n(1))^2 = c\,n^{\frac{3}{2}+2p}\mathbf{E}\int_{[t_{n-1},1]^2} S_u^2\gamma_u S_v^2\gamma_v\sqrt{f'(u)}\sqrt{f'(v)}dudv.$$

We use the Cauchy–Schwarz inequality, Inequalities (3) and (17). From the explicit formula of f', we obtain that

$$n^{\frac{1}{2}+2p}\mathbf{E}\,(R_{20}^n(1))^2 \le c\,n^{1+2p}\int_{[t_{n-1},1]^2}\frac{dudv}{(1-u)^{5/8-3/(8\mu)}(1-v)^{5/8-3/(8\mu)}},$$

$$\leq c\,\frac{n^{1+2p}}{n^{3/4+3/(4\mu)}}.$$

Since $\mu \in [1,2]$,

$$\frac{3}{4}+\frac{3}{4\mu}-(1+2p)=\frac{-3\mu^2+5\mu+3}{4\mu(\mu+1)}>0$$

and the claim follows. □

Lemma 11 $n^{\frac{1}{2}+2p}\mathbf{E}\,(\sup_t R_{21}^n(t))^2 \xrightarrow[n\to\infty]{} 0.$

Proof Let us consider $\Psi(t,x) := x^2\widehat{C}_{xx}(t,x)\sqrt{f'(t)}$. The Ito formula yields

$$\begin{aligned}\Psi(t,S_t) = {} & \Psi(t_{i-1},S_{t_{i-1}}) + \int_{t_{i-1}}^{t}\frac{\partial\Psi}{\partial x}(u,S_u)\sigma S_u dW_u + \int_{t_{i-1}}^{t}\frac{\partial\Psi}{\partial t}(u,S_u)du \\ & + \frac{1}{2}\int_{t_{i-1}}^{t}\frac{\partial^2\Psi}{\partial x^2}(u,S_u)\sigma^2 S_u^2 du,\end{aligned}$$

where

$$\begin{aligned}\frac{\partial\Psi}{\partial t}(t,x) &= x^2\left[\widehat{C}_{xxt}(t,x)\sqrt{f'(t)}+\widehat{C}_{xx}(t,x)\frac{f''(t)}{2\sqrt{f'(t)}}\right],\\ \frac{\partial\Psi}{\partial x}(t,x) &= \left[2x\widehat{C}_{xx}(t,x)+x^2\widehat{C}_{xxx}(t,x)\right]\sqrt{f'(t)},\\ \frac{\partial^2\Psi}{\partial x^2}(t,x) &= \left[2\widehat{C}_{xx}(t,x)+4x\widehat{C}_{xxx}(t,x)+x^2\widehat{C}_{xxxx}(t,x)\right]\sqrt{f'(t)}.\end{aligned}$$

If we set $X_t = S_t^2\widehat{C}_{xx}(t,x)\sqrt{f'(t)}$ then $dX_t = \zeta_t dt + \beta_t dW_t$, where

$$\zeta_t = \frac{\partial\Psi}{\partial t}(t,S_t)+\frac{1}{2}\frac{\partial^2\Psi}{\partial x^2}(t,S_t)\sigma^2 S_t^2, \qquad \beta_t = \frac{\partial\Psi}{\partial x}(t,S_t)\sigma S_t.$$

We write $n^{\frac{1}{4}+p}R_{21}^n(t) = A_t^n + B_t^n$ with

$$A_t^n := \sigma k_0 n^{\frac{3}{4}+p}\sqrt{\frac{2}{\pi}}\sum_{i\in J_1^n(t)}\int_{t_{i-1}}^{t_i}\left(\int_{t_{i-1}}^{t}\beta_u dW_u\right)dt,$$

$$B_t^n := \sigma k_0 n^{\frac{3}{4}+p}\sqrt{\frac{2}{\pi}}\sum_{i\in J_1^n(t)}\int_{t_{i-1}}^{t_i}\left(\int_{t_{i-1}}^{t}\zeta_u du\right)dt.$$

From (59), there exists a constant C such that

$$\mathbf{E}\,\beta_t^2 \le c\left(\mathbf{E}\,S_t^4\delta_t^2 + \mathbf{E}\,S_t^6\widehat{C}_{xxx}^2(t,S_t)\right)f'(t) \le \frac{Cf'(t)^{\frac{1}{4}}}{n^{\frac{3}{4}}(1-t)^{\frac{3}{2}}}.$$

Using Assumption (A1), we claim that there exists a constant $\widetilde{c}$ such that

$$\frac{|f''(t)|}{\sqrt{f'(t)}} = \frac{\widetilde{c}}{(1-t)^{\frac{3}{2}-1/(2/\mu)}}.$$

Thus, using (58)–(63), we find some constant C such that the following inequality holds:

$$\mathbf{E}\,\zeta_t^2 \le \frac{c\,(1-t)^{3/(4\mu)}}{n^{\frac{1}{4}}(1-t)^{13/4}} + \frac{c\,(1-t)^{3/(4\mu)}}{n^{3/4}(1-t)^{7/4}} + \frac{c}{n^{5/4}(1-t)^{9/4+1/(4\mu)}}. \tag{28}$$

By means of the stochastic Fubini theorem, we obtain that

$$A_t^n = \sigma k_0 n^{\frac{3}{4}+2p}\sqrt{\frac{2}{\pi}}\sum_{i\in J_1^n(t)}\int_{t_{i-1}}^{t_i}(t_i-u)\beta_u dW_u.$$

Since the Doob inequality $\mathbf{E}\,(\sup_t A_t^n)^2 \le 4\mathbf{E}\,(A_1^n)^2$ holds, it suffices to estimate $\mathbf{E}\,(A_1^n)^2$. From the boundedness of $\sqrt{(t_i-u)/(1-u)}$ and $f'(u)(t_i-u)n$ on $u \in [t_{i-1},t_i)$, we obtain the following estimates:

$$\begin{aligned}
\mathbf{E}\,\left(A_1^n\right)^2 &\le cn^{\frac{3}{2}+2p}\sum_{i=1}^{n-1}\int_{t_{i-1}}^{t_i}(t_i-u)^2\mathbf{E}\,\beta_u^2 du \\
&\le cn^{\frac{3}{2}+2p}\sum_{i=1}^{n-1}\int_{t_{i-1}}^{t_i}\frac{(t_i-u)^2 f'(u)^{\frac{1}{4}}}{n^{3/4}(1-u)^{\frac{3}{2}}}du, \\
&\le \frac{cn^{2p}}{n^{\frac{1}{4}}}\sum_{i=1}^{n-1}\int_{t_{i-1}}^{t_i}\frac{(t_i-u)}{(1-u)^{\frac{3}{2}}}du \le c\,\frac{n^{2p}\log n}{n^{3/4}} \xrightarrow[n\to\infty]{} 0.
\end{aligned}$$

We conclude from here that $\mathbf{E}\,(\sup_t A_t^n)^2 \xrightarrow[n\to\infty]{} 0$.

Secondly, we write:

$$B_t^n = cn^{3/4+p}\sum_{i\in J_1^n(t)}\int_{t_{i-1}}^{t_i}\zeta_u\int_{t_{i-1}}^{t_i}\mathbf{1}_{t\ge u}dt\,du = cn^{3/4+p}\sum_{i\in J_1^n(t)}\int_{t_{i-1}}^{t_i}(t_i-u)\zeta_u du.$$

Then,

$$\sup_t |B_t^n| \le cn^{3/4+p}\sum_{i=1}^{n-1}\int_{t_{i-1}}^{t_i}(t_i-u)|\zeta_u|du.$$

Thus, there exists a constant c such that $\mathbf{E}\ \sup_t |B_t^n|^2 \le c\, n^{\frac{3}{2}+2p}\Upsilon^n$, where

$$\Upsilon^n = \mathbf{E}\left(\int_0^1 \sum_{i=1}^{n-1}(t_i - u)|\zeta_u|\mathbf{1}_{(t_{i-1},t_i]}(u)du\right)^2,$$

$$= \mathbf{E}\int_0^1\int_0^1 \sum_{i,\,j=1}^{n-1}(t_i - u)(t_j - v)|\zeta_u||\zeta_v|\mathbf{1}_{(t_{i-1},t_i]}(u)\mathbf{1}_{(t_{j-1},t_j]}(v)du\,dv.$$

Using the Cauchy–Schwarz inequality and (28), we can bound Υ^n:

$$\Upsilon^n \le \int_0^1\int_0^1 \sum_{i,\,j=1}^{n-1}(t_i - u)(t_j - v)\left(\mathbf{E}\,\zeta_u^2\right)^{\frac{1}{2}}\left(\mathbf{E}\,\zeta_v^2\right)^{\frac{1}{2}}\mathbf{1}_{(t_{i-1},t_i]}(u)\mathbf{1}_{(t_{j-1},t_j]}(v)du\,dv,$$

$$\le \left(\int_0^1 \sum_{i=1}^{n-1}(t_i - u)\left(\mathbf{E}\,\zeta_u^2\right)^{\frac{1}{2}}\mathbf{1}_{(t_{i-1},t_i]}(u)du\right)^2 \le c\,(\Upsilon^{1n} + \Upsilon^{2n} + \Upsilon^{3n}),$$

where

$$\Upsilon^{1n} \le \left(\sum_{i\le n-1}\frac{(\Delta t_i)^2}{(1-t_i)}\frac{1}{n^{1/8}(1-t)^{5/8-3/(8\mu)}}\right)^2 \le \frac{c\,\log n}{n^{1+3/(4\mu)}}. \tag{29}$$

In a same way, we obtain the following inequalities:

$$\Upsilon^{2n} \le \left(\sum_{i\le n-1}\frac{(\Delta t_i)^2}{n^{3/8}(1-t_i)^{7/8}}\right)^2 \le \frac{C}{n^{\frac{5}{4}}}, \tag{30}$$

$$\Upsilon^{3n} \le \left(\sum_{i\le n-1}\frac{(\Delta t_i)^2}{n^{5/8}(1-t_i)^{1+(1/8+1/(8\mu))}}\right)^2 \le \frac{c\,\log n}{n^{7/2+1/(4\mu)}}. \tag{31}$$

Then, from inequalities (29), (30), and (31) we deduce that

$$\mathbf{E}\ \sup_t |B_t^n|^2 \le \frac{c\,n^{\frac{3}{2}+2p}\log n}{n^{1+3/(4\mu)}} \le \frac{c\,\log n}{n^{3/(4\mu)-\frac{1}{2}-2p}}$$

where

$$3/(4\mu) - \frac{1}{2} - 2p = \frac{-4\mu^2 + 3\mu + 3}{4\mu(\mu+1)}.$$

Assumption (A1) yields $-4\mu^2 + 3\mu + 3 > 0$. Hence the result follows. □

Lemma 12 $n^{\frac{1}{2}+2p}\mathbf{E}\,(\sup_t R_{22}^n(t))^2 \xrightarrow[n\to\infty]{} 0.$

Proof We write $-R^n_{22}(t) = k_n \sum_{i\in J^n_1(t)} \gamma_{t_{i-1}} S^2_{t_{i-1}} \{\mathscr{E}^{t_i}_{t_{i-1}}\}^2_s = U^n(t) + V^n(t)$ where U^n is a martingale defined by the formula

$$U^n(t) := k_0 \sum_{i\in J^n_1(t)} \gamma_{t_{i-1}} S^2_{t_{i-1}} \left(\{\mathscr{E}^{t_i}_{t_{i-1}}\}^2_s - \mathbf{E}\,\{\mathscr{E}^{t_i}_{t_{i-1}}\}^2_s\right),$$

and $V^n(t) := k_0 \sum_{i\in J^n_1(t)} \gamma_{t_{i-1}} S^2_{t_{i-1}} \mathbf{E}\,\{\mathscr{E}^{t_i}_{t_{i-1}}\}^2_s$. Recall that from Lemma 1

$$\mathbf{E}\,\{\mathscr{E}^{t_i}_{t_{i-1}}\}^2_s = k(\Delta t_j)^{\frac{3}{2}}\left(1 + o(n^{-\frac{1}{4}})\right).$$

We deduce that for n large enough, $0 \le \mathbf{E}\,\{\mathscr{E}^{t_i}_{t_{i-1}}\}^2_s \le c(\Delta t_i)^{\frac{3}{2}}$. Using the Doob inequality $\mathbf{E}\,(\sup_t U^n(t))^2 \le 4\mathbf{E}\,(U^n(1))^2$, it suffices to estimate $\mathbf{E}\,(U^n(1))^2$. The independence of the increments of the Brownian motion implies the equality

$$\mathbf{E}\,\left(U^n(1)\right)^2 = k_0^2 \sum_{i=1}^{n-1} \mathbf{E}\,\widehat{C}^2_{xx}(t_{i-1}, S_{t_{i-1}}) S^4_{t_{i-1}} \mathbf{E}\,\left(\{\mathscr{E}^{t_i}_{t_{i-1}}\}^2_s - \mathbf{E}\,\{\mathscr{E}^{t_i}_{t_{i-1}}\}^2_s\right)^2.$$

Then, there exists a constant C such that

$$n^{\frac{1}{2}+2p}\mathbf{E}\,\left(U^n(1)\right)^2 \le \frac{C\,n^{2p}}{n^{\frac{1}{4}}} \xrightarrow[n\to\infty]{} 0.$$

At last, for n large enough, $\mathbf{E}\,\{\mathscr{E}^{t_i}_{t_{i-1}}\}^2_s \ge 0$. Hence, $0 \le \sup_t V^n(t) \le N^n(1)$. In order to prove that $n^{\frac{1}{2}+2p}\mathbf{E}\,V^n(1)^2 \xrightarrow[n\to\infty]{} 0$, we first analyze the following sum

$$n^{\frac{1}{2}+2p} k_0^2 \sum_{i=1}^{n-1} \mathbf{E}\,\widehat{C}^2_{xx}(t_{i-1}, S_{t_{i-1}}) S^4_{t_{i-1}} (\mathbf{E}\,\{\mathscr{E}^{t_i}_{t_{i-1}}\}^2_s)^2 \le \frac{c\,n^{2p}}{n^{7/4}} \xrightarrow[n\to\infty]{} 0.$$

Using the Cauchy–Schwarz inequality, we also have that

$$n^{\frac{1}{2}+2p} \sum_{t_i<t_j\le t_{n-1}} \mathbf{E}\,\gamma_{t_{i-1}} S^2_{t_{i-1}} \gamma_{t_{j-1}} S^2_{t_{j-1}} \mathbf{E}\,\{\mathscr{E}^{t_i}_{t_{i-1}}\}^2_s \mathbf{E}\,\{\mathscr{E}^{t_j}_{t_{j-1}}\}^2_s \le \frac{c\,n^{2p}}{n} \xrightarrow[n\to\infty]{} 0.$$

We obtain that $n^{\frac{1}{2}+2p}\mathbf{E}\;V^n(1)^2 \xrightarrow[n\to\infty]{} 0$ and, finally, $n^{\frac{1}{2}+2p}\mathbf{E}\;(\sup_t R^n_{22}(t))^2 \xrightarrow[n\to\infty]{} 0$. □

Lemma 13 $n^{\frac{1}{2}+2p}\mathbf{E}\,(\sup_t R^n_{23}(t))^2 \xrightarrow[n\to\infty]{} 0$.

Proof We write $R^n_{23}(t) = R^n_{231}(t) + R^n_{232}(t)$, where

$$R^n_{231}(t) := k_0 \sum_{i\in J^n_1(t)} \Theta^1_i (S_{t_i} - S_{t_{i-1}}),$$

$$R^n_{232}(t) := k_0 \sum_{i \in J^n_1(t)} \left(\Theta_i - \Theta^1_i\right)(S_{t_i} - S_{t_{i-1}})$$

with $\Theta^1_i := \gamma_{t_{i-1}}|S_{t_i} - S_{t_{i-1}}| - |\widehat{C}_x(t_i, S_{t_i}) - \widehat{C}_x(t_{i-1}, S_{t_{i-1}})|$.

We note that $\sup_t |R^n_{231}(t)|$ is bounded by

$$k_0 \sum_{i=1}^{n-1} \left|\widehat{C}_x(t_i, S_{t_i}) - \widehat{C}_x(t_{i-1}, S_{t_{i-1}}) - \gamma_{t_{i-1}}(S_{t_i} - S_{t_{i-1}})\right| |S_{t_i} - S_{t_{i-1}}|.$$

Applying the Taylor formula to the difference $\widehat{C}_x(t_i, S_{t_i}) - \widehat{C}_x(t_{i-1}, S_{t_{i-1}})$ it is easy to see that it sufficient to estimate the sums (32), $\cdots$,(35). For the first one we have

$$n^{\frac{1}{4}+p} \left\| k_0 \sum_{i=1}^{n-1} \widehat{C}_{xt}(t_{i-1}, S_{t_{i-1}})(\Delta t_i)(S_{t_i} - S_{t_{i-1}}) \right\|_2 \le C \frac{n^p}{n^{1/8}} \to 0. \tag{32}$$

Indeed, from Corollary 5, we deduce that

$$\mathbf{E}\, \widehat{C}^2_{xt}(t_{i-1}, S_{t_{i-1}})(\Delta t_i)^2 (S_{t_i} - S_{t_{i-1}})^2 \le C \frac{(\Delta t_i)^3 n^{\frac{1}{4}} f'(t_{i-1})^{\frac{1}{4}}}{(1-t_i)^{\frac{3}{2}}}.$$

The second one verifies

$$n^{\frac{1}{4}+p} \left\| \sum_{i=1}^{n-1} \widehat{C}_{xxx}(\widetilde{t}_{i-1}, \widetilde{S}\_t_{i-1})(S_{t_i} - S_{t_{i-1}})^3 \right\|_2 \le C \frac{n^p \log n}{n^{\frac{1}{2}}} \to 0. \tag{33}$$

Thirdly, from (65), we deduce that

$$\mathbf{E}\, \widehat{C}^2_{xxt}(\widetilde{t}_{i-1}, \widetilde{S}_{t_{i-1}})(S_{t_i} - S_{t_{i-1}})^4 (\Delta t_i)^2 \le \frac{C(\Delta t_i)^4}{(1-t_i)^3}$$

and it follows that

$$n^{\frac{1}{4}+p} \left\| \sum_{i=1}^{n-1} \widehat{C}_{xxt}(\widetilde{t}_{i-1}, \widetilde{S}_{t_{i-1}})(S_{t_i} - S_{t_{i-1}})^2 \Delta t_i \right\|_2 \le C \frac{n^p \log n}{n^{\frac{1}{4}}} \xrightarrow[n\to\infty]{} 0. \tag{34}$$

Finally, from Lemma 28, we get that

$$\mathbf{E}\, \widehat{C}^2_{xtt}(\widetilde{t}_{i-1}, \widetilde{S}_{t_{i-1}})(S_{t_i} - S_{t_{i-1}})^2 (\Delta t_i)^4 \le \frac{C(\Delta t_i)^5}{(1-t_i)^4}$$

and

$$n^{\frac{1}{4}+p} \left\| \sum_{i=1}^{n-1} \widehat{C}_{xtt}(\widetilde{t}_{i-1}, \widetilde{S}_{t_{i-1}})(S_{t_i} - S_{t_{i-1}})(\Delta t_i)^2 \right\|_2 \le C \frac{n^p \log n}{n^{\frac{1}{4}}} \xrightarrow[n\to\infty]{} 0. \tag{35}$$

From above, we can conclude about that $n^{\frac{1}{2}+2p}\mathbf{E}\,(\sup_t R^n_{231}(t))^2 \xrightarrow[n\to\infty]{} 0$.

As for $R^n_{232}(t)$, we use the inequality $|\Theta_i - \Theta_i^1| \le |\Delta K^n_{t_i}|$ and we deduce from Definition (8) the bound

$$|R^n_{232}(t)| \le c \sup_i \left|\Delta S_{t_i}\right| \int_0^{t_{n-1}} |\widehat{C}_{xt}(t_u, S_u)| du,$$

with

$$|\widehat{C}_{xt}(t_u, S_u)| \le \frac{c}{\sqrt{S_u}(1-u)}$$

so that $|R^n_{232}(t)| \le c \log(n) \sup_t S_t^{-\frac{1}{2}} \sup_i |\Delta S_{t_i}|$.

Using the Cauchy–Schwarz inequality, the boundedness of $\mathbf{E}\,\sup_t S_t^{-2}$ implies that

$$\mathbf{E}\,\sup_t \left(R^n_{232}(t)\right)^2 \le C \log^2(n) \sqrt{\mathbf{E}\,\sup_i \left(\Delta S_{t_i}\right)^4}.$$

Moreover,

$$\begin{aligned}\mathbf{E}\,\sup_i \left(\Delta S_{t_i}\right)^4 &\le n^{-\frac{3}{2}} + \mathbf{E}\,\sup_i \left(\Delta S_{t_i}\right)^4 \mathbf{1}_{\sup_i (\Delta S_{t_i})^4 \ge n^{-\frac{3}{2}}} \\ &\le n^{-\frac{3}{2}} + C\sqrt{\mathbb{P}\left(\sup_i \left(\Delta S_{t_i}\right)^4 \ge n^{-\frac{3}{2}}\right)}.\end{aligned}$$

By virtue of the Bienaymé–Tchebytchev inequality $\mathbb{P}(|X| \ge k) \le k^{-8}\mathbf{E}\,X^8$,

$$\mathbb{P}\left(\sup_i \left(\Delta S_{t_i}\right)^4 \ge n^{-\frac{3}{2}}\right) \le n^{12} \sum_i \mathbf{E}\,\left(\Delta S_{t_i}\right)^{32} \le C n^{-3}.$$

We deduce that $\mathbf{E}\,\sup_i (\Delta S_{t_i})^4 \le C n^{-\frac{3}{2}}$ and, finally, $\mathbf{E}\,\sup_t (R^n_{232}(t))^2 \le C n^{-3/4} \log^2(n)$ and we obtain the claim of the lemma. □

Lemma 14 *We have* $n^{\frac{1}{2}+2p}\mathbf{E}\,(\sup_t R^n_{24}(t))^2 \xrightarrow[n\to\infty]{} 0$.

Proof Let us notice that $\sup_t |R^n_{24}(t)|$ is bounded by the random variable

$$k_0 \sum_{i=1}^{n-1} \left|\widehat{C}_x(t_i, S_{t_i}) - \widehat{C}_x(t_{i-1}, S_{t_{i-1}}) + \Delta K^n_{t_i} - \widehat{C}_{xx}(t_{i-1}, S_{t_{i-1}})\left(S_{t_i} - S_{t_{i-1}}\right)\right| S_{t_{i-1}}.$$

Using the Ito formula for the increments $\widehat{C}_x(t_i, S_{t_i}) - \widehat{C}_x(t_{i-1}, S_{t_{i-1}})$, we obtain that

$$\sup_t |R^n_{24}(t)| \le k_0 \sum_{i=1}^{n-1} S_{t_{i-1}} \left| \int_{t_{i-1}}^{t_i} \sigma S_u \left[\widehat{C}_{xx}(u, S_u) - \widehat{C}_{xx}(t_{i-1}, S_{t_{i-1}})\right] dW_u\right.$$

$$+\frac{1}{2}\int_{t_{i-1}}^{t_i}\sigma^2 S_u^2 \widehat{C}_{xxx}(u,S_u)du\Bigg|. \tag{36}$$

Thus $n^{\frac{1}{4}+p}\|\sup_t R_{24}^n(t)\|_2 \le T_n^1 + T_n^2$, where

$$T_n^1 = \sigma k_0 n^{\frac{1}{4}+p}\sum_{i=1}^{n-1}\left(\int_{t_{i-1}}^{t_i}\mathbf{E}\,S_{t_{i-1}}^2 S_u^2\left(\gamma_u-\gamma_{t_{i-1}}\right)^2 du\right)^{\frac{1}{2}}$$

and

$$T_n^2 = \frac{k_0 n^{\frac{1}{4}+p}\sigma^4}{4}\sum_{i=1}^{n-1}(\Delta t_i)^{\frac{1}{2}}\left(\int_{t_{i-1}}^{t_i}\mathbf{E}\,S_{t_{i-1}}^2 S_u^4 \widehat{C}_{xxx}^2(u,S_u)du\right)^{\frac{1}{2}}.$$

We first prove that $T_n^1 \xrightarrow[n\to\infty]{} 0$. Using the Taylor formula, we get that

$$\begin{aligned}\gamma_u-\gamma_{t_{i-1}} &= \gamma_u-\widehat{C}_{xx}(u,S_{t_{i-1}})+\widehat{C}_{xx}(u,S_{t_{i-1}})-\gamma_{t_{i-1}}\\ &= \widehat{C}_{xxx}(u,S_{t_{i-1}})(S_u-S_{t_{i-1}})+\frac{1}{2}\widehat{C}_{xxxx}(u,\widetilde{S}_{t_{i-1}})(S_u-S_{t_{i-1}})^2\\ &\quad+\widehat{C}_{xxt}(\widetilde{t}_{i-1},S_{t_{i-1}})(u-t_{i-1}).\end{aligned}$$

Using estimations from the Appendix, we then obtain that

$$\mathbf{E}\,S_{t_{i-1}}^2 S_u^2\left(\gamma_u-\gamma_{t_{i-1}}\right))^2 \le \frac{c\Delta t_i}{n^{\frac{7}{8}}(1-t_i)^{\frac{7}{4}}}+\frac{c(\Delta t_i)^2}{n^{\frac{3}{2}}(1-t_i)^3 f'(t_i)^{\frac{3}{2}}}+\frac{c(\Delta t_i)^2}{n^{3/4}(1-t_i)^{\frac{11}{4}}}.$$

The last estimate follows from Corollary 63. Indeed, the proof is the same since $\rho_{\widetilde{t}_{i-1}} \le \rho_{t_{i-1}}$. We can therefore deduce that $T_n^1 \xrightarrow[n\to\infty]{} 0$.

We then prove that $T_n^2 \xrightarrow[n\to\infty]{} 0$. We deduce from the Appendix the following inequality:

$$\mathbf{E}\,S_{t_{i-1}}^2 S_u^4 \widehat{C}_{xxx}^2(u,S_u) \le \frac{c}{n^{7/8}(1-t_i)^{7/4}}.$$

It suffices to obtain the convergence

$$n^{\frac{1}{4}+p}\sum_{i=1}^{n-1}\frac{\Delta t_i}{n^{7/16}(1-t_i)^{7/8}} \le \frac{c\,n^p}{n^{3/16}} \xrightarrow[n\to\infty]{} 0$$

and to conclude.

The last lemma completes the proof of Theorem 3. □

5.3 Step 3: Asymptotic Distribution

From the previous subsection, it turns out that the deviation between the "real world" terminal portfolio and the payoff $h(S_1)$ is essentially composed of a martingale as $n \to \infty$. To study the asymptotic distribution of $n^{\frac{1}{4}+p} M_1^n$, we consider it as terminal values of the following sequence of martingales $(N_j^n)_{j=0,\cdots,n}$ with respect to the filtration $\mathscr{F}^n = (\mathscr{F}_{t_i})_i$: $N_j^n := n^{\frac{1}{4}+p} M_j^n = \sum_{i=1}^j (\chi_i + \vartheta_i)$, where

$$\chi_i := k_0 n^{1/4+p} \gamma_{t_{i-1}} S_{t_{i-1}}^2 \left[\mathscr{E}_{t_{i-1}}^{t_i}\right]_c,$$

$$\vartheta_i := k_0 n^{1/4+p} K_{t_{i-1}}^n (S_{t_i} - S_{t_{i-1}}).$$

We achieve the proof of Theorem 2 by means of results in [6] recalled by Theorem 4 in the Appendix. We need some more results.

Lemma 15 *The sequence of martingales $(N_i^n)_{i=0,\cdots,n}$ satisfies the following property*:

$$\text{for all } \varepsilon > 0, \quad \sum_i \mathbf{E}\left((\chi_i + \vartheta_i)^2 \mathbf{1}_{|\chi_i+\vartheta_i|>\varepsilon} | \mathscr{F}_{t_{i-1}}\right) \xrightarrow[n\to\infty]{L^1} 0. \tag{37}$$

Proof We use the inequality $(\chi_i + \vartheta_i)^2 \le 2\chi_i^2 + 2\vartheta_i^2$ and we deduce the convergence in L^1. First, let us show that $\mathbf{E}\ (\vartheta_i^2 \mathbf{1}_{|\chi_i+\vartheta_i|>\varepsilon}) \xrightarrow[n\to\infty]{} 0$. By virtue of the Markov inequality, we obtain that

$$\mathbf{E}\,\vartheta_i^2 \mathbf{1}_{|\vartheta_i|>\varepsilon/2} \le \sqrt{\mathbf{E}\,\vartheta_i^4}\sqrt{\mathbb{P}(|\vartheta_i| > \varepsilon/2)} \le C\varepsilon^{-6}\sqrt{\mathbf{E}\,\vartheta_i^4}\sqrt{\mathbf{E}\,\vartheta_i^{12}}.$$

By independance, we have:

$$\mathbf{E}\,\vartheta_i^4 = k_0^2 n^{1+4p} \mathbf{E}\,(K_{t_{i-1}}^n)^4 S_{t_{i-1}}^4 \mathbf{E}\,(\mathscr{E}_{t_{i-1}}^{t_i})^4,$$

$$\mathbf{E}\,\vartheta_i^{12} = k_0^2 n^{3+12p} \mathbf{E}\,(K_{t_{i-1}}^n)^{12} S_{t_{i-1}}^{12} \mathbf{E}\,(\mathscr{E}_{t_{i-1}}^{t_i})^{12}.$$

By virtue of Lemma 22 there exists a constant C such that

$$|K_{t_{i-1}}^n|^4 \le C \sup_{0\le u\le T} S_u^{-2} \left(\int_0^{t_{n-1}} \frac{du}{1-u}\right)^4 \le C \sup_{0\le u\le T} S_u^{-2} \log^4 n.$$

We deduce that

$$\begin{aligned} \mathbf{E}\,\vartheta_i^4 &\le C \log^4(n) n^{4p-1}, \\ \mathbf{E}, \vartheta_i^{12} &\le C \log^{12}(n) n^{12p-3}, \\ \sum_i \mathbf{E}\,\vartheta_i^{12} &\le C \log^{12}(n) n^{12p-2} \to 0. \end{aligned} \tag{38}$$

Since $p < 1/8$, we infer that

$$\sum_i \mathbf{E}\,\vartheta_i^2 \mathbf{1}_{|\vartheta_i|>\varepsilon/2} \le C\,\varepsilon^{-6} n^{8p-1} \log^8 n \sum_{i\le n} n^{-1} \le C\,\varepsilon^{-6} n^{8p-1} \log^8(n) \xrightarrow[n\to\infty]{} 0.$$

Let us study $\mathbf{E}\,(\vartheta_i^2 \mathbf{1}_{|\chi_i|>\varepsilon/2})$. Again,

$$\mathbf{E}\,\vartheta_i^2 \mathbf{1}_{|\chi_i|>\varepsilon/2} \le \sqrt{\mathbf{E}\,\vartheta_i^4}\sqrt{\mathbb{P}(|\chi_i|>\varepsilon/2)} \le C\varepsilon^{-2}\sqrt{\mathbf{E}\,\vartheta_i^4}\sqrt{\mathbf{E}\,\chi_i^4}.$$

Again by independence, $\mathbf{E}\,\chi_i^4 = k_0^4 n^{1+4p} \mathbf{E}\,\gamma_{t_{i-1}}^4 S_{t_{i-1}}^8 \mathbf{E}\,[\mathscr{E}_{t_{i-1}}^{t_i}]_c^4$. We easily deduce from Lemma 1 the inequality $\mathbf{E}\,[\mathscr{E}_{t_{i-1}}^{t_i}]_c^4 \le C(\Delta t_i)^2$. Using the inequality (58) we obtain that

$$\mathbf{E}\,\chi_i^4 \le C\,n^{1+4p} \frac{(\Delta t_i)^2}{(n^{1/4}\sqrt{1-t_{i-1}})^3} \le C\,n^{4p-1/4} \tag{39}$$

$$\sum_i \mathbf{E}\,\chi_i^4 \le C\,n^{1+4p} \sum_i \frac{(\Delta t_i)^2}{(n^{1/4}\sqrt{1-t_{i-1}})^3} \le C\,\log(n) n^{4p-1/4}. \tag{40}$$

Since $p < \frac{1}{16} < \frac{3}{32}$, then

$$\begin{aligned}\sum_i \mathbf{E}\,\vartheta_i^2 \mathbf{1}_{|\chi_i|>\varepsilon/2} &\le C\,\varepsilon^{-2} n^{4p} \log^2(n) \sum_i \frac{\Delta t_i}{n^{3/8}(1-t_{i-1})^{3/4}}.\\ &\le C\,\varepsilon^{-2} n^{4p-3/8} \log^2(n) \sum_i \frac{\Delta t_i}{1-t_{i-1}}\\ &\le C\varepsilon^{-2}\, n^{-3/8+4p} \log^3(n) \xrightarrow[n\to\infty]{} 0.\end{aligned}$$

From the inequality $\mathbf{1}_{|\chi_i+\vartheta_i|>\varepsilon} \le \mathbf{1}_{|\chi_i|>\varepsilon/2} + \mathbf{1}_{|\vartheta_i|>\varepsilon/2}$ we then deduce that

$$\sum_i \mathbf{E}\,\vartheta_i^2 \mathbf{1}_{|\chi_i+\vartheta_i|>\varepsilon} \xrightarrow[n\to\infty]{} 0.$$

Second, let us show that $\mathbf{E}\,(\chi_i^2 \mathbf{1}_{|\chi_i+\vartheta_i|>\varepsilon}) \xrightarrow[n\to\infty]{} 0$. In the same way, we have:

$$\mathbf{E}\,\left(\chi_i^2 \mathbf{1}_{|\vartheta_i|>\varepsilon/2}\right) \le \sqrt{\mathbf{E}\,\chi_i^4}\sqrt{\mathbb{P}(|\vartheta_i|>\varepsilon/2)} \le C\varepsilon^{-6}\sqrt{\mathbf{E}\,\chi_i^4}\sqrt{\mathbf{E}\,\vartheta_i^{12}}.$$

From (39) we have $\mathbf{E}\,\chi_i^4 \le C\,n^{4p-1/4}$. Thus, using $p < \frac{1}{16} < \frac{5}{64}$,

$$\sum_i \mathbf{E}\,\left(\chi_i^2 \mathbf{1}_{|\vartheta_i|>\varepsilon/2}\right) \le C\varepsilon^{-6}\, n^{8p-5/8} \log^6(n) \xrightarrow[n\to\infty]{} 0. \tag{41}$$

Let us now study $\mathbf{E}\,(\chi_i^2 \mathbf{1}_{|\chi_i|>\varepsilon/2})$.

$$\mathbf{E}\,\left(\chi_i^2 \mathbf{1}_{|\chi_i|>\varepsilon/2}\right) \le \sqrt{\mathbf{E}\,\chi_i^4}\sqrt{\mathbb{P}(|\chi_i|>\varepsilon/2)} \le C\varepsilon^{-2}\mathbf{E}\,\chi_i^4.$$

Using the bound (39), we obtain $\sum_i \mathbf{E}\,(\chi_i^2 \mathbf{1}_{|\chi_i|>\varepsilon/2}) \le C n^{4p-1/4} \xrightarrow[n\to\infty]{} 0.$

This proves the lemma. □

Inspecting the proof above, we get the following:

Corollary 1 *Foe the sequence of martingales* $(N_i^n)_{i=0,\cdots,n}$ *we have*:

$$\max_i \mathbf{E}\,\left((\chi_i+\vartheta_i)^2|\mathscr{F}_{t_{i-1}}\right) \xrightarrow[n\to\infty]{P} 0.$$

Proof Indeed, by virtue of Inequalities (38) and (40), for a given $\varepsilon > 0$

$$\begin{aligned}
&P\Big(\max_i E\big((\chi_i+\vartheta_i)^2|\mathscr{F}_{t_{i-1}}\big) > \varepsilon\Big)\\
&\le P\Big(2\max_i E\big(\chi_i^2|\mathscr{F}_{t_{i-1}}\big) + 2\max_i E\left(\vartheta_i^2|\mathscr{F}_{t_{i-1}}\right) > \varepsilon\Big)\\
&\le P\Big(\max_i E\big(\chi_i^2|\mathscr{F}_{t_{i-1}}\big) > \varepsilon/4\Big) + P\Big(\max_i E\big(\vartheta_i^2|\mathscr{F}_{t_{i-1}}\big) > \varepsilon/4\Big)\\
&\le C\varepsilon^{-2}\sum_i E\chi_i^4 + C\varepsilon^{-6}\sum_i E\vartheta_i^{12} \to 0.
\end{aligned}$$

□

Lemma 16 *The sequence of martingales* $(M_i^n)_{i=0,\cdots,n}$ *satisfies the following convergence*

$$V_n^2 := \sum_i \mathbf{E}\,\left((\chi_i+\vartheta_i)^2|\mathscr{F}_{t_{i-1}}\right) \xrightarrow[n\to\infty]{P} \eta^2, \tag{42}$$

where

$$\eta := \mathscr{Q}(\mu)(k_0\sigma)^{1-4p} S_1^2 \int_0^\infty x^{4p}\left\{\left(\int_x^\infty J(y,S_1)dy\right)^2 + \left(1-\frac{2}{\pi}\right)\widetilde{J}(x,S_1)^2\right\}dx,$$

with

$$\begin{aligned}
J(x,S_1) &:= \frac{1}{2x}\int_{-\infty}^{\infty} h'(S_u e^{\sqrt{x}y+x/2})(-y^2-\sqrt{x}y+1)\varphi(y)dy,\\
\widetilde{J}(x,S_1) &:= \frac{1}{\sqrt{x}}\int_{-\infty}^{\infty} h'(S_u e^{\sqrt{x}y+x/2})y\varphi(y)dy.
\end{aligned}$$

Proof First, let us study the term $\xi_n^\vartheta := \sum_i \mathbf{E}\,(\vartheta_i^2|\mathscr{F}_{t_{i-1}})$. By independence, we obtain $\mathbf{E}\,(\vartheta_i^2|\mathscr{F}_{t_{i-1}}) = k_0^2 n^{1/2+2p}(K_{t_{i-1}}^n)^2 S_{t_{i-1}}^2 \mathbf{E}\,(\mathscr{E}_{t_{i-1}}^{t_i})^2$. Hence, using Lemma 1 and the change of variable $y=\rho_u^2$ and $x_i=\rho_{t_i}^2$,

$$\begin{aligned}
&\mathbf{E}\,\left(\vartheta_i^2|\mathscr{F}_{t_{i-1}}\right)\\
&= k_0^2 n^{1/2+2p} K_{t_{i-1}}^{n2} S_{t_{i-1}}^2 \sigma^2 \Delta t_i (1+O(n^{-1}))
\end{aligned}$$

$$= k_0^2\sigma^2 n^{1/2+2p} S_{t_{i-1}}^2 \left(\int_0^{t_{i-1}} \widehat{C}_{xt}(u, S_u)du\right)^2 \Delta t_i(1 + O(n^{-1}))$$

$$= k_0^2\sigma^2 S_{t_{i-1}}^2 \left(\int_0^{t_{i-1}} \widehat{C}_{xt}(u, S_u)du\right)^2 \frac{n^{1/2+2p}\Delta t_i}{x_{i-1} - x_i}\Delta x_i(1 + O(n^{-1}))$$

$$= k_0^2\sigma^2 S_{t_{i-1}}^2 \left(\int_{x_{i-1}}^{\rho_0^2} \widehat{C}_{xt}(u, S_u)\widehat{\sigma}_u^{-2}dy\right)^2 \frac{n^{1/2+2p}\Delta t_i}{x_{i-1} - x_i}\Delta x_i(1 + O(n^{-1})).$$

We then deduce that

$$\xi_n^{\vartheta} = (1 + O(n^{-1}))\int_0^{\infty} z_n^{\vartheta}(x)dx \tag{43}$$

where

$$z_n^{\vartheta}(x) := S_{t_{i-1}}^2 k_0^2\sigma^2 \sum_i \left(\int_{x_{i-1}}^{\rho_0^2} \widehat{C}_{xt}(u, S_u)\widehat{\sigma}_u^{-2}dx\right)^2 \frac{n^{1/2+2p}\Delta t_i}{x_{i-1} - x_i}\mathbf{1}_{(x_i, x_{i-1}]}(x).$$

Recall that $|\widehat{C}_{xt}(u, S_u)\widehat{\sigma}_u^{-2}|du \le c\, G_1(x, S_u)$, $x = \rho_u^2$, where

$$G_1(x, y) = \frac{1}{x}e^{-x/8}\left(\sum_{j=1}^{p}\frac{|\log(y/K_j)|}{\sqrt{x}}\exp\left\{-\frac{\log^2(y/K_j)}{2x}\right\} + \sqrt{x} + x\right).$$

In particular,

$$\sqrt{x}G_1(x, y) \le G(x), \tag{44}$$

where $G(x) = c\,x^{-\frac{1}{2}}e^{-x/16}$, $c > 0$ is a constant. Hence, a.s.,

$$\left|\int_{x_{i-1}}^{\rho_0^2} \widehat{C}_{xt}(u, S_u)\widehat{\sigma}_u^{-2}dy\right| \le \int_x^{\infty} G(x')dx' \le \int_0^{\infty} G(x')dx' < +\infty. \tag{45}$$

Therefore, using (20), we get that

$$|z_n^{\vartheta}(x)| \le C(1 + x)\left(\int_x^{\infty} G(x')dx'\right)^2 \sup_{u\in[0,1]} S_u^2. \tag{46}$$

But, due to Hölder's inequality,

$$\int_0^{\infty}(1 + x)\left(\int_x^{\infty} G(x')dx'\right)^2 dx < \infty.$$

Thus, we can apply the Lebesgue theorem, using Corollary 9 and (19):

$$\xi_n^{\vartheta} \xrightarrow[n\to\infty]{a.s.} \mathscr{Q}(\mu)(k_0\sigma)^{1-4p}S_1^2\int_0^{\infty} x^{4p}\left(\int_x^{\infty} J(y, S_1)dy\right)^2 dx. \tag{47}$$

Second, let us study the term $\xi_n^\chi = \sum_i \mathbf{E}\,(\chi_i^2|\mathscr{F}_{t_{i-1}})$. By independence, we obtain

$$\mathbf{E}\left(\chi_i^2|\mathscr{F}_{t_{i-1}}\right) = k_0^2 n^{1/2+2p}\gamma_{t_{i-1}}^2 S_{t_{i-1}}^4 \mathbf{E}\left[\mathscr{E}_{t_{i-1}}^{t_i}\right]_c^2.$$

Then $\mathbf{E}\,(\chi_i^2|\mathscr{F}_{t_{i-1}}) = k_0^2\sigma^2 n^{1/2+2p}\gamma_{t_{i-1}}^2 S_{t_{i-1}}^4(1-\frac{2}{\pi})\Delta t_i(1+o(1))$. We then deduce that

$$\sum_i \mathbf{E}\left(\chi_i^2|\mathscr{F}_{t_{i-1}}\right) = (1+O(n^{-1}))\int_0^\infty z_n^\chi(x)dx, \tag{48}$$

where

$$z_n^\chi(x) := S_{t_{i-1}}^4 k_0^2\sigma^2 \sum_i \gamma_{t_{i-1}}^2 \frac{n^{1/2+2p}\Delta t_i}{x_{i-1}-x_i}\mathbf{1}_{(x_i,x_{i-1}]}(x).$$

Let us obtain a suitable bound for $z_n^\chi(x)$, integrable in x. Recall that

$$\begin{aligned}\gamma_{t_{i-1}} &= \frac{1}{\rho_{t_{i-1}}S_{t_{i-1}}}\int_{-\infty}^\infty h'(S_{t_{i-1}}e^{\rho_{t_{i-1}}y+\rho_{t_{i-1}}^2/2})y\varphi(y)dy \\ &= \frac{1}{\sqrt{x_{i-1}}S_{t_{i-1}}}\int_{-\infty}^\infty h'(S_{t_{i-1}}e^{\sqrt{x_{i-1}}y+x_{i-1}/2})y\varphi(y)dy.\end{aligned}$$

Due to inequality (56), we claim that a.s. (in ω) for n large enough, there is a constant c_ω which does not depend on n such that

$$|\gamma_{t_{i-1}}| \le C\sup_{u\le 1} S_u^{-3/2}e^{-x/8}\left(\mathbf{1}_{x\ge 1}+\left(\frac{e^{-\frac{c_\omega}{x}}}{\sqrt{x}}+1\right)\mathbf{1}_{x\le 1}\right). \tag{49}$$

Indeed, this is obvious for $x\ge 1$. Otherwise, $1\ge x=\rho_u^2\ge cn^{1/2}(1-u^n(x))$ implies that $u=u^n(x)$ is close to 1 uniformly in $x\le 1$ as soon as n is large enough. It then suffices to choose S_1 out of the null-set $\{S_1=K_1,\cdots,K_{p_h}\}$ to obtain by continuity that $S_{u^n(x)}$ is also far enough from the points $K_1,\cdots,K_{p_h}$ if $x\ge 1$. We conclude that, for all j, there is a bound $\log^2(K_j/S_{u^n(x)})\ge c_{\omega,j}$ for some constants $c_{\omega,j}>0$.

Therefore,

$$|S_{t_{i-1}}|^4|\gamma_{t_{i-1}}|^2 \le C\xi e^{-x/4}\left(\mathbf{1}_{x\ge 1}+\left(\frac{e^{-\frac{c_\omega}{x}}}{\sqrt{x}}+1\right)^2\mathbf{1}_{x\le 1}\right),$$

where $\xi := \sup_{0\le u\le 1} S_u^4 \sup_{0\le u\le 1} S_u^{-3}$. Thus, due to (19)

$$|z_n^\chi(x)| \le C\xi(1+x)e^{-x/4}\left(\mathbf{1}_{x\ge 1}+\left(\frac{e^{-\frac{c_\omega}{x}}}{\sqrt{x}}+1\right)^2\mathbf{1}_{x\le 1}\right).$$

We can then apply the dominated convergence theorem using the limit (20). We obtain that

$$\xi_n^\chi \xrightarrow[n\to\infty]{a.s.} \left(1-\frac{2}{\pi}\right)\mathcal{Q}(\mu)(k_0\sigma)^{1-4p}S_1^2\int_0^\infty x^{4p}\widetilde{J}(x,S_1)^2dx.$$

Finally, let us study the term $\sum_i \mathbf{E}\,(\chi_i\vartheta_i|\mathcal{F}_{t_{i-1}})$. By independence, we have

$$\mathbf{E}\,\left(\chi_i\vartheta_i|\mathcal{F}_{t_{i-1}}\right)=k_0^2n^{1/2+2p}\gamma_{t_{i-1}}S_{t_{i-1}}^2K_{t_{i-1}}^nS_{t_{i-1}}\mathbf{E}\,\left(\left[\mathcal{E}_{t_{i-1}}^{t_i}\right]_c\mathcal{E}_{t_{i-1}}^{t_i}\right).$$

But

$$\mathbf{E}\,\left(\left[\mathcal{E}_{t_{i-1}}^{t_i}\right]_c\mathcal{E}_{t_{i-1}}^{t_i}\right)=\mathbf{E}\,\left[\mathcal{E}_{t_{i-1}}^{t_i}\right]_c^2\operatorname{sgn}\mathcal{E}_{t_{i-1}}^{t_i}=\left(1-\frac{2}{\pi}\right)\sigma^2(\Delta t_i)^{\frac{3}{2}}(1+o(1)).$$

Due to (19), we obtain

$$\frac{(\Delta t_i)^{\frac{3}{2}}n^{1/2+2p}}{x_{i-1}-x_i}\xrightarrow[n\to\infty]{}0.$$

From the bounds (20), (45), (49) and by applying again the Lebesgue theorem, we then obtain the following limit: $\sum_i \mathbf{E}\,(\chi_i\vartheta_i|\mathcal{F}_{t_{i-1}})\xrightarrow[n\to\infty]{a.s.}0$. □

Lemma 17 *We have that* $\mathbf{E}\,(N_1^n)^2\xrightarrow[n\to\infty]{}\mathbf{E}\,\eta^2$.

Proof Due to the independence of the increments of the Wiener process, we have $\mathbf{E}\,(\chi_i+\vartheta_i)(\chi_j+\vartheta_j)=0$ whenever $i\neq j$. We thus obtain that

$$\mathbf{E}\,(N_1^n)^2=\sum_i\mathbf{E}\,(\chi_i+\vartheta_i)^2=\mathbf{E}\,\sum_i\mathbf{E}\,\left((\chi_i+\vartheta_i)^2|\mathcal{F}_{t_{i-1}}\right).$$

But

$$\sum_i\mathbf{E}\,\left((\chi_i+\vartheta_i)^2|\mathcal{F}_{t_{i-1}}\right)\leq 2\sum_i\mathbf{E}\,\left(\chi_i^2+\vartheta_i^2|\mathcal{F}_{t_{i-1}}\right)=2(\xi_n^\chi+\xi_n^\vartheta).$$

Let us show that $\xi_n:=\xi_n^\chi+\xi_n^\vartheta$ is uniformly integrable. First, let us note that ξ_n is bounded in $L^1(\Omega)$. Indeed, from Corollary 3, inequalities (46) and (19), we obtain that for all n

$$\mathbf{E}\,|\xi_n|\leq C\int_0^\infty(1+x)\left((\mathbf{E}\,S_1^2)\frac{e^{-x}}{\sqrt{x}}+\left(\int_x^\infty G(x')dx'\right)\right)<\infty.$$

Now, using the Cauchy–Schwarz inequality and then the Markov inequality, we have:

$$\mathbf{E}\,\xi_n^\vartheta\mathbf{1}_{\xi_n\geq k}\leq C\int_0^\infty(1+x)\left(\int_x^\infty G(x')dx'\right)^2dx\sqrt{\mathbf{E}\,S_1^4}\sqrt{\mathbb{P}(\xi_n\geq k)}$$

$$\leq C\sqrt{\frac{\sup_n \mathbf{E}\,|\xi_n|}{k}} \xrightarrow[k\to\infty]{} 0.$$

Recall that

$$z_n^{\chi}(x)\mathbf{1}_{\xi^n\geq M_0} := k_0^2\sigma^2\sum_i S_{t_{i-1}}^4\gamma_{t_{i-1}}^2\mathbf{1}_{\xi^n\geq M_0}\frac{n^{1/2+2p}\Delta t_i}{x_{i-1}-x_i}\mathbf{1}_{(x_i,x_{i-1}]}(x).$$

Therefore, applying successively the Cauchy–Schwarz inequality, (19), Corollary 7, and the Markov inequality, we obtain that

$$\mathbf{E}\,\xi_n^{\chi}\mathbf{1}_{\xi_n\geq k} \leq C\sum_i\left(\mathbf{E}\,S_{t_{i-1}}^5\gamma_{t_{i-1}}^{5/2}\right)^{4/5}\left(\mathbb{P}(\xi^n\geq k)\right)^{1/5}\frac{n^{1/2+2p}\Delta t_i}{x_{i-1}-x_i}\mathbf{1}_{(x_i,x_{i-1}]}(x)$$

$$\leq C\int_0^{\infty}(1+x)\left(\frac{e^{-5x/32}}{x^{15/16}}\right)^{4/5}dx\left(\frac{\sup_n \mathbf{E}\,\xi_n}{k}\right)^{1/5} \xrightarrow[k\to\infty]{} 0.$$

Therefore, ξ_n is uniformly integrable, and so is $\sum_i \mathbf{E}\,((\chi_i+\vartheta_i)^2|\mathscr{F}_{t_{i-1}})$, which moreover converges to η a.s. This yields the conclusion of the lemma. □

We easily deduce the following:

Corollary 2 *We have* $\sup_n \mathbf{E}\,(\max_i(\chi_i+\vartheta_i)^2) < \infty$.

These last lemmas and corollaries complete the proof of Theorem 2.

5.4 Conclusion

Let us summarize the results of the previous theorems:

$$n^{\frac{1}{2}+2p}\mathbf{E}\,(\varepsilon_t^n)^2 \xrightarrow[n\to\infty]{} 0$$

and $n^{\frac{1}{4}+p}N_1^n \xrightarrow[n\to\infty]{d} Z$. Therefore, $n^{\frac{1}{4}+p}(V_1^n - h(S_1)) \xrightarrow[n\to\infty]{d} Z$ and

$$n^{\frac{1}{2}+2p}\mathbf{E}\,(V_1^n - h(S_1))^2 \xrightarrow[n\to\infty]{} \mathbf{E}\,\eta^2 = \mathbf{E}\,Z^2.$$

The proof of the limit theorem is then complete.

Acknowledgements The authors thanks the anonymous referees for constructive criticism and helpful suggestions which improved the presentation of the paper. The authors thank Yuri Kabanov and the other organizers of the Bachelier colloquium 2011 at Metabief.

The authors thank the "Chair Les Particuliers Face aux Risque" sponsored by Groupama for their support.

Appendix

The following limit result combines Theorem 3.4 (p. 67) and Theorem 3.5 (p. 71) in [6]:

Theorem 4 *Let $\{M_i^n, \mathscr{F}_{t_i}, 0 \le i \le n\}$ be a zero-mean square integrable martingale with increments $\Delta M_i^n = X_i^n$ and let η^2 be a finite r.v. Suppose that*

$$\text{for all}\, \varepsilon > 0, \quad \sum_i \mathbf{E}\left((X_i^n)^2 \mathbf{1}_{|X_i^n|>\varepsilon} | \mathscr{F}_{t_{i-1}} \right) \xrightarrow[n\to\infty]{L^1} 0, \tag{50}$$

$$V_n^2 = \sum_i \mathbf{E}\left((X_i^n)^2 | \mathscr{F}_{t_{i-1}} \right) \xrightarrow[n\to\infty]{P} \eta^2, \tag{51}$$

$$\max_i E\left((X_i^n)^2 | \mathscr{F}_{t_{i-1}} \right) \xrightarrow[n\to\infty]{P} 0, \tag{52}$$

$$\sup_n E\left(\max_i (X_i^n)^2 \right) < \infty. \tag{53}$$

Then $M_n^n \xrightarrow[n\to\infty]{d} Y$ where the r.v. Y has the characteristic function $\mathbf{E}\ \exp -\frac{1}{2}\eta^2 t^2$.

Proof Under conditions (50), (51) and (52), we deduce, by virtue of Theorem 3.5 (p. 71) in [6] that $U_n^2 \xrightarrow[n\to\infty]{L^1} \eta^2$ where $U_n^2 := \sum_i (X_i^n)^2$. Observe that the condition (50) implies that $\max_i |X_i^n| \xrightarrow[n\to\infty]{P} 0$. Applying Theorem 3.4 page 67 [6], we conclude. □

A.1 Explicit Formulae

We recall from [2] the following expressions for the successive derivatives. They are based on direct computations using the integration by parts formula under suitable assumptions on the payoff function h.

Lemma 18 *Let $\widehat{C}(t,x)$ is given by* (4). *Then*

$$\widehat{C}_x(t,x) = \int_{-\infty}^{\infty} h'(xe^{\rho y+\rho^2/2})\varphi(y)dy,$$

$$\widehat{C}_{xx}(t,x) = \frac{1}{\rho x}\int_{-\infty}^{\infty} h'(xe^{\rho y+\rho^2/2})y\varphi(y)dy,$$

$$\widehat{C}_{xxx}(t,x) = \frac{1}{\rho^2 x^2}\int_{-\infty}^{\infty} h'(xe^{\rho y+\rho^2/2})P_2(y)\varphi(y)dy,$$

$$\widehat{C}_{xxxx}(t,x) = \frac{1}{\rho^3 x^3}\int_{-\infty}^{\infty} h'(xe^{\rho y+\rho^2/2})P_3(y)\varphi(y)dy,$$

where

$$P_2(y) := y^2 - \rho y - 1,$$
$$P_3(y) := y^3 - 3\rho y^2 + (2\rho^2 - 3)y + 3\rho.$$

In particular, $|\widehat{C}_x(t,x)| \le \|h'\|_\infty$. Similarly, we obtain the following expressions for the successive derivatives in t:

Lemma 19 *Let* $\widehat{C}(t,x)$ *is given by* (4). *Then*

$$\widehat{C}_t(t,x) = \frac{-\widehat{\sigma}_t^2 x}{2\rho}\int_{-\infty}^{\infty} h'(xe^{\rho y+\rho^2/2})y\varphi(y)dy, \tag{54}$$

$$\widehat{C}_{tx}(t,x) = \frac{\widehat{\sigma}_t^2}{2\rho^2}\int_{-\infty}^{\infty} h'(xe^{\rho y+\rho^2/2})Q_2(y)\varphi(y)dy, \tag{55}$$

where

$$Q_2(y) := -y^2 - \rho y + 1.$$

Lemma 20 *We have*:

$$\widehat{C}_{xxt}(t,x) = \frac{\widehat{\sigma}_t^2}{2\rho_t^3 x}\int_{-\infty}^{\infty} h'(xe^{\rho_t y+\rho_t^2/2})P_1(\rho_t,y)\varphi(y)dy,$$

$$\widehat{C}_{xtt}(t,x) = -\frac{\rho_t''}{\rho_t}\int_{-\infty}^{\infty} h'(xe^{\rho_t y+\rho_t^2/2})P_2(\rho_t,y)\varphi(y)dy$$
$$+\frac{\widehat{\sigma}_t^4}{2\rho_t^4}\int_{-\infty}^{\infty} h'(xe^{\rho_t y+\rho_t^2/2})P_3(\rho_t,y)\varphi(y)dy,$$

$$\widehat{C}_{xxxt}(t,x) = \frac{\widehat{\sigma}_t^2}{2\rho_t^4 x^2}\int_{-\infty}^{\infty} h'(xe^{\rho_t y+\rho_t^2/2})P_4(\rho_t,y)\varphi(y)dy$$

where

$$P_1(x,y) := -y^3 - xy^2 + 3y + x,$$
$$P_2(x,y) := -y^2 - xy + 1,$$
$$P_3(x,y) := y^4 - (4+x^2)y^2 + 2xy + x^2 + 1,$$
$$P_4(x,y) := -y^4 + 2xy^3 + (6-x^2)y^2 - 8xy + x^2 - 3.$$

A.2 Estimates

To study the residual terms generated by the discretization of the theoretical portfolio $\widehat{C}(T, S_T)$, we use Taylor approximations. We then need to estimate some bounds of the successive derivatives of $\widehat{C}$.

Lemma 21 *There is a constant $C > 0$ such that*

$$|\widehat{C}_{xx}(t, x)| \leq C \frac{e^{-\rho^2/8}}{\rho x^{3/2}} \sum_{j=1}^{p} \exp\left\{-\frac{1}{2}\frac{\log^2(K_j/x)}{\rho^2}\right\} + c\frac{e^{-\rho^2/8}}{x^{3/2}}. \tag{56}$$

Corollary 3 *There exists a constant C such that for $t \in [0, 1[$*

$$\mathbf{E}\ S_t^4 \widehat{C}_{xx}^2(t, S_t) \leq \frac{C}{\rho} e^{-\rho^2/4}.$$

Corollary 4 *There exists a constant c such that for $t \in [0, 1[$*

$$\mathbf{E}\ S_t^2 \widehat{C}_{xx}^2(t, S_t) \leq c \left(\sum_{j=1}^{p} \frac{1}{\rho^2 \sqrt{2u^2 + 1}} \exp\left\{-\frac{v_j^2}{2u^2 + 1}\right\} + e^{-\rho^2/4} \right)$$

where c is a constant, $u = \alpha_t / \rho$ and

$$v_j := \frac{\log(S_0/K_j) - \alpha_t^2/2}{\rho} + \frac{\rho}{2}.$$

Lemma 22 *There exists a constant c such that*

$$|\widehat{C}_{xxx}(t, x)| \leq \frac{c e^{-\rho^2/8}}{\rho^2 x^{5/2}} \left(L(x, \rho) + \rho\right),$$

$$|\widehat{C}_{xxxx}(t, x)| \leq c e^{-\rho^2/8} x^{-7/2} P_3(\rho^{-1}),$$

$$|\widehat{C}_{tx}(t, x)| \leq \frac{c \widehat{\sigma}^2 e^{-\frac{\rho^2}{8}}}{x^{1/2} \rho^2} \left(L(x, \rho) + \rho + \rho^2\right),$$

$$|\widehat{C}_{xxt}(t, x)| \leq c \widehat{\sigma}^2 e^{-\rho^2/8} x^{-3/2} (\rho^{-1} + \rho^{-3}),$$

where P_3 is a polynomial of the third order and

$$L(x, \rho) := \sum_{j=1}^{p} \frac{|\log(x/K_j)|}{\rho} \exp\left\{-\frac{\log^2(x/K_j)}{2\rho^2}\right\}.$$

Lemma 23 *There exists a constant c and a polynomial Q of third order such that*

$$\mathbf{E}\ S_t^m \widehat{C}_{tx}^2(t, S_t) \leq c \widehat{\sigma}_t^4 Q(\rho^{-1}) e^{-\rho^2/4}.$$

Lemma 24 *The following bounds hold*:

$$|\widehat{C}_{xxt}(t,x)| \le c\frac{e^{-\rho_t^2/8}}{x^{3/2}}\frac{\widehat{\sigma}_t^2}{\rho_t^3}\left(\sum_{j=1}^{p}\Big(\rho_j(x)^2+\rho_t^2/4+1\Big)e^{-\rho_j(x)^2/2}+\rho_t+\rho_t^3\right),$$

$$|\widehat{C}_{xtt}(t,x)| \le \mathscr{X}^1(t,x)+\mathscr{X}^2(t,x),$$

where

$$\mathscr{X}^1(t,x) := c\frac{e^{-\rho_t^2/8}}{\sqrt{x}}\frac{|\rho_t''|}{\rho_t}\left(\sum_{j=1}^{p}\rho_j(x)e^{-\rho_j(x)^2/2}+\rho_t+\rho_t^2\right),$$

$$\mathscr{X}^2(t,x) := c\frac{e^{-\rho_t^2/8}}{\sqrt{x}}\frac{\widehat{\sigma}_t^4}{\rho_t^4}\left(\sum_{j=1}^{p}\Big(\rho_j(x)^3+\rho_j(x)\Big)e^{-\rho_j(x)^2/2}+\sum_{j=1}^{4}\rho_t^j\right),$$

and $\rho_j(x) := |\log(K_j/x)|/\rho_t$.

Lemma 25 *Assume that Assumption* (A1) *holds. Then there is a constant* c *such that* $\varepsilon_i := n^{1/2}\sqrt{\Delta t_i}\sqrt{f'(t_{i-1})}$, $i \le n-1$ *satisfies the inequality* $|\varepsilon_i - 1| \le c\Delta t_i/(1-t_i)$ *for* n *large enough.*

Proof We have obviously

$$|\varepsilon_i - 1| \le |n\Delta t_i f'(t_{i-1}) - 1|,$$

where $\Delta t_i = g'(\theta_i)n^{-1}$ and $\theta_i \in [(i-1)/n, i/n]$. Then, $d_i := g(\theta_i) - t_{i-1} \in [0, \Delta t_i]$. We deduce that:

$$|\varepsilon_i - 1| \le \left|\frac{f'(g(\theta_i)-h_i)}{f'(g(\theta_i))} - 1\right| \le c\frac{\Delta t_i}{1-t_i}.$$

Indeed, we use the first order Taylor expansion to estimate the difference

$$f'(g(\theta_i)-h_i) - f'(g(\theta_i)).$$

We obtain the claim by using the explicit expression of f, g and also the inequality $(1-t_{i-1})/(1-t_i) \le c$ for $i \le n-1$. □

The following lemma plays an important role to get estimations of expectations in several proofs.

Lemma 26 *Suppose that* $t \le u < 1$, $m \in \mathbb{R}$, $q \in 2\mathbb{N}$, *and* $K > 0$. *There exists a constant* $c = c(m,q)$ *such that*

$$\mathbf{E}\, S_u^m \log^q \frac{S_u}{K}\exp\left\{-\frac{\log^2(S_u/K)}{\rho_t^2}\right\} \le cP_q(\rho_t)$$

where

$$P_0(\rho_t) := \rho_t, \quad P_2(\rho_t) := \rho_t^3 + \rho_t^5,$$
$$P_4(\rho_t) := \rho_t^5 + \rho_t^7 + \rho_t^9,$$
$$P_{2q}(\rho_t) := \rho_t^{2q+1} + \rho_t^{2q+3} + \cdots + \rho_t^{4q+1}.$$

Proof We set $p = \log \frac{S_0}{K} - \sigma^2 u/2$, $\alpha = \sigma\sqrt{u}$, and

$$A(q) = \mathbf{E}\, S_u^m \log^q \frac{S_u}{K} \exp\left\{-\frac{\log^2(S_u/K)}{\rho_t^2}\right\}.$$

Then,

$$A(q) = \frac{S_0^m}{\sqrt{2\pi}} \int_{-\infty}^{\infty} (p+\alpha y)^q \exp\left\{\alpha m y - \alpha^2 m/2 - \frac{1}{\rho_t^2}(p+\alpha y)^2 - y^2/2\right\} dy,$$
$$A(q) = \frac{S_0^m e^{A_1}}{\sqrt{2\pi}} \int_{-\infty}^{\infty} (p+\alpha y)^q \exp\left\{-\frac{1}{2}\left(1+\frac{2\alpha^2}{\rho_t^2}\right) y^2 + \alpha\left(m - \frac{2p}{\rho_t^2}\right) y\right\} dy,$$

where

$$A_1 = -\frac{\alpha^2 m}{2} - \frac{p^2}{\rho_t^2}.$$

Let $y = z/A_2$ with $A_2 = \sqrt{1 + 2\alpha^2/\rho_t^2}$. Then

$$A(q) = \frac{S_0^m e^{A_4}}{\sqrt{2\pi}\, A_2} \int_{-\infty}^{\infty} \left(p + \frac{\alpha z}{A_2}\right)^q \exp\left\{-\frac{1}{2}\left[z^2 - 2(A_3/A_2)z + A_3^2/A_2^2\right]\right\} dz,$$

where $A_3 = \alpha(m - 2p/\rho_t^2)$ and $A_4 = A_1 + A_3^2/(2A_2^2)$. After the change of variable $y = z - A_3/A_2$, we obtain that

$$A(2) = \frac{S_0^m \rho_t e^{A_4}}{\sqrt{\rho_t^2 + 2\alpha^2}} \left[\left(p + \frac{\alpha\rho_t^2 A_3}{\rho_t^2 + 2\alpha^2}\right)^2 + \frac{\alpha^2\rho_t^2}{\rho_t^2 + 2\alpha^2}\right].$$

Moreover, if $u \geq t$, then the inequality $\rho_t^2 \geq \sigma^2(1-t)$ implies that

$$\rho_t^2 + 2\alpha^2 \geq \sigma^2(1-t) + \sigma^2 u \geq \sigma^2.$$

We have that

$$A_4 = -\frac{m\alpha^2}{2} - \frac{p^2}{\rho_t^2} + \frac{\alpha^2\rho_t^2}{2(\rho_t^2 + 2\alpha^2)}\left(m^2 + \frac{4p^2}{\rho_t^4} - \frac{4pm}{\rho_t^2}\right),$$

where p, α are bounded. But the term

$$\frac{\alpha^2\rho_t^2}{2(\rho_t^2+2\alpha^2)}m^2$$

is obviously bounded whereas we can establish the following inequality

$$\frac{\alpha^2\rho_t^2}{2(\rho_t^2+2\alpha^2)}\frac{4p^2}{\rho_t^4}\leq\frac{p^2}{\rho_t^2}.$$

The term

$$\left|\frac{\alpha^2\rho_t^2}{2(\rho_t^2+2\alpha^2)}\frac{4pm}{\rho_t^2}\right|$$

is also bounded. It follows that e^{A_4} is bounded and we conclude for $q=2$. In a similar way, we can conclude for any $q\in 2\mathbb{N}$ because we use, in particular, the property

$$\int_{-\infty}^{\infty}y^k\varphi(y)dy=0,\quad \text{if } k\in 2\mathbb{N}+1. \qquad \square$$

Corollary 5 *If $m\in\mathbb{R}$ and $u\geq t$, then there exists a constant $c_m>0$ such that*

$$\mathbf{E}\ S_u^m\widehat{C}_{xt}^2(t,S_u)\leq\frac{c_m\widehat{\sigma}_t^4}{\rho_t^3}e^{-\rho_t^2/8}.$$

Proof Indeed, it suffices to use Lemma 22 and apply the previous lemma. □

In a similar way, we have:

Corollary 6 *If $m\in\mathbb{R}$ and $u\geq t$, then there exists a constant $C_m>0$ such that*

$$\mathbf{E}\ S_u^m\widehat{C}_{xt}^4(t,S_u)\leq\frac{c_m\widehat{\sigma}_t^8}{\rho_t^7}e^{-\rho_t^2/8}, \tag{57}$$

$$\mathbf{E}\ S_u^m\widehat{C}_{xx}^4(t,S_u)\leq\frac{c_m}{\rho_t^3}e^{-\rho_t^2/4}. \tag{58}$$

Corollary 7 *If $m\in\mathbb{R}$, then there exists a constant $c_m>0$ such that*

$$\mathbf{E}\ S_t^m\widehat{C}_{xx}^{5/2}(t,S_t)\leq\frac{c_m}{\rho_t^{15/8}}e^{-5\rho_t^2/32}.$$

Proof We write $\mathbf{E}\ S_t^m\widehat{C}_{xx}^{5/2}(t,S_t)=\mathbf{E}\ S_t^m\widehat{C}_{xx}^{3/2}(t,S_t)\widehat{C}_{xx}(t,S_t)$ and apply the Cauchy–Schwarz inequality with $p=4/3$ and $q=4$ such that $p^{-1}+q^{-1}=1$. We obtain that

$$\mathbf{E}\ S_t^m\widehat{C}_{xx}^{5/2}(t,S_t)\leq\left(\mathbf{E}\ S_t^{4m/3}\widehat{C}_{xx}^2(t,S_t)\right)^{3/4}\left(\mathbf{E}\ \widehat{C}_{xx}^4(t,S_t)\right)^{1/4}$$

$$\leq \left(C_m \mathbf{E}\, \widehat{C}^4_{xx}(t, S_t)\right)^{3/8} \left(\mathbf{E}\, \widehat{C}^4_{xx}(t, S_t)\right)^{1/4}$$

$$\leq \left(C_m \frac{c}{\rho_t^3} e^{-\rho_t^2/4}\right)^{3/8} \left(\frac{c}{\rho_t^3} e^{-\rho_t^2/4}\right)^{1/4},$$

where the last inequality is deduced from (58). The claim follows. □

Corollary 8 *If $m \in \mathbb{R}$ and $u \geq t$, then there exists a constant $c_m > 0$ such that*

$$\mathbf{E}\, S_u^m \widehat{C}^2_{xxx}(t, S_u) \leq \frac{c_m}{\rho_t^3} e^{-\rho_t^2/8}, \tag{59}$$

$$\mathbf{E}\, S_u^m \widehat{C}^2_{xxt}(t, S_u) \leq \frac{c_m \widehat{\sigma}_t^4}{\rho_t^5} e^{-\rho_t^2/8}, \tag{60}$$

$$\mathbf{E}\, S_u^m \widehat{C}^4_{xxx}(t, S_u) \leq \frac{c_m}{\rho_t^7} e^{-\rho_t^2/8}, \tag{61}$$

$$\mathbf{E}\, S_u^m \widehat{C}^2_{xxxx}(t, S_u) \leq \frac{c_m}{\rho_t^5} e^{-\rho_t^2/8}, \tag{62}$$

$$\mathbf{E}\, S_u^m \widehat{C}^4_{xxt}(t, S_u) \leq \frac{c_m \widehat{\sigma}_t^8}{\rho_u^{11}} e^{-\rho_t^2/8}. \tag{63}$$

Let $\widetilde{S}_{t_{i-1}} \in [S_{t_{i-1}}, S_{t_i}]$ and $\widetilde{t}_{i-1} \in [t_{i-1}, t_i]$ be some random variables.

Lemma 27 *There exists a constant c such that*

$$\mathbf{E}\, \widehat{C}^4_{xt}(\widetilde{t}_{i-1}, \widetilde{S}_{t_{i-1}}) \leq \frac{c e^{-\rho_{t_i}^2/4}}{(1-t_i)^4}.$$

Proof We have $\widetilde{S}^m_{t_{i-1}} \leq S^m_{t_{i-1}} + S^m_{t_i}$, and $\rho_{\widetilde{t}_{i-1}} \geq \rho_{t_i}$. Furthermore, in virtue of Lemma 22,

$$|\widehat{C}_{xt}(t, x)| \leq c \frac{\widehat{\sigma}_t^2 e^{-\rho_t^2/8}}{x^{1/2} \rho_t^2}.$$

This implies the result. □

In the same way we can prove the following:

Lemma 28 *There exists a constant C such that*

$$\mathbf{E}\, \widehat{C}^4_{xtt}(\widetilde{t}_{i-1}, \widetilde{S}_{t_{i-1}}) \leq \frac{C e^{-\rho_{t_i}^2/4}}{(1-t_i)^8}.$$

Proof The arguments are similar to the previous ones but we also use the inequality

$$\frac{g''(u)}{g'(u)^2} \le \frac{C}{(1-g(u))^{3/2}}, \quad \forall u < 1$$

in order to get the bound $\frac{\rho_t''}{\rho_t} \le \frac{C}{(1-t)^2}$. □

Lemma 29 *There exists a constant C such that*

$$\mathbf{E}\,\widehat{C}^4_{xxx}(\widetilde{t}_{i-1}, \widetilde{S}_{t_{i-1}}) \le \frac{Ce^{-\rho^2_{t_i}/4}}{\rho^8_{t_i}}, \tag{64}$$

$$\mathbf{E}\,\widehat{C}^4_{xxt}(\widetilde{t}_{i-1}, \widetilde{S}_{t_{i-1}}) \le \frac{Ce^{-\rho^2_{t_i}/4}}{n(1-t_i)^6 f'(t_i)}, \tag{65}$$

$$\mathbf{E}\,\widehat{C}^4_{xxxx}(\widetilde{t}_{i-1}, \widetilde{S}_{t_{i-1}}) \le \frac{Ce^{-\rho^2_{t_i}/4}}{\rho^{12}_{t_i}}. \tag{66}$$

A.3 Technical Lemmas

Recall the two following lemmas (see [8]). These results ensures the convergence of the Leland scheme without any hedging error when using the modified Leland strategy. The change of variable $x = \rho^2_u$ appears to be as essential in the following proofs and points out the significative role of the revision dates near the maturity.

Lemma 30 *We have the following equality*

$$\int_s^t \widehat{C}_{xt}(u, S_u)du = \int_{\rho^2_t}^{\rho^2_s} \widehat{C}_{xt}(u, S_u)\widehat{\sigma}_u^{-2}dx,$$

where $u = u(x, n)$ *is defined by* $x = \rho^2_u$ *and verifies* $\lim_{n\to\infty} u(x, n) = 1$. *Moreover,*

$$C_{xt}(u, S_u)\widehat{\sigma}_u^{-2} = \frac{1}{2x}\int_{-\infty}^{\infty} h'(S_u e^{\sqrt{x}y + x/2})(-y^2 - \sqrt{x}y + 1)\varphi(y)dy$$

satisfies the inequality $|\widehat{C}_{xt}(u, S_u)\widehat{\sigma}_u^{-2}|du \le c\,G_1(x, S_u)$, *where*

$$G_1(x, S) := \frac{1}{x}e^{-x/8}\left(\sum_{j=1}^{p} \frac{|\log(S/K_j)|}{\sqrt{x}} \exp\left\{-\frac{\log^2(S/K_j)}{2x}\right\} + \sqrt{x} + x\right).$$

Corollary 9 *Assume that we have two sequences* $(t_{k^n})_{n\in\mathbf{N}}$ *and* $(s_{k^n})_{n\in\mathbf{N}}$ *in* $[0,1]$ *such that* $\rho_{t_{kn}}$ *and* $\rho_{s_{kn}}$ *converge to* $a \in [0,\infty]$ *and* $b \in [0,\infty]$, *respectively. Then*

$$\lim_{n\to\infty} \int_{s_{kn}}^{t_{kn}} \widehat{C}_{xt}(u, S_u)du = \int_a^b J(x, S_1)dx < \infty, \quad a.s.$$

Proof We apply Lemma 30 with the change of variable $x = \rho_u^2$. Recall that we have the bounds $0 \le 1 - u \le c\,x\,n^{-1/2}$, so that $u \to 1$ as $n \to \infty$ for a given $x \ge 0$. We can apply the Lebesgue theorem by dominating the function $G_1(x, S_u)$ whether $x \le 1$ or not because $x \le 1$ implies that u is sufficiently near from 1 independently of x for $n \ge n_0$. Indeed, outside of the null-set $\bigcup_i \{S_1 = K_i\}$, we have that

$$0 < a \le |\log(S_u/K_j)| \le b$$

for some constants a, b (depending on ω) provided that u is sufficiently near unit. □

References

1. Denis, E.: Marchés avec coûts de transaction: approximation de Leland et arbitrage. Thèse, Université de Franche-Comté (2008)
2. Denis, E.: Approximate hedging of contingent claims under transaction costs. Appl. Math. Finance **17**, 491–518 (2010)
3. Denis, E., Kabanov, Y.: Mean square error for the Leland-Lott hedging strategy: convex payoffs. Finance Stoch. (2009)
4. Gamys, M., Kabanov, Y.: Mean square error for the Leland–Lott hedging strategy. In: Recent Advances in Financial Engineering: Proceedings of the 2008 Daiwa International Workshop on Financial Engineering. World Scientific, Singapore (2009)
5. Granditz, P., Schachinger, W.: Leland's approach to option pricing: the evolution of discontinuity. Math. Finance **11**, 347–355 (2001)
6. Hall, P., Heyde, C.C.: Martingale limit theory and its application. In: Probability and Mathematical Statistics. Academic Press, Harcourt Brace Jovanovich, New York (1980). xii+308 pp.
7. Kabanov, Y., Safarian, M.: On Leland's strategy of option pricing with transaction costs. Finance Stoch. **1**, 239–250 (1997)
8. Lépinette, E.: Modified Leland's strategy for constant transaction costs rate. Math. Finance **22**(4), 741–752 (2012)
9. Lott, K.: Ein verfahren zur replikation von optionen unter transaktionkosten in stetiger zeit. Dissertation. Universität der Bundeswehr München, Institut für Mathematik und Datenverarbeitung (1993)
10. Pergamenshchikov, S.: Limit theorem for Leland's strategy. Ann. Appl. Probab. **13**, 1099–1118 (2003)
11. Sekine, J., Yano, J.: Hedging errors of Leland's strategies with time-inhomogeneous rebalancing. Preprint
12. Zhao, Y., Hedging, Z.W.T.: Errors with Leland's option model in the presence of transaction costs. Finance Res. Lett. **4**(1), 49–58 (2007)
13. Zhao, Y., Ziemba, W.T.: Comments on and corrigendum to "Hedging errors with Leland's option model in the presence of transaction costs". Finance Res. Lett. **4**(3), 196–199 (2007)

Conditional Default Probability and Density

N. El Karoui, M. Jeanblanc, Y. Jiao, and B. Zargari

Abstract We construct explicit models of conditional probability and density processes given a reference filtration for one or several default times. For this purpose, different methods are proposed such as the dynamic copula, change of time, change of probability measure and filtering.

This paper is dedicated to our friend Marek, for his birthday. Two of us know Marek since more than 20 years, when we embarked in the adventure of Mathematics for Finance. Our paths diverged, but we always kept strong ties. Thank you, Marek, for all the fruitful discussions we have had. We hope you will find some interest in this paper and the modeling of credit risk we present, and we are looking forward to sharing a enjoyable week in Métabief together, sipping Arbois wine, tasting Jura cheese, walking in the snow, and attending to nice talks.

N. El Karoui
Laboratoire de Probabilités et Modèles Aléatoires, Université Pierre et Marie Curie, Paris, France

N. El Karoui
Centre de Mathématiques Appliquées, École Polytechnique, Palaiseau cedex, France
e-mail: nicole.elkaroui@cmap.polytechnique.fr

M. Jeanblanc · B. Zargari
Laboratoire Analyse et Probabilités, Université d'Evry-Val-D'Essonne, Évry, France

M. Jeanblanc
e-mail: monique.jeanblanc@univ-evry.fr

B. Zargari
e-mail: behnaz.zargari@univ-evry.fr

M. Jeanblanc
Institut Europlace de Finance, Paris, France

Y. Jiao (✉)
ISFA, Université Claude Bernard-Lyon I, 50 avenue Tony Garnier, 69007 Lyon, France
e-mail: jiao@math.univ-paris-diderot.fr

B. Zargari
Sharif University of Technology, Tehran, Iran

Y. Kabanov et al. (eds.), *Inspired by Finance*, DOI 10.1007/978-3-319-02069-3_9,

Keywords Credit risk · Default models · Survival process · Brownian motion · Gaussian copula · Filtering

Mathematics Subject Classification (2010) 91G20 · 91G40

1 Introduction

The goal of this paper is to give examples of the conditional law of a random variable (or a random vector), given a reference filtration, and methods to construct dynamics of conditional laws, in order to model price processes with default risk. This methodology appears in some recent papers (El Karoui et al. [4], Filipovic et al. [7]) and it is important to present techniques to build concrete examples. We have chosen to characterize the (conditional) law of a random variable through its (conditional) survival probability or through its (conditional) density, if it exists.

In Sect. 2, we give the definition of martingale survival processes and density processes. In Sect. 3, we give standard examples of conditional laws, in particular a Gaussian model, and we give methods to construct other ones. In Sect. 4, we show that, in the case of random times (i.e., non-negative random variables), the density methodology can be seen as an extension of the Cox model, and we recall a result which allows to construct default times having the same intensity and different conditional laws. We build the change of probability framework in Sect. 5 and show how it can be applied to filtering theory for computing the conditional law of the random variable which represents the signal.

2 Definitions

Let $(\Omega, \mathscr{A}, \mathbf{F}, P)$ be a filtered probability space with a filtration $\mathbf{F} = (\mathscr{F}_t)_{t\geq 0}$ satisfying the usual conditions, $\mathscr{F}_\infty \subset \mathscr{A}$ and $\mathscr{F}_0$ is trivial. Let E be equal to one of the following spaces: $\mathbb{R}$, $\mathbb{R}^d$, $\mathbb{R}_+$, or $\mathbb{R}^d_+$.

A family of $(P, \mathbf{F})$-martingale survival processes on E is a family of $(P, \mathbf{F})$-martingales $G_.(\theta)$, $\theta \in E$ with values in [0, 1] such that $\theta \to G_t(\theta)$ is decreasing. We have used the standard convention for maps from $\mathbb{R}^d$ to $\mathbb{R}$: such a map G is decreasing if $\theta \leq \widetilde{\theta}$ implies $G(\theta) \geq G(\widetilde{\theta})$, where $\theta \leq \widetilde{\theta}$ means that $\theta_i \leq \widetilde{\theta}_i$ for $i = 1, \dots, d$.

A $(P, \mathbf{F})$-density process on E is a family $g_.(\theta)$, $\theta \in E$ of non-negative, $(P, \mathbf{F})$-martingales such that for all t

$$\int_E g_t(u)du = 1 \quad \text{a.s.} \tag{1}$$

where du denotes the Lebesgue measure on E. If there is no ambiguity, we shall simply say a martingale survival process and a density process.

If G is a family of martingale survival processes on E, absolutely continuous with respect to the Lebesgue measure, i.e., $G_t(\theta) = \int_\theta^\infty g_t(u)du$, the family g is a density process (see Jacod [9] for important regularity conditions).

The martingale survival process of an $\mathcal{A}$-measurable $\mathbb{R}^d$-valued random variable X is the family of càdlàg processes $G_t(\theta) = P(X > \theta|\mathcal{F}_t)$. Obviously, this is a martingale survival process (it is decreasing in θ). In particular, assuming regularity conditions, the non-negative function g_0 such that $G_0(\theta) = \int_\theta^\infty g_0(s)ds$ is the probability density of X.

If we are given a family of density processes $g_.(\theta)$, then there exists a random variable X (constructed on an extended probability space) such that

$$P(X > \theta|\mathcal{F}_t) = G_t(\theta) = \int_\theta^\infty g_t(u)du \quad \text{a.s.}$$

where (with an abuse of notation) P is a probability measure on the extended space, which coincides with the given probability measure on $\mathbf{F}$. For the construction, one starts with a random variable X on $\Omega \times E$ independent of $\mathbf{F}$, with probability density g_0 and one checks that $(g_t(X),\ t \geq 0)$ is an $\mathbf{F} \vee \sigma(X)$-martingale. Then, setting $dQ|_{\mathcal{F}_t \vee \sigma(X)} = \frac{g_t(X)}{g_0(X)} dP|_{\mathcal{F}_t \vee \sigma(X)}$, one obtains, from the Bayes formula that $Q(X > \theta|\mathcal{F}_t) = G_t(\theta)$. This construction was important in Grorud and Pontier [8] and in Amendinger [1] in an initial enlargement of filtration framework for application to insider trading.

In the specific case of random times (non-negative random variables), one has to consider martingale survival processes defined on $\mathbb{R}_+$. They can be deduced from martingale survival processes on $\mathbb{R}$ by a simple change of variable: if G is the martingale survival process on $\mathbb{R}$ of the real valued random variable X and h a strictly increasing function from $\mathbb{R}_+$ to $\mathbb{R}$, then $G_t^h(u) := G_t(h(u))$ defines a martingale survival process on $\mathbb{R}_+$ (corresponding to the change of variable $Y = h^{-1}(X)$). In the case where h is differentiable, the density process is $g^h(u) = g_t(h(u))h'(u)$.

It is important to note that, due to the martingale property, in order to characterize the family $g_t(\theta)$ for any pair $(t,\theta) \in \mathbb{R}_+ \times \mathbb{R}$, it suffices to know this family for any pair (t,θ) such that $\theta \leq t$. Hence, in what follows, we shall concentrate on the construction for $\theta \leq t$.

In the paper, the natural filtration of a process Y is denoted by $\mathbf{F}^Y$.

3 Examples of Martingale Survival Processes

We first present two specific examples of conditional law of an $\mathcal{F}_\infty^B$-measurable random variable, when $\mathbf{F}^B$ is the natural filtration of a Brownian motion B. Then we give two large classes of examples, based on Markov processes and diffusion processes.

The first example, despite its simplicity, will allow us to construct a dynamic copula, in a Gaussian framework; more precisely, we construct, for any t, the (conditional) copula of a family of random times $P(\tau_i > t_i,\ i = 1,\dots,n|\mathcal{F}_t)$ and we can choose the parameters so that $P(\tau_i > t_i,\ i = 1,\dots,n)$ equals a given (static) Gaussian copula. To the best of our knowledge, there are very few explicit constructions of such a model.

In [5] Fermanian and Vigneron apply a copula methodology, using a factor Y. However, the processes they use to fit the conditional probabilities

$$P\big(\tau_i > t_i,\ i = 1, \dots, n\,|\mathscr{F}_t \vee \sigma(Y)\big)$$

are not martingales. Using some adequate parametrization, they can produce a model such that $P(\tau_i > t_i,\ i = 1, \dots, n|\mathscr{F}_t)$ are martingales. Our model will satisfy both martingale conditions.

In [2] Carmona is interested in the dynamics of prices of assets corresponding to a payoff which is a Bernoulli random variable (with values 0 or 1). In other words, he is looking for examples of dynamics of [0, 1]-valued martingales with a given terminal condition. Surprisingly, the example he provides corresponds to the one we give below in Sect. 3.1, up to a particular choice of the parameters to satisfy the terminal constraint.

In a second example, we construct another dynamic copula, again in an explicit way, with a more complicated dependence.

Furthermore, we show that a class of examples can be obtained from a Markov model, where the decreasing property is introduced via a change of variable. In the second class of examples, the decreasing property is modeled via the dependence of a diffusion through its initial condition. To close the loop, we show that we can recover the Gaussian model of the first example within this framework.

3.1 A Dynamic Gaussian Copula Model

In this subsection Φ is the standard Gaussian distribution function, $\varphi = \Phi'$.

We consider the random variable $X := \int_0^\infty f(s)dB_s$ where f is a deterministic, square-integrable function. For any real number θ and any positive t

$$P\big(X > \theta|\mathscr{F}_t^B\big) = P\left(m_t > \theta - \int_t^\infty f(s)dB_s|\mathscr{F}_t^B\right)$$

where $m_t = \int_0^t f(s)dB_s$ is $\mathscr{F}_t^B$-measurable. The random variable $\int_t^\infty f(s)dB_s$ has a centered Gaussian law with variance $\sigma^2(t) = \int_t^\infty f^2(s)ds$ and is independent of $\mathscr{F}_t^B$. Assuming that $\sigma(t)$ does not vanish, one has

$$P\big(X > \theta|\mathscr{F}_t^B\big) = \Phi\left(\frac{m_t - \theta}{\sigma(t)}\right). \tag{2}$$

In other words, the conditional law of X given $\mathscr{F}_t^B$ is a Gaussian law with mean m_t and variance $\sigma^2(t)$. We summarize the result[1] in the following proposition, and we give the dynamics of the martingale survival process, obtained with a standard use of Itô's rule.

[1]More results on that model, in an enlargement of filtration setting, can be found in Chaleyat-Maurel and Jeulin [3] and Yor [17].

Proposition 1 *Let $B = (B_t)$ be a Brownian motion, f be a deterministic L^2-function, $m_t = \int_0^t f(s)dB_s$ and $\sigma^2(t) = \int_t^\infty f^2(s)ds$. The family*

$$G_t(\theta) = \Phi\left(\frac{m_t - \theta}{\sigma(t)}\right)$$

is a family of $\boldsymbol{F}^B$-martingales with values in $[0, 1]$, *and decreasing in θ. Moreover,*

$$dG_t(\theta) = \varphi\left(\frac{m_t - \theta}{\sigma(t)}\right)\frac{f(t)}{\sigma(t)}dB_t\,.$$

The dynamics of the martingale survival process can be written

$$dG_t(\theta) = \varphi\big(\Phi^{-1}\big(G_t(\theta)\big)\big)\frac{f(t)}{\sigma(t)}\,dB_t\,. \tag{3}$$

We obtain the associated density family by differentiating $G_t(\theta)$ with respect to θ,

$$g_t(\theta) = \frac{1}{\sqrt{2\pi}\,\sigma(t)}\exp\left(-\frac{(m_t - \theta)^2}{2\sigma^2(t)}\right)$$

and its dynamics

$$dg_t(\theta) = -g_t(\theta)\frac{m_t - \theta}{\sigma^2(t)}f(t)dB_t. \tag{4}$$

Let us emphasize that, starting from (3), it is not obvious to check that the solution is decreasing in θ, or, as it is done in [5] and [2], to find the solution. In the same way, the solution of (4) with initial condition a probability density g_0, is a density process if and only if $\int_{-\infty}^\infty g_t(u)du = 1$, or equivalently, $\int_{-\infty}^\infty g_t(\theta)\frac{m_t-\theta}{\sigma^2(t)}f(t)d\theta = 0$. This last equality reduces to

$$\int_{-\infty}^\infty g_t(\theta)(m_t - \theta)d\theta = m_t - \int_{-\infty}^\infty g_t(\theta)\theta d\theta = 0$$

and we do not see how to check this equality if one does not know the explicit solution.

In order to provide conditional survival probabilities for positive random variables, we consider $\widetilde{X} = \psi(X)$ where ψ is a differentiable, positive and strictly increasing function which inverse ψ^{-1} we denote by h. The conditional law of $\widetilde{X}$ is

$$\tilde{G}_t(\theta) = \Phi\left(\frac{m_t - h(\theta)}{\sigma(t)}\right).$$

We obtain that

$$\widetilde{g}_t(\theta) = \frac{1}{\sqrt{2\pi}\sigma(t)}h'(\theta)\exp\left\{-\frac{(m_t - h(\theta))^2}{2\sigma^2(t)}\right\}$$

and

$$d\widetilde{G}_t(\theta) = \varphi\left(\frac{m_t - h(\theta)}{\sigma(t)}\right)\frac{f(t)}{\sigma(t)}dB_t,$$

$$d\widetilde{g}_t(\theta) = -\widetilde{g}_t(\theta)\frac{m_t - h(\theta)}{\sigma(t)}\frac{f(t)}{\sigma(t)}dB_t.$$

Introducing an n-dimensional standard Brownian motion $B = (B^i,\ i = 1, \dots, n)$ and a factor Y, independent of $\mathbf{F}^B$, gives a dynamic copula approach, as we present now. For h_i an increasing function, mapping $\mathbb{R}^+$ into $\mathbb{R}$, and setting

$$\tau_i = (h_i)^{-1}\left(\sqrt{1-\rho_i^2}\int_0^\infty f_i(s)dB_s^i + \rho_i Y\right),$$

for $\rho_i \in (-1, 1)$, an immediate extension of the Gaussian model leads to

$$P\big(\tau_i > t_i,\ \forall i = 1, \dots, n \mid \mathscr{F}_t^B \vee \sigma(Y)\big) = \prod_{i=1}^n \Phi\left(\frac{1}{\sigma_i(t)}\left(m_t^i - \frac{h_i(t_i) - \rho_i Y}{\sqrt{1-\rho_i^2}}\right)\right)$$

where $m_t^i = \int_0^t f_i(s)dB_s^i$ and $\sigma_i^2(t) = \int_t^\infty f_i^2(s)ds$. It follows that

$$P\big(\tau_i > t_i, \forall i = 1, \dots, n \mid \mathscr{F}_t^B\big)$$
$$= \int_{-\infty}^{\infty}\prod_{i=1}^n \Phi\left(\frac{1}{\sigma_i(t)}\left(m_t^i - \frac{h_i(t_i) - \rho_i y}{\sqrt{1-\rho_i^2}}\right)\right) f_Y(y)dy.$$

Note that, in that setting, the random times $(\tau_i,\ i = 1, \dots, n)$ are conditionally independent given $\mathbf{F}^B \vee \sigma(Y)$, a useful property which is not satisfied in Fermanian and Vigneron model. For $t = 0$, choosing f_i so that $\sigma_i(0) = 1$, and Y with a standard Gaussian law, we obtain

$$P(\tau_i > t_i,\ \forall i = 1, \dots, n) = \int_{-\infty}^{\infty}\prod_{i=1}^n \Phi\left(-\frac{h_i(t_i) - \rho_i y}{\sqrt{1-\rho_i^2}}\right)\varphi(y)dy$$

which corresponds, by construction, to the standard Gaussian copula (because $h_i(\tau_i) = \sqrt{1-\rho_i^2}X_i + \rho_i Y$, where $X_i,\ Y$ are independent standard Gaussian variables).

Relaxing the independence condition on the components of the process B leads to more sophisticated examples.

3.2 A Gamma Model

Here, we present another model, where the processes involved are no more Gaussian ones. Consider $A_t^{(\mu)} := \int_0^t e^{2B_s^{(\mu)}} ds$ where $B_t^{(\mu)} = B_t + \mu t$, μ being a positive constant. Matsumoto and Yor [15] have established that $A_\infty^{(-\mu)} = A_t^{(-\mu)} + e^{2B_t^{(-\mu)}} \widetilde{A}_\infty^{(-\mu)}$ where $\widetilde{A}_\infty^{(-\mu)}$ is independent of $\mathscr{F}_t^B$, with the same law as $A_\infty^{(-\mu)}$. The law of $A_\infty^{(-\mu)}$ is proved to be the law of $1/(2\gamma_\mu)$, γ_μ being a Gamma random variable with parameter μ. The survival probability of $A_\infty^{(-\mu)}$ is

$$\Upsilon(x) = \frac{1}{\Gamma(\mu)} \int_0^{1/(2x)} y^{\mu-1} e^{-y} dy,$$

where Γ is the Gamma function. Then, one obtains

$$G_t(\theta) = P\big(A_\infty^{(-\mu)} > \theta | \mathscr{F}_t^B\big) = \Upsilon\left(\frac{\theta - A_t^{(-\mu)}}{e^{2B_t^{(-\mu)}}}\right) \mathbb{1}_{\theta > A_t^{(-\mu)}} + \mathbb{1}_{\theta \leq A_t^{(-\mu)}}.$$

This gives a family of martingale survival processes G, similar to (5), with gamma structure. It follows that, on $\{\theta > A_t^{(-\mu)}\}$,

$$dG_t(\theta) = \frac{1}{2^{\mu-1}\Gamma(\mu)} e^{-\frac{1}{2}Z_t(\theta)} \big(Z_t(\theta)\big)^\mu dB_t$$

where $Z_t(\theta) = \frac{e^{2B_t^{(-\mu)}}}{\theta - A_t^{(-\mu)}}$ (to simplify notation, we do not specify that this process Z depends on μ). One can check that $G_t(\cdot)$ is differentiable with respect to θ, so that $G_t(\theta) = \int_\theta^\infty g_t(u) du$, where

$$g_t(\theta) = \mathbb{1}_{\theta > A_t^{(-\mu)}} \frac{1}{2^\mu \Gamma(\mu)} \big(Z_t(\theta)\big)^{\mu+1} e^{-\frac{1}{2}Z_t(\theta) - 2B_t^{(-\mu)}}.$$

Again, introducing an n-dimensional Brownian motion, a factor Y and the random variables $\alpha_i A_\infty^{(-\mu,i)} + \rho_i Y$, where α_i and ρ_i are constants, will give an example of a dynamic copula.

3.3 Markov Processes

Let X be a real-valued Markov process with transition probability

$$p_T(t, x, y) dy = P(X_T \in dy | X_t = x),$$

and Ψ a family of functions $\mathbb{R} \times \mathbb{R} \to [0, 1]$, decreasing in the second variable and such that

$$\Psi(x, -\infty) = 1, \qquad \Psi(x, \infty) = 0.$$

Then, for any T,

$$G_t(\theta) := E\big(\Psi(X_T,\theta)|\mathscr{F}_t^X\big) = \int_{-\infty}^{\infty} p_T(t, X_t, y)\Psi(y,\theta)dy$$

is a family of martingale survival processes on $\mathbb{R}$. While modeling $(T;x)$-bond prices, Filipovic et al. [6] have used this approach in an affine process framework. See also Keller-Ressel et al. [13].

Example 1 Let X be a Brownian motion, and $\Psi(x,\theta) = e^{-\theta x^2}\mathbf{1}_{\theta\geq 0} + \mathbf{1}_{\theta\leq 0}$. We obtain a martingale survival process on $\mathbb{R}_+$, defined for $\theta \geq 0$ and $t < T$ as,

$$G_t(\theta) = E\big[\exp\{-\theta X_T^2\}|\mathscr{F}_t^X\big] = \frac{1}{\sqrt{1+2(T-t)\theta}}\exp\bigg\{-\frac{\theta X_t^2}{1+2(T-t)\theta}\bigg\}.$$

The construction given above provides a martingale survival process $G(\theta)$ on the time interval $[0, T]$. Using a (deterministic) change of time, one can easily deduce a martingale survival process on the whole interval $[0,\infty[$: setting

$$\hat{G}_t(\theta) = G_{h(t)}(\theta)$$

for a differentiable increasing function h from $[0,\infty]$ to $[0, T]$, and assuming that $dG_t(\theta) = G_t(\theta)K_t(\theta)dB_t, t < T$, one obtains

$$d\hat{G}_t(\theta) = \hat{G}_t(\theta)K_{h(t)}(\theta)\sqrt{h'(t)}dW_t$$

where W is a Brownian motion.

One can also randomize the terminal date and consider T as an exponential random variable independent of $\mathbf{F}$. Noting that the previous $G_t(\theta)$'s depend on T, one can write them as $G_t(\theta, T)$ and consider

$$\widetilde{G}_t(\theta) = \int_0^{\infty} G_t(\theta, z)e^{-z}dz$$

which is a martingale survival process. The same construction can be done with a random time T with any given density, independent of $\mathbf{F}$.

3.4 Diffusion-Based Model with Initial Value

Proposition 2 *Let Ψ be a probability distribution function of class C^2, and let Y be the solution of*

$$dY_t = a(t, Y_t)dt + \nu(t, Y_t)dB_t, \quad Y_0 = y_0,$$

where a and ν are deterministic functions smooth enough to ensure that the solution of the above SDE is unique. Then, the process $(\Psi(Y_t),\ t \geq 0)$ *is a martingale, valued in* $[0, 1]$, *if and only if*

$$a(t, y)\Psi'(y) + \frac{1}{2}\nu^2(t, y)\Psi''(y) = 0. \tag{5}$$

Proof The result follows by applying Itô's formula and noting that $\Psi(Y_t)$, being a (bounded) local martingale, is a martingale. □

We denote by $Y_t(y)$ the solution of the above SDE with initial condition $Y_0 = y$. Note that, from the uniqueness of the solution, $y \to Y_t(y)$ is increasing (i.e., $y_1 > y_2$ implies $Y_t(y_1) \geq Y_t(y_2)$). It follows that

$$G_t(\theta) := 1 - \Psi\big(Y_t(\theta)\big)$$

is a family of martingale survival processes.

Example 2 Let us reduce our attention to the case where Ψ is the cumulative distribution function of a standard Gaussian variable. Since $\Phi''(y) = -y\Phi'(y)$, the equation (5) reduces to

$$a(t, y) - \frac{1}{2}y\nu^2(t, y) = 0.$$

In the particular the case where $\nu(t, y) = \nu(t)$, straightforward computation leads to

$$Y_t(y) = e^{\frac{1}{2}\int_0^t \nu^2(s)ds}\left(y + \int_0^t e^{-\frac{1}{2}\int_0^s \nu^2(u)du}\nu(s)dB_s\right).$$

Setting $f(s) = -\nu(s)\exp(-\frac{1}{2}\int_0^s \nu^2(u)du)$, one deduces that $Y_t(y) = \frac{y-m_t}{\sigma(t)}$, where $\sigma^2(t) = \int_t^\infty f^2(s)ds$ and $m_t =: \int_0^t f(s)dB_s$, and we recover the Gaussian example of Sect. 3.1.

4 Density Models

In this section, we are interested in densities on $\mathbb{R}_+$ in order to give models for the conditional law of a random time τ. We recall the classical constructions of default times as first hitting time of a barrier, independent of the reference filtration, and we extend these constructions to the case where the barrier is no more independent of the reference filtration. It is then natural to characterize the dependence of this barrier and the filtration by means of its conditional law.

In the literature on credit risk modeling, the attention is mostly focused on the intensity process, i.e., to the process Λ such that $\mathbb{1}_{\tau \leq t} - \Lambda_{t\wedge\tau}$ is a $\mathbf{G} = \mathbf{F} \vee \mathbf{H}$-martingale, where $\mathscr{H}_t = \sigma(t \wedge \tau)$. We recall that the intensity process Λ is the only increasing predictable process such that the survival process $G_t := P(\tau > t|\mathscr{F}_t)$ admits the decomposition $G_t = N_t e^{-\Lambda_t}$ where N is a local martingale. We recall that the intensity process can be recovered form the density process as $d\Lambda_s = \frac{g_s(s)}{G_s(s)}ds$ (see [4]). We end the section giving an explicit example of two different martingale survival processes having the same survival processes (hence the intensities are equal).

4.1 Structural and Reduced-Form Models

In the literature, models for default times are often based on a threshold: the default occurs when some driving process X reaches a given barrier. Based on this observation, we consider the random time on $\mathbb{R}_+$ in a general threshold model. Let X be a stochastic process and Θ be a barrier which we shall precise later. Define the random time as the first passage time

$$\tau := \inf\{t \,:\, X_t \geq \Theta\}.$$

In classical structural models, the process X is an $\mathbf{F}$-adapted process associated with the value of a firm and the barrier Θ is a constant. So, τ is an $\mathbf{F}$-stopping time. In this case, the conditional distribution of τ does not have a density process, since $P(\tau > \theta|\mathscr{F}_t) = \mathbb{1}_{\theta<\tau}$ for $\theta \leq t$.

To obtain a density process, the model has to be changed, for example one can stipulate that the driving process X is not observable and that the observation is a filtration $\mathbf{F}$, smaller than the filtration $\mathbf{F}^X$, or a filtration including some noise. The goal is again to compute the conditional law of the default $P(\tau > \theta|\mathscr{F}_t)$, using for example filtering theory.

Another method is to consider a right-continuous $\mathbf{F}$-adapted increasing process Γ and to randomize the barrier. The easiest way is to take the barrier Θ as an $\mathscr{A}$-measurable random variable independent of $\mathbf{F}$, and to consider

$$\tau := \inf\{t \,: \Gamma_t \geq \Theta\}. \tag{6}$$

If Γ is continuous, τ is the inverse of Γ taken at Θ, and $\Gamma_\tau = \Theta$. The $\mathbf{F}$-conditional law of τ is

$$P(\tau > \theta|\mathscr{F}_t) = G^{\Theta}(\Gamma_\theta), \quad \theta \leq t,$$

where G^{Θ} is the survival probability of Θ given by $G^{\Theta}(t) = P(\Theta > t)$. We note that in this particular case, $P(\tau > \theta|\mathscr{F}_t) = P(\tau > \theta|\mathscr{F}_\infty)$ for any $\theta \leq t$, which means that the H-hypothesis is satisfied[2] and that the martingale survival processes remain constant after θ (i.e., $G_t(\theta) = G_\theta(\theta)$ for $t \geq \theta$). This result is stable by

[2] We recall that H-hypothesis stands for any $\mathbf{F}$-martingale is a $\mathbf{G} = \mathbf{F} \vee \mathbf{H}$ martingale.

increasing transformation of the barrier, so that we can assume without loss of generality that the barrier is the standard exponential random variable $-\log G^{\Theta}(\Theta)$.

If the increasing process Γ is assumed to be absolutely continuous with respect to the Lebesgue measure with Radon–Nikodym density γ and if G^{Θ} is differentiable, then the random time τ admits a density process given by

$$\begin{aligned} g_t(\theta) &= -\big(G^{\Theta}\big)'(\Gamma_\theta)\gamma_\theta = g_\theta(\theta), && \theta \le t, \\ &= E\big(g_\theta(\theta)|\mathscr{F}_t\big), && \theta > t. \end{aligned} \tag{7}$$

Example (Cox process model) In the widely used Cox process model, the independent barrier Θ follows the exponential law and $\Gamma_t = \int_0^t \gamma_s ds$ represents the default compensator process. As a direct consequence of (7),

$$g_t(\theta) = \gamma_\theta e^{-\Gamma_\theta}, \quad \theta \le t.$$

4.2 Generalized Threshold Models

In this subsection, we relax the assumption that the threshold Θ is independent of $\mathscr{F}_\infty$. We assume that the barrier Θ is a strictly positive random variable whose conditional distribution w.r.t. $\mathbf{F}$ admits a density process, i.e., there exists a family of $\mathscr{F}_t \otimes \mathscr{B}(\mathbb{R}_+)$-measurable functions $p_t(u)$ such that

$$G_t^{\Theta}(\theta) := P(\Theta > \theta|\mathscr{F}_t) = \int_\theta^\infty p_t(u)du\,. \tag{8}$$

We assume in addition that the process Γ is absolutely continuous w.r.t. the Lebesgue measure, i.e., $\Gamma_t = \int_0^t \gamma_s ds$. We still consider τ defined as in (6) by $\tau = \Gamma^{-1}(\Theta)$ and we say that a random time constructed in such a setting is given by *a generalized threshold*.

Proposition 3 *Let τ be given by a generalized threshold. Then τ admits the density process $g(\theta)$ where*

$$g_t(\theta) = \gamma_\theta p_t(\Gamma_\theta), \quad \theta \le t. \tag{9}$$

Proof By definition and by the fact that Γ is strictly increasing and absolutely continuous, we have for $t \ge \theta$,

$$\begin{aligned} G_t(\theta) &:= P(\tau > \theta|\mathscr{F}_t) = P(\Theta > \Gamma_\theta|\mathscr{F}_t) = G_t^{\Theta}(\Gamma_\theta) = \int_{\Gamma_\theta}^\infty p_t(u)du \\ &= \int_\theta^\infty p_t(\Gamma_u)\gamma_u du, \end{aligned}$$

which implies $g_t(\theta) = \gamma_\theta p_t(\Gamma_\theta)$ for $t \ge \theta$.

Obviously, in the particular case where the threshold Θ is independent of $\mathscr{F}_\infty$, we recover the classical results (7) recalled above.

Conversely, if we are given a density process g, then it is possible to construct a random time τ by a generalized threshold, that is, to find Θ such that the associated τ has g as density, as we show now. It suffices to define $\tau = \inf\{t : t \geq \Theta\}$ where Θ is a random variable with conditional density $p_t = g_t$. Of course, for any increasing process Γ, $\tau = \inf\{t : \Gamma_t \geq \Delta\}$ where $\Delta := \Gamma_\Theta$ is a different way to obtain a solution! □

4.3 An Example with Same Survival Processes

We recall that, starting with a survival martingale process $\widetilde{G}_t(\theta)$, one can construct other survival martingale processes $G_t(\theta)$ admitting the same survival process (i.e., $\widetilde{G}_t(t) = G_t(t)$), in particular, the same intensity. The construction is based on the general result obtained in Jeanblanc and Song [11]: for any supermartingale Z valued in $[0, 1[$, with multiplicative decomposition $Ne^{-\Lambda}$, where Λ is continuous, the family

$$G_t(\theta) = 1 - (1 - Z_t)\exp\left\{-\int_\theta^t \frac{Z_s}{1 - Z_s} d\Lambda_s\right\}, \quad 0 < \theta \leq t \leq \infty,$$

is a martingale survival process (called the basic martingale survival process) which satisfies $G_t(t) = Z_t$ and, if N is continuous, $dG_t(\theta) = \frac{1-G_t(\theta)}{1-Z_t} e^{-\Lambda_t} dN_t$. In particular, the associated intensity process is Λ (we emphasize that the intensity process does not contain enough information about the conditional law).

We illustrate this construction in the Gaussian example presented in Sect. 3.1 where we set $Y_t = \frac{m_t - h(t)}{\sigma(t)}$. The multiplicative decomposition of the supermartingale $\widetilde{G}_t = P(\tau > t | \mathscr{F}_t^B)$ is $\widetilde{G}_t = N_t \exp\{-\int_0^t \lambda_s ds\}$ where

$$dN_t = N_t \frac{\varphi(Y_t)}{\sigma(t)\Phi(Y_t)} dm_t, \qquad \lambda_t = \frac{h'(t)\,\varphi(Y_t)}{\sigma(t)\,\Phi(Y_t)}.$$

Using the fact that $\widetilde{G}_t(t) = \Phi(Y_t)$, one checks that the basic martingale survival process satisfies

$$dG_t(\theta) = \big(1 - G_t(\theta)\big)\frac{f(t)\varphi(Y_t)}{\sigma(t)\Phi(-Y_t)} dB_t, \quad t \geq \theta, \qquad G_\theta(\theta) = \Phi(Y_\theta),$$

providing a new example of martingale survival processes, with density process

$$g_t(\theta) = (1 - G_t) e^{-\int_\theta^t \frac{G_s}{1-G_s}\lambda_s ds} \frac{G_\theta \lambda_\theta}{1 - G_\theta}, \quad \theta \leq t.$$

Other constructions of martingale survival processes having a given survival process can be found in [12], as well as constructions of local-martingales N such that $Ne^{-\Lambda}$ is valued in $[0, 1]$ for a given increasing continuous process Λ.

5 Change of Probability Measure and Filtering

In this section, our goal is to show how, using a change of probability measure, one can construct density processes. The main idea is that, starting from the (unconditional) law of τ, we construct a conditional density in a dynamic way using a change of probability. This methodology is a very particular case of the general change of measure approach developed in [4]. Then, we apply the idea of change of probability framework to a filtering problem (due to Kallianpur and Striebel [10]), to obtain the Kallianpur–Striebel formula for the conditional density (see also Meyer [16]). Our results are established in a very simple way, in a general filtering model, when the signal is a random variable, and contain, in the simple case, the results of Filipovic et al. [7]. We end the section with an example of the traditional Gaussian filtering problem.

5.1 Change of Measure

One starts with the elementary model where, on the filtered probability space $(\Omega, \mathscr{A}, \mathbf{F}, P)$, an $\mathscr{A}$-measurable random variable X is independent from the reference filtration $\mathbf{F} = (\mathscr{F}_t)_{t\geq 0}$ and its law admits a density probability g_0, so that

$$P(X > \theta|\mathscr{F}_t) = P(X > \theta) = \int_\theta^\infty g_0(u)du\,.$$

We denote by $\mathbf{G}^X = \mathbf{F} \vee \sigma(X)$ the filtration generated by $\mathbf{F}$ and X.

Let $(\beta_t(u),\ t \in \mathbb{R}_+)$ be a family of positive $(P, \mathbf{F})$-martingales such that $\beta_0(u) = 1$ for all $u \in \mathbb{R}$. Note that, due to the assumed independence of X and $\mathbf{F}$, the process $(\beta_t(X),\ t \geq 0)$ is a $\mathbf{G}^X$-martingale and one can define a probability measure Q on $(\Omega, \mathscr{G}_t^X)$, by $dQ = \beta_t(X)dP$. Since $\mathbf{F}$ is a subfiltration of $\mathbf{G}^X$, the positive $\mathbf{F}$-martingale

$$m_t^\beta := E\big(\beta_t(X)|\mathscr{F}_t\big) = \int_0^\infty \beta_t(u)g_0(u)du$$

is the Radon–Nikodym density of the measure Q, restricted to $\mathscr{F}_t$ with respect to P (note that $m_0^\beta = 1$). Moreover, the Q-conditional density of X with respect to $\mathscr{F}_t$ can be computed, from the Bayes formula

$$Q(X \in B|\mathscr{F}_t) = \frac{1}{E(\beta_t(X)|\mathscr{F}_t)} E\big(\mathbf{1}_B(X)\beta_t(X)|\mathscr{F}_t\big) = \frac{1}{m_t^\beta}\int_B \beta_t(u)g_0(u)du$$

where we have used, in the last equality the independence between X and $\mathbf{F}$, under P. Let us summarize this simple but important result:

Proposition 4 *If X is a random variable with probability density g_0, independent from $\mathbf{F}$ under P, and if Q is a probability measure, equivalent to P on $\mathbf{F} \vee \sigma(X)$*

with Radon–Nikodym density $\beta_t(X)$, $t \geq 0$, then the $(Q, \boldsymbol{F})$-density process of X is

$$g_t^Q(u)du := Q(X \in du|\mathscr{F}_t) = \frac{1}{m_t^\beta}\, \beta_t(u)g_0(u)du \tag{10}$$

where m^β is the normalizing factor, $m_t^\beta = \int_{-\infty}^{\infty} \beta_t(u)g_0(u)du$. In particular

$$Q(\tau \in du) = P(\tau \in du) = g_0(u)du\,.$$

The right-hand side of (10) can be understood as the ratio of $\beta_t(u)g_0(u)$ (the change of probability times the P probability density) and a normalizing coefficient m_t^β. One can say that $(\beta_t(u)g_0(u),\ t \geq 0)$ is the unnormalized density, obtained by a linear transformation from the initial density. The normalization factor $m_t^\beta = \int \beta_t(u)g_0(u)du$ introduces a nonlinear dependence of $g_t^Q(u)$ with respect to the initial density. The example of the filtering theory provides an explicit form to this dependence when the martingales $\beta_t(u)$ are stochastic integrals with respect to a Brownian motion.

Remark 1 We present here some important remarks.

(1) If, for any t, $m_t^\beta = 1$, then the probability measures P and Q coincide on $\mathbf{F}$. In that case, the process $(\beta_t(u)g_0(u),\ t \geq 0)$ is a density process.

(2) Let $\mathbf{G} = (\mathscr{G}_t)_{t\geq 0}$ be the usual right-continuous and complete filtration in the default framework (i.e. when $X = \tau$ is a nonnegative random variable) generated by $\mathscr{F}_t \vee \sigma(\tau \wedge t)$. Similar calculation may be made with respect to $\mathscr{G}_t$. The only difference is that the conditional distribution of τ is a Dirac mass on the set $\{t \geq \tau\}$. On the set $\{\tau > t\}$, and under Q, the distribution of τ admits a density given by:

$$Q(\tau \in du|\mathscr{G}_t) = \beta_t(u)g_0(u)\frac{1}{\int_t^{\infty} \beta_t(\theta)g_0(\theta)d\theta}du.$$

(3) This methodology can be easily extended to a multivariate setting: one starts with an elementary model, where the τ_i, $i = 1, \ldots, d$, are independent from $\mathbf{F}$, with joint density $g(u_1, \ldots, u_d)$. With a family of non-negative martingales $\beta(\theta_1, \ldots, \theta_d)$, the associated change of probability provides a multidimensional density process.

5.2 *Filtering Theory*

The change of probability approach presented in the previous Sect. 5.1 is based on the idea that we can restrict our attention to the simple case where the random variable is independent from the filtration and use a change of probability. The same idea is the building block of filtering theory as we present now.

Let W be a Brownian motion on the probability space $(\Omega, \mathscr{A}, P)$, and X be a random variable independent of W, with probability density g_0. We denote by

$$dY_t = a(t, Y_t, X)dt + b(t, Y_t)dW_t \tag{11}$$

the observation process, where a and b are smooth enough to have a solution and where b does not vanish. The goal is to compute the conditional density of X with respect to the filtration $\mathbf{F}^Y$. The way we shall solve the problem is to construct a probability Q, equivalent to P, such that, under Q, the signal X and the observation $\mathbf{F}^Y$ are independent, and to compute the density of X under P by means of the change of probability approach of the previous section. It is known in nonlinear filtering theory as the Kallianpur–Striebel methodology [10], a way to linearize the problem.

Note that, from the independence assumption on X and W, we see that W is a $\mathbf{G}^X = \mathbf{F}^W \vee \sigma(X)$-martingale under P.

5.2.1 Simple Case

We start with the simple case where the dynamics of the observation is

$$dY_t = a(t, X)dt + dW_t\,.$$

We assume that a is smooth enough so that the solution of

$$d\beta_t(X) = -\beta_t(X)a(t, X)dW_t, \quad \beta_0(X) = 1,$$

is a $(P, \mathbf{G}^X)$-martingale, and we define a probability measure Q on $\mathscr{G}_t^X$ by putting $dQ = \beta_t(X)dP$. Then, by Girsanov's theorem, the process Y is a $(Q, \mathbf{G}^X)$-Brownian motion, hence is independent from $\mathscr{G}_0^X = \sigma(X)$, under Q. Then, we apply our change of probability methodology, writing

$$dP = \frac{1}{\beta_t(X)}dQ =: \ell_t(X)dQ$$

with

$$d\ell_t(X) = \ell_t(X)a(t, X)dY_t, \quad \ell_0(X) = 1,$$

(in other words, $\ell_t(u) = \frac{1}{\beta_t(u)} = \exp(\int_0^t a(s, u)dY_s - \frac{1}{2}\int_0^t a^2(s, u)ds)$) and we get from Proposition 4 that the density of X under P, with respect to $\mathbf{F}^Y$, is $g_t(u)$, given by

$$P\big(X \in du|\mathscr{F}_t^Y\big) = g_t(u)du = \frac{1}{m_t^\ell}g_0(u)\ell_t(u)du$$

where $m_t^\ell = E_Q(\ell_t(X)|\mathscr{F}_t^Y) = \int_{-\infty}^{\infty}\ell_t(u)g_0(u)du$. Since

$$dm_t^\ell = \left(\int_{-\infty}^{\infty}\ell_t(u)a(t, u)g_0(u)du\right)dY_t = m_t^\ell\left(\int_{-\infty}^{\infty}g_t(u)a(t, u)du\right)dY_t$$

and setting

$$\widehat{a}_t := E\big(a(t,X)|\mathscr{F}_t^Y\big) = \int_{-\infty}^{\infty} g_t(u)a(t,u)du\,,$$

Girsanov's theorem implies that the process B given by

$$dB_t = dY_t - \widehat{a}_t dt = dW_t + \big(a(t,X) - \widehat{a}_t\big)dt$$

is a $(P, \mathbf{F}^Y)$-Brownian motion (it is the innovation process). From Itô's calculus, it is easy to show that the density process satisfies the nonlinear filtering equation

$$\begin{aligned} dg_t(u) &= g_t(u)\left(a(t,u) - \frac{1}{m_t^\ell}\int_{-\infty}^{\infty} dy\, g_0(y)a(t,y)\ell_t(y)\right)dB_t \\ &= g_t(u)\big(a(t,u) - \widehat{a}_t\big)dB_t\,. \end{aligned} \tag{12}$$

Remark 2 Observe that conversely, given a solution $g_t(u)$ of (12), and the process μ solution of $d\mu_t = \mu_t\widehat{a}_t dY_t$, then $h_t(u) = \mu_t g_t(u)$ is solution of the linear equation $dh_t(u) = h_t(u)a(t,u)dY_t$.

5.2.2 General Case

Using the same ideas, we now solve the filtering problem in the case where the observation follows (11). Let $\beta(X)$ be the $\mathbf{G}^X$-local martingale, solution of

$$d\beta_t(X) = \beta_t(X)\sigma_t(X)dW_t, \quad \beta_0(X) = 1,$$

with $\sigma_t(X) = -\frac{a(t,Y_t,X)}{b(t,Y_t)}$. We assume that a and b are smooth enough so that β is a martingale. Let Q be defined on $\mathscr{G}_t^X$ by $dQ = \beta_t(X)dP$.

From Girsanov's theorem, the process $\widehat{W}$ defined as

$$d\widehat{W}_t = dW_t - \sigma_t(X)dt = \frac{1}{b(t,Y_t)}dY_t$$

is a $(Q, \mathbf{G}^X)$-Brownian motion, hence $\widehat{W}$ is independent from $\mathscr{G}_0^X = \sigma(X)$. Being $\mathbf{F}^Y$-adapted, the process $\widehat{W}$ is a $(Q, \mathbf{F}^Y)$-Brownian motion, X is independent from $\mathbf{F}^Y$ under Q, and, as mentioned in Proposition 4, admits, under Q, the probability density g_0.

We now assume that the natural filtrations of Y and $\widehat{W}$ are the same. To do so, note that it is obvious that $\mathbf{F}^{\widehat{W}} \subseteq \mathbf{F}^Y$. If the SDE $dY_t = b(t,Y_t)d\widehat{W}_t$ has a strong solution (e.g., if b is Lipschitz, with linear growth) then $\mathbf{F}^Y \subseteq \mathbf{F}^{\widehat{W}}$ and the equality between the two filtrations holds.

Then, we apply our change of probability methodology, with $\mathbf{F}^Y$ as the reference filtration, writing $dP = \ell_t(X)dQ$ with $d\ell_t(X) = -\ell_t(X)\sigma_t(X)d\widehat{W}_t$ (which follows from $\ell_t(X) = \frac{1}{\beta_t(X)}$) and we get that the density of X under P, with respect to $\mathbf{F}^Y$

is $g_t(u)$ given by

$$g_t(u) = \frac{1}{m_t^\ell} g_0(u)\ell_t(u)$$

with dynamics

$$\begin{aligned} dg_t(u) &= -g_t(u)\left(\sigma_t(u) - \frac{1}{m_t^\ell}\int_{-\infty}^{\infty} dy\, g_0(y)\sigma_t(y)\ell_t(y)\right)dB_t \\ &= g_t(u)\left(\frac{a(t,Y_t,u)}{b(t,Y_t)} - \frac{1}{b(t,Y_t)}\int_{-\infty}^{\infty} dy\, g_t(y)a(t,Y_t,y)\right)dB_t \\ &= g_t(u)\left(\frac{a(t,Y_t,u)}{b(t,Y_t)} - \frac{\widehat{a}_t}{b(t,Y_t)}\right)dB_t\,. \end{aligned} \tag{13}$$

Here B is a $(P, \mathbf{F}^Y)$-Brownian motion (the innovation process) given by

$$dB_t = dW_t + \left(\frac{a(t,Y_t,X)}{b(t,Y_t)} - \frac{\widehat{a}_t}{b(t,Y_t)}\right)dt\,,$$

where $\widehat{a}_t = E(a(t,Y_t,X)|\mathscr{F}_t^Y)$.

Proposition 5 *If the signal X has probability density $g_0(u)$ and is independent from the Brownian motion W, and if the observation process Y follows*

$$dY_t = a(t,Y_t,X)dt + b(t,Y_t)dW_t\,,$$

then the conditional density of X given $\mathscr{F}_t^Y$ is

$$P\big(X \in du|\mathscr{F}_t^Y\big) = g_t(u)du = \frac{1}{m_t^\ell} g_0(u)\ell_t(u)du \tag{14}$$

where

$$\begin{aligned} \ell_t(u) &= \exp\left\{\int_0^t \frac{a(s,Y_s,u)}{b^2(s,Y_s)}dY_s - \frac{1}{2}\int_0^t \frac{a^2(s,Y_s,u)}{b^2(s,Y_s)}ds\right\}, \\ m_t^\ell &= \int_{-\infty}^{\infty} \ell_t(u)g_0(u)du, \end{aligned}$$

and its dynamics is given by (13).

5.3 Gaussian Filter

We apply our results to the well-known case of Gaussian filter. Let W be a Brownian motion, X be a random variable (the signal) with Gaussian density g_0 with mean m_0

and variance γ_0, independent of the Brownian motion W, and let Y (the observation) be the solution of

$$dY_t = \big(a_0(t,Y_t) + a_1(t,Y_t)X\big)dt + b(t,Y_t)dW_t.$$

Then, from the previous results, the density process $g_t(u)$ is of the form

$$\frac{1}{m_t^\ell}\exp\left\{\int_0^t \frac{a_0(s,Y_s)+a_1(s,Y_s)u}{b^2(s,Y_s)}\,dY_t \right.$$
$$\left. - \frac{1}{2}\int_0^t \left(\frac{a_0(s,Y_s)+a_1(s,Y_s)u}{b(s,Y_s)}\right)^2 ds\right\} g_0(u).$$

The logarithm of $g_t(u)$ is a quadratic form in u with stochastic coefficient, so that $g_t(u)$ is a Gaussian density, with mean m_t and variance γ_t (as proved already by Liptser and Shiryaev [14]). A tedious computation, purely algebraic, shows that

$$\gamma_t = \frac{\gamma_0}{1+\gamma_0\int_0^t \frac{a_1^2(s,Y_s)}{b^2(s,Y_s)}ds}, \qquad m_t = m_0 + \int_0^t \gamma_s \frac{a_1(s,Y_s)}{b(s,Y_s)}\,dB_s$$

with

$$dB_t = dW_t + \frac{a_1(t,Y_t)}{b(t,Y_t)}\big(X - E\big(X|\mathscr{F}_t^Y\big)\big)dt.$$

Back to the Gaussian example in Sect. 3.1: In the case where the coefficients of the process Y are deterministic functions of time, i.e.

$$dY_t = \big(a_0(t) + a_1(t)X\big)dt + b(t)dW_t,$$

the variance $\gamma(t)$ is deterministic and the mean is an $\mathbf{F}^Y$-Gaussian martingale

$$\gamma(t) = \frac{\gamma_0}{1+\gamma_0\int_0^t \alpha^2(s)ds}, \qquad m_t = m_0 + \int_0^t \gamma(s)\alpha(s)dB_s$$

where $\alpha = a_1/b$. Furthermore, $\mathbf{F}^Y = \mathbf{F}^B$.

Choosing $f(s) = \frac{\gamma(s)a_1(s)}{b(s)}$ in the example of Sect. 3.1 leads to the same conditional law (with $m_0 = 0$); indeed, it is not difficult to check that this choice of parameter leads to $\int_t^\infty f^2(s)ds = \sigma^2(t) = \gamma(t)$ so that the two variances are equal.

The similarity between filtering and the example of Sect. 3.1 can be also explained as follows. Let us start from the setting of Sect. 3.1 where the random variable $X = \int_0^\infty f(s)dB_s$ and introduce $\mathbf{G}^X = \mathbf{F}^B \vee \sigma(X)$, where B is the given Brownian motion. Standard results of enlargement of filtration (see Jacod [9]) show that

$$W_t := B_t + \int_0^t \frac{m_s - X}{\sigma^2(s)} f(s)ds$$

is a $\mathbf{G}^X$-Brownian motion, hence is a $\mathbf{G}^W$-Brownian motion independent of X. So, the example presented in Sect. 3.1 is equivalent to the following filtering problem: the signal is X a Gaussian variable, centered, with variance $\gamma(0) = \int_0^\infty f^2(s)ds$ and the observation

$$dY_t = f(t)Xdt + \left(\int_t^\infty f^2(s)ds\right)dW_t = f(t)Xdt + \sigma^2(t)dW_t .$$

References

1. Amendinger, J.: Initial enlargement of filtrations and additional information in financial markets. PhD thesis, Technischen Universität Berlin (1999)
2. Carmona, R.: Emissions option pricing. Slides Heidelberg (2010)
3. Chaleyat-Maurel, M., Jeulin, T.: Grossissement Gaussien de la filtration Brownienne. Lecture Notes in Math., vol. 1118, pp. 59–109. Springer, Berlin (1985)
4. El Karoui, N., Jeanblanc, M., Jiao, Y.: What happens after a default: the conditional density approach. Stoch. Process. Appl. **120**, 1011–1032 (2010)
5. Fermanian, J.D., Vigneron, O.: 2010, On break-even correlation: the way to price structured credit derivatives by replication. Preprint
6. Filipović, D., Overbeck, L., Schmidt, T.: Dynamic CDO term structure modeling. Math. Finance (2009). Forthcoming
7. Filipovic, D., Hughston, L., Macrina, A.: Conditional density models for asset pricing. Preprint (2010)
8. Grorud, A., Pontier, M.: Asymmetrical information and incomplete markets. Int. J. Theor. Appl. Finance **4**, 285–302 (2001)
9. Jacod, J.: Grossissement initial, hypothèse (H') et théorème de Girsanov. Lecture Notes in Math., vol. 1118, pp. 15–35. Springer, Berlin (1985)
10. Kallianpur, G., Striebel, C.: Estimation of stochastic systems: arbitrary system process with additive white noise observation errors. Ann. Math. Stat. **39**(3), 785–801 (1968)
11. Jeanblanc, M., Song, S.: Explicit model of default time with given survival probability. Preprint (2010)
12. Jeanblanc, M., Song, S.: Default times with given survival probability and their **F**-martingale decomposition formula. Preprint (2010)
13. Keller-Ressel M., Papapantoleon, A., Teichman, J.: The Affine Libor Models. Preprint (2010)
14. Liptser, R.S., Shiryaev, A.N.: Statistics of Random Processes, II Applications, 2nd edn. Springer, Berlin (2001)
15. Matsumoto, H., Yor, M.: A relationship between Brownian motions with opposite drifts via certain enlargements of the Brownian filtration. Osaka J. Math. **38**, 383–398 (2001)
16. Meyer, P.-A.: Sur un problème de filtration. In: Séminaire de Probabilités VII. Lecture Notes in Math., vol. 321, pp. 223–247. Springer, Berlin (1973)
17. Yor, M.: Grossissement de filtrations et absolue continuité de noyaux. Lecture Notes in Math., vol. 1118, pp. 6–14. Springer, Berlin (1985)

Yield Curve Smoothing and Residual Variance of Fixed Income Positions

Raphaël Douady

Abstract We model the yield curve in any given country as an object lying in an infinite-dimensional Hilbert space, the evolution of which is driven by what is known as a *cylindrical Brownian motion*. We assume that volatilities and correlations do not depend on rates (which hence are Gaussian). We prove that a principal component analysis (PCA) can be made. These components are called *eigenmodes* or *principal deformations* of the yield curve in this space. We then proceed to provide the best approximation of the curve evolution by a Gaussian Heath–Jarrow–Morton model that has a given finite number of factors. Finally, we describe a method, based on finite elements, to compute the eigenmodes using historical interest rate data series and show how it can be used to compute approximate hedges which optimize a criterion depending on transaction costs and residual variance.

Keywords Cylindrical Brownian motion · Term structure of interest rates · Yield curve · Heath–Jarrow–Morton model · Fixed-income models · Asymptotic arbitrage

Mathematics Subject Classification (2010) 91G30 · 91G60

1 Introduction

Infinity is a word that economists usually do not like. Nothing, in economy, can be considered either as infinitely large, or as infinitely small. The size of worldwide markets is finite, as well as the total number of various stocks and bonds. Conversely, transactions cannot be infinitely close in time (a minimum time period is required between two transactions on the same asset) and price variations cannot be less than a "tick", neither can they be infinitely large. Nevertheless, two seminal articles

R. Douady (✉)
CES, Univ. Paris 1, 106 Bd de l'hôpital, 75647 Paris cedex 13, France
e-mail: rdouady@univ-paris1.fr

Y. Kabanov et al. (eds.), *Inspired by Finance*, DOI 10.1007/978-3-319-02069-3_10,

introduced, perhaps unwillingly, infinity into the finance literature.[1] The first one is Robert C. Merton's article on continuous time finance [26, 1973] (see also [27] and ref. cit.). Indeed, considering the possibility of continuous time trading replaces the setting of a finite set of random innovations by the infinite-dimensional Wiener space of Brownian motions. Although we just said that such trading strategies are physically impossible to execute, we consider this theory as financially extremely significant. The reason is that when one wants to study the results of a given trading strategy over a time period that is huge compared to the minimal trading interval, then any discrete time approach based on the maximal trading frequency would be, for an equivalent numerical precision, much more complex to implement than the corresponding continuous time limit. Consequently, though practical hedging should be performed with a view of optimizing a near future situation, according to all particularities of the market at the present time, pricing, which is based on an average of the resulting wealth of forecasted hedging strategy, is better handled in a continuous time framework.[2]

The second article is Heath–Jarrow–Morton interest rate model [21]. This model, which we call H.J.M. in the sequel, "summarizes" information about the interest rate market (Libor, Libor futures, swaps, fixed income assets, etc.) into a "yield curve" or, more precisely, a curve of "forward spot rates". The knowledge of this curve is equivalent, through a simple integration with respect to maturity, to the price of zero-coupon bonds of any maturity. Again, here the word "any" means a continuum of maturities. The set of possible curves is an infinite dimensional functional space and the market cannot be described by a finite set of "market variables", although only finitely many assets are traded. One could argue that the model can be reduced to a finite-dimensional subspace and that the knowledge of a finite number of variables is enough to describe the whole market. In fact, this argument does not hold. Even if, as explicated in Heath–Jarrow–Morton article, the infinitesimal evolution of the curve is given by a finite number of "factors", the support of the distribution of possible curves after an arbitrarily small, but finite, amount of time is in general equal to the whole functional space.[3] Economically speaking, one must understand that both the finite number of factors and the continuous curve are introduced for the sake of simplicity, but none of them corresponds to reality: the total number of asset prices is finite, and the number of sources of noise, though also finite, is much larger than what is currently implemented in most trading floors.

A few articles describe capital markets as a random field. The first one that came to our knowledge is Kennedy [24]. This model is a Gaussian random field which could be considered as a generalization of the Gaussian H.J.M. model with volatility factors which do not depend on the level of the rates. In the example he provides, the forward spot rate $f(t,T)$ is a "Brownian sheet", which is in contradiction with

[1]One should add that the theoretical justification of arbitrage theory is itself an—actually questionable—"infinity" argument: if some true arbitrage opportunity existed, it could be implemented with an infinite nominal amount, hence reducing it to zero.

[2]"Hedge in discrete time finance, price in continuous time finance" (N. Taleb, 1996).

[3]In [28], Musiela gives an example of a one-factor HJM model with this property.

statistical evidence. Indeed, the process $f(t, T)$ has a very asymmetrical behavior with respect to its two variables. For fixed T, it behaves as a stochastic process with respect to the current date t, but at a given date t the curve $T \longmapsto f(t, T)$ is generally smooth. In the random field framework, this can be achieved by requiring the correlation function of the field to be smooth along the diagonal in the transverse direction. See Bricio-Hernandez [6] for a theoretical study of this topic and Turner [32] where a statistical study of the correlation function, showing this smoothness along the diagonal, is performed.

In this article, we develop another approach towards an infinite dimensional model, based on so-called "cylindrical Brownian motions". These processes are a generalization multidimensional Brownian motions to infinite dimensional Hilbert spaces. They were introduced by Gaveau in 1953 (see [19]). We refer to Yor [34] and to Da Prato–Zapczyk [12, p. 96] for a complete presentation of this theory. Our model can be seen as a limit case of H.J.M. model in infinite dimensions. As in Brace–Gatarek–Musiela model (B.G.M.) (see [4]), we consider the term structure of interest rates as an object in a certain functional space. We then proceed to study the motion of the vector representing this object. For simplicity reasons, we chose to work in a framework where rates are Gaussian. However this theory can be easily generalized to B.G.M. log-normal setting, or to any specified diffusion process for the term structure in which rate volatilities depend on the global term structure. Obviously certain technical assumptions apply.

Under very natural hypothesis,[4] we show that this type of motion can always be decomposed into an (infinite) sum of one-dimensional Brownian motions, which we call *eigenmodes* or *principal deformations*. This turns out to be a principal component analysis (P.C.A.) of the motion. Listing all the works on the yield curve P.C.A. would be impossible. Let us mention the initial study (as it came to our knowledge) of Litterman and Scheinkman [25], the theoretical article of the Banque de France [18], and the statistical analysis cited in this article. It is shown in Sect. 8 that the n-factor H.J.M. model that best reproduces an infinite dimensional diffusion of the yield curve, in the sense of minimizing the variance of the error, is provided by the truncated P.C.A.

R.C. Merton said in his preface to *Continuous-time Finance: "The continuous time model is a watershed between the static and dynamic models of finance"*. Similarly, we could say that this "functional analysis" of term structure models—a less polemical term than "infinite dimensions", though representing the same thing—is a watershed between one-dimensional and multi-dimensional arbitrage pricing.

From the point of view of risk management, this approach allows a substantial reduction of the computational burden relative to the usual "bucketing" method, while not losing any precision on market data fitting and risk evaluation. For this purpose, performing a P.C.A. of a statistically estimated variance-covariance matrix

[4]We assume that the price of zero-coupons always depends continuously on the maturity, and that their variance is finite at all time.

of the yield curve movements is of little help because of the high instability of this matrix. We recommend to choose a fixed series of basic deformations of the yield curve, which could be inspired by Fourier analysis or wavelets. A thorough historical analysis has to be performed to find the minimum number of terms one needs in order to reproduce, up to a tightly controlled error, all possible variations, even in case of crisis. In [13], we provide an example of such basic deformations for which seven terms are sufficient to reproduce the variations of all exchange quoted Euro-dollar futures over a 10-years period with an error that never exceeds two basis points. The same number of terms applies to cash and swap rate variations from one month until thirty years.[5] In comparison, Basel committee recommends to use 13 buckets in yield curve deformations.

Option pricing theory faces an unexpected difficulty in the infinite dimensional setting. Even if the whole volatility structure of the yield curve — that is, rate volatilities and correlations—is deterministic and known, some options may not support perfect replication, although they may have an "arbitrage price". In fact one will seek a sequence of "almost replicating" strategies, with a wealth variance tending to 0 and converging sequence of initial price. This leads us to introduce the notion of "quasi-arbitrage", that is, a sequence of trading strategies with returns bounded from below and wealth variance tending to 0 (Kabanov and Kramkov [23] call it "asymptotic arbitrage"). Only in the absence of quasi-arbitrage (A.Q.A. assumption) will an equivalent risk-neutral probability exist. Then an option price is the risk-neutral expectation of its discounted pay-off. This theoretical impossibility to perfectly replicate options does not create more difficulty in practical dynamic hedging than the inability to implement of a purely continuous time dynamic hedging. In a sense, it induces even less risk for, as mentioned above, the "spatial uncertainty", which measures how rate interpolations can be inaccurate, is much less unpredictable than the time uncertainty, which measures rate variations between two dynamic hedging transactions. Actually, traders often use linear interpolations to evaluate, when necessary, the rate to apply on a period which corresponds to no standard products. This practice justifies our approach based on the yield curve regularity.

This article is organized as follows. After preliminaries and notations (Sect. 3), we first expose (Sect. 5) the infinite dimensional diffusion of the yield curve and study the existence of an absolutely continuous risk-neutral probability, introducing the notion of quasi-arbitrage. In Sect. 7, we show that such a model can be seen as a limit case of H.J.M. finite dimensional model (in fact an extended version of the strict H.J.M. framework). In particular, we show in Sect. 8 the possibility of performing an infinite dimensional P.C.A. In Sect. 10, we provide basic option pricing formulas. The last part (Sects. 11 and 12) is devoted to numerical methods for the

[5] However, it does not correctly shows bond yield variations because, for economical reasons that are not the topic of this article, each bond price is subject to its own individual source of noise and the "smooth curve principle" only applies up to a limit of 20–30 bp's error size that cannot be captured by smooth functions of the maturity. Note that usual one-year buckets face the same inaccuracy.

three following purposes: option pricing and hedging, calibration, P.C.A. computation. Hedges optimizing a cost vs. residual variance criterion are provided when transaction costs apply.

In this study, we assume that rates only follow (infinite dimension) diffusion processes. In particular, we exclude jumps and other processes not driven by Brownian motions.

2 History, Tribute and Recent Bibliography

A first version of this article was written in 1994, while the author was working at Société Générale in the fixed income and foreign exchange derivatives trading room. This version has been finalized in 2001, but was never published before. This work raised the interest of Marek Musiela, who at that time was teaching at the University of New South Wales in Sydney, and was at the origin of the long and fruitful scientific relationship between the author and Marek Musiela. Modeling yield curves with infinitely many risk factors raised a lot of skepticism from the mathematical finance community at this time, and Marek Musiela was one of the first person to perceive this deep fact that infinite dimensions do not come as an *increase in complexity* of the models but, on the contrary, as the necessary *path to the most parsimonious* models of a complex reality.

Since then, many authors wrote on infinite dimensions modeling of the term structure, and also of other aspects of financial markets, such as volatility surfaces. Let us cite here only major references, which the reader is invited to consult, as well as other articles cited in those references. One of the most complete study on this topic is Damir Filipovic's thesis [17]. A good statistical study of the regularity properties of the yield curve in the US has been performed by Bouchaud & al. [3]. Another approach to infinite dimensions modeling is through stochastic partial differential equations (SPDE), see Cont's article [9] on this matter. In [28], Musiela and Sondermann pointed out that even a one-factor model can lead to a yield curve lying in an infinite dimensional space.

As general references to interest rate modeling, we recommend books by Musiela and Rutkowski [29], Brigo and Mercurio [7] and Rebonato [30], who also wrote in 2003 a thorough survey of interest rate modeling, as it appeared at that time [31].

3 Notations and Definitions

The framework described here is the classical framework of Heath–Jarrow–Morton [21], Brace–Gatarek–Musiela [4], El Karoui–Myneni–Viswanathan [16] and Jamshidian [22]. It is extended in the sense that each individual forward spot rate or discount factor is driven by its own Brownian motion (possibly correlated to others).

3.1 Term Structure of Interest Rates

We denote by $z(t,T)$ the discount factor applying on the period $[t,T]$, that is, the price at time t of an asset delivering one unit of numeraire at time T and let

$$\ell(t,T) = -\log z(t,T), \qquad y(t,T) = \frac{1}{T-t}\,\ell(t,T), \tag{1}$$

so that $y(t,T)$ is the continuously-compounded *zero-coupon rate,* or *yield,* over the time period $[t,T]$. One has $z(t,t)=1$, $\ell(t,t)=0$, $z(t,T)>0$. We shall not make any assumption about positive interest rates, because our model will be of Gaussian type, which does not rates from being negative, although with a very low probability. Such an assumption may be achieved by changing the volatility structure (see for instance [2, 4] for a log-normal structure, and [10, 14] for a χ^2-type distribution).

The *spot rate* $r(t)$ is the value of the zero-coupon rate when $T=t$:

$$r(t) = y(t,t).$$

In the H.J.M. framework, the *forward spot rate* $f(t,T)$ is considered:

$$f(t,T) = \frac{\partial \ell(t,T)}{\partial T}.$$

It is linked to the zero-coupon price and yield by the formulas

$$z(t,T) = \exp\left\{-\int_t^T f(t,s)\,ds\right\}, \tag{2}$$

$$y(t,T) = \frac{1}{T-t}\int_t^T f(t,s)\,ds. \tag{3}$$

In this article, we only consider continuously compounded rates, namely, those defined via the *logarithm* of zero-coupon prices. The usual rates, with finite compounding periods, are computed from those by simple formulas.

The functions z, ℓ, y and f will be considered as various representations of the same term structure. They will always be linked by the formulas above.

We denote by $\beta(t,T)$ the savings account at date T initiated at time t:

$$\beta(t,T) = \exp\left\{\int_t^T r(s)\,ds\right\}.$$

The value of $\beta(t,T)$ is only known at date T.

3.2 Risk-Neutral Probability

In the sequel, for any random process X_t, we denote by $\mathbf{E}_{\mathbf{P}}[X_T\,|\,t]$ the conditional expectation of X_T knowing the past until date t under the probability measure $\mathbf{P}$. We shall simply write $\mathbf{E}[X_T]$ if there is no ambiguity about t and $\mathbf{P}$.

An origin of time $t_0 = 0$ is fixed once for all, as well as a maximum maturity date of assets $T_{\max}$. For any Ito process X_t, we set

$$X_t = X_0 + \int_0^t dX_u.$$

In particular, we shall write indifferently

$$dX_t = \mu(t)\,dt + \sigma(t)\,dW_t \quad \Longleftrightarrow \quad X_t = X_0 + \int_0^t \mu(u)\,du + \int_0^t \sigma(u)\,dW_u,$$

where W_t is Brownian motion under $\mathbf{P}$ or another probability and $\mu(t)$, $\sigma(t)$ are predictable processes.

We let now $\mathbf{P}$ denote the real, or historical, probability. In the absence of arbitrage opportunities, for any maturity T, there exists a "risk-neutral" probability $\mathbf{Q}_T$ equivalent to $\mathbf{P}$, such that the discount factor $z(t,T)$ is the expectation at date t of $\beta(t,T)^{-1}$:

$$z(t,T) = \mathbf{E}_{\mathbf{Q}_T}\big[\beta(t,T)^{-1}\,|\,t\big].$$

Under this probability, the price of any asset X_t *depending only on discount factors* $z(t',T)$, $t < t' < T$, is such that $\beta(t_0,t)^{-1}X_t$ is a $\mathbf{Q}_T$-martingale for any initial date t_0. We call such assets T-assets. If the market of T-assets is complete, then $\mathbf{Q}_T$ is unique in the sense that two such probabilities would coincide on the space of T-assets. In this case, $\mathbf{Q}_T$ is characterized by the fact that $\beta(t_0,t)^{-1}z(t,T)$ is a martingale. The *risk-neutral* probability $\mathbf{Q}_T$ should not be confused with the *forward-neutral* one, which we shall denote $\mathbf{Q}'_T$ and characterized by the fact that $z(t,T)^{-1}X_t$ is a martingale for any T-asset X_t.

The Radon–Nikodym density of $\mathbf{Q}_T$ is given with respect to $\mathbf{P}$ by the mean of the "market price of risk", which a priori depends on the asset $z(t,T)$, thus on T. In fact, for any given *finite* set of maturities $\mathbf{T} = (T_1, \ldots, T_n)$ there exists a probability $\mathbf{Q}_{\mathbf{T}}$ under which all the t_0-actualized discount factors $\beta(t_0,t)^{-1}z(t,T_i)$ are martingales. The existence, uniqueness and absolute continuity of a risk-neutral probability for an infinite set of maturities, for instance a whole interval, is in general not ensured. This will be the topic of Sect. 3.3.2. When it exists, we call $\mathbf{Q}$ the probability which is risk-neutral with respect to every maturity $T \in [0, T_{\max}]$.

3.3 *Diffusion of Discount Factors and Forward Rates*

3.3.1 Discount Factors and Yields

For every fixed T, the discount factor process (defined for $t \le T$)

$$t \longmapsto z(t,T)$$

follows an Ito process

$$dz(t,T)|_{T \text{ fixed}} = z(t,T)\big(\mu(t,T)\,dt - \sigma(t,T)\,d\tilde{W}_t^T\big), \tag{4}$$

where $\tilde{W}_t^T$ is a standard Wiener process in t under the probability $\mathbf{P}$, depending on the parameter T, $\mu(t,T)$ and $\sigma(t,T)$ are predictable processes for the drift and the volatility respectively (the "minus" sign is artificial and has been put for technical reasons). The identity $z(t,t)=1$ implies

$$\sigma(t,t) = 0,$$

and, because

$$z(t,t+dt) = 1 - r(t)\,dt + O(dt^2),$$

one has

$$\mu(t,t) = r(t).$$

Taking the logarithm of (4), we get from the Ito formula

$$d\ell(t,T)|_{T \text{ fixed}} = \left(-\mu(t,T) + \frac{1}{2}\sigma(t,T)^2\right)dt + \sigma(t,T)\,d\tilde{W}^T. \tag{5}$$

Assumption. We make the following non-degeneracy assumption, which is a strong version of the completeness of the market of T-assets:

$$t < T \quad \Longrightarrow \quad \sigma(t,T) > 0.$$

Let the process W^T be defined by

$$W_t^T = \tilde{W}_t^T + \int_0^t \lambda(u,T)\,du, \qquad \lambda(u,T) = \frac{r(u) - \mu(u,T)}{\sigma(u,T)}.$$

This is a standard Brownian motion under probability $\mathbf{Q}_T$ and under $\mathbf{Q}$ if it exists (see [11] or [30]). One has

$$dz(t,T)|_{T \text{ fixed}} = z(t,T)\big(r(t)\,dt - \sigma(t,T)\,dW_t^T\big), \tag{6}$$

and

$$d\ell(t,T)|_{T \text{ fixed}} = \left(-r(t) + \frac{1}{2}\sigma(t,T)^2\right)dt + \sigma(t,T)\,dW^T, \tag{7}$$

which can be written, in terms of the quadratic variation process of $\ell(t,T)$,

$$d\ell(t,T)|_{T \text{ fixed}} = -r(t)\,dt + \frac{1}{2}d\,\langle \ell(t,T)\rangle + \sigma(t,T)\,dW^T. \tag{8}$$

For the zero-coupon rates $y(t, T)$, we get

$$dy(t,T) = \left(\frac{y(t,T) - r(t)}{T-t} + \frac{T-t}{2}\sigma_y(t,T)^2 \right) dt + \sigma_y(t,T)\, dW^T , \tag{9}$$

where

$$\sigma_y(t,T) = \frac{1}{T-t}\,\sigma(t,T) .$$

3.3.2 Forward Rates

Assume now that, following Heath–Jarrow–Morton [21], forward spot rates themselves follow an Ito process

$$df(t,T)|_{T \text{ fixed}} = \mu_f(t,T)\, dt + \sigma_f(t,T)\, d\tilde{W}_f^T , \tag{10}$$

where $\tilde{W}_f^T$ are Brownian motions under $\mathbf{P}$ and $\mu_f(.,T)$, $\sigma_f(.,T)$ are predictable processes depending on the maturity T such that, for any $t_0 \leq t \leq T \leq T_{\max}$

$$\mathbf{E} \int_t^T \left|\mu_f(t,s)\right| ds < \infty, \qquad \mathbf{E} \left(\int_t^T \left|\sigma_f(t,s)\right| ds \right)^2 < \infty . \tag{11}$$

We also assume that the family $(\tilde{W}_f^T)_T$ has independent increments, that is, for any T, T' the increment $d\tilde{W}_f^{T'}(t)$ is independent of $\tilde{W}_f^T(t)$. The *instantaneous correlation function* $\rho(t, T, T')$ is defined by

$$\rho(t,T,T') = \mathrm{Corr}_{\mathbf{P}} \left(d\tilde{W}_f^T(t), d\tilde{W}_f^{T'}(t) \right) ,$$

or, in terms of cross-variation process,

$$d\left\langle \tilde{W}_f^T(t), \tilde{W}_f^{T'}(t) \right\rangle = \rho(t,T,T')\, dt .$$

Obviously, $\rho(t,T,T) = 1$, $|\rho(t,T,T')| \leq 1$ and $\rho(t,T,T') = \rho(t,T',T)$ for any (t,T,T'). Moreover, for any sequence of maturities $(T_1, \ldots, T_n)$, the matrix (ρ_{ij}) where $\rho_{ij} = \rho(t, T_i, T_j)$ is symmetric and positive.

Assumption. We shall assume that, for any $(t; T_1, \ldots, T_n)$ such that $t < T_i \neq T_j$ for any $i \neq j$, the matrix (ρ_{ij}) is positive definite, and

$$\sigma_f(t, T_i) > 0 ,$$

that is, no finite combination of forward spot rates has, at no time, zero volatility.

In order to define "risk-neutral" Brownian motions dW_f^T, we shall make heuristic calculations that will be made rigorous below. We wish to write

$$dW_f^T = \lambda_f(t,T)\,dt + d\tilde{W}_f^T\,.$$

Identifying the **P**-martingale component of $\ell(t, T+\delta T) - \ell(t,T)$ with that of $f(t,T)\delta T$, we get

$$\sigma(t, T+\delta T)\,d\tilde{W}^{T+\delta T} - \sigma(t,T)\,d\tilde{W}^T = \sigma_f(t,T)\,\delta T\,d\tilde{W}_f^T + O(\delta T^2)\,.$$

Assuming that the risk-neutral probability **Q** exists, one must have, by taking **Q**-expectation

$$\mu(t,T) - \mu(t,T+\delta T) = \sigma_f(t,T)\,\delta T\,\lambda_f(t,T) + O(\delta T^2)\,,$$

which leads, when $\delta T \to 0$, to

$$\lambda_f = -\frac{1}{\sigma_f}\frac{\partial \mu}{\partial T}\,.$$

On the other hand, by taking the **P**-expectation in (5) and (10), we get

$$\mu_f(t,T)\,\delta T = \mu(t,T) - \mu(t,T+\delta T) + \frac{1}{2}\sigma(t,T+\delta T)^2 - \frac{1}{2}\sigma(t,T)^2 + O(\delta T^2)\,,$$

therefore, letting $\delta T \to 0$, one gets

$$\frac{\partial \mu}{\partial T} = -\mu_f + \sigma\,\frac{\partial \sigma}{\partial T}\,,$$

and, finally,

$$df(t,T)|_{T\text{ fixed}} = \left(\mu_f + \frac{\partial \mu}{\partial T}\right)dt + \sigma_f\,dW_f^T = \sigma\,\frac{\partial \sigma}{\partial T}\,dt + \sigma_f\,dW_f^T\,. \tag{12}$$

The link between volatilities σ and σ_f is as follows. From

$$\ell(t,T) = \int_t^T f(t,s)\,ds$$

we deduce

$$d\,\langle \ell(t,T)\rangle = \iint_{(u,v)\in[t,T]^2} d\,\langle f(t,u), f(t,v)\rangle\,du\,dv\,,$$

that is

$$\sigma(t,T)^2 = \iint_{(u,v)\in[t,T]^2} \sigma_f(t,u)\sigma_f(t,v)\rho(t,u,v)\,du\,dv\,.$$

Again, this formula, which generalizes H.J.M., will be rigorously proved later on, after having set the formalism of function valued random processes. It provides another expression of the risk-neutral drift of forward spot rates

$$\sigma(t,T)\frac{\partial\sigma}{\partial T}(t,T)=\sigma_f(t,T)\int_t^T \sigma_f(t,u)\rho(t,u,T)\,du\,, \tag{13}$$

from which we deduce the following

Proposition 1 *One has*

$$\left|\frac{\partial\sigma}{\partial T}(t,T)\right|\leq (T-t)\,\sigma_f(t,T)$$

with equality in the case of a local one-factor Ho–Lee model, that is, $\sigma_f(t,u)$ *doesn't depend on* u *and* $\rho(t,u,v)=1$ *for any* (u,v).

Proof Consider the scalar product on functions $\phi:[t,T]\to\mathbb{R}$

$$\phi.\psi=\iint_{(u,v)\in[t,T]^2}\phi(u)\psi(v)\rho(t,u,v)\,dudv\,.$$

Schwarz inequality between functions σ_f and 1 implies

$$\begin{aligned}\left(\int_t^T \sigma_f(t,u)\rho(t,u,T)\,du\right)^2 &\leq \sigma(t,T)^2\iint_{(u,v)\in[t,T]^2}\rho(t,u,v)\,dudv\\ &\leq (T-t)^2\sigma(t,T)^2\,.\end{aligned}$$

Equality occurs only if $\sigma_f(t,.)$ is constant and $\rho(t,.,.)\equiv 1$. □

It is worthy of note that Eqs. (12) and (13), which generalize results obtained by Heath–Jarrow–Morton [21] and by Brace–Gatarek–Musiela [4], only assume that $\mathbf{Q}$ is a risk-neutral probability, but do not require that the whole yield curve evolution is driven by a finite number of Brownian motions.

3.4 Function Valued Random Processes

The yield curve at time t is defined as the function

$$\mathbf{y}_t : x\longmapsto y(t,t+x)\,.$$

It is defined on an interval $I=[0,M]$ which we shall suppose fixed, for instance $M=30$ years. The variable t therefore spans the interval $[0,T_{\max}-M]$. The yield curve $\mathbf{y}_t$ belongs to some functional space H^y over I. We shall define H^y more precisely later on. For the moment, we only assume it is a Banach space, equipped with a norm $\|.\|_{H^y}$ and contained in the space of continuous functions. This defines a random process $\mathbf{y}_t$ with values in H^y.

Similarly, we define functions $\mathbf{z}_t \in H^z$, $\mathbf{l}_t \in H^\ell$, $\mathbf{f}_t \in H^f$. These functions are linked to $\mathbf{y}_t$ and between themselves by Eqs. (1) and (2). The spaces H^z, H^ℓ and H^f should also be linked in a similar way. For example, if $H^f = C^0(I)$ then H^ℓ is the space of C^1 functions vanishing at 0, H^y is the space of continuous functions on I, of class C^1 on $(0, M]$ and whose derivative (with respect to x) is $O(\frac{1}{x})$ at 0, and $H^z = C^1(I)$, or the affine subspace of functions taking value 1 at 0. Note that the correspondence between $\mathbf{y}_t$, $\mathbf{l}_t$ and $\mathbf{f}_t$ are linear, unlike that with $\mathbf{z}_t$.

In order to define function valued processes $\mathbf{y}_t$, $\mathbf{z}_t$, $\mathbf{l}_t$ and $\mathbf{f}_t$, we shall use the formalism of so-called *cylindrical Brownian motions* which appears to be best suited for our purposes. A static portfolio made only of linear assets—bonds, swaps, F.R.A., but not options—can be seen as a finite combination of Dirac masses on H^z, corresponding to payment dates and amounts.

In fact, in [35], Yor proved that, in order to define a cylindrical Brownian motion in the infinite dimensional space H^y (this will be our theoretical setting), one needs to choose a Hilbert space, for instance $L^2(I, \mu)$, where μ is a measure on I, or a Sobolev space with respect to a measure on I.

Remark 1 The choice of the space H^y or, equivalently, of its norm, that is, of the Sobolev exponent and of the measure μ is one of the most important issues. Indeed, this norm measures the risk and should be in accordance with the most probable moves of the yield curve. Generally speaking, we shall see that the most appropriate choice for μ is linked to the distribution in maturities of the significant quoted rates, while the Sobolev exponent, which stands for the curve smoothness, results from market practice and can be deduced in a rather reliable way from the statistics.

Although we have not yet defined processes in H^y, we see that its drift will depend on the diffusion with fixed $x = T - t$, that is, with slipping maturity T. One has

$$\begin{aligned} dy(t, t+x)|_{x \text{ fixed}} &= dy(t, t+x)|_{t+x \text{ fixed}} + \frac{\partial y}{\partial T}(t, t+x)\, dt \\ &= \left(\frac{1}{x}(f(t, t+x) - r(t)) + \frac{x}{2}\sigma_y(t, t+x)^2\right) dt \\ &\quad + \sigma_y(t, t+x)\, dW^T. \end{aligned}$$

The same holds for z, ℓ and f (provided the function f is differentiable):

$$df|_{x \text{ fixed}} = df|_{T \text{ fixed}} + \frac{\partial f}{\partial T}\, dt,$$

$$d\ell|_{x \text{ fixed}} = \left(f(t, t+x) - r(t) + \frac{1}{2}\sigma(t, t+x)^2\right) dt + \sigma(t, t+x)\, dW^T,$$

$$dz|_{x \text{ fixed}} = z(t, t+x)\Big((r(t) - f(t, t+x))\, dt - \sigma(t, t+x)\, dW^T\Big).$$

4 Market Data on the Term Structure

Before setting an abstract framework for the term structure of interest rates, we shall first make a short presentation of the market data that provides it. This will either justify certain modeling and assumptions, or show their limits. We mean by market data the prices of assets which are interest rate dependent, or quoted rates which are linked to these prices by a standard formula. Although they are well known, we shall list these assets in order to examine the features of each one from the point of view of their incidence on the yield curve smoothness. For a complete study of market data on the term structure and on their interrelations, we refer to Anderson–Breedon–Deacon–Derry–Murphy [1]. The reader who is familiar with the fixed income market is advised to jump directly to the conclusions of this section.

4.1 Bonds

A bond delivers a "coupon" C at dates $T_1, \ldots, T_n$ (where $T_k = T_0 + k\,\delta T$, $\delta T =$ 3, 6 or 12 months) and the principal N at T_n. The *coupon rate* R is defined by the formula

$$C = R\, N\, \delta T\,.$$

Hence, its price at time $t < T_1$ is, or should be

$$P_{\text{bond}}(t) = N\left(z(t, T_n) + R\,\delta T \sum_{k=1}^{n} z(t, T_k)\right),$$

where $z(t, T)$ is the discount factor between dates t and T.

Remark 2 In practice, some bonds have prices trading above or below the theoretically determined price, owing to taxation, institutional factors or simply liquidity reasons.

The market also quotes very liquid bond futures, which cannot be theoretically perfectly linked to discount factors for two reasons. First, there is a system of margin calls, which make them (slightly) sensitive to the covariance between the bond price and the short rate. Second, they quote the value of the cheapest bond in a given pool, hence they involve an option associated to the possible change in the "cheapest-to-deliver" (most of the time, this option is far from being at the money).

4.2 Swaps

As it is well known, a swap is an exchange of a fixed interest rate loan with a variable rate one, both of the same principal and the same maturity. The variable leg can be replicated by a rolling loan of the principal over the whole period. The following formula gives the price that should be paid at the beginning by the side paying the fixed rate in order to enter an asset swap[6] with fixed rate R and settlement dates $(T_1, \ldots, T_n)$, $T_k = T_0 + k\,\delta T$:

$$P_{\text{swap}}(t) = N\left(z(t, T_0) - z(t, T_n) - R\,\delta T \sum_{k=1}^{n} z(t, T_k)\right).$$

The *swap rate* is the value of R that cancels the price (for other fixed rates, this is an asset swap)

$$R(t, n\delta T) = \frac{z(t, T_0) - z(t, T_n)}{\delta T \sum_{k=1}^{n} z(t, T_k)}.$$

Bond prices and swap rates provide an information on the value of a given discount factor with respect to an average of others with a shorter maturity. One usually uses a "boot strapping" method to compute discount factors, indeed, errors are at each step multiplied by the coupon rate and do not accumulate.

4.3 Cash and Future Short Rates

A cash rate is the rate of a loan without coupon (a "zero-coupon"). All cash rates from the "overnight" to one year[7] are permanently quoted on market screens.

Future contracts are *forward rate agreements* (FRA) on a 3 months loan on prescribed periods,[8] with a system of *margin calls*. Their quotation has been recently extended to a four-years forward period in Europe (France, GB, Germany) and in Japan, and 10 years in the USA. The price of futures is slightly different from that of the corresponding FRA (no more than a few basis points) because of the convexity induced by the margin calls.

An FRA rate on the period $[T, T']$ evaluated at date $t < T$ is given by

$$\text{FRA}(t, T, T') = \frac{1}{T' - T}\big((T' - t)\, y(t, T') - (T - t)\, y(t, T)\big).$$

---

[6] As usual, settled in advance, paid in arrears; there are also swaps paid in advance.

[7] Precisely O/N, next day, 1 and 2 weeks, 1, 2, 3, 6, 9 and 12 month (all intermediaries are immediately given by market makers on request).

[8] So-called IMM dates, around mid-March, June, September and December.

If T and T' are close, the FRA gives an estimate of the forward spot rate, that is, of the *derivative*[9] of the yield curve with respect to maturity. Practically speaking, if T is worth several years, then $T' = T + 3$ months can be considered as close.

4.4 STRIP, or the Decomposition of Bonds

In the USA, T-bonds can be "stripped", that is, fragmented into the principal and the coupons, which can be negotiated separately. The same holds (since a more recent period) in other countries for government bonds. The coupons are called *Strips.* Theoretically, this gives the level of every zero-coupon rate. However, they are much less liquid than the rest of the market and, in fact, market makers use that other information to price them. Furthermore, their price is a little affected by the fact that, in order to reconstruct a full T-bond, one is not allowed to replace the principal by an accumulation of coupons at the same date. Because of the "reconstruction" opportunity that it provides, the principal is slightly more expensive than its "theoretical" price. The stripping generates large amounts of data, but it must be used with care.

The behavior of coupon prices is more surprising. For short maturities, they are below the theoretical price, but, at some point (between 10 and 20 years) they pass to the other side. To explain this fact, let $\tilde{z}_P(t, T_0, T_n, \delta T, R)$ be the price at date t of the principal of a bond that has coupons of size R at dates $T_0,\ T_0 + \delta T, \dots, T_n = T_0 + n\delta T$, and let $\tilde{z}_C(t, T_k)$ be the price of a strip of maturity T_k. Reconstructing the bond provides a relation between these prices and the theoretical prices

$$\tilde{z}_N(t, T_0, T_n, \delta T, R) - z(t, T_n) = R\,\delta T \sum_k^n (z(t, T_k) - \tilde{z}_C(t, T_k)).$$

This difference should always be positive, but not too big. If one sees the right hand side as a discrete approximation of an integral, we see that the "liquidity spread" on the principal price is equal to the algebraic area[10] between the theoretical zero-coupon price curve and the strip curve (counted negatively if the strip curve goes above). When one tries to fit a constant liquidity spread for the principal, then the accumulation effect of the strips must be compensated on the long term part. Of course, the behavior of rates is symmetrical.

Remark 3 An important observation is that market makers on futures and on strip markets have a tendency to *smooth* the curve with respect to T, as if some kind of elasticity tried to erase possible angles.

[9]In the sense of the mathematical derivative of a function.

[10]That is, counted with signs.

4.5 Conclusion

After these observations, we look at the yield curve as an object lying in some functional space H that has either infinite dimensions or at least a large one. The discount factors and the yields are *implicit variables,* in the sense that explicit data are not reliable (see comment on strips). We must therefore take into account other linear and nonlinear functions of the rates. Remark 3 tends to indicate that the space H should consist of differentiable functions.

We tend to see the term structure of interest rates as a "smooth skeleton" given by averaging the available information, with some noise due partly to rounding to the nearest basis point (or to the bid/ask spread), partly to the particularities of each market. What will be described in a theoretical framework is the evolution of the "smooth skeleton", for the "noise" can be considered as bounded and does not represent a risk that should be hedged according to the usual Black-Scholes theory based on diffusion processes. In practice, cash and swap rates are extremely close to a smooth curve (about 1–2 bp), while each government bond has its own spread (on the positive side) over the cash-swap curve, and this series of spreads cannot be modeled as a curve.

As we mentioned in the introduction, we shall see how assuming first that H is infinite-dimensional allows us to find very good approximation subspaces of rather low dimension, through standard finite element techniques.

5 Brownian Motions in a Hilbert Space

Defining a Brownian motion in an infinite dimensional space is not a simple task. Indeed, such a space is not locally compact and there is no Lebesgue measure[11] on it, thus no Gaussian density (although Gaussian probability measures exist). To overcome this difficulty, we shall refer to the formalism developed in 1973 by Gaveau [19] and well explained in Da Prato–Zapczyk [12, p. 96] and in Yor [34].

Yor gives three different definitions of a Brownian motion in a Hilbert space H and shows that these three frameworks are equivalent. In our situation, the most natural one is the so-called *cylindrical Brownian motion*: for any $h \in H$, a real valued centered Brownian motion $B_t(h)$ of volatility $\|h\|_H^2$ is given.[12]

Intuitively, if $B_t \in H$, then

$$B_t(h) = h.B_t\,.$$

In fact, it is not difficult to see that, if H is infinite dimensional, B_t cannot belong to H for every t (see Sect. 7) and that

$$\mathbf{E}\|B_t\|_H^2 = \infty\,.$$

[11]That is, a uniformly distributed measure, invariant by translation.

[12]I.e., its variance at date t is $t\,\|h\|_H^2$.

The process B_t will have to lie in a super-space[13] of H in which H will be dense.

In our situation, bond prices, swap rates, etc., and, generally speaking, portfolios containing the assets described above, are of the type $\mu(\mathbf{z}_t)$, where μ is a linear form on H^z. This corresponds to a stochastic integral with respect to a Brownian motion $B_t(\mu)$.

Although they are Hilbert spaces, we shall from now on make a distinction between the space $H = H^z$ of curves[14] and its dual, the space H^* of portfolios, to which the linear forms μ belong.

Of course, in order to define a cylindrical Brownian motion, we need to know $h.B_t$ for *every* $h \in H$ (or equivalently every $\mu \in H^*$). Here, only *some* μ are given, but we shall assume that they span a dense subspace, so that an entire cylindrical Brownian motion can be uniquely defined by extension.

We shall also assume that the price process of any static portfolio is an Ito process, in particular, it has a finite variance.

Remark 4 The linear forms of the above type (Sect. 3.4) contain Dirac masses. Hence they do not belong to the space L^2 but to the Sobolev space of distributions H^{-1} (or even to $H^{-\frac{1}{2}-\varepsilon}$, $\varepsilon > 0$)[15].

We shall now give a more formal definition to these two assumptions. We fix $s > \frac{1}{2}$ and we assume that $H = H^s$ and hence, that $H^* = H^{-s}$. The choice of the regularity parameter s will be discussed in Sect. 9.

6 Assumptions

6.1 Almost Complete Market

ACM. We assume that the set of traded assets is dense in H^* for the weak topology and that if the sequence μ_n of traded assets weakly tends to $\mu \in H^*$, then the price processes of assets μ_n converge in L^2.

For instance, if $H = H^s, s > \frac{1}{2}$, then the space of finite combinations of Dirac masses is dense in H^*.

---

[13] I.e. a space containing H.

[14] We drop the superscript z to ease notations. If a process is defined in H^z, we get the corresponding processes in H^y and H^f by applying formulas (2) and (3).

[15] When s is a positive integer, H^s is the space of functions whose s-th derivative belongs to L^2 and H^{-s} is the dual of H^s for the L^2 dot product. For non integer s, this space is defined by the mean of Fourier transform. If $s > \frac{1}{2}$ then functions lying in H^s are continuous, and Dirac masses belong to H^{-s}.

6.2 Finite Variance

FV For every $\mu \in H^*$, we assume that $\mu(\mathbf{z}_t)$ follows an Ito process driven by a Brownian motion $\tilde{B}_t(\mu)$

$$d\mu(\mathbf{z}_t) = \mu\left(\mathbf{a}(t,\mathbf{z})\right) dt + dB_t(\mu),$$
$$dB_t(\mu) = \sigma_\mu(t,\mathbf{z})\, d\tilde{B}_t(\mu), \qquad \mathbf{a}(t,\mathbf{y}) : x \longmapsto a(t, t+x, \mathbf{z}). \tag{14}$$

In particular $\frac{1}{dt} \operatorname{Var} \mu(d\mathbf{z}_t) < \infty$ for any $\mu \in H^*$.

The relevance of these two hypotheses has been discussed at the end of the introduction.

6.3 Gaussian Rates

We shall also add the (less natural but convenient) assumption that volatility does not depend on the level of rates.

Gauss. $\operatorname{Var} dB_t(\mu)$ does not depends on the yield curve $\mathbf{y}$.

This implies that for any $\mu \in H^*$, the distribution of $\mu(\mathbf{y}_t)$ is Gaussian.[16]

7 Principal Component Analysis

7.1 The Volatility Operator

On the space H^* of portfolios, there is a natural time dependent bilinear form $\tilde{Q}_t$ induced by the cross-variation process of the stochastic part of two portfolios

$$\tilde{Q}_t(\mu,\nu) = \frac{\langle dB_t(\mu), dB_t(\nu)\rangle}{dt}.$$

According to the CV hypothesis, $\tilde{Q}$ only depends on μ and ν. We shall denote by Q_t the quadratic form associated with $\tilde{Q}_t$

$$Q_t(\mu) = \frac{\langle dB_t(\mu)\rangle}{dt}.$$

Obviously, the quadratic form Q is positive. If its rank is finite, then we find the usual Gaussian H.J.M. model with a finite number of factors. On the contrary, we

[16] $\mu(\mathbf{y}_t)$ linearly depends on rates. If their distribution is Gaussian, then so is that of $\mu(\mathbf{y}_t)$ and it has a (very low) probability of becoming negative. But a bond the price of which is given by $\mu(\mathbf{z}_t)$, where μ is a positive measure, will always have a positive price.

shall assume that it is non degenerate (any portfolio moves, even slightly, none is rigorously hedged). This assumption allows us to consider Q as a new *norm*[17] on H^*. When completing the space H^* with respect to this norm (we do not change notations), we get a cylindrical Brownian motion $B_t(\mu)$.

In such a situation, Yor [34, prop. I.4.2] shows that this motion can be realized: one can find a super-space[18] V of the dual H of H^* and a process B_t with values in V, almost surely continuous for the norm of V, such that

$$\mathbf{E}\,\|B_t\|_V^2 < \infty\,,$$

and

$$\mu.dB_t = \mu(d\mathbf{y}_t - \mathbf{b}(t, \mathbf{y}_t)dt)$$

for any $t > 0$ and any $\mu \in V^* \subset H^*$.

It should be reminded that V is the space where lie the *yield curves* $\mathbf{y}_t$ and H^* that of *linear forms* (or portfolios) μ, and that

$$V^* \subset H^*\,, \qquad H \subset V\,.$$

We shall see in Sect. 9 the variety of possible spaces V.

It is now possible, in a rigorous mathematical language, to say that the random process $\mathbf{y}_t$ is determined by the following stochastic differential equation, driven by the cylindrical Brownian motion B_t

$$d\mathbf{y}_t = \mathbf{b}(t, \mathbf{y}_t)\,dt + dB_t\,.$$

Remark 5 There are different Hilbert spaces, with different dot products. If $\mu \in V^*$ is a linear form, thus a measure[19] on the time interval $[0, T_{\max}]$ that has a density χ_μ (possibly generalized, that is, with Dirac masses), and if $B_t \in V$, then one has

$$(\mu.B_t)_{V^* \times V} = \int_0^M \chi_\mu(x)\,B_t(x)\,dx\,.$$

However, H^* (hence also H) is a space which is specifically adapted to the quadratic form Q, for it has been completed with respect to it.

---

[17]If Q is degenerate, then one may slightly modify it to get a new norm on H^*:

$$\|\mu\|^2 = \varepsilon\|\mu_1\|_{L^2}^2 + Q(\mu_2)\,, \quad \mu = \mu_1 + \mu_2\,, \quad \mu_1 \in \ker Q\,, \quad h_2 \in \ker Q^\perp\,, \quad \varepsilon > 0\,.$$

[18]Note that if V is a super space of H, then its norm is dominated by that of H

$$\|h\|_V \le \text{cst}\ \|h\|_H$$

for any vector $h \in H$ should be able to be measured in V (the inclusion $H \hookrightarrow V$ is continuous).

[19]Riesz representation theorem.

7.2 Principal Component Analysis

We shall now look at Q_t as a quadratic form on V^*. There exists a positive symmetric operator A_t on V^* such that for any $\mu \in V^*$

$$Q_t(\mu) = \mu \,.\, A_t \mu \quad \text{(dot product in } V^*\text{)}.$$

Proposition 2 *The operator A_t has a finite trace. In particular, it is compact.*

Proof Let $(\mu_n)_{n\in\mathbb{N}}$ be an orthonormal basis of V^*. Then

$$Q_t(\mu_n) = \frac{1}{dt}\mathbf{E}\,(\mu_n \,.\, dB_t)^2,$$

hence

$$\operatorname{Tr} Q_t = \sum_{n=0}^{\infty} Q_t(\mu_n) = \frac{1}{dt}\mathbf{E}\sum_{n=0}^{\infty}(\mu_n \,.\, dB_t)^2 = \frac{1}{dt}\mathbf{E}\,\|dB_t\|_V^2 \;<\infty\,.$$ □

Corollary 1 *The operator A_t is diagonalizable in an orthonormal basis, because its spectrum is discrete.*

One can thus find an orthonormal basis $(\bar{\mu}_n)_{n\in\mathbb{N}}$ of V^* and a sequence of eigenvalues $(\lambda_n)_{n\in\mathbb{N}}$ such that, if $\mu \in V^*$ and $\mu = \sum a_n \mu_n$, then

$$Q_t(\mu) = \sum_{n=0}^{\infty} \lambda_n \, a_n^2\,.$$

Because A_t is positive with a finite trace, one has

$$\forall n \in \mathbb{N}\,, \quad \lambda_n > 0 \quad \text{and} \quad \sum_{n=0}^{\infty} \lambda_n < \infty\,.$$

We assume that the eigenvalues λ_n are decreasingly ordered: $\lambda_{n+1} \le \lambda_n$ (possibly repeated if they are multiple).

We go back to the space V of yield curves and denote by $(\psi_n)_{n\in\mathbb{N}}$ the dual basis[20] of $(\mu_n)_{n\in\mathbb{N}}$. By definition

$$dB_t(x) = \sum_{n=0}^{\infty} dv_n(t)\, \psi_n(x)\,,$$

where

$$dv_n(t) = \mu_n \,.\, dB_t\,.$$

[20] $\mu_n(\psi_p) = 1$, if $n = p$, 0 otherwise.

Consequently

$$d\langle B_t(x)\,,\; B_t(x')\rangle = dt\sum_{n=0}^{\infty}\lambda_n\,\psi_n(x)\,\psi_n(x')\,.$$

Definition 1 The functions

$$\varphi_n = \sqrt{\lambda_n}\,\psi_n$$

will be called *eigenmodes*, or *principal deformations* of the yield curve $\mathbf{y}_t$.

7.3 Infinite Dimensional H.J.M. Representation

The covariance of two zero-coupon rates is given by the formula

$$\langle d\mathbf{y}_t(x)\,, d\,\mathbf{y}_t(x')\rangle = dt\sum_{n=0}^{\infty}\varphi_n(x)\,\varphi_n(x')\,.$$

When setting $x = x'$, we get

$$\sigma(t, t+x)^2 = \sum_{n=0}^{\infty}\varphi_n(x)^2\,. \tag{15}$$

Let

$$w_n(t) = \frac{1}{\sqrt{\lambda_n}}\,v_n(t)\,.$$

The w_n are independent standard Brownian motions (i.e. with volatility 1) and

$$d\mathbf{y}_t(x) = \frac{1}{x}\left(f(t, t+x) - r(t) + \frac{1}{2}\sum_{n=0}^{\infty}\varphi_n(x)^2\right) + \sum_{n=0}^{\infty}\varphi_n(x)\,dw_n(t)\,. \tag{16}$$

Under this form, we clearly see the P.C.A. of the yield curve process.

Multiplying this equation by x then deriving it with respect to x yields the Brace–Musiela equation on forward spot rates, generalized to an infinite summation. Let

$$\mathbf{f}_t(x) = f(t, t+x) = \mathbf{y}_t(x) + x\,\frac{d\mathbf{y}_t}{dx}(x)\,, \qquad \mathbf{f}_t'(x) = \frac{d\mathbf{f}_t}{dx}(x) = \frac{\partial f}{\partial T}(t, t+x)\,,$$

$$\Phi_n(x) = x\,\varphi_n(x)\,, \qquad \varphi_n^f(x) = \frac{d\Phi_n}{dx}(x) = \varphi_n(x) + x\,\frac{d\varphi_n}{dx}(x)\,.$$

Then we have (see Brace–Musiela [5])

$$d\mathbf{f}_t(x) = \left(\mathbf{f}_t'(x) + \sum_{n=0}^{\infty}\Phi_n(x)\,\varphi_n^f(x)\right)dt + \sum_{n=0}^{\infty}\varphi_n^f(x)\,dw_n(t)\,. \tag{17}$$

From these equations, we deduce the initial Brownian motions W^T and W^{fT}

$$dW^T = \frac{1}{\sigma(x)} \sum_{n=0}^{\infty} \varphi_n(x)\, dw_n(t), \qquad dW^{fT} = \frac{1}{\sigma^f(x)} \sum_{n=0}^{\infty} \varphi_n^f(x)\, dw_n(t),$$

with

$$\sigma^f(x)^2 = \sum_{n=0}^{\infty} \varphi_n^f(x)^2, \qquad \sigma(x)\sigma^f(x)\,\mathrm{Corr}\big[dW^T, dW^{fT}\big] = \sum_{n=0}^{\infty} \varphi_n(x)\,\varphi_n^f(x).$$

Remark 6 Notice the double orthogonality:

1. The eigenmodes φ_n are orthogonal in V,
2. The Brownian motions w_n are independent.

Remark 7 The P.C.A. (the eigenmodes, etc.) depends on the space V, that is in fact on its norm, which, according to Da Prato–Zapczyk [12], can be any norm such that Q has a finite trace in V (see Sect. 9).

8 Optimal Representation with an N-Factor Model

We fix an integer N and we define $\mathbf{y}_t^N \in V$ by $\mathbf{y}_0^N = \mathbf{y}_0$ and by the stochastic differential equation

$$d\mathbf{y}_t^N(x) = \frac{1}{x}\left(\mathbf{f}_t^N(x) - \mathbf{y}_t^N(0) + \frac{1}{2}\sum_{n=0}^{N-1} \varphi_n(x)^2\right) + \sum_{n=0}^{N-1} \varphi_n(x)\, dw_n(t).$$

Theorem 1 *The curve $\mathbf{y}_t^N(x)$ is the best approximation of $\mathbf{y}_t(x)$ in the sense of the norm of V the evolution of which is described by N Brownian motions. More precisely, if $\mathbf{u}_t$ is the solution of an SDE with values in V, driven by N real valued Brownian motions, then*[21]

$$\mathbf{E}\,\|d\mathbf{y}_t - d\mathbf{u}_t\|_V^2 \geq \mathbf{E}\,\|d\mathbf{y}_t - d\mathbf{y}_t^N\|_V^2 = dt \sum_{n=N}^{\infty} \lambda_n$$

and

$$\max_{\|\mu\|_{V*}=1} \mathrm{Var}\big[\mu.(d\mathbf{y}_t - d\mathbf{u}_t)\big] \geq \max_{\|\mu\|_{V*}=1} \mathrm{Var}\Big[\mu.(d\mathbf{y}_t - d\mathbf{y}_t^N)\Big] = \lambda_N\, dt.$$

[21]For a random process $X(t)$ in $\mathbb{R}$, with a Meyer decomposition $X = \langle X\rangle + \tilde{X}$ into a process with finite variation and a martingale, we set

$$\mathbf{E}[dX] = d\langle X\rangle, \qquad \mathrm{Var}\, dX = d\langle \tilde{X}^2\rangle, \qquad dX^{\text{stoch.}} = d\tilde{X}.$$

Proof The first thing to do is to characterize the linear combinations of N Brownian motions. A one-dimensional Brownian motion in V is a process

$$(\omega, t) \in \Omega \times \mathbb{R}_+ \longmapsto \theta_t(\omega) = X_t(\omega)\,\xi\,,$$

where Ω is the space of randomness, $X_t(\omega)$ is a real valued Brownian motion and $\xi \in V$.

We fix a date t and consider a time interval $\delta t > 0$ that will later tend to 0. One can see that $\delta\theta = \theta_{t+\delta t} - \theta_t$ as an element of a tensor product

$$\delta\theta = \delta X \otimes \xi \in L^2(\Omega) \otimes V\,.$$

Let

$$\Theta = L^2(\Omega)\hat{\otimes} V$$

be the *Hilbert-Schmidt completion* of this tensor product. If $(e_n)_{n\in\mathbb{N}}$ is an orthonormal basis of $L^2(\Omega)$ and if $(f_n)_{n\in\mathbb{N}}$ is one of V, then $L^2(\Omega) \otimes V$ is endowed with the norm for which $(e_p \otimes f_q)_{(p,q)\in\mathbb{N}^2}$ is an orthonormal basis (this norm does not depend on the chosen bases), and completed with respect to this norm.[22] Let $\zeta \in \Theta$ be a *random curve,* then

$$\|\zeta\|_{\Theta}^2 = \mathbf{E}\,\|\zeta\|_V^2.$$

In particular, $\delta B \in \Theta$ and one has

$$\|\delta B\|_{\Theta}^2 = \sum_{n=0}^{\infty} \mathrm{Var}\,[\delta w_n]\,\|\varphi_n\|_V^2 = \delta t \sum_{n=0}^{\infty} \lambda_n\,.$$

Denote by $\Gamma_N \subset \Theta$ the cone of N-tensors, that is the sum of N simple tensors

$$\Gamma_N = \left\{ \sum_{i=1}^{N} X_i \otimes \xi_i \mid X_i \in L^2(\Omega)\,,\ \xi_i \in V\,,\ i = 1, \ldots, N \right\}.$$

In order to show the first inequality, it is enough to prove that, for any sufficiently small $\delta t > 0$, the distance between δB and the set Γ_N is reached at the point $\delta B^N = B^N(t + \delta t) - B^N(t)$, where B^N is the Brownian motion defined by

$$B^N(t) = \sum_{n=0}^{N-1} \varphi_n(x)\, w_n(t)\,.$$

Indeed, by definition

$$\frac{1}{dt}\mathbf{E}\,\|d\mathbf{y}_t - d\mathbf{u}_t\|_V^2 = \lim_{\delta t \to 0} \frac{1}{\delta t}\mathbf{E}\,\|\delta\mathbf{y}_t^{\text{stoch.}} - \delta\mathbf{u}_t^{\text{stoch.}}\|_V^2,$$

[22] For instance, when $V = L^2(I)$, then the elements of the tensor product $L^2(\Omega) \otimes V$ are functions defined on $\Omega \times I$ and the Hilbert-Schmidt completion is nothing else but $L^2(\Omega \times I)$.

$$\frac{1}{dt}\mathbf{E}\,\|d\mathbf{y}_t - d\mathbf{y}_t^N\|_V^2 = \lim_{\delta t \to 0} \frac{1}{\delta t}\mathbf{E}\,\|\delta\mathbf{y}_t^{\text{stoch.}} - \delta\mathbf{y}_t^{N\,\text{stoch.}}\|_V^2 .$$

To show that δB^N is the closest point of Γ_N to δB, we identify elements of Θ with linear operators from $L^2(\Omega)$ to V by setting

$$(X \otimes \xi) \, . \, Y = \mathrm{Cov}\,[X, Y]\;\xi \, .$$

The Hilbert-Schmidt norm is then given by

$$\|u\|_\Theta^2 = \mathrm{Tr}\left({}^t u\, u\right) ,$$

where ${}^t u$ is the transposed of u, and Γ_N is made of operators whose rank is less than or equal to N.

Lemma 1 *Assume that there exists $\bar{u} \in \Gamma_N$ such that*

$$\|\delta B - \bar{u}\|_\Theta = \mathrm{dist}\,(\delta B \, , \, \Gamma_N) \, .$$

Then the image $\mathrm{Im}\,\bar{u}$ *is stable under* $\delta B\, {}^t\delta B$.

Corollary 2 $\mathrm{Im}\,\bar{u}$ *is spanned by eigenvectors of $\delta B\, {}^t\delta B$ (that is, the φ_n or linear combinations between φ_n's corresponding to the same eigenvalue if it is multiple).*

Proof of lemma. We know that $\delta B - \bar{u}$ is orthogonal in $\bar{u}$ to Γ_N (with respect to the dot product in Θ). For any endomorphism α of $L^2(\Omega)$, the rank of $\bar{u} + \bar{u}\alpha$ remains bounded by N, thus

$$\mathrm{Tr}\left({}^t\alpha\, {}^t\bar{u}\,(\delta B - \bar{u})\right) = 0$$

and

$${}^t\bar{u}\,(\delta B - \bar{u}) = 0 \, .$$

Symmetrically, using endomorphisms of V, we see that

$$\bar{u}\,({}^t\delta B - {}^t\bar{u}) = 0 \, .$$

Combining these two identities, we get

$$\delta B\, {}^t\delta B\, \bar{u} = \bar{u}\, \delta B\, {}^t\delta B = \bar{u}\, {}^t\bar{u}\, \bar{u} \, .$$

The lemma follows.

End of proof. Going back to the tensor product, if $\bar{u}$ exists, it can be written as

$$\bar{u} = \sum_{i=1}^{N} X_i \otimes \varphi_{n_i}$$

(if some eigenvalues are multiple, one might have to change the corresponding φ_n into another orthonormal basis of the eigenspace; this does not affects the decomposition of δB). Let

$$J = \{n_1, \ldots, n_N\}.$$

One has

$$\delta B - \bar{u} = \sum_{i=1}^{N} (w_{n_i} - X_i) \otimes \varphi_{n_i} + \sum_{n \notin J} w_n \otimes \varphi_n$$

and

$$\|\delta B - \bar{u}\|_{\Theta}^2 \geq \sum_{n \notin J} \lambda_n \geq \sum_{n=N}^{\infty} \lambda_n = \|\delta B - \delta B^N\|_{\Theta}^2 .$$

This would end the proof if we knew that $\bar{u}$ exists. It is the case if Θ is finite dimensional. Let q be an integer which, later, will tend to infinity. We set

$$E_q = \mathrm{Vect}(w_0, \ldots, w_q), \qquad V_q = \mathrm{Vect}(\varphi_0, \ldots, \varphi_q),$$

and let

$$\pi_q : L^2(\Omega) \longrightarrow E_q ,$$
$$\pi'_q : V \longrightarrow V_q ,$$

be the orthogonal projections. It is easy to check that

$$\pi_q \otimes \pi'_q : \Theta \longrightarrow \Theta_q = E_q \otimes V_q ,$$
$$X \otimes \xi \longmapsto \pi_q(X) \otimes \pi'_q(\xi) ,$$

is the orthogonal projection of Θ onto Θ_q and that

$$\Gamma_{N,q} = \pi_q \otimes \pi'_q (\Gamma_N) = \left\{ \sum_{i=1}^{N} X_i \otimes \xi_i \mid X_i \in E_q , \ \xi_i \in V_q , \ i = 1, \ldots, N \right\} .$$

Therefore, if $\bar{u}_q \in \Gamma_{N,q}$ realizes the distance from $\pi_q \otimes \pi'_q (\delta B)$ to $\Gamma_{N,q}$ then

$$\mathrm{dist}\,(\delta B , \Gamma_N)^2 \geq \|\pi_q \otimes \pi'_q (\delta B) - \bar{u}_q\|_{\Theta_q}^2 \geq \sum_{n=N}^{q} \lambda_n$$

(the first inequality comes from the fact that an orthogonal projection does not increases distances). This lower bound is valid for every q, hence

$$\mathrm{dist}\,(\delta B , \Gamma_N)^2 \geq \sum_{n=N}^{\infty} \lambda_n = \|\delta B - \delta B^N\|_{\Theta}^2 .$$

Second inequality. We notice that the quantity to minimize is the usual operator norm of the transpose of $\delta B - \bar{u}$, as an operator from V to $L^2(\Omega)$, that is

$$\max_{\|\zeta\|_V=1} \|({}^t\delta B - {}^t\bar{u})(\zeta)\|_{L^2(\Omega)} .$$

As the rank of ${}^t\bar{u}$ is at most N, its kernel has a co-dimension greater than or equal to N and

$$\ker {}^t\bar{u} \cap \mathrm{Vect}(\varphi_0, \ldots, \varphi_N) \neq \{0\} .$$

Let ζ be an element of this intersection such that $\|\zeta\|_V = 1$. One has

$$\|({}^t\delta B - {}^t\bar{u}) . \zeta\|_{L^2(\Omega)} = \|{}^t\delta B . \zeta\|_{L^2(\Omega)} \geq \lambda_N .$$

When $\bar{u} = \delta B^N$ the equality is implied the orthogonality of the φ_n.

9 Possible Choice in the Hilbert Space V

In the statement of the previous theorem, we mentioned: *in the sense of the norm of* V. Indeed, Yor's construction leaves out some latitude on the choice of this space. According to Da Prato–Zapczyk [12], V^* can be any subspace of H^* provided with a Hilbert norm with respect to which Q has a finite trace. Then V is the super-space of H which is the dual of V^*. In particular, if V fits, any super-space of V fits. Nevertheless, a smallest acceptable space does not exist.

Remark 8 In all these kinds of considerations, defining such or such space where such object lies is somewhat abstract, for, in the reality, these objects are finite dimensional and lie in any reasonable space. The very "physical" meaning of this type of statement relies in the evaluation of the corresponding norms. It has a meaning to say that such norm has a reasonable value or is extremely large, that such measurement of an error is bounded, while we have no clue of another evaluation. We understand the previous analysis in this context.

The linear forms μ which intervene when differentiating the prices of bonds and swaps are all of the kind "*integral over an interval + Dirac mass*",[23] thus their principal singularity is a Dirac mass. In other term, the price of a zero-coupon is well defined. We deduce that V is always contained in $H^{\frac{1}{2}}$ (since $V^* \supset H^{-\frac{1}{2}}$, see Sect. 5, Remark 4, Footnote 15). In countries in which a large number of futures is traded, like the United States (over a range of 10 years), we know better: a future contract evaluates in fact a forward short rate (3 months), that is an approximation

[23] Here, we see that we are again more concerned with the general profile of μ, rather than with details like knowing whether the distribution of coupons is continuous or discrete. This really makes a small difference in their value.

of the derivative of the zero-coupon rate. As these also follow an Ito process with finite variance, we conclude that $V \subset H^{\frac{3}{2}}$. Yet one may choose any wider space.

The reality of markets is, on the contrary, oriented towards more smoothness: an tight analysis on US future curve shows that the $H^{\frac{3}{2}}$-norm of this curve, that is the $H^{\frac{5}{2}}$-norm of the yield curve, is almost always bounded (with however a slight difficulty due to the tic discretization and to a regular shift on the value of December contracts). A similar observation can be made on the French data series on OAT and BTAN bonds.

Remark 9 The choice of the V-norm should be made carefully, in particular according to the profile of one's portfolio, indeed, as we already said, the P.C.A. is optimal with respect to this norm, and eigenmodes depend on its choice.

In practice, we shall minimize a *least square criterion,* possibly weighted, on the prices of assets we are dealing. This criterion provides a quadratic form on the space of yield curve, which is the most natural choice as a norm for V. For instance, assume that we are dealing a series of bonds $B_1, \ldots, B_n$ the price of which is, at the first order, approximated by the measures $\mu_1, \ldots, \mu_n$ and that the least square criterion weights the bond B_i with a coefficient α_i (to take into account an unequal distribution of the portfolio). The norm on V can be set to

$$\|\mathbf{y}\|_V^2 = \sum_{i=1}^{n} \alpha_i \, \mu_i(\mathbf{y})^2 .$$

In fact, this can be only a semi-norm (it may vanish for $\mathbf{y} \neq 0$). If the number of bonds is sufficient, and if their duration is well distributed, such a drawback will be avoided. Otherwise, one has to combine this sum with an L^2-like norm, directly on zero-coupon rates

$$\|\mathbf{y}\|_V^2 = \int_0^M \mathbf{y}(x)^2 \, m(x) \, dx ,$$

the weight $m(x) > 0$ being again adapted to the portfolio profile.

When dealing with futures, calendar spreads, etc., one should rather choose a Sobolev norm, that is a norm (still of Hilbert type) involving the derivative $\mathbf{y}'(x) = \frac{d\mathbf{y}}{dx}(x)$.

10 Option Pricing

Jamshidian's [22] and Brace–Musiela's [5] formulae can easily be generalized to an infinite number of factors. The results match those of Kennedy [24]. We give in this section the expectation and the variance of any zero-coupon, as well as the covariance of any pair of such. In our model, rates are Gaussian and the zero-coupons have a log-normal distribution. Therefore, these data are sufficient to evaluate the price of any plain vanilla option (put or call) on any portfolio which is a linear combination of zero-coupons. This includes caplets, floorlets, options on bonds, swaps, and even on Forward Rate Agreements (options on "yield curve spreads").

Our model being a limit of Gaussian H.J.M. models with a finite number of factors, expectations, variances and covariances provided the N-factor H.J.M. model (see [5]) tend, when N tends to infinity, to a limit which corresponds to the model driven by the cylindrical Brownian motion B_t.

The option prices computed this way are of course arbitrage prices (provided the model fits the reality), but there is a little difficulty. Assume that, in the reality, interest rates satisfy the diffusion equation (16). If we try to hedge a cap against N modes of deformation using an approximation of the reality by an N-factor model, we get a price C_N and, as the hedge is not perfect, also a variance v_N. When N tends to infinity, v_N tends to 0 and the price C_N has a limit C, which is the price we propose. However, although there is a theoretically infinite number of hedging instruments P_i , $i = 1, 2, \ldots$, the N-factor model will use only N of them to cancel N hedge ratios Δ_i^N , $i = 1, \ldots, N$. When N tends to infinity, the hedge ratios tend to a well defined limit Δ_i , $i = 1, \ldots$ but it may happen that

$$\sum_{i=1}^{\infty} |\Delta_i|\, P_i = \infty,$$

while, because of high correlations (due to the fact that the variance of Q is finite), the management cost (theoretical, that is transaction cost free) of this "infinite" portfolio remains finite. In practice, "infinite" means a prohibitive high value. Besides, the presence of transaction costs makes a rigorous replication strategy impossible (but this remark is valid even for an option on a single asset).

Equations (7), then (15) and (16) provide the diffusion of logarithms of "forward zero-coupon" $z^F(t, T, T') = z(t, T')/z(t, T)$ when T and T' are fixed

$$\begin{aligned} d \log z^F(t, T, T') = {} & \frac{1}{2} \sum_{n=0}^{\infty} \Big((T-t)^2 \varphi_n(T-t)^2 - (T'-t)^2 \varphi_n(T'-t)^2 \Big)\, dt \\ & + \sum_{n=0}^{\infty} \big((T-t)\, \varphi_n(T-t) - \big(T'-t\big)\, \varphi_n(T'-t) \big)\, dw_n\,. \end{aligned}$$

The first series converges absolutely, while the second one converges absolutely as a function of t with values in $L^2(\Omega)$. If we set $\delta T = T' - t$ and $\Phi_n(x) = x\, \varphi_n(x)$, we get

$$\begin{aligned} & \mathbf{E}\,[z(T, T+\delta T)\,|\,t] \\ & \quad = \frac{z(t, T+\delta T)}{z(t, T)} \exp\left(\sum_{n=0}^{\infty} \int_0^{T-t} \Big(\Phi_n(s)\, \Phi_n(s+\delta T) - \Phi_n(s+\delta T)^2 \Big)\, ds \right), \end{aligned} \tag{18}$$

$$\mathrm{Var}\left[\log z(T, T+\delta T)\,|\,t\right] = \sum_{n=0}^{\infty} \int_0^{T-t} (\Phi_n(s) - \Phi_n(s+\delta T))^2\, ds. \tag{19}$$

$$\mathrm{Cov}\left[\log z(T, T+\delta T_1), \log z(T, T+\delta T_2)\,|\,t\right]$$
$$= \sum_{n=0}^{\infty} \int_0^{T-t} (\Phi_n(s) - \Phi_n(s+\delta T_1))\,(\Phi_n(s) - \Phi_n(s+\delta T_2))\, ds\,. \tag{20}$$

By truncating these summations at rank $N-1$, we obtain the expectations, variances and covariances of zero-coupons as if the yield curve were the approximate one $\mathbf{y}_t^N$.

11 Computation of Eigenmodes

11.1 Reconstruction and Smoothing of the Yield Curve

In order to perform the P.C.A. of the yield curve out of historical data series, we first need to restrict ourselves to the finite dimension through a finite element method. We shall thus approximate the yield curve by a function depending on a finite number n of parameters: polynomial, spline, piecewise linear, etc. The main point is that, because of the usually high heteroskedasticity, one needs a fixed type of approximation, and not an approximation that depends itself on historical data. This second kind of simplification will namely be provided by the P.C.A. we are going to undertake.

Let $(\mathscr{E}^n)_{n\in\mathbb{N}}$ be a *Galerkin decomposition* of V. Each $\mathscr{E}^n$ is an n-dimensional subspace of V contained in the next one $\mathscr{E}^{n+1}$ and the union of all the $\mathscr{E}^n$ is dense in V. By the Schmidt orthonormalization procedure,[24] one can find an orthonormal basis $(L_n)_{n\in\mathbb{N}}$ of V adapted to this decomposition: for any n, $(L_1, \ldots, L_n)$ is a basis of $\mathscr{E}^n$. The subspaces $\mathscr{E}^n$ are endowed with the same norm as V.

From now on, the dimension n is fixed. In practice, if we consider the bid/offer spread as a limit for precision, then most of the time, one can find an acceptable 6 to 8 dimensional Galerkin subspace. Each yield curve $\mathbf{y}$ will then be approximated by its orthogonal projection $\mathbf{y}^n$ onto the subspace $\mathscr{E}^n$. As the basis $(L_1, \ldots, L_n)$ is orthonormal, the approximate (or smoothen, see Remark 10 below) curve is given by the simple formula

$$\mathbf{y}^n = \sum_{i=1}^{n} (\mathbf{y}.L_i)_V\, L_i\,.$$

[24] If $(J_1, \ldots)$ is a basis adapted to the Galerkin decomposition, that is, $(J_1, \ldots, J_n)$ spans $\mathscr{E}^n$, but not necessarily orthonormal, the orthonormal $(L_1, \ldots)$ basis is built by first normalizing J_1, then moving J_2 parallel to J_1 to make it orthogonal, and normalizing, and so forth.

This allows to identify the movement of the yield curve $\mathbf{y}_t$ with an n-dimensional random process

$$\mathbf{a}(t) = ((\mathbf{y}_t.L_1)_V, \dots, (\mathbf{y}_t.L_n)_V) \in \mathbb{R}^n .$$

An important issue is that the norm and dot product of the space V should be easily computable out of explicit available data.

Remark 10 If elements of $\mathcal{E}^n$ are smooth yield curves for any n, then the approximation of a curve $\mathbf{y}$ by an element $\mathbf{y}^n \in \mathcal{E}^n$ is by construction a smoothing of the yield curve.

11.2 Eigenmode Computation from the Historical Series

After this first dimension reduction we made, we shall only be able to compute the first n eigenmodes. Moreover, the last ones will lack precision. Nevertheless, this does not rise a big problem, because most of the time, the quasi-totality of the variance of the motion (more than 99 %) is borne by the first three modes. Anyway, we shall have to leave out some variance because of the impossibility to perform a strict time continuous hedge. Our aim will therefore to compute only the first three or four eigenmodes.

The previous section shows how to identify the yield curve with a vector evolving in $\mathbb{R}^n$:

$$\mathbf{a}(t) = (a_1(t), \dots, a_n(t)) \in \mathbb{R}^n .$$

Remark 11 The identification

$$\mathbf{y}^n = \sum_{i=1}^{n} a_i L_i \in \mathcal{E}^n \quad \longmapsto \quad \mathbf{a} = (a_1, \dots, a_n) \in \mathbb{R}^n$$

is an isometry when $\mathcal{E}^n$ is provided with the norm $\| \, \|_V$ and $\mathbb{R}^n$ with its usual Euclidean norm, for base functions L_i are orthonormal with respect to $\| \, \|_V$. Therefore, eigenmodes in $\mathcal{E}^n$ and in $\mathbb{R}^n$ are identical.

We now fix $\delta t > 0$. When t varies, the vectors

$$\mathbf{a}'(t) = \frac{1}{\sqrt{\delta t}} (\mathbf{a}(t + \delta t) - \mathbf{a}(t))$$

form a cloud of points in $\mathbb{R}^n$ the principal axes of which are the "historical" eigenmodes. Indeed, let π_n be the orthogonal projection of V onto $\mathcal{E}^n$. The quadratic form Q^n defined on the dual $\mathcal{E}^{n*}$ of $\mathcal{E}^n$ by

$$Q^n(\mu) = \frac{1}{\delta t} \operatorname{Var}\big[\, \mu(\mathbf{y}^n_{t+\delta t}) \,|\, t\big]$$

is obtained from Q by the transpose of the projection π_n

$$Q^n = Q \circ {}^t\pi_n\,, \qquad {}^t\pi_n : \mathscr{E}^{n*} \longrightarrow V^*\,.$$

It is a well known result that, in this situation, the eigenspaces of Q^n (that is its P.C.A.) tend, in the weak sense (that is index wise[25]), to those of Q. In practice, taking $n = 7$ or 8 gives a very good approximation of the first four modes.

Definition 2 The matrix S of Q^n in the basis $(L_1, \ldots, L_n)$ is called the *covariance matrix* of the process $\mathbf{a}(t)$ (or $\mathbf{y}_t$). It is defined by

$$S_{ij} = \frac{1}{\delta t}\,\mathrm{Cov}\left[\mathbf{y}^n_{t+\delta t}.L_i\,,\ \mathbf{y}^n_{t+\delta t}.L_j \,|\, t\right].$$

Because we took an orthonormal basis, this matrix represents the quadratic form Q^n and its diagonalization provides the eigenmodes: if $\mathbf{u} = (u_1, \ldots, u_n)$ is an eigenvector of S associated with the eigenvalue λ then $\sum u_i L_i$ is the eigenmode (in the $\mathscr{E}^n$ approximation) associated with the same eigenvalue λ.

Remark 12 This is a purely historical evaluation of the covariance matrix and of the eigenmodes. If one is concerned with Vega hedging, he should rather try to perform an implicit evaluation of the factors out of the market prices of options, or mix the two methods.

12 Dimension Reduction

The previous analysis provides two opportunities to reduce the dimension of the overall space of yield curves. The first one relies in the projection onto the Galerkin subspace $\mathscr{E}^n$. It corresponds, as we said in Remark 10, to smoothing the yield curve. This reduction should not depend on the movement of the yield curve. Indeed, the hedges we are going to compute do not take into account the errors made at this step, hence only a serious statistical analysis can insure that these errors are bounded in any market state, even catastrophic.

The second reduction is performed after the principal component analysis of the move of the approximated curve $\mathbf{y}^n_t$. Once principal deformations have been determined, we just keep the first d of them, $d = 2$ or 3. This way, we get an H.J.M. model with d factors. The space to which the curve belongs is still $\mathscr{E}^n$ but, infinitesimally, there are only d types of possible deformations. Nevertheless, as it is very

---

[25] Let $(\lambda_1, \ldots)$ be the eigenvalues of Q and $(\lambda^n_1, \ldots, \lambda^n_n)$ be those of Q^n. For fixed k, then λ^n_k tends to λ_k as n tends to ∞ and, if λ_k is not multiple, then the corresponding eigenvector φ^n_k tends to φ_k. When there is some multiplicity, then the whole eigenspace corresponding to λ^n_k tends to that corresponding to λ_k (they have the same dimension if n is sufficiently large).

unlikely that these deformations form the basis of a Markovian model,[26] the movement, even reduced to d sources of randomness, may explore the whole space $\mathcal{E}^n$. Consequently, its complexity (for instance to price a swaption) is the dimension n and not d (keeping only d factors still simplifies computations, but not as much as one could have hoped).

12.1 The Drift Term and the Real Option Pricing

When projecting the whole yield curve movement on one of the Galerkin subspaces $\mathcal{E}^n$, another problem arises. The real statistical drift of the diffusion could belong to this subspace, or be also projected onto it. But in order to fulfill the AAO hypothesis, the risk-neutral drift is imposed and there is no reason why it should lie in $\mathcal{E}^n$. A first solution would be to choose special subspaces, such that if a yield curve $\mathbf{y}^n \in \mathcal{E}^n$ then the corresponding drift $\mathbf{b}(t, \mathbf{y}^n)$ (see Eq. (14) in Sect. 5) also belongs to $\mathcal{E}^n$. This solution is that adopted by N. El Karoui in number of papers (see, e.g. [8] and [15] to [16]), and by other authors, beginning with Vasiček [33]. Although we respect the relevance of this approach, which is well adapted to implicit evaluation of the volatility structure, our own experience shows that one would rather take spaces $\mathcal{E}^n$ that really fit well the data, whatever they look like (we give in Sect. 12.3.1 examples of Galerkin spaces that appeared to be efficient).

We now have two solutions. Either we consider that the space $\mathcal{E}^n$ is here only to size the deformation factors, but we keep the entire risk-neutral drift, and the formulae (18) to (20) compute options on zero-coupons and on portfolios of such. This is the approach of Brace–Musiela [5] and Jamshidian [22]. For instance, an option on a 10 years quarterly swap leads to a 40-dimensional integral, hard to compute. One solution would be Monte-Carlo technique, or deterministic low discrepancy sequences (Sobol, etc.). Another is again to diagonalize the 40-dimensional covariance matrix of the zero-coupons involved in the swap, and to compute the integral only on the three dimensional subspace spanned by the eigenvectors corresponding to the three biggest eigenvalues, by a Gauss–Legendre interpolation.

Remark 13 The faster an option evaluation, the better it allows to compute implicit deformation factors out of option market prices (see Remark 12).

The second solution is to give up the strict AAO assumption. Indeed, the theoretical arbitrages that one could achieve in such a setting are impossible to realize in practice because of transaction costs. In other words, even if the vector $\mathbf{b}(t, \mathbf{y}^n)$ does not belong to $\mathcal{E}^n$, it is so close to its projection $\mathbf{b}^n(t, \mathbf{y}^n) = \pi_n(\mathbf{b}(t, \mathbf{y}^n))$ that the difference cannot be made into a real free lunch. Therefore, it is possible to let the curve evolve with the "almost risk-neutral" drift $\mathbf{b}^n(t, \mathbf{y}^n)$ and compute as well

[26]For this, they should show an exponential or polynomial shape with respect to the maturity x (see [15]).

vanilla options, as exotic ones (barriers, etc.) by Monte-Carlo techniques or PDE discretization inside the space $\mathscr{E}^n$.

12.2 Practical Option Hedging

Practical option hedging is always an optimization between costs and residual risks. In practice, dynamic hedging will never be able to totally offset risks, therefore some level of residual risk must be accepted. Moreover, in an optimal hedging strategy, the various sources of risk are comparable. There are mostly four sources of such residual risks:

- Discrete time dynamic hedging,
- Uncertain volatility and correlations,
- Shocks and non-diffusion processes,
- Hedging only finitely many risk factors.

Comparing the size of residual risk stemming from these sources leads to the optimal choice of number of risk factors to choose. In practice, one first assess the minimum amount of residual risk one cannot avoid by any dynamic hedging, then the level of acceptable residual risk with respect to the corresponding transaction costs. Finally, equally splitting this acceptable level of risk across the four sources above, implement the appropriate dynamic hedging strategy in order to achieve the targeted level of risk.

12.3 Difficulties

12.3.1 Galerkin Space

The first difficulty is to find good Galerkin subspaces $\mathscr{E}^n$ in order to optimize the *computation/accuracy* ratio of the model. Let us mention the following series, with their advantages and drawbacks.

- *Polynomials of degree* $n-1$. Arbitrage free, but not performing: it cannot at the same time the variety of short term rates and the barely changing behavior of long term rates.
- *Decreasing exponential* $e^{-\alpha_k x}$, $\alpha_k \geq 0$. So-called generalized Vasiček (see [15]). Arbitrage free and better than the previous one. Good for implicit evaluation of factors, because of possibility of rather fast evaluation of swaptions.
- *Cubic splines* (piecewise third degree polynomials we C^2 fit at junction). Good for fitting the prices of assets with rather small number of parameters, but not arbitrage free. Dimension n equals $3+$ *number of splines*. Most common: three splines (see Turner [32]).

- *Polynomials of degree $n-1$ with a change of variable on the maturity*. Also non arbitrage free. One of the most efficient is the *Log* change for it fits the fast changing shape of the short term part and the very regular one of the long term part. This idea was first suggested by P. Gaye, and appears to have several theoretical justifications, one of them being that the yield curve and the curve of forward spot rates belong to the same space (the forward spot rate is the derivative of the zero-coupon rate with respect to the *Log* of the maturity). It is also better than cubic splines because it does not particularize any maturity and thus gives a "nicer" smoothing.

12.3.2 Instability of Eigenmodes

In this first formulation, our model assumes that volatilities and correlations are constant. This fact is obviously denied by most statistical analysis. Studies we made on the French curve show that the plane spanned by the first two modes almost follows a Brownian path in the manifold of 2-dimensional planes of $\mathscr{E}^n$ for $n=6$ (polynomials in *Log*). Short term options, or even European ones (options on futures, swaps, bonds, etc.) can afford a homoskedastic model, provided we don't mix the maturity of the various options, whereas caps and floors really need a heteroskedastic model (see Brace–Gatarek–Musiela [4]).

Similarly, it is of importance to detect "changes of regime", that is situations where we need to take into account a larger number of factors of a bigger Galerkin subspace to keep the "noise" under control.

12.3.3 Statistical Evaluation of Drift and Volatility

Evaluating the diffusion coefficients (drifts, volatilities and correlations) of a multi-dimensional process, especially when data are not always of good quality, can easily become a challenge, see Genon-Catalot & Jacod [20]. One again needs to optimize the value of n in order to guaranty a correct estimation of the coefficients, as well as a good control on the noise.

ARCH and GARCH models should also be considered.

Note that the "tick" discretization introduces its own noise for large n, and without it, the move of prices inside the bid/offer spread is rather erratic.

12.3.4 Mixing Historical and Implicit Data

Historical data are useful to price illiquid vanilla options (options on bonds, etc.), for we seldom have a Vega hedging and we need to forecast the behavior of volatility. For any other option (liquid, exotic when liquid vanilla exist, etc.) one need an implicit evaluation of the factors out of the market data on the prices of liquid options. The problem is that the number of parameters to estimate can be larger than the number of reliable data. This means that we necessarily need to mix implicit data with historical ones, through the optimization of some penalty function that weights both.

Incoherence between historical and implicit data, or even among implicit data, can sometimes give rise to (quasi) arbitrage opportunities, provided transaction costs allow to enter the setting up of such position.

Acknowledgements The author wishes to thank the Société Générale Research and Development Team on Interest Rate and Forex Markets, and especially Pierre Gaye, who asked all the questions that initiated this work, and Jean-Michel Fayolle whose programming skills have been of great help. He is also grateful to Nicole El Karoui, Marek Musiela, Marc Yor, Marco Avellaneda and Albert Shiryaev for helpful discussions and comments.

References

1. Anderson, N., Breedon, F., Deacon, M., Derry, A., Murphy, G.: Estimating and Interpreting the Yield Curve. Wiley, New York (1996)
2. Black, F., Derman, E., Toy, W.: A one-factor model of interest rates and its application to treasury bond options. Financ. Anal. J. **46**(1), 33–39 (1990)
3. Bouchaud, J.-P., Sagna, N., Cont, R., El Karoui, N., Potters, M.: Phenomenology pf the Interest rate curve. Preprint SSRN (1997)
4. Brace, A., Gatarek, D., Musiela, M.: The market model of interest rate dynamics. Preprint, Univ. of New South Wales, Sydney (1995)
5. Brace, A., Musiela, M.: A multifactor Gauss-Markov implementation of Heath, Jarrow and Morton. Math. Finance **4**(3), 259–283 (1994)
6. Bricio-Hernandez, D.: Lectures on Probability and Second Order Random Fields. Series on Adv. in Math. for Appl. Sci., vol. 30. World Scientific, Singapore (1995)
7. Brigo, D., Mercurio, F.: Interest Rate Models: Theory and Practice, 2nd edn. Springer, Heidelberg (2006)
8. Cherif, T., El Karoui, N., Myneni, R., Viswanathan, R.: Arbitrage pricing and hedging of quanto options and interest rate claims with quadratic Gaussian state variables. Preprint, Lab. of proba., Univ. of Paris VI, (1995)
9. Cont, R.: Modeling term structure dynamics: an infinite dimensional approach. Preprint, École Polytechnique, Paris (2002)
10. Cox, J.C., Ingersoll, J.E., Ross, S.A.: A theory of the term structure of interest rates. Econometrica **53**(2) (1985)
11. Dana, R.A., Jeanblanc-Piqué, M.: Marchés financiers en temps continu. Economica, coll. Recherche en Gestion, Paris (1994). In English: Financial Markets in Continuous Time. Springer Finance (2007)
12. Da Prato, G., Zapczyk, J.: Stochastic Equations in Infinite Dimensions. Cambridge University Press, Cambridge (1992)
13. Douady, R.: A Fourier-log analysis of US euro-dollar futures. Working paper (1997)
14. El Karoui, N., Durand, P.: Quadratic Gaussian model of interest rates and quanto options. Preprint, Labo. de proba, de Paris VI (1997)
15. El Karoui, N., Lacoste, V.: Multifactor model of the term structure of interest rates. Preprint, Labo. de proba, de Paris VI (1992)
16. El Karoui, N., Myneni, R., Visnawathan, R.: Arbitrage pricing and hedging of interest rate claim with state variables. Preprint, Labo. de proba, de Paris VI et Univ. de Stanford, février (1992)
17. Filipovic, D.: Consistency Problems for Heath-Jarrow-Morton Interest Rate Models. Springer Lecture Notes in Mathematics, vol. 1760. Springer, Berlin (2004)

18. Frachot, A., Janci, D., Lacoste, V.: Factor Analysis of the Term Structure: a Probabilistic Approach. de la Banque de France, Labo (1992). Preprint
19. Gaveau, B.: Intégrale stochastique radonifiante. C. R. Acad. Sci. (Paris) **276**, mai (1973)
20. Genon-Catalot, V., Jacod, J.: On the estimation of the diffusion coefficients for multidimensional diffusion processes. Ann. Inst. Henri Poincaré **29**, 119–152 (1993)
21. Heath, D., Jarrow, R., Morton, A.: Bond pricing and the term structure of interest rates: a new methodology for contingent claims valuation. Econometrica **60**, 77–106 (1992)
22. Jamshidian, F.: Option and future evaluation with deterministic volatilities. Math. Finance **3**(2), 149–159 (1993)
23. Kabanov, Yu., Kramkov, D.: Large financial markets: asymptotic arbitrage and contiguity. Theory Probab. Appl. **39**(1), 182–187 (1994)
24. Kennedy, D.P.: The term structure of interest rates as a Gaussian random field. Math. Finance **4**(3), 247–258 (1994)
25. Litterman, R., Scheinkman, J.: Common factors affecting bond returns. Technical Report 62, Goldman-Sachs Financial Strategies Group, septembre (1988)
26. Merton, R.: Theory of rational option pricing. Bell J. Econ. Manag. Sci. **4**, 141–183 (1973)
27. Merton, R.: Continuous Time Finance. Blackwell, Oxford (1991)
28. Musiela, M., Sondermann, D.: Different Dynamical Specifications of the Term Structure of Interest Rates and their Implications. Preprint, Univ. Bonn (1994)
29. Musiela, M., Rutkowski, M.: Martingale Methods in Financial Modeling, 2nd edn. Springer, Berlin (2005)
30. Rebonato, R.: Interest-Rate Option Models, 2nd edn. Wiley, Chichester (1998)
31. Rebonato, R.: Interest-rate term-structure pricing models: a review. Preprint (2003)
32. Turner, S.R.E.: Modeling interest rates and assessing risk of derivative securities. Preprint, Univ. of Cambridge, Statistical Lab. (1993)
33. Vasiček, O.A.: An equilibrium characterisation of the term structure. J. Financ. Econ. **5**, 177–188 (1977)
34. Yor, M.: Existence et unicité de diffusions à valeur dans un espace de Hilbert. Ann. Inst. Henri Poincaré **X**(1), 55–58 (1974)
35. Yor, M.: Sur les intégrales stochastiques à valeur dans un espace de Banach. Ann. Inst. Henri Poincaré **X**(1), 31–36 (1974)

Maximally Acceptable Portfolios

Ernst Eberlein and Dilip B. Madan

Abstract Portfolios are selected in non-Gaussian contexts to maximize a Cherny and Madan index of acceptability. Analytical gradients are developed for the purpose of optimizing portfolio searches on the unit sphere. It is shown that though an acceptability index is not a preference ordering, many utilities will concur with acceptability maximization. A stylized economy illustrates the advantages from the perspective of acceptability of nonlinear securities and options. In sample results for the year 2008 indicate that maximizing the acceptability index can lead to portfolios that second order stochastically dominate their Gaussian counterparts. Backtests over the period 1997 to 2008 reflect gains to maximizing acceptability over holding a maximal Sharpe ratio portfolio.

Keywords Acceptability index · Distorsion · Risk measure · Sharp ratio · Maximally acceptable portfolio

Mathematics Subject Classification (2010) 91G10

Portfolio rebalancing is now possible and is being executed at much higher frequencies than has been possible in the past. Some algorithms trade every five to fifteen minutes a fairly large number of stocks ranging from a thousand stocks upward. It has been known for some time now that at such short horizons returns are extremely non-Gaussian displaying significant levels of skewness and excess kurtosis. Additionally modern economies directly provide access to nonlinear cash flows via the markets for options and variance swaps. Optimal portfolio selection in such non-Gaussian contexts is expected to diverge from the multivariate Gaussian model that essentially focuses on maximizing the Sharpe ratio. This is primarily due to the

E. Eberlein
Department of Mathematical Stochastics, University of Freiburg, Freiburg, Germany
e-mail: eberlein@stochastik.uni-freiburg.de

D.B. Madan (✉)
Robert H. Smith School of Business, University of Maryland, College Park, MD 20742, USA
e-mail: dbm@rhsmith.umd.edu

Y. Kabanov et al. (eds.), *Inspired by Finance*, DOI 10.1007/978-3-319-02069-3_11,

recognition that investors are not indifferent to other aspects of a return distribution and ceteris paribus they prefer positive skewness and peakedness and dislike tailweightedness. Kurtosis, as noted in [8], is a preferentially confused statistic as it combines both peakedness and tailweightedness.

The choice of criterion on which to base portfolio selection is then a critical issue and many alternatives have been formulated in the literature. We refer to Biglova, Ortobelli, Rachev and Stoyanov [4] for a survey and application of a number of these criteria that all take the form of ratios using the expected return in the numerator and a suitably chosen risk measure in the denominator. Here we propose to follow the generalization of Sharpe ratios to arbitrage consistent performance measures developed in [7]. These measures are not based on ratios and they do not separate risk from reward. Instead they attempt to directly measure the quality of cash flow distributions accessed at zero cost. First, by construction nonnegative cash flows accessed at zero cost are considered to be infinitely good as they are arbitrages. For other cash flows that are exposed to losses, one computes a stressed expectation and the quality of the cash flow is proportional to the level of stress that it can withstand. The measures are termed acceptability indices and the higher the index the smaller is the set of cash flow distributions acceptable at this level. At all levels the set of acceptable cash flows forms, as random variables, a convex set containing all the non-negative cash flows.

We develop in this paper fast algorithms for maximizing the acceptability index attained by a portfolio and show how to operationalize and implement the optimization procedure. When working with a single underlier we explicitly introduce nonlinear payoffs and options. For multiple assets the exercise requires the specification of the non-Gaussian joint law of asset returns and we recognize that there are numerous ways to do this. The algorithm we develop requires that one be able to simulate the joint law for the assets of interest and many researchers would like to work with their favorite specifications in this domain. Since our focus is on explaining and operationalizing the maximization of an index of acceptability we adopt a fairly simple yet adequate formulation of the joint law for our purposes. We thereby leave refinements in this direction to future investigations. In formulating the joint law we follow the suggestions of Malevergne and Sornette [13] and merely compute a covariance matrix after transformation of marginals to a standard normal variate by passing through the composition of the distribution function and inverse normal cumulative distribution function. Malevergne and Sornette [13] estimate marginals in a modified Weibull family but as we construct samples from the joint law with some frequency on 50 stocks we just employ the empirical distribution function of our samples. Refinements associated with estimating and simulating more general and more complicated densities preferably associated with limit laws or self-decomposable random variables can easily be entertained in extensions. One may also reestimate a few parameters at each rebalance while reestimating the whole set at a lower frequency. These are considerations that must be analyzed in developing an industrial strategy but are not essential for the initial exposition of procedures devoted to designing maximally acceptable trades.

The outline of the rest of the paper is as follows. Section 1 provides basic details on indices of acceptability. The algorithm for constructing maximally acceptable portfolios is developed in Sect. 2. Section 3 presents a stylized economy in which we study the advantages offered by nonlinear cash flows and options from the perspective of enhancing acceptability. Section 4 applies this algorithm to recent data covering the volatile period of the year ending in December 5, 2008. Section 5 presents a backtest rebalancing maximally acceptable portfolios every 5 days compared with a maximal Sharpe ratio investment. Section 6 concludes.

1 Acceptability Indices

We present here the essential details leading to the operational indices of acceptability defined in [7]. For this purpose, we model the financial outcomes of trading as zero-cost terminal cash flows seen as random variables on a probability space $(\Omega, \mathcal{F}, P)$. A short review of the development of acceptability indices and its links to more classical ideas may be helpful. For an expected utility maximizing investor, with utility function u, with a given random initial position W the set of zero cost random variables acceptable to this investor is given by the set of all random variables X such that $E[u(W+X)] \geq E[u(W)]$, or the classical better than set. This is typically a convex set containing the nonnegative cash flows. If one is interested in cash flows acceptable to many investors then one must intersect all such convex sets, but the result will remain a convex set containing the nonnegative cash flows. If we now shift attention to cash flows that move marginally in the direction X by taking the position $W + \varepsilon X$ for a small number ε, thereby leaving issues of size to other considerations like market depth or impact, then one may model the acceptable cash flows by the smallest convex cone containing all the classical better than convex sets.

Such a formulation for acceptable cash flows was axiomatized and adopted in [1] and studied further for its asset pricing implications in [5]. Such cones of acceptable cash flows are supported by a set of probability measures and cash flows are acceptable just if they have a positive expectation under all the supporting probability measures. It follows that the larger is the set of supporting measures the smaller is the cone of acceptability. Cherny and Madan [7] went on to index a decreasing sequence of cones by a real valued level of acceptability with the property that the higher the level of acceptability, the larger the set of supporting measures. Cash flows with a positive expectation are acceptable at level zero while arbitrages are infinitely acceptable. They then constructed a performance measure for cash flows as the highest level of acceptability attained by a potential cash flow. Such performance measures based on indices of acceptability are a generalization of the Sharpe ratio and the Gain-Loss ratio of Bernardo and Ledoit [3] and like them are scale invariant, but improve on the associated economic properties.

The construction of operational cones of acceptability led Cherny and Madan [7] to consider law invariant cones of acceptability. Here the decision on the acceptability of a cash flow depends only on the distribution function. This property, though not ideal, is shared with expected utility, and all the various ratios used in risk analysis and mentioned earlier by reference to Biglova, Ortobelli, Rachev and Stoyanov [4]. Such law invariant operational cones of acceptability are related to a sequence of concave distortions $\Psi^{\gamma}(y)$ also studied in [8]. Each function $\Psi^{\gamma}(y)$ is a concave distribution function defined on the unit interval with values in the unit interval that is pointwise increasing in the level of the distortion γ. A random variable X with distribution function $F(x)$ is acceptable at level γ just if its expectation under such a distortion is nonnegative or that

$$\int_{-\infty}^{\infty} x d\Psi^{\gamma}(F(x)) \geq 0.$$

The acceptability index of X, $\gamma^*(X)$ is then given by

$$\gamma^*(X) = \sup\left\{\gamma : \int_{-\infty}^{\infty} x d\Psi^{\gamma}(F(x)) \geq 0\right\}.$$

It may be tempting to think of the level of acceptability as a degree of risk aversion but this is not correct. A few remarks address the important differences. First, risk aversions may be increased to arbitrarily high levels depending on the preferences being represented. Levels of acceptability can not be increased in the same way as there is a market determined limit to the highest level possibly attainable. Essentially in markets prohibiting arbitrage the highest level of γ attainable in the market is bounded by a finite positive real number. Second, we observe that increases in γ amount to a further distortion of probability and do not introduce greater concavity in utility. In fact there is no distortion of wealth, comparable to its utility, occurring in the definition of an acceptability index. We refer the reader to [12] for a deeper discussion of all these distortions. However, we note by way of comparison to utility considerations that expectations under concave distortions are also expectations under a change of measure as, supposing the existence of the density $f(x)$ of the distribution function $F(x)$, we have that

$$\begin{aligned}\int_{-\infty}^{\infty} x d\Psi^{\gamma}(F(x)) &= \int_{-\infty}^{\infty} x \Psi^{\gamma\prime}(F(x)) f(x) dx \\ &= E^{Q}[X]\end{aligned}$$

where the change of measure is

$$\frac{dQ}{dP} = \Psi^{\gamma\prime}(F(X)). \tag{1}$$

Note that the measure change depends explicitly on the cash flow X as indicated in expression (1). We note that increased risk aversion introduces greater concavity and nonlinearity in the measure change and the same applies to increasing γ but as already noted, there are market determined limits to how far γ may be increased but no such limits apply to risk aversion.

Critical to the various levels of acceptability are the measures supporting acceptability at this level. Fortunately there is a clear understanding of these measures provided in [6]. One has to first construct the conjugate dual Φ^{γ} to the distortion defined by

$$\Phi^{\gamma}(x) = \sup_{0 \le y \le 1} \left(\Psi^{\gamma}(y) - xy \right)$$

and the supporting set of measures has densities Z with respect to P satisfying

$$E\,(Z - c)^{+} \le \Phi^{\gamma}(c), \quad c \ge 0.$$

Cherny and Madan [7] provide four examples of useful concave distortions. The first termed *MINVAR* is given by

$$\Psi^{\gamma}(y) = 1 - (1 - y)^{1+\gamma}.$$

An expectation under this distortion for integral γ is easily seen to be the expectation of the minimum of $(1+\gamma)$ independent draws from the distribution function. Hence more generally we say that X is *MINVAR* acceptable at level γ if the minimum of $1+\gamma$ independent draws has a positive expectation. A simple computation shows that the measure change (1) does not reweight large losses, when $F(x)$ is near zero, to arbitrarily high levels and hence the economic dissatisfication with this distortion. A similar critique accompanies the Gain-Loss ratio.

The second distortion termed *MAXVAR* is given by

$$\Psi^{\gamma}(y) = y^{\frac{1}{1+\gamma}}.$$

Here large losses are reweighted up to infinity but the gains are not discounted to zero. Expectation under this distortion is from the distribution function of a random variable that is so bad that one has to make $1+\gamma$ independent draws and take the maximum outcome to get to the original distribution being evaluated. The other two combine these in two ways. We shall here work with *MINMAXVAR* for which

$$\Psi^{\gamma}(y) = 1 - (1 - y^{\frac{1}{1+\gamma}})^{1+\gamma}$$

and we note that in this case both, large losses and large gains, are respectively reweighted up to infinity and down to zero. This property also holds for the distortion *MAXMINVAR* for which

$$\Psi^{\gamma}(y) = \left(1 - (1 - y)^{1+\gamma}\right)^{\frac{1}{1+\gamma}}.$$

When γ is an integer we may interpret both *MINMAXVAR* and *MAXMINVAR* for example as first drawing from a distribution so bad that we take maximum of γ draws and then we repeat this procedure another γ times and take the minimum outcome.

Given that an index of acceptability is a performance measure, like the Sharpe ratio, and not a preference ordering for an investor, the question arises as to why one should consider maximizing this index of acceptability. We recognize that though

Sharpe ratios have been maximized in practice, we have been forewarned in numerous studies and we cite Goetzmann, Ingersoll, Spiegel and Welch [10] and Agarwal and Naik [2] about how such strategies may be preferentially inferior. It is well recognized that outside a Gaussian framework, one may for example increase the Sharpe ratio by accessing negative skewness on selling downside puts but actually take positions that decrease expected utilities.

When managing money for a single investor, expected utility is a well established and sound criterion, notwithstanding its more modern critique from the considerations of behavioral finance. One of the motivations behind acceptability is the recognition that money is often managed on behalf of large groups of individuals and here one would like to maximize the consent of a sizable set of economically sensible supporting kernels. Certainly an arbitrage would have the full consent of all rational kernels. We also recognize that if a random variable X second order stochastically dominates Y then it has a higher acceptability level. This is not true for many performance measures but it does hold for an index of acceptability. However, the implication does not go in the reverse direction though we shall encounter occasions where we are able to associate with a higher acceptability level a situation of second order stochastic dominance, in which case we have carried all preference orderings along.

Unlike the situation with Sharpe ratios, one has a much clearer understanding of all the preference orderings that will concur with a particular trade in a direction enhancing an index of acceptability. If the random variable X with distribution function F is acceptable at level γ for a distortion Ψ then we have that

$$\int_{-\infty}^{\infty} x d\Psi^{\gamma}(F(x)) \geq 0. \tag{2}$$

We also know that such a trade is marginally acceptable to a utility function u at a random initial wealth W provided

$$E\left[u'(W)X\right] \geq 0. \tag{3}$$

Now define by

$$\Lambda(x) = E\left[u'(W)|X = x\right]$$

and write

$$Eu'(W)X = \int_{-\infty}^{\infty} x\Lambda(x) f(x)dx.$$

We now note that on the provision

$$\left(\Lambda(x) - \Psi^{\gamma\prime}(F(x))\right)x \geq 0$$

we have that (2) implies (3). Hence for investors whose expected marginal utility does not rise on losses beyond $\Psi^{\gamma\prime}(F(x))$ and does not fall on gains beyond $\Psi^{\gamma\prime}(F(x))$ a positive acceptability receives their consent. Equivalently one requires that $(\Psi^{\gamma'})^{-1}(U'(x)) \leq F(x)$ and the condition involves all three entities, the distribution function, the distortion and the utility function and so a general statement involving two of these entities is not possible. The importance of having $\Psi^{\gamma'}$ go to infinity and zero at the two extremes of zero and unity is now even clearer as we do expect marginal utilities to behave this way for a wide class of utility functions. We recognize that we will not necessarily carry all utilities but there is a large class that comes along. As mentioned earlier we shall have occasion to associate with a particular enhancement in acceptability a second order stochastic dominance and then we do carry all utility functions.

Acceptability is thus considerably differentiated from utility and in particular one does not have to specify a degree of risk aversion in working with acceptability as an objective. The acceptability level γ^* will be endogenously determined through the optimization and unlike risk aversion, it is not an input that needs to be specified. One may then wonder what happens to investor preferences in this approach. They essentially go into the choice of distortions. For example the distortion *MINVAR* is relatively lenient towards large losses with a maximal reweighting of losses capped at $1 + \gamma$. Such a distortion will not carry many utility functions along with its decisions as the expected marginal utility $\Lambda(x)$ for losses will easily rise above this bound of $1 + \gamma$. This is why the use of *MINMAXVAR* is more conservative. However, once one has chosen a distortion that has a derivative rising sufficiently fast for losses and falling sufficiently fast for gains, its decisions will satisfy a sufficiently large number of utilities and one can concentrate on improving the quality of cash flows for wide collections of investors simultaneously, by maximizing acceptability and leaving issues of risk aversion aside.

2 Constructing Maximally Acceptable Portfolios

We develop in this section an efficient algorithm for constructing portfolios that are maximally acceptable over a prespecified finite set of potential stock investments. We envisage the investment as being on day t to be unwound either the next day or a few days later. The use of such a short horizon is predicated on the belief that we are unable to describe adequately multivariate return possibilities over long horizons using statistical data on recent daily returns. We may not be able to describe the possibilities over the short horizon either but we suspend our disbelief in this proposition and entertain a statistical approach to such short term investment.

Our first task is to describe the joint law for daily returns on n selected assets that we denote by $R = (R_1, \ldots, R_n)$. We suppose the marginal distribution function of the ith return is $F_i(r)$. In constructing the joint law we follow Malevergne and Sornette [13] and define standard Gaussian random variates Z_i by

$$Z_i = N^{-1}(F_i(R_i))$$

where $N(x)$ is the distribution function of a standard normal variate. This is a particularly simple structure to implement. For other possibilities we reference Eberlein and Madan [9] and Khanna and Madan [11]. We postulate that the variables Z_i are correlated with a correlation matrix C. They have unit variance and zero means by construction. The non-Gaussian nature of our returns is captured in the nonlinear transformation back with

$$R_i = F_i^{-1}(N(Z_i)).$$

We wish to construct a portfolio with h_i dollars invested long or short in asset i with the portfolio return

$$Y = h'R.$$

We wish to find the portfolio weights h with a view to maximizing the level of acceptability of the cash flow Y. The optimization will be conducted on a simulated sample space where we generate M readings on the n joint returns that are stored in the n by M matrix A. The portfolio returns on this sample space are then given by the vector

$$c = h'A.$$

We sort the vector c in increasing order to construct

$$s_i = c_{k(i)}$$

where s_1 is the smallest element and s_M is the largest element of the vector c. The acceptability index for the vector c, $\gamma(c)$ is implicitly defined by the equation

$$\sum_{i=1}^{M} s_i \left(\Psi^{\gamma}\left(\frac{i}{M}\right) - \Psi^{\gamma}\left(\frac{i-1}{M}\right)\right) = 0. \tag{4}$$

The summation in Eq. (4) is an estimate for the distorted expectation at level γ and the acceptability index is the value of γ for which this distorted expectation is zero. Given that acceptability indices are scale invariant by construction, the search for the optimal h may be restricted to the surface of the sphere in dimension n defined by $h'h = 1$. The search algorithm is then fairly simple once we have the gradient

$$\gamma_h = \frac{\partial \gamma}{\partial h}.$$

We merely follow the gradient to the point $h + \gamma_h$ which we renormalize to unit length and stop when the renormalized point equals the original point h. Hence for implementation we need an explicit gradient computation of γ_h.

Taking the total differential of (4) we get that

$$\sum_{i=1}^{N} ds_i \left(\Psi^{\gamma}\left(\frac{i}{N}\right) - \Psi^{\gamma}\left(\frac{i-1}{N}\right)\right) + \sum_{i=1}^{N} s_i \frac{\partial}{\partial \gamma}\left[\Psi^{\gamma}\left(\frac{i}{N}\right) - \Psi^{\gamma}\left(\frac{i-1}{N}\right)\right] d\gamma = 0.$$

It follows that

$$\frac{d\gamma}{ds_i} = -\frac{\Psi^\gamma(\frac{i}{N}) - \Psi^\gamma(\frac{i-1}{N})}{\sum_{i=1}^N s_i \frac{\partial}{\partial\gamma}[\Psi^\gamma(\frac{i}{N}) - \Psi^\gamma(\frac{i-1}{N})]}.$$

For *MINMAXVAR* we have that

$$\Psi^\gamma(y) = 1 - (1 - y^{\frac{1}{1+\gamma}})^{1+\gamma}$$

$$\frac{\partial}{\partial\gamma}\Psi^\gamma(y) = -\left(1 - y^{\frac{1}{1+\gamma}}\right)^{1+\gamma} \ln\left(1 - y^{\frac{1}{1+\gamma}}\right) - \left(1 - y^{\frac{1}{1+\gamma}}\right)^\gamma y^{\frac{1}{1+\gamma}} \frac{\ln(y)}{1+\gamma}.$$

For the other distortions we have: for *MINVAR*

$$\frac{\partial}{\partial\gamma}\Psi^\gamma(y) = -(1-y)^{1+\gamma}\ln(1-y)$$

and for *MAXVAR*

$$\frac{\partial}{\partial\gamma}\Psi^\gamma(y) = -y^{\frac{1}{1+\gamma}}\frac{\ln(y)}{(1+\gamma)^2}.$$

Finally, for *MAXMINVAR* we have

$$\begin{aligned}\frac{\partial}{\partial\gamma}\Psi^\gamma(y) = &-\left(1-(1-y)^{1+\gamma}\right)^{\frac{1}{1+\gamma}} \frac{\ln(1-(1-y)^{1+\gamma})}{(1+\gamma)^2} \\ &- \frac{1}{1+\gamma}\left(1-(1-y)^{1+\gamma}\right)^{-\frac{\gamma}{1+\gamma}} (1-y)^{1+\gamma}\ln(1-y)\end{aligned}$$

To construct the partial of the acceptability index γ with respect to h_j we must evaluate

$$\frac{\partial\gamma}{\partial h_j} = \sum_i \frac{d\gamma}{ds_i} R_{jk(i)}$$

where $R_{jk(i)}$ is the ith largest element of the jth row and $k(i)$ is the column number for this i^{th} largest element. We employ this gradient computation in a search restricted to the surface of the sphere $h'h = 1$ to find the portfolio that maximizes the acceptability index.

3 Nonlinearity and Acceptability in Economies

We consider in this section a stylized economy and the role played by nonlinear securities like variance swaps and options in enhancing the acceptability of cash flows that may be accessed in markets. The distortion employed is *MINMAXVAR*. Consider a two date one period economy with a single risky asset and a zero interest rate. The risky asset is assumed to be lognormally distributed with a mean rate of return of $\mu = .15$ and a volatility $\sigma = .35$. The final asset value is

$$S = \exp\left(\mu + \sigma Z - \frac{\sigma^2}{2}\right)$$

where Z is a standard normal variate. The initial price of this risky asset is unity and the pricing kernel or measure change is given by the measure change for the Black–Scholes economy with

$$\frac{dQ}{dP} = \exp\left(\alpha Z - \frac{\alpha^2}{2}\right)$$

for $\alpha = -\mu/\sigma$.

The first zero cost cash flow available to investors is the risky return

$$R = S - 1.$$

The level of acceptability of this cash flow using *MINMAXVAR* is .2624.

We now successively introduce nonlinear securities into this economy with cash flows given by R^2, R^3 and two out of the money options, a put on S struck at the 5 % level and a call, struck at the 95 % level. The specific strikes are .6141 and 1.9715. We price these securities using the measure change and the zero discount rate to get the prices .1724, .1258, .0056 and .0108 respectively.

Now on just introducing the squared return the level of acceptability rises to .2946 and the trade direction on the unit circle is .9186 shares and $-.3951$ units of the squared return. If we now introduce the claim paying R^3 the acceptability rises to .2971 and the trade direction is $(.9003, -.4346, .0226)$.

We next introduce the put option and then the call option. The levels of acceptability rise to .3001 and .3021 respectively. The final trade direction is

$$(.8528, -.5160, .0728, .0314, .00013)$$

reflecting investment in the risky asset, shorting the squared return and buying skewness and some out of the money puts and calls. We present in Fig. 1 the cash flow accessed with squared and cubic assets, and then the final cash flow including the options.

4 In Sample Application to Portfolios Constructed for the Year 2008

It is well recognized that the year 2008 was very volatile with significant possibilities for departure from Gaussian returns. In the next section we shall consider backtesting over a much longer period starting in 1997 and finishing in December 2008. For this longer period we obtained data on 771 stocks that were continuously

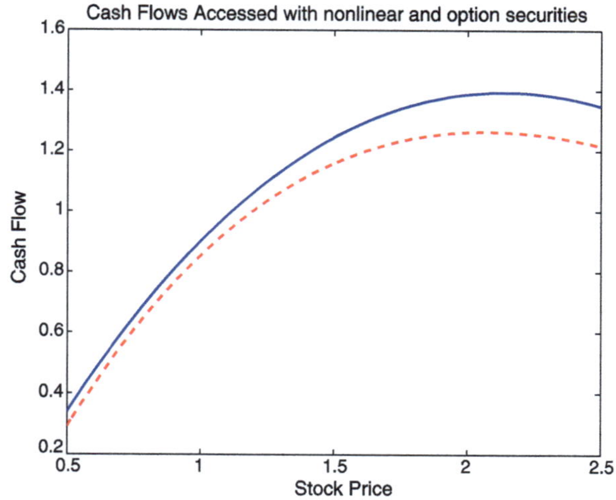

Fig. 1 Cash flows accessed using squared and cubic securities in a *solid line*. Cash flows with options and nonlinear securities as a *dashed line*

quoted among the top 1500 names over the whole period. In this section we consider three portfolios of 50 stocks made up of those with the top 50 realized means over the year, the second 50 and third 50 realized means. For each of these three sets of 50 stocks we first construct the benchmark Gaussian investment by normalizing to the unit sphere the vector

$$a = V^{-1}m, \qquad g = \frac{a}{\sqrt{a'a}}$$

where V is the covariance matrix of the 50 returns over the year and m is vector of realized means over the year.

Next we transform to standard Gaussian variates using the empirical distribution function constructed from daily returns over the past year, (252 observations), we then compute the correlation matrix of these transformed variates. Finally we generate 10000 draws from a multivariate Gaussian model with this correlation matrix and transform back via $F_i^{-1}(N(x))$ to get 10000 joint readings on our 50 stocks. This gives us three sets of 50 by 10000, potential A matrices for which we implement the search procedure to find the maximally acceptable portfolio h for the distortion *MINMAXVAR*.

We then construct, for each of the three sets separately, the returns $g'A$ and $h'A$ and present in Figs. 2, 3 and 4 the empirical densities for the Gaussian and maximally acceptable portfolios.

We observe that for the top 50 means there is a clear domination by the maximally acceptable portfolio of the Gaussian portfolio. To investigate this further we constructed the double integral of the empirical density or the integral of the distribution function to find that the Gaussian distribution function integral lies above the maximally acceptable distribution function integral for both, the top 50 and second 50, sets of portfolios. This suggests that the maximally acceptable portfolios second order stochastically dominate the Gaussian portfolios in these two cases. In this case

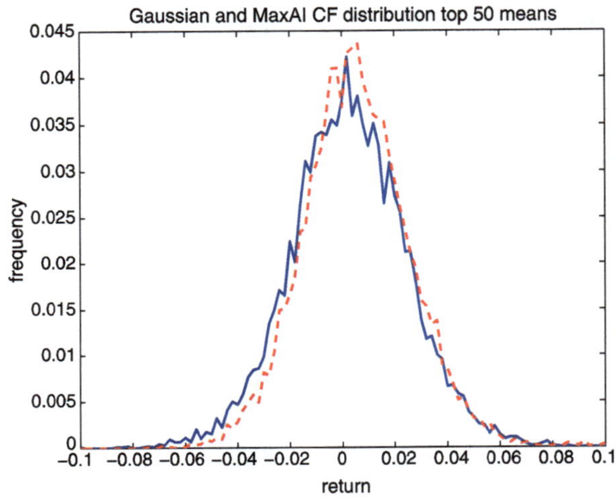

Fig. 2 Gaussian empirical density as a *solid line* and maximally acceptable density as a *dashed line* for the stocks with the top 50 realized annual mean returns

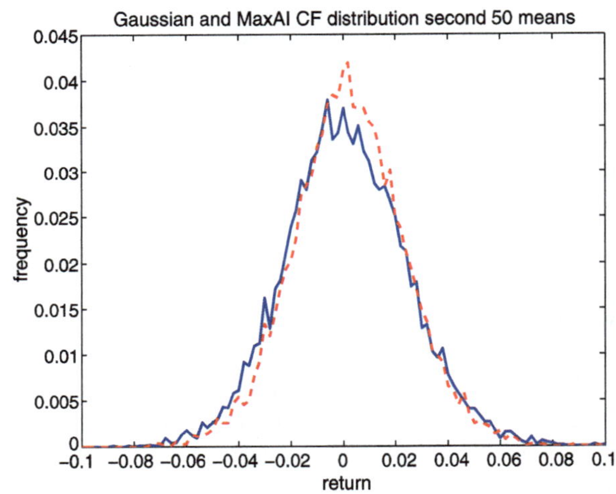

Fig. 3 Gaussian empirical density as a *solid line* and maximally acceptable density as a *dashed line* for the stocks with the second 50 realized annual mean returns

all utility functions would prefer the maximally acceptable portfolio to the Gaussian one. We present in Figs. 5, 6 and 7 the integrals of these distribution functions.

We see clearly that for the third 50 stocks this domination is lost and we dominate only for utility functions that are strictly concave for large positive returns but are linear for small positive and negative returns.

5 Backtesting Portfolio Rebalancing from 1997 to 2008

We report in this section the results of a backtest where we start on March 10 1997 and end on November 28 2008, rebalancing portfolios every five days on the stocks with top 50, second 50 and third 50 realized mean returns over the past year. For

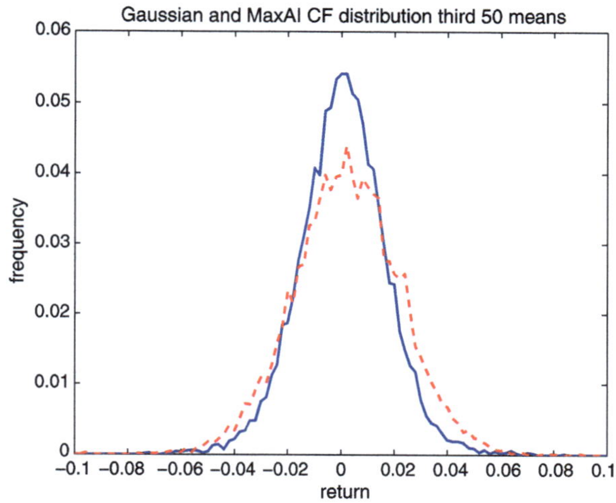

Fig. 4 Gaussian empirical density as a *solid line* and maximally acceptable density as a *dashed line* for the stocks with the third 50 realized annual mean returns

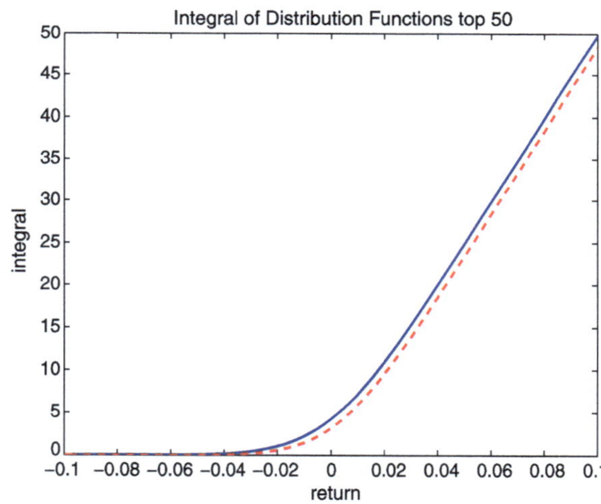

Fig. 5 Distribution function integrals, Gaussian as a *solid line* and maximally acceptable as a *dashed line*, top 50

each of these three sets of stocks we construct two portfolios, the straight Gaussian portfolio normalized to the unit sphere and the maximally acceptable one optimized on the unit sphere as per the construction described in Sect. 2. Every five days we transform to standard Gaussians, draw from a suitably correlated Gaussian model 10000 joint return possibilities and maximize over the sphere for the portfolio h. Both the Gaussian and maximally acceptable portfolios are held for five days when they are unwound and a new portfolio is formed for the next five days.

There are in all six cash flows of length 591 for the 591 rebalancings that occurred over this period. They are the maximally acceptable and Gaussian results for the top, second and third 50 stocks for each rebalance day. We present in Fig. 8 the backtested cumulated cash flows from these strategies.

We observe a clear domination of the top 50 over the second 50 and the third 50 for both strategies and a domination of the Gaussian by the maximally acceptable.

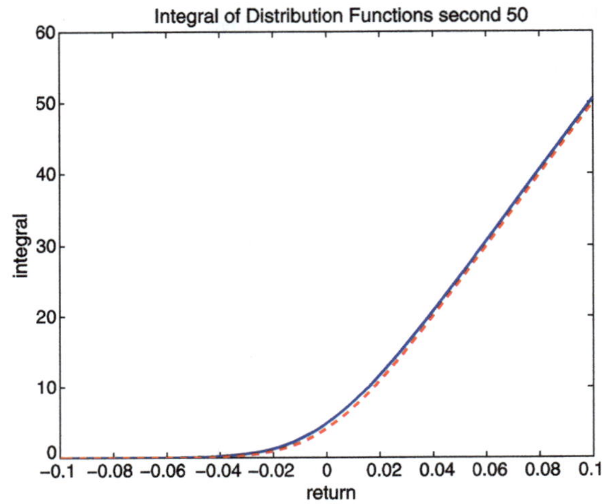

Fig. 6 Distribution function integrals, Gaussian as a *solid line* and maximally acceptable as a *dashed line*, second 50

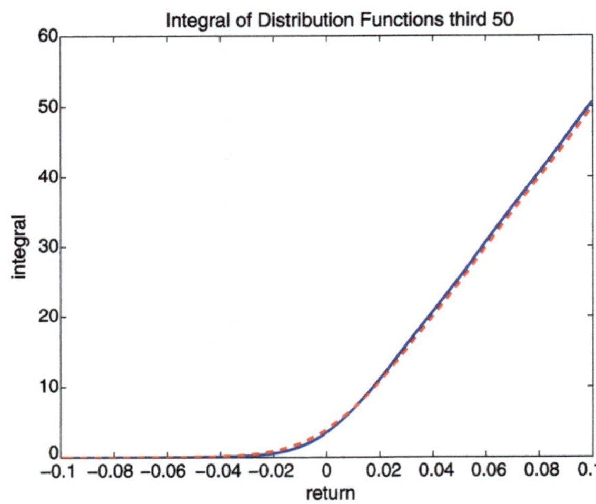

Fig. 7 Distribution function integrals, Gaussian as a *solid line* and maximally acceptable as a *dashed line*, third 50

The strategies took considerable losses towards the end of 2008, a phenomenon experienced by many strategies.

6 Conclusion

Portfolio selection in non-Gaussian environments is studied with a view towards maximizing an index of acceptability as defined in [7]. As the indices are scale invariant, optimal long short portfolios may be constructed by maximizing over the unit sphere. Analytical gradients are developed for the purpose of enhancing this search. The indices of acceptability are heuristically described as the maximum level of stress a potential cash flow can be subjected to before its stress distorted

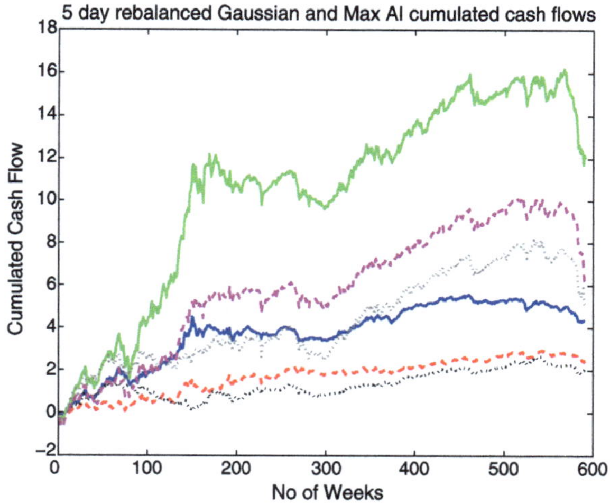

Fig. 8 Cumulated cash flows maximally acceptable as the *upper solid, dashed and dotted lines* and Gaussian as *lower solid, dashed and dotted lines* for the top 50, second 50 and third 50 respectively

expectation turns negative. It is shown that though an acceptability index is not a preference ordering, it is related to preferences and certain well understood classes of utilities concur with its decisions. In fact, conditionally expected marginal utilities, conditional on the outcome, that rise less for losses and fall more for gains, than the derivative of the distortion taken at the cash flow quantile, agree with acceptability.

A stylized economy illustrates the acceptability enhancing features of nonlinear securities and options. In sample results for the year 2008 indicate that some portfolios maximizing the acceptability index in fact second order stochastically dominate their Gaussian counterparts. Backtests over the period 1997 to 2008 reflect gains to maximizing acceptability over holding a maximal Sharpe ratio portfolio.

Acknowledgements Dilip Madan acknowledges support from the Humboldt foundation as a Research Award Winner.

References

1. Artzner, P., Delbaen, F., Eber, J.M., Heath, D.: Coherent measures of risk. Math. Finance **9**, 203–228 (1999)
2. Agarwal, V., Naik, N.: Risk and portfolio decisions involving hedge funds. Rev. Financ. Stud. **17**, 63–98 (2004)
3. Bernardo, A., Ledoit, O.: Gain, loss, and asset pricing. J. Polit. Econ. **108**, 144–172 (2000)
4. Biglova, A., Ortobelli, S., Rachev, S., Stoyanov, S.: Different approaches to risk estimation in portfolio theory. J. Portf. Manag. **31**, 103–112 (2004)
5. Carr, P., Geman, H., Madan, D.: Pricing and hedging in incomplete markets. J. Financ. Econ. **62**, 131–167 (2001)
6. Cherny, A.: Weighted VAR and its properties. Finance Stoch. **10**, 367–393 (2006)
7. Cherny, A., Madan, D.B.: New measures for performance evaluation. Rev. Financ. Stud. **22**, 2571–2606 (2009)

8. Eberlein, E., Madan, D.B.: Hedge fund performance: sources and measures. Int. J. Theor. Appl. Finance **12**, 267–282 (2009)
9. Eberlein, E., Madan, D.B.: On correlating Lévy processes. J. Risk **13**, 1 (2010)
10. Goetzmann, W., Ingersoll, J., Spiegel, M., Welch, I.: Sharpening Sharpe ratios. NBER Working Paper 9116, Cambridge (2002)
11. Khanna, A., Madan, D.V.: Non-Gaussian models of dependence in returns. SSRN 1540875 (2010)
12. Jin, H., Zhou, X.Y.: Behavioral portfolio selection in continuous time. Math. Finance **18**, 385–426 (2008)
13. Malevergne, Y., Sornette, D.: High-order moments and cumulants of multivariate Weibull asset return distributions: analytical theory and empirical tests: II. Finance Lett. **3**, 54–63 (2005)

Some Extensions of Norros' Lemma in Models with Several Defaults

Pavel V. Gapeev

Abstract We provide some extensions of Norros' lemma for a model with several default times and nontrivial reference filtrations. These results allow a characterization of the filtration immersion properties in terms of the terminal values of compensators of the associated default processes. The method of proof is based on the analysis of properties of exponential martingales associated with the default times.

Keywords Default times · Default processes and their compensators · Intensity processes · Reference filtration · Filtration immersions

Mathematics Subject Classification (2010) 91B70 · 60G44 · 60G40

1 Introduction

It is known that a filtration is said to be immersed in a larger one whenever every martingale with respect to the former filtration keeps the martingale property with respect to the latter one (see, e.g. Mansuy and Yor [10, Chap. I] or Bielecki and Rutkowski [2, Chap. VIII]). For the first time, such a situation was described in Brémaud and Yor [3] and referred to as the (H)-hypothesis. Kusuoka [8] introduced that hypothesis for the credit risk setting and considered the case in which the given (nontrivial) reference filtration is immersed in the filtration progressively enlarged with that generated by the associated default process. In models of reliability theory, where the reference filtration is trivial, the so-called Norros' lemma states the following assertion. If the failure times are finite and neither two of them can occur at the same time almost surely, then the continuous compensator processes evaluated at the failure times are independent random variables having standard exponential law (see, e.g. Norros [11]).

In this paper, we extend Norros' lemma for the case of credit risk models in which the reference filtration is no longer trivial. We show that if the reference filtration is

P.V. Gapeev (✉)
Department of Mathematics, London School of Economics, Houghton Street, London WC2A 2AE, UK
e-mail: p.v.gapeev@lse.ac.uk

Y. Kabanov et al. (eds.), *Inspired by Finance*, DOI 10.1007/978-3-319-02069-3_12,

immersed into every filtration progressively enlarged by any particular default time, then the terminal values of the compensators of the associated default processes are independent of the observations. Moreover, we provide the links between various immersion properties and (conditional) independence of the terminal values of the compensators (with respect to the reference filtration). The results of the paper can naturally be derived for a model with finitely many default times, and we restrict our consideration to the case of two defaults, for simplicity of exposition.

The paper is organized as follows. In Sect. 2, we formulate a credit risk model with two default times and recall the notion of filtration immersions. In Sect. 3, we provide extensions of Norros' lemma for the case of two defaults, under non-trivial reference filtrations. The main result of the paper is stated in Proposition 2.

2 Default Times and Filtration Immersions

In this section, we introduce a credit risk model with two default times and recall the notion of filtration immersions.

2.1 The Setting

Let us suppose that on a probability space $(\Omega, \mathscr{G}, P)$ there exist (nonnegative) finite random variables τ_i, $i = 1, 2$, which we call the *default* times. For every $i = 1, 2$ fixed, let $H^i = (H^i_t)_{t \ge 0}$ be the *default* process associated with the time τ_i and defined by $H^i_t = I(\tau_i \le t)$, where $I(\cdot)$ is the indicator function. Let $(\mathscr{H}^i_t)_{t \ge 0}$ be the natural filtration of H^i, so that $\mathscr{H}^i_t = \sigma(H^i_s : 0 \le s \le t)$ for all $t \ge 0$. Let us denote by $(\mathscr{F}_t)_{t \ge 0}$ the *reference* filtration and define the filtrations $(\mathscr{G}^{i*}_t)_{t \ge 0}$ and $(\mathscr{G}^*_t)_{t \ge 0}$ by $\mathscr{G}^{i*}_t = \mathscr{F}_t \vee \mathscr{H}^i_t$ and $\mathscr{G}^*_t = \mathscr{F}_t \vee \mathscr{H}^1_t \vee \mathscr{H}^2_t$ for all $t \ge 0$, respectively. Let us denote by $(\mathscr{G}^i_t)_{t \ge 0}$ and $(\mathscr{G}_t)_{t \ge 0}$ the right-continuous versions of $(\mathscr{G}^{i*}_t)_{t \ge 0}$ and $(\mathscr{G}^*_t)_{t \ge 0}$, so that $\mathscr{G}^i_t = \mathscr{G}^{i*}_{t+}$ and $\mathscr{G}_t = \mathscr{G}^*_{t+}$ for all $t \ge 0$. To simplify further notations, we shall assume that $(\mathscr{G}^i_t)_{t \ge 0}$ and $(\mathscr{G}_t)_{t \ge 0}$ are right-continuous and completed by all P-negligible sets.

For every $i = 1, 2$ fixed, let $G^i = (G^i_t)_{t \ge 0}$ be the $(\mathscr{F}_t)_{t \ge 0}$*-conditional survival probability* process of the default time τ_i defined by $G^i_t = P(\tau_i > t \,|\, \mathscr{F}_t)$ for all $t \ge 0$.

Hypothesis 1 *Assume that the process* $G^i = (G^i_t)_{t \ge 0}$ *is continuous and satisfies the condition* $0 < G^i_t \le 1$, *for all* $t > 0$ *and every* $i = 1, 2$.

Note that the assumption stated above yields the fact that τ_i fails to be an $(\mathscr{F}_t)_{t \ge 0}$-stopping time. Being a continuous $(\mathscr{F}_t)_{t \ge 0}$-supermartingale, the process G^i admits the continuous compensator $C^i = (C^i_t)_{t \ge 0}$ such that $C^i_0 = 1 - G^i_0 = 0$ and $G^i + C^i$ forms an $(\mathscr{F}_t)_{t \ge 0}$-martingale, for every $i = 1, 2$. In the same way, there exists a $(\mathscr{G}_t^i)_{t\geq 0}$-predictable increasing process $A^i = (A_t^i)_{t\geq 0}$ such that the process $M^i = (M_t^i)_{t\geq 0}$ defined by:

$$M_t^i = H_t^i - A_t^i \tag{1}$$

is a $(\mathscr{G}_t^i)_{t\geq 0}$-martingale. It is well known (see, e.g. [6]) that $A_t^i = A_{t\wedge\tau_i}^i$ and also $A_t^i I(\tau_i \geq t) = \Lambda_t^i I(\tau_i \geq t)$ holds, where $\Lambda^i = (\Lambda_t^i)_{t\geq 0}$ is an $(\mathscr{F}_t)_{t\geq 0}$-predictable continuous increasing process given by:

$$\Lambda_t^i = \int_0^t \frac{dC_s^i}{G_s^i}, \tag{2}$$

the integral is supposed to be convergent, for all $t \geq 0$ and every $i = 1, 2$. Hence, the default time τ_i turns out to be a $(\mathscr{G}_t^i)_{t\geq 0}$-totally inaccessible stopping time (see, e.g. [12, Chap. VI, Sect. 13]). In the credit risk literature, A^i is called the $(\mathscr{G}_t^i)_{t\geq 0}$-*intensity* process and Λ^i is called the $(\mathscr{F}_t)_{t\geq 0}$-*intensity* process of the default time τ_i (see, e.g. [2, Chap. V]).

2.2 Immersion Properties

Let $(\mathscr{F}_t')_{t\geq 0}$ and $(\mathscr{F}_t'')_{t\geq 0}$ be two right-continuous completed filtrations such that $\mathscr{F}_t' \subseteq \mathscr{F}_t''$ for all $t \geq 0$. The filtration $(\mathscr{F}_t')_{t\geq 0}$ is said to be *immersed* in the filtration $(\mathscr{F}_t'')_{t\geq 0}$ if any $(\mathscr{F}_t')_{t\geq 0}$-martingale remains an $(\mathscr{F}_t'')_{t\geq 0}$-martingale. This notion is also known as the (H)-*hypothesis* for the filtrations $(\mathscr{F}_t')_{t\geq 0}$ and $(\mathscr{F}_t'')_{t\geq 0}$ in the literature (see, e.g. [3], [10, Chap. V, Sect. 4] or [2, Chap. VIII, Sect. 3]), and that is equivalent to the conditional independence of $\mathscr{F}_t''$ and $\mathscr{F}_\infty'$ with respect to $\mathscr{F}_t'$, for any $t \geq 0$ (see, e.g. [4]). Recall that, in the particular case in which $\mathscr{G}_t^i = \mathscr{F}_t \vee \mathscr{H}_t^i$, the filtration $(\mathscr{F}_t)_{t\geq 0}$ is immersed in the filtration $(\mathscr{G}_t^i)_{t\geq 0}$ if and only if the equality:

$$P(\tau_i > t \,|\, \mathscr{F}_t) = P(\tau_i > t \,|\, \mathscr{F}_\infty) \tag{3}$$

holds for all $t \geq 0$ (see, e.g. [3] or [6]). Note that, in the case in which $(\mathscr{F}_t)_{t\geq 0}$ is a trivial filtration (like in models of reliability theory), the (H)-hypothesis holds for $(\mathscr{F}_t)_{t\geq 0}$ and $(\mathscr{G}_t^i)_{t\geq 0}$ automatically. Observe that the condition of (3) necessarily implies the fact that the process G^i is decreasing, and thus, because of the assumption of continuity of G^i, we have $C_t^i = 1 - G_t^i$ for all $t \geq 0$. We will further study the case in which the equality:

$$P(\tau_i > t \,|\, \mathscr{F}_t) = P(\tau_i > t \,|\, \mathscr{G}_t^{3-i}) \tag{4}$$

is satisfied, that is equivalent to the fact that $\mathscr{G}_t^i$ and $\mathscr{G}_t^{3-i}$ are conditionally independent with respect to $\mathscr{F}_t$, for any $t \geq 0$. Here, for $i = 1$ we have $3 - i = 2$, and for $i = 2$ we have $3 - i = 1$, respectively. We also add that $(\mathscr{G}_t^i)_{t\geq 0}$ is immersed in the filtration $(\mathscr{G}_t)_{t\geq 0}$ if and only if the equality:

$$P(\tau_i > t \,|\, \mathscr{G}_t^{3-i}) = P(\tau_i > t \,|\, \mathscr{G}_\infty^{3-i}) \tag{5}$$

holds, signifying the fact that $\mathscr{G}_t$ and $\mathscr{G}_\infty^{3-i}$ are conditionally independent with respect to $\mathscr{G}_t^{3-i}$, for any $t \geq 0$ and every $i = 1, 2$.

3 Extensions of Norros' Lemma

In this section, we provide some links between the filtration immersions and the properties of the terminal values of compensators of the default processes.

3.1 The Case of One Default Time

We begin with an appropriate assertion for a model with one default time. In the case of trivial reference filtration, part (i) was obtained in [11, Theorem 2.1]. In the case of general reference filtration, the assertion of part (ii) and its inverse (see Remark 1 below) can be found as an exercise in [10, p. 99, Example 38]. We keep a proof of that result for completeness.

For this, we assume that on $(\Omega, \mathscr{G}, P)$ there exists a default time τ. We define by $H = (H_t)_{t\geq 0}$ the compensator process of the time τ by $H_t = I(\tau \leq t)$ and denote by $(\mathscr{H}_t)_{t\geq 0}$ its natural filtration $\mathscr{H}_t = \sigma(H_s :\ 0 \leq s \leq t)$ for all $t \geq 0$. In this case, the filtration $(\mathscr{G}_t)_{t\geq 0}$ defined by $\mathscr{G}_t = \mathscr{F}_t \vee \mathscr{H}_t$ for $t \geq 0$ is assumed to be right-continuous and completed by all P-negligible sets. It is further assumed that the $(\mathscr{F}_t)_{t\geq 0}$-*conditional survival probability* process $G = (G_t)_{t\geq 0}$ defined by $G_t = P(\tau > t \,|\, \mathscr{F}_t)$ is *continuous* and satisfies the condition $0 < G_t \leq 1$ for all $t > 0$. We also define the corresponding $(\mathscr{F}_t)_{t\geq 0}$-compensator process $C = (C_t)_{t\geq 0}$ such that $C_0 = 1 - G_0 = 0$ and $G + C$ forms an $(\mathscr{F}_t)_{t\geq 0}$-martingale. In the same way, there exists a $(\mathscr{G}_t)_{t\geq 0}$-predictable increasing process $A = (A_t)_{t\geq 0}$ such that the process $M = (M_t)_{t\geq 0}$ defined by:

$$M_t = H_t - A_t \tag{6}$$

is a $(\mathscr{G}_t)_{t\geq 0}$-martingale. Moreover, we assume that the $(\mathscr{F}_t)_{t\geq 0}$-predictable continuous increasing process $\Lambda = (\Lambda_t)_{t\geq 0}$ by:

$$\Lambda_t = \int_0^t \frac{dC_s}{G_s} \tag{7}$$

where the integral is supposed to be convergent, and note that $A_t = A_{t\wedge\tau}$ and also $A_t\, I(\tau \geq t) = \Lambda_t\, I(\tau \geq t)$ holds for all $t \geq 0$ according to [6].

Proposition 1 *Let the process* $G = (G_t)_{t\geq 0}$ *be continuous and such that* $G_0 = 1$. *Then, the following conclusions hold*:

(i) *the variable* $A_\tau \equiv \Lambda_\tau$, *defined in* (6)–(7), *has a standard exponential law* (*with parameter* 1);

(ii) *if* $(\mathscr{F}_t)_{t\geq 0}$ *is immersed in* $(\mathscr{G}_t)_{t\geq 0}$ (*i.e. if* (3) *holds for* τ *and all* $t \geq 0$), *then the variable* $A_\tau \equiv \Lambda_\tau$ *is independent of* $\mathscr{F}_\infty$.

Proof (i) In this part, we reproduce the arguments from [11] for the reader convenience. Consider the process $L = (L_t)_{t \geq 0}$ defined by:

$$L_t = (1+z)^{H_t} \, e^{-zA_t} \tag{8}$$

for all $t \geq 0$ and any $z > 0$ fixed. Then, applying the integration-by-parts formula to (8), we get:

$$dL_t = z \, e^{-zA_t} \, dM_t \tag{9}$$

where the process M, defined in (6), is a $(\mathscr{G}_t)_{t \geq 0}$-martingale. Hence, by virtue of the assumption that $z > 0$, it follows from (9) that L is a $(\mathscr{G}_t)_{t \geq 0}$-martingale too, so that:

$$E\big[(1+z)^{H_t} \, e^{-zA_t} \,\big|\, \mathscr{G}_s\big] = (1+z)^{H_s} \, e^{-zA_s} \tag{10}$$

holds for all $0 \leq s \leq t$. In view of the implied by $z > 0$ uniform integrability of L, we may let t go to infinity in (10). Setting s equal to zero in (10) and using the fact that $A_\infty = A_\tau$, we therefore obtain:

$$E\big[(1+z) \, e^{-zA_\tau}\big] = 1. \tag{11}$$

This means that the Laplace transform of A_τ is the same as that of a standard exponential variable and thus proves the claim. This property was also proved in [1] (see also [7, Chap. IV]).

(ii) Applying the change-of-variable formula, we get:

$$\begin{aligned}
e^{-zA_{t\wedge\tau}} = e^{-zA_t} &= 1 - z \int_0^t e^{-zA_s} \, dA_s \\
&= 1 - z \int_0^t \exp\left(-z \int_0^s \frac{I(\tau > u)}{G_u} \, dC_u\right) I(\tau > s) \, d\Lambda_s \\
&= 1 - z \int_0^t \exp\left(-z \int_0^s \frac{dC_u}{G_u}\right) I(\tau > s) \, d\Lambda_s \\
&= 1 - z \int_0^t e^{-z\Lambda_s} \, \frac{I(\tau > s)}{G_s} \, dC_s
\end{aligned} \tag{12}$$

for all $t \geq 0$ and any $z > 0$ fixed. Then, taking the conditional expectations under $\mathscr{F}_t$ from both parts of the expression in (12) and applying Fubini's theorem, we obtain from the immersion of $(\mathscr{F}_t)_{t \geq 0}$ in $(\mathscr{G}_t)_{t \geq 0}$ that:

$$\begin{aligned}
E\big[e^{-zA_t} \,\big|\, \mathscr{F}_t\big] &= 1 - z \int_0^t E\left[e^{-z\Lambda_s} \, \frac{I(\tau > s)}{G_s} \,\middle|\, \mathscr{F}_t\right] dC_s \\
&= 1 - z \int_0^t e^{-z\Lambda_s} \, \frac{P(\tau > s \,|\, \mathscr{F}_t)}{G_s} \, dC_s \\
&= 1 - z \int_0^t e^{-z\Lambda_s} \, dC_s
\end{aligned} \tag{13}$$

holds for all $t \ge 0$. Hence, using the fact that the immersion of $(\mathscr{F}_t)_{t\ge0}$ in $(\mathscr{G}_t)_{t\ge0}$ implies the decrease of the process G, so that $C_t = 1 - G_t$ and $\Lambda_t = -\ln G_t$, we see from (13) that:

$$E\big[e^{-zA_{t\wedge\tau}} \,|\, \mathscr{F}_t\big] = 1 + \frac{z}{1+z}\Big((G_t)^{1+z} - (G_0)^{1+z}\Big) \tag{14}$$

is satisfied for all $t \ge 0$. Letting t go to infinity and using the assumption $G_0 = 1$, as well as the fact that $G_\infty = 0$ (P-a.s.), we therefore obtain from (14) by virtue of the uniform integrability of L that:

$$E\big[e^{-zA_\tau} \,|\, \mathscr{F}_\infty\big] = \frac{1}{1+z} \tag{15}$$

holds, that signifies the desired assertion. Note that a similar result was obtained in [5], by means of the time-change technique and under the assumption of strict decrease of the process G. □

Remark 1 To show that an assertion inverse to part (ii) of Proposition 1 holds true, we use the fact that the process A is continuous. Then, the default time τ can obviously be represented in the form:

$$\tau = \inf\{t \ge 0 : \; A_t \ge A_\tau\}.$$

Hence, if A_τ is independent of $\mathscr{F}_\infty$, then we obtain:

$$P(\tau > t \,|\, \mathscr{F}_t) = P(A_\tau > A_t \,|\, \mathscr{F}_t) = P(A_\tau > A_t \,|\, \mathscr{F}_\infty) = P(\tau > t \,|\, \mathscr{F}_\infty)$$

for all $t \ge 0$, so that the condition of (3) holds with τ, signifying that $(\mathscr{F}_t)_{t\ge0}$ is immersed in $(\mathscr{G}_t)_{t\ge0}$ (see also [10, p. 99, Example 38]).

3.2 The Case of Two Default Times

Let us now formulate and prove the related result for the two-defaults setting.

Proposition 2 *Let the processes* $G^i = (G^i_t)_{t\ge0}$, $i = 1, 2$, *be continuous and such that* $G^i_0 = 1$, *and assume that* $P(\tau_1 = \tau_2) = 0$ *is satisfied. Then, the following conclusions hold*:

(i) *if* $(\mathscr{G}^i_t)_{t\ge0}$, $i = 1, 2$, *are immersed in* $(\mathscr{G}_t)_{t\ge0}$ (*i.e. if* (5) *holds for all* $t \ge 0$), *then the variables* $A^i_{\tau_i} \equiv \Lambda^i_{\tau_i}$, $i = 1, 2$, *defined in* (1)–(2), *are independent*;

(ii) *if* $(\mathscr{F}_t)_{t\ge0}$ *is immersed in* $(\mathscr{G}^i_t)_{t\ge0}$ (*i.e. if* (3) *holds for all* $t \ge 0$) *and* (4) *holds for all* $t \ge 0$ *and every* $i = 1, 2$, *then the variables* $A^i_{\tau_i} \equiv \Lambda^i_{\tau_i}$, $i = 1, 2$, *are conditionally independent with respect to* $\mathscr{F}_\infty$.

Proof (i) Observe that the condition of (5) yields that, for every $i = 1, 2$, the process $L^i = (L^i_t)_{t \ge 0}$ defined by:

$$L^i_t = (1 + z_i)^{H^i_t} \, e^{-z_i A^i_t} \tag{16}$$

is $(\mathscr{G}^i_t)_{t \ge 0}$ as well as $(\mathscr{G}_t)_{t \ge 0}$-martingale. Then, following the arguments from [11] and applying the implied by $P(\tau_1 = \tau_2) = 0$ orthogonality of the pure jump processes L^i, $i = 1, 2$, in (16), we obtain:

$$E\big[(1+z_1)^{H^1_t} \, e^{-z_1 A^1_t} \, (1+z_2)^{H^2_t} \, e^{-z_2 A^2_t} \,\big|\, \mathscr{G}_s\big] = (1+z_1)^{H^1_s} \, e^{-z_1 A^1_s} \, (1+z_2)^{H^2_s} \, e^{-z_2 A^2_s} \tag{17}$$

for all $0 \le s \le t$. Hence, letting t go to infinity and setting s equal to zero in (17), we get that:

$$E\big[(1+z_1) \, e^{-z_1 A^1_{\tau_1}} \, (1+z_2) \, e^{-z_2 A^2_{\tau_2}}\big] = 1 \tag{18}$$

holds. Upon recalling the expression in (11) applied for every $i = 1, 2$, we see from (18) that:

$$E\big[e^{-z_1 A^1_{\tau_1}} \, e^{-z_2 A^2_{\tau_2}}\big] = E \, e^{-z_1 A^1_{\tau_1}} \;\; E e^{-z_2 A^2_{\tau_2}}$$

is satisfied, thus proving the claim.

(ii) Using the arguments from the part (ii) of Proposition 1 above, we see that the expression in (12) applied for every $i = 1, 2$ implies:

$$\begin{aligned} e^{-z_1 A^1_t} \, e^{-z_2 A^2_t} &= 1 - z_1 \int_0^t e^{-z_1 A^1_u} \frac{I(\tau_1 > u)}{G^1_u} \, dC^1_u - z_2 \int_0^t e^{-z_2 A^2_v} \frac{I(\tau_2 > v)}{G^2_v} \, dC^2_v \\ &\quad + z_1 z_2 \int_0^t \int_0^t e^{-z_1 A^1_u} \, e^{-z_2 A^2_v} \, \frac{I(\tau_1 > u, \tau_2 > v)}{G^1_u G^2_v} \, dC^1_u \, dC^2_v \end{aligned} \tag{19}$$

for all $t \ge 0$. Then, taking the conditional expectations under $\mathscr{F}_t$ from both parts of the expression in (19) and applying Fubini's theorem, we have:

$$\begin{aligned} &E\big[e^{-z_1 A^1_t} \, e^{-z_2 A^2_t} \,\big|\, \mathscr{F}_t\big] \\ &\quad = 1 - z_1 \int_0^t e^{-z_1 A^1_u} \, dC^1_u - z_2 \int_0^t e^{-z_2 A^2_v} \, dC^2_v \\ &\qquad + z_1 z_2 \int_0^t \int_0^t e^{-z_1 A^1_u} \, e^{-z_2 A^2_v} \, \frac{P(\tau_1 > u, \tau_2 > v \,|\, \mathscr{F}_t)}{G^1_u G^2_v} \, dC^1_u \, dC^2_v \end{aligned} \tag{20}$$

for all $t \ge 0$. Observe that it follows from the assumptions of (3) and (4) that:

$$\begin{aligned} P(\tau_i > u, \tau_{3-i} > v \,|\, \mathscr{F}_t) &= P(\tau_i > u \,|\, \mathscr{F}_t) \, P(\tau_{3-i} > v \,|\, \mathscr{F}_t) \\ &= P(\tau_i > u \,|\, \mathscr{F}_u) \, P(\tau_{3-i} > v \,|\, \mathscr{F}_v) = G^i_u G^{3-i}_v \end{aligned} \tag{21}$$

holds for all $0 \le u, v \le t$ and every $i = 1, 2$. Hence, using the expression in (14) applied for every $i = 1, 2$ and the fact that the assumption of (3) implies the decrease of the process G^i, so that $C_t^i = 1 - G_t^i$ and $\Lambda_t^i = -\ln G_t^i$, we get from (20) and (21) that:

$$E\big[e^{-z_1 \Lambda_t^1}\, e^{-z_2 \Lambda_t^2} \,\big|\, \mathscr{F}_t\big] = \left(1 + \frac{z_1}{1+z_1}\left((G_t^1)^{1+z_1} - (G_0^1)^{1+z_1}\right)\right) \times \left(1 + \frac{z_2}{1+z_2}\left((G_t^2)^{1+z_2} - (G_0^2)^{1+z_2}\right)\right) \tag{22}$$

for all $t \geq 0$. Therefore, letting t go to infinity and using the assumption that $G_0^i = 1$ as well as the fact that $G_\infty^i = 0$ (P-a.s.), we obtain from (22) by virtue of the uniform integrability of L^i, $i = 1, 2$, that:

$$E\big[e^{-z_1 \Lambda_{\tau_1}^1}\, e^{-z_2 \Lambda_{\tau_2}^2} \,\big|\, \mathscr{F}_\infty\big] = \frac{1}{1+z_1}\,\frac{1}{1+z_2} \tag{23}$$

holds. Upon recalling the expression in (15) applied for every $i = 1, 2$, we thus conclude from (23) that the equality

$$E\big[e^{-z_1 \Lambda_{\tau_1}^1}\, e^{-z_2 \Lambda_{\tau_2}^2} \,\big|\, \mathscr{F}_\infty\big] = E\big[e^{-z_1 \Lambda_{\tau_1}^1} \,\big|\, \mathscr{F}_\infty\big]\, E\big[e^{-z_2 \Lambda_{\tau_2}^2} \,\big|\, \mathscr{F}_\infty\big]$$

is satisfied, signifying the desired assertion. □

Remark 2 Following the approach of [9], we finally suppose that on the initial probability space $(\Omega, \mathscr{G}, P)$ there exists random variables U_i, $i = 1, 2$, being uniformly distributed on the interval (0, 1). For every $i = 1, 2$, let us define the random time $\widehat{\tau}_i$ by:

$$\widehat{\tau}_i = \inf\{t \geq 0 :\ \lambda_i t \geq -\ln U_i\}$$

where $\lambda_i > 0$ is fixed. Let us set the corresponding default process $\widehat{H}^i = (\widehat{H}_t^i)_{t \geq 0}$ by $\widehat{H}_t^i = I(\widehat{\tau}_i \leq t)$ and its natural filtration $(\widehat{\mathscr{H}_t^i})_{t \geq 0}$ by $\widehat{\mathscr{H}_t^i} = \sigma(\widehat{H}_s^i : 0 \leq s \leq t)$, for all $t \geq 0$ and every $i = 1, 2$. Assume that the variables U_i, $i = 1, 2$, are independent of $\mathscr{F}_\infty$, so that the condition of (3) holds for $\widehat{\tau}_i$ signifying that $(\mathscr{F}_t)_{t \geq 0}$ is immersed in $(\widehat{\mathscr{G}_t^i})_{t \geq 0}$ with $\widehat{\mathscr{G}_t^i} = \mathscr{F}_t \vee \widehat{\mathscr{H}_t^i}$ for $t \geq 0$. It is shown directly that the $(\widehat{\mathscr{G}_t^i})_{t \geq 0}$-compensator of the default process $\widehat{H}^i$ is given by $\widehat{A}_t^i = \lambda_i (t \wedge \widehat{\tau}_i)$, so that the corresponding $(\mathscr{F}_t)_{t \geq 0}$-intensity process takes the form $\widehat{\Lambda}_t^i = \lambda_i t$ for all $t \geq 0$. Observe that, since U_i, $i = 1, 2$, are independent of $\mathscr{F}_\infty$, the conditions of (4) and (5) do not hold in this case, unless the variables U_i, $i = 1, 2$, are conditionally independent with respect to $\mathscr{F}_\infty$ and thus independent. This fact means that the corresponding enlarged filtration $(\widehat{\mathscr{G}_t^i})_{t \geq 0}$ is not generally immersed in the full filtration $(\widehat{\mathscr{G}_t})_{t \geq 0}$ with $\widehat{\mathscr{G}_t} = \mathscr{F}_t \vee \widehat{\mathscr{H}_t^1} \vee \widehat{\mathscr{H}_t^2}$ for all $t \geq 0$, even when $(\mathscr{F}_t)_{t \geq 0}$ is immersed in $(\widehat{\mathscr{G}_t^i})_{t \geq 0}$ for every $i = 1, 2$.

Acknowledgements The paper was initiated when the author was visiting the Université d'Evry-Val-d'Essonne in November 2008. He is grateful to Monique Jeanblanc and the Département de Mathématiques for helpful discussions and warm hospitality. Financial support from the Europlace Institute of Finance and the European Science Foundation (ESF) through the grant number 2500 of the program Advanced Mathematical Methods for Finance (AMaMeF) are gratefully acknowledged. The author thanks Ashkan Nikeghbali for his comments and references to the literature. It is also a pleasure to thank Ilkka Norros for his encouragement to extend his result to the case of credit risk models.

This research also benefited from the support of the 'Chaire Risque de Crédit', Fédération Bancaire Française.

References

1. Azéma, J.: Quelques applications de la théorie générale des processus. Invent. Math. **18**, 293–336 (1972)
2. Bielecki, T.R., Rutkowski, M.: Credit Risk: Modeling, Valuation and Hedging. Springer, Berlin (2002)
3. Brémaud, P., Yor, M.: Changes of filtrations and of probability measures. Z. Wahrscheinlichkeitstheor. Verw. Geb. **45**, 269–295 (1978)
4. Dellacherie, C., Meyer, P.A.: Probabilités et Potentiel. Hermann, Paris (1975). Chapitres I–IV; English translation: Probabilities and Potential, Chapters I–IV. North-Holland (1978)
5. El Karoui, N.: Modélisation de l'information. CEA-EDF-INRIA, École d'été (1999)
6. Elliott, R.J., Jeanblanc, M., Yor, M.: On models of default risk. Math. Finance **10**, 179–195 (2000)
7. Jeulin, T.: Semi-Martingales et Grossissement d'une Filtration. Lecture Notes in Mathematics, vol. 833. Springer, Berlin (1980)
8. Kusuoka, S.: A remark on default risk models. Adv. Math. Econ. **1**, 69–82 (1999)
9. Lando, D.: On Cox processes and credit risky securities. Rev. Deriv. Res. **2**, 99–120 (1998)
10. Mansuy, R., Yor, M.: Random Times and Enlargements of Filtrations in a Brownian Setting. Lecture Notes in Mathematics, vol. 1873. Springer, Berlin (2004)
11. Norros, I.: A compensator representation of multivariate life length distributions, with applications. Scand. J. Stat. **13**, 99–112 (1986)
12. Rogers, L.C.G., Williams, D.: Diffusions, Markov Processes and Martingales II. Itô Calculus. Wiley, New York (1987)

On the Pricing of Perpetual American Compound Options

Pavel V. Gapeev and Neofytos Rodosthenous

Abstract We present explicit solutions to the perpetual American compound option pricing problems in the Black-Merton-Scholes model.

The method of proof is based on the reduction of the initial two-step optimal stopping problems for the underlying geometric Brownian motion to appropriate sequences of ordinary one-step problems. The latter are solved through their associated one-sided free-boundary problems and the subsequent martingale verification. We also obtain a closed form solution to the perpetual American chooser option pricing problem, by means of the analysis of the equivalent two-sided free-boundary problem.

Keywords Perpetual American compound options · The Black–Merton–Scholes model · Geometric Brownian motion · Multi-step optimal stopping problem · First hitting time · Free-boundary problem · Local time-space formula

Mathematics Subject Classification (2010) 91B28 · 60G40 · 34K10

1 Introduction

Compound options are financial contracts which give their holders the right (but not the obligation) to buy or sell some other options at certain times in the future by the strike prices given. Such contingent claims are widely used in currency, stock, and fixed income markets, for the sake of risk protection (see, e.g. Geske [10, 11] and Hodges and Selby [12] for the first financial applications of compound options of European type). In the real financial world, a common application of such contracts is the hedging of bids for business opportunities which may or may not be accepted in the future, and which become available only after the previous ones are

---

P.V. Gapeev (✉) · N. Rodosthenous
Department of Mathematics, London School of Economics, Houghton Street, London WC2A 2AE, UK
e-mail: p.v.gapeev@lse.ac.uk

N. Rodosthenous
e-mail: n.rodosthenous@lse.ac.uk

Y. Kabanov et al. (eds.), *Inspired by Finance*, DOI 10.1007/978-3-319-02069-3_13,

undertaken. This fact makes compound options an important example of using real options to undertake business decisions which can be expressed in the presented perspective (see Dixit and Pindyck [5] for an extensive introduction). Other important modifications of such contracts are compound contingent claims of American type in which both the initial and underlying options can be exercised at any (random) times up to maturity. The rational pricing problems for such contracts can thus be embedded into two-step optimal stopping problems for the underlying asset price processes. The latter are decomposed into appropriate sequences of ordinary one-step optimal stopping problems which are then solved sequentially.

Apart from the extensive literature on optimal switching as well as impulse and singular stochastic control, the multi-step optimal stopping problems for underlying one-dimensional diffusion processes have recently drawn a considerable attention. Duckworth and Zervos [7] studied an investment model with entry and exit decisions alongside a choice of the production rate for a single commodity. The initial valuation problem was reduced to a two-step optimal stopping problem which was solved through its associated dynamic programming differential equation. Carmona and Touzi [2] derived a constructive solution to the problem of pricing of perpetual swing contracts, the recall components of which could be viewed as contingent claims with multiple exercises of American type, using the connection between optimal stopping problems and their associated Snell envelopes. Carmona and Dayanik [1] then obtained a closed form solution of a multi-step optimal stopping problem for a general linear regular diffusion process and a general payoff function. Algorithmic constructions of the related exercise boundaries were also proposed and illustrated with several examples of such optimal stopping problems for several linear and mean-reverting diffusions. Other infinite horizon optimal stopping problems with finite sequences of stopping times are being sought. Some of them are related to hiring and firing options and were recently considered by Egami and Xu [6] among others.

In the present paper, we derive explicit solutions to the problems of pricing of the perpetual American standard compound options in the Black-Merton-Scholes model, something which has not been done so far, to the best of our knowledge. For this, we follow the approach described profoundly in the monograph of Peskir and Shiryaev [18], which is based on the reduction of the resulting optimal stopping problems to their associated one-sided ordinary differential free-boundary problems (see also Dayanik and Karatzas [4]). It turns out that the payoff functions of some compound options are concave and the resulting value functions may have different structure, depending on the relations between the strike prices given. Moreover, we obtain a closed form solution to the problem of pricing of the perpetual American chooser option through its associated two-sided ordinary differential free-boundary problem. It is shown that the admissible intervals for the resulting exercise boundaries are smaller than the ones of the related strangle option recently studied by Gapeev and Lerche [9]. Note that the problem of pricing of American compound options was recently studied by Chiarella and Kang [3] in more general stochastic volatility framework. The associated two-step free-boundary problems for partial differential equations were solved numerically, by means of a modified sparse grid approach.

The paper is organized as follows. In Sect. 2, we formulate the perpetual American compound option problems and then specify the decompositions of the initial two-step optimal stopping problems into sequences of ordinary one-step problems for the underlying geometric Brownian motion. In Sect. 3, we derive explicit solutions of the four resulting one-sided ordinary differential free-boundary problems. In Sect. 4, we verify that the solution of the free-boundary problem related to the most informative put-on-call case provides the solution of the initial two-step optimal stopping problem. In Sect. 5, we present a closed form solution to the two-sided free-boundary problem associated with the perpetual American chooser option.

2 Preliminaries

In this section, we give a formulation of the perpetual American compound option optimal stopping problems and the associated ordinary differential free-boundary problems.

2.1 Formulation of the Problem

For a precise formulation of the problem, let us consider a probability space $(\Omega, \mathcal{F}, P)$ carrying a standard one-dimensional Brownian motion $B = (B_t)_{t \geq 0}$. Let us define the process $S = (S_t)_{t \geq 0}$ by

$$S_t = s \exp\left(\left(r - \delta - \frac{\sigma^2}{2}\right) t + \sigma\, B_t\right), \tag{1}$$

which solves the stochastic differential equation

$$dS_t = (r - \delta)\, S_t\, dt + \sigma\, S_t\, dB_t \quad (S_0 = s) \tag{2}$$

for $s > 0$, where $\sigma > 0$ and $0 < \delta < r$. Assume that the process S describes the risk-neutral dynamics of the price of a risky asset paying dividends, where r represents the riskless interest rate and δS is the dividend rate paid to stockholders.

We further consider the problem of pricing of the *initial* perpetual American standard *compound* options which are contracts giving their holders the right to buy or sell some other *underlying* (perpetual American) call or put options at certain (random) exercise times by the (positive) strike prices given. More precisely, the call-on-call (call-on-put) option gives its holder the right to buy at an exercise time τ for the price of K_1 a call (put) option with the strike K_2 (L_2) and exercise time ζ. Furthermore, the put-on-call (put-on-put) option gives its holder the right to sell at an exercise time τ for the price of L_1 a call (put) option with the strike K_2 (L_2) and exercise time ζ. Then, the *rational* (or *no-arbitrage*) prices of such perpetual American contingent claims are given by the values of the optimal stopping problems

$$V_1^*(s) = \sup_\tau \sup_\zeta E\Big[e^{-r\tau}\big(e^{-r(\zeta-\tau)}\,(S_\zeta - K_2)^+ - K_1\big)^+\Big], \tag{3}$$

$$V_2^*(s) = \sup_\tau \sup_\zeta E\Big[e^{-r\tau}\big(e^{-r(\zeta-\tau)}\,(L_2 - S_\zeta)^+ - K_1\big)^+\Big], \tag{4}$$

$$V_3^*(s) = \sup_\tau \inf_\zeta E\Big[e^{-r\tau}\big(L_1 - e^{-r(\zeta-\tau)}\,(S_\zeta - K_2)^+\big)^+\Big], \tag{5}$$

$$V_4^*(s) = \sup_\tau \inf_\zeta E\Big[e^{-r\tau}\big(L_1 - e^{-r(\zeta-\tau)}\,(L_2 - S_\zeta)^+\big)^+\Big], \tag{6}$$

where the suprema and infima are taken over the sets of stopping times $0 \le \tau \le \zeta$ with respect to the natural filtration $(\mathscr{F}_t)_{t\ge 0}$ of the asset price process S, that is $\mathscr{F}_t = \sigma(S_u \mid 0 \le u \le t)$, for all $t \ge 0$. Here, the expectations are taken with respect to the equivalent martingale measure under which the dynamics of S started at $s > 0$ are given by (1)–(2), and z^+ denotes the positive part $\max\{z, 0\}$ of any $z \in \mathbb{R}$. Note that the payoff of the call-on-call option in (3) is unbounded, while the payoffs, and thus the related rational prices of the other options in (4)–(6), are bounded by L_2 and L_1, respectively. Moreover, it is easily seen from (4) and will be shown for (6) below that the optimal exercise times of the related options are trivial whenever $K_1 \ge L_2$ and $L_1 \ge L_2$ holds, respectively.

Observe that the value functions in (3)–(4) are given by the optimal sequential choices of τ and ζ, that results in the suprema over both such stopping times, since the holders of the initial compound options can buy the underlying calls or puts at the time τ and then control the exercise time ζ. This is not the case for the value functions in (5)–(6), due to the fact that, in the case in which the holders of the compound options exercise the initial puts at the time τ by selling the underlying calls or puts, they cannot control the subsequent exercise time ζ of the latter options. We should then assume that the holders of the underlying options exercise them optimally. This turns out to be the worst case scenario for the holders of the initial compound options, resulting in the infima over ζ in the expressions of (5)–(6).

2.2 The Structure of the Optimal Stopping Times

The optimal stopping problems formulated above involve the sequential choice of the stopping times τ and ζ. Hence, the initial two-step optimal stopping problems can then be decomposed into sequences of two one-step optimal stopping problems which can then be solved separately. More precisely, using the strong Markov property of the process S, we further show that the expressions for $V_i^*(s)$, $i = 1, \ldots, 4$, in (3)–(6) can be reduced to the values of the optimal stopping problems

$$V_i^*(s) = \sup_\tau E\big[e^{-r\tau}\, H_i^+(S_\tau)\big], \tag{7}$$

where the payoff functions $H_i(s)$, $i = 1, \dots, 4$, are given by

$$H_1(s) = W(s) - K_1; \qquad H_2(s) = U(s) - K_1; \qquad H_3(s) = L_1 - W(s); \\ H_4(s) = L_1 - U(s) \tag{8}$$

for all $s > 0$. Here we denote the rational prices of the underlying perpetual American put and call options by $U(s)$ and $W(s)$ with strike prices L_2 and K_2, respectively. These are given by

$$U(s) = \sup_{\eta} E\big[e^{-r\eta}(L_2 - S_\eta)^+\big] \quad \text{and} \quad W(s) = \sup_{\eta} E\big[e^{-r\eta}(S_\eta - K_2)^+\big], \tag{9}$$

where the suprema are taken over the stopping times η of the process S started at $s > 0$. It is well known (see, e.g. [15] and [20, Chap. VIII, Sect. 2a]) that the value functions in (9) are continuously differentiable and have the form

$$U(s) = \begin{cases} -(g_*/\gamma_-)(s/g_*)^{\gamma_-}, & s > g_*, \\ L_2 - s, & s \le g_*, \end{cases} \qquad W(s) = \begin{cases} (h_*/\gamma_+)(s/h_*)^{\gamma_+}, & s < h_*, \\ s - K_2, & s \ge h_*. \end{cases} \tag{10}$$

The optimal exercise times have the structure

$$\eta_g^* = \inf\{t \ge 0 : \ S_t \le g_*\} \quad \text{and} \quad \eta_h^* = \inf\{t \ge 0 : \ S_t \ge h_*\}, \tag{11}$$

and the hitting boundaries are given by

$$g_* = \frac{\gamma_- L_2}{\gamma_- - 1} \quad \text{and} \quad h_* = \frac{\gamma_+ K_2}{\gamma_+ - 1} \tag{12}$$

with

$$\gamma_\pm = \frac{1}{2} - \frac{r-\delta}{\sigma^2} \pm \sqrt{\left(\frac{1}{2} - \frac{r-\delta}{\sigma^2}\right)^2 + \frac{2r}{\sigma^2}}, \tag{13}$$

so that $\gamma_- < 0 < 1 < \gamma_+$ holds.

It follows from the general theory of optimal stopping for Markov processes (see, e.g. [18, Chap. I, Sect. 2.2]) that the optimal stopping times in the problems of (7)-(8) are given by

$$\tau_i^* = \inf\{t \ge 0 : \ V_i^*(S_t) = H_i^+(S_t)\}$$

whenever they exist. Analyzing the structure of the outer and inner payoffs in (3)–(6), we observe that the call-on-call and put-on-put options should be exercised at the first time at which the price of the underlying risky asset rises to some upper levels b_i^*, while the call-on-put and put-on-call options should be exercised at the first time at which the asset price falls to some lower levels a_i^*. Hence, we need further to search for optimal stopping times in the problems of (7)–(8) in the form

$$\tau_i^* = \inf\{t \ge 0 : \ S_t \le a_i^*\} \quad \text{or} \quad \tau_i^* = \inf\{t \ge 0 : \ S_t \ge b_i^*\} \tag{14}$$

for some $a_i^* > 0$ and $b_i^* > 0$ to be determined, where the left-hand stopping time in (14) is optimal for the cases of $i = 2, 3$, and the right-hand one is optimal for the cases of $i = 1, 4$. Taking into account the structure of the stopping times in (11), we then further assume that the optimal stopping times ζ_i^* in (3)–(6) have the form

$$\zeta_i^* = \inf\{t \geq \tau_i^* : S_t \leq g_*\} \quad \text{or} \quad \zeta_i^* = \inf\{t \geq \tau_i^* : S_t \geq h_*\} \tag{15}$$

depending on the view of the payoff functions of the underlying options.

2.3 *The Free-Boundary Problem*

It can be shown by means of standard arguments (see, e.g. [13, Chap. V, Sect. 5.1] or [16, Chap. VII, Sect. 7.3]) that the infinitesimal operator $\mathbb{L}$ of the process S acts on an arbitrary twice continuously differentiable locally bounded function $F(s)$ according to the rule

$$(\mathbb{L}F)(s) = (r - \delta)\, s\, F'(s) + \frac{\sigma^2}{2}\, s^2\, F''(s)$$

for all $s > 0$. In order to find explicit expressions for the unknown value functions $V_i^*(s)$, $i = 1, \ldots, 4$, from (7)–(8) and the unknown boundaries a_i^* and b_i^* from (14), we may use the results of the general theory of optimal stopping problems for continuous time Markov processes (see, e.g. [19, Chap. III, Sect. 8] and [18, Chap. IV, Sect. 8]). We formulate the associated free-boundary problems

$$(\mathbb{L}V_i)(s) = r\, V_i(s) \quad \text{for} \quad s > a_i \quad \text{or} \quad s < b_i, \tag{16}$$

$$V_i(a_i+) = H_i^+(a_i) \quad \text{or} \quad V_i(b_i-) = H_i^+(b_i) \quad \text{(instantaneous stopping)}, \tag{17}$$

$$V_i'(a_i+) = H_i^{+\prime}(a_i) \quad \text{or} \quad V_i'(b_i-) = H_i^{+\prime}(b_i) \quad \text{(smooth fit)}, \tag{18}$$

$$V_i(s) = H_i^+(s) \quad \text{for} \quad s < a_i \quad \text{or} \quad s > b_i, \tag{19}$$

$$V_i(s) > H_i^+(s) \quad \text{for} \quad s > a_i \quad \text{or} \quad s < b_i, \tag{20}$$

$$(\mathbb{L}V_i)(s) < r\, V_i(s) \quad \text{for} \quad s < a_i \quad \text{or} \quad s > b_i, \tag{21}$$

for some $a_i > 0$ and $b_i > 0$ fixed, depending on the structure of the payoff $H_i^+(s)$ in (8), for every $i = 1, \ldots, 4$.

3 Solutions of the Free-Boundary Problems

We further derive solutions of the free-boundary problems related to the optimal stopping problems in (7)–(8), by specifying whether the left-hand or the right-hand part of the system in (16)–(21) is realized in every case of $i = 1, \ldots, 4$. For this we first note that the general solution of the second order ordinary differential equation in (16) is given by

$$V_i(s) = C_{+,i}\, s^{\gamma_+} + C_{-,i}\, s^{\gamma_-}, \tag{22}$$

where $C_{+,i}$ and $C_{-,i}$ are some arbitrary constants, and $\gamma_- < 0 < 1 < \gamma_+$ are defined in (13). Observe that we should have $C_{-,i} = 0$ in (22) when the right-hand part of the system in (16)–(21) is realized, since otherwise $V_i(s) \to \pm\infty$, which must be excluded because the value functions in (7) are bounded under $s \downarrow 0$. Similarly, we should also have $C_{+,i} = 0$ in (22) when the left-hand part of the system in (16)–(21) is realized, since otherwise $V_i(s) \to \pm\infty$, which must be excluded because the value functions in (7) are less than s under $s \uparrow \infty$.

3.1 The Call-on-Call Option

Let us first consider the case of $i = 1$ in which the right-hand stopping time from (14) is optimal in (3) and (7)–(8), so that the right-hand part of the free-boundary problem is realized in (16)–(21). Applying the conditions of the right-hand parts of the equations in (17) and (18) to the function in (22) with $C_{-,1} = 0$, we obtain after some rearrangements that if $b_1 < h_*$ then the equalities

$$C_{+,1}\, b_1^{\gamma_+} = \frac{h_*}{\gamma_+}\Big(\frac{b_1}{h_*}\Big)^{\gamma_+} - K_1 \quad \text{and} \quad C_{+,1}\, \gamma_+\, b_1^{\gamma_+} = h_*\Big(\frac{b_1}{h_*}\Big)^{\gamma_+} \tag{23}$$

should hold, and if $b_1 \geq h_*$ then the equalities

$$C_{+,1}\, b_1^{\gamma_+} = b_1 - K_2 - K_1 \quad \text{and} \quad C_{+,1}\, \gamma_+\, b_1^{\gamma_+} = b_1 \tag{24}$$

are satisfied for some $C_{+,1}$ and $b_1 > 0$, where h_* is given by (12). Multiplying the first equation in (23) by γ_+, we conclude from the second one there that the system in (16)–(18) does not have solutions, so that the subcase $b_1^* < h_*$ cannot be realized. Solving the system in (24), we obtain the solution of the right-hand part of the system in (16)–(18) having the form

$$V_1(s; b_1^*) = \frac{b_1^*}{\gamma_+}\Big(\frac{s}{b_1^*}\Big)^{\gamma_+} \quad \text{with} \quad b_1^* = \frac{\gamma_+(K_1 + K_2)}{\gamma_+ - 1} \equiv \frac{\gamma_+ K_1}{\gamma_+ - 1} + h_*. \tag{25}$$

3.2 The Call-on-Put Option

Let us then proceed with the case of $i = 2$ in which the left-hand stopping time from (14) is optimal in (4) and (7)–(8), so that the left-hand part of the free-boundary problem is realized in (16)–(21). Applying the conditions of the left-hand parts of the equations in (17) and (18) to the function in (22) with $C_{+,2}=0$, we obtain after some rearrangements that if $a_2 > g_*$ then the equalities

$$C_{-,2}\, a_2^{\gamma_-} = -\frac{g_*}{\gamma_-}\Big(\frac{a_2}{g_*}\Big)^{\gamma_-} - K_1 \quad \text{and} \quad C_{-,2}\, \gamma_- \, a_2^{\gamma_-} = -g_* \Big(\frac{a_2}{g_*}\Big)^{\gamma_-} \tag{26}$$

should hold, and if $a_2 \le g_*$ then the equalities

$$C_{-,2}\, a_2^{\gamma_-} = L_2 - a_2 - K_1 \quad \text{and} \quad C_{-,2}\, \gamma_- \, a_2^{\gamma_-} = -a_2 \tag{27}$$

are satisfied for some $C_{-,2}$ and $a_2 > 0$, where g_* is given by (12). Multiplying the first equation in (26) by γ_-, we conclude from the second one there that the system in (16)–(18) does not have solutions, so that the subcase $a_2^* > g_*$ cannot be realized. Solving the system in (27), we obtain the solution of the left-hand part of the system in (16)–(18) having the form

$$V_2(s; a_2^*) = -\frac{a_2^*}{\gamma_-}\Big(\frac{s}{a_2^*}\Big)^{\gamma_-} \quad \text{with} \quad a_2^* = \frac{\gamma_-(L_2 - K_1)}{\gamma_- - 1} \equiv g_* - \frac{\gamma_- K_1}{\gamma_- - 1}, \tag{28}$$

where the number a_2^* is strictly positive if and only if $L_2 > K_1$.

3.3 The Put-on-Call Option

Let us now continue with the case of $i = 3$ in which the left-hand stopping time from (14) is optimal in (5) and (7)–(8), so that the left-hand part of the free-boundary problem is realized in (16)–(21). Applying the conditions of the left-hand parts of the equations in (17) and (18) to the function in (22) with $C_{+,3} = 0$, we get after some rearrangements that if $a_3 < h_*$ then the equalities

$$C_{-,3}\, a_3^{\gamma_-} = L_1 - \frac{h_*}{\gamma_+}\Big(\frac{a_3}{h_*}\Big)^{\gamma_+} \quad \text{and} \quad C_{-,3}\, \gamma_- \, a_3^{\gamma_-} = -h_* \Big(\frac{a_3}{h_*}\Big)^{\gamma_+} \tag{29}$$

hold, and if $a_3 \ge h_*$ then the equalities

$$C_{-,3}\, a_3^{\gamma_-} = L_1 - a_3 + K_2 \quad \text{and} \quad C_{-,3}\, \gamma_- \, a_3^{\gamma_-} = -a_3 \tag{30}$$

are satisfied for some $C_{-,3}$ and $a_3 > 0$, where h_* is given by (12). Solving the systems in (29) and (30), we conclude that the two regions for L_1 and K_2, with qualitatively different solutions of the free-boundary problem, can be distinguished. By means of straightforward computations, if the condition

$$L_1 < \frac{\gamma_- - \gamma_+}{\gamma_+ \gamma_-} h_* \equiv \frac{(\gamma_- - \gamma_+) K_2}{\gamma_-(\gamma_+ - 1)} \tag{31}$$

is satisfied, then $a_3^* < h_*$ holds and the solution of the left-hand part of the system in (16)–(18) has the form

$$V_3(s; a_3^*, h_*) = -\frac{h_*}{\gamma_-}\Big(\frac{a_3^*}{h_*}\Big)^{\gamma_+}\Big(\frac{s}{a_3^*}\Big)^{\gamma_-} \tag{32}$$

with

$$a_3^* = h_*\Big(\frac{\gamma_+\gamma_- L_1}{(\gamma_- - \gamma_+)h_*}\Big)^{1/\gamma_+} \equiv \frac{\gamma_+ K_2}{\gamma_+ - 1}\Big(\frac{\gamma_-(\gamma_+ - 1)L_1}{(\gamma_- - \gamma_+)K_2}\Big)^{1/\gamma_+}. \tag{33}$$

Using similar arguments, if the condition

$$L_1 \geq \frac{\gamma_- - \gamma_+}{\gamma_+\gamma_-}\, h_* \equiv \frac{(\gamma_- - \gamma_+)K_2}{\gamma_-(\gamma_+ - 1)} \tag{34}$$

is satisfied, then $a_3^* \geq h_*$ holds and the solution of the left-hand part of the system in (16)–(18) has the form

$$V_3(s; a_3^*) = -\frac{a_3^*}{\gamma_-}\Big(\frac{s}{a_3^*}\Big)^{\gamma_-} \quad \text{with} \quad a_3^* = \frac{\gamma_-(L_1 + K_2)}{\gamma_- - 1}. \tag{35}$$

3.4 The Put-on-Put Option

Let us finally consider the case of $i = 4$ in which the right-hand stopping time from (14) is optimal in (6) and (7)–(8), so that the right-hand part of the free-boundary problem is realized in (16)–(21). Applying the conditions of the right-hand parts of the equations in (17) and (18) to the function in (22) with $C_{-,4} = 0$, we get after some rearrangements that if $b_4 > g_*$ then the equalities

$$C_{+,4}\, b_4^{\gamma_+} = L_1 + \frac{g_*}{\gamma_-}\Big(\frac{b_4}{g_*}\Big)^{\gamma_-} \quad \text{and} \quad C_{+,4}\, \gamma_+\, b_4^{\gamma_+} = g_*\Big(\frac{b_4}{g_*}\Big)^{\gamma_-} \tag{36}$$

hold, and if $b_4 \leq g_*$ then the equalities

$$C_{+,4}\, b_4^{\gamma_+} = L_1 - L_2 + b_4 \quad \text{and} \quad C_{+,4}\, \gamma_+\, b_4^{\gamma_+} = b_4 \tag{37}$$

are satisfied for some $C_{+,4}$ and $b_4 > 0$. Solving the systems in (36) and (37), we conclude that the two regions for L_1 and L_2, with qualitatively different solutions of the free-boundary problem (besides the trivial solution in the case $L_1 \geq L_2$), can be distinguished. By means of straightforward computations, if the condition

$$L_1 < \frac{\gamma_- - \gamma_+}{\gamma_+\gamma_-}\, g_* \equiv \frac{(\gamma_- - \gamma_+)L_2}{\gamma_+(\gamma_- - 1)} \tag{38}$$

is satisfied, then $b_4^* > g_*$ holds and the solution of the left-hand part of the system in (16)–(18) has the form

$$V_4(s; b_4^*, g_*) = \frac{g_*}{\gamma_+}\left(\frac{b_4^*}{g_*}\right)^{\gamma_-}\left(\frac{s}{b_4^*}\right)^{\gamma_+} \tag{39}$$

with

$$b_4^* = g_*\left(\frac{\gamma_+\gamma_- L_1}{(\gamma_- - \gamma_+)g_*}\right)^{1/\gamma_-} \equiv \frac{\gamma_- L_2}{\gamma_- - 1}\left(\frac{\gamma_+(\gamma_- - 1)L_1}{(\gamma_- - \gamma_+)L_2}\right)^{1/\gamma_-}. \tag{40}$$

Using similar arguments, if the condition

$$L_1 \geq \frac{\gamma_- - \gamma_+}{\gamma_+\gamma_-}\, g_* \equiv \frac{(\gamma_- - \gamma_+)L_2}{\gamma_+(\gamma_- - 1)} \tag{41}$$

is satisfied, then $b_4^* \leq g_*$ holds and the solution of the left-hand part of the system in (16)–(18) has the form

$$V_4(s; b_4^*) = \frac{b_4^*}{\gamma_+}\left(\frac{s}{b_4^*}\right)^{\gamma_+} \quad \text{with} \quad b_4^* = \frac{\gamma_+(L_2 - L_1)}{\gamma_+ - 1}, \tag{42}$$

where the number b_4^* is strictly positive if and only if $L_2 > L_1$.

4 Main Results and Proofs

Taking into account the facts proved above, let us now formulate the main assertions of the paper. We recall that the price process S of the underlying risky asset is defined in (1)–(2), and the exercise boundaries g_* and h_* for the underlying perpetual American put and call options are given by (12).

Proposition 1 *In the optimal stopping problem of* (3), *related to the perpetual American call-on-call option with strike prices $K_1 > 0$ and $K_2 > 0$ of the outer and inner payoffs, respectively, the value function has the form*

$$V_1^*(s) = \begin{cases} V_1(s; b_1^*), & \text{if } s < b_1^*, \\ (s - K_2) - K_1, & \text{if } s \geq b_1^*, \end{cases}$$

where the function $V_1(s; b_1^*)$ and the hitting boundary $b_1^* \geq h_*$ for the right-hand optimal exercise time τ_1^* in (14) *are given by* (25) (*see Fig.* 1).

Proposition 2 *In the optimal stopping problem of* (4), *related to the perpetual American call-on-put option with strike prices $0 < K_1 < L_2$ of the outer and inner payoffs, respectively, the value function has the form*

$$V_2^*(s) = \begin{cases} V_2(s; a_2^*), & \text{if } s > a_2^*, \\ (L_2 - s) - K_1, & \text{if } s \leq a_2^*, \end{cases}$$

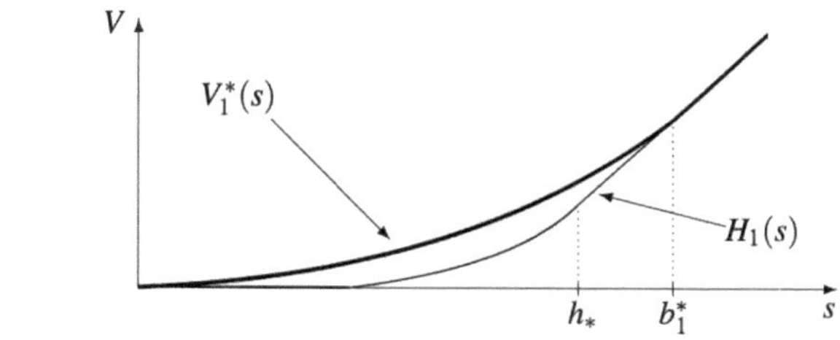

Fig. 1 A computer drawing of the payoff function $H_1(s)$ and the resulting value function $V_1^*(s)$

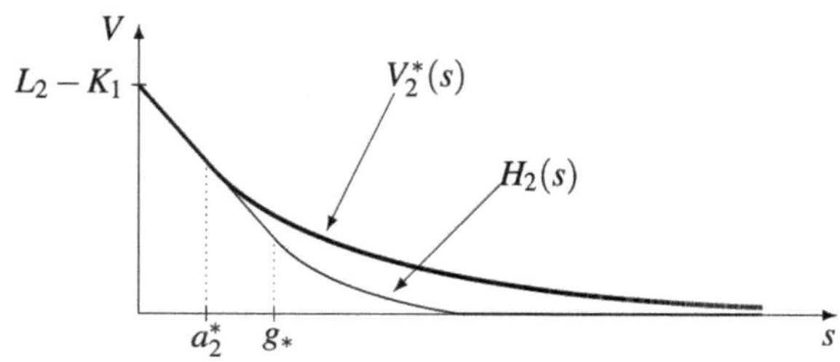

Fig. 2 A computer drawing of the payoff function $H_2(s)$ and the resulting value function $V_2^*(s)$

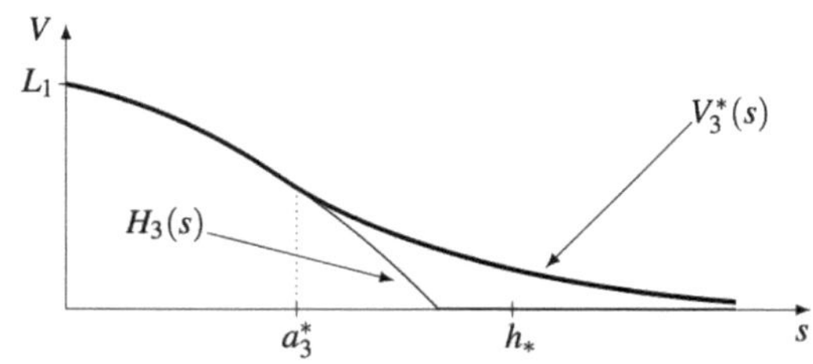

Fig. 3 A computer drawing of the payoff function $H_3(s)$ and the value function $V_3^*(s)$, when (31) holds for L_1 and K_2

where the function $V_2(s; a_2^*)$ and the hitting boundary $a_2^* \le g_*$ for the left-hand optimal exercise time τ_2^* in (14) *are given by* (28) (*see Fig.* 2), *while $V_2^*(s) = 0$ and $\tau_2^* = 0$ whenever $K_1 \ge L_2$.*

Proposition 3 *In the optimal stopping problem of* (5), *related to the perpetual American put-on-call option with strike prices $L_1 > 0$ and $K_2 > 0$ of the outer and inner payoffs, respectively, the following assertions hold*:

(i) *if* (31) *holds for L_1 and K_2 then the value function has the form*:

$$V_3^*(s) = \begin{cases} V_3(s; a_3^*, h_*), & \text{if } s > a_3^*, \\ L_1 - (h_*/\gamma_+)(s/h_*)^{\gamma_+}, & \text{if } s \le a_3^*, \end{cases} \tag{43}$$

where the function $V_3(s; a_3^*, h_*)$ and the hitting boundary $a_3^* < h_*$ for the left-hand optimal exercise time τ_3^* in (14) *are given by* (32) *and* (33), *respectively* (*see Fig.* 3);

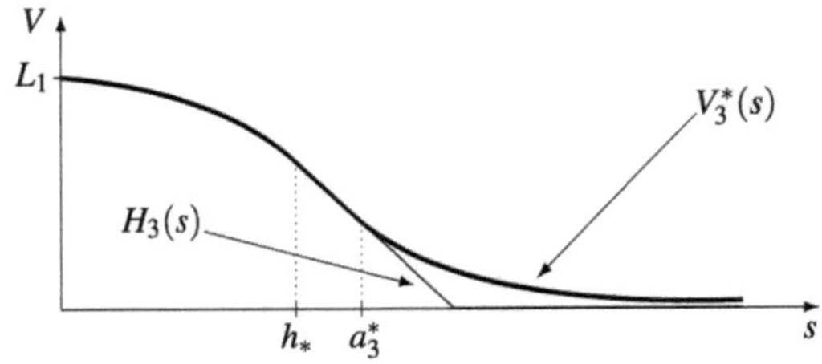

Fig. 4 A computer drawing of the payoff function $H_3(s)$ and the value function $V_3^*(s)$, when (34) holds for L_1 and K_2

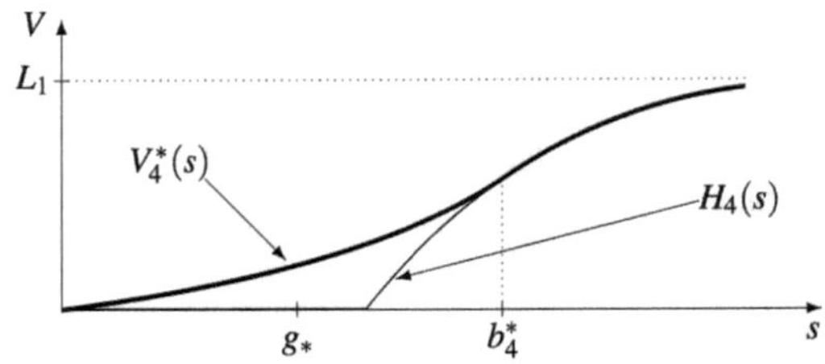

Fig. 5 A computer drawing of the payoff function $H_4(s)$ and the value function $V_4^*(s)$, when (38) holds for L_1 and L_2

(ii) *if* (34) *holds for* L_1 *and* K_2 *then the value function has the form*:

$$V_3^*(s) = \begin{cases} V_3(s; a_3^*), & \text{if } s > a_3^*, \\ L_1 - (s - K_2), & \text{if } h_* \le s \le a_3^*, \\ L_1 - (h_*/\gamma_+)(s/h_*)^{\gamma_+}, & \text{if } s < h_*, \end{cases} \tag{44}$$

where the function $V_3(s; a_3^*)$ *and the hitting boundary* a_3^* *for the left-hand optimal exercise time* τ_3^* *in* (14) *are given by* (35) *(see Fig.* 4*).*

Proposition 4 *In the optimal stopping problem of* (6)*, related to the perpetual American put-on-put option with strike prices* $L_1 > 0$ *and* $L_2 > 0$ *of the outer and inner payoffs, respectively, the following assertions hold*:

(i) *if* (38) *holds for* L_1 *and* L_2*, then the value function has the form*

$$V_4^*(s) = \begin{cases} V_4(s; b_4^*, g_*), & \text{if } s < b_4^*, \\ L_1 + (g_*/\gamma_-)(s/g_*)^{\gamma_-}, & \text{if } s \ge b_4^*, \end{cases}$$

where the function $V_4(s; b_4^*, g_*)$ *and the hitting boundary* $b_4^* > g_*$ *for the right-hand optimal exercise time* τ_4^* *in* (14) *are given by* (39) *and* (40)*, respectively (see Fig.* 5*)*;

(ii) *if* (41) *holds with* $L_1 < L_2$*, then the value function has the form*

$$V_4^*(s) = \begin{cases} V_4(s; b_4^*), & \text{if } s < b_4^*, \\ L_1 - (L_2 - s), & \text{if } b_4^* \le s \le g_*, \\ L_1 + (g_*/\gamma_-)(s/g_*)^{\gamma_-}, & \text{if } s > g_*, \end{cases}$$

where the function $V_4(s; b_4^*)$ *and the hitting boundary* b_4^* *for the right-hand optimal exercise time* τ_4^* *in* (14) *are given by* (42) *(see Fig.* 6*), while the value function has the form* $V_4^*(s) = L_1 - (L_2 - s)$ *and* $\tau_4^* = 0$ *whenever* $L_1 \ge L_2$.

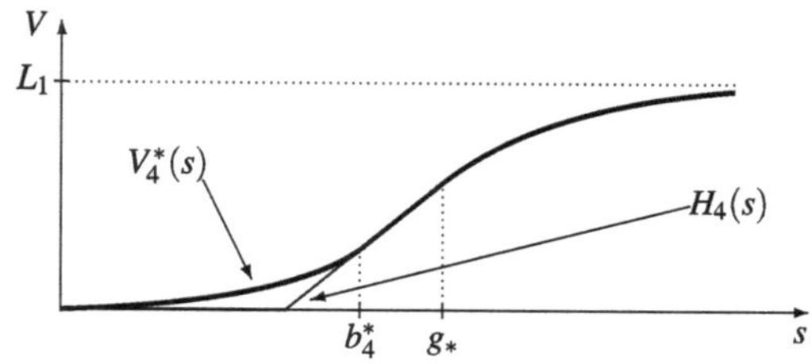

Fig. 6 A computer drawing of the payoff function $H_4(s)$ and the value function $V_4^*(s)$, when (41) holds for L_1 and L_2

Since all the assertions formulated above are proved using similar arguments, we only give a proof for the problem related to the perpetual American put-on-call option, which represents the most complicated and informative case.

Proof In order to verify the assertion of Proposition 3 stated above, it remains to show that the function $V_3^*(s)$ defined in either (43) or (44) coincides with the value function in (5), and that the stopping time τ_3^* in the left-hand side of (14) is optimal with a_3^* given by either (33) or (35). Let us denote by $V_3(s)$ the right-hand side of the expression in (43) or (44). Applying the local time-space formula from [17] (see also [18, Chap. II, Sect. 3.5] for a summary of the related results as well as further references) and taking into account the smooth-fit condition in (18) and the smoothness of the functions in (10), the following expressions

$$e^{-rt}\, V_3(S_t) = V_3(s) + \int_0^t e^{-ru}\, (\mathbb{L} V_3 - r\, V_3)(S_u)\, I(S_u \neq a_3^*)\, du + M_t \tag{45}$$

and

$$e^{-rt}\, W(S_t) = W(s) + \int_0^t e^{-ru}\, (\mathbb{L} W - r\, W)(S_u)\, I(S_u \neq h_*)\, du + N_t \tag{46}$$

hold, where $I(\cdot)$ denotes the indicator function and the processes $M = (M_t)_{t \geq 0}$ and $N = (N_t)_{t \geq 0}$ defined by

$$M_t = \int_0^t e^{-ru}\, V_3'(S_u)\, \sigma\, S_u\, dB_u \quad \text{and} \quad N_t = \int_0^t e^{-ru}\, W'(S_u)\, \sigma\, S_u\, dB_u \tag{47}$$

are continuous square integrable martingales with respect to the probability measure P. The latter fact can easily be observed, since the derivatives $V_3'(s)$ and $W'(s)$ are bounded functions.

By means of straightforward calculations similar to those of the previous section, it can be verified that the conditions of (20) and (21) hold with a_3^* given by either (33) or (35). These facts together with the conditions in (16)–(17) and (19) yield that $(\mathbb{L} V_3 - r\, V_3)(s) \leq 0$ holds for all $s \neq a_3^*$, and $V_3(s) \geq (L_1 - W(s))^+$ is satisfied for all $s > 0$. It is well known (see, e.g. [20, Chap. VIII, Sect. 2a]) that $(\mathbb{L} W - r W)(s) \leq 0$ holds for all $s \neq h_*$, and $W(s) \geq (s - K_2)^+$ is satisfied for all $s > 0$. Moreover, since the time spent by the process S at the boundaries a_3^* and h_* is of

Lebesgue measure zero, the indicators which appear in the integrals of (45)–(46) can be ignored. Hence, it follows from the expressions in (45)–(46) that the inequalities

$$e^{-r(\tau\wedge t)}\,(L_1 - W(S_{\tau\wedge t}))^+ \le e^{-r(\tau\wedge t)}\,V_3(S_{\tau\wedge t}) \le V_3(s) + M_{\tau\wedge t} \tag{48}$$

and

$$e^{-r(\zeta\wedge u)}\,(S_{\zeta\wedge u} - K_2)^+ \le e^{-r(\zeta\wedge u)}\,W(S_{\zeta\wedge u}) \le e^{-r(\tau\wedge t)}\,W(S_{\tau\wedge t}) + N_{\zeta\wedge u} - N_{\tau\wedge t} \tag{49}$$

hold for all $0 \le t \le u$ and any stopping times $0 \le \tau \le \zeta$ of the process S started at $s > 0$. Then, taking the (conditional) expectations with respect to P in (48)–(49), by means of Doob's optional sampling theorem (see, e.g. [14, Theorem 3.6] or [13, Chap. I, Theorem 3.22]), we get that the inequalities

$$E\big[e^{-r(\tau\wedge t)}\,(L_1 - W(S_{\tau\wedge t}))^+\big] \le E\big[e^{-r(\tau\wedge t)}\,V_3(S_{\tau\wedge t})\big] \le V_3(s) + E\,M_{\tau\wedge t} = V_3(s)$$

and

$$\begin{aligned} &E\big[e^{-r(\zeta\wedge u)}\,(S_{\zeta\wedge u} - K_2)^+ \,\big|\, \mathscr{F}_{\tau\wedge t}\big] \\ &\quad\le E\big[e^{-r(\zeta\wedge u)}\,W(S_{\zeta\wedge u}) \,\big|\, \mathscr{F}_{\tau\wedge t}\big] \\ &\quad\le e^{-r(\tau\wedge t)}\,W(S_{\tau\wedge t}) + E\big[N_{\zeta\wedge u} - N_{\tau\wedge t} \,\big|\, \mathscr{F}_{\tau\wedge t}\big] = e^{-r(\tau\wedge t)}\,W(S_{\tau\wedge t}) \quad (P\text{-a.s.}) \end{aligned}$$

hold for all $s > 0$. Thus, letting u and then t go to infinity and using (conditional) Fatou's lemma, we obtain

$$E\big[e^{-r\tau}\,(L_1 - W(S_\tau))\big] \le E\big[e^{-r\tau}\,(L_1 - W(S_\tau))^+\big] \le E\big[e^{-r\tau}\,V_3(S_\tau)\big] \le V_3(s) \tag{50}$$

and

$$E\big[e^{-r\zeta}\,(S_\zeta - K_2)^+ \,\big|\, \mathscr{F}_\tau\big] \le E\big[e^{-r\zeta}\,W(S_\zeta) \,\big|\, \mathscr{F}_\tau\big] \le e^{-r\tau}\,W(S_\tau) \quad (P\text{-a.s.}) \tag{51}$$

for any stopping times $0 \le \tau \le \zeta$ and all $s > 0$. By virtue of the structure of the stopping times in (14) and (15), it is readily seen that the equalities in (50)-(51) hold with τ_3^* and ζ_3^* instead of τ and ζ, when $s \le a_3^*$ and $S_{\tau_3^*} \ge h_*$ (P-a.s.).

It remains to be shown that the equalities are attained in (50)–(51) when τ_3^* and ζ_3^* replace τ and ζ, respectively, when $s > a_3^*$ and $S_{\tau_3^*} < h_*$ (P-a.s.). By virtue of the fact that the function $V_3(s; a_3^*, h_*)$ and the boundary a_3^* satisfy the conditions in (16) and (17) as well as for the function $W(s)$ and the boundary h_* the condition $(\mathbb{L}W - rW)(s) = 0$ is satisfied for $s < h_*$ and $W(h_*-) = h_* - K_2$ holds, it follows from the expressions in (45)-(46) and the structure of the stopping times τ_3^* and ζ_3^* in (14) and (15) that the equalities

$$e^{-r(\tau_3^*\wedge t)}\,V_3(S_{\tau_3^*\wedge t}) = V_3(s) + M_{\tau_3^*\wedge t} \tag{52}$$

and

$$e^{-r(\zeta_3^*\wedge u)}\,W(S_{\zeta_3^*\wedge u}) = e^{-r(\tau_3^*\wedge t)}\,W(S_{\tau_3^*\wedge t}) + N_{\zeta_3^*\wedge u} - N_{\tau_3^*\wedge t} \tag{53}$$

are satisfied for all $0 \le t \le u$, when $s > a_3^*$ and $S_{\tau_3^*} < h_*$ (P-a.s.), and where the processes M and N are defined in (47). Taking into account the fact that $V_3(s)$ is bounded by L_1 from above and the properties of the function $W(s)$ in (10) (see, e.g. [20, Chap. VIII, Sect. 2a]), we conclude from (52)–(53) that the variables $e^{-r\tau_3^*} V_3(S_{\tau_3^*})$ and $e^{-r\zeta_3^*} W(S_{\zeta_3^*})$ are equal to zero on the sets $\{\tau_3^* = \infty\}$ and $\{\zeta_3^* = \infty\}$ (P-a.s.), respectively, and the processes $(M_{\tau_3^* \wedge t})_{t \ge 0}$ and $(N_{\zeta_3^* \wedge t})_{t \ge 0}$ are uniformly integrable martingales. Therefore, taking the (conditional) expectations with respect to P and letting u and then t go to infinity, we apply the (conditional) Lebesgue dominated convergence theorem to obtain the equalities

$$E\big[e^{-r\tau_3^*} (L_1 - W(S_{\tau_3^*}))\big] = E\big[e^{-r\tau_3^*} (L_1 - W(S_{\tau_3^*}))^+\big] = E\big[e^{-r\tau_3^*} V_3(S_{\tau_3^*})\big] = V_3(s)$$

and

$$E\big[e^{-r\zeta_3^*} (S_{\zeta_3^*} - K_2)^+ \,\big|\, \mathscr{F}_{\tau_3^*}\big] = E\big[e^{-r\zeta_3^*} W(S_{\zeta_3^*}) \,\big|\, \mathscr{F}_{\tau_3^*}\big] = e^{-r\tau_3^*} W(S_{\tau_3^*}) \quad (P\text{-a.s.})$$

for all $s > a_3^*$ and $S_{\tau_3^*} < h_*$ (P-a.s.). The latter, together with the inequalities in (50)–(51), imply the fact that $V_3(s)$ coincides with the function $V_3^*(s)$ from (5), and τ_3^* and ζ_3^* from (14) and (15) are the optimal stopping times. □

Remark 1 Note that in the cases of call-on-call and call-on-put options in Propositions 1 and 2 above, one should not stop the underlying process S when $s < b_1^*$ and $s > a_2^*$, respectively. However, both the initial and underlying options should be exercised immediately when $s \ge b_1^*$ and $s \le a_2^*$, accordingly. Moreover, in the case of put-on-call option in Proposition 3 above, one should not stop the underlying process when $s > a_3^*$ holds, one should exercise the initial option only when either $s \le a_3^*$ under (31) or $s < h_*$ under (34) is satisfied, while both the initial and underlying options should be exercised immediately when $h_* \le s \le a_3^*$ holds under (34). Similarly, in the case of put-on-put option in Proposition 4 above, one should not stop the underlying process when $s < b_4^*$, one should exercise the initial option only when either $s \ge b_4^*$ under (38) or $s > g_*$ under (41) is satisfied with $L_1 < L_2$, while both the initial and underlying options should be exercised immediately when $b_4^* \le s \le g_*$ holds under (41) with $L_1 < L_2$.

5 Chooser Options

In this section, we give a formulation of the perpetual American chooser option optimal stopping problem and prove the uniqueness of solution of the associated free-boundary problem.

5.1 Formulation of the Problem

Let us finally consider the perpetual American *chooser* option which is a contract giving its holder the right to decide at an exercise time τ whether the initial compound option acts further as the underlying perpetual American put or call option. Then, according to the arguments above, the rational price of such a contingent claim is given by the value of the optimal stopping problem

$$V^*(s) = \sup_\tau E\big[e^{-r\tau}\,\big(U(S_\tau) \vee W(S_\tau)\big)\big], \tag{54}$$

where the supremum is taken over the stopping times τ of the process S started at $s > 0$, and $x \vee y$ denotes the maximum $\max\{x, y\}$ of any $x, y \in \mathbb{R}$. Recall that the functions $U(s)$ and $W(s)$ represent the rational prices of the underlying perpetual American put and call options defined in (9), respectively. By virtue of the structure of the resulting convex and strictly monotone value functions in (10), we further search for an optimal stopping time in the problem of (54) of the form

$$\tau^* = \inf\{t \geq 0 : \; S_t \notin (p_*, q_*)\} \tag{55}$$

for some numbers $0 < p_* < c < q_* < \infty$ to be determined, where c denotes the point of intersection of the curves associated with the functions $U(s)$ and $W(s)$ (see Fig. 8). Note that the latter inequalities always hold, since we have the inequalities $U'(c-) < 0 < W'(c+)$, so that it is never optimal to exercise the option at $s = c$ (see, e.g. [4, Sect. 4] or [9, Sect. 3]).

In order to find explicit expressions for the unknown value function $V^*(s)$ from (54) and the unknown boundaries p_* and q_* from (55), we follow the schema of arguments above and formulate the free-boundary problem

$$(\mathbb{L}V)(s) = rV(s) \quad \text{for} \quad p < s < q, \tag{56}$$

$$V(p+) = U(p) \quad \text{and} \quad V(q-) = W(q) \quad \text{(instantaneous stopping)}, \tag{57}$$

$$V'(p+) = U'(p) \quad \text{and} \quad V'(q-) = W'(q) \quad \text{(smooth fit)}, \tag{58}$$

$$V(s) = U(s) \vee W(s) \quad \text{for} \quad s < p \quad \text{and} \quad s > q, \tag{59}$$

$$V(s) > U(s) \vee W(s) \quad \text{for} \quad p < s < q, \tag{60}$$

$$(\mathbb{L}V)(s) < rV(s) \quad \text{for} \quad s < p \quad \text{and} \quad s > q, \tag{61}$$

for some $0 < p < c < q < \infty$ fixed.

5.2 Solution of the Free-Boundary Problem

In order to solve the free-boundary problem in (56)–(61), we first recall that the general solution of the differential equation in (56) has the form of (22) with some arbitrary constants C_+ and C_-. Hence, applying the instantaneous stopping conditions from (57) to the function in (22), we obtain the equalities

$$C_+\, p^{\gamma_+} + C_-\, p^{\gamma_-} = U(p) \quad \text{and} \quad C_+\, q^{\gamma_+} + C_-\, q^{\gamma_-} = W(q), \tag{62}$$

which hold for some $0 < p < c < q < \infty$, where c is uniquely determined by the equation $U(c) = W(c)$. Solving the system of equations in (62), we obtain the function

$$V(s; p, q) = C_+(p, q)\, s^{\gamma_+} + C_-(p, q)\, s^{\gamma_-}, \tag{63}$$

which satisfies the system in (56)–(57) with

$$C_+(p, q) = \frac{U(p) q^{\gamma_-} - W(q) p^{\gamma_-}}{p^{\gamma_+} q^{\gamma_-} - q^{\gamma_+} p^{\gamma_-}} \quad \text{and} \quad C_-(p, q) = \frac{W(q) p^{\gamma_+} - U(p) q^{\gamma_+}}{p^{\gamma_+} q^{\gamma_-} - q^{\gamma_+} p^{\gamma_-}}, \tag{64}$$

for $0 < p < c < q < \infty$. Applying the smooth-fit conditions from (58) to the function in (63), we obtain the equalities

$$C_+(p, q)\, \gamma_+\, p^{\gamma_+} + C_-(p, q)\, \gamma_-\, p^{\gamma_-} = p\, U'(p), \tag{65}$$

$$C_+(p, q)\, \gamma_+\, q^{\gamma_+} + C_-(p, q)\, \gamma_-\, q^{\gamma_-} = q\, W'(q), \tag{66}$$

which hold with $C_+(p, q)$ and $C_-(p, q)$ given by (64). It is shown by means of standard arguments that the system in (65)–(66) is equivalent to

$$I_+(p) = J_+(q) \quad \text{and} \quad I_-(p) = J_-(q) \tag{67}$$

with

$$I_+(p) = \frac{pU'(p) - \gamma_- U(p)}{p^{\gamma_+}} \quad \text{and} \quad J_+(q) = \frac{qW'(q) - \gamma_- W(q)}{q^{\gamma_+}}, \tag{68}$$

$$I_-(p) = \frac{\gamma_+ U(p) - pU'(p)}{p^{\gamma_-}} \quad \text{and} \quad J_-(q) = \frac{\gamma_+ W(q) - qW'(q)}{q^{\gamma_-}}, \tag{69}$$

for all $0 < p < c < q < \infty$.

In order to show the existence and uniqueness of a solution of the system of equations in (67), we follow the schema of arguments from [9, Sect. 4] which are based on the idea of the proof of the existence and uniqueness of solutions applied to the systems of equations in (4.73)–(4.74) from [19, Chap. IV, Sect. 2] and (3.16)–(3.17) from [8, Sect. 3]. For this, we observe that, for the derivatives of the functions in (68)–(69), the expressions

$$I'_+(p) = -\frac{(\gamma_+ - 1)(\gamma_- - 1)p - \gamma_+\gamma_- L_2}{p^{\gamma_+ + 1}} \equiv -\frac{(\gamma_+ - 1)(\gamma_- - 1)(p - \overline{L}_2)}{p^{\gamma_+ + 1}} < 0,$$

$$J'_+(q) = \frac{(\gamma_+ - 1)(\gamma_- - 1)q - \gamma_+\gamma_- K_2}{q^{\gamma_+ + 1}} \equiv \frac{(\gamma_+ - 1)(\gamma_- - 1)(q - \overline{K}_2)}{q^{\gamma_+ + 1}} < 0,$$

$$I'_-(p) = \frac{(\gamma_+ - 1)(\gamma_- - 1)p - \gamma_+\gamma_- L_2}{p^{\gamma_- + 1}} \equiv \frac{(\gamma_+ - 1)(\gamma_- - 1)(p - \overline{L}_2)}{p^{\gamma_- + 1}} > 0,$$

$$J'_-(q) = -\frac{(\gamma_+ - 1)(\gamma_- - 1)q - \gamma_+\gamma_- K_2}{q^{\gamma_- + 1}} \equiv -\frac{(\gamma_+ - 1)(\gamma_- - 1)(q - \overline{K}_2)}{q^{\gamma_- + 1}} > 0$$

hold under $0 < p < g_* < \overline{L}_2$ and $\overline{K}_2 < h_* < q < \infty$, and are equal to zero otherwise, where we set

$$\overline{L}_2 = \frac{\gamma_+ \gamma_- L_2}{(\gamma_+ - 1)(\gamma_- - 1)} \equiv \frac{rL_2}{\delta} \quad \text{and} \quad \overline{K}_2 = \frac{\gamma_+ \gamma_- K_2}{(\gamma_+ - 1)(\gamma_- - 1)} \equiv \frac{rK_2}{\delta}. \tag{70}$$

Hence, the function $I_+(p)$ decreases on the interval $(0, g_*)$ from $I_+(0+) = \infty$ to $I_+(g_*) = 0$, and then remains equal to zero on the interval (g_*, ∞), so that the range of its values is given by the interval $(0, \infty)$. The function $J_+(q)$ is equal to the value $J_+(h_*) = (\gamma_+ - \gamma_-)h_*^{1-\gamma_+}/\gamma_+ > 0$ on the interval $(0, h_*)$, and then decreases to zero on the interval (h_*, ∞), so that the range is $(0, J_+(h_*))$. The function $I_-(p)$ increases from zero to $I_-(g_*) = (\gamma_- - \gamma_+)g_*^{1-\gamma_-}/\gamma_- > 0$ on the interval $(0, g_*)$, and then remains equal to $I_-(g_*)$ on the interval (g_*, ∞), so that the range is $(0, I_-(g_*))$.

The function $J_-(q)$ is equal to zero on the interval $(0, h_*)$, and then increases from $J_-(h_*) = 0$ to infinity on the interval (h_*, ∞), so that the range is $(0, \infty)$. It is shown by means of straightforward computations that the bounds $I_+(g_* \wedge c) < J_+(h_* \vee c)$ and $I_-(g_* \wedge c) > J_-(h_* \vee c)$ holds. This fact guarantees that the ranges of values of the left- and right-hand sides of the equations in (67) have nontrivial intersections.

It thus follows from the left-hand equation in (67) that, for each $q \in (h_* \vee c, \infty)$, there exists a unique number $p \in (\widehat{p}, g_* \wedge c)$, where $\widehat{p}$ is uniquely determined by the equation $I_+(\widehat{p}) = J_+(h_* \vee c)$. It also follows from the right-hand equation in (67) that, for each $p \in (0, g_* \wedge c)$, there exists a unique number $q \in (h_* \vee c, \widehat{q})$, where $\widehat{q}$ is uniquely determined by the equation $I_-(g_* \wedge c) = J_-(\widehat{q})$ (see Fig. 7). We may therefore conclude that the equations in (67) uniquely define the function $q_+(p)$ on $(\widehat{p}, g_* \wedge c)$ with the range $(h_* \vee c, \infty)$ and the function $q_-(p)$ on $(0, g_* \wedge c)$ with the range $(h_* \vee c, \widehat{q})$, respectively. This fact directly implies that, for every point $p \in (\widehat{p}, g_* \wedge c)$, there are unique values $q_+(p)$ and $q_-(p)$ belonging to $(h_* \vee c, \infty)$, that together with the inequalities $h_* \vee c \equiv q_+(\hat{p}) \equiv q_-(0+) < q_-(g_* \wedge c) < \infty \equiv q_+(g_*)$ guarantees the existence of exactly one intersection point with the coordinates p_* and q_* of the curves associated with the functions $q_+(p)$ and $q_-(p)$ on the interval $(\widehat{p}, g_* \wedge c)$ such that $h_* \vee c < q_+(p_*) \equiv q_* \equiv q_-(p_*) < \widehat{q}$ holds (see Fig. 7). This completes the proof of the claim.

Summarizing the facts proved above, we are now ready to formulate the following result.

Proposition 5 *Let the process S be given by* (1)–(2), *the functions $U(s)$ and $W(s)$ be defined in* (9)–(10), *and the number c be uniquely determined by $U(c) = W(c)$. Hence, in the optimal stopping problem of* (54), *related to the perpetual American chooser option with the inner put and call payoffs with strike prices $L_2 > 0$ and $K_2 > 0$, respectively, the value function has the form*

$$V^*(s) = \begin{cases} V(s; p_*, q_*), & \text{if } p_* < s < q_*, \\ U(s) \vee W(s), & \text{if } s \le p_* \text{ or } s \ge q_*, \end{cases} \tag{71}$$

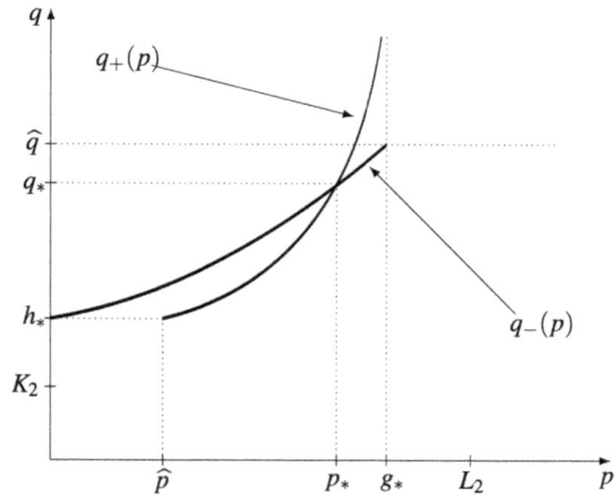

Fig. 7 A computer drawing of the functions $q_+(p)$ and $q_-(p)$

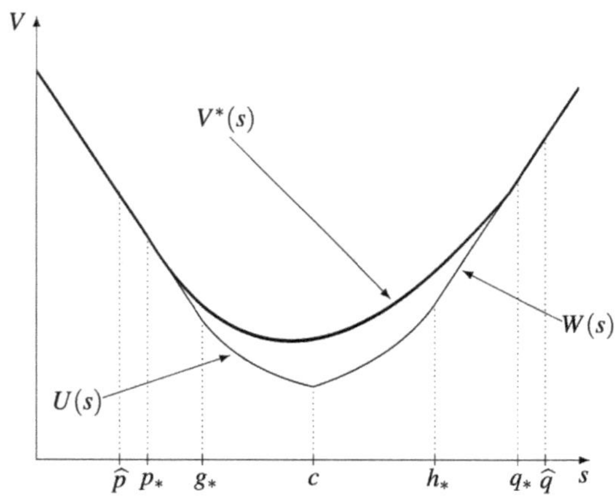

Fig. 8 A computer drawing of the value function $V^*(s)$ for the case $g_* < c < h_*$ for the payoff function $U(s) \vee W(s)$

where the function $V(s; p, q)$ *is given by* (63)–(64), *and the exit boundaries* p_* *and* q_* *such that* $0 < p_* < g_* \wedge c \leq h_* \vee c < q_* < \infty$ *for the optimal exercise time* τ^* *in* (55) *are uniquely determined by the system of* (67) (*see Fig.* 8). *The underlying perpetual American put or call option should then be exercised at the same time* τ^*.

Proof In order to verify the assertion stated above, let us follow the schema of arguments from [9, Theorem 3.1] and show that the function defined in (71) coincides with the value function in (54), and that the stopping time τ^* in (55) is optimal with the boundaries p_* and q_* specified above. Let us denote by $V(s)$ the right-hand side of the expression in (71). Applying the local time-space formula from [17] and taking into account the smooth-fit conditions in (58), the following expression

$$e^{-rt}\, V(S_t) = V(s) + \int_0^t e^{-ru}\, (\mathbb{L}V - rV)(S_u)\, I(S_u \neq p_*, S_u \neq q_*)\, du + M_t^* \quad (72)$$

holds for all $t \geq 0$, where the process $M^* = (M_t^*)_{t\geq 0}$ defined by

$$M_t^* = \int_0^t e^{-ru}\, V'(S_u)\, \sigma\, S_u\, dB_u \tag{73}$$

is a continuous square integrable martingale with respect to P. The latter fact can be easily observed, since the derivative $V'(s)$ is a bounded function.

By means of straightforward computations, it can be verified that the conditions of (60) and (61) hold with p_* and q_* being a unique solution of the system in (67). These facts together with the conditions in (56)–(57) and (59) yield that $(\mathbb{L}V - rV)(s) \leq 0$ holds for any $s > 0$ such that $s \neq p_*$ and $s \neq q_*$, and $V(s) \geq U(s) \vee W(s)$ is satisfied for all $s > 0$. Moreover, since the time spent by the process S at the boundaries p_* and q_* is of Lebesgue measure zero, the indicator which appear in the integral of (72) can be ignored. Hence, it follows from the expression in (72) that the inequalities

$$e^{-r(\tau\wedge t)}\, \big(U(S_{\tau\wedge t}) \vee W(S_{\tau\wedge t})\big) \leq e^{-r(\tau\wedge t)}\, V(S_{\tau\wedge t}) \leq V(s) + M_{\tau\wedge t}^* \tag{74}$$

hold for any stopping time τ of the process S started at $s > 0$. Then, taking the expectations with respect to P in (74), by means of Doob's optional sampling theorem, we get that the inequalities

$$E\big[e^{-r(\tau\wedge t)}\, \big(U(S_{\tau\wedge t}) \vee W(S_{\tau\wedge t})\big)\big] \leq E\big[e^{-r(\tau\wedge t)}\, V(S_{\tau\wedge t})\big] \leq V(s) + E\, M_{\tau\wedge t}^* = V(s)$$

hold for all $s > 0$. Hence, letting t go to infinity and using Fatou's lemma, we obtain

$$E\big[e^{-r\tau}\, \big(U(S_\tau) \vee W(S_\tau)\big)\big] \leq E\big[e^{-r\tau}\, V(S_\tau)\big] \leq V(s) \tag{75}$$

for any stopping time τ and all $s > 0$. By virtue of the structure of the stopping time in (55), it is readily seen that the equalities in (75) hold with τ^* instead of τ when either $s \leq p_*$ or $s \geq q_*$.

It remains to be shown that the equalities are attained in (75) when τ^* replaces τ for $p_* < s < q_*$. By virtue of the fact that the function $V(s; p_*, q_*)$ and the boundaries p_* and q_* satisfy the conditions in (56) and (57), it follows from the expression in (72) and the structure of the stopping time in (55) that the equality

$$e^{-r(\tau^*\wedge t)}\, V(S_{\tau^*\wedge t}; p_*, q_*) = V(s) + M_{\tau^*\wedge t}^* \tag{76}$$

is satisfied for all $s \in (p_*, q_*)$, where the process M^* is defined in (73). Observe that the explicit form of the function in (63) and (64) yields that the condition

$$E\Big[\sup_{t\geq 0} e^{-r(\tau^*\wedge t)}\, V(S_{\tau^*\wedge t}; p_*, q_*)\Big] < \infty \tag{77}$$

holds for all $s \in (p_*, q_*)$, as well as the variable $e^{-r\tau^*} V(S_{\tau^*}; p_*, q_*)$ is equal to zero on the event $\{\tau^* = \infty\}$ (P-a.s.). Hence, taking into account the property in (77), we conclude from the expression in (76) that the process $(M_{\tau^*\wedge t}^*)_{t\geq 0}$ is a uniformly integrable martingale. Therefore, taking the expectation in (76) and letting t

go to infinity, we apply the Lebesgue dominated convergence theorem to obtain the equalities

$$E\big[e^{-r\tau^*}\big(U(S_{\tau^*}) \vee W(S_{\tau^*})\big)\big] = E\big[e^{-r\tau^*}\, V(S_{\tau^*}; p_*, q_*)\big] = V(s)$$

for all $s \in (p_*, q_*)$. The latter, together with the inequalities in (75), implies the fact that $V(s)$ coincides with the value function $V^*(s)$ from (54) and τ^* from (55) is the optimal stopping time. □

Remark 2 Note that the system (67) is equivalent to the system (4.5) from [9] with the only difference that $(\widehat{p}, g_* \wedge c)$ and $(h_* \vee c, \widehat{q})$ are allowed for p_* and q_*, respectively, which are eventually smaller than the corresponding ones $(\overline{p}, g_* \wedge c)$ and $(h_* \vee c, \overline{q})$ from [9, Sect. 4]. Here, the numbers g_* and h_* are given by (12), and the boundaries $\overline{p} < \widehat{p}$ and $\overline{q} > \widehat{q}$ are uniquely determined by the equations $I_+(\overline{p}) = J_+(\overline{K}_2)$ and $I_-(\overline{L}_2) = J_-(\overline{q})$ with $\overline{L}_2$ and $\overline{K}_2$ defined in (70). It follows from the arguments above that the rational price $V^*(s)$ of the perpetual American chooser option in (54) coincides with the one of the perpetual American strangle option in [9, Example 4.2].

Acknowledgements The authors are grateful to Mihail Zervos for many useful discussions. The authors thank the Editor and two anonymous Referees for their careful reading of the manuscript and helpful suggestions. The second author gratefully acknowledges the scholarship of the Alexander Onassis Public Benefit Foundation for his doctoral studies at the London School of Economics and Political Science.

References

1. Carmona, R., Dayanik, S.: Optimal multiple-stopping of linear diffusions. Math. Oper. Res. **33**(2), 446–460 (2008)
2. Carmona, R., Touzi, N.: Optimal multiple stopping and valuation of swing options. Math. Finance **18**(2), 239–268 (2008)
3. Chiarella, C., Kang, B.: The evaluation of American compound option prices under stochastic volatility using the sparse grid approach. In: Research Centre Research Paper No. 245. University of Technology, Sydney, Quantitative Finance (2009). http://www.business.uts.edu.au/qfrc/research/research_papers/rp245.pdf
4. Dayanik, S., Karatzas, I.: On the optimal stopping problem for one-dimensional diffusions. Stoch. Process. Appl. **107**, 173–212 (2003)
5. Dixit, A.K., Pindyck, R.S.: Investment Under Uncertainty. Princeton University Press, Princeton (1994)
6. Egami, M., Xu, M.: A continuous-time search model with job switch and jumps. Math. Methods Oper. Res. **70**(2), 241–267 (2008)
7. Duckworth, J.K., Zervos, M.: An investment model with entry and exit decisions. J. Appl. Probab. **37**, 547–559 (2000)
8. Gapeev, P.V.: The spread option optimal stopping game. In: Kyprianou, A., Schoutens, W., Wilmott, P. (eds.) Exotic Option Pricing and Advanced Levy Models, pp. 293–305. Wiley, Chichester (2005)
9. Gapeev, P.V., Lerche, H.R.: On the structure of discounted optimal stopping problems for one-dimensional diffusions. In: Stochastics: An International Journal of Probability and Stochastic

Processes. CDAM Research Report LSE-CDAM-2009-03 (2010). http://www.maths.lse.ac.uk/Personal/pavel/PDF/Publications/Gapeev-Lerche-SSR.pdf
10. Geske, R.: The valuation of corporate liabilities as compound options. J. Financ. Quant. Anal. **12**, 541–552 (1977)
11. Geske, R.: The valuation of compound options. J. Financ. Econ. **7**, 63–81 (1979)
12. Hodges, S.D., Selby, M.J.P.: On the evaluation of compound options. Manag. Sci. **33**, 347–355 (1987)
13. Karatzas, I., Shreve, S.E.: Brownian Motion and Stochastic Calculus, 2nd edn. Springer, New York (1991)
14. Liptser, R.S., Shiryaev, A.N.: Statistics of Random Processes I, 2nd edn. Springer, Berlin (2001)
15. McKean, H.P.: Appendix: a free boundary problem for the heat equation arising from a problem of mathematical economics. Ind. Manage. Rev. **6**, 32–39 (1965)
16. Øksendal, B.: Stochastic Differential Equations: An Introduction with Applications, 5th edn. Springer, Berlin (1998)
17. Peskir, G.: A change-of-variable formula with local time on curves. J. Theor. Probab. **18**, 499–535 (2005)
18. Peskir, G., Shiryaev, A.N.: Optimal Stopping and Free-Boundary Problems. Birkhäuser, Basel (2006)
19. Shiryaev, A.N.: Optimal Stopping Rules. Springer, Berlin (1978)
20. Shiryaev, A.N.: Essentials of Stochastic Finance. World Scientific, Singapore (1999)

New Approximations in Local Volatility Models

E. Gobet and A. Suleiman

Abstract For general time-dependent local volatility models, we propose new approximation formulas for the price of call options. This extends previous results of Benhamou et al. (Int. J. Theor. Appl. Finance 13(4):603–634, 2010) where stochastic expansions combined with Malliavin calculus were performed to obtain approximation formulas based on the local volatility At The Money. Here, we derive alternative expansions involving the local volatility at strike. Averaging both expansions give even more accurate results. Approximations of the implied volatility are provided as well.

Keywords Option pricing · Local volatility model · Stochastic expansion · Malliavin calculus

Mathematics Subject Classification (2010) 91G20 · 91G60

1 Introduction

1.1 Framework

We consider a linear Brownian motion $(W_t)_{t\ \leq T}$ defined on a filtered probability space $(\Omega, \mathscr{F}_T, (\mathscr{F}_t)_{t\leq T}, \mathbb{P})$ where $T > 0$ is a fixed terminal time. Here, $(\mathscr{F}_t)_{t\leq T}$ is the completion of the natural filtration of W. This is used to model the dynamics of a risky asset S (e.g. a stock or an index), which price process is $(S_t)_{t\leq T}$. We are mainly interested in valuing European-style financial contracts written on S, exercised at maturity T, which related payoff is of the form $\Phi(S_T)$. We especially pay attention

E. Gobet (✉)
CMAP, École Polytechnique, Route de Saclay, 91128 Palaiseau Cedex, France
e-mail: emmanuel.gobet@polytechnique.edu

A. Suleiman
Ensimag, Domaine Universitaire, 681 rue de la passerelle, 38402 St Martin d'Hères, France
e-mail: ali.suleiman9@gmail.com

Y. Kabanov et al. (eds.), *Inspired by Finance*, DOI 10.1007/978-3-319-02069-3_14,

to vanilla options, i.e. $\Phi(S) = (S - K)^+$ (call options) and $\Phi(S) = (K - S)^+$ (put options).

We consider the standard framework of complete market (see for instance [10]), and more specifically, we assume that

1. the short-term interest rate $(r_t)_{t \le T}$ is deterministic and bounded;
2. the risky asset pays a continuous dividend $(q_t)_{t \le T}$, which is deterministic and bounded;
3. S follows a local volatility model, which dynamics is defined by the solution of the following stochastic differential equation:

$$\frac{dS_t}{S_t} = (r_t - q_t)dt + \sigma(t, S_t)dW_t. \tag{1}$$

We denote the compound factor by

$$C_t = \exp\left\{ \int_0^t (r_s - q_s)ds \right\}. \tag{2}$$

Thus, we have

$$S_t = C_t e^{X_t}, \tag{3}$$

$$X_t = \log(S_0) + \int_0^t \sigma(s, S_s)dW_s - \frac{1}{2}\int_0^t \sigma^2(s, S_s)ds. \tag{4}$$

Note that the above dynamics are directly given under the risk-neutral measure, since we only focus on pricing formulas. Then, the option price at time 0 is given by $\mathbb{E}(e^{-\int_0^T r_s ds}\Phi(S_T))$. Of course, due to the general form of the local volatility function $\sigma(t, S)$, it is hopeless to derive exact closed formulas for such option prices. The aim of this work is to obtain accurate approximations.

1.2 Literature Background

The interest in local volatility models probably dates back to the work by Dupire [5] among others, who shows that such models are able to fit all call and put option prices at a given observation date (the calibration date). However, except in a few cases, analytical pricing formulas are available (for instance, in the CEV model $\sigma(t, S) = \nu S^{\beta - 1}$, see [12]). As alternative numerical methods, one could use a PDE approach but to achieve real-time pricing and calibration routines, it is better to search for approximative formulas, quicker to evaluate. Hagan et al. [7] use singular perturbation techniques to obtain an implied volatility expansion, in the case of separable volatility $\sigma(t, S) = \alpha(t)A(S)$. Henry-Labordère [8] transfers heat kernel expansions on price expansions. To tackle the case of non-separable volatility, Piterbarg [11] suggests the use of parameter averaging for some choices of $\sigma(t, S)$.

A different approach has been developed in [1]: first a model proxy is chosen, then a smart expansion around this proxy is performed, involving Malliavin calculus to determine explicitly the expansion terms. This approach appears to be quite flexible since it naturally handles time-dependent coefficients and various modeling situations including so far jumps, discrete dividends or stochastic interest rates. More precisely, applications to local volatility model including jumps have been developed in [1] and deeply investigated along further directions in [3]. Allowing the interest rates to be stochastic is achieved in [2], while in [4] the case of time-dependent Heston model is considered. In [6], the authors investigate the case of assets paying discrete dividends. Within this approach, we are able to prove explicit error estimates that depend on σ and its derivatives, on the maturity and on the payoff. It helps to better understand the roles of each parameter. In addition, the regularity of the payoff is crucial in order to design the expansion and to establish error estimates. These features are extensively discussed in [3] and [2].

Nevertheless, regarding the results in [3], one could legitimately formulate the criticism that we use the local volatility only At The Money (ATM in short) when we take the model proxy as Black–Scholes model and when we compute the expansions. For arbitrary payoffs, this is natural, but for call/put options, this may be strange since the spot and strike variables play somewhat symmetric roles.

Here, we correct this drawback by providing new expansion formulas based on the local volatility at strike (and we even mix the expansions). This article is organized as follows: in the next paragraphs, we define the assumptions and notations used throughout the paper. Then, in the next section, main results are stated. The main proofs are postponed at the end of this article (Sects. 4, 5, 6). Numerical experiments are presented in Sect. 3.

As suggested before, this kind of approximations is mainly useful for the calibration of a local volatility model using market call/put option prices. The calibration is known to be an ill-posed inverse problem, which makes it a challenging issue. Although we do not discuss these aspects, our analytical formulas potentially speed up any calibration routines.

1.3 Standing Assumptions for the Approximations

Throughout the paper, we assume the following:

- **Assumption** (E). The function σ is bounded and positive (i.e. $\sigma_{inf} = \inf_{(t,x)\in[0,T]\times\mathbb{R}_+}\sigma(t,S) > 0$). We denote by $c_E \geq 1$ the smallest constant such that

$$\sup_{(t,S)\in[0,T]\times\mathbb{R}_+}\sigma(t,S) \leq c_E \inf_{(t,S)\in[0,T]\times\mathbb{R}_+}\sigma(t,S).$$

- **Assumption** (R). The function σ is seven-times continuously differentiable in the S-variable and

$$M_1 := \max_{1\leq i\leq 7} \sup_{(t,x)\in[0,T]\times\mathbb{R}} \left|\frac{\partial^i}{\partial x^i}[\sigma(t,\exp(x))]\right| < \infty, \tag{5}$$

$$M_0 := \max\Big(M_1, \sup_{(t,S)\in[0,T]\times\mathbb{R}_+} \sigma(t,S)\Big) < \infty. \tag{6}$$

The assumption (R) is used in at least two respects: it allows for differentiating coefficients to obtain an expansion formula; it is used to derive error estimates. The assumption (E) is an ellipticity-type condition that enables us to handle the error analysis for non-smooth payoffs Φ (such as call/put options). This is the standard framework developed in [1].

Note that for deterministic volatility functions, one has $M_1 = 0$.

1.4 Definitions and Other Notations

In the representation of the expansion formulas, we repeatedly use the following integral operator.

Definition 1 (Integral Operator) The integral operator ω^T is defined as follows: for any integrable function l, we set

$$\omega(l)_t^T := \int_t^T l_u du$$

for $t \in [0, T]$. Similarly, for integrable functions (l_1, l_2), we put for $t \in [0, T]$

$$\omega(l_1, l_2)_t^T := \omega(l_1\omega(l_2)_{\cdot}^T)_t^T = \int_t^T l_{1,r}\Big(\int_r^T l_{2,s}ds\Big)dr.$$

The n-times iteration is defined analogously: for any integrable functions $(l_1, \ldots, l_n)$, we set

$$\omega(l_1, \ldots, l_n)_t^T := \omega(l_1\omega(l_2, \ldots, l_n)_{\cdot}^T)_t^T$$

for $t \in [0, T]$.

We also use a short notation for Greeks.

Definition 2 (Greeks) Let Z be a random variable and let h be a payoff function. We define the ith Greek for the variable Z by the quantity (if it has a meaning)

$$\text{Greek}_i^h(Z) := \frac{\partial^i \mathbb{E}[h(Z+x)]}{\partial x^i}\Big|_{x=0}.$$

Definition 3 (Black–Scholes formula and related Greeks) Using usual notation, the Black–Scholes formula for call option and constant parameters (σ, r, q) writes

$$\mathrm{Call}^{\mathrm{BS}}(t, S; T, K; \sigma, r, q) = Se^{-q(T-t)}\mathcal{N}(d_1) - Ke^{-r(T-t)}\mathcal{N}(d_2),$$

where $\mathcal{N}(d) = \frac{1}{\sqrt{2\pi}}\int_{-\infty}^{d} e^{-u^2/2}du$ and

$$d_1 = d_1(t, S; T, K; \sigma, r, q) = \frac{1}{\sigma\sqrt{T-t}}\log\frac{Se^{-q(T-t)}}{Ke^{-r(T-t)}} + \frac{1}{2}\sigma\sqrt{T-t},$$

$$d_2 = d_2(t, S; T, K; \sigma, r, q) = d_1 - \sigma\sqrt{T-t}.$$

For time dependent coefficients $(\sigma_s, r_s, q_s)_{s\leq T}$, the call price formula is deduced from the Black–Scholes formula by replacing the arguments σ^2, r and q by their time-average on the interval $[t, T]$. The resulting formula will be denoted by

$$\mathrm{Call}^{\mathrm{BS}}(t, S; T, K; (\sigma_s)_s, (r_s)_s, (q_s)_s).$$

For $t < T$ and $\sigma > 0$, the function $(S, K) \mapsto \mathrm{Call}^{\mathrm{BS}}(t, S; T, K; \sigma, r, q)$ is smooth and its sensitivities $\frac{\partial^i}{\partial S^i}\mathrm{Call}^{\mathrm{BS}}(t, S; T, K; \sigma, r, q)$ and $\frac{\partial^i}{\partial K^i}\mathrm{Call}^{\mathrm{BS}}(t, S; T, K; \sigma, r, q)$ are given explicitly in Proposition 1 (see Sect. 6), for $i = 1, \ldots, 6$. They will be used in our expansion formulas (see Theorems 2 and 3).

2 Expansion Formulas

In this section, we give several expansion formulas, with a second and third order accuracy. The general principle for deriving such approximations is to choose a relevant proxy and to expand the quantities of interest around this proxy. First, we recall the general results from [1], where the proxy is obtained by freezing the local volatility at the initial spot value (ATM). Second we apply these expansions to call options. Third, using the Dupire forward PDE satisfied by the call price as a function of maturity and strike, we propose a new proxy where the volatility is frozen at the strike value K (instead of S_0). We then derive new second and third approximation formulas. Finally, some expansions of implied volatility are provided.

2.1 A General Result

We first state two expansion results in a quite general form, so that we can apply it later to various situations. Let $(Y_t)_{t\leq T}$ the solution of

$$dY_t = -\frac{1}{2}a^2(t, Y_t)dt + a(t, Y_t)dW_t, \quad Y_0 \text{ given.} \tag{7}$$

Theorem 1 (Second and third order approximations [3, Theorems 2.1 and 2.3]) *Assume that*

- *the function a is bounded and positive ($a_{inf} = \inf_{(t,y)\in[0,T]\times\mathbb{R}} a(t,y) > 0$). We denote by $c_E \geq 1$ the smallest constant such that*

$$\sup_{(t,y)\in[0,T]\times\mathbb{R}} a(t,y) \leq c_E \inf_{(t,y)\in[0,T]\times\mathbb{R}} a(t,y).$$

- *the function a is seven-times continuously differentiable in the y-variable and*

$$M_{Y,1} = \max_{1\leq i\leq 7} \sup_{(t,y)\in[0,T]\times\mathbb{R}} \left|\partial_y^i a(t,y)\right| < \infty, \tag{8}$$

$$M_{Y,0} = \max\left(M_{Y,1}, \sup_{(t,y)\in[0,T]\times\mathbb{R}} a(t,y)\right) < \infty. \tag{9}$$

- *the function $h : \mathbb{R} \mapsto \mathbb{R}$ is a.e. differentiable. In addition, h and h' have at most an exponential growth: $|h(x)| + |h'(x)| \leq c_h e^{c_h|x|}$ for any x, for a constant c_h.*

Define

- *the Gaussian process $(Y_t^P)_{t\leq T}$ by*

$$Y_t^P = Y_0 - \frac{1}{2}\int_0^t a^2(s,Y_0)ds + \int_0^t a(s,Y_0)dW_s;$$

- *$a(t) := a(t,Y_0)$, $a^{(1)}(t) := \partial_y^1 a(t,Y_0)$ and $a^{(2)}(t) := \partial_y^2 a(t,Y_0)$;*
- *the expansion coefficients computed using the function $a(t,.)$ at Y_0:*

$$c_{1,T} = \omega(a^2, aa^{(1)})_0^T, \qquad c_{2,T} = \omega(a^2, (a^{(1)})^2)_0^T,$$
$$c_{3,T} = \omega(a^2, aa^{(2)})_0^T, \qquad c_{4,T} = \omega(a^2, a^2, (a^{(1)})^2)_0^T,$$
$$c_{5,T} = \omega(a^2, a^2, aa^{(2)})_0^T, \qquad c_{6,T} = \omega(a^2, aa^{(1)}, aa^{(1)})_0^T,$$
$$c_{7,T} = \omega(a^2, a^2, aa^{(1)}, aa^{(1)})_0^T, \qquad c_{8,T} = \omega(a^2, aa^{(1)}, a^2, aa^{(1)})_0^T.$$

Then, the following expansion formulas hold.

a) Second order approximation. *One has*

$$\mathbb{E}h(Y_T) = \mathbb{E}h(Y_T^P) + c_{1,T}\left(\frac{1}{2}\text{Greek}_1^h(Y_T^P) - \frac{3}{2}\text{Greek}_2^h(Y_T^P) + \text{Greek}_3^h(Y_T^P)\right) + \text{Error}_2, \tag{10}$$

where

$$|\text{Error}_2| \leq C \sup_{v\in[0,1]} \|h^{(1)}(vY_T + (1-v)Y_T^P)\|_2 \frac{M_{Y,0}}{a_{inf}} M_{Y,1} M_{Y,0}^2 T^{3/2}$$

and the constant C depends (in an increasing way) only on the upper bounds of the model parameters, on c_E and on the maturity.

b) Third order approximation. *One has*

$$\mathbb{E}h(Y_T) = \mathbb{E}h(Y_T^P) + \sum_{i=1}^{6} \eta_{i,T}\text{Greek}_i^h(Y_T^P) + \text{Error}_3, \tag{11}$$

where

$$\begin{aligned}
\eta_{1,T} =& \frac{c_{1,T}}{2} - \frac{c_{2,T}}{2} - \frac{c_{3,T}}{2} - \frac{c_{4,T}}{4} - \frac{c_{5,T}}{4} - \frac{c_{6,T}}{2},\\
\eta_{2,T} =& -\frac{3c_{1,T}}{2} + \frac{c_{2,T}}{2} + \frac{c_{3,T}}{2} + \frac{5c_{4,T}}{4} + \frac{5c_{5,T}}{4} + \frac{7c_{6,T}}{2} + \frac{c_{7,T}}{2} + \frac{c_{8,T}}{4},\\
\eta_{3,T} =& c_{1,T} - 2c_{4,T} - 2c_{5,T} - 6c_{6,T} - 3c_{7,T} - \frac{3c_{8,T}}{2},\\
\eta_{4,T} =& c_{4,T} + c_{5,T} + 3c_{6,T} + \frac{13c_{7,T}}{2} + \frac{13c_{8,T}}{4},\\
\eta_{5,T} =& -6c_{7,T} - 3c_{8,T},\\
\eta_{6,T} =& 2c_{7,T} + c_{8,T},
\end{aligned}$$

and

$$|\text{Error}_3| \le C \sup_{v\in[0,1]} \|h^{(1)}(vY_T + (1-v)Y_T^P)\|_2 \Big(\frac{M_{Y,0}}{a_{inf}}\Big)^2 M_{Y,1}M_{Y,0}^3 T^2.$$

As before, the constant C depends (in an increasing way) only on the upper bounds of the model parameters, on c_E and on the maturity.

As explained in [3], the approximation order is related to the power m in the error upper bounds $M_{Y,1}M_{Y,0}^m(\sqrt{T})^{m+1}$. The smaller the volatility ($M_{Y,0} \to 0$) or its variations ($M_{Y,1} \to 0$) or the maturity ($T \to 0$), the more accurate the approximations. See Sect. 2.5 for the explicit bounds in a time-dependent CEV model. Since the proxy is Gaussian, the computation of $\mathbb{E}h(Y_T^P)$ and $\text{Greek}_i^h(Y_T^P)$ can be performed in closed forms for usual functions h (such as call/put payoffs), or by using efficient numerical integration techniques in other cases.

An interesting property of these expansion formulas is that they are exact for $h(x) = e^x$ (indeed $\mathbb{E}h(Y_T) = \mathbb{E}h(Y_T^P) = \text{Greek}_i^h(Y_T^P) = e^{Y_0}$, and the sum of expansion coefficients is equal to zero). In particular, when further applied to the local volatility model (1), this implies that the call/put parity will be preserved within these approximations.

When the function $(t,x) \mapsto a(t,x)$ is piecewise constant w.r.t. the time variable, the coefficients $(c_{i,T})_{1\le i\le 8}$ can be quickly and simultaneously computed for different maturities T, using recursion (see [1, Proposition 4.1]). In other situations, numerical integration is likely needed.

2.2 Application to Expansion Formulas for Call Price

We go back to the local volatility model (1) and to the evaluation of call options. In view of (4), the call price at time 0 is equal to

$$\text{Call}\,(T,K) = \mathbb{E}e^{-\int_0^T r_s ds}(S_T - K)^+ = \mathbb{E}h(X_T)$$

where $h(x) = e^{-\int_0^T r_s ds}(C_T e^x - K)^+$. In order to apply previous expansion results, it remains to identify the function $a(\cdot)$ in the dynamics of

$$dX_t = a(t,X_t)dW_t - \frac{1}{2}a^2(t,X_t)dt.$$

Comparing with (4), it follows that

$$a(t,x) = \sigma(t, C_t e^x).$$

Owing to the assumptions (R) and (E) on σ, one can apply Theorem 1 to $Y = X$ and to $h(x) = e^{-\int_0^T r_s ds}(C_T e^x - K)^+$, in order to obtain expansion formulas for call prices in local volatility models. The next step consists in transforming the Greeks in the X-variable in the (usual) Greeks in the S-variable, and in expressing the coefficients $c_{i,T}$ using the derivatives of σ. These computations are detailed in Sect. 4. We obtain the following

Theorem 2 (Second and third order approximations for call options, based on the ATM local volatility) *Assume (E) and (R). Set $\sigma_t := \sigma(t, C_t S_0)$, $\sigma_t^{(1)} := \partial_S \sigma(t, C_t S_0)$, $\sigma_t^{(2)} := \partial_S^2 \sigma(t, C_t S_0)$ and*

$$\alpha_{1,T} = \omega(\sigma^2, S_0 C\sigma\sigma^{(1)})_0^T,$$

$$\alpha_{2,T} = \omega(\sigma^2, (S_0 C\sigma^{(1)})^2)_0^T,$$

$$\alpha_{3,T} = \omega(\sigma^2, S_0^2 C^2 \sigma\sigma^{(2)} + S_0 C\sigma\sigma^{(1)})_0^T,$$

$$\alpha_{4,T} = \omega(\sigma^2, \sigma^2, (S_0 C\sigma^{(1)})^2)_0^T,$$

$$\alpha_{5,T} = \omega(\sigma^2, \sigma^2, S_0^2 C^2 \sigma\sigma^{(2)} + S_0 C\sigma\sigma^{(1)})_0^T,$$

$$\alpha_{6,T} = \omega(\sigma^2, S_0 C\sigma\sigma^{(1)}, S_0 C\sigma\sigma^{(1)})_0^T,$$

$$\alpha_{7,T} = \omega(\sigma^2, \sigma^2, S_0 C\sigma\sigma^{(1)}, S_0 C\sigma\sigma^{(1)})_0^T,$$

$$\alpha_{8,T} = \omega(\sigma^2, S_0 C\sigma\sigma^{(1)}, \sigma^2, S_0 C\sigma\sigma^{(1)})_0^T.$$

a) Second order approximation. *One has*

$$\begin{aligned}\mathrm{Call}\,(T,K) &= \mathrm{Call}^{\mathrm{BS}}(0,S_0;T,K)\\ &\quad+\alpha_{1,T}\Big(\frac{3}{2}S_0^2\partial_S^2\mathrm{Call}^{\mathrm{BS}}(0,S_0;T,K)+S_0^3\partial_S^3\mathrm{Call}^{\mathrm{BS}}(0,S_0;T,K)\Big)\\ &\quad+\mathrm{Error}_2, \qquad (12)\end{aligned}$$

$$|\mathrm{Error}_2|\le C S_0\exp\Big(-\frac{\log^2(S_0C_T/K)}{8|\sigma|_\infty^2 T}\Big)\frac{M_0}{\sigma_{inf}}M_1M_0^2T^{3/2}$$

where the Black–Scholes price and Greeks are computed using the time dependent parameters $(\sigma_t, r_t, q_t)_{t\le T}$.

b) Third order approximation. *One has*

$$\mathrm{Call}\,(T,K)=\mathrm{Call}^{\mathrm{BS}}(0,S_0;T,K)+\sum_{i=2}^{6}\pi_{i,T}S_0^i\partial_S^i\mathrm{Call}^{\mathrm{BS}}(0,S_0;T,K)+\mathrm{Error}_3, \qquad (13)$$

where

$$\begin{aligned}\pi_{2,T}&=\frac{3}{2}\alpha_{1,T}+\frac{1}{2}\alpha_{2,T}+\frac{1}{2}\alpha_{3,T}+\frac{9}{4}\alpha_{4,T}+\frac{9}{4}\alpha_{5,T}+\frac{13}{2}\alpha_{6,T}+9\alpha_{7,T}+\frac{9}{2}\alpha_{8,T},\\ \pi_{3,T}&=\alpha_{1,T}+4\alpha_{4,T}+4\alpha_{5,T}+12\alpha_{6,T}+66\alpha_{7,T}+33\alpha_{8,T},\\ \pi_{4,T}&=\alpha_{4,T}+\alpha_{5,T}+3\alpha_{6,T}+\frac{153}{2}\alpha_{7,T}+\frac{153}{4}\alpha_{8,T},\\ \pi_{5,T}&=24\alpha_{7,T}+12\alpha_{8,T},\\ \pi_{6,T}&=2\alpha_{7,T}+\alpha_{8,T},\end{aligned}$$

$$|\mathrm{Error}_3|\le C S_0\exp\Big\{-\frac{\log^2(S_0C_T/K)}{8|\sigma|_\infty^2 T}\Big\}\Big(\frac{M_0}{\sigma_{inf}}\Big)^2 M_1M_0^3T^2.$$

In the above expansions, the constant C depends (in an increasing way) only on the upper bounds of the model parameters, on c_E and on the maturity.

Note that the local volatility and its derivatives are computed along the ATM forward curve $(S_0C_t)_{0\le t\le T}$.

2.3 Other Expansions Based on the Local Volatility at Strike

In the previous approximation formulas, the ATM local volatility plays a central role. This is quite natural for arbitrary functions h, like in the general form of Theorem 1. But when dealing with call-put options, the local volatility at strike presumably plays a similarly important role. The aim of this paragraph is to derive similar expansion formulas, but using the volatility at strike. To achieve this goal, we follow the Dupire approach [5], which writes a PDE satisfied by the call price function $(T,K)\mapsto \mathrm{Call}(T,K)=\mathbb{E}(e^{-\int_0^T r_s ds}(S_T-K)^+)$. Indeed, we know that

$$\begin{aligned}\frac{\partial \mathrm{Call}\,(T,K)}{\partial T} &= -q_T\mathrm{Call}\,(T,K)-(r_T-q_T)K\frac{\partial \mathrm{Call}\,(T,K)}{\partial K}\\ &\quad+\frac{1}{2}\sigma^2(T,K)K^2\frac{\partial^2 \mathrm{Call}\,(T,K)}{\partial K^2},\\ \mathrm{Call}\,(0,K) &= (S_0-K)^+.\end{aligned}$$

In other words, instead of handling a PDE in the backward variables (t,S) with a call payoff as a terminal condition, we now deal with a PDE in the forward variables (T,K), with a put payoff as an initial condition. This latter has a probabilistic Feynman–Kac representation

$$\mathrm{Call}\,(T,K)=e^{-\int_0^T q_{T-t}dt}\mathbb{E}(S_0-K_T)^+$$

using the following diffusion process $(K_t)_{t\le T}$:

$$\frac{dK_t}{K_t}=-(r_{T-t}-q_{T-t})dt+\sigma(T-t,K_t)dW_t,\qquad K_0=K. \tag{14}$$

Define the process $(Y_t)_{t\,\le T}$ as follows:

$$K_t=e^{-\int_0^t(r_{T-s}-q_{T-s})ds}e^{Y_t}=\frac{C_{T-t}}{C_T}e^{Y_t}. \tag{15}$$

Then, Y has a dynamics of the form (7) with $a(t,y)=\sigma\big(T-t,\frac{C_{T-t}}{C_T}e^y\big)$. Thus, we are in a position to apply the general Theorem 1, to Y and to the function $h(y)=e^{-\int_0^T q_{T-t}dt}\big(S_0-\frac{C_0}{C_T}e^y\big)^+$. Retransforming the Greeks with respect to the Y-variable into usual Greeks with respect to K, we obtain the new following expansion formulas (see Sect. 5 for the proof).

Theorem 3 (Second and third order approximations for call options, based on the local volatility at strike) *Assume* (E) *and* (R). *Set* $\tilde C_t=\frac{C_{T-t}}{C_T}$, $\tilde\sigma_t:=\sigma(T-t,\tilde C_tK)$, $\tilde\sigma_t^{(1)}:=\partial_S\sigma(T-t,\tilde C_tK)$, $\tilde\sigma_t^{(2)}=\partial^2_{S^2}\sigma(T-t,\tilde C_tK)$ *and*

$$\begin{aligned}\tilde\alpha_{1,T}&=\omega(\tilde\sigma^2,K\tilde C\tilde\sigma\tilde\sigma^{(1)})_0^T,\\ \tilde\alpha_{2,T}&=\omega(\tilde\sigma^2,(K\tilde C\tilde\sigma^{(1)})^2)_0^T,\\ \tilde\alpha_{3,T}&=\omega(\tilde\sigma^2,K^2\tilde C^2\tilde\sigma\tilde\sigma^{(2)}+K\tilde C\tilde\sigma\tilde\sigma^{(1)})_0^T,\\ \tilde\alpha_{4,T}&=\omega(\tilde\sigma^2,\tilde\sigma^2,(K\tilde C\tilde\sigma^{(1)})^2)_0^T,\\ \tilde\alpha_{5,T}&=\omega(\tilde\sigma^2,\tilde\sigma^2,K^2\tilde C^2\tilde\sigma\tilde\sigma^{(2)}+K\tilde C\tilde\sigma\tilde\sigma^{(1)})_0^T,\end{aligned}$$

$$\alpha_{6,T} = \omega(\tilde{\sigma}^2, K\tilde{C}\tilde{\sigma}\tilde{\sigma}^{(1)}, K\tilde{C}\tilde{\sigma}\tilde{\sigma}^{(1)})_0^T,$$
$$\tilde{\alpha}_{7,T} = \omega(\tilde{\sigma}^2, \tilde{\sigma}^2, K\tilde{C}\tilde{\sigma}\tilde{\sigma}^{(1)}, K\tilde{C}\tilde{\sigma}\tilde{\sigma}^{(1)})_0^T,$$
$$\tilde{\alpha}_{8,T} = \omega(\tilde{\sigma}^2, K\tilde{C}\tilde{\sigma}\tilde{\sigma}^{(1)}, \tilde{\sigma}^2, K\tilde{C}\tilde{\sigma}\tilde{\sigma}^{(1)})_0^T.$$

a*) *Second order approximation. *One has*

$$\begin{aligned}\mathrm{Call}\,(T,K) = {} & \mathrm{Call}^{\mathrm{BS}}(0,S_0;T,K) \\ & + \tilde{\alpha}_{1,T}\Big(\frac{3}{2}K^2\partial_K^2\mathrm{Call}^{\mathrm{BS}}(0,S_0;T,K) + K^3\partial_K^3\mathrm{Call}^{\mathrm{BS}}(0,S_0;T,K)\Big) \\ & + \mathrm{Error}_2, \qquad (16)\end{aligned}$$

$$|\mathrm{Error}_2| \le CK\exp\Big\{-\frac{\log^2(S_0C_T/K)}{8|\sigma|_\infty^2T}\Big\}\frac{M_0}{\sigma_{inf}}M_1M_0^2T^{3/2}$$

where the Black–Scholes price and Greeks are computed using the time dependent parameters $(\tilde{\sigma}_t, r_t, q_t)_{t\le T}$.
b*) *Third order approximation. *One has*

$$\mathrm{Call}\,(T,K) = \mathrm{Call}^{\mathrm{BS}}\,(0,S_0;T,K) + \sum_{i=2}^{6}\tilde{\pi}_{i,T}K^i\partial_K^i\mathrm{Call}^{\mathrm{BS}}(0,S_0;T,K) + \mathrm{Error}_3, \qquad (17)$$

where

$$\begin{aligned}\tilde{\pi}_{2,T} &= \frac{3}{2}\tilde{\alpha}_{1,T} + \frac{1}{2}\tilde{\alpha}_{2,T} + \frac{1}{2}\tilde{\alpha}_{3,T} + \frac{9}{4}\tilde{\alpha}_{4,T} + \frac{9}{4}\tilde{\alpha}_{5,T} + \frac{13}{2}\tilde{\alpha}_{6,T} + 9\tilde{\alpha}_{7,T} + \frac{9}{2}\tilde{\alpha}_{8,T},\\ \tilde{\pi}_{3,T} &= \tilde{\alpha}_{1,T} + 4\tilde{\alpha}_{4,T} + 4\tilde{\alpha}_{5,T} + 12\tilde{\alpha}_{6,T} + 66\tilde{\alpha}_{7,T} + 33\tilde{\alpha}_{8,T},\\ \tilde{\pi}_{4,T} &= \tilde{\alpha}_{4,T} + \tilde{\alpha}_{5,T} + 3\tilde{\alpha}_{6,T} + \frac{153}{2}\tilde{\alpha}_{7,T} + \frac{153}{4}\tilde{\alpha}_{8,T},\\ \tilde{\pi}_{5,T} &= 24\tilde{\alpha}_{7,T} + 12\tilde{\alpha}_{8,T},\\ \tilde{\pi}_{6,T} &= 2\tilde{\alpha}_{7,T} + \tilde{\alpha}_{8,T},\end{aligned}$$

$$|\mathrm{Error}_3| \le CK\exp\Big\{-\frac{[\log(S_0C_T/K)]^2}{8|\sigma|_\infty^2T}\Big\}\Big(\frac{M_0}{\sigma_{inf}}\Big)^2M_1M_0^3T^2.$$

In the above expansions, the constant C *depends (in an increasing way) only on the upper bounds of the model parameters, on* c_E *and on the maturity.*

2.4 Expansion Formulas for Implied Volatility

Interestingly, the previous expansions of call price can be turned into expansions of implied volatility $\sigma^I(0, S_0; T, K)$ defined by

$$\text{Call}\,(T, K) = \text{Call}^{\text{BS}}(0, S_0; T, K; \sigma^I(0, S_0; T, K), (r_t)_{t\le T}, (q_t)_{t\le T}).$$

To achieve this, we use nice relations between Greeks (see below and Proposition 1), omitting to indicate all the parameters whenever unambiguous:

$$\begin{aligned}
\text{Vega} &= \frac{\partial}{\partial\sigma}\text{Call}^{\text{BS}}(0, S; T, K) = Se^{-qT}\mathcal{N}'(d_1)\sqrt{T} = Ke^{-rT}\mathcal{N}'(d_2)\sqrt{T},\\
S^2\Gamma_{\text{S}} &= S^2\frac{\partial^2}{\partial S^2}\text{Call}^{\text{BS}}(0, S; T, K) = \frac{Se^{-qT}\mathcal{N}'(d_1)}{\sigma\sqrt{T}} = \frac{\text{Vega}}{\sigma T},\\
S^3\text{Speed}_{\text{S}} &= S^3\frac{\partial^3}{\partial S^3}\text{Call}^{\text{BS}}(0, S; T, K) = -S^2\Gamma_{\text{S}}\Big(\frac{d_1}{\sigma\sqrt{T}} + 1\Big)\\
&= -\frac{\text{Vega}}{\sigma T}\Big(\frac{d_1}{\sigma\sqrt{T}} + 1\Big),\\
K^2\Gamma_{\text{K}} &= \frac{\partial^2}{\partial K^2}\text{Call}^{\text{BS}}(0, S; T, K) = \frac{Ke^{-rT}\mathcal{N}'(d_2)}{\sigma\sqrt{T}} = \frac{\text{Vega}}{\sigma T},\\
K^3\text{Speed}_{\text{K}} &= K^3\frac{\partial^3}{\partial K^3}\text{Call}^{\text{BS}}(0, S; T, K) = -K^2\Gamma_{\text{K}}\Big(1 - \frac{d_2}{\sigma\sqrt{T}}\Big)\\
&= -\frac{\text{Vega}}{\sigma T}\Big(1 - \frac{d_2}{\sigma\sqrt{T}}\Big).
\end{aligned}$$

Now, consider the second order expansion formula based on the ATM local volatility: it becomes

$$\text{Call}\,(T, K) = \text{Call}^{\text{BS}}(0, S_0; T, K) - \text{Vega}\frac{\alpha_{1,T}}{\sqrt{T}(\int_0^T \sigma_s^2 ds)^{3/2}}\log\frac{S_0 C_T}{K} + \text{Error}_2.$$

Since $\text{Vega} = \frac{\partial}{\partial\sigma}\text{Call}^{\text{BS}}$, this directly reads as an expansion of the implied volatility. The derivation is similar for the second order expansion formula based on the local volatility at strike. We have proved the following

Theorem 4 (Second order approximations on implied volatilities) *Assume* (E) *and* (R). *Using the notations of Theorems* 2 *and* 3, *we have*

$$\sigma^I(0, S_0; T, K) = \Big(\frac{1}{T}\int_0^T \sigma_s^2 ds\Big)^{\frac{1}{2}} - \frac{\alpha_{1,T}}{T^{\frac{1}{2}}(\int_0^T \sigma_s^2 ds)^{\frac{3}{2}}}\log\frac{S_0 C_T}{K} + \text{Error}_2^I, \quad (18)$$

$$\sigma^I(0, S_0; T, K) = \Big(\frac{1}{T}\int_0^T \tilde{\sigma}_s^2 ds\Big)^{\frac{1}{2}} + \frac{\tilde{\alpha}_{1,T}}{T^{\frac{1}{2}}(\int_0^T \tilde{\sigma}_s^2 ds)^{\frac{3}{2}}}\log\frac{S_0 C_T}{K} + \widetilde{\text{Error}}_2^I. \quad (19)$$

Note that in the first case (18), the local volatility is computed ATM, while in the second one (19), it is computed at strike.

In addition to these direct implied volatility approximations, one can upper bound the residual terms Error_2^I and $\widetilde{\mathrm{Error}}_2^I$, simply applying the error estimates from Theorems 2 and 3. We do not give the details of this derivation. As it can be expected, the error estimates depend on the ratio $\frac{\log(S_0 C_T/K)}{|\sigma|_\infty \sqrt{T}}$, but actually, they are locally uniform w.r.t. this ratio. More precisely, for any $\xi > 0$, there is a constant C_ξ which depends (in an increasing way) on ξ, on the upper bounds of the model parameters, on c_E, on the maturity and on the ratio M_0/σ_{inf} such that for any S_0 and K satisfying $|\log(S_0 C_T/K)| \le \xi |\sigma|_\infty \sqrt{T}$ we have

$$|\mathrm{Error}_2^I| + |\widetilde{\mathrm{Error}}_2^I| \le C_\xi M_1 M_0^2 T.$$

Thus, inaccuracies may occur for very small or very large strikes, a feature which is confirmed by the further numerical experiments. In view of the above upper bounds, the *relative* errors on implied volatility are locally of order $M_1 M_0 T$, justifying the label of second order approximations.

This paves the way for the derivation of a third order expansion of implied volatility, but unfortunately, we have not been able to simplify the computations in order to get a sufficiently nice expression. This will be further investigated.

2.5 Applications to Time-Dependent CEV Model

To conclude this section, we specify the results when the volatility has the form

$$\sigma(t, S) = \nu_t S^{\beta_t - 1}, \tag{20}$$

i.e. a CEV-type volatility with a time-dependent level $(\nu_t)_{t \le T}$ and a time-dependent skew $(\beta_t)_{t \le T}$. To force the volatility function to fulfill the assumptions (E) and (R), we could alternatively set $\sigma(t, S) = \nu_t [\varphi(S)]^{\beta_t - 1}$, where $\varphi(.)$ is a C_b^∞-function such that $\varepsilon \le \varphi(S) \le \frac{1}{\varepsilon}$ (for a small positive parameter ε) and $\varphi(S) = S$ for $S \in [2\varepsilon, \frac{1}{2\varepsilon}]$. The related expansion coefficients coincide with those computed from (20) provided S_0 and K are in the interval $[2\varepsilon, \frac{1}{2\varepsilon}]$.

Expansion coefficients In order to apply Theorems 2 and 3, all what is needed is to give the expressions for the coefficients $(\alpha_{i,T}, \tilde{\alpha}_{i,T})_{1 \le i \le 6}$. First, the proxy volatilities are given by $\sigma_t = \nu_t (C_t S_0)^{\beta_t - 1}$ and $\tilde{\sigma}_t = \nu_{T-t} (\tilde{C}_t K)^{\beta_{T-t} - 1}$, where $\tilde{C}_t = C_{T-t}/C_T$; then, we have

$$\alpha_{1,T} = \omega(\sigma^2, (\beta - 1)\sigma^2)_0^T,$$

$$\alpha_{2,T} = \alpha_{3,T} = \omega(\sigma^2, (\beta - 1)^2 \sigma^2)_0^T,$$

$$\alpha_{4,T} = \alpha_{5,T} = \omega(\sigma^2, \sigma^2, (\beta - 1)^2 \sigma^2)_0^T,$$

$$\alpha_{6,T} = \omega(\sigma^2, (\beta-1)\sigma^2, (\beta-1)\sigma^2)_0^T,$$
$$\alpha_{7,T} = \omega(\sigma^2, \sigma^2, (\beta-1)\sigma^2, (\beta-1)\sigma^2)_0^T,$$
$$\alpha_{8,T} = \omega(\sigma^2, (\beta-1)\sigma^2, \sigma^2, (\beta-1)\sigma^2)_0^T.$$

The expressions are similar for $(\tilde{\alpha}_{i,T})_{1\le i\le 6}$, by replacing σ_t by $\tilde{\sigma}_t$ and $(\beta_t - 1)$ by $(\beta_{T-t} - 1)$ in the above formulas. In the case of constant parameters $\nu_t = \nu$, $\beta_t = \beta$ and $\mu = r - q$, all the previous quantities can be expressed in closed forms (the values of the integral operator $\omega(.)_0^T$ are given by iterated integrals of exponential functions). We give them in the simple case $\mu = 0$. By setting $\sigma = \nu S_0^{\beta-1}$ and $\tilde{\sigma} = \nu K^{\beta-1}$, we obtain

$$\alpha_{1,T} = (\beta-1)\sigma^4\frac{T^2}{2}, \qquad \alpha_{2,T} = \alpha_{3,T} = (\beta-1)^2\sigma^4\frac{T^2}{2},$$
$$\alpha_{4,T} = \alpha_{5,T} = \alpha_{6,T} = (\beta-1)^2\sigma^6\frac{T^3}{6}, \qquad \alpha_{7,T} = \alpha_{8,T} = (\beta-1)^2\sigma^8\frac{T^4}{24}.$$

Replacing σ by $\tilde{\sigma}$ gives the values for $(\tilde{\alpha}_{i,T})_{1\le i\le 6}$.

Error estimates The errors are related to the coefficients M_0 and M_1 that are given by

$$M_0 \le c|\nu_.|_\infty \quad \text{and} \quad M_1 \le c|\nu_.|_\infty|\beta_. - 1|_\infty.$$

This easily follows from $|\partial_x^i\sigma(.,.)|_\infty \le c_i|\nu_.|_\infty|\beta_. - 1|_\infty^i$. Thus, a *small volatility level* ($|\nu_.|_\infty \to 0$) gives both small M_0 and M_1. A *small volatility slope* ($|\beta_. - 1|_\infty \to 0$) gives small M_1. In view of Theorem 2 (and this is analogous for Theorem 3), the error estimates are respectively of order

$$S_0 \exp\left\{-\frac{\log^2(S_0C_T/K)}{8|\sigma|_\infty^2 T}\right\} |\nu_.|_\infty^3|\beta_. - 1|_\infty T^{3/2}$$

and

$$S_0 \exp\left\{-\frac{\log^2(S_0C_T/K)}{8|\sigma|_\infty^2 T}\right\} |\nu_.|_\infty^4|\beta_. - 1|_\infty T^2$$

for the second and the third order approximations. Consequently, the formulas are expected to be more accurate for small volatility levels ($|\nu_.|_\infty \to 0$), or small maturities ($T \to 0$), or small volatility slopes ($|\beta_. - 1|_\infty \to 0$); note that these asymptotics can hold simultaneously, so that the approximations may be even more accurate. We illustrate the features related to T and β in the next section.

3 Numerical Results

In the numerical tests we report here, we take $r = q = 0$ and we consider a CEV model (20) for the volatility, with constant parameters ν and β. For additional tests

Table 1 Set of maturities and strikes used for the numerical tests

T	K												
3M	0.70	0.75	0.80	0.85	0.90	0.95	1.00	1.05	1.10	1.20	1.25	1.30	1.35
6M	0.65	0.75	0.80	0.85	0.90	0.95	1.00	1.05	1.10	1.15	1.25	1.35	1.50
1Y	0.55	0.65	0.75	0.80	0.90	0.95	1.00	1.05	1.15	1.25	1.40	1.50	1.80
1.5Y	0.50	0.60	0.70	0.75	0.85	0.95	1.00	1.10	1.15	1.30	1.50	1.65	2.00
2Y	0.45	0.55	0.65	0.75	0.85	0.90	1.00	1.10	1.20	1.35	1.55	1.80	2.30
3Y	0.35	0.50	0.55	0.70	0.80	0.90	1.00	1.10	1.25	1.45	1.75	2.05	2.70
5Y	0.25	0.40	0.50	0.60	0.75	0.85	1.00	1.15	1.35	1.60	2.05	2.50	3.60
10Y	0.15	0.25	0.35	0.50	0.65	0.80	1.00	1.20	1.50	1.95	2.75	3.65	6.30

Table 2 CEV model ($\beta = 0.8$): implied volatilities in %

3M	25.908	25.728	25.563	25.409	25.265	25.129	25.001	24.879	24.763	24.548	24.447	24.350	24.258
6M	26.096	25.728	25.564	25.410	25.266	25.130	25.001	24.880	24.764	24.654	24.448	24.258	24.001
1Y	26.530	26.096	25.729	25.565	25.267	25.131	25.003	24.881	24.655	24.449	24.171	24.002	23.562
1.5Y	26.780	26.304	25.907	25.731	25.413	25.133	25.004	24.766	24.656	24.353	24.003	23.772	23.311
2Y	27.058	26.531	26.099	25.732	25.414	25.270	25.005	24.768	24.552	24.262	23.925	23.564	22.980
3Y	27.729	26.783	26.534	25.911	25.570	25.272	25.008	24.770	24.453	24.089	23.633	23.254	22.605
5Y	28.646	27.377	26.788	26.313	25.739	25.421	25.012	24.664	24.268	23.854	23.258	22.788	21.943
10Y	30.079	28.658	27.746	26.800	26.118	25.586	25.022	24.568	24.020	23.386	22.573	21.918	20.694

with time-dependent parameters, see [3]. We choose $S_0 = 1$, $\nu = 25$ % and we allow β to vary. Actually, we consider two values: $\beta = 0.8$ which is not far from the log-normal case, and $\beta = 0.2$ which is rather different. We test the accuracy of different approximations, for various maturities (3–6 months, 1-1.5-2-3-5-10 years) and various strikes. The range of strikes depends on the maturity: the tested values are reported in Table 1. Essentially, the strikes are roughly equal to $S_0 \exp(\xi \nu \sqrt{T})$ where ξ is taken as various quantiles of the standard Gaussian law (we take the quantiles 1 % − 5 % − 10 % − 20 % − 30 % − 40 % − 50 % − 60 % − 70 % − 80 % − 90 % − 95 % − 99 %): this means that the first and last columns of strikes are associated to very ITM options or very OTM options.

For the sake of completeness, in Table 2 and 3 we report the implied volatilities related to the (exact) call price in CEV model with constant parameters (our computations are based on the work by Schroder [12]). We aim at comparing the following different approximations.

1. `ImpVol(AppPrice(2,S0))`: this is the implied volatility of the second order expansion based on the ATM local volatility (see (12) in Theorem 2).
2. `AppImpVol(2,S0)`: this is the second order implied volatility expansion based on the ATM local volatility (see (18) in Theorem 4).
3. `ImpVol(AppPrice(2,K))`: this is the implied volatility of the second order expansion based on the local volatility at strike (see (16) in Theorem 3).

Table 3 CEV model ($\beta = 0.2$): implied volatilities in %

3M	28.755	28.003	27.312	26.673	26.080	25.528	25.010	24.535	24.074	23.232	22.845	22.477	22.128
6M	29.590	28.017	27.325	26.686	26.092	25.539	25.021	24.535	24.078	23.646	22.851	22.133	21.177
1Y	31.537	29.624	28.046	27.352	26.116	25.561	25.042	24.555	23.664	22.867	21.814	21.189	19.602
1.5Y	32.706	30.568	28.831	28.075	26.736	25.583	25.062	24.115	23.681	22.513	21.202	20.359	18.733
2Y	34.034	31.618	29.692	28.103	26.761	26.163	25.083	24.133	23.288	22.177	20.921	19.621	17.619
3Y	37.339	32.840	31.698	28.924	27.459	26.209	25.124	24.170	22.930	21.547	19.882	18.555	16.406
5Y	42.069	35.797	33.000	30.816	28.271	26.908	25.205	23.802	22.262	20.709	18.589	17.011	14.382
10Y	47.850	41.604	37.460	33.144	30.082	27.758	25.378	23.535	21.407	19.089	16.346	14.325	10.993

4. `AppImpVol(2,K)`: this is the second order implied volatility expansion based on the local volatility at strike (see (19) in Theorem 4).
5. `ImpVol(AppPrice(3,S0))`: this is the implied volatility of the third order expansion based on the ATM local volatility (see (13) in Theorem 2).
6. `ImpVol(AppPrice(3,K))`: this is the implied volatility of the third order expansion based on the local volatility at strike (see (17) in Theorem 3).
7. `Av.ImpVol(AppPrice(3,.))`: this is the average of `ImpVol(AppPrice (3,S0))` and `ImpVol(AppPrice(3,K))`. The interest in this approximation is explained later.

In Table 4 (resp. Table 5), we report the errors on implied volatility using the six first aforementioned approximations, for $\beta = 0.8$ (resp. $\beta = 0.2$). The errors are expressed in bps (basis points): an implied volatility of 25.01 % instead of 25 % yields 1 bp error. For instance, on the first row of Table 4, the value -12.3 is associated to the approximation error of `ImpVol(AppPrice(2,S0))` for the first strike of maturity $T = 3M$ (i.e. $K = 0.70$); on the fourth row of Table 4, the value -0.9 refers to the approximation error of `AppImpVol(2,K)` for the second strike of maturity $T = 3M$ (i.e. $K = 0.75$), and so one. Sometimes (especially for very small and very large strikes), the price approximation is out of the non-arbitrage interval for call options: in this case, one can not define a value for the implied volatility and we report ND in the tabular. For all these results, a medium (or large) error on implied volatility may yield a small (or reasonable) error on prices: this is especially true for ITM or OTM options, for which the Vega is small (see the discussion in [4]).

Influence of β and T Generally speaking, we observe that for $\beta = 0.8$, the errors are smaller compared to $\beta = 0.2$: it is not surprising since the lognormal proxy suits better in the first case. This can also be explained by our error estimates, since M_1 is essentially proportional to $|\beta - 1|$. Errors are increasing w.r.t. T, which is also coherent with our error estimates.

Influence of K For usual values of strike (essentially in the Gaussian quantile range $[10\,\%, 90\,\%]$), errors are small (or very small, depending on the approximation that is used), usually smaller than 10 bps for $\beta = 0.8$ up to 10Y maturity, and smaller than 20 bps for $\beta = 0.2$ up to maturity 5Y. Error approximations on implied

Table 4 CEV model ($\beta = 0.8$): errors in bps on the implied volatility using the 6 approximations `ImpVol(AppPrice(2,S0))`, `AppImpVol(2,S0)`, `ImpVol(AppPrice(2,K))`, `AppImpVol(2,K)`, `ImpVol(AppPrice(3,S0))` and `ImpVol(AppPrice(3,K))`

3M	−12.3	−5.8	−2.4	−0.9	−0.3	−0.1	−0.1	−0.1	−0.2	−1.3	−2.6	−4.9	−8.5
	−1.7	−0.9	−0.5	−0.3	−0.2	−0.1	−0.1	−0.1	−0.1	−0.3	−0.5	−0.6	−0.8
	−17.1	−6.8	−2.6	−0.9	−0.3	−0.1	−0.1	−0.1	−0.2	−1.2	−2.4	−4.2	−6.8
	−1.7	−0.9	−0.5	−0.3	−0.2	−0.1	−0.1	−0.1	−0.1	−0.3	−0.5	−0.6	−0.8
	−1.4	−0.4	−0.1	0.0	0.0	0.0	0.0	0.0	0.0	0.0	0.1	0.1	0.4
	0.6	0.1	0.0	0.0	0.0	0.0	0.0	0.0	0.0	0.0	0.0	−0.1	−0.4
6M	−13.3	−3.4	−1.5	−0.6	−0.3	−0.2	−0.1	−0.2	−0.2	−0.4	−1.6	−4.4	−14.8
	−1.9	−0.9	−0.6	−0.4	−0.2	−0.2	−0.1	−0.2	−0.2	−0.3	−0.5	−0.9	−1.5
	−17.7	−3.7	−1.6	−0.7	−0.3	−0.2	−0.1	−0.2	−0.2	−0.4	−1.5	−4.0	−11.2
	−2.1	−0.9	−0.6	−0.4	−0.2	−0.2	−0.1	−0.2	−0.2	−0.3	−0.5	−0.8	−1.4
	−1.1	−0.1	0.0	0.0	0.0	0.0	0.0	0.0	0.0	0.0	0.0	0.1	0.8
	0.7	0.0	0.0	0.0	0.0	0.0	0.0	0.0	0.0	0.0	0.0	−0.1	−0.7
1Y	−23.5	−8.0	−2.3	−1.2	−0.4	−0.3	−0.3	−0.3	−0.5	−1.2	−3.9	−7.7	−36.8
	−3.5	−1.9	−1.0	−0.7	−0.4	−0.3	−0.3	−0.3	−0.4	−0.7	−1.2	−1.6	−3.1
	−34.1	−9.2	−2.4	−1.2	−0.4	−0.3	−0.3	−0.3	−0.5	−1.1	−3.6	−6.7	−23.2
	−3.9	−2.0	−1.0	−0.7	−0.4	−0.3	−0.3	−0.3	−0.4	−0.6	−1.1	−1.5	−2.8
	−2.1	−0.3	−0.1	0.0	0.0	0.0	0.0	0.0	0.0	0.0	0.1	0.2	3.0
	2.0	0.2	0.0	0.0	0.0	0.0	0.0	0.0	0.0	0.0	−0.1	−0.2	−2.3
1.5Y	−28.4	−10.6	−3.5	−2.0	−0.7	−0.4	−0.4	−0.5	−0.6	−1.6	−5.6	−12.1	−50.0
	−4.7	−2.7	−1.5	−1.1	−0.6	−0.4	−0.4	−0.5	−0.5	−0.9	−1.7	−2.4	−4.3
	−41.3	−12.2	−3.7	−2.1	−0.7	−0.4	−0.4	−0.5	−0.6	−1.5	−5.1	−10.2	−29.7
	−5.3	−2.9	−1.6	−1.2	−0.6	−0.4	−0.4	−0.5	−0.5	−0.9	−1.6	−2.2	−3.8
	−2.5	−0.4	−0.1	−0.1	0.0	0.0	0.0	0.0	0.0	0.0	0.1	0.4	4.4
	2.5	0.3	0.0	0.0	0.0	0.0	0.0	0.0	0.0	0.0	−0.1	−0.4	−3.3
2Y	−36.5	−14.5	−5.3	−1.9	−0.8	−0.6	−0.5	−0.6	−0.9	−2.0	−6.0	−17.6	−91.0
	−6.2	−3.7	−2.2	−1.3	−0.8	−0.6	−0.5	−0.6	−0.8	−1.2	−2.1	−3.3	−6.2
	−55.7	−17.2	−5.6	−1.9	−0.8	−0.6	−0.5	−0.6	−0.9	−2.0	−5.5	−14.2	−43.6
	−7.1	−4.0	−2.3	−1.3	−0.8	−0.6	−0.5	−0.6	−0.8	−1.2	−1.9	−3.0	−5.3
	−3.5	−0.6	−0.2	−0.1	0.0	0.0	0.0	0.0	0.0	0.0	0.1	0.7	10.0
	3.6	0.5	0.1	0.0	0.0	0.0	0.0	0.0	0.0	−0.1	−0.1	−0.6	−6.6
3Y	−64.7	−17.8	−11.3	−2.9	−1.4	−0.9	−0.8	−0.8	−1.3	−3.1	−10.5	−27.0	−140.9
	−10.5	−5.0	−3.9	−1.9	−1.2	−0.9	−0.8	−0.8	−1.1	−1.8	−3.2	−4.9	−8.8
	−122.7	−21.1	−12.6	−3.0	−1.4	−0.9	−0.8	−0.8	−1.2	−3.0	−9.3	−20.7	−57.5
	−12.6	−5.6	−4.3	−2.0	−1.3	−0.9	−0.8	−0.8	−1.1	−1.8	−3.0	−4.3	−7.3
	−8.9	−0.8	−0.4	−0.1	−0.1	0.0	0.0	0.0	0.0	0.1	0.3	1.2	16.0
	10.7	0.6	0.3	0.1	0.0	0.0	0.0	0.0	−0.1	−0.1	−0.3	−1.2	−10.1
5Y	−106.7	−30.6	−13.2	−5.9	−2.2	−1.5	−1.2	−1.3	−2.1	−4.8	−17.3	−45.5	−471.9
	−18.1	−8.6	−5.5	−3.6	−2.0	−1.5	−1.2	−1.3	−1.9	−2.9	−5.3	−7.9	−14.5
	−256.0	−38.2	−14.5	−6.1	−2.2	−1.5	−1.2	−1.3	−2.1	−4.6	−14.9	−32.2	−88.5
	−23.2	−10.0	−6.1	−3.8	−2.1	−1.5	−1.2	−1.3	−1.8	−2.7	−4.7	−6.7	−11.5
	−18.8	−1.6	−0.5	−0.3	−0.1	−0.1	0.0	0.0	0.1	0.1	0.5	2.4	38.9
	23.1	1.3	0.3	0.2	0.1	0.0	0.0	−0.1	−0.1	−0.2	−0.5	−2.4	−20.9
10Y	−172.3	−69.5	−30.2	−10.0	−4.4	−2.7	−2.2	−2.4	−3.6	−9.1	−34.6	−103.0	ND
	−33.7	−19.2	−12.1	−6.7	−4.1	−2.8	−2.2	−2.4	−3.4	−5.6	−10.2	−15.5	−29.6
	−472.8	−94.7	−34.0	−10.3	−4.4	−2.8	−2.2	−2.4	−3.6	−8.7	−28.2	−60.5	−159.3
	−47.5	−24.3	−14.3	−7.3	−4.2	−2.8	−2.2	−2.4	−3.3	−5.1	−8.7	−12.3	−20.9
	−33.9	−5.0	−1.1	−0.5	−0.3	−0.2	0.0	0.1	0.2	0.3	1.1	6.7	146.5
	27.4	2.5	0.6	0.4	0.2	0.1	0.0	−0.1	−0.2	−0.3	−1.3	−7.0	−58.7

Table 5 CEV model ($\beta = 0.2$): errors in bps on the implied volatility using the 6 approximations `ImpVol(AppPrice(2,S0))`, `AppImpVol(2,S0)`, `ImpVol(AppPrice(2,K))`, `AppImpVol(2,K)`, `ImpVol(AppPrice(3,S0))` and `ImpVol(AppPrice(3,K))`

3M	−131.8	−71.5	−33.0	−12.7	−4.2	−1.5	−1.0	−2.3	−3.9	−22.8	−52.9	−124.1	−525.3
	−18.8	−12.6	−8.0	−4.8	−2.7	−1.5	−1.0	−2.3	−2.7	−5.5	−7.6	−10.1	−12.9
	ND	−134.0	−43.9	−14.2	−4.3	−1.6	−1.0	−2.3	−3.7	−18.5	−34.8	−57.1	−84.0
	−24.4	−15.4	−9.3	−5.3	−2.8	−1.5	−1.0	−2.2	−2.7	−4.9	−6.5	−8.4	−10.3
	−31.8	−9.5	−2.1	−0.5	−0.2	−0.1	0.0	−0.9	−0.5	0.6	3.4	12.8	40.5
	57.2	11.7	2.1	0.5	0.2	0.1	0.0	−1.0	−0.7	−1.1	−3.3	−9.1	−20.3
6M	−152.3	−47.5	−22.8	−10.1	−4.6	−2.6	−2.1	−2.3	−3.6	−7.0	−27.8	−97.9	ND
	−28.2	−14.0	−9.4	−6.1	−3.8	−2.6	−2.1	−2.3	−3.1	−4.4	−8.2	−13.4	−23.2
	−466.9	−63.4	−26.3	−10.8	−4.7	−2.6	−2.1	−2.3	−3.5	−6.6	−22.7	−56.0	−130.1
	−38.4	−16.8	−10.7	−6.6	−4.0	−2.6	−2.1	−2.3	−3.0	−4.1	−7.1	−10.9	−17.1
	−31.0	−3.3	−1.1	−0.6	−0.3	−0.2	0.0	0.1	0.2	0.4	1.2	7.8	94.9
	41.7	3.4	1.2	0.6	0.3	0.1	0.0	−0.1	−0.2	−0.3	−1.1	−6.7	−41.0
1Y	−257.5	−105.8	−34.8	−18.9	−6.5	−4.8	−4.2	−4.3	−7.4	−18.9	−77.7	−203.6	ND
	−55.8	−31.6	−16.9	−12.1	−6.2	−4.8	−4.2	−4.3	−6.1	−9.8	−17.8	−24.4	−48.0
	ND	−164.6	−40.5	−20.7	−6.7	−4.8	−4.2	−4.3	−7.1	−16.9	−52.8	−90.9	−227.4
	−84.9	−41.8	−19.7	−13.4	−6.3	−4.8	−4.2	−4.2	−5.8	−8.7	−14.3	−18.3	−30.8
	−63.4	−11.2	−2.4	−1.5	−0.6	−0.3	−0.1	0.2	0.6	1.2	4.9	18.0	343.5
	77.5	12.0	3.0	1.8	0.6	0.2	−0.1	−0.3	−0.6	−1.0	−4.7	−15.0	−111.1
1.5Y	−313.2	−139.6	−53.3	−31.8	−12.3	−7.0	−6.2	−6.9	−8.6	−25.4	−118.7	−574.7	ND
	−77.5	−46.0	−26.4	−19.8	−11.1	−7.0	−6.2	−6.8	−7.9	−13.7	−25.7	−36.7	−66.4
	ND	−221.0	−63.9	−35.7	−12.8	−7.0	−6.2	−6.8	−8.4	−22.4	−73.0	−131.2	−273.3
	−124.7	−63.5	−32.0	−22.6	−11.6	−7.0	−6.2	−6.7	−7.6	−12.0	−19.6	−25.7	−39.3
	−76.9	−16.4	−4.4	−3.0	−1.5	−0.5	−0.1	0.6	0.9	2.0	8.6	39.6	470.0
	81.1	18.1	6.0	3.8	1.5	0.3	−0.1	−0.7	−0.9	−1.6	−8.1	−29.8	−150.5
2Y	−395.3	−187.6	−79.0	−31.6	−14.5	−11.0	−8.3	−8.7	−12.8	−33.0	−125.5	ND	ND
	−104.9	−63.9	−38.4	−22.7	−13.6	−10.9	−8.3	−8.6	−11.1	−17.8	−30.4	−49.9	−94.8
	ND	−314.7	−98.3	−34.7	−14.9	−11.1	−8.3	−8.6	−12.2	−28.6	−78.5	−172.3	−342.2
	−180.4	−93.1	−48.5	−25.5	−14.1	−11.0	−8.3	−8.5	−10.5	−15.2	−22.8	−32.7	−50.2
	−105.8	−25.5	−7.6	−3.8	−2.0	−1.4	−0.2	0.7	1.5	2.9	9.3	76.9	770.3
	99.9	28.0	10.5	4.6	1.8	0.9	−0.2	−1.0	−1.5	−2.2	−8.7	−50.8	−228.6
3Y	−651.4	−234.4	−160.1	−48.9	−24.7	−15.6	−12.4	−12.3	−18.6	−51.2	−279.0	ND	ND
	−184.1	−90.8	−72.0	−35.7	−22.8	−15.5	−12.4	−12.3	−16.1	−26.3	−47.8	−73.4	−133.8
	ND	−375.7	−219.9	−54.3	−25.7	−15.7	−12.4	−12.2	−17.6	−42.6	−125.7	−231.1	−397.4
	−375.3	−138.1	−101.1	−41.3	−24.0	−15.7	−12.4	−12.2	−15.0	−21.5	−32.8	−43.5	−62.4
	−228.9	−33.9	−19.4	−7.2	−4.3	−2.1	−0.5	0.8	2.6	5.1	22.6	156.9	1012.9
	129.8	37.9	25.9	8.8	3.9	1.1	−0.5	−1.5	−2.5	−3.7	−21.6	−89.8	−295.9
5Y	ND	−392.5	−198.4	−100.5	−42.0	−28.2	−20.5	−19.9	−30.9	−80.9	ND	ND	ND
	−320.6	−163.4	−106.9	−70.8	−39.5	−28.2	−20.5	−19.9	−26.3	−40.9	−76.7	−117.4	−219.1
	ND	−618.2	−247.5	−113.0	−43.8	−28.6	−20.5	−19.7	−28.7	−64.3	−188.0	−314.1	−468.3
	−830.8	−283.4	−154.1	−88.3	−42.3	−28.7	−20.5	−19.7	−23.7	−31.7	−46.8	−59.8	−81.3
	−414.2	−68.3	−27.9	−17.0	−9.3	−5.5	−1.3	1.8	5.2	9.4	49.9	341.0	1498.8
	−666.6	40.4	35.9	21.7	7.9	2.8	−1.3	−3.4	−4.8	−6.7	−47.8	−165.5	−400.1
10Y	ND	−731.7	−386.6	−160.2	−80.6	−51.7	−37.8	−35.2	−51.8	−169.5	ND	ND	ND
	−387.9	−274.1	−196.2	−121.2	−77.4	−52.6	−37.8	−35.8	−46.2	−76.8	−146.2	−227.2	−439.9
	1545.1	−250.4	−303.5	−155.1	−81.7	−52.1	−37.8	−34.9	−48.1	−118.9	−304.8	−431.9	−406.7
	ND	−784.3	−387.4	−168.5	−87.5	−53.9	−37.8	−35.2	−40.1	−52.3	−71.3	−85.6	−103.8
	−447.8	−67.5	−13.5	−17.4	−16.2	−10.3	−2.6	3.5	10.6	19.8	151.4	855.1	2397.6
	ND	−1411.2	−136.9	34.6	23.9	8.4	−2.6	−7.4	−9.8	−15.8	−135.4	−323.5	−406.7

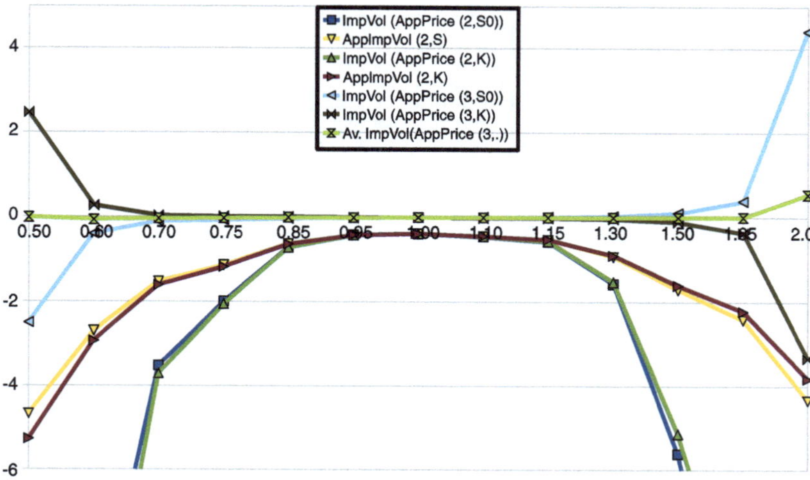

Fig. 1 CEV model ($\beta = 0.8$): errors in bps on the implied volatility using the 7 approximations `ImpVol(AppPrice(2,S0))`, `AppImpVol(2,S0)`, `ImpVol(AppPrice(2,K))`, `AppImpVol(2,K)`, `ImpVol(AppPrice(3,S0))`, `ImpVol(AppPrice(3,K))` and `Av. ImpVol(AppPrice(3,.))`

volatility are much larger for very ITM or very OTM options. For these situations, it may be a good idea to incorporate known asymptotic on the implied volatility (see for instance [9]).

Influence of the type of approximation Regarding the second order approximations, within this model it gives lower bounds on implied volatility (and on price). This systematic underestimation is a drawback of these approximations. Notice that it is usually much better to use the direct approximation on implied volatility (Theorem 4) compared to the implied volatility of the price approximation. However, these implied volatility expansions underestimate the true value as well.

As expected, third order approximations are more accurate than second order ones. The improvement is more significant for $\beta = 0.2$. In Figures 1 and 2, we plot the errors on implied volatility for the maturity $T = 1.5Y$ (this choice is unimportant) for both values of β. We first observe that `ImpVol(AppPrice(3,S0))` overestimates the true value for $K \gg S_0$ and yields an underestimation for $K \ll S_0$. This is the converse regarding `ImpVol(AppPrice(3,K))`. On Tables 4 and 5, we can check that this is generally satisfied for any maturity. Thus, an heuristic rule may be to consider the following confidence interval for the exact implied volatility:

$$\sigma^I(0, S_0; T, K) \in \big(\mathtt{ImpVol(AppPrice(3,K))}, \mathtt{ImpVol(AppPrice(3,S0))}\big).$$

If the width of this interval is too large, it somehow indicates an inaccuracy in our approximations.

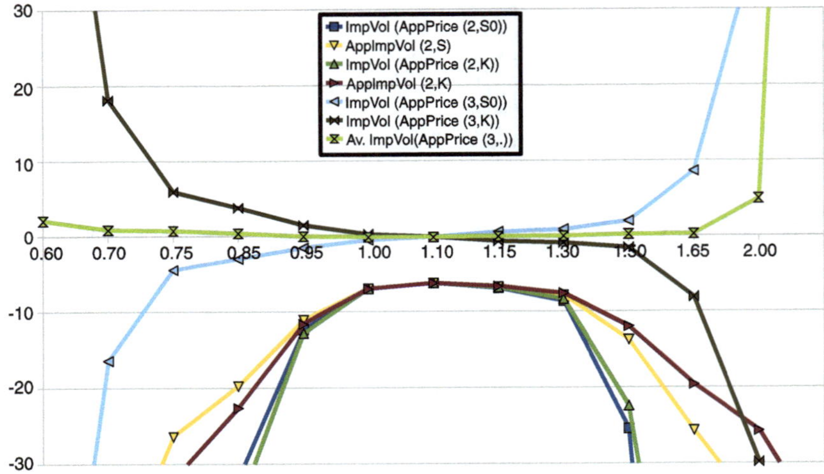

Fig. 2 CEV model ($\beta = 0.2$): errors in bps on the implied volatility using the 7 approximations `ImpVol(AppPrice(2,S0))`, `AppImpVol(2,S0)`, `ImpVol(AppPrice(2,K))`, `AppImpVol(2,K)`, `ImpVol(AppPrice(3,S0))`, `ImpVol(AppPrice(3,K))` and `Av. ImpVol(AppPrice(3,.))`

Table 6 CEV model ($\beta = 0.8$): errors in bps on the implied volatility using `Av. ImpVol(AppPrice(3,.))`

3M	−0.41	−0.13	−0.04	−0.02	−0.02	−0.01	0.00	0.00	0.00	0.00	0.01	0.00	−0.01
6M	−0.16	−0.04	−0.03	−0.02	−0.01	−0.01	0.00	−0.01	0.00	−0.01	0.00	0.01	0.04
1Y	−0.05	−0.06	−0.04	−0.03	−0.01	−0.02	−0.01	−0.01	0.00	−0.01	0.00	0.00	0.34
1.5Y	−0.02	−0.07	−0.04	−0.03	−0.02	−0.02	−0.01	−0.01	−0.01	−0.01	0.00	0.01	0.54
2Y	0.06	−0.08	−0.05	−0.03	−0.02	−0.02	−0.02	−0.01	−0.01	−0.01	−0.01	0.01	1.69
3Y	0.89	−0.09	−0.06	−0.03	−0.02	−0.02	−0.02	−0.02	−0.01	−0.01	−0.01	0.02	2.98
5Y	2.17	−0.16	−0.06	−0.04	−0.03	−0.03	−0.03	−0.02	−0.01	−0.01	−0.01	0.01	8.99
10Y	−3.24	−1.24	−0.23	−0.05	−0.05	−0.05	−0.04	−0.03	−0.02	−0.01	−0.12	−0.13	43.89

Secondly, we observe that the errors using `ImpVol(AppPrice(3,S0))` and `ImpVol(AppPrice(3,K))` have roughly the same magnitude (but with opposite signs). Then, if we define the average

$$\begin{aligned} &\texttt{Av. ImpVol(AppPrice(3,.))} \\ &\quad = \frac{1}{2}\big(\texttt{ImpVol(AppPrice(3,S0))} + \texttt{ImpVol(AppPrice(3,K))}\big), \qquad (21) \end{aligned}$$

we expect to obtain a much better implied volatility estimate. The errors for `Av. ImpVol(AppPrice(3,.))` for $\beta = 0.8$ and $\beta = 0.2$ are reported in Tables 6 and 7. Observe that for maturities smaller than 5Y, the accuracy is truly excellent (i.e. smaller than few bps) for a widened range of strikes. We have compared our approximations with the known implied volatility approximation in the CEV model

Table 7 CEV model ($\beta = 0.2$): errors in bps on the implied volatility using `Av. ImpVol (AppPrice(3,.))`

3M	12.69	1.08	0.01	0.00	−0.01	−0.01	0.00	−0.95	−0.60	−0.27	0.01	1.86	10.08
6M	5.36	0.07	0.07	0.03	0.00	−0.02	−0.02	−0.02	0.00	0.02	0.04	0.54	26.98
1Y	7.01	0.37	0.31	0.14	−0.04	−0.07	−0.06	−0.06	0.01	0.11	0.13	1.52	116.22
1.5Y	2.09	0.87	0.76	0.40	−0.03	−0.14	−0.14	−0.07	−0.02	0.21	0.23	4.94	159.77
2Y	−2.93	1.24	1.46	0.40	−0.12	−0.21	−0.23	−0.14	0.02	0.34	0.28	13.05	270.84
3Y	−49.57	2.02	3.24	0.82	−0.19	−0.50	−0.50	−0.34	0.05	0.69	0.49	33.54	358.51
5Y	−540.41	−13.94	3.99	2.33	−0.69	−1.31	−1.27	−0.77	0.22	1.37	1.04	87.77	549.35
10Y	ND	−739.31	−75.21	8.58	3.84	−0.94	−2.64	−1.96	0.38	2.00	7.99	265.78	995.47

(with zero interest rates and zero dividend) (see [8, formula (5.41) p.141]):

$$\sigma^I(0, S_0; T, K) \approx \nu \frac{(1-\beta)\ln(K/S_0)}{K^{1-\beta} - S_0^{1-\beta}}\left(1 + \frac{(\beta-1)^2\nu^2 T}{24}\left(\frac{S_0+K}{2}\right)^{2\beta-2}\right).$$

This latter approximation yields a slightly better numerical accuracy compared to ours (and it is quicker to evaluate). However, our approximations are also able to deal naturally with general time-dependent local volatility (with piecewise continuity in time), as a difference with [8, Chap. 5] for instance, or with stochastic interest rates [2]. This may be a significant advantage compared to other approaches, while maintaining tight error estimates.

4 Proof of Theorem 2

We apply Theorem 1, by taking $h(x) = e^{-\int_0^T r_s ds}(C_T e^x - K)^+$ and $a(t,x) = \sigma(t, C_t e^x)$. The required assumptions on h and a are satisfied owing to assumptions (E) and (R). By simple computations, we easily check that $M_{Y,0} = M_0$ and $M_{Y,1} = M_1$. The proxy of X used in Theorem 1 now writes

$$X_t^P = \log S_0 - \frac{1}{2}\int_0^t \sigma_s^2 ds + \int_0^t \sigma_s dW_s.$$

Main term and correction terms From this, we deduce that the main term $\mathbb{E}(h(X_T^P))$ in the expansion is equal to

$$\mathbb{E}e^{-\int_0^T r_s ds}(C_T e^{X_T^P} - K)^+ = \text{Call}^{\text{BS}}(0, S_0; T, K; (\sigma_t)_{t\le T}, (r_t)_{t\le T}, (q_t)_{t\le T}).$$

In the following, for the sake of brevity, we omit to indicate in the Black-Scholes formula the dependence w.r.t. $(\sigma_t, r_t, q_t)_{t\le T}$.

For computing the sensitivities $\text{Greek}_i^h(X_T^P) = \partial_x^i \mathbb{E}h(X_T^P + x)|_{x=0}$, we proceed similarly to the main term. First, we have $\mathbb{E}h(X_T^P + x) = \text{Call}^{\text{BS}}(0, S_0 e^x; T, K)$. By successive differentiations, we obtain (using matrix notation)

$$\begin{pmatrix} \mathrm{Greek}_1^h(X_T^P) \\ \mathrm{Greek}_2^h(X_T^P) \\ \mathrm{Greek}_3^h(X_T^P) \\ \mathrm{Greek}_4^h(X_T^P) \\ \mathrm{Greek}_5^h(X_T^P) \\ \mathrm{Greek}_6^h(X_T^P) \end{pmatrix} = \begin{pmatrix} 1 & 0 & 0 & 0 & 0 & 0 \\ 1 & 1 & 0 & 0 & 0 & 0 \\ 1 & 3 & 1 & 0 & 0 & 0 \\ 1 & 7 & 6 & 1 & 0 & 0 \\ 1 & 15 & 25 & 10 & 1 & 0 \\ 1 & 31 & 90 & 65 & 15 & 1 \end{pmatrix} \begin{pmatrix} S_0 \partial_S \mathrm{Call}^{\mathrm{BS}}(0, S_0; T, K) \\ S_0^2 \partial_S^2 \mathrm{Call}^{\mathrm{BS}}(0, S_0; T, K) \\ S_0^3 \partial_S^3 \mathrm{Call}^{\mathrm{BS}}(0, S_0; T, K) \\ S_0^4 \partial_S^4 \mathrm{Call}^{\mathrm{BS}}(0, S_0; T, K) \\ S_0^5 \partial_S^5 \mathrm{Call}^{\mathrm{BS}}(0, S_0; T, K) \\ S_0^6 \partial_S^6 \mathrm{Call}^{\mathrm{BS}}(0, S_0; T, K) \end{pmatrix}. \tag{22}$$

Regarding the summation of the correction terms, it implies that $\sum_{i=1}^6 \eta_{i,T} \mathrm{Greek}_i^h(X_T^P) = \sum_{i=1}^6 \hat{\eta}_{i,T} S_0^i \partial_S^i \mathrm{Call}^{\mathrm{BS}}(0, S_0; T, K)$ where

$$\hat{\eta}_{1,T} = 0,$$

$$\hat{\eta}_{2,T} = \frac{3}{2} c_{1,T} + \frac{1}{2} c_{2,T} + \frac{1}{2} c_{3,T} + \frac{9}{4} c_{4,T} + \frac{9}{4} c_{5,T} + \frac{13}{2} c_{6,T} + 9 c_{7,T} + \frac{9}{2} c_{8,T},$$

$$\hat{\eta}_{3,T} = c_{1,T} + 4 c_{4,T} + 4 c_{5,T} + 12 c_{6,T} + 66 c_{7,T} + 33 c_{8,T},$$

$$\hat{\eta}_{4,T} = c_{4,T} + c_{5,T} + 3 c_{6,T} + \frac{153}{2} c_{7,T} + \frac{153}{4} c_{8,T},$$

$$\hat{\eta}_{5,T} = 24 c_{7,T} + 12 c_{8,T},$$

$$\hat{\eta}_{6,T} = 2 c_{7,T} + c_{8,T}.$$

The expressions of the coefficients $(c_{i,T})_{1 \le i \le 8}$ are given in Theorem 1, but in order to specify them in the current case $a(t, x) = \sigma(t, C_t \exp(x))$, we denote them by $(\alpha_{i,T})_{1 \le i \le 8}$ instead of $(c_{i,T})_{1 \le i \le 8}$. Easy computations show that these definitions coincide with those given in Theorem 2. Then, the second order expansion formula is obtained by keeping only the first coefficient $\alpha_{1,T}$, while all the coefficients are taken for the third order expansion formula.

Error estimates We have already observed that $M_{Y,0} = M_0$ and $M_{Y,1} = M_1$. It remains to estimate the factor $\|h^{(1)}(vX_T + (1-v)X_T^P)\|_2$ arising in the error bounds of Theorem 1. For $v \in [0, 1]$, define $\sigma_t^v := v\sigma(t, X_t) + (1-v)\sigma_t \in [\sigma_{inf}, |\sigma|_\infty]$ and $\sigma_t^{2,v} := v\sigma^2(t, X_t) + (1-v)\sigma_t^2 \in [\sigma_{inf}^2, |\sigma|_\infty^2]$: clearly we have

$$d(vX_t + (1-v)X_t^P) = \sigma_t^v dW_t - \frac{1}{2}\sigma_t^{2,v} dt.$$

We denote by $\mathbb{P}^v$ the probability measure under which $W_t^v = W_t - 2\int_0^t \sigma_s^v ds$ is a Brownian motion. Then, using $h'(x) = e^{-\int_0^T q_s ds} e^x \mathbf{1}_{x - \log S_0 > -d_0}$ where $d_0 = \log(S_0 C_T / K)$, we obtain

$$\mathbb{E}[h'(vX_T + (1-v)X_T^P)]^2$$

$$= S_0^2 e^{-2\int_0^T q_s ds}\mathbb{E} e^{2\int_0^T \sigma_s^v dW_s - \int_0^T \sigma_s^{2,v} ds}\mathbf{1}_{\int_0^T \sigma_s^v dW_s - \frac{1}{2}\int_0^T \sigma_s^{2,v} ds > -d_0}$$
$$= S_0^2 e^{-2\int_0^T q_s ds}\mathbb{E}^v e^{2\int_0^T [\sigma_s^v]^2 ds - \int_0^T \sigma_s^{2,v} ds}\mathbf{1}_{\int_0^T \sigma_s^v dW_s^v + 2\int_0^T [\sigma_s^v]^2 ds - \frac{1}{2}\int_0^T \sigma_s^{2,v} ds > -d_0}$$
$$\leq S_0^2 e^{-2\int_0^T q_s ds + 2|\sigma|_\infty^2 T}\mathbb{P}^v\Big(\int_0^T \sigma_s^v dW_s^v + 2|\sigma|_\infty^2 T > -d_0\Big). \quad (23)$$

If $-d_0 > 2|\sigma|_\infty^2 T$, one can apply the Bernstein exponential inequality to show that the above probability is bounded by $\exp\{-\frac{(-d_0-2|\sigma|_\infty^2 T)^2}{2|\sigma|_\infty^2 T}\}$. Using the inequality $(a-b)^2 \geq \frac{1}{2}a^2 - b^2$, it follows that

$$\sup_{v\in[0,1]} \mathbb{E}[h'(vX_T + (1-v)X_T^P)]^2 \leq C S_0^2 \exp\Big\{-\frac{d_0^2}{4|\sigma|_\infty^2 T}\Big\} \quad (24)$$

where the constant C depends in an increasing way on the bounds on the coefficients and on the maturity. Note that the inequality (24) is also valid if $0 \leq -d_0 \leq 2|\sigma|_\infty^2 T$: indeed, from (23), we write

$$\begin{aligned}
&\mathbb{E}[h'(vX_T + (1-v)X_T^P)]^2 \\
&\leq S_0^2 e^{-2\int_0^T q_s ds + 2|\sigma|_\infty^2 T} \\
&\leq S_0^2 e^{-2\int_0^T q_s ds + 2|\sigma|_\infty^2 T} \exp\Big\{\frac{(2|\sigma|_\infty^2 T)^2}{4|\sigma|_\infty^2 T}\Big\}\exp\Big\{-\frac{d_0^2}{4|\sigma|_\infty^2 T}\Big\} \\
&\leq C S_0^2 \exp\Big\{-\frac{d_0^2}{4|\sigma|_\infty^2 T}\Big\}.
\end{aligned}$$

To sum up we obtain

$$\sup_{v\in[0,1]} \|[h'(vX_T + (1-v)X_T^P)\|_2 \leq C S_0 \exp\Big\{-\frac{[\log(S_0 C_T/K)]^2}{8|\sigma|_\infty^2 T}\Big\} \quad (25)$$

for any $d_0 \leq 0$, or equivalently for any $K \geq S_0 C_T$. Thus, the announced estimates on Error_2 and Error_3 are valid for any Out of The Money calls. Using a similar analysis, the same estimates hold for Out of The Money puts ($K \leq S_0 C_T$). But, since the call/put parity is preserved within these expansions, error estimates are equal for call/put with the same characteristics. Thus, estimates for Out of The Money puts transfer to In The Money calls. This completes the proof. □

A careful inspection of the current proof and that of Theorem 1 reveals that the factor 8 in the exponential (25) can be improved and actually, it can be taken strictly larger than 2: this gives presumably better error estimates for $K \ll S_0$ or $K \gg S_0$.

5 Proof of Theorem 3

The derivation of the expansion is obtained following the same lines as those for Theorem 2. We detail only the main arguments. The proxy for the process $(Y_t)_{t\le T}$ is defined by

$$Y_t^P = \log K + \int_0^t \tilde{\sigma}_s dW_s - \frac{1}{2}\int_0^t \tilde{\sigma}_s^2 ds.$$

We interpret $e^{Y_T^P}/K$ as the Radon–Nikodym derivative of a new measure $\tilde{\mathbb{P}}$ with respect to $\mathbb{P}$ on $\mathscr{F}_T$, under which $\tilde{W}_t = W_t - \int_0^t \tilde{\sigma}_s ds$ is a standard Brownian motion; then we obtain

$$\begin{aligned}
&\mathbb{E}e^{-\int_0^T q_{T-t}dt}\Big(S_0 - \frac{C_0}{C_T}e^{Y_T^P}\Big)^+\\
&= \mathbb{E}e^{-\int_0^T r_s ds}\frac{e^{Y_T^P}}{K}\Big(S_0 C_T \frac{K}{e^{Y_T^P}} - K\Big)^+\\
&= \tilde{\mathbb{E}}e^{-\int_0^T r_s ds}\Big(S_0 e^{\int_0^T (r_s - q_s)ds} e^{-\int_0^T \tilde{\sigma}_s d\tilde{W}_s - \frac{1}{2}\int_0^T \tilde{\sigma}_s^2 ds} - K\Big)^+\\
&= \mathrm{Call}^{\mathrm{BS}}(0, S_0; T, K; (\tilde{\sigma}_t)_{t\le T}, (r_t)_{t\le T}, (q_t)_{t\le T}).
\end{aligned}$$

This gives the main term in the expansion. Regarding the computation of the sensitivities $\mathrm{Greek}_i^h(Y_T^P)$, observe that $\mathbb{E}h(Y_T^P + x) = \mathrm{Call}^{\mathrm{BS}}(0, S_0; T, Ke^x)$, omitting the last parameters $(\tilde{\sigma}_t, r_t, q_t)_{t\le T}$. Thus, we easily relate the sensitivities $\mathrm{Greek}_i^h(Y_T^P)$ to the Greeks of $\mathrm{Call}^{\mathrm{BS}}(0, S_0; T, K)$ with respect to K (instead of S_0 in the Theorem 2). The relation is affine and is similar to (22). The other steps of the proof are analogous to that of Theorem 2, replacing S_0 and σ by K and $\tilde{\sigma}$ in most places. □

6 Computations of Derivatives of the Black–Scholes Price Function with Respect to *S* and *K*

In the following proposition, we make explicit the formulas for the six first derivatives of $\mathrm{Call}^{\mathrm{BS}}(0, S; T, K; \sigma, r, q)$ (in short $\mathrm{Call}^{\mathrm{BS}}(0, S; T, K)$) w.r.t. S and K, leaving the proofs to the reader. These formulas are necessary to implement the expansions of Theorems 2 and 3.

Proposition 1 (Black–Scholes Greeks) *Using the notation from Definition* 3, *the sensitivities w.r.t. S are given by*

$$\Delta_S(t, S; T, K) = \frac{\partial}{\partial S}\mathrm{Call}^{\mathrm{BS}}(t, S; T, K) = e^{-q(T-t)}\mathcal{N}(d_1),$$

$$\Gamma_{\rm S}(t,S;T,K)=\frac{\partial^2}{\partial S^2}{\rm Call}^{\rm BS}(t,S;T,K)=e^{-q(T-t)}\frac{\mathcal{N}'(d_1)}{S\sigma\sqrt{T-t}},$$

$$\text{Speed}_{\rm S}(t,S;T,K)=\frac{\partial^3}{\partial S^3}{\rm Call}^{\rm BS}(t,S;T,K)=-\frac{\Gamma_{\rm S}}{S}\Big(\frac{d_1}{\sigma\sqrt{T-t}}+1\Big),$$

$$\frac{\partial^4}{\partial S^4}{\rm Call}^{\rm BS}(t,S;T,K)=\frac{\Gamma_{\rm S}}{S^2}\Big(2+\frac{3d_1}{\sigma\sqrt{T-t}}+\frac{d_1^2-1}{\sigma^2(T-t)}\Big),$$

$$\frac{\partial^5}{\partial S^5}{\rm Call}^{\rm BS}(t,S;T,K)=\frac{\Gamma_{\rm S}}{S^3}\Big(-6-\frac{11d_1}{\sigma\sqrt{T-t}}+\frac{6(1-d_1^2)}{\sigma^2(T-t)}+\frac{3d_1-d_1^3}{\sigma^3(T-t)^{\frac{3}{2}}}\Big),$$

$$\frac{\partial^6}{\partial S^6}{\rm Call}^{\rm BS}(t,S;T,K)=\frac{\Gamma_{\rm S}}{S^4}\Big(24+\frac{50d_1}{\sigma\sqrt{T-t}}+\frac{35(d_1^2-1)}{\sigma^2(T-t)}+\frac{10d_1(d_1^2-3)}{\sigma^3(T-t)^{\frac{3}{2}}}+\frac{3(1-2d_1^2)+d_1^4}{\sigma^4(T-t)^2}\Big).$$

The sensitivities with respect to K are given by

$$\Delta_{\rm K}(t,S;T,K)=\frac{\partial}{\partial K}{\rm Call}^{\rm BS}(t,S;T,K)=-e^{-r(T-t)}\mathcal{N}(d_2),$$

$$\Gamma_{\rm K}(t,S;T,K)=\frac{\partial^2}{\partial K^2}{\rm Call}^{\rm BS}(t,S;T,K)=e^{-r(T-t)}\frac{\mathcal{N}'(d_2)}{K\sigma\sqrt{T-t}},$$

$$\text{Speed}_{\rm K}(t,S;T,K)=\frac{\partial^3}{\partial K^3}{\rm Call}^{\rm BS}(t,S;T,K)=-\frac{\Gamma_{\rm K}}{K}\Big(1-\frac{d_2}{\sigma\sqrt{T-t}}\Big),$$

$$\frac{\partial^4}{\partial K^4}{\rm Call}^{\rm BS}(t,S;T,K)=\frac{\Gamma_{\rm K}}{K^2}\Big(2-\frac{3d_2}{\sigma\sqrt{T-t}}+\frac{d_2^2-1}{\sigma^2(T-t)}\Big),$$

$$\frac{\partial^5}{\partial K^5}{\rm Call}^{\rm BS}(t,S;T,K)=\frac{\Gamma_{\rm K}}{K^3}\Big(-6+\frac{11d_2}{\sigma\sqrt{T-t}}+\frac{6(1-d_2^2)}{\sigma^2(T-t)}+\frac{d_2^3-3d_2}{\sigma^3(T-t)^{\frac{3}{2}}}\Big),$$

$$\frac{\partial^6}{\partial K^6}{\rm Call}^{\rm BS}(t,S;T,K)=\frac{\Gamma_{\rm K}}{K^4}\Big(24-\frac{50d_2}{\sigma\sqrt{T-t}}+\frac{35(d_2^2-1)}{\sigma^2(T-t)}-\frac{10d_2(d_2^2-3)}{\sigma^3(T-t)^{\frac{3}{2}}}+\frac{3(1-2d_2^2)+d_2^4}{\sigma^4(T-t)^2}\Big).$$

Acknowledgements The first author is grateful to Chair *Financial Risks* of the *Risk Foundation* for its financial support. This work has been partly done when the first author was affiliated to Grenoble INP–Ensimag.

References

1. Benhamou, E., Gobet, E., Miri, M.: Smart expansion and fast calibration for jump diffusion. Finance Stoch. **13**(4), 563–589 (2009)

2. Benhamou, E., Gobet, E., Miri, M.: Analytical formulas for local volatility model with stochastic rates. Quant. Finance **12**(2), 185–198 (2012)
3. Benhamou, E., Gobet, E., Miri, M.: Expansion formulas for European options in a local volatility model. Int. J. Theor. Appl. Finance **13**(4), 603–634 (2010)
4. Benhamou, E., Gobet, E., Miri, M.: Time dependent Heston model. SIAM J. Financ. Math. **1**, 289–325 (2010)
5. Dupire, B.: Pricing with a smile. Risk **7**(1), 18–20 (1994)
6. Etore, P., Gobet, E.: Stochastic expansion for the pricing of call options with discrete dividends. Appl. Math. Finance **19**(3), 233–264 (2012)
7. Hagan, P., Woodward, D.: Equivalent Black volatilities. Appl. Math. Finance **6**, 147–157 (1999)
8. Henry-Labordère, P.: Analysis, Geometry, and Modeling in Finance: Advanced Methods in Option Pricing. Chapman and Hall, London (2008)
9. Lee, R.W.: The moment formula for implied volatility at extreme strikes. Math. Finance **14**(3), 469–480 (2004)
10. Musiela, M., Rutkowski, M.: Martingale Methods in Financial Modelling. Springer, Berlin (1998)
11. Piterbarg, V.V.: Stochastic volatility model with time-dependent skew. Appl. Math. Finance **12**(2), 147–185 (2005)
12. Schroder, M.: Computing the constant elasticity of variance option pricing formula. J. Finance **44**, 211–219 (1989)

Low-Dimensional Partial Integro-differential Equations for High-Dimensional Asian Options

Peter Hepperger

Abstract Asian options on a single asset under a jump-diffusion model can be priced by solving a partial integro-differential equation (PIDE). We consider the more challenging case of an option whose payoff depends on a large number (or even a continuum) of assets. Possible applications include options on a stock basket index and electricity contracts with a delivery period. Both of these can be modeled with an exponential, time-inhomogeneous, Hilbert space valued jump-diffusion process. We derive the corresponding high- or even infinite-dimensional PIDE for Asian option prices in this setting and show how to approximate it with a low-dimensional PIDE. To this end, we employ proper orthogonal decomposition (POD) to reduce the dimension. We generalize the convergence results known for European options to the case of Asian options and give an estimate for the approximation error.

Keywords Hilbert space valued jump-diffusion · Asian option · Partial integro-differential equation · Dimension reduction · Proper orthogonal decomposition

Mathematics Subject Classification (2010) 91B25 · 60H35 · 35R15 · 35R09

1 Introduction

It is well known that the price of an Asian option on a single asset driven by a geometric Brownian motion is the solution of a partial differential equation [15]. This equation depends on two space variables, the value of the underlying and its average up to the current time. If we add jumps to the model, we obtain an additional integral term which yields a partial integro-differential equation (PIDE). In fact, there are several ways to derive such a PIDE. Using clever parametrizations, it is possible to obtain a PIDE with only one space variable [18].

P. Hepperger (✉)
Mathematische Statistik M4, Technische Universität München, Boltzmannstr. 3, 85748 Garching, Germany
e-mail: peter.hepperger@tum.de

Y. Kabanov et al. (eds.), *Inspired by Finance*, DOI 10.1007/978-3-319-02069-3_15,

The PIDEs corresponding to Asian options in general cannot be solved analytically. They are, however, the basis for numerical pricing methods. Using appropriate algorithms, the PIDEs can be solved in a numerically stable way, see [6, 19] and the references therein. For an overview of methods for pricing Asian options, we refer to [17].

In the present article, we consider arithmetic average Asian options depending on more than one underlying asset. More precisely, we will use the time-inhomogeneous, Hilbert space valued jump-diffusion model introduced in [10]. This is a quite general approach suitable for a wide range of applications. We may, e.g., price Asian options written on an index depending on a large basket of stocks. In this case, we would choose the Hilbert space to be finite-dimensional, the dimension equals to the number of stocks. There are, however, also markets in which the option depends on a continuum of assets. This happens, among others, in electricity markets. Electricity option payoffs depend on the forward curve of prices which can be modeled with a function-valued process [8]. We discuss our model and the driving stochastic process, which is applicable to both stock baskets and electricity contracts, in Sect. 2.

Introducing the arithmetic average as an additional space variable, the option price can be written as a function of time, the average value, and the Hilbert space valued variable describing the state of the underlying assets. This is a high-dimensional (possibly infinite-dimensional) object. The main objective of this article is to derive a low-dimensional PIDE which approximates the option price. To this end, we generalize the dimension reduction method for European options presented in [9] to Asian options. The reduction is based on proper orthogonal decomposition (POD) and uses a similar idea as principal component analysis. In Sect. 3, we first describe the POD method for Asian options in detail. Then, we derive the low-dimensional PIDE satisfied by the approximated price process. We show convergence of the PIDE solution to the true value of the Asian option in Theorem 4, which is the main result of this paper. The numerical solution of the PIDE is beyond the scope of this article. This will be a topic for future research. All the results presented here are also applicable to European options as a special case.

2 Hilbert Space Valued Jump-Diffusion

In this section, we state our market model. We first define the driving stochastic process, a time-inhomogeneous Hilbert space valued jump-diffusion. Then, we construct the exponential of this process, which we will use to model the underlying assets. Finally, we discuss the payoff of an Asian option.

2.1 Driving Stochastic Process

Since we consider Hilbert space valued processes, we will make use of infinite-dimensional stochastic analysis. For a definition of integrals with respect to Hilbert space valued Brownian motion see, e.g., [5, 11]. An overview of Poisson random measures in Hilbert spaces can be found in [7], the case of Lévy processes is treated in [13].

Let $(D, \mathcal{F}_D, \mu_D)$ be a finite measure space. We consider the separable Hilbert space

$$H := L^2(D; \mu_D). \tag{1}$$

For every $h \in H$, we denote the corresponding norm by

$$\|h\|_H := \sqrt{\int_D [h(u)]^2 \, \mu_D(u)}. \tag{2}$$

This is the state space for the underlying assets of the Asian option. To model, e.g., a basket of stocks, we could choose a discrete set D, with $\|\cdot\|_H$ denoting the Euclidean norm. For a continuum of assets, on the other hand, we may consider a compact interval $D \subset \mathbb{R}$ and the Lebesgue measure μ_D.

We assume that our model is stated under the risk neutral measure. The driving stochastic process for our model is the H-valued process

$$X_t := \int_0^t \gamma_s \, ds + \int_0^t \sigma_s \, dW(s) + \int_0^t \int_H \eta_s(\xi) \, \widetilde{M}(d\xi, ds), \quad t \geq 0. \tag{3}$$

The diffusion part is driven by an H-valued Wiener process W whose covariance is a symmetric nonnegative definite trace class operator Q. The jumps are characterized by $\widetilde{M}$, the compensated random measure of an H-valued compound Poisson process

$$J_t = \sum_{i=1}^{N_t} Y_i, \quad t \geq 0, \tag{4}$$

which is independent of W. Here, N denotes a Poisson process with intensity λ and $Y_i \sim P^Y$ $(i = 1, 2, \ldots)$ are iid on H (and independent of N). The corresponding Lévy measure is denoted by $\nu = \lambda P^Y$. We denote by $\mathrm{L}(H, H)$ the space of all bounded linear operators on H. We shall assume the drift $\gamma : [0, T] \to H$, the volatility $\sigma : [0, T] \to \mathrm{L}(H, H)$, and the jump dampening factor $\eta : [0, T] \to \mathrm{L}(H, H)$ to be deterministic functions. Let further $(\Omega, (\mathcal{F}_t)_{t \in [0,T]})$ be the filtered measurable space on which the risk neutral measure is defined, with the natural filtration $(\mathcal{F}_t)_{t \in [0,T]}$ generated by X. We make the following assumption, which is similar to the finite-dimensional moment conditions in [16, Sect. 25].

Assumption 4 *We assume that the second exponential moment of the jump distribution Y exists*:

$$E[e^{2\|Y\|_H}] = \int_H e^{2\|\xi\|_H} \, P^Y(d\xi) < \infty. \tag{5}$$

We assume further $\gamma \in \mathrm{L}^2(0, T; H)$, $\sigma \in \mathrm{L}^2(0, T; \mathrm{L}(H, H))$, *and*

$$\|\eta_t\|_{\mathrm{L}(H,H)} \leq 1 \quad \textit{for every } t \in [0, T]. \tag{6}$$

In a finite-dimensional setting ($\dim H < \infty$), the value of each underlying asset at time $t \geq 0$ is modeled by the exponential of one component of the driving process X,

$$S_i(t) = S_i(0)\, e^{X_i(t)} \in \mathbb{R}, \quad i = 1, \dots, \dim H, \tag{7}$$

where $S_i(0) \in \mathbb{R}$ denotes the initial value. For a generalization of the exponential to an infinite-dimensional Hilbert space, let $\{e_k\}_{k \in \mathbb{N}}$ be an orthonormal basis of H. We then define

$$S_t := \sum_{k \in \mathbb{N}} \langle S_0, e_k \rangle_H \, e^{\langle X_t, e_k \rangle_H} \, e_k \in H, \tag{8}$$

for $t > 0$, with the initial value $S_0 \in H$. While it might not be obvious that S_t is an element of H again, this is indeed a consequence of Assumption 4, see [8, Thm. 2.2]. Note that this definition reproduces (7) in the finite-dimensional case, if we choose e_i to be standard unit vectors.

2.2 Value of an Asian Option

Before we can define the value of an arithmetic average Asian option, we need to clarify what exactly *average* is supposed to mean in our Hilbert space valued setting. Consider the application of our model to a basket of stocks. An index on such a basket is basically a weighted sum of the individual stock values. The Asian option is then written on the time-average of this sum. The weight factors are nothing more than a linear mapping working on the vector of asset prices. More generally, we consider an arbitrary bounded linear mapping $w : H \to \mathbb{R}$, which we identify with $w \in H$ by the representation theorem of Fréchet–Riesz. The arithmetic average up to time $t > 0$ is then given by

$$A_t := \frac{1}{t} \int_0^t \langle w, S_u \rangle_H du \in \mathbb{R}. \tag{9}$$

Using the Jensen inequality, the Cauchy–Schwarz inequality, and Fubini's theorem, we obtain

$$E\left[A_t^2\right] = \frac{1}{t^2} E\left(\int_0^t \langle w, S_u \rangle_H du \right)^2 \leq \frac{1}{t^2} \|w\|_H^2 \int_0^t E \|S_u\|_H^2 \, du. \tag{10}$$

This expression is finite by [8, Thm. 2.2]. Hence, the average is a well defined random variable in $L^2(\Omega)$ for $t > 0$. The defining equation (9) is, however, not valid for $t = 0$. Intuitively,

$$A_0 := \langle w, S_0 \rangle_H \tag{11}$$

is the obvious continuation for A. The following theorem shows that this is indeed the correct choice.

Proposition 1 *The following convergence holds almost surely:*

$$\lim_{t\to 0} A_t = \langle w, S_0\rangle_H. \tag{12}$$

Proof Using the definition of A, we find

$$|A_t - \langle w, S_0\rangle_H| \le \frac{1}{t}\int_0^t |\langle w, S_u - S_0\rangle_H|\, du. \tag{13}$$

In order to find a bound for $\langle w, S_u - S_0\rangle_H$, we consider the driving process X. From the proof of [10, Thm. 2.2], we know that

$$E\,\|X_t\|_H^2 \le \int_0^t \Big(\|\gamma_s\|_H^2 + (\mathrm{tr}\,Q)\ \|\sigma_s\|_{\mathrm{L}(H,H)}^2 + C\int_H \|\eta_s(\xi)\|_H^2\ \nu(d\xi)\Big) ds. \tag{14}$$

Thus, $\lim_{t\to 0}\|X_t\|_H = 0$ in $\mathrm{L}^2(\Omega)$. Consequently, there is a sequence $\{t_n\}_{n\in\mathbb{N}} \subset \mathbb{R}^+$ satisfying $\lim_{n\to\infty} t_n = 0$ such that almost surely

$$\lim_{n\to\infty} \left\|X_{t_n}\right\|_H = 0. \tag{15}$$

Moreover, almost surely there exists $\delta > 0$ such that the path of X is continuous in $[0, \delta)$. Consequently, we have almost surely

$$\lim_{t\to 0} \|X_t\|_H = 0. \tag{16}$$

Due to the Cauchy–Schwarz inequality, this yields almost surely $\lim_{t\to 0}\langle X_t, e_k\rangle_H = 0$ and thus

$$\lim_{t\to 0} e^{\langle X_t, e_k\rangle_H} = 1 \tag{17}$$

uniformly in k. Hence, we have almost surely

$$|\langle w, S_t - S_0\rangle_H| = \sum_{k\in\mathbb{N}} \langle S_0, e_k\rangle_H\ \langle w, e_k\rangle_H \left(e^{\langle X_t, e_k\rangle_H} - 1\right) \to 0 \quad \text{for } t\to 0. \tag{18}$$

We apply this limit to (13) and the proof is complete. □

Let $T > 0$ be the maturity of an Asian option. By definition, the value of the option depends on A_T. In addition, it may depend on the state S_T of the underlying at maturity, e.g., in the case of a floating strike. The state S_T in turn is a function of the driving process X_T, defined in (8). It turns out that in view of the dimension reduction methods which we will discuss in Sect. 3 it is useful to introduce the centered process

$$Z_t := X_t - E[X_t], \quad t \ge 0. \tag{19}$$

Hence, $S_t = S_t(Z_t)$ is completely determined by Z_t. We can write it as the function

$$S_t : \begin{cases} H & \to H, \\ z & \mapsto \sum_{k\in\mathbb{N}} \langle S_0, e_k\rangle_H \, e^{\langle \int_0^t \gamma(u)\,du + z,\, e_k\rangle_H} \, e_k. \end{cases} \tag{20}$$

We denote the value of the option at time $t \in [0, T]$, discounted to time 0, by

$$\widehat{V}(t, z, a) := e^{-rT} E[G(Z_T, A_T) \big| Z_t = z, A_t = a] \quad \text{for every } z \in H, a \in \mathbb{R}. \tag{21}$$

This is the conditional expectation of the payoff $G : H \times \mathbb{R} \to \mathbb{R}$ at maturity T given the current state $z \in H$ of the underlying assets and the average $a \in \mathbb{R}$. We make the following assumption concerning the payoff.

Assumption 5 *We assume that there are constants L_z^G and L_a^G such that the payoff function G satisfies the Lipschitz conditions*

$$|G(z_1, a) - G(z_2, a)| \le L_z^G \, \|z_1 - z_2\|_H \quad \textit{for every } z_1, z_2 \in H, \; a \in \mathbb{R}, \tag{22}$$

$$|G(z, a_1) - G(z, a_2)| \le L_a^G \, |a_1 - a_2| \quad \textit{for every } z \in H, \; a_1, a_2 \in \mathbb{R}. \tag{23}$$

Note that this assumption is satisfied, e.g., for Asian call and put options on A_T with fixed or floating strike.

Similar to the finite-dimensional case, the option value $\widehat{V}$ satisfies a PIDE. In order to derive this PIDE in the Hilbert space valued setting, we need H-valued generalizations of two concepts: covariances and derivatives. Covariance matrices are replaced by covariance operators which can be interpreted as possibly infinite dimensional matrices. By [10, Thm. 2.4],

$$\mathcal{C}_{X_T} : \begin{cases} H & \to H', \\ h & \mapsto E\big[\langle X_T - E[X_T], h\rangle_H \langle X_T - E[X_T], \cdot\rangle_H\big] \end{cases} \tag{24}$$

is a well defined, symmetric, nonnegative definite trace class operator (and thus compact). We are particularly interested in the subspace of H where $\mathcal{C}_{X_T}$ is strictly positive definite, i.e., the orthogonal complement of its kernel. We denote this space by $E_0(\mathcal{C}_{X_T})^{\perp}$ (E_0 denoting the eigenspace corresponding to eigenvalue 0).

Furthermore, we denote by $D_z\widehat{V}(t, z, a) \in \mathrm{L}(H, \mathbb{R})$ the Fréchet derivative of $\widehat{V}$ at $(t, z, a) \in [0, T] \times H \times \mathbb{R}$ with respect to z. The second derivative is $D_z^2\widehat{V}(t, z, a) \in \mathrm{L}(H, H)$. The derivatives are continuous linear operators such that for every $t \in [0, T]$, $z \in H$, and $a \in \mathbb{R}$ we have

$$\widehat{V}(t, z + \zeta, a) = \widehat{V}(t, z, a) + [D_z\widehat{V}(t, z, a)](\zeta) + \frac{1}{2}\big\langle [D_z^2\widehat{V}(t, z, a)](\zeta), \zeta\big\rangle_H + o(\|\zeta\|_H^2) \tag{25}$$

for every $\zeta \in H$. It is often convenient to identify $D_z^2\widehat{V}(t,z,a)$ with a bilinear form on $H \times H$, setting

$$[D_z^2\widehat{V}(t,z,a)](\zeta_1,\zeta_2) := \big\langle [D_z^2\widehat{V}(t,z,a)](\zeta_1), \zeta_2 \big\rangle_H \quad \text{for every } \zeta_1, \zeta_2 \in H. \tag{26}$$

The one-dimensional partial derivatives of $\widehat{V}$ with respect to time and average are denoted by $\partial_t \widehat{V}$ and $\partial_a \widehat{V}$, respectively. We can now state the Hilbert space valued PIDE for $\widehat{V}$. We denote the trace operator by $\operatorname{tr}(\cdot)$, and the adjoint operator of $\sigma_t \in \mathrm{L}(H,H)$ by σ_t^*.

Theorem 1 *Suppose that the discounted price $\widehat{V}$ defined in* (21) *is continuously differentiable with respect to t and twice continuously differentiable with respect to z and a. Moreover, assume that the second derivative with respect to z restricted to an arbitrary bounded subset of H is a uniformly continuous mapping to the Hilbert–Schmidt space $\mathrm{L_{HS}}(H,H)$. Then $\widehat{V}$ is a classical solution of the PIDE*

$$\begin{aligned}
-\partial_t \widehat{V}(t,z,a) &= \frac{1}{2}\operatorname{tr}\left(D_z^2\widehat{V}(t,z,a)\sigma_t Q \sigma_t^*\right) + \frac{1}{t}\left(\langle w, S_t(z)\rangle_H - a\right)\partial_a \widehat{V}(t,z,a) \\
&\quad + \int_H \Big[\widehat{V}(t,z+\eta_t(\zeta),a) - \widehat{V}(t,z,a) - \big[D_z\widehat{V}(t,z,a)\big]\eta_t(\zeta)\Big]\nu(d\zeta)
\end{aligned} \tag{27}$$

with terminal condition

$$\widehat{V}(T,z,a) = e^{-rT} G(z,a) \tag{28}$$

for a.e. $t \in (0,T)$, $z \in E_0(\mathcal{C}_{X_T})^{\perp}$, and $a \in \mathbb{R}$.

Proof The proof is very similar to the one of [8, Thm. 4.5]. Applying Itô's formula for Hilbert space valued processes [13, Thm. D.2] to $\widehat{V}(t, Z_t, A_t)$, $t > 0$, yields

$$\begin{aligned}
&\widehat{V}(t, Z_t, A_t) \\
&\quad = \widehat{V}(0, Z_0, A_0) + \int_0^t \partial_t \widehat{V}(u-, Z_{u-}, A_{u-})\,du + \int_0^t D_z\widehat{V}(u-, Z_{u-}, A_{u-})\,dZ_u \\
&\quad + \int_0^t \partial_a \widehat{V}(u-, Z_{u-}, A_{u-})\,dA_u + \frac{1}{2}\int_0^t D_z^2\widehat{V}(u-, Z_{u-}, A_{u-})\,d[Z,Z]_u^c \\
&\quad + \sum_{0\le u\le t} \Big[\widehat{V}(u, Z_u, A_u) - \widehat{V}(u-, Z_{u-}, A_{u-}) \\
&\quad - \big[D_z\widehat{V}(u-, Z_{u-}, A_{u-})\big](Z_u - Z_{u-})\Big],
\end{aligned} \tag{29}$$

where $[Z,Z]^c$ denotes the continuous part of the square bracket process as defined in [13]. Note that the average process A is continuous and of finite variation. Hence, the jump part of the equation does not contain the partial derivative $\partial_a \widehat{V}$. For the same reason, the square bracket processes $[A, A]$ and $[A, Z]$ do not occur in the equation.

We first simplify the covariation term. By the properties of quadratic variations for real-valued processes and [5, Cor. 4.14], we obtain

$$
\begin{aligned}
[Z, Z]_t^c &= \sum_{i,j\in\mathbb{N}} e_i \otimes e_j \big[X_i^c, X_j^c\big]_t \\
&= \sum_{i,j\in\mathbb{N}} e_i \otimes e_j \bigg[\Big\langle \int_0^{\cdot} \sigma_u \, dW_u, e_i \Big\rangle_H, \Big\langle \int_0^{\cdot} \sigma_u \, dW_u, e_j \Big\rangle_H \bigg]_t \\
&= \sum_{i,j\in\mathbb{N}} e_i \otimes e_j \Big(\int_0^t \langle [\sigma_u Q \sigma_u^*] e_j, e_i \rangle_H \, du \Big),
\end{aligned} \tag{30}
$$

where $e_i \otimes e_j$ denotes the tensor product of the two basis elements (compare also the proof of [8, Lemma 4.4]). Thus, we get

$$
\begin{aligned}
&\int_0^t D_z^2 \widehat{V}(u-, Z_{u-}, A_{u-}) \, d[Z, Z]_u^c \\
&\quad = \int_0^t \sum_{i,j\in\mathbb{N}} \big[D_z^2 \widehat{V}(u-, Z_{u-}, A_{u-})\big](e_i, e_j) \, \langle [\sigma_u Q \sigma_u^*] e_j, e_i \rangle_H \, du \\
&\quad = \int_0^t \sum_{j\in\mathbb{N}} \big[D_z^2 \widehat{V}(u-, Z_{u-}, A_{u-})\big]\Big([\sigma_u Q \sigma_u^*] e_j, e_j \Big) \, du \\
&\quad = \int_0^t \operatorname{tr}\Big(D_z^2 \widehat{V}(u-, Z_{u-}, A_{u-}) \sigma_u Q \sigma_u^* \Big) \, du.
\end{aligned} \tag{31}
$$

Next we calculate dA_u. By definition (9) of A we have

$$
\langle w, S_u \rangle_H du = d(uA_u) = A_u du + u dA_u. \tag{32}
$$

Hence, we obtain

$$
dA_u = \frac{1}{u} \left(\langle w, S_u(Z_u) \rangle_H - A_u \right) du. \tag{33}
$$

Finally, we reorganize the jump terms in (29) exactly in the same way as in the proof of [8, Lemma 4.4]. The result is

$$
\begin{aligned}
&d\widehat{V}(t, Z_t, A_t) \\
&\quad = \partial_t \widehat{V}(t-, Z_{t-}, A_{t-}) dt + \frac{1}{2} \operatorname{tr}\Big(D_z^2 \widehat{V}(t-, Z_{t-}, A_{t-}) \sigma_t Q \sigma_t^* \Big) dt \\
&\qquad + \frac{1}{t} \big(\langle w, S_t(Z_{t-}) \rangle_H - A_{t-} \big) \, \partial_a \widehat{V}(t-, Z_{t-}, A_{t-}) \, dt
\end{aligned}
$$

$$+\int_H \Big[\widehat{V}(t, Z_{t-} + \eta_t(\zeta), A_{t-}) - \widehat{V}(t-, Z_{t-}, A_{t-})$$
$$- D_z\widehat{V}(t-, Z_{t-}, A_{t-})\eta_t(\zeta)\Big] \nu(d\zeta)\,dt + D_z\widehat{V}(t-, Z_{t-}, A_{t-})\sigma_t\,dW_t$$
$$+\int_H \Big[\widehat{V}(t, Z_{t-} + \eta_t(\zeta), A_{t-}) - \widehat{V}(t-, Z_{t-}, A_{t-})\Big] \widetilde{M}(d\zeta, dt). \tag{34}$$

The last two summands in this equation are local martingales by definition of the stochastic integral [13, Thms. 8.7, 8.23]. Due to the fact that continuous local martingales of finite variation are almost surely constant [14, Ch. II, Thm. 27], the sum of the remaining integral terms must equal 0. This yields the PIDE. □

3 Approximate Pricing with POD

The PIDE derived in the previous section depends on H-valued objects. In order to obtain a lower-dimensional equation which allows for a numerical solution, we reduce the dimension using POD. The basic idea is to find a small set of orthonormal vectors in H which allow for an accurate approximation of the state S_t of the underlying assets for every $t \in [0, T]$. The POD method has been discussed in [9] in the context of European options. We generalize the approach to Asian options. In particular, we state an error estimate for the solution of the approximating equation.

3.1 POD for the Driving Process

We start with an approximation of the centered driving process Z at maturity $T > 0$.

Definition 1 A sequence of orthonormal elements $\{p_l\}_{l\in\mathbb{N}} \subset H$ is called a *POD-basis* for Z_T, if it solves the minimization problem

$$\min_{\langle p_i, p_j\rangle_H = \delta_{ij}} E\left\| Z_T - \sum_{l=1}^{d} p_l \langle Z_T, p_l\rangle_H \right\|_H^2 \tag{35}$$

for every $d \in \mathbb{N}$.

In other words, a POD basis is a set of deterministic orthonormal functions such that we expect the projection of the random vector $Z_T = X_T - E[X_T] \in H$ onto the first d elements of this basis to be a good approximation. Projecting to a POD basis is equivalent to using the partial sum of the first d elements of a Karhunen–Loève expansion, which itself is closely connected to the eigenvector problem of the covariance operator $\mathcal{C}_{X_T}$ defined in (24). The following proposition is quoted from [10, Thm. 3.3]. It shows that the eigenvectors of $\mathcal{C}_{X_T}$ are indeed a POD basis.

Proposition 2 *Every sequence of orthonormal eigenvectors* $(p_l)_{l\in\mathbb{N}}$ *of the operator* $\mathcal{C}_{X_T}$, *ordered by descending size of the corresponding eigenvalues* $\mu_1 \geq \mu_2 \geq \cdots \geq 0$, *solves the maximization problem*

$$\max_{\langle p_i, p_j\rangle_H=\delta_{ij}} \sum_{l=1}^{d} \langle \mathcal{C}_{X_T} p_l, p_l\rangle_H \tag{36}$$

for every $d \in \{1, 2, \ldots, \dim H\}$. *The maximum value is*

$$\sum_{l=1}^{d} \langle \mathcal{C}_{X_T} p_l, p_l\rangle_H = \sum_{l=1}^{d} \mu_l. \tag{37}$$

Moreover, the eigenvectors are a POD basis in the sense of Definition 1, *and the expectation of the projection error is*

$$E\left\| Z_T - \sum_{l=1}^{d} p_l \langle Z_T, p_l\rangle_H \right\|_H^2 = \sum_{l=d+1}^{\dim H} \mu_l. \tag{38}$$

Subsequently, let $(p_l)_{l\in\mathbb{N}}$ and $(\mu_l)_{l\in\mathbb{N}}$ denote the orthonormal basis and eigenvalues from Proposition 2. Further, let

$$U_d := \operatorname{span}\{p_1, p_2, \ldots, p_d\} \subset H \tag{39}$$

be the d-dimensional subspace spanned by the eigenvectors corresponding to the largest eigenvalues. We will assume that $\mu_1 \geq \cdots \geq \mu_d > 0$, as there is no need to include eigenvectors of the covariance operator corresponding to eigenvalue 0. We define the projection operator

$$\mathcal{P}_d : \begin{cases} H & \to U_d \cong \mathbb{R}^d, \\ z & \mapsto \sum_{l=1}^{d} \langle z, p_l\rangle_H \, p_l. \end{cases} \tag{40}$$

Hence, we can rewrite (38) as

$$E\left\| Z_T - \mathcal{P}_d Z_T \right\|_H^2 = \sum_{l=d+1}^{\dim H} \mu_l. \tag{41}$$

Whenever necessary, we will identify U_d with $\mathbb{R}^d$ via the isometry

$$\iota : \begin{cases} \left(U_d, \|\cdot\|_H\right) & \to \left(\mathbb{R}^d, \|\cdot\|\right), \\ x & \mapsto \left(\langle x, p_l\rangle_H\right)_{l=1}^{d}. \end{cases} \tag{42}$$

So far, we have approximated the value of Z only at time T. It turns out, however, that this is indeed sufficient to obtain small projection errors for arbitrary $t \in [0, T]$.

Proposition 3 *Let Z be the centered jump-diffusion defined in* (19). *For every $t \in [0, T]$, we have*

$$E\left[\|Z_t - \mathcal{P}_d Z_t\|_H^2\right] \leq \sum_{l=d+1}^{\dim H} \mu_l. \tag{43}$$

Proof This is a direct consequence of the independent increments of Z. Using the Pythagorean theorem, we obtain

$$\begin{aligned} E\,\|Z_T - \mathcal{P}_d Z_T\|_H^2 &= E\,\|Z_t - \mathcal{P}_d Z_t + (Z_T - Z_t) - \mathcal{P}_d(Z_T - Z_t)\|_H^2 \\ &= E\,\|Z_t - \mathcal{P}_d Z_t\|_H^2 + E\,\|(Z_T - Z_t) - \mathcal{P}_d(Z_T - Z_t)\|_H^2 \\ &\geq E\,\|Z_t - \mathcal{P}_d Z_t\|_H^2 \,. \end{aligned} \tag{44}$$

Applying Proposition 2 yields (43). □

Consequently, it is not necessary to change Definition 1 in order to approximate the whole path Z_t, $t \in [0, T]$. This is due to the fact that by approximating Z_T, we obviously capture also the events up to time T. In the time-homogeneous case, we even obtain the following t-dependent equality.

Proposition 4 *Let Z be the centered jump-diffusion defined in* (19). *Suppose Z is a time-homogeneous jump-diffusion process, i.e., σ and η in* (3) *do not depend on t. For every $t \in [0, T]$, we then have*

$$E\,\|Z_t - \mathcal{P}_d Z_t\|_H^2 = \frac{t}{T} \sum_{l=d+1}^{\dim H} \mu_l. \tag{45}$$

Proof Due to i.i.d. increments, the covariance operator of $Z(t)$ is given by

$$\mathcal{C}_{X_t} = \frac{t}{T}\mathcal{C}_{X_T}. \tag{46}$$

Hence, the eigenpairs of $\mathcal{C}_{X_t}$ are given by $(\frac{t}{T}\mu_l, p_l)$, $l \in \mathbb{N}$. Applying Proposition 2 (setting $T = t$) yields (45). □

3.2 POD for the Average

Besides the centered driving process Z, the payoff G of the Asian option also depends on the average process A which is a function of the exponential S. Thus, to approximate (27) with a low-dimensional PIDE, we need to show that A and S can be accurately represented with the POD basis as well. To this end, recall that S is defined as a deterministic function of Z by (20). If we apply this function to $\mathcal{P}_d Z_t$ for arbitrary $t \in [0, T]$, we obtain

$$S_t(\mathcal{P}_d Z_t) = \sum_{k \in \mathbb{N}} \langle S_0, e_k \rangle_H \, e^{\langle \int_0^t \gamma(u)\,du + \mathcal{P}_d Z_t,\, e_k \rangle_H} \, e_k \in H. \tag{47}$$

The following theorem is the central part of generalizing the POD method to Asian options.

Theorem 2 *There is a constant $C > 0$ (depending on T) such that*

$$E\left| \langle w, S_t(Z_t) \rangle_H - \langle w, S_t(\mathcal{P}_d Z_t) \rangle_H \right| \leq C \, \|w\|_H \left(\sum_{l=d+1}^{\dim H} \mu_l \right)^{\frac{1}{2}} \tag{48}$$

for every $t \in [0, T]$.

Proof From the definition of S_t, we get

$$\begin{aligned}
&E\left| \langle w, S_t(Z_t) \rangle_H - \langle w, S_t(\mathcal{P}_d Z_t) \rangle_H \right| \\
&\quad = E\left| \sum_{k \in \mathbb{N}} \langle w, e_k \rangle_H \, \langle S_0, e_k \rangle_H \left(e^{\langle \int_0^t \gamma(u)\,du + Z_t,\, e_k \rangle_H} - e^{\langle \int_0^t \gamma(u)\,du + \mathcal{P}_d Z_t,\, e_k \rangle_H} \right) \right| \\
&\quad \leq E \sum_{k \in \mathbb{N}} \left| \langle w, e_k \rangle_H \, \langle S_0, e_k \rangle_H \, e^{\int_0^t \langle \gamma(u), e_k \rangle_H du} \left(e^{\langle Z_t, e_k \rangle_H} - e^{\langle \mathcal{P}_d Z_t, e_k \rangle_H} \right) \right|.
\end{aligned} \tag{49}$$

For the term depending on γ, we use Assumption 4 and obtain

$$\left| \int_0^t \langle \gamma(u), e_k \rangle_H du \right| \leq \int_0^t \|\gamma(u)\|_H \, du \leq C_1 \left(\int_0^t \|\gamma(u)\|_H^2 \, du \right)^{\frac{1}{2}} \leq C_2, \tag{50}$$

with positive constants C_1, C_2 depending on T but not on t. Next, we apply the mean-value theorem to the exponential function and make use of the self-adjointness of the projection operator $\mathcal{P}_d$ for the estimate

$$\begin{aligned}
\left| e^{\langle Z_t, e_k \rangle_H} - e^{\langle \mathcal{P}_d Z_t, e_k \rangle_H} \right| &\leq e^{\max\{\langle Z_t, e_k \rangle_H, \langle \mathcal{P}_d Z_t, e_k \rangle_H\}} \, |\langle Z_t - \mathcal{P}_d Z_t,\, e_k \rangle_H| \\
&\leq e^{\max\{\langle Z_t, e_k \rangle_H, \langle Z_t, \mathcal{P}_d e_k \rangle_H\}} \, \|Z_t - \mathcal{P}_d Z_t\|_H
\end{aligned} \tag{51}$$

for every $k \in \mathbb{N}$. Inserting these results into (49) and using the monotone convergence theorem yields

$$\begin{aligned}
&E\left| \langle w, S_t(Z_t) \rangle_H - \langle w, S_t(\mathcal{P}_d Z_t) \rangle_H \right| \\
&\quad \leq C \sum_{k \in \mathbb{N}} |\langle w, e_k \rangle_H \, \langle S_0, e_k \rangle_H| \; E\left[e^{\max\{\langle Z_t, e_k \rangle_H, \langle Z_t, \mathcal{P}_d e_k \rangle_H\}} \, \|Z_t - \mathcal{P}_d Z_t\|_H \right].
\end{aligned} \tag{52}$$

With the Cauchy–Schwarz inequality, we find

$$\begin{aligned} E\left|\langle w, S_t(Z_t)\rangle_H - \langle w, S_t(\mathcal{P}_d Z_t)\rangle_H\right| \\ \leq C \sum_{k\in\mathbb{N}} \left|\langle w, e_k\rangle_H \langle S_0, e_k\rangle_H\right| \left(E\left[e^{2\max\{\langle Z_t, e_k\rangle_H, \langle Z_t, \mathcal{P}_d e_k\rangle_H\}}\right]\right)^{\frac{1}{2}} \\ \times \left(E\left\|Z_t - \mathcal{P}_d Z_t\right\|_H^2\right)^{\frac{1}{2}}. \end{aligned} \tag{53}$$

For the first expectation, we use [8, Proposition. 2.3]:

$$\begin{aligned} E\left[e^{2\max\{\langle Z_t, e_k\rangle_H, \langle Z_t, \mathcal{P}_d e_k\rangle_H\}}\right] &= E\left[\max\{e^{\langle Z_t, 2e_k\rangle_H}, e^{\langle Z_t, 2\mathcal{P}_d e_k\rangle_H}\}\right] \\ &\leq E\left[e^{\langle Z_t, 2e_k\rangle_H} + e^{\langle Z_t, 2\mathcal{P}_d e_k\rangle_H}\right] \\ &\leq C_3 e^{C_4 T} \end{aligned} \tag{54}$$

with constants C_3, C_4. The Cauchy–Schwarz inequality in $l^2(\mathbb{N})$ yields the following bound for the remaining sum in k:

$$\sum_{k\in\mathbb{N}} \left|\langle w, e_k\rangle_H \langle S_0, e_k\rangle_H\right| \leq \|w\|_H \|S_0\|_H. \tag{55}$$

By Proposition 3, we thus get

$$\begin{aligned} E\left|\langle w, S_t(Z_t)\rangle_H - \langle w, S_t(\mathcal{P}_d Z_t)\rangle_H\right| &\leq C\|w\|_H \|S_0\|_H \left(E\|Z_t - \mathcal{P}_d Z_t\|_H^2\right)^{\frac{1}{2}} \\ &\leq C\|w\|_H \|S_0\|_H \left(\sum_{l=d+1}^{\dim H} \mu_l\right)^{\frac{1}{2}}. \end{aligned} \tag{56}$$

□

Although $S_t(\mathcal{P}_d Z_t)$ is still an element of the possibly infinite-dimensional Hilbert space H, it can be computed from the d-dimensional object $\mathcal{P}_d Z_t$. This makes the approximation suitable for numerical computations. Similar to (9), we define the arithmetic average corresponding to $S_t(\mathcal{P}_d Z_t)$ by

$$A_t^d := \frac{1}{t}\int_0^t \langle w, S_u(\mathcal{P}_d Z_u)\rangle_H du \in \mathbb{R} \tag{57}$$

for $t > 0$. Similar to (11), we set

$$A_0^d := \langle w, S_0(\mathcal{P}_d Z_0)\rangle_H = \langle w, S_0\rangle_H. \tag{58}$$

We find the following estimate for the approximation error.

Corollary 1 *There is a constant $C > 0$ (depending on T) such that*

$$E\left|A_t - A_t^d\right| \leq C\,\|w\|_H \left(\sum_{l=d+1}^{\dim H} \mu_l\right)^{\frac{1}{2}} \tag{59}$$

for every $t \in [0, T]$.

Proof By definition, $A_0^d = A_0$. For $t > 0$, we have

$$\begin{aligned} E\left|A_t - A_t^d\right| &= \frac{1}{t} E\left|\int_0^t \langle w, S_u(Z_u) - S_u(\mathcal{P}_d Z_u)\rangle_H du\right| \\ &\leq \frac{1}{t} E\left[\int_0^t |\langle w, S_u(Z_u) - S_u(\mathcal{P}_d Z_u)\rangle_H|\, du\right]. \end{aligned} \tag{60}$$

Using Fubini's theorem and applying Theorem 2 yields

$$\begin{aligned} E\left|A_t - A_t^d\right| &\leq \frac{1}{t}\int_0^t E\,|\langle w, S_u(Z_u) - S_u(\mathcal{P}_d Z_u)\rangle_H|\, du \\ &\leq \frac{1}{t}\int_0^t C\,\|w\|_H \left(\sum_{l=d+1}^{\dim H} \mu_l\right)^{\frac{1}{2}} du. \end{aligned} \tag{61}$$

Since the integrand does no longer depend on the integration variable u, the proof is complete. □

As before, we obtain an t-dependent estimate for the approximation error in the time-homogeneous case.

Corollary 2 *Suppose that Z is a time-homogeneous jump-diffusion process. Then there is a constant $C > 0$ (depending on T) such that*

$$E\left|A_t - A_t^d\right| \leq C\,\|w\|_H \sqrt{\frac{t}{T}} \left(\sum_{l=d+1}^{\dim H} \mu_l\right)^{\frac{1}{2}} \tag{62}$$

for every $t \in [0, T]$.

Proof We apply Proposition 4 to Eq. (56) in the proof of 2 to obtain

$$E\,|\langle w, S_t(Z_t)\rangle_H - \langle w, S_t(\mathcal{P}_d Z_t)\rangle_H| \leq C\,\|w\|_H\, \|S_0\|_H \sqrt{\frac{t}{T}} \left(\sum_{l=d+1}^{\dim H} \mu_l\right)^{\frac{1}{2}}. \tag{63}$$

We proceed as in the proof of Corollary 1 and find

$$E\left|A_t - A_t^d\right| \leq \frac{1}{t}\int_0^t \sqrt{\frac{u}{T}} C \|w\|_H \left(\sum_{l=d+1}^{\dim H} \mu_l\right)^{\frac{1}{2}} du. \tag{64}$$

Since

$$\frac{1}{t}\int_0^t \sqrt{\frac{u}{T}}\, du = \frac{2}{3}\sqrt{\frac{t}{T}}, \tag{65}$$

the proof is complete. □

3.3 Approximate Pricing

In the previous sections, we have seen how to approximate the processes on which the payoff G of the Asian option depends, the centered process Z, and the average A. Now, we use these results to find a finite-dimensional approximation of the discounted option value $\widehat{V}$. For $t \in [0, T]$, we define

$$\widehat{V}_d(t, z, a) := e^{-rT} E[G(\mathcal{P}_d Z_T, A_T^d) | \mathcal{P}_d Z_t = z, A_t^d = a] \quad \text{for every } z \in U_d, a \in \mathbb{R}. \tag{66}$$

In contrast to the definition of $\widehat{V}$ in (21), the payoff is applied to the projected random variables $\mathcal{P}_d Z_T$ and A_T^d here instead of Z_T and A_T. Thus, $\widehat{V}_d$ is defined on the finite dimensional domain $[0, T] \times U_d \times \mathbb{R}$ which allows for numerical discretization. Similar to Theorem 27, we find that $\widehat{V}_d$ satisfies a PIDE. The PIDE is finite-dimensional.

Theorem 3 *Suppose that the approximated option value $\widehat{V}_d$ defined in* (66) *is continuously differentiable with respect to t and twice continuously differentiable with respect to z and a. Then $\widehat{V}_d$ is a classical solution of the PIDE*

$$\begin{aligned}
-\partial_t \widehat{V}_d(t, z, a) = {} & \frac{1}{2}\sum_{i,j=1}^{d} c_{ij}(t)\, \partial_i \partial_j \widehat{V}_d(t, z, a) + \frac{1}{t}\left(\langle w, S_t(z)\rangle_H - a\right) \partial_a \widehat{V}(t, z, a) \\
& + \int_H \Big[\widehat{V}_d(t, z + \mathcal{P}_d \eta_t(\zeta), a) - \widehat{V}_d(t, z, a) \\
& - \sum_{i=1}^{d} \langle \eta_t(\zeta), p_i\rangle_H \partial_i \widehat{V}_d(t, z, a)\Big] \nu(d\zeta),
\end{aligned} \tag{67}$$

with time-dependent coefficients

$$c_{ij}(t) := \left\langle \sigma_t Q \sigma_t^* p_i, p_j \right\rangle_H, \quad i, j = 1, \ldots, d, \tag{68}$$

and terminal condition

$$\widehat{V}_d(T, z, a) = e^{-rT} G(z, a) \tag{69}$$

for a.e. $t \in (0, T)$, $z \in U_d$, *and* $a \in \mathbb{R}$.

Proof This can be shown along the very same lines as in the proof of Theorem 1. The main difference is that we make use of a finite-dimensional version of Itô's formula (see, e.g., [4, Prop. 8.19]). This yields finite sums of second derivatives instead of the trace operator. □

The value of the Asian option at time $t = 0$ is given by $\widehat{V}(0, 0, \langle w, S_0\rangle_H)$, since $Z_0 = 0 \in H$ and $A_0 = \langle w, S_0\rangle_H \in \mathbb{R}$ by definition. The solution of the finite-dimensional PIDE yields $\widehat{V}_d(0, 0, \langle w, S_0\rangle_H)$. The following theorem states an upper bound of the approximation error for the option value.

Theorem 4 *There is a constant $C > 0$ (depending on T) such that the difference of the true Asian option price and its finite dimensional approximation satisfies*

$$\left|\widehat{V}(0, 0, \langle w, S_0\rangle_H) - \widehat{V}_d(0, 0, \langle w, S_0\rangle_H)\right| \le C \left(\sum_{l=d+1}^{\dim H} \mu_l\right)^{\frac{1}{2}}. \tag{70}$$

Proof We start with the definition of $\widehat{V}$ and $\widehat{V}_d$ and make use of Assumption 5 to find

$$\begin{aligned}
&\left|\widehat{V}(0, 0, A_0) - \widehat{V}_d(0, 0, A_0^d)\right| \\
&\quad = e^{-rT} \left|E[G(Z_T, A_T) - E[G(\mathcal{P}_d Z_T, A_T^d)]\right| \\
&\quad \le e^{-rT} E\left[L_z^G \|Z_T - \mathcal{P}_d Z_T\|_H + L_a^G \left|A_T - A_T^d\right|\right] \\
&\quad \le e^{-rT} \max\{L_z^G, L_a^G\} \left(E \|Z_T - \mathcal{P}_d Z_T\|_H + E\left|A_T - A_T^d\right|\right).
\end{aligned} \tag{71}$$

With the Cauchy–Schwarz inequality, we get

$$\left|\widehat{V}(0, 0, A_0) - \widehat{V}_d(0, 0, A_0^d)\right| \le C \left(\left(E \|Z_T - \mathcal{P}_d Z_T\|_H^2\right)^{\frac{1}{2}} + E\left|A_T - A_T^d\right|\right). \tag{72}$$

Applying Proposition 2 to $E \|Z_T - \mathcal{P}_d Z_T\|_H^2$ and Corollary 1 to $E\left|A_T - A_T^d\right|$ completes the proof. □

The theorem shows that we can achieve a good approximation, if the right-hand side of (70) is small. In practice, we can first compute the eigenvalues μ_l, $l = 1, 2, \ldots$, and then decide how many POD components we have to include in the projection in order to satisfy a given absolute tolerance.

For the discretization of the PIDE (67), sparse grid methods similar to those presented in [10] may be suitable. The nonlocal integral terms, which are due to the jumps in the model, can be discretized using a Galerkin approach with a wavelet basis [12]. The POD method in combination with sparse grids was already shown to be a promising approach to break the "curse of dimension" in the case of European options.

There is, however, an additional numerical difficulty when dealing with Asian options. The fact that there is no diffusion in the variable a representing the average requires special attention. Equations of this kind are often termed "degenerate parabolic" PIDEs. A large number of authors has dealt with such problems, see, e.g., [1–3, 19] and the references therein. Since the dimension reduced equation is finite-dimensional, the numerical schemes and convergence result presented there can be applied directly. These include, e.g., flux limiting methods, operator splitting, and difference-quadrature methods. Numerical experiments concerning the presented PIDE for Asian options will be a topic for future research.

References

1. Barles, G.: Convergence of numerical schemes for degenerate parabolic equations arising in finance theory. In: Numerical Methods in Finance, Publ. Newton Inst, pp. 1–21. Cambridge Univ. Press, Cambridge (1997)
2. Biswas, I.H., Jakobsen, E.R., Karlsen, K.H.: Difference-quadrature schemes for nonlinear degenerate parabolic integro-PDE. SIAM J. Numer. Anal. **48**(3), 1110–1135 (2010)
3. Briani, M., Chioma, C.L., Natalini, R.: Convergence of numerical schemes for viscosity solutions to integro-differential degenerate parabolic problems arising in financial theory. Numer. Math. **98**, 607–646 (2004)
4. Cont, R., Tankov, P.: Financial Modelling with Jump Processes. Chapman & Hall, Boca Raton (2004)
5. Da Prato, G., Zabczyk, J.: Stochastic Equations in Infinite Dimensions. Cambridge University Press, Cambridge (1992)
6. d'Halluin, Y., Forsyth, P., Labahn, G.: A semi-Lagrangian approach for American Asian options under jump diffusion. SIAM J. Sci. Comput. **27**, 315–345 (2005)
7. Hausenblas, E.: A note on the Itô formula of stochastic integrals in Banach spaces. Random Oper. Stoch. Equ. **14**(1), 45–58 (2006)
8. Hepperger, P.: Hedging electricity swaptions using partial integro-differential equations. Stoch. Process. Appl. **122**, 600–622 (2012)
9. Hepperger, P.: Numerical hedging of electricity swaptions using dimension reduction. Int. J. Theor. Appl. Finance **15**(6), 1250042 (2012). (pp. 26)
10. Hepperger, P.: Option pricing in Hilbert space valued jump-diffusion models using partial integro-differential equations. SIAM J. Financ. Math. **1**, 454–489 (2010)
11. Kunita, H.: Stochastic integrals based on martingales taking values in Hilbert space. Nagoya Math. J. **38**, 41–52 (1970)
12. Matache, A.M., von Petersdorff, T., Schwab, C.: Fast deterministic pricing of options on Lévy-driven assets. Math. Model. Numer. Analysis **38**(1), 37–71 (2004)
13. Peszat, S., Zabczyk, J.: Stochastic Partial Differential Equations with Lévy Noise: An Evolution Equation Approach. Cambridge University Press, Cambridge (2007)
14. Protter, P.E.: Stochastic Integration and Differential Equations, 2nd edn. Springer, Berlin (2005)
15. Rogers, L.C.G., Shi, Z.: The value of an Asian option. J. Appl. Probab. **32**, 1077–1088 (1995)

16. Sato, K.: Lévy Processes and Infinitely Divisible Distributions. Cambridge University Press, Cambridge (1999)
17. Schoutens, W.: Exotic options under Lévy models: An overview. Tech. rep., UCS 2004-06, Leuven, Belgium. http://perswww.kuleuven.ac.be/~u0009713/Schout04.pdf
18. Věcěr, J., Xu, M.: Pricing Asian options in a semimartingale model. Quant. Finance **4**, 170–175 (2004)
19. Zvan, R., Forsyth, P., Vetzal, K.: Robust numerical methods for PDE models of Asian options. J. Comput. Finance **1**, 39–78 (1998)

A Time Before Which Insiders Would not Undertake Risk

Constantinos Kardaras

Abstract A continuous-path semimartingale market model with wealth processes discounted by a riskless asset is considered. The numéraire portfolio is the unique strictly positive wealth process that, when used as a benchmark to denominate all other wealth, makes all wealth processes local martingales. It is assumed that the numéraire portfolio exists and that its wealth increases to infinity as time goes to infinity. Under this setting, an initial enlargement of the filtration is performed, by including the overall minimum of the numéraire portfolio. It is established that all nonnegative wealth processes, when stopped at the time of the overall minimum of the numéraire portfolio, become local martingales in the enlarged filtration. This implies that risk-averse insider traders would refrain from investing in the risky assets before that time. A partial converse to the previous result is also established in the case of complete markets, showing that the time of the overall minimum of the numéraire portfolio is in a certain sense unique in rendering undesirable the act of undertaking risky positions before it. The aforementioned results shed light to the importance of the numéraire portfolio as an indicator of overall market performance.

Keywords Semimartingale market model · Initial enlargement of the filtration · Numéraire portfolio

Mathematics Subject Classification (2010) 90G20

1 Introduction

When modeling insider trading, one usually enlarges the "public" information flow by including knowledge of a non-trivial random variable, which represents the extra information of the insider, from the very beginning. (This method called *initial filtration enlargement*, as opposed to *progressive filtration enlargement*—for more details, see [11, Chap. VI].) It is then of interest to explore the effect that the extra

C. Kardaras (✉)
Department of Statistics, London School of Economics, 10 Houghton street, London, WC2A 2AE, UK
e-mail: k.kardaras@lse.ac.uk

Y. Kabanov et al. (eds.), *Inspired by Finance*, DOI 10.1007/978-3-319-02069-3_16,

information has on the trading behavior of the insider—for an example, see [1]. Under this light, the topic of the present paper may be considered slightly unorthodox, as we identify an initial filtration enlargement and a stopping time of the enlarged filtration (which is *not* a stopping time of the original filtration) with the property that risk-averse insider traders would refrain from taking risky positions before that time. As will be revealed, this apparently "negative" result, though not helpful in the theory of insider trading, sheds more light to the importance of a specific investment opportunity, namely, the numéraire portfolio.

Our setting is a continuous-path semimartingale market model with d asset-price processes $S^1, \dots, S^d$. All wealth is discounted with respect to some locally riskless asset. Natural structural assumptions are imposed—in particular, we only enforce a mild market viability condition, and allow for the existence of some discounted wealth process that will grow unconditionally as time goes to infinity. Such assumptions are satisfied in every reasonable infinite time-horizon model. In such an environment, the numéraire portfolio—an appellation coined in [9]—is the unique nonnegative wealth process $\widehat{X}$ with unit initial capital such that all processes $S^i / \widehat{X}$ become local martingales. The numéraire portfolio has several interesting optimality properties. For instance, it maximizes expected logarithmic utility for all time-horizons and achieves maximal long-term growth—for more information, check [6]. The goal of the present paper is to add yet one more to the remarkable list of properties of the numéraire portfolio.

The original filtration $\mathbf{F}$ is enlarged to $\mathbf{G}$, which further contains information on the overall minimum level $\min_{t \in \mathbb{R}_+} \widehat{X}(t)$ of the numéraire portfolio. In particular, the time ρ that this overall minimum is achieved (which can be shown to be almost surely unique) becomes a stopping time with respect to $\mathbf{G}$. Our first main result states that all S^i become local martingales up to time ρ under the *enlarged filtration* $\mathbf{G}$ and *original probability* $\mathbb{P}$. Note that the asset-price processes are discounted by the locally riskless wealth process, and not by the numéraire portfolio. (The latter discounting makes asset price-processes local martingales under $(\mathbf{F}, \mathbb{P})$, while the former discounting makes asset price-processes, when stopped at ρ, local martingales under $(\mathbf{G}, \mathbb{P})$.) In essence, $\mathbb{P}$ becomes a risk-neutral measure for the model with enlarged filtration up to time ρ. An immediate consequence of this fact is that a risk-averse investor would refrain from taking risky positions up to time ρ, since they would result in no compensation for the risk that is being undertaken, in terms of excess return relative to the riskless account. (Note, however, that an insider can arbitrage unconditionally after time ρ with *no* downside risk whatsoever involved, simply by taking arbitrarily large long positions in the numéraire portfolio immediately after ρ.) In effect, trading in the market occurs simply because traders do not have information about the time of the overall minimum of the numéraire portfolio. In fact, until time ρ, not only the numéraire portfolio, but the whole market performs badly, since the expected outcome of any portfolio at time ρ is necessarily less or equal than the initial capital used to set it up.

A partial converse to the previous result is also presented. Under an extra completeness assumption on the market, it is shown that if a random time ϕ (satisfying a couple of technical properties) is such that $\mathbb{E}X(\phi) \leq X(0)$ holds for any nonnegative wealth process X formed by trading with information $\mathbf{F}$, then ϕ is necessarily equal to the time of the overall minimum of the numéraire portfolio. Combined with our first main result, this clarifies the unique role of the numéraire portfolio as an indicator of overall market performance.

The structure of the remainder of the paper is simple. In Sect. 2 the results are presented, while Sect. 3 contains the proofs.

2 Results

2.1 The Set-up

Let $(\Omega, \mathscr{F}, \mathbf{F}, \mathbb{P})$ be a filtered probability space—here, $(\Omega, \mathscr{F}, \mathbb{P})$ is a complete probability space and $\mathbf{F} = (\mathscr{F}(t))_{t \in \mathbb{R}_+}$ is a right-continuous filtration such that $\mathscr{F}(t) \subseteq \mathscr{F}$ and $\mathscr{F}(t)$ contains all $\mathbb{P}$-null sets of $\mathscr{F}$—in other words, $\mathbf{F}$ satisfies the *usual conditions*. Without affecting in any way the generality of our discussion, we shall be assuming that $\mathscr{F}(0)$ is trivial modulo $\mathbb{P}$. Relationships involving random variables are to be understood in the $\mathbb{P}$-a.s. sense; relationships involving processes hold modulo evanescence.

On $(\Omega, \mathbf{F}, \mathbb{P})$, let $S = (S^i)_{i=1,\ldots,d}$ be a vector-valued semimartingale with continuous paths. The component S^i represents the discounted, with respect to some baseline security, price of the ith liquid asset in the market. The baseline security, which we shall simply call *discounting process*, should be thought as a locally riskless account. In contrast, the other assets are supposed to represent riskier investments. We also set $S^0 := 1$ to denote the wealth accumulated by the baseline locally riskless security, discounted by itself.

Starting with initial capital $x \in \mathbb{R}_+$, and investing according to some d-dimensional, $\mathbf{F}$-predictable and S-integrable strategy ϑ modeling the number of liquid assets held in the portfolio, an economic agent's discounted wealth is given by $X^{x,\vartheta} = x + \int_0^{\cdot} \vartheta^\top(t) \mathrm{d}S(t)$. Define $\mathscr{X}_{\mathbf{F}}(x)$ as the set of all processes $X^{x,\vartheta}$ in the previous notation that remain nonnegative at all times. Furthermore, we set $\mathscr{X}_{\mathbf{F}} := \bigcup_{x \in \mathbb{R}_+} \mathscr{X}_{\mathbf{F}}(x)$.

Below, we gather some definitions and results that have appeared previously in the literature. More information about them can be found in [6] and, for the special case of continuous-path semimartingales that is considered here, in [7, Sect. 4].

Definition 1 We shall say that the market allows for *arbitrage of the first kind* if there exists $T \in \mathbb{R}_+$ and an $\mathscr{F}(T)$-measurable random variable $\xi \geq 0$, $\xi \neq 0$, such that for all $x > 0$ one can find $X \in \mathscr{X}(x)$ with $X(T) \geq \xi$. If the market does not allow for any arbitrage of the first kind, we say that condition NA_1 holds.

Condition NA_1 is weaker than the "No Free Lunch with Vanishing Risk" market viability condition of [2], and is actually equivalent to the requirement that $\lim_{\ell \to \infty} \sup_{X \in \mathscr{X}^{\mathbf{F}}(x)} \mathbb{P}[X(T) > \ell] = 0$ holds for all $x \in \mathbb{R}_+$ and $T \in \mathbb{R}_+$—see [7,

Proposition 1]. The latter boundedness-in-probability requirement is coined condition BK in [5] and condition "No Unbounded Profit with Bounded Risk" (NUPBR) in [6].

Definition 2 A strictly positive local martingale deflator is a strictly positive process Y with $Y(0) = 1$ such that YS^i is a local martingale on $(\Omega, \mathbf{F}, \mathbb{P})$ for all $i \in \{0, \dots, d\}$. (The last requirement is equivalent to asking that YX is a local martingale on $(\Omega, \mathbf{F}, \mathbb{P})$ for all $X \in \mathscr{X}_{\mathbf{F}}$.) A strictly positive process $\widehat{X} \in \mathscr{X}_{\mathbf{F}}(1)$ will be called *the numéraire portfolio* if $\widehat{Y} := 1/\widehat{X}$ is a (necessarily, strictly positive) local martingale deflator.

By Jensen's inequality, it is straightforward to see that if the numéraire portfolio $\widehat{X}$ exists, then it is unique. Obviously, if the numéraire portfolio exists then at least one strictly positive local martingale deflator exists in the market. Interestingly, the converse also holds, i.e., existence of the numéraire portfolio is equivalent to existence of at least one strictly positive local martingale deflator. Furthermore, the previous are also equivalent to condition NA_1 holding in the market.

Condition NA_1 can also be described in terms of the asset-prices process drifts and volatilities. More precisely, let $A = (A^1, \dots, A^d)$ be the continuous-path finite-variation process appearing in the Doob-Meyer decomposition of the continuous-path semimartingale S. For $i, k \in \{1, \dots, d\}$, denote by $[S^i, S^k]$ the quadratic (co)variation of S^i and S^k. Also, let $[S, S]$ be the $d \times d$ nonnegative-definite symmetric matrix-valued process whose (i, k)-entry is $[S^i, S^k]$. Call now $G := \mathsf{trace}[S, S]$, where trace is the operator returning the trace of a matrix. Observe that G is an increasing, adapted, continuous process, and that there exists a $d \times d$ nonnegative-definite symmetric matrix-valued process c such that $[S^i, S^k] = \int_0^\cdot c^{i,k}(t)\mathrm{d}G(t)$ for $i \in \{1, \dots, d\}$ and $k \in \{1, \dots, d\}$; $[S, S] = \int_0^\cdot c(t)\mathrm{d}G(t)$ in short. Then, condition NA_1 is equivalent to the existence of a d-dimensional, predictable process ξ such that $A = \int_0^\cdot (c(t)\xi(t))\mathrm{d}G(t)$, satisfying $\int_0^T (\xi^\top(t)c(t)\xi(t))\mathrm{d}G(t) < \infty$ for all $T \in \mathbb{R}_+$. In fact, with the previous notation, it can be checked that the numéraire portfolio is given by $\widehat{X} = \mathscr{E}(\int_0^\cdot \xi^\top(t)\mathrm{d}S(t))$, where "$\mathscr{E}$" denotes the stochastic exponential operator.

Definition 3 We shall say that *the discounting process is asymptotically suboptimal* if there exists $X \in \mathscr{X}_{\mathbf{F}}$ such that $X(t) \to \infty$ $\mathbb{P}$-a.s. as $t \to \infty$.

The previous definition is self-explanatory—the locally riskless discounting process (which is used as a baseline to denominate all other wealth) is asymptotically suboptimal if it can be beaten unconditionally in the long run by some other wealth process in the market. As a simple example where the discounting process is asymptotically suboptimal, we mention any multi-dimensional Black-Scholes model such that the probability $\mathbb{P}$ is not a risk-neutral one.

Given the existence of the numéraire portfolio $\widehat{X}$ (i.e., under the validity of condition NA_1), the discounting process is asymptotically suboptimal if and only if

$\widehat{X}(t) \to \infty$ $\mathbb{P}$-a.s. Indeed, if there exists some $X \in \mathscr{X}_{\mathbf{F}}$ such that $X(t) \to \infty$ $\mathbb{P}$-a.s., the supermartingale property of $X/\widehat{X}$ and Doob's nonnegative supermartingale convergence theorem give $\widehat{X}(t) \to \infty$ $\mathbb{P}$-a.s. Furthermore, under condition NA_1, and with the notation used in the paragraph right before Definition 3, it can be checked that the discounting process is asymptotically suboptimal if and only if $\int_0^\infty (\xi^\top(t)c(t)\xi(t))\mathrm{d}G(t) = \infty$.

2.2 The First Result

For the purposes of Sect. 2.2, assume that condition NA_1 holds in the market and the discounting process is asymptotically suboptimal. Recall that this is equivalent to existence of the numéraire portfolio $\widehat{X}$, which satisfies $\widehat{X}(t) \to \infty$ $\mathbb{P}$-a.s.

Define the nonincreasing process $I := \inf_{t \in [0,\cdot]} \widehat{X}(t)$; then the random variable $I(\infty) = \inf_{t \in \mathbb{R}_+} \widehat{X}(t)$ is the overall minimum of $\widehat{X}$. Let $\mathbf{G} = (\mathscr{G}(t))_{t \in \mathbb{R}_+}$ be the smallest filtration satisfying the usual hypotheses, containing $\mathbf{F}$, and making $I(\infty)$ a $\mathscr{G}(0)$-measurable random variable. Consider any random time ρ such that

$$\widehat{X}(\rho) = \inf_{t \in \mathbb{R}_+} \widehat{X}(t) = I(\infty).$$

This means that $\widehat{X}$ achieves at ρ its overall minimum. Since $\widehat{X}(t) \to \infty$ $\mathbb{P}$-a.s., such a time is $\mathbb{P}$-a.s. finite. In fact, it is also $\mathbb{P}$-a.s. unique, as will be revealed in Theorem 1 below. Therefore, $\mathbb{P}$-a.s., $\rho = \inf\{t \in \mathbb{R}_+ \mid \widehat{X}(t) = I(\infty)\}$, the latter being a stopping time on $(\Omega, \mathbf{G})$; since $\mathscr{G}(0)$ contains all $\mathbb{P}$-null sets of $\mathscr{F}$, it follows that ρ is a stopping time on $(\Omega, \mathbf{G})$. Therefore, $\mathbf{G}$ is strictly larger than the smallest filtration that satisfies the usual hypotheses, contains $\mathbf{F}$, and makes ρ a stopping time.

Theorem 1 *Assume that condition* NA_1 *holds and that the discounting process is asymptotically suboptimal. Then, the time of minimum of* $\widehat{X}$ *is* $\mathbb{P}$*-a.s. unique. With* ρ *denoting such a time, the process* $S^\rho = (S(\rho \wedge t))_{t \in \mathbb{R}_+}$ *is a local martingale on* $(\Omega, \mathbf{G}, \mathbb{P})$.

Remark 1 The result of Theorem 1 does not appear to follow directly from well known results in the theory of filtration enlargements. In particular:

- A widely used sufficient condition that enables the use of the theory of initial filtration enlargements is the so-called *Jacod's criterion* [3], which states that the conditional law of the random variable $I(\infty)$ given $\mathscr{F}(t)$ is absolutely continuous with respect to its unconditional law for all $t \in \mathbb{R}_+$. However, the conditional law of $I(\infty)$ given $\mathscr{F}(t)$ has a Dirac component of mass $1 - I(t)\widehat{Y}(t)$ at the point $I(t)$, as follows from Doob's maximal identity ([10, Lemma 2.1]—see also the beginning of Sect. 3), while the unconditional law of $I(\infty)$ is standard uniform (this is proved in Sect. 3). Therefore, Jacod's criterion fails.

- The Jeulin–Yor semimartingale decomposition result (see [4]) cannot be utilized, because this is not a case of progressive filtration enlargement. Furthermore, as already noted, the filtration $\mathbf{G}$ is strictly larger than the smallest filtration that satisfies the usual hypotheses, contains $\mathbf{F}$, and makes ρ a stopping time.

One could use the general results of [10, Sect. 3] in order to establish the validity of Theorem 1. Here, we provide a simple, self-contained alternative proof, in the course of which the concepts of local martingale deflators and martingale measures will play an important role.

Remark 2 Theorem 1 justifies the title of the paper. With the insider information flow $\mathbf{G}$, investing in the risky assets before time ρ gives the same instantaneous return as the locally riskless asset, but entails (locally) higher risk; therefore, before ρ an insider would not be willing to take any position on the risky assets. One can make the point more precise. Let $\mathscr{X}_{\mathbf{G}}^{\rho}$ be the class of nonnegative processes of the form $x + \int_0^\cdot \vartheta^\top(t) \mathrm{d}S^\rho(t)$, where now x is $\mathscr{G}(0)$-measurable and ϑ is $\mathbf{G}$-predictable and S^ρ-integrable. By Theorem 1, all processes in $\mathscr{X}_{\mathbf{G}}^{\rho}$ are nonnegative local martingales on $(\Omega, \mathbf{G}, \mathbb{P})$, which implies that they are nonnegative supermartingales on $(\Omega, \mathbf{G}, \mathbb{P})$. Therefore, $\mathbb{E}[X(\rho) \mid I(\infty)] \le X(0)$ holds for all $X \in \mathscr{X}_{\mathbf{G}}^{\rho}$. (In particular, $\mathbb{E}X(\rho) \le X(0)$ holds for all $X \in \mathscr{X}_{\mathbf{F}}$, which sharpens the conclusion of [8, Theorem 2.15] for continuous-path semimartingale models.) Jensen's inequality then implies that any expected utility maximizer having an increasing and concave utility function, information flow $\mathbf{G}$, and time-horizon before ρ, would not take any position in the risky assets.

Remark 3 At first sight, Theorem 1 appears counterintuitive. If the overall minimum of $\widehat{X}$ is known from the outset exactly, and especially if it is going to be extremely low, taking an opposite (short) position in it should ensure particularly good performance at the time of the overall minimum of $\widehat{X}$. Of course, admissibility constraints prevent one from taking an *absolute* short position on the numéraire portfolio; still, one can imagine that a *relative* short position on the numéraire portfolio should result in something substantial. To understand better why this intuition fails, remember that $\widehat{X} = \mathscr{E}(\int_0^\cdot \xi^\top(t) \mathrm{d}S(t))$ in the notation of Sect. 2.1, which was noted in the discussion before Definition 3. A relative short position would result in the wealth $X = \mathscr{E}(-\int_0^\cdot \xi^\top(t) \mathrm{d}S(t))$. Straightforward computations show that

$$X(\rho) = \frac{1}{\widehat{X}(\rho)} \exp\left\{ -\int_0^\rho \left(\xi^\top(t) c(t) \xi(t) \right) \mathrm{d}G(t) \right\}.$$

From $\widehat{X} = \mathscr{E}(\int_0^\cdot \xi^\top(t) \mathrm{d}S(t))$, it follows that the term $\int_0^\rho (\xi^\top(t) c(t) \xi(t)) \mathrm{d}G(t)$ is the integrated squared volatility of the numéraire portfolio up to time ρ. Even when $\widehat{X}(\rho)$ is close to zero, the term $\exp\{-\int_0^\rho (\xi^\top(t) c(t) \xi(t)) \mathrm{d}G(t)\}$ will compensate for the very small values of $\widehat{X}(\rho)$. In effect, the integrated squared volatility of the numéraire portfolio up to the time of its overall minimum will eliminate any chance of profit by taking short positions in it.

2.3 A Partial Converse to Theorem 1

In Remark 2, it was argued that $\mathbb{E}X(\rho) \le X(0)$ holds for all $X \in \mathscr{X}_{\mathbf{F}}$. A partial converse of the previous result will be presented now. Before stating the result, some definitions are needed.

Definition 4 Consider a market as described in Sect. 2.1, satisfying condition NA_1. The market will be called *complete* if for any stopping time τ and any $\mathscr{F}_\tau$-measurable nonnegative random variable H_τ with $\mathbb{E}\widehat{Y}_\tau H_\tau < \infty$, there exists $X \in \mathscr{X}_{\mathbf{F}}$ such that $X_\tau = H_\tau$.

Remark 4 A market as described in Sect. 2.1 satisfies condition NA_1 *if and only if* there exists at least one strictly positive supermartingale deflator. It can be actually shown that the market is further complete in the sense of Definition 4 *if and only if* there exists a unique strictly positive supermartingale deflator. The proof is similar to the one for the case where an equivalent martingale measure exists in the market—one has to utilize results on optional decomposition under the assumption that a strictly positive local martingale deflator (but not necessarily an equivalent martingale measure) exists in the market; such results are presented in [12]. In fact, it can be further shown that in a complete market, for any stopping time τ and $\mathscr{F}_\tau$-measurable nonnegative random variable H_τ, one has

$$\mathbb{E}\widehat{Y}_\tau H_\tau = \min\{x \in \mathbb{R}_+ \mid \text{ there exists } X \in \mathscr{X}_{\mathbf{F}}(x) \text{ with } X_\tau = H_\tau\},$$

which gives a formula for the minimal hedging price of the payoff H_τ delivered at time τ.

Definition 5 Let ϕ be a random time on $(\Omega, \mathbf{F}, \mathbb{P})$. If $\mathbb{P}[\phi = \tau] = 0$ holds for all stopping times τ on $(\Omega, \mathbf{F})$, we shall say that ϕ *avoids all stopping times on* $(\Omega, \mathbf{F}, \mathbb{P})$. Furthermore, ϕ will be called an honest time on $(\Omega, \mathbf{F})$ if for all $t \in \mathbb{R}_+$ there exists an $\mathscr{F}_t$-measurable random variable ϕ_t such that $\phi = \phi_t$ holds on $\{\phi \le t\}$.

As it turns out (and will come as an immediate consequence of Theorem 2 below), the random time ρ defined in Sect. 2.2 is an honest time that avoids all stopping times on $(\Omega, \mathbf{F}, \mathbb{P})$. The next result states that, if the market is viable and complete, ρ is the *unique* honest time that avoids all stopping times on $(\Omega, \mathbf{F}, \mathbb{P})$, with the property that a wealth processes sampled at this random time has expectation dominated by its initial capital.

Theorem 2 *Assume that condition* NA_1 *holds and that the market is complete. Let* ϕ *be an honest time that avoids all stopping times on* $(\Omega, \mathbf{F}, \mathbb{P})$, *such that* $\mathbb{E}X(\phi) \le X(0)$ *holds for all* $X \in \mathscr{X}_{\mathbf{F}}$. *Then, the discounting process is asymptotically suboptimal and* $\phi = \rho$.

Remark 5 An inspection of the proof of Theorem 1 shows that, under its assumptions, whenever ϕ is the time of maximum of a continuous-path local martingale deflator (which is an honest time that avoids all stopping times), $\mathbb{E}X(\phi) \leq X(0)$ holds for all $X \in \mathscr{X}_{\mathbf{F}}$. Therefore, if the market is incomplete, in which case there exist more than one local martingale deflators, the result of Theorem 2 is no longer valid.

Furthermore, since the honest time $\phi = 0$ is such that $\mathbb{E}X(\phi) \leq X(0)$ trivially holds for all $X \in \mathscr{X}_{\mathbf{F}}$, the assumption that ϕ avoids all stopping times on $(\Omega, \mathbf{F}, \mathbb{P})$ cannot be avoided in the statement of Theorem 2. It is less clear how essential the assumption that ϕ is an honest time is. No immediate counterexample comes to mind, although it is quite possible that one exists. Note, however, that ϕ being an honest time is instrumental in the proof of Theorem 2; therefore, further investigation of this issue is not undertaken.

3 Proofs

In the course of the proofs below, we shall use the so-called Doob's maximal identity, which we briefly recall for the reader's convenience. If M is a continuous-path nonnegative local martingale on $(\Omega, \mathbf{F}, \mathbb{P})$ such that $\lim_{t \to \infty} M_t = 0$, $\mathbb{P}$-a.s. holds, then, with $M^* := \max_{t \in [0, \cdot]} M_t$ and ρ^M denoting *any* time of maximum of M, one has the equality $\mathbb{P}[\rho^M > \tau \mid \mathscr{F}_\tau] = M_\tau / M^*_\tau$ whenever τ is a finite stopping time on $(\Omega, \mathbf{F})$. Doob's maximal identity can be shown by applying Doob's optional sampling theorem. For a proof of the identity presented above, see [10, Lemma 2.1].

3.1 Proof of Theorem 1

We shall first show that ρ is $\mathbb{P}$-a.s. unique. Define the random times

$$\rho' := \inf\{t \in \mathbb{R}_+ \mid \widehat{X}(t) = I(\infty)\}$$

and

$$\rho'' := \sup\{t \in \mathbb{R}_+ \mid \widehat{X}(t) = I(\infty)\}.$$

Since $\widehat{Y} := 1/\widehat{X}$ a nonnegative local martingale that vanishes at infinity on $(\Omega, \mathbf{F}, \mathbb{P})$, Doob's maximal identity implies that

$$\mathbb{P}[\rho' > t \mid \mathscr{F}(t)] = \mathbb{P}[\rho'' > t \mid \mathscr{F}(t)] = I(t)\widehat{Y}(t) \quad \forall t \in \mathbb{R}_+.$$

The previous imply that ρ' and ρ'' have the same law under $\mathbb{P}$. Since $\rho' \leq \rho''$, it follows that $\rho' = \rho''$. Furthermore, since for any time ρ of minimum of $\widehat{X}$ (which is a time of maximum of $\widehat{Y}$) we have $\rho' \leq \rho \leq \rho''$, it follows that the time of minimum of $\widehat{X}$ is $\mathbb{P}$-a.s. unique.

For all $u \in [0, 1)$ define $\eta_u := \inf\{t \in \mathbb{R}_+ \mid \widehat{Y}(t) = 1/(1-u)\}$; then, $(\eta_u)_{u \in [0,1)}$ is a nondecreasing collection of stopping times on $(\Omega, \mathbf{F})$. Recall that $\widehat{Y}$ is a nonnegative local martingale on $(\Omega, \mathbf{F}, \mathbb{P})$ such that $\widehat{Y}(0) = 1$ and $\widehat{Y}(t) \to 0$ $\mathbb{P}$-a.s., $t \to \infty$. Also, $1/I = \sup_{t \in [0,\cdot]} \widehat{Y}(t)$ and $I(\rho) = I(\infty)$. By the definition of $(\eta_u)_{u \in [0,1)}$, $\widehat{Y}^{\eta_u}$ is a uniformly bounded martingale on $(\Omega, \mathbf{F}, \mathbb{P})$ with terminal value

$$\widehat{Y}^{\eta_u}_\infty = \widehat{Y}_{\eta_u} = 1/(1-u)\mathbb{I}_{\{\eta_u < \infty\}}.$$

In particular, Doob's optional sampling theorem gives $\mathbb{P}[\eta_u < \infty] = 1 - u$; therefore, $I(\infty)$ has the standard uniform distribution under $\mathbb{P}$ since

$$\mathbb{P}[I(\infty) \leq 1-u] = \mathbb{P}\left[\eta_u < \infty\right] = 1-u, \quad u \in [0, 1).$$

For $u \in [0, 1)$, let $\mathbb{P}_u$ be the probability $\mathbb{P}$ on $(\Omega, \mathscr{F})$ conditioned on $\{\eta_u < \infty\}$; of course, $\mathbb{P}_u$ is absolutely continuous with respect to $\mathbb{P}$. From the discussion above, $\mathrm{d}\mathbb{P}_u/\mathrm{d}\mathbb{P} = (1/(1-u))\mathbb{I}_{\{\eta_u<\infty\}} = \widehat{Y}(\eta_u)$ holds for all $u \in [0, 1)$. We use "$\mathbb{E}_u$" to denote expectation under $\mathbb{P}_u$ for $u \in [0, 1)$ and "$\mathbb{E}$" to denote expectation under $\mathbb{P} = \mathbb{P}_0$.

Remark 6 Since $\widehat{Y}S = (\widehat{Y}S^i)_{i \in \{1,\dots,d\}}$ is a local martingale on $(\Omega, \mathbf{F}, \mathbb{P})$, it follows that S^{η_u} is a local martingale in $(\Omega, \mathbf{F}, \mathbb{P}_u)$ for all $u \in [0, 1)$. In other words, $\mathbb{P}_u$ is an absolutely continuous local martingale measure for S^{η_u} for all $u \in [0, 1)$.

A key step towards the proof of Theorem 1 will be Lemma 3 below. Loosely interpreted, it states that taking the expectation of an $(\Omega, \mathbf{F})$-optional process sampled at ρ is tantamount to taking the expectation of the same process sampled at η_u under $\mathbb{P}_u$, where "u has standard uniform distribution, independent of everything else". Combined with the fact that $\mathbb{P}_u$ is an absolutely continuous local martingale measure for S^{η_u} for all $u \in [0, 1)$, this immediately connects to the statement of Theorem 1.

Before stating and proving Lemma 3, define also the nonnegative nondecreasing process $U = 1 - I$. Of course, $U(\infty) = U(\rho) = 1 - I(\infty)$ has the standard uniform distribution under $\mathbb{P}$.

Lemma 3 *For all $u \in [0, 1)$ we have that $\mathbb{P}_u[\eta_u < \infty] = 1$ holds; in particular, $\mathbb{P}_u[U(\eta_u) = u] = 1$. Furthermore, for any bounded and d-dimensional process V that is optional on $(\Omega, \mathbf{F})$, we have*

$$\mathbb{E}V(\rho) = \mathbb{E}\int_{\mathbb{R}_+} V(t)\widehat{Y}(t)\mathrm{d}U(t) = \int_{[0,1)} \mathbb{E}_u V(\eta_u)\mathrm{d}u. \quad (1)$$

Proof First of all, note that

$$\mathbb{P}_u[\eta_u < \infty] = \mathbb{E}(1/(1-u))\mathbb{I}_{\{\eta_u<\infty\}} = (1/(1-u))\mathbb{P}[\eta_u < \infty] = 1$$

holds for all $u \in [0, 1)$.

In order to establish (1), start by observing that $\mathbb{P}[\rho > t \mid \mathscr{F}(t)] = I(t)\widehat{Y}(t)$ holds for all $t \in \mathbb{R}_+$, in view of Doob's maximal identity. (Recall that $\widehat{Y}$ is a continuous-path nonnegative local martingale on $(\Omega, \mathbf{F}, \mathbb{P})$ and that $\mathbb{P}[\lim_{t\to\infty} \widehat{Y}_t = 0] = 1$.) Fix $s \in \mathbb{R}_+$ and $t \in \mathbb{R}_+$ with $s \leq t$. The definition of I and the integration-by-parts formula give

$$\begin{aligned} I(s)\widehat{Y}(s) - I(t)\widehat{Y}(t) &= -\int_s^t \widehat{Y}(v)\mathrm{d}I(v) - \int_s^t I(v)\mathrm{d}\widehat{Y}(v) \\ &= -\int_s^t \frac{1}{I(v)}\mathrm{d}I(v) - \int_s^t I(v)\mathrm{d}\widehat{Y}(v) \\ &= \log I(s) - \log I(t) - \int_s^t I(v)\mathrm{d}\widehat{Y}(v), \end{aligned}$$

the second equality following from the fact that $\int_{\mathbb{R}_+} \mathbb{I}_{\{\widehat{Y}(t) \neq 1/I(t)\}}\mathrm{d}I(t) = 0$. Note that $0 \leq I\widehat{Y} \leq 1$. With $(\tau_n)_{n\in\mathbb{N}}$ denoting a localizing sequence for $\int_0^\cdot I(v)\mathrm{d}\widehat{Y}(v)$, which is a local martingale on $(\Omega, \mathbf{F}, \mathbb{P})$, it follows that

$$\begin{aligned} &\mathbb{P}[s \wedge \tau_n < \rho \leq t \wedge \tau_n \mid \mathscr{F}(s \wedge \tau_n)] \\ &\quad = \mathbb{E}\left[I(s \wedge \tau_n)\widehat{Y}(s \wedge \tau_n) - I(t \wedge \tau_n)\widehat{Y}(t \wedge \tau_n) \mid \mathscr{F}(s \wedge \tau_n)\right] \\ &\quad = \mathbb{E}\left[\log I(s \wedge \tau_n) - \log I(t \wedge \tau_n) \mid \mathscr{F}(s \wedge \tau_n)\right]. \end{aligned}$$

Upon sending n to infinity, appropriate versions of the bounded and monotone convergence theorem applied to the first and last sides of the above equality will give

$$\mathbb{P}[s < \rho \leq t \mid \mathscr{F}(s)] = \mathbb{E}\left[\log I(s) - \log I(t) \mid \mathscr{F}(s)\right].$$

As $-\log I$ is non-decreasing and adapted, it coincides with the optional compensator (dual optional projection) of $\mathbb{I}_{[\![\rho,\infty[\![}$ on $(\Omega, \mathbf{F}, \mathbb{P})$. In other words,

$$\begin{aligned} \mathbb{E}V(\rho) &= -\mathbb{E}\int_{\mathbb{R}_+} V(t)\frac{\mathrm{d}I(t)}{I(t)} \\ &= -\mathbb{E}\int_{\mathbb{R}_+} V(t)\widehat{Y}(t)\mathrm{d}I(t) \\ &= \mathbb{E}\int_{\mathbb{R}_+} V(t)\widehat{Y}(t)\mathrm{d}U(t) \\ &= \mathbb{E}\int_{[0,1)} V(\eta_u)\widehat{Y}(\eta_u)\mathbb{I}_{\{\eta_u<\infty\}}\mathrm{d}u \\ &= \int_{[0,1)} \mathbb{E}\widehat{Y}(\eta_u)V(\eta_u)\mathrm{d}u \\ &= \int_{[0,1)} \mathbb{E}_u V(\eta_u)\mathrm{d}u, \end{aligned}$$

the 2nd equality following from the fact that $\int_{\mathbb{R}_+} \mathbb{I}_{\{\widehat{Y}(t) \neq 1/I(t)\}} \mathrm{d}I(t) = 0$ and the 4th by a simple time-change. The above establishes (1) and completes the proof of Lemma 3. □

Continuing with the proof of Theorem 1, we may assume that S is actually bounded via a simple localization argument. In all that follows, fix arbitrary $s, t \in \mathbb{R}_+$ with $s \leq t$, $B \in \mathscr{F}_s$, as well as a bounded deterministic function $f : [0, 1) \mapsto \mathbb{R}_+$. The π-λ-theorem implies that one only needs to show that

$$\mathbb{E} S^{\rho}(t) f(U(\infty)) \mathbb{I}_B = \mathbb{E} S^{\rho}(s) f(U(\infty)) \mathbb{I}_B.$$

Further noticing that $U(\infty) = U(\rho)$, and using the obvious equality

$$S^{\rho}(t) f(U(\rho)) \mathbb{I}_B = S^{\rho}(s) f(U(\rho)) \mathbb{I}_B \mathbb{I}_{\{\rho \leq s\}} + S^{\rho}(t) f(U(\rho)) \mathbb{I}_B \mathbb{I}_{\{\rho > s\}},$$

one needs to establish the identity

$$\mathbb{E} S^{\rho}(t) f(U(\rho)) \mathbb{I}_B \mathbb{I}_{\{\rho > s\}} = \mathbb{E} S^{\rho}(s) f(U(\rho)) \mathbb{I}_B \mathbb{I}_{\{\rho > s\}}. \tag{2}$$

Since S is assumed bounded, Remark 6 implies that S^{η_u} is a martingale on $(\Omega, \mathbf{F}, \mathbb{P}_u)$ for all $u \in [0, 1)$. The process $V := S^t f(U) \mathbb{I}_B \mathbb{I}_{]\!]s, \infty[\![}$ is optional on $(\Omega, \mathbf{F})$; furthermore, $V(\rho) = S^{\rho}(t) f(U(\rho)) \mathbb{I}_B \mathbb{I}_{\{\rho > s\}}$. Therefore, from Lemma 3, recalling that $\mathbb{P}_u[U(\eta_u) = u]$ for all $u \in [0, 1)$, we obtain that

$$\begin{aligned}
\mathbb{E} S^{\rho}(t) f(U(\rho)) \mathbb{I}_B \mathbb{I}_{\{\rho > s\}} &= \int_{[0,1)} f(u) \mathbb{E}_u S^{\eta_u}(t) \mathbb{I}_B \mathbb{I}_{\{\eta_u > s\}} \mathrm{d}u \\
&= \int_{[0,1)} f(u) \mathbb{E}_u S^{\eta_u}(s) \mathbb{I}_B \mathbb{I}_{\{\eta_u > s\}} \mathrm{d}u \\
&= \mathbb{E} S^{\rho}(s) f(U(\rho)) \mathbb{I}_B \mathbb{I}_{\{\rho > s\}},
\end{aligned}$$

which is exactly (2) and completes the proof of Theorem 1.

3.2 *Proof of Theorem 2*

To begin with, note that $(\Omega, \mathbf{F}, \mathbb{P})$ supports only continuous local martingales. Indeed, otherwise there would exist a nontrivial strictly positive process N with $N(0) = 1$, such that N is a purely discontinuous local martingale on $(\Omega, \mathbf{F}, \mathbb{P})$; but then, $N\widehat{Y}$ would be a strictly positive local martingale deflator in the market, which contradicts the uniqueness of the strictly positive local martingale deflator $\widehat{Y}$.

Since all local martingales on $(\Omega, \mathbf{F})$ are continuous and ϕ is an honest time that avoids all stopping times on $(\Omega, \mathbf{F}, \mathbb{P})$, [10, Theorem 4.1] implies that ϕ is the time of overall maximum of a nonnegative continuous local martingale L on $(\Omega, \mathbf{F}, \mathbb{P})$ with $L(0) = 1$ and $\mathbb{P}[\lim_{t \to \infty} L(t) = 0] = 1$. We shall show below that $L = \widehat{Y}$; this shows at the same time that $\phi = \rho$ and that the discounting process is asymptotically suboptimal, the latter following from $\mathbb{P}[\lim_{t\to\infty} L(t) = 0] = 1$.

As in the proof of Theorem 1, with L replacing $\widehat{Y}$ and ϕ replacing ρ, for all $u \in [0, 1)$ define $\eta_u := \inf\{t \in \mathbb{R}_+ \mid L(t) = 1/(1-u)\}$ and $\mathbb{P}_u$ via

$$\mathrm{d}\mathbb{P}_u = L(\eta_u)\mathrm{d}\mathbb{P} = (1/(1-u))\,\mathbb{I}_{\{\eta_u<\infty\}}.$$

Define the nondecreasing processes $L^* := \sup_{t\in[0,\cdot]} L(t)$ and $K := 1 - 1/L^*$. Following the reasoning of Lemma 3 (replacing $\widehat{Y}$ and U there by L and K respectively—note that in the proof of Lemma 3, we only use the facts that $\widehat{Y}$ is a nonnegative continuous local martingale on $(\Omega, \mathbf{F}, \mathbb{P})$ with $\widehat{Y}(0) = 1$ and $\widehat{Y}(t) \to 0$ $\mathbb{P}$-a.s. as $t \to \infty$, properties that $\widehat{Y}$ shares with L), we obtain that

$$\mathbb{E}V(\phi) = \mathbb{E}\int_{\mathbb{R}_+} V(t)L(t)\mathrm{d}K(t), \tag{3}$$

holding for all nonnegative optional process V on $(\Omega, \mathbf{F})$.

Lemma 4 *For uniformly bounded $X \in \mathscr{X}_{\mathbf{F}}$, we have*:

$$\mathbb{E}_u \int_0^{\eta_u} (1 - K(t))\mathrm{d}X(t) \le 0, \quad \textit{for all } u \in [0, 1). \tag{4}$$

Proof Let $B := \int_{[0,\cdot]} X(t)\mathrm{d}K(t)$; clearly, B is a uniformly bounded nondecreasing continuous and adapted process on $(\Omega, \mathbf{F})$. Fix $u \in [0, 1)$. Using integration-by-parts, write

$$\begin{aligned}
&\int_{\mathbb{R}_+} X^{\eta_u}(t)L(t)\mathrm{d}K(t)\\
&\quad= \int_0^{\eta_u} L(t)\mathrm{d}B(t) + X(\eta_u)\int_{\eta_u}^{\infty} L(t)\mathrm{d}K(t)\\
&\quad= \int_0^{\eta_u} L(t)\mathrm{d}B(t) + X(\eta_u)\int_{\eta_u}^{\infty} L(t)\frac{1}{(L^*(t))^2}\mathrm{d}L^*(t)\\
&\quad= \int_0^{\eta_u} L(t)\mathrm{d}B(t) + X(\eta_u)\int_{\eta_u}^{\infty} \frac{1}{L^*(t)}\mathrm{d}L^*(t)\\
&\quad= L(\eta_u)B(\eta_u) - \int_0^{\eta_u} B(t)\mathrm{d}L(t) + X(\eta_u)\big(\log L^*(\infty) + \log(1-u)\big)\,\mathbb{I}_{\{\eta_u<\infty\}},
\end{aligned}$$

the 3rd equality following because $\int_{\mathbb{R}_+} \mathbb{I}_{\{L(t)\neq L^*(t)\}}\mathrm{d}L^*(t) = 0$. Now, observe that

$$\mathbb{E}L(\eta_u)B(\eta_u) = \mathbb{E}_u B(\eta_u) = \mathbb{E}_u \int_0^{\eta_u} X(t)\mathrm{d}K(t)$$

and $\mathbb{E}\int_0^{\eta_u} B(t)\mathrm{d}L(t) = 0$, the latter following from the facts that B is uniformly bounded and L^{η_u} is a uniformly bounded martingale on $(\Omega, \mathbf{F}, \mathbb{P})$. Furthermore, using Doob's maximal identity we obtain that

$$\mathbb{E}\left[\log L^*(\infty) + \log(1-u) \mid \mathscr{F}(\eta_u)\right] = 1 \text{ holds on } \{\eta_u < \infty\}.$$

Therefore,

$$\begin{aligned}\mathbb{E}\left[X(\eta_u)\left(\log L^*(\infty) + \log(1-u)\right)\mathbb{I}_{\{\eta_u<\infty\}}\right] &= \mathbb{E}\left[X(\eta_u)\mathbb{I}_{\{\eta_u<\infty\}}\right] \\ &= (1-u)\mathbb{E}_u[X(\eta_u)].\end{aligned}$$

In view of the fact that $\mathbb{E}\int_{\mathbb{R}_+} X^{\eta_u}(t)L(t)\mathrm{d}K(t) = \mathbb{E}X^{\eta_u}(\phi) \le X(0)$, as follows from (3) and the assumptions of Theorem 2, all the previous give

$$\mathbb{E}_u\left[\int_0^{\eta_u} X(t)\mathrm{d}K(t) + (1-u)X(\eta_u)\right] \le X(0).$$

On $\{\eta_u < \infty\}$, it holds that

$$\int_0^{\eta_u} X(t)\mathrm{d}K(t) = K(\eta_u)X(\eta_u) - \int_0^{\eta_u} K(t)\mathrm{d}X(t) = uX(\eta_u) - \int_0^{\eta_u} K(t)\mathrm{d}X(t);$$

since $\mathbb{P}_u[\eta_u < \infty] = 1$, we furthermore obtain

$$\mathbb{E}_u\left[X(\eta_u) - \int_0^{\eta_u} K(t)\mathrm{d}X(t)\right] \le X(0),$$

which is the same as (4) and proves Lemma 4. □

Continuing, for each $i \in \{1, \dots, d\}$ and $n \in \mathbb{N}$, define

$$\tau_n^i := \inf\left\{t \in \mathbb{R}_+ \mid |S^i(t) - S^i(0)| \ge n\right\},$$

which is a stopping time on $(\Omega, \mathbf{F})$. Furthermore, define $X_n^i \in \mathscr{X}_{\mathbf{F}}(1)$ via

$$X_n^i := 1 + n^{-1}(S^i - S^i(0))^{\tau_n^i}.$$

It is clear that $0 \le X_n^i \le 2$. For an arbitrary stopping time τ on $(\Omega, \mathbf{F})$, apply (4) with $(X_n^i)^\tau$ replacing X; one then obtains the bound $\mathbb{E}_u\int_0^{\eta_u\wedge\tau_n^i\wedge\tau}(1-K(t))\mathrm{d}S^i(t) \le 0$. Performing exactly the previous work by redefining $X_n^i := 1 - n^{-1}(S^i - S^i(0))^{\tau_n^i}$, one obtains $\mathbb{E}_u\int_0^{\eta_u\wedge\tau_n^i\wedge\tau}(1-K(t))\mathrm{d}S^i(t) \ge 0$. Therefore, for all $i \in \{1, \dots, d\}$, $n \in \mathbb{N}$, and any stopping time τ on $(\Omega, \mathbf{F})$, $\mathbb{E}_u\int_0^{\eta_u\wedge\tau_n^i\wedge\tau}(1-K(t))\mathrm{d}S^i(t) = 0$ holds. This implies that each process $\int_0^{\eta_u\wedge\cdot}(1-K(t))\mathrm{d}S^i(t)$ is a local martingale on $(\Omega, \mathbf{F}, \mathbb{P}_u)$. Since $1-K > 0$, we further obtain that each process $(S^i)^{\eta_u}$ is a local martingale on $(\Omega, \mathbf{F}, \mathbb{P}_u)$. By the definition of the collection $(\mathbb{P}_u)_{u\in[0,1)}$, we conclude that LS^i is a local martingale on $(\Omega, \mathbf{F}, \mathbb{P})$ for all $i \in \{1, \dots, d\}$. This implies that L is a local martingale deflator. Since $1/\widehat{X}$ is the unique local martingale deflator, we finally conclude that $L = 1/\widehat{X}$, which proves Theorem 2.

References

1. Ankirchner, S., Dereich, S., Imkeller, P.: The Shannon information of filtrations and the additional logarithmic utility of insiders. Ann. Probab. **34**, 743–778 (2006)
2. Delbaen, F., Schachermayer, W.: A general version of the fundamental theorem of asset pricing. Math. Ann. **300**, 463–520 (1994)
3. Jacod, J.: Grossissement Initial, Hypothèse (H) et Théorème de Girsanov. Lecture Notes in Mathematics, vol. 1118. Springer, Berlin (1985). Jeulin, T. and Yor, M., Grossissements de filtrations: exemples et applications
4. Jeulin, T., Yor, M.: Grossissement d'une filtration et semi-martingales: formules explicites. In: Séminaire de Probabilités, XII, Univ. Strasbourg, Strasbourg, 1976/1977. Lecture Notes in Math., vol. 649, pp. 78–97. Springer, Berlin (1978)
5. Kabanov, Y.M.: On the FTAP of Kreps–Delbaen-Schachermayer. In: Statistics and Control of Stochastic Processes, pp. 191–203. World Scientific, River Edge (1997)
6. Karatzas, I., Kardaras, C.: The numéraire portfolio in semimartingale financial models. Finance Stoch. **11**, 447–493 (2007)
7. Kardaras, C.: Finitely additive probabilities and the fundamental theorem of asset pricing. In: Contemporary Quantitative Finance: Essays in Honour of Eckhard Platen, pp. 19–34. Springer, Berlin Heidelberg (2010)
8. Kardaras, C.: Numéraire-invariant preferences in financial modeling. Ann. Appl. Probab. **20**, 1697–1728 (2010)
9. Long, J.B. Jr.: The numéraire portfolio. J. Financ. Econ. **26**, 29–69 (1990)
10. Nikeghbali, A., Yor, M.: Doob's maximal identity, multiplicative decompositions and enlargements of filtrations. Ill. J. Math. **50**, 791–814 (2006) (electronic)
11. Protter, P.: Stochastic Integration and Differential Equations. Springer, Berlin (1990)
12. Stricker, C., Yan, J.A.: Some remarks on the optional decomposition theorem. In: Séminaire de Probabilités, XXXII. Lecture Notes in Math., vol. 1686, pp. 56–66. Springer, Berlin (1998)

Sensitivity with Respect to the Yield Curve: Duration in a Stochastic Setting

Paul C. Kettler, Frank Proske, and Mark Rubtsov

Abstract Bond duration in its basic deterministic meaning form is a concept well understood. Its meaning in the context of a yield curve on a stochastic path is less well developed. In this paper we extend the basic idea to a stochastic setting. More precisely, we introduce the concept of stochastic duration as a Malliavin derivative in the direction of a stochastic yield surface modeled by the Musiela equation. Further, using this concept we also propose a mathematical framework for the construction of immunization strategies (or delta hedges) of portfolios of interest rate securities with respect to the fluctuation of the whole yield surface.

Keywords Bond duration · Malliavin derivative · Yield surface · Immunization strategies · Delta hedges

Mathematics Subject Classification (2010) 91G30 · 60H07 · 91B02

1 Introduction

The concept of bond duration dates to a foundational book defining the idea [32]. Through the years there have been many presentations on the idea. One of note is [27]. Other tracts obtain, most frequently addressing the bond with periodic coupons and a terminal payment of principal. Such discussions tend to concentrate on the idea of an annuity as the sum of a geometric series, presented in a variety of flavors. We eschew these notions as being of scant academic interest, and focus on the continuously compounded zero coupon bond as a building block, leaving the con-

P.C. Kettler (✉) · F. Proske · M. Rubtsov
CMA, Department of Mathematics, University of Oslo, P.O. Box 1053, Blindern, 316 Oslo, Norway
e-mail: mail@paulcarlislekettler.net

F. Proske
e-mail: proske@math.uio.no

M. Rubtsov
e-mail: rubtsov@math.uio.no

Y. Kabanov et al. (eds.), *Inspired by Finance*, DOI 10.1007/978-3-319-02069-3_17,

struction of instruments with component payments to others. See the Appendix for a brief discussion of Macaulay duration in context.

The bond market worldwide has about \$82 trillion outstanding, with about \$1 trillion trading on a typical day. Other than price, the most widely quoted parameter in the market, without question, is duration. It appears on quotation screens, on traders' lips, and in all manner of literature on the market. Yet the concept, which dates back 70 years, addresses the sensitivity of a bond's price with respect to changes in yield, assumes a uniform rate of interest through the life of a bond, an unrealistic posture.

In basic bond analysis one considers a zero coupon bond with present value (or price) v given as a function of a level interest rate r, maturing to future value 1 at time T. The relationship of variables is this

$$v = e^{-rT}. \tag{1}$$

The quantity

$$d := \frac{1}{v}\frac{\partial v}{\partial r} = \frac{\partial}{\partial r}\log v = -T \tag{2}$$

is known as the *duration*, and the quantity

$$c := \frac{1}{2v}\frac{\partial^2 v}{\partial r^2} = \frac{1}{2}T^2$$

is known as convexity. Note that d and c are the coefficients, respectively, of r and r^2 in the Taylor series expansion of v:

$$v = 1 - Tr + \frac{1}{2}T^2 r^2 - \cdots \tag{3}$$

Bond traders routinely employ duration and convexity in market analysis to estimate the effects of rate changes.

An important fact about duration, which makes it useful for portfolio analysis, is that the duration of a portfolio is the average of the component durations weighted by present values. A two security case is sufficient to illustrate. Let

$$v = \alpha_1 v_1 + \alpha_2 v_2 = \alpha_1 e^{-rT_1} + \alpha_2 e^{-rT_2}.$$

Then

$$d = -\frac{\alpha_1 v_1}{\alpha_1 v_1 + \alpha_2 v_2}T_1 - \frac{\alpha_2 v_2}{\alpha_1 v_1 + \alpha_2 v_2}T_2\,.$$

One may generalize this concept of bond to incorporate a piecewise constant interest rate $r(s)$, where

$$r(s) = \begin{cases} r_1 & \text{if } 0 =: s_0 \le s < s_1, \\ r_2 & \text{if } s_1 \le s < s_2, \\ \cdots & \\ r_n & \text{if } s_{n-1} \le s \le s_n =: T. \end{cases}$$

Then Eq. (1) becomes

$$v = \exp\left\{-\sum_{i=1}^{n} r_i(s_i - s_{i-1})\right\}. \tag{4}$$

From this expression we obtain the ith *partial duration*

$$d_i := \frac{\partial}{\partial r_i}\log v = -(s_i - s_{i-1}), \quad 1 \le i \le n,$$

and the ith *partial convexity*

$$c_i = \frac{1}{2}(s_i - s_{i-1})^2, \quad 1 \le i \le n.$$

Observe that the partial durations add to the total duration, whereas the partial convexities (and higher order related partial terms) do not.

One may elaborate further on the themes of Eqs. (1) and (4) by putting r and the $\{r_i\}$ on stochastic paths. To start, denote by $P(t,T)$ the price at time t of a zero coupon bond, which pays 1$ at maturity T. Then one can define instantaneous forward rates as

$$f(t;T) = -\frac{\partial \log P(t,T)}{\partial T}, \quad 0 \le t \le T \tag{5}$$

for each maturity T (see [23]). So we can recast (1) as

$$v = P(t,T) = \exp\left\{-\int_t^T f(t,s)\,\mathrm{d}s\right\}. \tag{6}$$

Since the outcome of future interest rates are not known in advance it is reasonable to model instantaneous forward rates $\{f(t,s)\}_{0\le s\le t}$ as stochastic processes. In this context we may interpret $f(t,s)$ as the overnight interest rate at (future) time t as seen from time s. The case $f(t,t) =: r(t)$ is simply the overnight rate or short rate.

The literature is replete with examples on stochastic interest rates. A small sample of papers, not otherwise cited in the text, is this [1, 4–6, 12–14, 18, 25, 30, 34, 41, 42, 44, 45] and [19]. All address stochastic interest rates in financial modeling. Of interest within are these references including co-author Marek Musiela: [8, 9, 35] and [21].

As mentioned above the classical duration is based on the assumption that interest rates are flat or piecewise flat. This assumption is quite unrealistic and only applies to sensitivity measurements with respect to (piecewise) parallel shifts of interest rates. The latter is especially unsatisfying for a trader who manages a complex portfolio of interest rate sensitive securities (as e.g. caps, swaps, bond options, ...). In this case it would be desirable to measure the interest rate risk of the portfolio with respect to the *stochastic fluctuations* of the *whole* term structure or even the *yield surface*, that is

$$(t,x) \mapsto Y(t,t+x), \tag{7}$$

where $Y(t,T)$ is the yield given by

$$Y(t,T) = -\frac{1}{T-t}\log P(t,T).$$

Here x in (7) stands for the time-to-maturity.

Using relation (6) we can represent the yield surface $Y_t(x) := Y(t,t+x)$ as

$$Y_t(x) = \frac{1}{x}\int_0^x f_t(s)\,\mathrm{d}s, \tag{8}$$

where $f_t(s) := f(t,t+s)$. Because of the "linear" correspondence (8) between the yield curves $Y_t(\cdot)$ and the forward curves $f_t(\cdot)$ we can and will refer to

$$(t,x) \mapsto f_t(x) \tag{9}$$

as the yield surface in this paper.

Assuming e.g. the Heath–Jarrow–Morton model (see [23]) for the dynamics of instantaneous interest rates, one shows (under certain conditions) that the yield surface (9) is described by a stochastic partial differential equation, called Musiela equation (see e.g. [10]).

In this paper we wish to develop an analogous concept to the classical duration of Macaulay in the Heath–Jarrow–Morton setting. More precisely, we want to measure the sensitivity of interest rate claims with respect to the Musiela dynamics of the yield surface (9).

An apparently analogous way to the classical case would be to define the duration of an interest rate security by means of the Fréchet derivative for each interest rate scenario. However, interest rate securities or even dynamically hedged portfolios composed of those are in general complicated functionals of the yield surface, which are usually not even continuous.

In order to overcome this problem one may think of weaker forms of derivatives than the Fréchet derivative to measure interest rate sensitivities. A possible candidate could be the Malliavin derivative, which can be considered a stochastic Gateaux derivative. See Sect. 2.

In this paper we want to base the stochastic duration concept on this stochastic Gateaux derivative. This concept is analogous to the classical one in the sense that it relies on a derivative of an infinite dimensional version of the Taylor expansion (3). Using this concept we also define stochastic convexity as a measure for the curvature of yield surface movements.

The paper is organized as follows: In Sect. 2 we define the concept of stochastic duration by using Malliavin calculus for general Gaussian random fields. In Sect. 3 we propose a mathematical framework for the construction of immunization strategies of portfolios, which are composed of interest rate instruments.

2 An Expanded Concept of Duration via Malliavin Calculus

In this section we want to elaborate a duration concept for stochastic yield curves. This definition extends the classical duration of Macaulay to a stochastic setting.

Denote by $P(t,T)$ the price at time t of a zero coupon bond, which pays 1$ at maturity T. Suppose that the bond prices are modeled by non-negative adapted processes $\{P(t,T)\}_{0\le t\le T}$ for each $T>0$ on a filtered probability space

$$\left(\Omega, \mathscr{F}, \{\mathscr{F}_t\}_{t\ge 0}, P\right). \tag{10}$$

In the following we assume that the bond prices $P(t,T)$ are described by the Heath–Jarrow–Morton model (HJM-model) [23], that is the bond prices take the form

$$P(t,T) = \exp\left\{-\int_t^T f(t,s)\,\mathrm{d}s\right\}, \tag{11}$$

where $f(t,T),\ \ 0\le t\le T<\infty$, are instantaneous forward rates modeled by the SDE

$$\mathrm{d}f(t,T) = \alpha(t,T)\,\mathrm{d}t + \sigma(t,T)\,\mathrm{d}B_t, \quad 0\le t\le T<\infty. \tag{12}$$

Here we require that $\sigma(\cdot,T)$ be a deterministic Borel measurable function and $\alpha(\cdot,T)$ a predictable process for all T w.r.t. the P-completed filtration $(\mathscr{F}_t)$ generated by a (1-dimensional) Brownian motion $B_t,\ t\ge 0$.

Now let us reparametrize the forward rates by the *time-to-maturity* $x=T-t$, that is let us consider the forward curves

$$f_t(x) := f(t,t+x). \tag{13}$$

Then an application of the generalized Itô formula (see Theorem 3.3.1 in [29]) shows that under certain conditions on $\alpha(\cdot,T),\sigma(\cdot,T)$ the forward curves $f_t(x)$ satisfy the first order SPDE

$$\mathrm{d}f_t(x) = \frac{\mathrm{d}}{\mathrm{d}x} f_t(x) + \alpha_t(x)\,\mathrm{d}t + \sigma_t(x)\,\mathrm{d}B_t. \tag{14}$$

Here we use the notation $\alpha_t(x) := \alpha(t,t+x), \sigma_t(x) := \sigma(t,t+x)$. Note that (14) is referred to as *Musiela equation* in the literature (see e.g. [10]). See also [15] and the references therein for more information about SPDE's.

A deficiency of the model (14) is that it does not capture the feature of *maturity-specific risk*. A model with such a property would enable hedging of bond options with unique portfolio strategies. On the other hand, it would meet the intuitive requirement that maturities of the bonds underlying the bond option are used in the hedging portfolio.

A more realistic model than (14) which takes into account maturity-specific risk would consequently have the (formal) form

$$\mathrm{d}f_t(x) = \left(\frac{\mathrm{d}}{\mathrm{d}x} f_t(x) + \alpha_t(x)\right)\mathrm{d}t + \sigma_t(x)\,\mathrm{d}B_t(x), \tag{15}$$

where each noise $B_t(x)$ stands for the risk arising from the time-to-maturity x. Here we may think of $B_t(x)$ to be a Brownian sheet in t and x. So (15) can be recast as

$$\mathrm{d} f_t(x) = \Big(\frac{\mathrm{d}}{\mathrm{d}x} f_t(x) + \alpha_t(x)\Big)\, \mathrm{d}t + \sum_{k\geq 1} \sigma_t^{(k)}(x)\, \mathrm{d}B_t^{(k)}, \tag{16}$$

where $\sigma_\cdot^{(k)}(\cdot),\ k \geq 1$ are deterministic measurable functions and $B_t^{(k)},\ k \geq 1$ independent 1-dimensional Brownian motions.

In what follows we want to assume that the forward curves are modeled by functions of a Hilbert space H. This space should exhibit the natural feature that evaluation functionals on it are continuous, that is

$$\delta_x : H \to \mathbb{R}; \quad f \mapsto f(x), \tag{17}$$

is continuous on H for all x. Further it is desirable that the generator $A = \frac{\mathrm{d}}{\mathrm{d}x}$ in (16) admits a strongly continuous semigroup S_t on H. The semigroup S_t is the left shift operator given by

$$(S_t f)(x) = f(t + x). \tag{18}$$

The following family of Hilbert spaces H_w of Sobolev type introduced by [17] fulfills the above mentioned conditions: Let $w : [0,\infty) \longrightarrow (0,\infty)$ be a non-decreasing function such that

$$\int_0^\infty \frac{1}{w(x)}\, \mathrm{d}x < \infty.$$

Then H_w is defined as

$$H_w = \left\{ f : \mathbb{R}^+ \to \mathbb{R} : f \text{ absolutely continuous}, \int_0^\infty (f'(x))^2\, w(x)\, \mathrm{d}x < \infty \right\} \tag{19}$$

and equipped with the scalar product

$$\langle f, g\rangle_{H_w} = f(0)g(0) + \int_0^\infty f'(x) g'(x)\, w(x)\, \mathrm{d}x.$$

In the sequel we require that

$$\alpha_t(\cdot), \sigma_t^{(k)}(\cdot) \in H \text{ a.e.} \tag{20}$$

for all $t \geq 0$.

Consider the special case that $\alpha_t(x) = \alpha_t^*(x) f_t(x)$ for a deterministic function $\alpha_t^*(x)$. Then, using integrating factors we observe that the (mild) solution of (16) is explicitly given by the Gaussian random field

$$f_t(x) = e^{\int_0^t \alpha^*(s,t+x)\,\mathrm{d}s} f(0,t+x) + \sum_{k\geq 1} \int_0^t e^{\int_s^t \alpha^*(u,t+x)\,\mathrm{d}u} \sigma^{(k)}(s,t+x)\, \mathrm{d}B_t^{(k)}. \tag{21}$$

Now, let W_t be a Q-Wiener process , where Q is a symmetric non-negative operator on a separable Hilbert space U with Trace $Q < \infty$. Set $U_0 = Q^{1/2}(U)$, which is a Hilbert space with norm

$$\|h\|_0 := \left\| Q^{-1/2}(h) \right\|, \quad u \in U_0.$$

Denote by $L_2(U,H)$ the space of Hilbert–Schmidt operators from U to H with the operator norm $\|\cdot\|_{L_2}$. Further, let u_k, $k \geq 1$, be an orthonormal basis of U, and suppose that there exists a Borel measurable map

$$\sigma : [0,T] \to L(U_0, H)$$

such that

$$\sigma_t[Q^{1/2}(u_k)] = \sigma_t^{(k)}(\cdot)$$

and

$$\sigma_t \circ Q^{1/2} \in L_2(U,H)$$

for all t, k in (16), where $\circ$ stands for the composition of operators. Then we can view $\{B_t^{(k)}\}_{0 \leq t \leq T}$, $k \geq 1$, in (16) as a Wiener process B_t cylindrically defined on U and rewrite (16) as

$$\mathrm{d} f_t = (A f_t + \alpha_t)\,\mathrm{d}t + \sigma_t\,\mathrm{d}W_t. \tag{22}$$

In the sequel we assume that there exists a predictable unique strong solution

$$(t \mapsto f_t(\cdot)) \in C([0,T]; H)$$

to (22).

Remark 1 Suppose that $\alpha_t = b(t, f_t)$ in (22), where $b : [0,T] \times H \longrightarrow H$ is a Borel measurable map. Then the following set of conditions provide sufficient criteria for the existence of a unique strong solution of (22):

1. f_t is a unique mild solution of (22);
2. $f_0 \in D(A)$ (domain of A), $S_{t-s} b(s,x) \in D(A)$, $S_{t-s}\sigma_s u \in D(A)$ for all $u \in U_0$, $t \geq s$;
3. $\|A S_{t-s} b(s,x)\|_H \leq q(t-s)\,\|x\|_H$ for $q \in L^1([0,T]; \mathbb{R}_+)$;
4. $\|A S_{t-s} \sigma_s\|_H = g(t-s)$ for $g \in L^2([0,T]; \mathbb{R}_+)$.

See, e.g., [28].

Assume that σ is invertible for all $t \leq T$ a.e. and that the integrability condition

$$\sup_{t \in [0,T]} E \exp\left\{ \delta \left\| \sigma_t^{-1} \left[A f_t + \alpha_t \right] \right\|_0^2 \right\} < \infty \tag{23}$$

holds for some $\delta > 0$. Then Girsanov's theorem (see e.g. [2]) applied to (22) entails that

$$\mathrm{d} f_t = \sigma_t \, \mathrm{d}\widehat{W}_t, \tag{24}$$

where

$$\widehat{W}_t = W_t - \int_0^t \psi(s) \, \mathrm{d}s$$

is a Q-Wiener process under the change of measure $\widehat{P}$ given by

$$\widehat{P}(A) = E\chi_A \exp\left\{\int_0^T < \psi(s), \, \mathrm{d}W_s >_0 - \frac{1}{2}\int_0^T \|\psi(s)\|_0^2 \, \mathrm{d}s\right\}$$

with

$$\psi(t) := \sigma_t^{-1}\left[A f_t + \alpha_t\right].$$

Consequently f_t is a Gaussian $\mathscr{F}_t$-martingale with respect to $\widehat{P}$. Define

$$\widehat{f_t} = f_t - f_0 = \int_0^t \sigma_s \, \mathrm{d}\widehat{W}_s. \tag{25}$$

Thus $\widehat{f_t}(x)$ is a centered Gaussian random field with respect to time and time-to-maturity under $\widehat{P}$. We wish to use these forward curves to define an expanded concept of duration which serves as a tool to measure interest rate sensitivities of bond options or bond portfolios with respect to the *whole* yield surface

$$((t,x) \mapsto f_t(x)).$$

In view of the relation between Malliavin derivatives and Gateaux derivatives it is reasonable to define the duration of an interest rate instrument as the Malliavin derivative of a square integrable functional of $\widehat{f_t}(x)$.

To this end we have to introduce a Malliavin calculus with respect to $\widehat{f_t}(x)$ which is the centered forward curve in the "risk neutral world". For this purpose let

$$(\Omega, \widehat{\mathscr{F}}, \widehat{P}) \tag{26}$$

be our reference probability space, where $\widehat{\mathscr{F}}$ generated by $\widehat{f_t}(x)$. In the following we denote by I index set with respect to the tuples (t,x) and set $\widehat{f}(u) = \widehat{f_t}(x)$, if $u = (t,x) \in I$. Let

$$C(u,r) = E\widehat{f}(u)\widehat{f}(r) \tag{27}$$

be the covariance function of $\widehat{f}$. Further let us consider the reproducing kernel Hilbert space (RKHS) K of C (see e.g. [11]) with norm $\|\cdot\|_K$. Then K is isometrically isomorphic to the closure of the linear span of $\widehat{f}(u), u \in I$ in $L^2(\Omega, \widehat{\mathscr{F}}, \widehat{P})$.

Using in addition the continuity of evaluation functionals on H and the theorem of Banach–Steinhaus we find that K is isometrically isomorphic to the space

$$H(\widehat{f}) := \left\{ \lambda : [0,T] \to H^* \text{ Borel measurable}, \quad \int_0^T \|\lambda_s \circ \sigma_s\|^2_{L_2^0} \, \mathrm{d}s < \infty \right\}, \quad (28)$$

where $\|B\|_{L_2^0} := \|B \circ Q^{1/2}\|_{L_2} < \infty$ for $B \in L(H, H)$. Here H^* stands for the (topological) dual of H.

By [11] we obtain the following chaos decomposition:

$$L^2(\Omega, \widehat{\mathscr{F}}, \widehat{P}) = \sum_{p \geq 0} \oplus I_p(K^{\widehat{\otimes} p}), \quad (29)$$

where $K^{\widehat{\otimes} p}$ is the p-fold symmetric tensor product of K and where $I_p : K^{\otimes p} \longrightarrow L^2(\Omega, \widehat{\mathscr{F}}, \widehat{P})$ are linear operators such that the following properties hold:

$$E I_p(f) = 0 \quad (30)$$

$$E I_p(f) I_q(g) = \begin{cases} 0, & p \neq q, \\ p! < \widetilde{f}, \widetilde{g} >_K, & p = q, \end{cases} \quad (31)$$

for $f \in K^{\otimes p}$, $g \in K^{\otimes q}$, where $\widetilde{f}$ is the symmetrization of f. Here I_p is recursively defined by

$$I_{p+1}(gh) = I_p(g) I_1(h) - \sum_{k=1}^{p} I_{p-1}(g \underset{k}{\otimes} h) \quad (32)$$

for $g \in K^{\otimes p}$, $h \in K$, where

$$I_1(h) := \int_0^T h_s \, d(\sigma_s \widehat{W}_s) = \int_0^T h_s \circ \sigma_s \, \mathrm{d}\widehat{W}_s$$

for $h \in H(\widehat{f})$. See [33].

Now let $u \in L^2(\Omega; K)$ and let u_t have the chaos representation

$$u_t = \sum_{p \geq 0} I_p(f_p^t)$$

for unique $f_p^t \in K^{\widehat{\otimes} p}$ and each $t \in I$. Denote by $\widetilde{f_p}$ the symmetrization of an appropriate version of $f_p^t(t_1, \ldots, t_p)$ w.r.t. $t_1, \ldots, t_p$ and t.Then the Skorohod integral of the process u_t is defined as

$$\delta(u.) = \sum_{p \geq 1} I_{p+1}(\widetilde{f_p}) \quad (33)$$

if

$$\sum_{p\geq 1}(p+1)!\|\widetilde{f}_p\|^2_{K^{\widehat{\otimes}p+1}}<\infty \tag{34}$$

is fulfilled.

The *Malliavin derivative* $\mathrm{D}_u F \in L^2(\Omega; K)$ of a square integrable functional F of the forward curve $\widehat{f}$ can be defined as the adjoint operator of δ in (33). In the sequel we shall denote by $\mathbb{D}^{1,2} \subset L^2(\Omega, \widehat{\mathcal{F}}, \widehat{P})$ the domain of the Malliavin derivative D.

In view of the financial applications we have in mind it is important to note that the Malliavin derivative can be regarded as a sensitivity measure with respect to the fluctuations of the yield surface $((t,x) \mapsto f_t(x))$. The latter can be justified by the following relationship between the Malliavin derivative and the stochastic Gateaux K-derivative:

Let X be the support of the image measure μ of $\widehat{f}$ under $\widehat{P}$ in $C([0,T]; H)$. Then by [7] we find that X is the closure of K in $C([0,T]; H)$. Further Proposition 4.1 in [20] shows that if for $F \in L^2(\mu)$

$$\frac{F(x+\varepsilon k)-F(x)}{\varepsilon} \tag{35}$$

converges in $L^2(\mu)$ as $\varepsilon \to 0$ for $k \in K$, then $\mathrm{D}.F \in L^2(\mu; K)$ exists and the above limit equals $(\mathrm{D}.F, k)_K$.

Since the measure P in (12) is equivalent to $\widehat{P}$ we see that the convergence of (35) to $\langle \mathrm{D}.F, k\rangle_K$ also holds in probability with respect to the image measure of the forward curves under the original measure P. Therefore, if $F = \xi_T$ is the terminal value of a bond portfolio, we may interpret the Malliavin derivative $\mathrm{D}.F$ as a sensitivity measure of the fluctuations of the whole yield surface in this portfolio. The latter observation gives rise to introduce an expanded concept of duration as follows:

Definition 1 (Stochastic duration) Let F be a square integrable functional of the forward curve $\widehat{f}$ w.r.t. $\widehat{P}$. Assume that F is Malliavin differentiable with respect $\widehat{f}$. Then the *stochastic duration of* F is the stochastic process

$$\mathrm{D}.F \in L^2(\Omega, \widehat{\mathcal{F}}, \widehat{P}; K).$$

Remark 2 We shall mention that we could also have introduced our concept of stochastic duration w.r.t. mild solutions f_t of (22). In this case one can replace condition (23) by assuming that

$$\sup_{t\in[0,T]} E\exp\Bigl(\delta\left\|\sigma_t^{-1}[\alpha_t]\right\|_0^2\Bigr)<\infty$$

for some $\delta > 0$. Compared to mild solutions strong solutions are rather rare. However from the viewpoint of applications we have in mind (see Sect. 3) it is (technically) more convenient to deal with strong solutions.

We want to illustrate this concept by calculating the generalized duration of certain interest rate claims. For this purpose we need the following auxiliary results:

The first result gives a chain rule for the Malliavin derivative D.

Lemma 1 *Let F be Malliavin differentiable with respect to $\widehat{f}$, i.e. $F \in \mathbb{D}^{1,2}$. Further suppose that $g : \mathbb{R} \longrightarrow \mathbb{R}$ is continuously differentiable with bounded derivative. Then $g(F) \in \mathbb{D}^{1,2}$ and*

$$\mathrm{D}_u g(F) = g'(F)\,\mathrm{D}_u F$$

for each $u \in K$. Here g' stands for the derivative of g.

Proof The proof follows from arguments in the Brownian motion case. See Theorem 3.5 in [16] or Proposition 1.2.2 in [37]. □

The next lemma pertains to the closability of the Malliavin derivative.

Lemma 2 *Let $F \in L^2(\widehat{P})$ and $(F_k)_{k \geq 1} \subset \mathbb{D}^{1,2}$ such that*

$$F_k \underset{k \to \infty}{\to} F \quad \text{in } L^2(\widehat{P})$$

and $\mathrm{D}.F_k$ converges in $L^2(\widehat{P}; K)$. Then $F \in \mathbb{D}^{1,2}$ and

$$\mathrm{D}.F_k \underset{k \to \infty}{\to} \mathrm{D}.F \quad \text{in } L^2(\widehat{P}; K).$$

Proof See the arguments of Theorem 3.3 in [16]. □

Example 1 (Zero coupon bond) As before let $P(t, T)$ be the price at time t of a zero coupon bond, which pays 1$ at maturity T. Then using the instantaneous forward rates $f(t, s), 0 \leq t \leq s$ we have that

$$P(t, T) = \exp\left\{- \int_t^T f(t, s)\, ds\right\} = \exp\left\{- \int_0^{T-t} f_t(x)\, dx\right\}.$$

We find that

$$\begin{aligned} \mathrm{D}_{r,y}\left(\int_0^{T-t} f_t(x)\, dx\right) &= \int_0^{T-t} \mathrm{D}_{r,y}(f_t(x))\, dx \\ &= \int_0^{T-t} \chi_{[0,t]}(r)\, \mathrm{d}x = (T-t)\chi_{[0,t]}(r), \end{aligned}$$

where $\chi_{[0,t]}$ is the indicator function of $[0, t]$. Then the chain rule of Lemma 1 (in connection with Lemma 2) shows that the stochastic duration $\mathrm{D}.P(t, T)$ of $P(t, T)$ in the HJM-model is given by

$$\mathrm{D}_{r,y} P(t, T) = \begin{cases} -(T-t)P(t, T), & \text{if } 0 \leq r \leq t, \\ 0, & \text{otherwise.} \end{cases} \tag{36}$$

So $\mathrm{D}_{r,y}P(t,T)/P(t,T)$, $0 \le r \le t$, has the form of the classical duration in Sect. 1. The latter expression seems to suggest that we should rather use $\mathrm{D}.F/F$ as a generalized duration than $\mathrm{D}.F$. However, general interest rate claims F may be zero for a positive probability. Therefore it is reasonable to introduce $\mathrm{D}.F$ as an expanded concept of duration. Note that our definition does not generalize Macaulay's duration in the sense that $\mathrm{D}.F$ gives the classical duration, if the interest rate claim F is deterministic, that is a functional of a deterministic (flat) yield surface. The explanation for this is that the duration concepts are based on different interest rate models. The classical duration presumes yield surfaces which are flat or piecewise flat. Such a model is fundamentally different from a stochastic interest rate model. For example, under our conditions, yield surfaces in a risk-neutral HJM-model only assume a certain constant value with probability zero. In view of this we may therefore consider the stochastic duration a concept, which is analogous to the classical one in the HJM-setting.

Example 2 (Interest rate cap) Consider a cap of the form

$$F = (R(t,T) - K)^+ ,$$

where K is the cap rate and $R(t,T)$ the average interest rate given by

$$R(t,T) = \frac{1}{T-t}\int_t^T r(s)\,\mathrm{d}s .$$

Here $r(t) = f(t,t)$ is the overnight interest rate, also known as "short rate". We observe that

$$\begin{aligned}\mathrm{D}_{r,y}\left(\frac{1}{T-t}\int_t^T r(s)\,\mathrm{d}s\right) &= \frac{1}{T-t}\int_t^T \mathrm{D}_{r,y}(r(s))\,\mathrm{d}s\\ &= \frac{1}{T-t}\int_t^T \mathrm{D}_{r,y}(f_s(0))\,\mathrm{d}s = \chi_{[0,t]}(r).\end{aligned}$$

Now let us approximate the function $\varphi(x) := (x-K)^+$ by functions φ_n with

$$\varphi_n(x) = \varphi(x) \quad \text{for } |x-K| \ge \frac{1}{n}$$

and

$$0 \le \varphi_n'(x) \le 1 \quad \text{for all } x .$$

Then it follows from Lemma 1 and Lemma 2 that

$$\mathrm{D}_{r,y}F = \chi_{[K,\infty)}(R(t,T)) \cdot \chi_{[0,t]}(r).$$

Example 3 (Asian option) Let us also have a look at the following Asian type of option defined as

$$F = \frac{1}{(\overline{x}_2 - \overline{x}_1)(T_2 - T_1)} \int_{\overline{x}_1}^{\overline{x}_2} \int_{T_1}^{T_2} f_t(x)\,\mathrm{d}t\,\mathrm{d}x.$$

Then

$$\mathrm{D}_{r,y}F = \frac{1}{(\overline{x}_2 - \overline{x}_1)(T_2 - T_1)} \int_{\overline{x}_1}^{\overline{x}_2} \int_{T_1}^{T_2} \chi_{[0,t]}(r)\,\mathrm{d}t\,\mathrm{d}x = \chi_{[0,t]}(r).$$

3 Estimation of Stochastic Duration and the Construction of Immunization Strategies

In the previous section we introduced the concept of stochastic duration $\mathrm{D}_{t,y}F$ and gave examples of interest rate derivatives F, whose stochastic duration can be computed, explicitly. In general, the stochastic duration of an interest claim or a complex bond portfolio cannot be determined, explicitly. The latter is also due to the fact that e.g. a dynamically hedged bond portfolio is a stochastically weighted sum of interest rate claims. The weights of the portfolio or hedging strategy at any time point are usually complicated functionals of the stochastic forward curve. In order to overcome this deficiency, we aim at resorting to an estimate of $\mathrm{D}_{t,y}F$. A reasonable estimate of $\mathrm{D}_{t,y}F$ could be the expected stochastic duration of F given the "observed" forward curves $\widehat{f}_s$, $0 \leq s \leq t$. This estimate naturally appears in the Clark–Ocone formula or as a solution of a backward stochastic differential equation (BSDE).

Using the fact that the set

$$\left\{\exp\left\{I_1(h) - \frac{1}{2}\|h\|_K^2\right\} : h \in K\right\}$$

is total in $L^2(\Omega, \widehat{\mathscr{F}}, \widehat{P})$ (see also [15]), one finds in connection with (28) that the Clark–Ocone formula w.r.t. the forward curves $\widehat{f}_t$ takes the following form:

$$F = E_{\widehat{P}}F + \int_0^T E\big[\mathrm{D}_s^*(F)|\widehat{\mathscr{F}}_s\big]\,\mathrm{d}\widehat{f}_s, \tag{37}$$

where the $\mathscr{B}([0,T]) \otimes \widehat{\mathscr{F}}, \mathscr{B}(H^*)$-measurable map $\mathrm{D}_\cdot^*(F) : [0,T] \times \Omega \to H^*$ can be linear isometrically identified with the Malliavin derivative (i.e. stochastic duration) $\mathrm{D}_\cdot F$ in Definition 1. Further $F \in L^2(\Omega, \widehat{\mathscr{F}}, \widehat{P})$ is in the domain of D^* and $\widehat{\mathscr{F}}_t$ is the $\widehat{P}$-completed filtration generated by $\widehat{f}_s$, $0 \leq s \leq t$.

The H^*-valued conditional expectation

$$E\big[\mathrm{D}_t^*(F)|\widehat{\mathscr{F}}_t\big], \quad 0 \leq t \leq T, \tag{38}$$

can be regarded as an estimate of $\mathrm{D}.F$. Now let us have a look at the BSDE

$$Y_t = Y_T - \int_t^T Z_s \,\mathrm{d}\widehat{f}_s, \tag{39}$$

where $Y_T = F$. Then we observe that

$$Z_t = E\big[\mathrm{D}_t^*(F)|\widehat{\mathscr{F}}_t\big] \quad \widehat{P}\text{-a.e.}$$

for $t \le T$ a.e.

We wish to recast the dynamics of the solution (Y_t, Z_t) in (39) with respect to the original measure P. Since σ_t is invertible t-a.e. we see that the natural filtration of $\widehat{W}_t$ coincides with the filtration $\widehat{\mathscr{F}}_t$. Assume that there exists a unique strong solution f_t^* of the SPDE

$$f_t^* = \int_0^t \sigma_s^{-1}\big[Af_s^* + \alpha_s(s,\cdot)\big]\,\mathrm{d}s + W_t, \quad 0 \le t \le T, \tag{40}$$

where W_t is the Q-cylindrical Wiener process in (24). See e.g. [40] for criteria about the existence and uniqueness of solutions of (non-linear) SPDE's.

Remark 3 Let $\alpha_t = b(t, f_t)$ in (40) for a Borel measurable map $b : [0,T] \times H \to H$. Impose on A the (rather strong) condition to be a bounded operator on H. Further assume that the drift coefficient $F(t,x) := \sigma_t^{-1}[Ax + b(t,x)]$ satisfies a linear growth and Lipschitz condition w.r.t. x (uniformly in t). Then Picard iteration gives a unique strong solution of (40).

Assumption (40) entails that the natural filtration of W_t is given by $\widehat{\mathscr{F}}_t$. Then it follows from (24) that the solution (Y_t, Z_t) in (39) has the following BSDE dynamics under P:

$$\begin{aligned} Y_t &= Y_T + \int_t^T Z_s\big[Af_s + \alpha_s(s,\cdot)\big]\,\mathrm{d}s - \int_t^T Z_s\,\mathrm{d}W_s^*, \\ Y_T &= F, \end{aligned} \tag{41}$$

where W^* is the square integrable H-valued martingale given by

$$W_t^* = \int_0^t \sigma_s \,\mathrm{d}W_s. \tag{42}$$

So we see that the estimate Z_t of the stochastic duration of F satisfies the forward-backward stochastic partial differential equation (FBSPDE)

$$\begin{aligned} \mathrm{d}f_t &= Af_t + \alpha_t\,\mathrm{d}t + \sigma_t\,\mathrm{d}W_t, \\ Y_t &= Y_T + \int_t^T Z_s\big[Af_s + \alpha_s(s,\cdot)\big]\,\mathrm{d}s - \int_t^T Z_s\,\mathrm{d}W_s^*, \end{aligned} \tag{43}$$

$$Y_T = F,$$

where F is a measurable functional of the solution of the forward SPDE, i.e of the forward curves f_t. For more information about (linear) forward-backward S(P)DE's the reader may consult the book of [31]. See also [38].

Remark 4 In view of financial applications it would be desirable to develop a numerical approximation scheme for solutions (Y_t, Z_t) of FBSPDE's of the type (43). In general, this is a challenging task. A possible ansatz to this problem (in some special cases) would be to employ the results in [46] or [36] in connection with Galerkin approximation. Another approach could be based on finite element or finite difference schemes in a BSPDE setting. In the framework of the linear Gaussian model (21) for the forward curves one can simplify further the numerical analysis by using dimension reduction techniques as e.g. principal component analysis of interest rate data. See e.g. [10].

Remark 5 Using stochastic distribution theory (see e.g. [43] or [16]) the concept of stochastic duration for interest rate claims $F \in \mathbb{D}^{1,2}$ can be extended to the case of claims contained in a space of generalized random variables which comprises the space of square integrable functionals of the forward curves (w.r.t. $\widehat{P}$). As a consequence we may still interpret Z_t in (43) as an estimate of the stochastic duration of a claim F, when $F \in L^2(P) \cap L^2(\widehat{P})$.

Finally, we want to discuss an extension of the concept of delta hedge of interest rate sensitive securities developed by [26] to a stochastic setting, which involves the fluctuations of the whole yield surface. The purpose of delta hedge is to immunize bond portfolios of interest rate sensitive securities under Ho's interest rate scenario [24]. In other words the idea devised by [26] is to neutralize given financial positions in interest rate derivatives against parallel shifts of i-years spot rates (or key rates).

We want to propose a mathematical framework which facilitates the construction of immunization strategies of interest rate sensitive portfolios in the sense of [26] with respect to stochastic fluctuations of the yield surface. In fact, we aim at minimizing the exposure of given financial positions to interest risk by going short in bonds of a generalized bond portfolio, that is of self-financing portfolios composed of infinitely many bonds of any maturity.

To this end we need some notions and conditions. Suppose that the generalized HJM-model (22) for the forward curves f_t fulfills the HJM no-arbitrage condition

$$\alpha_t(x) = \sum_{k \geq 1} \sigma_t^{(k)}(x) \left(I_x(\sigma_t^{(k)}) + \lambda_t^{(k)} \right), \tag{44}$$

where the processes $\lambda_t^{(k)}$, $k \geq 1$, are the Fourier coefficients of a predictable H-valued process

$$\lambda_t = \sum_{k \geq 1} \lambda_t^{(k)} e_k \, .$$

Here $\{e_k\}$ is an ONB of H. Further $\sigma_t^{(k)}$, $k \geq 1$ is given as in (16) and I_x is an linear functional in H^* defined by

$$I_x(f) = \int_0^x f(u)\,\mathrm{d}u. \tag{45}$$

We remark that the processes $\lambda_t^{(k)}$, $k \geq 1$ admit the interpretation of market prices of risk w.r.t. different bond maturities.

Now let us consider the discounted bond price curve $\widetilde{P}_t(\cdot)$ given by

$$\widetilde{P}_t(x) = \exp\left\{-\int_0^t f_s(0)\,\mathrm{d}s - \int_0^x f_s(x)\,\mathrm{d}s\right\}. \tag{46}$$

We require that the conditions

$$E\exp\left\{\int_0^t \langle\lambda_s, \mathrm{d}W_s\rangle_0 - \frac{1}{2}\int_0^t \|\lambda_s\|_0^2\,\mathrm{d}s\right\} = 1 \tag{47}$$

and

$$\int_0^t \left(\int_0^s \|\delta_{t-u} \circ \sigma_s\|_{L_2^0}^2\,\mathrm{d}u\right)^{1/2}\mathrm{d}s < \infty \tag{48}$$

hold for all $t \geq 0$.

Then, using Itô's formula and Girsanov's theorem one finds that

$$\widetilde{P}(t,T) = P(0,T) - \int_0^t P(s,T) I_{T-s} \circ \sigma_s\,\mathrm{d}\widetilde{W}_s, \tag{49}$$

where

$$\widetilde{W}_t = W_t + \int_0^t \lambda_s\,\mathrm{d}s$$

is a Q-Wiener process under a local martingale measure $\widetilde{P}$.

Define

$$\widetilde{\sigma}_t(\omega, x) = P_t(x) I_x \circ \sigma_t. \tag{50}$$

Let G be a separable Hilbert space in $C([0,\infty))$ such that evaluation functionals δ_x on G are continuous and the semigroup S_t of left shift operators is strongly continuous on G (see (17), (18)). From now on we assume that $\widetilde{\sigma}_t$ in (50) is a predictable $L(U_0, G)$-valued process such that $\int_0^T \|\widetilde{\sigma}_s\|_{L_2^0}^2\,\mathrm{d}s < \infty$ a.e. The latter implies that the bond price curves $\widetilde{P}_t$ are G-valued and satisfy

$$\mathrm{d}\widetilde{P}_t = A\widetilde{P}_t\,\mathrm{d}t - \widetilde{\sigma}_t\,\mathrm{d}\widetilde{W}_t \tag{51}$$

or

$$\mathrm{d}\widetilde{P}_t = (A\widetilde{P}_t - \widetilde{\sigma}_t[\lambda_t])\,\mathrm{d}t - \widetilde{\sigma}_t\,\mathrm{d}W_t \tag{52}$$

in the mild sense.

Now let consider generalized bond portfolios [3], that is the wealth process V_t of such portfolios is given by

$$V_t := V_t(\phi) := \phi_t[P_t(\cdot)], \quad t \geq 0, \tag{53}$$

where ϕ_t is a predictable G^*-valued process. The process ϕ_t can be regarded the trading strategy of an investor who manages a portfolio with infinitely many bonds of any maturity. For example, the strategy $\phi_t = \delta_{T-t}$ stands for buying and holding a zero-coupon bond with maturity T, since $\phi_t[P_t(\cdot)] = P(0,T)$.

Assume that

$$E_{\widetilde{P}} \int_0^t \|\phi_s \circ \widetilde{\sigma}_s\|^2_{L_2^0} \, \mathrm{d}s < \infty$$

for all $t \geq 0$. Then we shall say that a trading strategy ϕ_t, $t \geq 0$, is *self-financing* if there is $V_0 \in \mathbb{R}$ such that

$$\widetilde{V}_t(\phi) - \int_0^t \phi_s \circ \widetilde{\sigma}_s \, \mathrm{d}\widetilde{W}_s = V_0 \tag{54}$$

for all $t \geq 0$ a.e., where $\widetilde{V}_t(\phi)$ is the discounted wealth process given by

$$\widetilde{V}_t(\phi) = \phi_t[\widetilde{P}_t(\cdot)].$$

See e.g. [3]. We denote the set of all self-financing strategies by $\mathscr{A}$.

Remark 6 In the infinite-dimensional HJM-framework the existence of a unique martingale measure does not imply in general that the bond market given by (49) is complete. The latter is a deficiency not shared finite-rank models. However, since the kernels of $\widetilde{\sigma}_t$ (50) are zero t-a.e. our bond market is approximatively complete in the following sense: For all $\varepsilon > 0$ there exists a strategy $\phi^{\varepsilon} \in \mathscr{A}$

$$E_{\widetilde{P}} \left(E_{\widetilde{P}}[\widetilde{h}] + \int_0^T \phi_s^{\varepsilon} \circ \widetilde{\sigma}_s \, \mathrm{d}\widetilde{W}_s - \widetilde{h} \right)^2 < \varepsilon,$$

where $\widetilde{h}$ a discounted contingent claim. See e.g. [3].

Suppose that a trader is long in interest rate securities at time $t \geq 0$ whose price process is L_t. In order to neutralize the risk coming from the fluctuations of the yield surface the trader wishes to go short in the generalized bond portfolio (54) for a self-financing strategy $\phi^* \in \mathscr{A}$ such that ϕ^* minimizes at any time point the worst scenario interest rate sensitivity of the resulting portfolio. More precisely, the trader tries to find a $\phi^* \in \mathscr{A}$ such that

$$\inf_{\phi^* \in \mathscr{A}} E \int_0^T \|\mathrm{D}.(L_t - V_t(\phi))\|_K^2 \, \mathrm{d}t = E \int_0^T \left\|\mathrm{D}.(L_t - V_t(\phi^*))\right\|_K^2 \, \mathrm{d}t < \infty, \tag{55}$$

where K is the RKHS of the forward curves. Note that

$$\sup_{\|k\|_K=1} \langle \mathrm{D}.F, k\rangle_K = \|\mathrm{D}.F\|_K$$

for an interest claim $F \in \mathbb{D}^{1,2}$. So (35) admits the interpretation that $\|D.F\|_K$ is the "worst scenario" sensitivity with respect to all directional interest changes $k \in K$.

Using the estimate $Z. = Z.(F)$ for the stochastic duration $\mathrm{D}.(F)$ in the FBSPDE (43) for $F = L_t - V_t(\phi)$ (see Remark 5) and relation (28) the optimization problem (55) then takes the form

$$\begin{aligned} &\inf_{\phi^* \in \mathscr{A}} E \int_0^T \int_0^T \|Z_u(L_t - V_t(\phi)) \circ \sigma_u\|^2_{L_2^0} \, \mathrm{d}u \, \mathrm{d}t \\ &= E \int_0^T \int_0^T \left\|Z_u(L_t - V_t(\phi^*)) \circ \sigma_u\right\|^2_{L_2^0} \mathrm{d}u \, \mathrm{d}t < \infty \end{aligned} \tag{56}$$

for $\phi^* \in \mathscr{A}$.

We see that the construction of an immunization strategy boils down to an optimal control problem of the FBSPDE (43) or the FBSPDE

$$\begin{aligned} \widetilde{V}_t(\phi) &= \widetilde{V}_0(\phi) - \int_0^t \phi_s \circ \widetilde{\sigma}_s \, \mathrm{d}\widetilde{W}_s, \\ Y_t &= Y_T + \int_t^T Z_s \left[Af_s + \alpha_s(s,\cdot)\right] \mathrm{d}s - \int_t^T Z_s \, \mathrm{d}W_s^*, \\ Y_T &= F, \end{aligned} \tag{57}$$

where $F = L_t - V_t(\phi)$ for each t, if L_t is a measurable functional of $\widetilde{V}.(\phi)$.

An approach to tackle this problem could be based on a stochastic maximum principle for FBSPDE's. See [22]. From a practical point of view it would be important to find numerical approximation schemes for a delta hedge $\phi^* \in \mathscr{A}$.

Remark 7

1. It is conceivable that the concept of g-expectation by [39] for BSDE's can be generalized to FBSPDE's of the type (43). The latter would enable the construction of risk measures of functionals of forward curves. Such a construction would reveal the role of the stochastic duration as a building block for general interest rate risk measures.
2. We point out that our framework also allows for the definition of *stochastic convexity*, that is a measure of "curvature" w.r.t. to the fluctuations of the yield surface. It makes sense to define the stochastic convexity of a twice Malliavin differentiable interest rate claim F as

$$\mathrm{D.D.}(F) \in L^2(\Omega, \widehat{\mathscr{F}}, \widehat{P}; K \otimes K). \tag{58}$$

Acknowledgements We thank Professor V. Mandrekar for his valuable comments on this work.

Appendix: Macaulay Duration and Portfolio Immunization

A.1 Discrete Case

In Macaulay's original concept duration was the weighted average by present value of the number of periods to maturity for a series of cash flows, typically those of interest and principal payments for a bond, normalized by the total present value [32]. For notation, let V be the present value (or price) of the bond, $r > 0$ be the (constant) rate of interest, and n be the number of periods to maturity. The expression

$$\mathscr{A}(r,n) = \frac{1-(1+r)^{-n}}{r}$$

is the closed form for the present value of an annuity in arrears for n periods at rate r, reflecting the typical payment scheme of a bond, e.g. a United States Treasury bond. Therefore the *Macaulay duration* d_{Mac} has the following definition for equally spaced cash flows of size C and return of principal P:

$$d_{\mathrm{Mac}} := \frac{C\sum_{k=1}^{n} k(1+r)^{-k} + nP(1+r)^{-n}}{C\sum_{k=1}^{n}(1+r)^{-k} + P(1+r)^{-n}}$$

or

$$d_{\mathrm{Mac}} = -(1+r)\frac{\partial}{\partial r}\log\left[C\,\mathscr{A}(r,n) + P(1+r)^{-n}\right]. \tag{A.1}$$

In the simple case of single cash flow—a zero coupon bond—Macaulay duration reduces to the number of periods n to that payment, justifying the name.

Soon, however, practitioners began preferring a version of duration as the simple negative of the derivative of V with respect to r, dropping the factor $(1+r)$. This version became known as the *modified duration* d_{mod}, with this definition:

$$d_{\mathrm{mod}} := -\frac{\partial}{\partial r}\log\left[C\,\mathscr{A}(r,n) + P(1+r)^{-n}\right]. \tag{A.2}$$

Such redefinition provides the relationship

$$d_{\mathrm{Mac}} = (1+r)\,d_{\mathrm{mod}},$$

so that the modified duration of a zero coupon bond is $(1+r)n$.

In ordinary parlance, either form of duration is stated as a positive number, e.g., "The duration of this bond is ten years", as indicated. A rationale exists, however, for stating the duration as a negative number, reflecting the inverse relationship between changes in the level of interest and changes in price. Such versions, inverting the minus signs of (A.1) and (A.2), more typically appear in Taylor series expansions of bond price, and in more developed mathematical expositions. The latter approach is assumed in this paper.

A.2 Continuous Case

The continuous case is a straightforward extension of the discrete case. Let C, as previously, be the cash flow assigned to a single period, but consider it divided equally into j parts flowing at the ends of j equally spaced *sub-periods*. As well, consider the interest rate r as that assigned to the entire period, but let it be divided by j providing a *sub-rate* for compounding across the sub-periods. Then term $C\mathscr{A}(r,n)$ of (A.1) then becomes

$$\begin{aligned} C\mathscr{A}(r,n) &:= \lim_{j\to\infty} \frac{C}{j} \frac{1-(1+r/j)^{-jn}}{r/j} \\ &= C\,\frac{1-e^{-rn}}{r}. \end{aligned}$$

So, if

$$\widehat{\mathscr{A}}(r,n) := \frac{1-e^{-rn}}{r},$$

then (A.1) and (A.2), respectively, become

$$\widehat{d}_{\text{Mac}} = -\frac{\partial}{\partial r} \log\left[C\widehat{\mathscr{A}}(r,n) + Pe^{-rn}\right]$$

and

$$\widehat{d}_{\text{mod}} = -\frac{\partial}{\partial r} \log\left[C\widehat{\mathscr{A}}(r,n) + Pe^{-rn}\right]$$

in the latter case because $\lim_{j\to\infty}(1+r/j) = 1$. So

$$\widehat{d}_{\text{Mac}} = \widehat{d}_{\text{mod}}, \tag{A.3}$$

justifying the use of the combined name *continuous duration* for both versions. As in the case of discrete Macaulay duration, in the simple case of a zero coupon bond continuous duration reduces to the number of periods n to that payment.

An alternative description of this result is that the modified duration is a continuous approximation to the Macaulay duration, or conversely, the Macaulay duration is a discrete approximation to the modified duration. As $n \to \infty$ with rn constant the two definitions merge.

It is stated without proof that the other common form of annuity timing, payments in advance, i.e., at the beginnings of the compounding periods rather than at the ends, results in the same continuous forms of (A.3).

A.3 Portfolio Immunization

An active part of portfolio management is the targeting of a specific duration. For example, a pension fund manager may wish to have a value certain at some future time $t = T$, starting at $t = 0$ now. Consider two portfolios A and B, with respective durations d_A and d_B, and present values (prices) of v_A and v_B. If these portfolios are combined, then the new portfolio $A + B$ has duration

$$d_{A+B} = \frac{v_A}{v_A + v_B} d_A + \frac{v_B}{v_A + v_B} d_B.$$

If A be the portfolio to be immunized to desired duration d_{A+B}, then one can solve for v_B knowing all other quantities. Specifically,

$$v_B = \frac{d_{A+B} - d_A}{d_B - d_{A+B}} v_A,$$

which may be positive or negative. If negative one can interpret the result as an amount proportioned to portfolio B to be sold from portfolio A to achieve the objective, or alternatively, the amount to sell short of portfolio B.

Bond immunization is a very big business. In recent years Japanese banking interests have been heavy buyers of 30-year United States Treasury Bond strips—having a duration of 30 years—in order to extend the durations of portfolios. The activity has been so significant as to keep the longest-term yields below those of somewhat shorter-term yields for extended periods of time, even in strongly positive yield curve environments otherwise.

References

1. Aihara, S.I., Bagchi, A.: Stochastic hyperbolic dynamics for infinite-dimensional forward rates and option pricing. Math. Finance **15**(1), 27–47 (2005)
2. Bensoussan, A.: Filtrage Optimal des Systèmes Linéaires. Dunod, Paris (1971)
3. Björk, T., Di Masi, G., Kabanov, Y., Runggaldier, W.: Towards a general theory of bond markets. Finance Stoch. **1**, 141–174 (1997)
4. Björk, T., Christensen, B.J., Gombani, A.: Some system theoretic aspects of interest rate theory. Insur. Math. Econ. **22**, 17–23 (1998)
5. Björk, T., Gombani, A.: Minimal realizations of interest rate models. Finance Stoch. **3**, 413–432 (1999)
6. Black, F., Derman, E., Toy, W.: A one-factor model of interest rates and its application to treasury bond options. Financ. Anal. J. **46**(1), 33–39 (1990)
7. Borel, C.: Gaussian random measures on locally convex space. Math. Scand. **38**, 265–284 (1976)
8. Brace, A., Musiela, M.: A multifactor Gauss Markov implementation of Heath, Jarrow, and Morton. Math. Finance **4**(3), 259–283 (1994)
9. Brace, A., Gatarek, D., Musiela, M.: The market model of interest rate dynamics. Math. Finance **7**(2), 127–155 (1997)
10. Carmona, R., Tehranchi, M.: Interest Rate Models: an Infinite Dimensional Stochastic Analysis Perspective. Springer Finance. Springer, Berlin (2006)
11. Chatterji, S.D., Mandrekar, V.: Equivalence and singularity of Gaussian measure and applications. In: Barucha-Reid, A.T. (ed.) Probabilistic Analysis and Related Topics, vol. 1, pp. 169–197. Academic Press, New York (1978)
12. Chen, L.: Interest Rate Dynamics, Derivatives Pricing, and Risk Management. Lecture Notes in Economics and Mathematical Systems, vol. 435. Springer, Berlin (1996)

13. Chen, L.: A three factor model of the term structure of interest rates and its applications in derivatives pricing and risk management. Financ. Mark. Inst. Instrum. **5**(1), 1–89 (1996)
14. Cox, J.C., Ingersoll, J.E. Jr., Ross, S.A.: A theory of the term structure of interest rates. Econometrica **53**(2), 385–407 (1985)
15. Da Prato, G., Zabczyk, J.: Stochastic Equations in Infinite Dimensions. Cambridge University Press, Cambridge (1992)
16. Di Nunno, G., Øksendal, B., Proske, F.: Malliavin Calculus for Lévy Processes with Applications to Finance. Universitext. Springer, Berlin (2009)
17. Filipović, D.: Consistency Problems for Heath–Jarrow–Morton Interest Rate Models. LNM, vol. 1760. Springer, Berlin (2001)
18. Filipović, D., Zabczyk, J.: Markovian term structure models in discrete time. Ann. Appl. Probab. **12**(2), 710–729 (2002)
19. Filipović, D., Tappe, S.: Existence of Lévy term structure models. Finance Stoch. **12**, 83–115 (2008)
20. Gawarecki, L., Mandrekar, V.: Itô-Ramer, Skorohod and Ogawa integrals with respect to Gaussian processes and their interrelationship. In: Perez-Abreu, V., Houdre, C. (eds.) Chaos Expansions, Multiple Wiener-Ito Integrals, and Their Applications, pp. 349–373. CRC Press, London (1993)
21. Goldys, B., Musiela, M., Sondermann, D.: Lognormality of rates and term structure models. Stoch. Anal. Appl. **18**(3), 375–396 (2000)
22. Haadem, S., Mandrekar, V.: A stochastic maximum principle for forward-backward SPDE's. Manuscript in preparation, University of Oslo (2010)
23. Heath, D., Jarrow, R., Morton, A.: Bond pricing and the term structure of interest rates: a new methodology for contingent claims valuation. Econometrica **60**, 77–105 (1992)
24. Ho, T.S.Y.: Key rate durations: measures of interest rate risk. J. Fixed Income **2**(2), 29–44 (1992)
25. Hull, J., White, A.: Pricing interest-rate-derivative securities. Rev. Financ. Stud. **3**(2), 573–592 (1990)
26. Hull, J., White, A.: The optimal hedge of interest rate sensitive securities. Research note, University of Toronto (1994)
27. Jarrow, R.A.: The relationship between yield, risk, and return of corporate bonds. J. Finance **33**(4), 1235–1240 (1978)
28. Kai, L.: Stability of Infinite Dimensional Stochastic Differential Equations with Applications. Monographs and Surveys in Pure and Applied Mathematics, vol. 135. Chapman & Hall/CRC Press, London (2006)
29. Kunita, H.: Stochastic Flows and Stochastic Differential Equations. Cambridge University Press, Cambridge (1990)
30. Lee, S.B., Ho, T.S.Y.: Term structure movements and pricing interest rate contingent claims. J. Finance **41**(5), 1011–1029 (1986)
31. Ma, J., Yong, J.: Forward-Backward Stochastic Differential Equations and Their Applications. LNM, vol. 1702. Springer, Berlin (1999)
32. Macaulay, F.R.: Some theoretical problems suggested by the movements of interest rates, bond yields and stock prices in The United States Since 1856. Columbia University Press, New York (1938)
33. Mandrekar, V., Zhang, S.: Skorohod integral and differentiation for Gaussian processes. In: R. R. Bahadur Festschrift. Wiley, New Delhi (1993)
34. Musiela, M.: General framework for pricing derivative securities. Stoch. Process. Appl. **55**, 227–251 (1995)
35. Musiela, M., Rutkowski, M.: Continuous-time term structure models: forward measure approach. Finance Stoch. **1**, 261–291 (1997)
36. Nakayama, T.: Approximation of BSDE's by stochastic difference equation's. J. Math. Sci. Univ. Tokyo **9**, 257–277 (2002)
37. Nualart, D.: The Malliavin Calculus and Related Topics. Springer, Berlin (1995)

38. Øksendal, B., Proske, F., Zhang, T.: Backward stochastic partial differential equations with jumps and application to optimal control of random jump fields. Stochastics **77**(5), 381–399 (2005)
39. Peng, S.: Backward SDE and related $g-$expectation. In: El Karoui, N., Mazliak, L. (eds.) Backward Stochastic Differential Equations. Pitman Research Notes in Math. Series, vol. 364. Springer, Berlin (1997)
40. Prévot, C., Röckner, M.: A Concise Course on Stochastic Partial Differential Equations. LNM, vol. 1905. Springer, Berlin (2007)
41. Rendleman, R.J. Jr., Bartter, B.J.: The pricing of options on debt securities. J. Financ. Quant. Anal. **15**(1), 11–24 (1980)
42. Ritchken, P.H., Sankarasubramanian, L.: Volatility structures of forward rates and the dynamics of the term structure. Math. Finance **5**(1), 55–72 (1995)
43. Üstünel, A.S.: An Introduction to Analysis on Wiener Space. LNM, vol. 1610. Springer, Berlin (1995)
44. Vargiolu, T.: Invariant measures for the Musiela equation with deterministic diffusion term. Finance Stoch. **3**, 483–492 (1999)
45. Vašíček, O.A.: An equilibrium characterization of the term structure. J. Financ. Econ. **5**, 177–188 (1977)
46. Zhang, J.: A numerical scheme for BSDE's. Ann. Appl. Probab. **14**(1), 459–488 (2004)

On the First Passage Time Under Regime-Switching with Jumps

Masaaki Kijima and Chi Chung Siu

Abstract In this paper, we present the analytical solution for the Laplace transform of the joint distribution of the first passage time and undershoot/overshoot value under a regime-switching jump-diffusion model. With the help of some martingale technique, the Laplace transform of the first passage time becomes the solution of a system of linear equations. The methodology discussed here is fairly elementary and can be applied to many stopping-time problems under a regime-switching model with jump risks. Some numerical examples are given to demonstrate the usefulness of our method.

Keywords First passage time · Regime-switching jump-diffusion model

Mathematics Subject Classification (2010) 90G20

1 Introduction

The first-passage-time problem has been one of the recurrent themes in the theory of stochastic processes. Closed-form expressions of the first passage time distribution prove to be vital in solving of many stochastic modeling problems. In the theory of option pricing, for example, studies on path-dependent options often reduce to the problem involving the first-passage-time distribution of the underlying processes; see Shreve [23]. In real options literature, optimal investment decisions are formulated as the first-passage-time problem; see, for example, Guo et al. [11]. Consequently, a systematic treatment of such problems can yield a wide variety of

M. Kijima
Graduate School of Social Sciences, Tokyo Metropolitan University, 1-1 Minami-Ohsawa, Hachiohji, Tokyo 192-0397, Japan
e-mail: kijima@tmu.ac.jp

C.C. Siu (✉)
UTS Business School, City Campus, 15 Broadway, Ultimo, NSW 2007, Australia
e-mail: chichung.siu@uts.edu.au

Y. Kabanov et al. (eds.), *Inspired by Finance*, DOI 10.1007/978-3-319-02069-3_18,

applications in finance. For this reason, there are by now many analytical expressions for the first-passage-time distributions for many stochastic processes, in both discrete- and continuous-time cases.

In many situations, close observation reveals that the analytical tractability is due to the Markovian structure of the underlying stochastic processes. A classical example is the geometric Brownian motion. Such Markovian structure enables one to recover, e.g., the density function of the maximum of Brownian motion. Another prominent class of Markov processes are Lévy processes. It is well-known that Lévy processes also possess the strong Markov property and many studies have been devoted to finding the analytical expressions for the stopping time problem for general Lévy processes.

However, difficulties immediately arise due to the inherent jump structures found in the Lévy models. Unlike the Brownian motions, where one knows exactly the location of a process at the first passage time, the overshoot/undershoot problem poses a great challenge to the study of the first passage time problem in the general Lévy processes. It is also for this reason that the mathematical machinery of stopping time problems becomes immensely involved when one makes a transition from the Brownian motion to the general Lévy processes.

One common tool used in the study of first passage times under general Lévy processes is the Wiener–Hopf factorization, which makes use of the fluctuation identities of Lévy processes. For the complete overview of the first passage times under general Lévy processes through this technique, the reader is referred to Kyprianou [17].

Besides the technicality, it has been shown that the undershoot and overshoot problems cannot be handled simultaneously under the general Lévy framework. Restriction to one-sided jumps (i.e., the case when either upward or downward jumps are allowed) is often made to retain some tractability of the stopping time distributions. However, in the option pricing literature, one-sided Lévy processes prove to be of limited use when one works with problems involving first exit time from an interval, in which both undershoot and overshoot features emerge concurrently.

Although for a general Lévy process one cannot simultaneously handle the undershoot and overshoot problems, there exists one special subclass of Lévy processes for which this problem can be easily solved. Kou and Wang [16] seems to be the first who solved the first-passage-time problem with two-sided jumps whose jump sizes follow a double-exponential distribution without making use of fluctuation theory.[1] Sepp [22] and Cai et al. [6] apply the double-exponential jump model for the pricing of different types of double barrier options. Asmussen et al. [4] generalize the results further by assuming that jump sizes follow phase-type distributions.[2]

[1]Mordecki [19, 20] solved the first-passage-time problem using one-sided jumps with exponential distribution without making use of fluctuation theory.

[2]Asmussen et al. [4] prove that the set of all phase-type jump-diffusion models is dense in the Lévy family, making it a suitable candidate to approximate any Lévy model with those of the phase-type distributions.

The important characteristics of the first-passage-time problem under this special class of Lévy processes is the conditional memoryless and independence properties. When jump sizes are exponentially distributed, the overshoot/undershoot values are independent of the underlying stochastic process before and at the stopping time. These properties greatly reduce the mathematical machinery required to solve the stopping time problem. Moreover, many stopping time problems under this subclass of Lévy processes possess closed-form expressions up to the Laplace transforms. As a result, the only numerical method needed is to perform a numerical Laplace inversion.[3]

In the asset pricing literature, the motivation of using Lévy models is to capture the short-run behavior of security prices. Although Lévy models successfully provide many important characteristics of the short-run behavior, the long-run phenomena remain unsolved by the ordinary Lévy models. In particular, close examination of many financial time series reveals that high volatility environment is persistent for some period of time, followed by a low volatility environment. Such feature is commonly known as the *volatility clustering*. Because Lévy processes have independent increments by definition, they cannot capture the volatility clustering feature at all.

A prominent class of stochastic processes that are used to explain the volatility clustering is collectively known as the stochastic volatility model, i.e., modeling volatility dynamics as another stochastic process. When one studies the time series of a security price, it often reveals that the price seems to follow some kind of business cycle, i.e., high volatility case can be seen as a busy period, whereas low volatility case is considered to be an idle period. The business cycle can be modeled through regime-switching models in many financial econometrics studies.

An analytical way of solving the first passage time under regime switching was first approached by Guo [9], who provided the analytical expression of the stopping time problem under geometric Brownian motion with two regimes. Since then, this framework has been employed and expanded toward the pricing of perpetual American options (see Guo and Zhang [10]; Jobert and Rogers [13]), in which multiple regimes are considered. The regime-switching Brownian motion also enables one to obtain closed-form expressions for the optimal investment time problem, subject to the business cycle; see Guo et al. [11].

If the Lévy models can handle the short-run behavior, while the regime-switching can tackle the long-run fluctuation, it makes sense to consider a regime-switching Lévy model as a prominent candidate that can capture both the long-run and short-run behaviors of the underlying securities simultaneously. In particular, it is natural to ask if certain subclass of the regime-switching Lévy models provides an analytical solution to the first-passage-time problem.

Recent papers (Jiang and Pistorius [12]; Mijatović and Pistorius [18]) provide the analytical expressions for the perpetual American options and other exotic options. Their technique mainly relies on the judicious use of matrix Wiener–Hopf

[3]Of course, the same conclusion holds for the case of Fourier transform. In fact, Borovkov and Novikov [5] illustrate the explicit relationship between moment generating function of the underlying process and its path-dependent payoff through pricing discrete lookback options under a general Lévy framework.

factorization. Despite being mathematically elegant, the matrix Wiener–Hopf factorization poses to be difficult to obtain numerical solutions without making use of advanced numerical techniques. Moreover, the complexity of Wiener–Hopf factorization makes the first-passage-time problem looks rather unrevealing.

For the sake of computation, we aim to provide simpler characterization of the first-passage-time problem under the regime-switching Lévy model. As explained later, under some situation, we can retain the analytical tractability of some regime-switching Lévy models. Moreover, our methodology involves only solving a system of linear equations and only numerical method needed is a Laplace inversion. Such simplicity and efficiency are essential when one wants to price derivatives under the regime-switching Lévy models.

Finally, very recently, Carr and Crosby [7] derived semi-closed form solution of the first-passage-time problem for a particular regime-switching Lévy model. Yet, our methodology is different from theirs in the sense that we appeal to pure probabilistic tools instead of guessing a solution to some ordinary integral-differential equation (OIDE for short) as considered in Carr and Crosby [7].[4] Although our approach and Carr and Crosby's approach can serve as alternative ways to compute the first-passage-time distribution under regime-switching Lévy processes, we believe that the probabilistic approach provides more insights into the first-passage-time problems.

The rest of the paper is organized as follows. Section 2 provides the background of the regime-switching Lévy model that is used throughout the paper. Section 3 discusses the conditional independence and memoryless properties of the regime-switching Lévy process and the corresponding first-passage-time problem. Section 4 provides numerical illustrations through computation of first passage probabilities. Section 5 concludes the paper. A brief discussion of Laplace inversion method is provided in the Appendix.

2 Regime-Switching Jump-Diffusion Process

In an attempt to capture both the long-run and short-run behaviors of financial securities, we assume that the process follows a regime-switching Lévy process. In this section, we provide the necessary background for the study of our problem. For a complete treatment, the reader is referred to, e.g., Asmussen [2].

Denote by X_t an ordinary Lévy process with $X_0 = 0$. For the case of practical interest, we shall confine to a class of finite-activity jump diffusion processes. Hence, X_t can be described by the stochastic differential equation (SDE for short)

$$\mathrm{d}X_t = \mu \mathrm{d}t + \sigma \mathrm{d}W_t + \mathrm{d}\left(\sum_{i=1}^{N_t} V_i\right), \tag{1}$$

[4] In their paper, they first take the Laplace transform with respect to time to convert the partial integral-differential equation into the corresponding OIDE. Hence, the desired results are subsequently obtained through the inversion of Laplace transform of the solution to the OIDE.

where μ and σ are constants describing the drift and volatility, respectively, W_t is the standard Brownian motion, N_t denotes a Poisson process with constant arrival rate λ, and $\{V_i,\ i = 1, 2, \dots\}$ is the sequence of independent and identically distributed random variables. Each V_i represents a random jump size. All the random quantities are mutually independent. The distribution of V_i is denoted by $\nu(\mathrm{d}x)$. Since the coefficients of the SDE (1) are constants, it has a strong solution. The solution of the SDE (1) is given by

$$X_t = \mu t + \sigma W_t + \sum_{i=1}^{N_t} V_i, \tag{2}$$

where $\sum_{i=1}^{0} V_i = 0$.

Let $\{J_t\}$ be a Markov chain with state space E. For simplicity, we assume that E is finite and contains n elements, i.e., $E = \{1, 2, \dots, n\}$. Let $\mathbf{Q}$ be the intensity matrix of J_t with respect to Lebesgue measure, i.e.

$$\mathbf{Q} = \left\{q_{ij}\right\}_{i,j\in E},$$

where

$$q_{ii} = -\sum_{i\neq j} q_{ij}.$$

The regime-switching Lévy process (X_t^J, J_t) is constructed as

$$\mathrm{d}X_t^J = \mu(J_t)\mathrm{d}t + \sigma(J_t)\mathrm{d}W_t + \mathrm{d}\left(\sum_{i=1}^{N_t(J_t)} V_i(J_t)\right), \tag{3}$$

where W_t denotes the standard Brownian motion and where, given $J_t = j \in E$, $\mu(j)$ and $\sigma(j)$ are constants, $N_t(j)$ denotes a Poisson process with constant arrival rate $\lambda(j)$, and $V(j)$ represents a random jump size with distribution $\nu_j(\mathrm{d}x)$. Let $J_t = j$ and assume that J_t does not jump during the time interval $[t, t+\delta]$, i.e., $J_s = j$ for all $s \in [t, t+\delta]$. Then X_s^J evolves as a Lévy process given by (1) with parameters $\mu(j)$, $\sigma(j)$, $\lambda(j)$ and jump distribution $\nu_j(\mathrm{d}x)$. Note that, under the current setting, the bivariate process (X_t^J, J_t) is jointly Markovian, although X_t^J itself is not Markovian with respect to the filtration generated by the underlying process. Hence, the SDE (3) does not admit such a simple solution as (2), although the integral representation is possible.

Similar to the case of ordinary Lévy processes, the moment generating function (MGF for short) represents the dynamics of the regime-switching Lévy process X_t^J. We denote by $\mathbf{F}_t[u]$ the $n \times n$ matrix with the ijth element $\mathbb{E}_i[\mathrm{e}^{uX_t^J};\ J_t = j]$, where

$$\mathbb{E}_i[\cdot] \equiv \mathbb{E}[\cdot|J_0 = i].$$

Then, from Proposition 5.2 in Asmussen [2], we have $\mathbf{F}_t[u] = e^{t\mathbf{K}[u]}$, where[5]

$$\mathbf{K}[u] = \mathbf{Q} + \{\kappa_j(u)\}_{diag} \tag{4}$$

and

$$\kappa_j(u) = \mu(j)u + \frac{\sigma^2(j)}{2}u^2 + \int_{-\infty}^{\infty} (e^{uy} - 1)\nu_j(dy). \tag{5}$$

Moreover, when $\nu_j(\cdot)$ is double exponentially distributed, i.e.,

$$\nu_j(dy) = \lambda(j)(p_j \eta_{j1} e^{-\eta_{j1}y} 1_{\{y\geq 0\}} + (1 - p_j)\eta_{j2} e^{\eta_{j2}y} 1_{\{y<0\}})dy, \tag{6}$$

where $\eta_{j1} > 1$, $\eta_{j2} > 0$ and $0 \leq p_j \leq 1$, it follows from Asmussen et al. [4] that we can remove all the jumps from the original process X_t^J through a transformation, called *fluidization*. Fluidization is possible mainly due to the independence and memoryless properties when X_t^J makes jump. Assuming these properties for the moment,[6] we will work with the fluid counterpart of X_t^J, which is denoted by $\tilde{X}_t^J$ throughout the paper.

In simple terms, the fluid model $\tilde{X}_t^J$ replaces the upward jump by a linear segment with slope of 1 and downward jump by a linear segment with slope of -1. To move from the original regime-switching Lévy model to its fluid counterpart, we shall augment the state space. Denote by $E_{(j,0)}$, $E_{(j,+)}$ and $E_{(j,-)}$ the states in which the process behaves as a pure diffusion, an upward jump and a downward jump, respectively, when the state is j. Hence, with $J_t = j$ fixed, we have turned the Lévy process to a process with positively-sloped segment as one state, negatively-sloped segment as another state, and the Brownian motion as the non-jump state. Under such characterization, the transformed process no longer possesses jumps, whence it has continuous sample paths. The state space of the regime-switching fluid model is denoted by

$$\tilde{E} = \{E_{(1,0)}, E_{(1,+)}, E_{(1,-)}, E_{(2,0)}, E_{(2,+)}, E_{(2,-)}, \ldots, E_{(n,0)}, E_{(n,+)}, E_{(n,-)}\},$$

and the process indicating the underlying state by $\tilde{J}_t$. Figure 1 provides a graphical representation of such transformation.

Note that the time frame under the fluid model is different from that of the original model, as we stretch the time when the process makes jumps. In other words, the time frame of the fluid model distorts the original time. In order to study the original stopping-time problem using the fluid model, we must restrict the elongated time so that the stopping time under the fluid model has the same distribution as the stopping time under the original model. Intuitively, before the stopping time of the

---

[5]More generally, we can consider a possibility that the Markov chain J_t changes the state at the same time as X_t^J makes a jump due to the Lévy component. However, for the sake of tractability, we rule out such simultaneous jumps in this paper.

[6]We shall show rigorously the independence and memoryless properties of X_t^J in Sect. 3.1.

Fig. 1 Fluidization

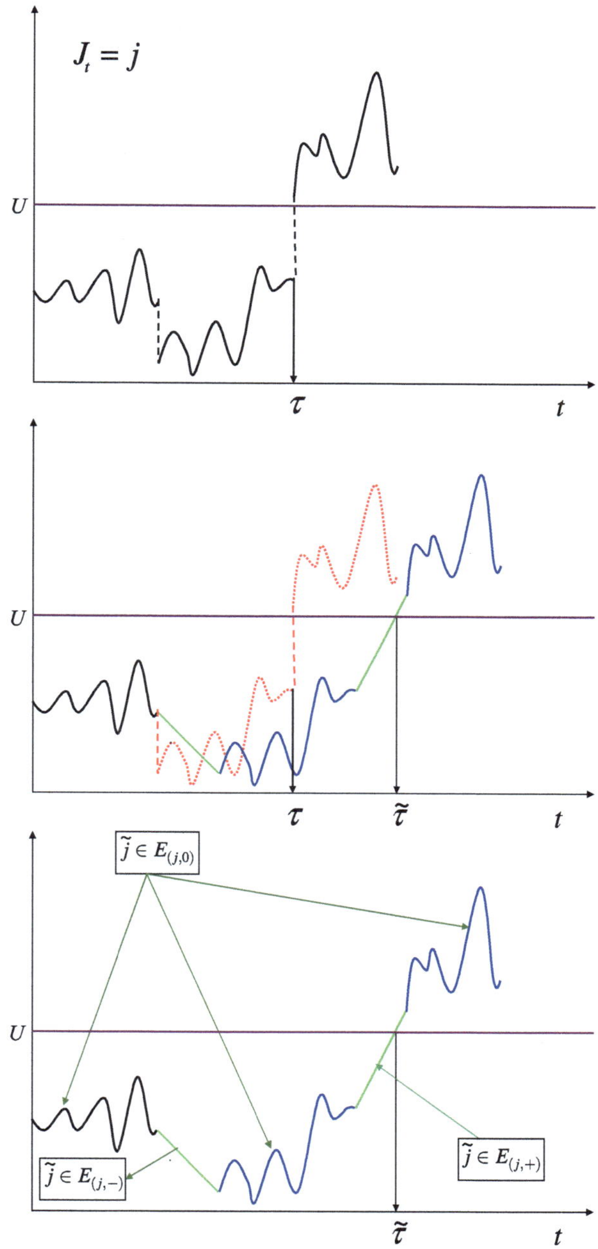

fluid model, we need to account only for the time in which the process behaves as a Brownian motion. To invoke such time restriction, we shall follow the concepts adopted in Jiang and Pistorius [12] to define *virtual time* and its right-continuous inverse. Throughout the paper, we denote by 1_A as the indicator function, meaning that $1_A = 1$ if A is true and $1_A = 0$ otherwise.

Definition 1 A function $T:\mathbb{R}\to\mathbb{R}$ is called a virtual time and is defined as, for every $t\geq 0$,

$$T(t)=\int_0^t 1_{\{\tilde{J}_s\in E_0\}}\mathrm{d}s,$$

where $E_0=\bigcup_j E_{(j,0)}$. The right-continuous inverse of T is defined as

$$T^{-1}(s)=\inf\{t\geq 0: T(t)>s\}.$$

From Definition 1, the virtual time $T(t)$ takes out all the elongated time due to jumps. Furthermore, by the definition of inverse $T^{-1}(s)$ of the virtual time, it follows that $(\tilde{X}^J_{T^{-1}(t)},\tilde{J}_{T^{-1}(t)})$ and (X^J_t, J_t) have the same distribution.

Note that the restriction also applies to the stopping times, and thus one can conclude that $T(\tilde{\tau})$ and τ agree almost surely, where τ is a stopping time of the original model and $\tilde{\tau}$ is the corresponding stopping time of the fluid model. See Fig. 1 for example. Because this observation plays a key role later when the stopping-time problem is considered, we state it formally as the following lemma.

Lemma 1 *Let $T(t)$ be the virtual time of the fluid model. For a stopping time τ of the original jump model, we have $T(\tilde{\tau})=\tau$ almost surely, where $\tilde{\tau}$ is the corresponding stopping time of the fluid model.*

2.1 A Special Case: Two Regimes

Consider for simplicity a two-state regime-switching Lévy model with double-exponential jumps. That is, we assume that $E=\{1,2\}$ and the jumps follow the distributions given by (6).

Denote the fluid counterpart of X^J_t by $\tilde{X}^J_t$. The state space of the fluid model is given by

$$\tilde{E}=\{E_{(1,0)},E_{(1,+)},E_{(1,-)},E_{(2,0)},E_{(2,+)},E_{(2,-)}\}.$$

The state indicator process is denoted by $\tilde{J}_t$.

The corresponding MGF is given by

$$\tilde{\mathbf{K}}[u]=\begin{pmatrix}\tilde{\mathbf{K}}^1[u] & \mathbf{O}\\ \mathbf{O} & \tilde{\mathbf{K}}^2[u]\end{pmatrix}+\begin{pmatrix}\tilde{\mathbf{Q}}_{11} & \tilde{\mathbf{Q}}_{12}\\ \tilde{\mathbf{Q}}_{21} & \tilde{\mathbf{Q}}_{22}\end{pmatrix}, \tag{7}$$

where

$$\tilde{\mathbf{K}}^j[u]=\begin{pmatrix}\mu(j)u+\frac{\sigma^2(j)}{2}u^2 & \lambda(j)(1-p_j) & \lambda(j)p_j\\ -\eta_{j2} & \eta_{j2}+u & 0\\ -\eta_{j1} & 0 & \eta_{j1}-u\end{pmatrix},\quad j=1,2, \tag{8}$$

and

$$\tilde{\mathbf{Q}}_{ii} = \begin{pmatrix} -q_{ii} & 0 & 0 \\ 0 & 0 & 0 \\ 0 & 0 & 0 \end{pmatrix}, \qquad \tilde{\mathbf{Q}}_{ij} = \begin{pmatrix} q_{ii} & 0 & 0 \\ 0 & 0 & 0 \\ 0 & 0 & 0 \end{pmatrix}. \tag{9}$$

Now, consider the equation

$$\det(\tilde{\mathbf{K}}[u] - a\mathbb{I}^+) = 0, \tag{10}$$

where $\mathbb{I}^+$ denotes the diagonal matrix with 1 on positions 1 and 4 and 0 elsewhere. It is readily seen after some algebra that Eq. (10) is equivalent to the equation

$$q_1 q_2 = (\kappa_1(u) - a - q_1)(\kappa_2(u) - a - q_2), \tag{11}$$

where $q_j = q_{jj}$ and

$$\kappa_j(u) = \mu(j)u + \frac{\sigma^2(j)}{2}u^2 + \lambda(j)\left(\frac{p_j\eta_{j1}}{\eta_{j1} - u} + \frac{(1-p_j)\eta_{j2}}{\eta_{j2} + u} - 1\right).$$

After some algebraic manipulation, it can be shown that Equation (11) is equivalent to the polynomial of degree 8. By the *Fundamental Theorem of Algebra*, we know that such a polynomial can have at most eight complex roots. As in the single regime Kou model [15], close observation reveals that, for any $a > 0$, we can get something more.

Lemma 2 *Suppose that*

$$-\infty < -\eta_{22} < -\eta_{12} < 0 < \eta_{11} < \eta_{21} < \infty.$$

Then, for any $a > 0$, the equation

$$f(u) = (\kappa_1(u) - a - q_1)(\kappa_2(u) - a - q_2) - q_1 q_2$$

has eight distinct real roots. Moreover, let $\rho_{1,a} < \cdots < \rho_{8,a}$ be the roots. Then, these roots are located as

$$\begin{aligned} -\infty < \rho_{1,a} < -\eta_{22} < \rho_{2,a} < -\eta_{12} < \rho_{3,a} < \beta_{2,a} < \rho_{4,a} < 0 \\ < \rho_{5,a} < \beta_{1,a} < \rho_{6,a} < \eta_{11} < \rho_{7,a} < \eta_{21} < \rho_{8,a} < \infty, \end{aligned}$$

where $\beta_{1,a}$ and $\beta_{2,a}$ are the roots of $g_1(s) \equiv \kappa_1(u) - a - q_1 = 0$ such that

$$-\eta_{12} < \beta_{2,a} < 0 < \beta_{1,a} < \eta_{11}.$$

Proof Let $g_j(u) = \kappa_j(u) - a - q_j$ so that $f(u) = g_1(u)g_2(u) - q_1q_2$. Under the given assumption, we observe that

$$g_j(\eta_{j1}-) = g_j(\eta_{j2}+) = +\infty, \qquad g_j(\eta_{j1}+) = g_j(-\eta_{j2}-) = -\infty,$$

which immediately implies that

$$f(\eta_{j1}-) = f(\eta_{j2}+) = +\infty, \qquad f(\eta_{j1}+) = f(-\eta_{j2}-) = -\infty.$$

In addition, we also see that $f(+\infty) = +\infty$ and $f(-\infty) = +\infty$. Hence, since $f(u)$ is continuous except at the singularities η_{11}, η_{21}, $-\eta_{12}$ and η_{22}, there exists at least one root at each of the intervals, $(-\infty, -\eta_{22})$, $(-\eta_{22}, -\eta_{12})$, (η_{11}, η_{21}) and (η_{21}, ∞).

To obtain the remaining roots, since $\beta_{1,a}$ and $\beta_{2,a}$ are the roots of $g_1(u) = 0$, we have $f(\beta_{1,a}) = f(-\beta_{2,a}) = -q_1 q_2 < 0$. Furthermore, observe that

$$f(0) = (a + q_1)(a + q_2) - q_1 q_2 > 0.$$

Thus, since $f(u)$ is continuous on the interval $(-\eta_{12}, \eta_{11})$, there exists at least one root at each of the intervals, $(-\eta_{12}, \beta_{2,a})$, $(\beta_{2,a}, 0)$, $(0, \beta_{1,a})$ and $(\beta_{1,a}, \eta_{11})$.

So far, we have found eight distinct real roots for $f(u) = 0$. However, since $f(u)$ is a polynomial of degree 8, we have at most 8 complex roots, and thus the proof is completed. □

In the following, we denote the roots of the determinant (10) by $\rho_{r,a}$, $r = 1, \ldots, 8$, $a > 0$, and assume that the roots $\rho_{r,a}$ are ordered as in Lemma 2.

With these roots at hand, we define

$$\mathbf{h}^r[a] = \begin{pmatrix} \gamma_r \mathbf{k}_1^{r,a} \\ -\mathbf{k}_2^{r,a} \end{pmatrix}, \qquad r = 1, 2, \ldots, 8, \tag{12}$$

where

$$\mathbf{k}_i^{r,a} = \begin{pmatrix} 1 \\ \frac{\eta_{i2}}{\eta_{i2}+\rho_{r,a}} \\ \frac{\eta_{i1}}{\eta_{i1}-\rho_{r,a}} \end{pmatrix}, \qquad \gamma_r = \frac{\kappa_2(\rho_{r,a}) - a - q_2}{q_2}. \tag{13}$$

After a bit of algebra, one can easily prove the following.

Lemma 3 *Let* $\mathbf{h}^r[a]$ *be given by* (12). *Then,*

$$(\tilde{\mathbf{K}}[\rho_{r,a}] - a\mathbb{I}^+)\mathbf{h}^r[a] = \mathbf{0}$$

for each $r = 1, 2, \ldots, 8$, *where* $\mathbf{0}$ *is the zero vector.*

3 First Passage Time Under Regime-Switching Double-Exponential Jump Model

For the rest of the paper, we continue to work on the two-state regime-switching Lévy model with double exponential jumps considered in the previous section. Extension to the general finite case is straightforward.

3.1 Conditional Independence and Memoryless Properties

As mentioned in the introductory section, one of the main reasons behind the popularity of Kou's double exponential model is due to its conditional independence and conditional memoryless properties. These properties greatly simplify the overshoot/undershoot problem since one can immediately know that, if upward/downward jump occurs, it must be exponentially distributed. Moreover, the conditional independence under the Kou [15] model enables one to handle the first passage time and its overshoot/undershoot separately. Such separation makes the calculation of the first-passage-time distributions much more efficient.

Now, in our regime-switching setting, observe that, when $J_t = j$, the process X_t^J is the Kou double exponential jump-diffusion process. Assume that $X_0^J = 0$ and define

$$\tau_U = \inf\{t > 0 : \ X_t^J \geq U\}, \quad U > 0.$$

Of interest is that similar conditional independence and memoryless properties are satisfied in the regime-switching framework. The following results show that such properties are still retained under some additional conditions.

Lemma 4 *For any $x > 0$, we have*

$$\begin{aligned} &\mathbb{P}\left(\tau_U \leq t,\ X_{\tau_U}^J - U > x,\ J_{\tau_U} = j\right) \\ &= \mathrm{e}^{-\eta_{j1} x}\, \mathbb{P}\left(\tau_U \leq t,\ X_{\tau_U}^J - U > 0,\ J_{\tau_U} = j\right). \end{aligned}$$

Proof First, note that, for any $x > 0$, the event $\{X_{\tau_U}^J - U > x\}$ occurs only by an upward jump. Hence, denoting by T_n, $n = 1, 2, \ldots$, the arrival times of jumps with $J_{T_n} = j$, we obtain

$$\mathbb{P}\left(\tau_U \leq t,\ X_{\tau_U}^J - U > x,\ J_{\tau_U} = j\right) = \sum_{n=1}^{\infty} \mathbb{P}_n,$$

where

$$\mathbb{P}_n \equiv \mathbb{P}\left(T_n = \tau_U \leq t,\ X_{\tau_U}^J - U > x,\ J_{\tau_U} = j\right).$$

Now, due to the conditional independence and the memoryless property of exponential distributions, we have

$$\mathbb{P}\left(X_{T_n}^J - U > x |\ X_{T_n -}^J < U, T_n \leq t, J_{T_n} = j\right) = \mathrm{e}^{-\eta_{j1} x}.$$

Since

$$\mathbb{P}\left(X_{T_n}^J - U > 0 |\ X_{T_n -}^J < U,\ T_n \leq t,\ J_{T_n} = j\right) = 1,$$

it follows that

$$\begin{aligned}\mathbb{P}_n &= \mathbb{P}\Big(\max_{s<T_n} X_s^J < U,\ X_{T_n}^J - U > x,\ T_n \le t,\ J_{T_n} = j\Big)\\ &= \mathbb{P}\Big(X_{T_n}^J - U > x \,\Big|\, \max_{s<T_n} X_s^J < U,\ T_n \le t,\ J_{T_n} = j\Big)\\ &\quad \times \mathbb{P}\Big(\max_{s<T_n} X_s^J < U,\ T_n \le t, J_{T_n} = j\Big)\end{aligned}$$

and thus

$$\begin{aligned}\mathbb{P}_n &= \mathrm{e}^{-\eta_{j1}x}\, \mathbb{P}\Big(X_{T_n}^J - U > 0 \,\Big|\, \max_{s<T_n} X_s^J < U,\ T_n \le t,\ J_{T_n} = j\Big)\\ &\quad \times \mathbb{P}\Big(\max_{s<T_n} X_s^J < U,\ T_n \le t,\ J_{T_n} = j\Big)\\ &= \mathrm{e}^{-\eta_{j1}x}\, \mathbb{P}\Big(\max_{s<T_n} X_s^J < U,\ X_{T_n}^J - U > 0,\ T_n \le t,\ J_{T_n} = j\Big)\\ &= \mathrm{e}^{-\eta_{j1}x}\, \mathbb{P}\Big(T_n = \tau_U \le t,\ X_{\tau_U}^J - U > 0,\ J_{\tau_U} = j\Big).\end{aligned}$$

The lemma now follows easily by taking the summation over n. □

The next result follows immediately by letting $t \to \infty$ and observing that, on the event $\{X_{\tau_U}^J > U\}$, the first passage time τ_U is finite almost surely by definition. That is, we obtain the conditional memoryless property of exponential jumps.

Corollary 1 *For any $x > 0$, we have*

$$\mathbb{P}\Big(X_{\tau_U}^J - U > x \mid X_{\tau_U}^J - U > 0,\ J_{\tau_U} = j\Big) = \mathrm{e}^{-\eta_{j1}x}.$$

Similarly, for downward jumps, we have

$$\mathbb{P}\Big(X_{\tau_L}^J - L < -x \mid X_{\tau_L}^J - L < 0,\ J_{\tau_L} = j\Big) = \mathrm{e}^{-\eta_{j2}x},$$

where $\tau_L = \inf\{t > 0:\ X_t^J \le L\}$, $L < 0$.

We note that the proof provided above is very much similar to the one given by Kou and Wang [16]. However, the conditional independence and memoryless properties in our setting are satisfied under the additional conditions that $J_{\tau_U} = j$ for each $j \in E$. As it will be shown in the next subsection, the conditional independence and memoryless properties result in a more transparent formulation of the first-passage-time problem.

3.2 The First-Passage-Time Problem

With the conditional independence and memoryless properties fitting nicely into our framework, we now come to the central theme of this paper, i.e., to solving the first-passage-time problem in an efficient way. Specifically, our goal is to study the moment when the regime-switching process X_t^J leaves the interval $[L, U]$ for the first time.

To this end, we assume that $X_0^J = y$, $L < y < U$, and we define

$$\tau = \inf\{t > 0 : \; X_t^J \notin [L, U]\}. \tag{14}$$

The definition of τ entails that it is a stopping time with respect to the σ-algebra $\sigma(X_t^J, J_t)$ generated by (X_t^J, J_t), i.e., for any $t \geq 0$, we have $\{\tau < t\} \in \sigma(X_t^J, J_t)$. In addition, this first-passage-time problem also includes the single barrier passage times, since one can obtain the solutions from the double-barrier solution immediately by taking $U \to \infty$ for the first passage time to the lower barrier and by $L \to -\infty$ for the first passage time to the upper barrier.

In the following, we consider the Laplace transform

$$\begin{aligned} &\mathbb{E}_{y,i}\left[\exp(-aT(\tilde{\tau}) + b\tilde{X}_{\tilde{\tau}}^J); \; \tilde{J}_{\tilde{\tau}}\right], \\ &a > 0, \; b \in \mathbb{R}\backslash\{\eta_{i1}, -\eta_{i2}, \; i = 1, 2\}, \end{aligned} \tag{15}$$

where $\mathbb{E}_{y,i}[\cdot] \equiv \mathbb{E}[\cdot|\tilde{X}_0^J = y, \; \tilde{J}_0 = (i, 0)]$, $\tilde{X}_t^J$ is the fluid version of X_t^J, $\tilde{\tau}$ is the first passage time of $\tilde{X}_t^J$ (see Fig. 1) defined by

$$\tilde{\tau} = \inf\{t > 0 : \; \tilde{X}_t^J \notin [L, U]\},$$

and $T(t)$ is the virtual time of the fluid model defined in Definition 1. As in the case of τ, the definition of $\tilde{\tau}$ also implies that it is a stopping time with respect to the filtration $\sigma(\tilde{X}_t^J, \tilde{J}_t)$. In addition to the given restriction on b, we shall also assume that $b > 0$ in order to study the joint distribution of $T(\tilde{\tau})$ and $\tilde{X}_{\tilde{\tau}}^J$ within our framework.

The motivation of focusing on the fluid model will be apparent:

(1) The fluid process $\tilde{X}_t^J$ has continuous sample paths, whence either $\tilde{X}_{\tilde{\tau}}^J = U$ or $\tilde{X}_{\tilde{\tau}}^J = L$ possibly occurs. Therefore, from (15), it makes sense to define, for $i, j \in \{1, 2\}$,

$$\begin{aligned} \pi_{(i,j)}^{(+,U)}[a] &\equiv \mathbb{E}_{y,i}\left[\exp(-aT(\tilde{\tau}))1_{\{\tilde{J}_{\tilde{\tau}}=(j,+),\; \tilde{X}_{\tilde{\tau}}^J=U\}}\right], \\ \pi_{(i,j)}^{(0,U)}[a] &\equiv \mathbb{E}_{y,i}\left[\exp(-aT(\tilde{\tau}))1_{\{\tilde{J}_{\tilde{\tau}}=(j,0),\; \tilde{X}_{\tilde{\tau}}^J=U\}}\right], \\ \pi_{(i,j)}^{(-,L)}[a] &\equiv \mathbb{E}_{y,i}\left[\exp(-aT(\tilde{\tau}))1_{\{\tilde{J}_{\tilde{\tau}}=(j,-),\; \tilde{X}_{\tilde{\tau}}^J=L\}}\right], \\ \pi_{(i,j)}^{(0,L)}[a] &\equiv \mathbb{E}_{y,i}\left[\exp(-aT(\tilde{\tau}))1_{\{\tilde{J}_{\tilde{\tau}}=(j,0),\; \tilde{X}_{\tilde{\tau}}^J=L\}}\right]. \end{aligned}$$

Note that there are $4+4=8$ quantities to be determined. Also, the event $\{\tilde{J}_{\tilde{\tau}} = (j,+),\ \tilde{X}^J_{\tilde{\tau}} = U\}$ corresponds to the case of overshoot, while the event $\{\tilde{J}_{\tilde{\tau}} = (j,0),\ \tilde{X}^J_{\tilde{\tau}} = U\}$ corresponds to the situation that the process diffuses to upper barrier U with no overshoot problem. The downward case is similar.

(2) From Lemma 1, we have $\tau = T(\tilde{\tau})$ almost surely. Hence, corresponding to the above definition, we obtain

$$\begin{aligned}
\pi^{(+,U)}_{(i,j)}[a] &= \mathbb{E}_{y,i}\left[\mathrm{e}^{-a\tau} 1_{\{\tilde{J}_{\tilde{\tau}}=(j,+),\ \tilde{X}^J_{\tilde{\tau}}=U\}}\right], \\
\pi^{(0,U)}_{(i,j)}[a] &= \mathbb{E}_{y,i}\left[\mathrm{e}^{-a\tau} 1_{\{\tilde{J}_{\tilde{\tau}}=(j,0),\ \tilde{X}^J_{\tilde{\tau}}=U\}}\right], \\
\pi^{(-,L)}_{(i,j)}[a] &= \mathbb{E}_{y,i}\left[\mathrm{e}^{-a\tau} 1_{\{\tilde{J}_{\tilde{\tau}}=(j,-),\ \tilde{X}^J_{\tilde{\tau}}=L\}}\right], \\
\pi^{(0,L)}_{(i,j)}[a] &= \mathbb{E}_{y,i}\left[\mathrm{e}^{-a\tau} 1_{\{\tilde{J}_{\tilde{\tau}}=(j,0),\ \tilde{X}^J_{\tilde{\tau}}=L\}}\right].
\end{aligned} \tag{16}$$

We are now in a position to state the main theorem of this paper.

Theorem 1 *The quantities defined in* (16) *are the solutions to the following system of linear equations*:

$$\begin{aligned}
\mathrm{e}^{\rho_{r,a} y} h^r_{(i,0)}[a] &= \sum_{j=1}^{2}\left[\pi^{(+,U)}_{(i,j)}[a]\mathrm{e}^{\rho_{r,a}U} h^r_{(j,+)}[a] + \pi^{(0,U)}_{(i,j)}[a]\mathrm{e}^{\rho_{r,a}U} h^r_{(j,0)}[a]\right] \\
&\quad + \sum_{j=1}^{2}\left[\pi^{(-,L)}_{(i,j)}[a]\mathrm{e}^{\rho_{r,a}L} h^r_{(j,-)}[a] + \pi^{(0,L)}_{(i,j)}[a]\mathrm{e}^{\rho_{r,a}L} h^r_{(j,0)}[a]\right]
\end{aligned}$$

for $r = 1,\dots,8$ *and* $i = 1,2$, *where* $\rho_{r,a}$ *are the roots defined in Lemma* 2 *and* $h^r_{\tilde{j}}[a]$, $\tilde{j} \in \tilde{E}$, *are the components of the vector* $\mathbf{h}^r[a]$ *defined by* (12).

Proof Let $Y_t = -aT(t)/b$ and $Z_t = \tilde{X}^J_t + Y_t$, and define the matrix-valued process

$$\mathbf{M}(a,t) \equiv \int_0^t \mathrm{e}^{bZ_s}\mathbf{1}_{\tilde{J}_s}\mathrm{d}s\tilde{\mathbf{K}}[b] + \mathrm{e}^{bZ_0}\mathbf{1}_{\tilde{J}_0} - \mathrm{e}^{bZ_t}\mathbf{1}_{\tilde{J}_t} + b\int_0^t \mathrm{e}^{bZ_s}\mathbf{1}_{\tilde{J}_s}\mathrm{d}Y_s,$$

where $\mathbf{1}_j$ denotes a 1×6 row vector with j^{th} entry equal to 1 and all other entries being 0. The matrix $\tilde{\mathbf{K}}[b]$ is defined by (7). Since the sample paths of $\tilde{X}^J_t$ are continuous, we can apply Theorem 2(d) of Asmussen and Kella [3] to conclude that $\mathbf{M}(a,t)$ is a zero-mean martingale. Note that

$$b\int_0^t \mathrm{e}^{bZ_s}\mathbf{1}_{\tilde{J}_s}\mathrm{d}Y_s = -a\int_0^t \mathrm{e}^{bZ_s}\mathbf{1}_{\tilde{J}_s}1_{\{\tilde{J}_s\in E_0\}}\mathrm{d}s = -a\int_0^t \mathrm{e}^{bZ_s}\mathbf{1}_{\tilde{J}_s}\mathbb{I}^+\mathrm{d}s,$$

where $\mathbb{I}^+$ is the diagonal matrix with 1 on positions $E_{(j,0)}$ and 0 elsewhere. It follows that

$$\mathbf{M}(a,t) = \int_0^t \exp(b\tilde{X}_t^J - aT(t))\mathbf{1}_{\tilde{J}_s} \mathrm{d}s(\tilde{\mathbf{K}}[b] - a\mathbb{I}^+) \\ + \mathrm{e}^{by}\mathbf{1}_{\tilde{J}_0} - \exp(b\tilde{X}_t^J - aT(t))\mathbf{1}_{\tilde{J}_t}.$$

In particular, post-multiplying the vector $\mathbf{h}^r[a]$ defined by (12), and then utilizing Lemma 3 for $\rho_{r,a}$, $r = 1, \dots, 8$, we obtain the zero-mean vector-valued martingale

$$\tilde{\mathbf{M}}(a,t) = \mathrm{e}^{\rho_{r,a} y}\mathbf{1}_{\tilde{J}_0}\mathbf{h}^r[a] - \exp(\rho_{r,a}\tilde{X}_t^J - aT(t))\mathbf{1}_{\tilde{J}_t}\mathbf{h}^r[a],$$

which, together with Doob's optional sampling theorem $\mathbb{E}[\tilde{\mathbf{M}}(a, t\wedge\tilde{\tau})] = 0$, yields

$$\mathrm{e}^{\rho_{r,a} y}\mathbf{1}_{\tilde{J}_0}\mathbf{h}^r[a] = \mathbb{E}[\exp(\rho_{r,a}\tilde{X}_{\tilde{\tau}}^J - aT(\tilde{\tau}))\mathbf{1}_{\tilde{J}_{\tilde{\tau}}}]\mathbf{h}^r[a]. \tag{17}$$

The theorem now follows by decomposing the expectation in (17) with respect to $\pi[a]$'s given above. □

Theorem 1 provides a solution to the Laplace transform of the first-passage-time distribution in the original regime-switching model X_τ^J with double exponential jumps. Let τ be the first passage time defined by (14). In order to obtain the Laplace transform $\mathbb{E}_{y,i}[\exp(-a\tau + bX_\tau^J)1_{\{J_\tau = j\}}]$, we need to consider the overshoot/undershoot problem at the first passage time τ. However, this problem is resolved by the conditional independence and memoryless properties when jump sizes follow double exponential distributions.

More specifically, first note that

$$1_{\{J_\tau = j\}} = 1_{\{\tilde{J}_{T(\tilde{\tau})} = (j,0)\}} + 1_{\{\tilde{J}_{T(\tilde{\tau})} = (j,+)\}} + 1_{\{\tilde{J}_{T(\tilde{\tau})} = (j,-)\}}.$$

Recall that the event $\{\tilde{J}_{T(\tilde{\tau})} = (j,0)\}$ corresponds to the situation that the process diffuses to either upper barrier U or lower barrier L when $J_\tau = j$, resulting no overshoot/undershoot problem. The Laplace transforms of X_τ^J for these cases are simply given by

$$\hat{f}_{(j,0)}^{(u)}(U) \equiv \mathrm{e}^{bU}, \qquad \hat{f}_{(j,0)}^{(d)}(L) \equiv \mathrm{e}^{bL}, \tag{18}$$

respectively. It follows that

$$\mathbb{E}_{y,i}\left[\exp(-a\tau + bX_\tau^J)1_{\{\tilde{J}_{T(\tilde{\tau})} = (j,0)\}}\right] \\ = \mathbb{E}_{y,i}\left[\mathrm{e}^{-a\tau + bU}1_{\{\tilde{J}_{\tilde{\tau}} = (j,0),\ \tilde{X}_{\tilde{\tau}}^J = U\}}\right] + \mathbb{E}_{y,i}\left[\mathrm{e}^{-a\tau + bL}1_{\{\tilde{J}_{\tilde{\tau}} = (j,0),\ \tilde{X}_{\tilde{\tau}}^J = L\}}\right] \\ = \pi_{(i,j)}^{(0,U)}\mathrm{e}^{bU} + \pi_{(i,j)}^{(0,L)}\mathrm{e}^{bL},$$

where $\pi[a]$'s are given by (16).

Next, the event $\{\tilde{J}_{T(\tilde{\tau})} = (j,+)\}$ corresponds to the case of overshoot where $J_\tau = j$. From Corollary 1, we know that the overshoot $X_\tau^J - U$ is independent of τ and exponentially distributed. Hence, we obtain

$$\begin{aligned}
&\mathbb{E}_{y,i}\left[\exp(-a\tau + bX_\tau^J)1_{\{\tilde{J}_{T(\tilde{\tau})}=(j,+)\}}\right] \\
&= \mathrm{e}^{bU}\mathbb{E}_{y,i}\left[\mathrm{e}^{-a\tau+b(X_\tau^J-U)}1_{\{\tilde{J}_{\tilde{\tau}}=(j,+)\}}\right] \\
&= \mathrm{e}^{bU}\mathbb{E}_{y,i}\left[\mathrm{e}^{-a\tau}1_{\{\tilde{J}_{\tilde{\tau}}=(j,+)\}}\right]\mathbb{E}_{y,i}\left[\mathrm{e}^{b(X_\tau^J-U)}1_{\{\tilde{J}_{\tilde{\tau}}=(j,+)\}}\right] \\
&= \pi_{(i,j)}^{(+,U)}\hat{f}_{(j,+)}^{(u)}(U),
\end{aligned}$$

where we define

$$\hat{f}_{(j,+)}^{(u)}(U) \equiv \mathrm{e}^{bU}\int_0^\infty \mathrm{e}^{by}\eta_{j1}\mathrm{e}^{-\eta_{j1}y}\mathrm{d}y = \frac{\eta_{j1}}{\eta_{j1}-b}\mathrm{e}^{bU}. \tag{19}$$

Similarly, for the case of undershoot, we have

$$\mathbb{E}_{y,i}\left[\exp(-a\tau + bX_\tau^J)1_{\{\tilde{J}_{T(\tilde{\tau})}=(j,+)\}}\right] = \pi_{(i,j)}^{(-,L)}\hat{f}_{(j,-)}^{(d)}(L),$$

where we define

$$\hat{f}_{(j,-)}^{(d)}(L) \equiv \mathrm{e}^{bL}\int_{-\infty}^0 \mathrm{e}^{by}\eta_{j2}\mathrm{e}^{\eta_{j2}y}\mathrm{d}y = \frac{\eta_{j2}}{\eta_{j2}+b}\mathrm{e}^{bL}. \tag{20}$$

Summarizing, we now have the following result.

Corollary 2 *For the original regime-switching Lévy process X_t^J, let τ be the first passage time defined by* (14)*. If the jump sizes follow double exponential distributions, the Laplace transform $\mathbb{E}_{y,i}[\exp(-a\tau + bX_\tau^J)]$ is given by*

$$\begin{aligned}
\mathbb{E}_{y,i}\left[\mathrm{e}^{-a\tau+bX_\tau^J}\right] = \sum_{j=1}^{2}\Big(&\pi_{(i,j)}^{(+,U)}[a]\hat{f}_{(j,+)}^{(u)}(U) + \pi_{(i,j)}^{(-,L)}[a]\hat{f}_{(j,-)}^{(d)}(L) \\
&+ \pi_{(i,j)}^{(0,U)}[a]\hat{f}_{(j,0)}^{(u)}(U) + \pi_{(i,j)}^{(0,L)}[a]\hat{f}_{(j,0)}^{(d)}(L)\Big),
\end{aligned}$$

where $J_0 = i$ and $X_0^J = y$.

Remark 1 Consider the single-barrier case that the process X_t^J crosses an upper level U from below, i.e.

$$\tau_U = \inf\{t > 0 : X_t^J \ge U\}, \quad X_0^J < U. \tag{21}$$

Under the assumption that $\mathbb{P}(\tau_U < \infty) = 1$, we have

$$\pi_{(i,j)}^{(-,L)}[a] = \pi_{(i,j)}^{(0,L)}[a] = 0,$$

in Eq. (16). Hence, the Laplace transform of $(\tau_U, X^J_{\tau_U})$ is obtained from Theorem 1 by simply setting $\pi^{(-,L)}_{(i,j)}[a] = \pi^{(0,L)}_{(i,j)}[a] = 0$.

Remark 2 We have just shown that, in spite of the complexity of our framework, the Laplace transform of the joint distribution of first passage time and overshoot/undershoot level is given in terms of the solution of a system of linear equations. By incorporating the Euler-inversion, first proposed by Abate and Whitt [1] and later extended to multidimensional case by Choudhury et al. [8] and to the two-sided case by Petrella [21], one can then obtain the desired joint distribution efficiently. See the Appendix for a brief discussion of Abate-Whitt method.

4 Numerical Examples

In the previous section, we provide a comprehensive scheme to solve the Laplace transform of the joint distribution of τ and X^J_τ in the form of a system of linear equations. In this section, for practical illustrations, we shall focus on the first passage time τ_U defined in (21) for an upper barrier $U > 0$, and seek to obtain the first passage probability and joint probability of X^J_t and its running maxima. That is, we demonstrate how to calculate $\mathbb{P}_{y,i}(\tau_U \le t)$ and $\mathbb{P}_{y,i}(\tau_U \le t,\ X^J_t > k)$ numerically. To this end, we first provide the Laplace transforms of these probabilities.

Corollary 3 *The Laplace transform of* $\mathbb{P}_{y,i}(\tau_U \le t)$ *is given by, for* $\alpha > 0$,

$$\mathscr{L}_\alpha[\mathbb{P}_{i,y}(\tau_U \le t)] = \sum_j \frac{1}{\alpha}\mathbb{E}_{i,y}\left[\mathrm{e}^{-\alpha\tau_U} 1_{\{J_{\tau_U}=j\}}\right].$$

Proof By virtue of Fubini's theorem, we obtain

$$\begin{aligned}\mathscr{L}_\alpha[\mathbb{P}_{i,y}(\tau_U \le t)] &= \mathscr{L}_\alpha\left[\mathbb{E}_{i,y}\left[1_{\{\tau_U\le t\}}\right]\right]\\ &= \sum_j \mathbb{E}_{i,y}\left[\int_{\tau_U}^\infty \mathrm{e}^{-\alpha t}\mathrm{d}t\, 1_{\{J_{\tau_U}=j\}}\right]\\ &= \sum_j \mathbb{E}_{i,y}\left[\int_0^\infty \mathrm{e}^{-\alpha(s+\tau_U)}\mathrm{d}s\, 1_{\{J_{\tau_U}=j\}}\right].\end{aligned}$$

The result follows by integrating with respect to s. □

Corollary 4 *Suppose that*

$$\lim_{s\to\infty} \mathrm{e}^{-\alpha s+\mathbf{K}[\xi]s} = 0$$

and that the matrix $(\mathbf{K}[\xi]-\alpha\mathbb{I})$ *is invertible. Then, the Laplace transform of* $\mathbb{P}_{y,i}(\tau_U \le t,\ X_t^J > k)$ *is given by*

$$\mathscr{L}_{\alpha,\xi}[\mathbb{P}_{i,y}(\tau_U \le t,\ X_t^J \ge k)] = \sum_{j,n} \frac{1}{\xi}\mathbb{E}_{i,y}[\mathrm{e}^{-\alpha\tau_U+\xi X_{\tau_U}^J} 1_{\{J_{\tau_U}=j\}}]\mathbf{A}_{jn},$$

where $\mathbf{A}=(\mathbf{K}[\xi]-\alpha\mathbb{I})^{-1}$ *and* $\mathbf{A}_{jn}$ *denotes the* jn*th element of matrix* $\mathbf{A}$.

Proof Under the present assumptions, the Laplace transform exists and we can interchange the order of integrations due to Fubini's theorem. Hence, we have

$$\begin{aligned}
&\mathscr{L}_{\alpha,\xi}[\mathbb{P}_{i,y}(\tau_U \le t,\ X_t^J \ge k_0)]\\
&= \sum_n \mathbb{E}_{i,y}\left[\int_{\tau_U}^{\infty}\int_{-X_t^J}^{\infty} \mathrm{e}^{-\alpha t-\xi k}\mathrm{d}k 1_{\{J_t=n\}}\mathrm{d}t\right]\\
&= \sum_n \frac{1}{\xi}\mathbb{E}_{i,y}\left[\int_0^{\infty} \mathrm{e}^{-\alpha(s+\tau_U)+\xi X_{s+\tau_U}^J} 1_{\{J_{s+\tau_U}=n\}}\mathrm{d}s\right]\\
&= \sum_j \frac{1}{\xi}\mathbb{E}_{i,y}\Bigg[\mathrm{e}^{-\alpha\tau_U+\xi X_{\tau_U}^J} 1_{\{J_{\tau_U}=j\}}\\
&\qquad \times \int_0^{\infty} \mathrm{e}^{-\alpha s}\sum_n \mathbb{E}_j\left[\mathrm{e}^{\xi X_{s+\tau_U}^J-\xi X_{\tau_U}^J} 1_{\{J_{s+\tau_U}=n\}}\Big|\mathscr{F}_{\tau_U}\right]\mathrm{d}s\Bigg]\\
&= \sum_{j,n} \frac{1}{\xi}\mathbb{E}_{i,y}\left[\mathrm{e}^{-\alpha\tau_U+\xi X_{\tau_U}^J} 1_{\{J_{\tau_U}=j\}}\left(\int_0^{\infty} \mathrm{e}^{-\alpha s+\mathbf{K}[\xi]s}\right)_{jn}\mathrm{d}s\right],
\end{aligned}$$

where the third equality follows from the strong Markov property of X_t^J. The result holds under the invertibility assumption of the matrix $(\mathbf{K}[\xi]-\alpha\mathbb{I})$. □

Remark 3 The assumptions in Corollary 4 are proved as the results of Lemmas 2 and 5 under more general conditions in Mijatović and Pistorius [18].

Using Corollaries 3 and 4, the probabilities $\mathbb{P}_{i,y}(\tau_U \le t)$ and $\mathbb{P}_{i,y}(\tau_U \le t,\ X_t^J \ge k)$ are obtained by applying the scheme we developed in the previous section together with the numerical inversion technique mentioned in Remark 2.

In the following, unless stated otherwise, the parameters of the model are set to be $y=\log 100$, $k=\log 105$, $U=\log 105$, $p_1=0.4$, $p_2=0.6$, $\eta_{11}=\eta_{12}=40$, $\eta_{21}=\eta_{22}=60$, $\sigma_1=0.1$, $\sigma_2=0.5$ and $t=1$. The validity of our numerical approach was checked by comparing our scheme with that of Kou [15] by restricting $J_0=1$ and $q_1=0$, because such restriction makes the two models identical.

First, we investigate the effect of upper barrier U on the probabilities $\mathbb{P}_{i,y}(\tau_U \le t)$ and $\mathbb{P}_{i,y}(\tau_U \le t, X_t^J \ge k)$. The results are shown in Table 1 for the two cases $q_1 =$

Table 1 First Passage Probabilities with respect to U

$(q_1 = q_2 = 100)$	$\mathbb{P}_y(\tau_U \le t)$		$\mathbb{P}_y(\tau_U \le t, X_t^J \ge k)$	
$\exp(U)$	$J_0 = 1$	$J_0 = 2$	$J_0 = 1$	$J_0 = 2$
105	0.8878	0.8873	0.4296	0.4293
110	0.7831	0.7831	0.4278	0.4275
115	0.6872	0.6876	0.4209	0.4206
120	0.6004	0.6011	0.4063	0.4062
125	0.5224	0.5233	0.3844	0.3846
130	0.4529	0.4541	0.3573	0.3577
135	0.3914	0.3928	0.3271	0.3277

$(q_1 = 50, q_2 = 200)$	$\mathbb{P}_y(\tau_U \le t)$		$\mathbb{P}_y(\tau_U \le t, X_t^J \ge k)$	
$\exp(U)$	$J_0 = 1$	$J_0 = 2$	$J_0 = 1$	$J_0 = 2$
105	0.8549	0.8538	0.4536	0.4532
110	0.7165	0.7166	0.4477	0.4473
115	0.5905	0.5917	0.4263	0.4263
120	0.4794	0.4813	0.3860	0.3867
125	0.3839	0.3864	0.3333	0.3348
130	0.3038	0.3065	0.2779	0.2800
135	0.2378	0.2407	0.2261	0.2285

$q_2 = 100$ (upper half) and $q_1 = 50$, $q_2 = 200$ (lower half). It is observed that, as U increases, the two probabilities decrease gradually, because the chance of hitting the upper level decreases, with everything else unchanged. Note that the impact of the initial regime on these probabilities are negligible. This is so, because the transition intensities are very high and the underlying Markov chain settles quickly.[7] However, in the lower half of Table 1, the probabilities starting from $J_0 = 1$ are smaller than those with $J_0 = 2$. Recall that the volatility in regime 1 ($\sigma_1 = 0.1$) is much smaller than that in regime 2 ($\sigma_2 = 0.5$) and the departing intensity from state 1 ($q_1 = 50$) is much smaller than that of state 2 ($q_1 = 200$),[8] which makes the chance of hitting the upper barrier starting from regime 1 smaller than that from regime 2, before the Markov chain settles down.

Next, Fig. 2 illustrates the differences in $\mathbb{P}_y(\tau_U \le t)$ under a regime-switching Brownian motion and under a regime-switching jump-diffusion process. It is observed that, as the upper barrier U increases, the probability $\mathbb{P}_y(\tau_U \le t)$ under the regime-switching Brownian motion decays faster than that under the regime-

[7] See, e.g., Kijima [14] for the speed of convergence of Markov chains.

[8] Since the volatility is the dominant term in this case, this models the situation that the process stays in a low-volatility environment more often than a high-volatility regime.

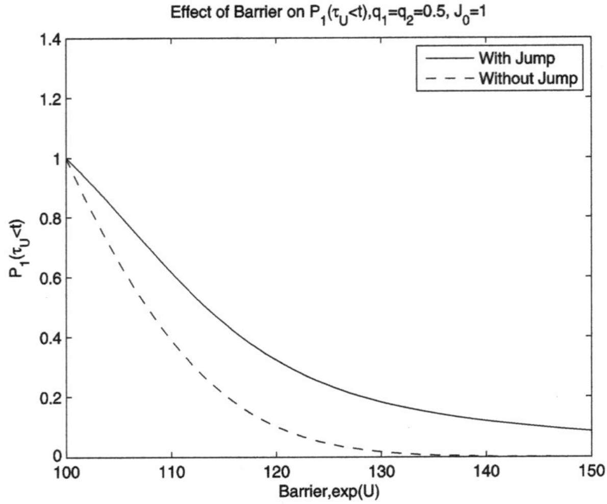

Fig. 2 Effect of Barrier on $\mathbb{P}_y(\tau_U \leq t)$

Table 2 First Passage Probabilities with respect to $q_1 = q_2$

$\exp(U) = 105, \exp(k) = 105$	$\mathbb{P}_y(\tau_U \leq t)$		$\mathbb{P}_y(\tau_U \leq t, X_t^J \geq k)$	
$q_1 = q_2$	$J_0 = 1$	$J_0 = 2$	$J_0 = 1$	$J_0 = 2$
0.5	0.8119	0.8936	0.4629	0.4130
10	0.8897	0.8860	0.4319	0.4284
50	0.8882	0.8872	0.4298	0.4292
100	0.8878	0.8873	0.4296	0.4293
200	0.8876	0.8874	0.4295	0.4293
500	0.8875	0.8874	0.4294	0.4293
1000	0.8875	0.8874	0.4294	0.4293
5000	0.8874	0.8874	0.4293	0.4293

switching jump-diffusion process.[9] This fits our intuition because, even without the regime-switching, the probability $\mathbb{P}_y(\tau_U \leq t)$ under the jump-diffusion process is always greater than that under the Brownian motion. Recall that the jump-diffusion model contains both diffusion and jump components that can contribute the overshoot of the upper barrier, resulting in a higher first passage probability. This intuition remains valid for the regime-switching case, too.

Table 2 investigates the speed of convergence of the regime-switching effect. As the switching intensities get large, switching between the two regimes occurs more

[9]The spread between the two curves reveals that the derivative prices obtained from the models with and without jumps may be significantly different. Similar caution should apply for the calculation of Value-at-Risk (VAR).

Table 3 $\mathbb{P}_y(\tau_U \leq t)$ with respect to σ_1

$(q_1 = q_2 = 0.5, \exp(U) = 105)$	$\lambda_1 = 1, \lambda_2 = 3$		$\lambda_1 = \lambda_2 = 0$	
σ_1	$J_0 = 1$	$J_0 = 2$	$J_0 = 1$	$J_0 = 2$
0.1	0.8224	0.9185	0.6517	0.9138
0.2	0.8680	0.9209	0.7700	0.9170
0.3	0.8907	0.9229	0.8253	0.9197
0.4	0.9044	0.9246	0.8569	0.9219
0.5	0.9137	0.9261	0.8772	0.9238
0.6	0.9204	0.9274	0.8914	0.9254
0.7	0.9255	0.9285	0.9017	0.9267
0.8	0.9295	0.9295	0.9096	0.9279

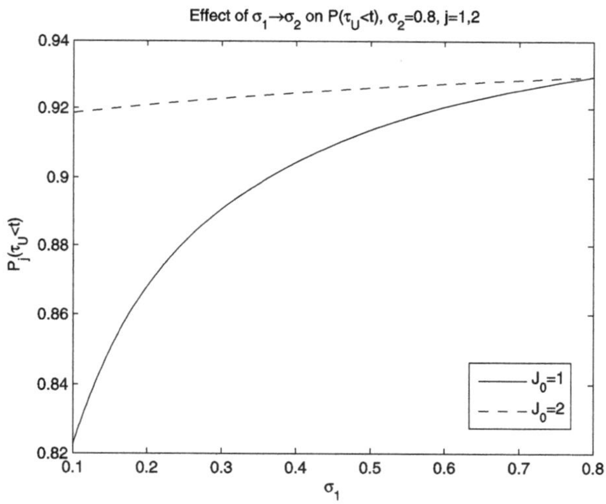

Fig. 3 Effect of Regime-Switching Intensity on $\mathbb{P}_y(\tau_U \leq t)$

frequently and the process converges to the steady state rapidly. The probabilities seem to converge after $q_1 = q_2 \geq 100$.

As the purpose of studying regime-switching models is to capture low and high volatility environments, it is of great interest to study the first passage probabilities against different values of σ_1. For this purpose, we set $\sigma_1 \leq \sigma_2$ and $\sigma_2 = 0.8$ to indicate that regime 1 is a low volatility environment, whereas regime 2 is a high volatility counterpart. The results are summarized in Table 3 and Fig. 3. To demonstrate the versatility of the model, we also provide the results for the regime-switching Brownian motion case, i.e. $\lambda_1 = \lambda_2 = 0$. It is explicitly observed from Table 3 and Fig. 3 that, as σ_1 increases, the two models get closer and the effect of the initial state disappears. These results are parallel with our intuition, since increasing σ_1 diminishes the effect of high and low volatility environments, i.e. the effect of regime switching disappears.

5 Conclusion

In this paper, we study the first-passage-time problem under a regime-switching double exponential jump-diffusion process. With the characterization of the fluid model, we can turn the original model into an augmented regime-switching diffusion model whose sample paths are continuous. Such characterization proves to have a significant advantage when one studies the problem of first exit time from an interval. With the help of the special Kella-Asmussen martingale (Asmussen and Kella [3]), the first-passage-time problem can be formulated as a system of linear equations. The methodology proves to be fairly elementary and one can obtain the Laplace transform of the first passage time by simply solving the linear equations. The numerical examples illustrate the efficiency of computing the first passage probabilities and the joint probabilities through the numerical Laplace inversion.

From a recent paper by Cai et al. [6], one can see that the regime-switching jump-diffusion model studied in this paper has a close resemblance to the hyper-exponential jump-diffusion model. In fact, an immediate generalization would be an extension to the case where jump sizes follow a phase-type distribution, as in the case of Asmussen et al. [4].

Furthermore, the regime-switching Lévy model discussed here possesses nice features as a security-price model, because it includes the short-run behavior captured by the jump-diffusion component and the long-run market cycle by the Markov chain component. Yet, such rich structure remains analytically tractable when one studies the first-passage-time problem. The results developed in this paper stimulates one to use this regime-switching Lévy structure for the option pricing. These are the subjects of our future research.

Appendix

There are many ways to perform numerical Laplace inversion. The one we adopt is the Fourier-series method developed by Abate and Whitt [1]. The benefits of Fourier-series method are that the methodology provides error bounds and converges rapidly.

Denote by $\widehat{F}(\alpha)$ the Laplace transform of function $F(t)$ with respect to t. Then $F(t)$ can be recovered from $\widehat{F}(\alpha)$ by the following formula:

$$F^{AW}(t) = \frac{e^{A/2}}{2t}\mathrm{Re}\left(\widehat{F}\left(\frac{A}{2t}\right)\right) + \frac{e^{A/2}}{t}\sum_{k=1}^{\infty}\mathrm{Re}\left(\widehat{F}\left(\frac{A+2k\pi\mathrm{i}}{2t}\right)\right),$$

where $F^{AW}(t)$ denotes the Laplace inversion by Abate–Whitt (AW) Fourier-series method, $\mathrm{i}=\sqrt{-1}$, and $\mathrm{Re}\,(a)$ denotes the real part of complex number a.

Abate and Whitt [1] also provide the error bound of such inversion. Assuming that F is bounded, i.e., $F(t) < C$ for some C, the discretization error of the Abate-

Whitt method can be bounded by

$$|F(t) - F^{AW}(t)| < C\frac{e^{-A}}{1-e^{-A}} \simeq Ce^{-A}.$$

Thus, we should set A large enough to make the error small. However, because of roundoff errors, increasing A would make inversion harder. In practice, Abate and Whitt [1] suggest that the choice of $A = 18.4$ should produce stable and accurate results.

Note that the Abate-Whitt algorithm is an infinite-series representation. To obtain high degree of accuracy, we need to add large number of terms. Large number of summation would certainly hinder the speed of inversion. Fortunately, close inspection of the Abate–Whitt algorithm reveals that it is in terms of an *alternating* series, which can be well approximated by an appropriate binomial expansion. To speed up the inversion procedure, we can modify $F^{AW}(t)$ by using the *Euler* algorithm

$$F^{AW}(t) \approx \sum_{k=0}^{m} C_m^k 2^{-m} s_{n+k}(t),$$

where

$$s_n(t) = \frac{e^{A/2}}{2t}\mathrm{Re}\left(\widehat{F}\left(\frac{A}{2t}\right)\right) + \frac{e^{A/2}}{t}\sum_{k=1}^{n}\mathrm{Re}\left(\widehat{F}\left(\frac{A+2k\pi \mathrm{i}}{2t}\right)\right).$$

By employing the Euler algorithm, we find that any $n > 40$ and $m > 15$ would produce stable results. Since the summation involves less than 100 terms, the algorithm is very efficient.

References

1. Abate, J., Whitt, W.: The Fourier-series method for inverting transforms of probability distributions. Queueing Syst. **10**, 5–88 (1992)
2. Asmussen, S.: Ruin Probabilities. World Scientific, Singapore (2000)
3. Asmussen, S., Kella, O.: A multi-dimensional martingale for Markov additive processes and its applications. Adv. Appl. Probab. **32**, 376–393 (2000)
4. Asmussen, S., Avram, F., Pistorius, M.: Russian and American put options under exponential phase-type Lévy models. Stoch. Process. Appl. **109**(1), 79–111 (2004)
5. Borovkov, K., Novikov, A.: On a new approach to calculating expectations for option pricing. J. Appl. Probab. **39**(4), 889–895 (2002)
6. Cai, N., Chen, N., Wan, X.: Pricing double-barrier options under a flexible jump diffusion model. Oper. Res. Lett. **37**, 163–167 (2009)
7. Carr, P., Crosby, J.: A class of Lévy process models with almost exact calibration to both barrier and vanilla FX options. Quant. Finance **10**(10), 1115–1136 (2010)
8. Choudhury, G.L., Lucantoni, D.M., Whitt, W.: Multidimensional transform inversion with applications to the transient M/M/1 queue. Ann. Appl. Probab. **4**, 719–740 (1994)
9. Guo, X.: An explicit solution to an optimal stopping problem with regime switching. J. Appl. Probab. **38**(2), 464–481 (2001)

10. Guo, X., Zhang, Q.: Closed-form solutions for perpetual American put options with regime switching. SIAM J. Appl. Math. **64**(6), 2034–2049 (2004)
11. Guo, X., Miao, J.J., Morellec, E.: Irreversible investment with regime shifts. J. Econ. Theory **122**(1), 37–59 (2005)
12. Jiang, Z., Pistorius, M.: On perpetual American put valuation and first passage in a regime-switching model with jumps. Finance Stoch. **12**, 331–355 (2008)
13. Jobert, A., Rogers, L.C.G.: Option pricing with Markov-modulated dynamics. SIAM J. Control Optim. **44**(6), 2063–2078 (2006)
14. Kijima, M.: Markov Processes for Stochastic Modeling. Chapman & Hall, London (1997)
15. Kou, S.G.: A jump-diffusion model for option pricing. Manag. Sci. **48**, 1086–1101 (2002)
16. Kou, S.G., Wang, H.: First passage times for a jump-diffusion process. Adv. Appl. Probab. **35**, 504–531 (2003)
17. Kyprianou, A.E.: Introductory Lectures on Fluctuations of Lévy Processes with Applications. Universitext. Springer, Berlin (2006)
18. Mijatović, A., Pistorius, M.: Exotic derivatives in a dense class of stochastic volatility models with jumps. In: Di Nuno, G., Øksendal, B. (eds.) Advanced Mathematical Methods for Finance, pp. 455–508. Springer, Berlin (2011)
19. Mordecki, E.: Optimal stopping for a diffusion with jumps. Finance Stoch. **3**, 227–236 (1999)
20. Mordecki, E.: Optimal stopping and perpetual for Lévy processes. Finance Stoch. **6**, 273–293 (2002)
21. Petrella, G.: An extension of the Euler Laplace transform inversion algorithm with applications in option pricing. Oper. Res. Lett. **32**, 380–389 (2004)
22. Sepp, A.: Analytical pricing of double-barrier options under a double-exponential jump diffusion process: applications of Laplace transform. Int. J. Theor. Appl. Finance **2**, 151–175 (2004)
23. Shreve, S.: Stochastic Calculus for Finance II: Continuous-Time Models. Springer Finance. Springer, New York (2004)

Strong Consistency of the Bayesian Estimator for the Ornstein–Uhlenbeck Process

Arturo Kohatsu-Higa, Nicolas Vayatis, and Kazuhiro Yasuda

Abstract In the accompanying paper Kohatsu-Higa et al. (submitted, 2013), we have done a theoretical study of the consistency of a computational intensive parameter estimation method for Markovian models. This method could be considered as an approximate Bayesian estimator method or a filtering problem approximated using particle methods. We showed in Kohatsu-Higa (submitted, 2013) that under certain conditions, which explicitly relate the number of data, the amount of simulations and the size of the kernel window, one obtains the rate of convergence of the method. In this first study, the conditions do not seem easy to verify and for this reason, we show in this paper how to verify these conditions in the toy example of the Ornstein–Uhlenbeck processes. We hope that this article will help the reader understand the theoretical background of our previous studies and how to interpret the required hypotheses.

Keywords Bayesian estimator · Computational intensive parameter estimation · Ornstein–Uhlenbeck process · Filtering problem · Particle method

Mathematics Subject Classification (2010) 62F15 · 91G70

A. Kohatsu-Higa (✉)
Ritsumeikan University and Japan Science and Technology Agency, 1-1-1 Nojihigashi, Kusatsu, Shiga, 525-8577, Japan
e-mail: khts00@fc.ritsumei.ac.jp

N. Vayatis
Centre de Mathématiques et de Leurs Applications (CMLA) UMR CNRS 8536, École Normale Supérieure de Cachan, 61, avenue du Président Wilson, 94 235 Cachan cedex, France
e-mail: vayatis@cmla.ens-cachan.fr

K. Yasuda
Hosei University, 3-7-2, Kajino-cho, Koganei-shi, Tokyo, 184-8584, Japan
e-mail: k_yasuda@hosei.ac.jp

Y. Kabanov et al. (eds.), *Inspired by Finance*, DOI 10.1007/978-3-319-02069-3_19,

1 Introduction

One method to estimate parameters in a Markovian model is to use a filtering method (also known as the Bayesian method). In such a framework, the estimation is carried out using a least-square principle, which leads to the calculation of the conditional expectation of the unknown density given the available data.

This expression is somehow theoretical, so one option is to use simulation to approximate the value of the unknown transition density if some theoretical model is proposed. This simulation procedure requires the choice of a variety of parameters. The procedure of choosing these parameters "correctly" is called tuning.

Recently, many computational statisticians have successfully proposed and studied several algorithms related to this idea, for example, using the Markov Chain Monte Carlo method (Roberts et al. [10]) between others. Many papers have confirmed the rate of convergence of the proposed method to the desired value using numerical experiments, but usually no mathematical proof is provided. In an accompanying paper [9], we adopt a particle method (details and other comments about this method can be found in Bain et al. [2]) to approximate the conditional expectation and study theoretically the rate of convergence and the proper tuning needed. This kind of filtering problem under discrete observations was studied by Del Moral et al. [4] who proved weak consistency and L^2-convergence. More recently, Cano et al. [3] studied the convergence of an approximated posterior distribution, which used the Euler–Maruyama approximation for stochastic differential equations (SDE). In Kohatsu-Higa et al. [9], we gave the rate of convergence of the approximated Bayesian estimator. In that set-up, the transition density function of an observation process is usually unknown, so that one approximates it by using the kernel density estimation method (KDE). As mentioned before, there are several new algorithms, which may work well in applications, but our objective was to provide a sound mathematical framework. Therefore, we choose the most basic method available within particle methods. Our method of analysis uses the Laplace method to obtain the rate of convergence $1/\sqrt{N}$, where N is a number of data under a strong hypothesis of convergence rate for the approximating average of likelihoods (see Assumption (A) (6)-(a)).

In the second part of Kohatsu-Higa et al. [9], we gave an explicit relationship between the number of data and approximation parameters, as to ensure that Assumption (A) (6)-(a) is satisfied. Here, we have three approximation parameters: (i) the first one is used to approximate the theoretical stochastic processes, (ii) the second one is to express the number of the Monte-Carlo simulations used for the approximating process, (iii) the last one is a bandwidth size of the KDE. We connect these three approximation parameters and the number of data. We believe that our study is the first that provides an explicit theoretical relationship between these parameters in order to achieve a certain rate of convergence. It also shows why a bad choice of tunning parameters may lead to unreliable estimation results.

Assumption (A) below states the hypotheses that are needed to achieve the rate of convergence announced previously. These hypotheses are not necessarily easy to understand and/or interpret. The objective of the present article is to consider an easy toy example where the reader may see how these conditions could be verified and, most importantly, what do they mean. In this paper, we consider the following Ornstein-Uhlenbeck process (OU process) as the parametrized observation process:

$$dX_t = -\theta X_t\, dt + dW_t,$$

where W_t is a Brownian motion and θ is a parameter, which we want to estimate. Then, we check the assumptions that give the strong consistency and the convergence rate. Clearly this is a toy example, as many elements can be directly computed and thus there is no need to use simulations. Furthermore, in that setting many other competing statistical methods exist (see, e.g., [1, 6–8, 11]).

We would like to emphasize again that the main objective here is to show that the general theory is applicable to a basic example. Clearly, there are still open problems to be considered—in particular, how to apply these results to other examples. We hope that with this article the reader may understand when a model satisfies the assumptions, although verifying them may still require a long procedure.

This paper is organized as follows: In Sect. 2, we recall the general theorem and the assumptions of Kohatsu-Higa et al. [9]. In Sect. 3, we check the assumptions with respect to the OU process and the Euler–Maruyama approximation of the OU process. Finally, in the Appendix, we give some properties of the mean and variance of the OU process and its Euler–Maruyama approximation.

2 Framework and General Theorem

2.1 Framework

In this article, we consider the following problem: Let $\theta_0 \in \Theta := [\theta^l, \theta^u]$, where $\theta^l < \theta^u$, be a parameter that we wish to estimate $\theta_0 \in \dot{\Theta}$, where $\dot{\Theta}$ denotes the interior of the set Θ. Let $(\Omega, \mathscr{F}, P_{\theta_0})$, $(\bar{\Omega}, \bar{\mathscr{F}}, \bar{P})$ and $(\hat{\Omega}, \hat{\mathscr{F}}, \hat{P})$ be three probability spaces, where the probability measure P_{θ_0} is parametrized by θ_0. A number $\Delta > 0$ is a fixed parameter that represents the time between observations. The observed Markov chain is defined on the probability space $(\Omega, \mathscr{F}, P_{\theta_0})$. The theoretical Markov chain (with the law P_θ) and its approximation are defined on $(\bar{\Omega}, \bar{\mathscr{F}}, \bar{P})$. Finally, simulations are defined on $(\hat{\Omega}, \hat{\mathscr{F}}, \hat{P})$, which will be used in estimating the transition density of the theoretical Markov chain.

(i). (*Observation process*) Let $\{Y_{i\Delta}\}_{i=0,1,\dots,N}$ be a sequence of $N+1$ observations of a Markov chain having the transition density $p_{\theta_0}(y,z)$, $y, z \in \mathbf{R}$ and an invariant measure μ_{θ_0}. This sequence is defined on the probability space $(\Omega, \mathscr{F}, P_{\theta_0})$. We write $Y_i := Y_{i\Delta}$ for $i = 0, 1, \dots, N$.

(ii). (*Model process*) Denote by $X^y(\theta)$ a random variable defined on the probability space $(\bar{\Omega}, \bar{\mathscr{F}}, \bar{P})$ such that its law is given by $p_\theta(y, z)$.

(iii). Denote by $(\hat{\Omega}, \hat{\mathscr{F}}, \hat{P})$ the probability space on which the simulation of the approximation to the process X^y is generated.

(iv). (*Approximating process*) Denote by $X^y_{(m)}(\theta)$ the approximation to $X^y(\theta)$, which is defined on $(\bar{\Omega}, \bar{\mathscr{F}}, \bar{P})$ and where $m = m(N)$ is the parameter that determines the quality of the approximation. Denote by $\tilde{p}^N_\theta(y,z) = \tilde{p}^N_\theta(y,z;m(N))$ the transition density for the process $X^y_{(m)}(\theta)$.

(v). (*Approximated transition density*) Let $K \in C^2(\mathbf{R};\mathbf{R}_+)$ (usually called kernel), which satisfies $\int K(x)\,dx = 1$. Denote by $\hat{p}^N_\theta(y,z)$ the kernel density estimate of $\tilde{p}^N_\theta(y,z)$ based on $n \equiv n(N)$ simulated i.i.d. copies of $X^y_{(m)}(\theta)$, which are defined on $(\hat{\Omega}, \hat{\mathscr{F}}, \hat{P})$ and denoted by $X^{y,(k)}_{(m)}(\theta,\cdot)$ for $k = 1,\ldots,n$. For $h \equiv h(N) > 0$,

$$\hat{p}^N_\theta(y,z) := \frac{1}{n(N)h(N)} \sum_{k=1}^{n(N)} K\left(\frac{X^{y,(k)}_{(m(N))}(\theta,\hat{\omega}) - z}{h(N)}\right).$$

(vi). For a given m, we introduce the "average" approximated transition density over all trajectories with respect to the kernel K by setting

$$\bar{p}^N_\theta(y,z) := \bar{p}^N_\theta(y,z;m(N),h(N)) := \hat{E}\big[\hat{p}^N_\theta(y,z)\big],$$

where $\hat{E}$ means the expectation with respect to $\hat{P}$.

As it can be deduced from the above set-up, we have preferred to state our problem in abstract terms without explicitly defining the dynamics that generate $X^y(\theta)$ or how the approximation $X^y_{(m)}(\theta)$ is defined. All the properties that will be required for p_θ and $\tilde{p}^N_\theta$ will be satisfied for an appropriate subclass of diffusion processes. Our objective in this article is to show that OU processes are in this class.

Remark 1 Without loss of generality, we can consider the product of the above three probability spaces so that all random variables are defined on the same probability space. We do this without any further mentioning.

Our purpose is to estimate the posterior expectation for some function $f \in C^1(\Theta)$ given the data

$$E_N[f] := E_\theta[f|Y_0,\ldots,Y_N] := \frac{\int f(\theta)\phi_\theta(Y_0^N)\pi(\theta)d\theta}{\int \phi_\theta(Y_0^N)\pi(\theta)d\theta},$$

where

$$\phi_\theta(Y_0^N) = \phi_\theta(Y_0,\ldots,Y_N) = \mu_\theta(Y_0)\prod_{j=1}^N p_\theta(Y_{j-1},Y_j)$$

is the joint density of $(Y_0, Y_1, \ldots, Y_N)$.

We propose to estimate this quantity on the basis of simulated instances of the process

$$\hat{E}^n_{N,m}[f] := \frac{\int f(\theta)\hat{\phi}^N_\theta(Y_0^N)\pi(\theta)d\theta}{\int \hat{\phi}^N_\theta(Y_0^N)\pi(\theta)d\theta},$$

where $\hat{\phi}^N_\theta(Y_0^N) := \mu_\theta(Y_0)\prod_{j=1}^N \hat{p}^N_\theta(Y_{j-1}, Y_j)$.

2.2 General Theorem of Kohatsu-Higa et al. [9]

Assumption (A): We assume the following:

(1). (*Observation process*) $\{Y_i\}_{i=0,1,\dots,N}$ is an α-mixing process with $\alpha_n = O(n^{-5})$.
(2). (*The prior distribution*) The prior distribution $\pi \in C(\Theta)$ and, for all $\theta \in \Theta$, $\pi(\theta) > 0$.
(3). (*Density regularity*) The transition densities $p, \bar{p}^N \in C^{2,0,0}(\Theta \times \mathbf{R}^2; \mathbf{R}_+)$, and for all $\theta \in \Theta$, $y, z \in \mathbf{R}$, we have that $\min\left\{p_\theta(y,z), \bar{p}^N_\theta(y,z)\right\} > 0$. Moreover, p_θ admits an invariant measure $\mu \in C_b^{0,0}(\Theta \times \mathbf{R}; \mathbf{R}_+)$ and, for all $\theta \in \Theta$, $\mu_\theta(y) > 0$ for every $y \in \mathbf{R}$.
(4). (*Identifiability*) We assume that there exist $c_1, c_2 : \mathbf{R} \to (0, \infty)$ such that, for all $\theta \in \Theta$,

$$\inf_N \int |q^i_\theta(y,z) - q^i_{\theta_0}(y,z)|dz \geq c_i(y)|\theta - \theta_0|,$$

$C_i(\theta_0) := \int c_i(y)^2 \mu_{\theta_0}(y)dy \in (0, +\infty)$ for $i = 1, 2$ and $q^1_\theta = p_\theta$, $q^2_\theta = \bar{p}^N_\theta$.
(5). (*Regularity of the log-density*) We assume that for $q_\theta = p_\theta$ and $\bar{p}^N_\theta$,

$$\sup_N \sup_{\theta\in\Theta} \int\int \left(\frac{\partial^i}{\partial\theta^i}\ln q_\theta(y,z)\right)^{12} p_{\theta_0}(y,z)\mu_{\theta_0}(y)dydz < \infty, \quad \text{for } i = 0, 1, 2, \tag{1}$$

$$\sup_N \sup_{\theta\in\Theta} \left|\frac{\partial^2}{\partial\theta^2}\int\int (\ln q_\theta(y,z))\, \bar{p}^N_{\theta_0}(y,z)\mu_{\theta_0}(y)\, dydz\right| < \infty, \tag{2}$$

$$\sup_N \sup_{\theta\in\Theta} \int\int \left|\frac{\partial^i}{\partial\theta^i}\ln q_\theta(y,z)\right| \bar{p}^N_{\theta_0}(y,z)\mu_{\theta_0}(y)\, dydz < \infty, \quad \text{for } i = 0, 1, \tag{3}$$

where $\frac{\partial^0}{\partial\theta^0} q_\theta = q_\theta$.
(6). (*Parameter tuning*)

(a). We assume the following boundedness

$$\sup_N \sup_{\theta\in\Theta} \left|\frac{1}{\sqrt{N}}\sum_{i=0}^{N-1}\left(\frac{\partial}{\partial\theta}\ln\hat{p}^N_\theta(Y_i, Y_{i+1}) - \frac{\partial}{\partial\theta}\ln\bar{p}^N_\theta(Y_i, Y_{i+1})\right)\right| < \infty,$$

a.s. (4)

(b). We assume that for each $y, z \in \mathbf{R}$, there exist functions $C_1^N(y,z)$ and $c_1(y,z)$ such that $|p_{\theta_0}(y,z) - \bar{p}_{\theta_0}^N(y,z)| \le C_1^N(y,z)a_1(N)$, where $\sup_N C_1^N(y,z) < +\infty$ and $a_1(N) \to 0$ as $N \to \infty$. Moreover, $C_1^N(y,z)a_1(N)\sqrt{N} < c_1(y,z)$, where c_1 in turn satisfies

$$\sup_N \sup_{\theta \in \Theta} \int\int \left|\frac{\partial}{\partial\theta} \ln \bar{p}_\theta^N(y,z)\right| c_1(y,z)\mu_{\theta_0}(y)\,dydz < \infty.$$

(c). There exist some function $g^N : \mathbf{R}^2 \to \mathbf{R}$ and constant $a_2(N)$, which depends on N, such that for all $y, z \in \mathbf{R}$,

$$\sup_{\theta\in\Theta}\left|\frac{\partial}{\partial\theta}\ln \bar{p}_\theta^N(y,z) - \frac{\partial}{\partial\theta}\ln p_\theta(y,z)\right| \le |g^N(y,z)|a_2(N),$$

where $\sup_N E_{\theta_0}[|g^N(Y_0,Y_1)|^4] < +\infty$ and $a_2(N) \to 0$ as $N \to \infty$.

Now we state the main result of [9].

Theorem 1 (Kohatsu-Higa et al. [9]) *Under Assumption* (A), *there exists some positive finite random variables* Ξ_1 *and* Ξ_2 *such that*

$$|E_N[f] - f(\theta_0)| \le \frac{\Xi_1}{\sqrt{N}} \quad a.s., \quad \text{and} \quad \left|\hat{E}_{N,m}^n[f] - f(\theta_0)\right| \le \frac{\Xi_2}{\sqrt{N}} \quad a.s.,$$

and thus

$$|E_N[f] - \hat{E}_{N,m}^n[f]| \le \frac{\Xi_1 + \Xi_2}{\sqrt{N}} \quad a.s.$$

2.3 Parameter Tuning for Assumption (A) (6)-(a)

All the conditions in Assumption (A) will be directly verified with the exception of Assumption (A) (6)-(a), which requires a special treatment. This section is devoted to show that Assumption (A) (6)-(a) is satisfied under sufficient smoothness hypothesis on the random variables and processes that appear in the problem, as well as a certain parameter tuning. We recall that the objective is to find conditions that assure that Assumption (A) (6)-(a) in Sect. 2.2 is satisfied.

Now $m \equiv m(N)$, $n \equiv n(N)$ and $h \equiv h(N)$ are parameters that depend on N. Let n be the number of Monte Carlo simulations used in order to estimate the density, m the generated random numbers used in the simulation of $X_{(m)}^{y,(1)}(\theta,\cdot)$ and h the window associated to the kernel density estimation method. In this sense, we will always think of hypotheses in terms of N although we will drop them from the notation and we will simply write m, n and h. The goal of this section is to prove that, under certain hypotheses, there is a choice of m, n and h that ensures that condition (4) is satisfied.

We work in this section under the following hypotheses:

(H1). There exist some positive constants φ_1, φ_2, where φ_1 is independent of N and φ_2 is independent of N and Δ, such that the following holds

$$\inf_{(\mathbf{x},\theta)\in B^N} \bar{p}_\theta^N(x, y) \geq \varphi_1 \exp\left(-\frac{\varphi_2 a_N^2}{\Delta}\right),$$

where the sequence a_N and the set B^N are defined in condition (ii) below.

(H2). The kernel K is the Gaussian kernel; $K(z) := \frac{1}{\sqrt{2\pi}} \exp(-\frac{1}{2}z^2)$.

(H5). There exists some positive constant $C_5 > 0$ such that

$$\left|\partial_x \bar{p}_\theta^N(x, y)\right|, \quad \left|\partial_y \bar{p}_\theta^N(x, y)\right|, \quad \left|\partial_\theta \bar{p}_\theta^N(x, y)\right| \leq C_5 < \infty,$$

for all $x, y \in \mathbf{R}$, $m \in \mathbf{N}$ and $\theta \in \Theta$.

(H5'). There exists some positive constant $\dot{C}_5 > 0$ such that

$$\left|\partial_x \partial_\theta \bar{p}_\theta^N(x, y)\right|, \quad \left|\partial_y \partial_\theta \bar{p}_\theta^N(x, y)\right|, \quad \left|\partial_\theta^2 \bar{p}_\theta^N(x, y)\right| \leq \dot{C}_5 < \infty,$$

for all $x, y \in \mathbf{R}$, $m \in \mathbf{N}$ and $\theta \in \Theta$.

Remark 2 For the ease of reference, we here use the same numbering of hypotheses as in Kohatsu-Higa et al. [9]. Note, however, that some of the intermediate hypotheses do not appear here. For the detailed explanations, we refer to Kohatsu-Higa et al. [9].

We need to find now a sequence of values for n and h such that all the hypotheses in Theorem 1 are satisfied and the upper bound is uniformly bounded in N. Now, we rewrite the needed conditions that are related to the parameters n and h. We assume stronger hypothesis that may help to better understand the existence of the right choice of parameters n and h.

The proof of Assumption (A) (6)-(a) uses a series of Borel–Cantelli lemmas for which we need the following hypotheses. We will assume the existence of some sequences of strictly positive numbers, which are assumed, without loss of generality, to be bigger than 1.

(ii). (Borel–Cantelli for Y_i)

For some constant $c_1 > 0$ and some sequence $\{a_N\}_{N\in\mathbf{N}} \subset [\theta^u - \theta^l, \infty)$, we have $m_{c_1} := E[e^{c_1|Y_1|^2}] < +\infty$ and

$$\sum_{N=1}^{\infty} \frac{N}{\exp(c_1 a_N^2)} < \infty.$$

We define $B^N = \{(\mathbf{x}, \theta) \in \mathbf{R}^2 \times \Theta;\ \|\mathbf{x}\| < a_N\}$, where $\|\cdot\|$ denotes the max-norm.

(iii). (Bore-l-Cantelli for $Z_{3,N}(\omega)$)

For some $r_3 > 0$ and some sequence $b_{3,N} \geq 1$, $N \in \mathbf{N}$,

$$\sum_{N=1}^{\infty} \frac{n a_N^{2r_3}}{(h^2 b_{3,N})^{r_3}} < +\infty$$

and $\sup_{N\in\mathbf{N}} E[|Z_{3,N}(\cdot)|^{r_3}] < \infty$ for each fixed $m \in \mathbf{N}$, where

$$Z_{3,N}(\omega) := a_N^{-2}\left(\sup_{(\mathbf{x},\theta)\in B^N} |X_{(m)}^x(\theta;\omega)| + 1\right) \sup_{(\mathbf{x},\theta)\in B^N} \left|\partial_\theta X_{(m)}^x(\theta;\omega)\right|.$$

(iv). (Borel–Cantelli for $Z_{4,N}(\omega)$)
For some $r_4 > 0$ and some sequence $b_{4,N} \geq 1$, $N \in \mathbf{N}$,

$$\sum_{N=1}^{\infty} \frac{n}{(b_{4,N})^{r_4}} < +\infty$$

and $\sup_{N\in\mathbf{N}} E[|Z_{4,N}(\cdot)|^{r_4}] < \infty$ for each fixed $m \in \mathbf{N}$, where

$$Z_{4,N}(\omega) := a_N^{-1}\left(\sup_{(\mathbf{x},\theta)\in B^N} |\partial_x X_{(m)}^x(\theta;\omega)| + \sup_{(\mathbf{x},\theta)\in B^N} |\partial_\theta X_{(m)}^x(\theta;\omega)|\right).$$

(vi). (Borel–Cantelli for $\dot{Z}_{4,N}(\omega)$)
For some $\dot{r}_4 > 0$ and some sequence $\dot{b}_{4,N} \geq 1$, $N \in \mathbf{N}$,

$$\sum_{N=1}^{\infty} \frac{n}{(\dot{b}_{4,N})^{\dot{r}_4}} < +\infty$$

and $\sup_{N\in\mathbf{N}} E[|\dot{Z}_{4,N}(\cdot)|^{\dot{r}_4}] < +\infty$ for each fixed $m \in \mathbf{N}$, where

$$\begin{aligned}\dot{Z}_{4,N}(\omega) := a_N^{-1}\Big(&h \sup_{(\mathbf{x},\theta)\in B^N} \left|\partial_x \partial_\theta X_{(m)}^x(\theta;\omega)\right| \\ &+ h \sup_{(\mathbf{x},\theta)\in B^N} \left|\partial_\theta \partial_\theta X_{(m)}^x(\theta;\omega)\right| \\ &+ (Z_{4,N}+1) \sup_{(\mathbf{x},\theta)\in B^N} \left|\partial_\theta X_{(m)}^x(\theta;\omega)\right|\Big).\end{aligned}$$

(viii). (Borel–Cantelli for $\dot{Z}_{6,N}(\omega)$)
For some $\dot{r}_6 > 0$ and some sequence of positive numbers $\dot{b}_{6,N}$,

$$\sum_{N=1}^{\infty} \frac{n}{(\dot{b}_{6,N})^{\dot{r}_6}} < +\infty$$

and $\sup_N E[|\dot{Z}_{6,N}(\cdot)|^{\dot{r}_6}] < +\infty$ for each fixed $m \in \mathbf{N}$, where

$$\dot{Z}_{6,N}(\omega) := a_N^{-1} \sup_{(\mathbf{x},\theta)\in B^N} \left\{ \left|\partial_\theta X_{(m)}^x(\theta;\omega)\right| + E\left[\left|\partial_\theta X_{(m)}^x(\theta;\cdot)\right|\right] \right\}.$$

(ix). For some $\dot{\alpha}_6 > 0$, $\dot{q}_6 > 1$ and $\dot{C}_6 > 0$, and some positive sequence η_N,

$$\left(\frac{\eta_N h^2}{(\|K'\|_\infty \dot{b}_{6,N})^2 a_N} \exp\left(-\frac{(\eta_N)^2}{2(\frac{\|K'\|_\infty}{h^2}\dot{b}_{6,N} a_N)^2} \right) \right)^{\dot{q}_6} \leq \frac{\dot{C}_6}{n^{1+\dot{\alpha}_6}}$$

and $\sup_{N\in\mathbf{N}} E[|\dot{Z}_{6,N}(\cdot)|^{\dot{q}_6}] < +\infty$.

Set $a_N := \sqrt{c_2 \ln N}$ for some positive constant c_2. Set $n = C_1 N^{\alpha_1}$ for $\alpha_1, C_1 > 0$ and $h = C_2 N^{-\alpha_2}$ for $\alpha_2, C_2 > 0$. Also, we set

$$b_{3,N} = \frac{C_3 (N^{\gamma_3} n)^{\frac{1}{r_3}} c_2 \ln N}{h^2}$$

for $\gamma_3 > 1$, and $\dot{b}_{6,N} = (\dot{C}_6 n N^{\dot{\gamma}_6})^{\frac{1}{\dot{r}_6}}$. Then we obtain the following result.

Theorem 2 (Kohatsu-Higa et al. [9]) *Assume that the constants are chosen so as to satisfy* $c_1 > \frac{2}{c_2}$,

$$\left(4\alpha_2 + 2\frac{\alpha_1 + \dot{\gamma}_6}{\dot{r}_6} + \frac{\varphi_2 c_2}{\Delta} + \frac{1}{2} + \frac{\gamma_3}{r_3} + \frac{\alpha_1}{r_3} \right) \dot{q}_6 > \alpha_1, \tag{5}$$

$$\alpha_1 \left(1 - \frac{2}{r_3} - \frac{2}{\dot{r}_6} \right) > 8\alpha_2 + 1 + \frac{2\varphi_2 c_2}{\Delta} + \frac{2\gamma_3}{r_3} + 2\frac{\dot{\gamma}_6}{\dot{r}_6}. \tag{6}$$

Furthermore, assume that the moment conditions stated in (ii), (iii), (iv), (vi), (viii) *and* (ix) *above are satisfied. If additionally, we assume* (H1), (H2), (H5), (H5′), *then Assumption* (A) (6)-(a) *is satisfied.*

Furthermore, if all other conditions in Assumption (A) *are satisfied then there exist some positive finite random variables* Ξ_1 *and* Ξ_2 *such that*

$$|E_N[f] - f(\theta_0)| \leq \frac{\Xi_1}{\sqrt{N}} \ a.s. \quad \text{and} \quad \left|E_{N,m}^n[f] - f(\theta_0)\right| \leq \frac{\Xi_2}{\sqrt{N}} \ a.s.,$$

and thus

$$|E_N[f] - E_{N,m}^n[f]| \leq \frac{\Xi_1 + \Xi_2}{\sqrt{N}} \ a.s.$$

Remark 3

(i). In (6), r_3 and $\dot{r}_6$ represent moment conditions on the derivatives of $X_{(m)}^x(\theta)$, φ_2^{-1} represents the variance of $X_{(m)}^x(\theta)$, Δ represents the length of the time interval between observations. Finally, $c_2 > 2c_1^{-1}$ expresses a moment condition on Y_i. In (5), recall that $\dot{q}_6$ determines a moment condition on $X^x_{(m)}(\theta)$.

(ii). Roughly speaking, if $r_3, \dot{r}_6$ and $\dot{q}_6$ are big enough (which implies a restriction on n) and we choose $\alpha_1 > 8\alpha_2 + 1 + 2\varphi_2 c_2 \Delta^{-1}$, $m = \sqrt{N}$, $h = C_2 N^{-\alpha_2}$ and $n = C_1 N^{\alpha_1}$, then Assumptions (A) (6)-(a) and (A) (6)-(b) are satisfied. Then conditions contain the main tuning requirements (see Proposition 10).

3 The Ornstein–Uhlenbeck Process

We consider the following Ornstein–Uhlenbeck process; without loss of generality for $\theta \in [\alpha, \beta]$, where $0 < \alpha < \beta < 2$,

$$dX_t = -\theta X_t dt + dW_t, \quad X_0 = x, \tag{7}$$

where W_t is a one-dimensional Brownian motion. Then we can write the solution explicitly as

$$X_t = X_s e^{-\theta(t-s)} + \int_s^t e^{-\theta(t-u)} dW_u.$$

It is well known that the OU process has the following expectation, variance and covariance, for $s < t$,

$$\mu(X_s, t-s, \theta) := X_s \mu(t-s, \theta) := E[X_t | X_s] = X_s e^{-\theta(t-s)},$$

$$\sigma^2_{t-s}(\theta) := \mathrm{Var}(X_t | X_s) = \frac{1}{2\theta} - \frac{1}{2\theta} e^{-2\theta(t-s)},$$

$$\mathrm{Cov}\,(X_t, X_s) := \frac{1}{2\theta} e^{-\theta(t-s)} - \frac{1}{2\theta} e^{-\theta(t+s)}.$$

From moment results for the Gaussian distribution, the moments of the OU process can also be bounded as follows

$$E\left[\left|X_t - X_s e^{-\theta(t-s)}\right|^{2k}\right] = E\left[\left|\int_s^t e^{-\theta(t-u)} dW_u\right|^{2k}\right] = \frac{(2k)!}{2^{2k} k!} \left(\frac{1 - e^{-2\theta(t-s)}}{\theta}\right)^k.$$

In particular, for $s = 0$, using Minkowski's inequality, we obtain

$$E\left[X_t^{2k}\right] \le C_k \left(\left(\frac{1 - e^{-2\theta t}}{\theta}\right)^k + E\left[X_0^{2k}\right]\right). \tag{8}$$

The conditional density of X_t given X_s is given by

$$p_\theta(X_s, x; s, t) := q\left(X_s, x; \mu(t-s, \theta), \sigma^2_{t-s}(\theta)\right), \tag{9}$$

where

$$q(y,z;\mu,\sigma^2) = \frac{1}{\sqrt{2\pi\sigma^2}} e^{-\frac{(z-y\mu)^2}{2\sigma^2}}.$$

Note that $p_\theta(X_s, y) = p_\theta(X_s, y; s, s+\Delta)$.

3.1 The Euler–Maruyama Approximation of the OU Process

For $m \in \mathbf{N}$ and $i = 1, \ldots, m$, we set

$$X^x_{i,m}(\theta) := X^x_{i-1,m}(\theta) - \theta X^x_{i-1,m}(\theta)\Delta t + \Delta_{i-1}W,$$

where $X^x_0(\theta) = x$, $\Delta t = t_i - t_{i-1} = \frac{\Delta}{m}$ and $\Delta_i W = W_{t_{i+1}} - W_{t_i}$. We denote $X^x_{(m)}(\theta) = X^x_{m,m}(\theta)$. We will find an explicit expression for this approximation by induction. First,

$$\bar{X}_{t_1} = x(1-\theta\Delta t) + W_{t_1},$$

for $\Delta t = \frac{\Delta}{m}$. Similarly, for $\Delta_i W = W(t_{i+1}) - W(t_i)$,

$$\bar{X}_{t_2} = (x(1-\theta\Delta t) + \Delta_0 W)(1-\theta\Delta t) + \Delta_1 W.$$

Therefore, in general, we have that

$$X^x_{(m)}(\theta) = \bar{X}_{t_m} = x(1-\theta\Delta t)^m + \sum_{i=0}^{m-1} \Delta_i W (1-\theta\Delta t)^{m-1-i}. \tag{10}$$

From the above expression, we can easily find that $X^x_{(m)}(\theta)$ has the Gaussian distribution with mean $\mu(x,m,\theta)$ and variance $\sigma^2(m,\theta)$ where

$$\mu(x,m,\theta) = x\mu(m,\theta) = x\Big(1 - \frac{\theta\Delta}{m}\Big)^m,$$

$$\sigma^2(m,\theta) = \frac{(1-\frac{\theta\Delta}{m})^{2m} - 1}{\theta(\frac{\theta}{m} - 2)},$$

where we exclude $\frac{\theta}{m} = 2$. For example, if we take $\beta < 2$ then, since $m \in \mathbf{N}$, we always have $\frac{\theta}{m} < 2$ for $\theta \in [\alpha, \beta]$, where $0 < \alpha < \beta < 2$. Then the transition density $\tilde{p}^{(m)}_\theta(x,y) \equiv \tilde{p}^N_\theta(x,y)$ is given as follows

$$\tilde{p}^{(m)}_\theta(x,y) = q\left(x, y, \mu(m,\theta), \sigma^2(m,\theta)\right).$$

Next, we can represent $\bar{p}_\theta^N(x, y)$ as follows

$$\bar{p}_\theta^N(x, y) = E\left[\tilde{p}_\theta^{(m)}(x, hX + y)\right] = \frac{d}{dy} P\left(X_{(m)}^x(\theta; \cdot) - hX \le y\right),$$

where X is a random variable with the standard Gaussian distribution. Now

$$X_{(m)}^x(\theta, \cdot) \sim N\left(x\left(1 - \frac{\theta\Delta}{m}\right)^m, \frac{(1 - \frac{\theta\Delta}{m})^{2m} - 1}{\theta(\frac{\theta}{m} - 2)}\right)$$

and is independent of X. Then

$$X_{(m)}^x(\theta; \cdot) - hX \sim N\left(x\left(1 - \frac{\theta\Delta}{m}\right)^m, \frac{(1 - \frac{\theta\Delta}{m})^{2m} - 1}{\theta(\frac{\theta}{m} - 2)} + h^2\right).$$

Therefore,

$$\bar{p}_\theta^N(x, y) = q\left(x, y, \mu(m, \theta), \sigma^2(m, \theta, h)\right), \tag{11}$$

where $\sigma^2(m, \theta, h) = \sigma^2(m, \theta) + h^2$.

Proposition 1 (Density conditions for $\bar{p}_\theta^N(x, y)$) *The function* $\bar{p}_\theta^N(x, y)$ *satisfies the hypotheses* (H1), (H5) *and* (H5′).

The proof follows directly from Lemma 10 in Appendix. In fact, in the OU process case, we can take $\varphi_2 = \frac{6\beta}{\alpha}$ if $0 < 2\alpha\Delta \le \ln 2$ and $\varphi_1 = \frac{1}{\sqrt{2\pi(C(0,\Delta,\alpha)+1)}}$, where $C(0, \Delta, \alpha)$ is given in Lemma 10.

3.2 About Assumptions (A) (1)–(5)

In this section, we will examine the validity of Assumptions (A) (1)–(5) for the OU process and its Euler–Maruyama approximation. Assumption (A) (6) will be discussed in the next section.

Proposition 2 *The OU process satisfies Assumption* (A) (1).

Proof From Proposition 3 on page 115 of Doukhan [5], we obtain that the OU process has the geometrically strong mixing property. The OU process satisfies Assumption (A) (1). □

Once we take a prior distribution $\pi(\theta)$ as the uniform distribution on Θ then it satisfies Assumption (A) (2).

Set $\mu_\theta(x) := \sqrt{\frac{\theta}{\pi}}\exp(-\theta x^2)$.

Lemma 1 *The density $\mu_\theta(x)$ is the probability density function of the invariant measure for the OU process* (7).

Proposition 3 *The OU process and its Euler–Maruyama approximation satisfy Assumption* (A) (3).

Proof From the expression (9) for the transition density $p_\theta(y,z) = p_\theta(y,z;s,s+\Delta)$ of the OU process and, in addition, from the assumption of the kernel K, we see that $p_\theta(y,z)$ and $\bar{p}_\theta^N(x,y)$ clearly satisfy Assumption (A) (3), that is, it is continuous in x, y and twice continuously differentiable in θ.[1] Also, from Lemma 1, the OU process satisfies Assumption (A) (3). □

Now we consider the identifiability condition for p in Assumption (A) (4).

Proposition 4 *The OU process satisfies Assumption* (A) (4) *for p.*

Proof First note that the identifiability condition for p is equivalent to

$$\infty > \int \left(\inf_{\theta\in\Theta} \int \frac{|p_\theta(x,y) - p_{\theta_0}(x,y)|}{|\theta-\theta_0|} dy \right)^2 \mu_{\theta_0}(x)dx \geq \int c(x)^2 \mu_{\theta_0}(x)dx > 0.$$

By using the fundamental theorem of calculus and changing variables and setting $\beta = \alpha\theta + (1-\alpha)\theta_0$, we obtain

$$\begin{aligned}\infty &> \int \left(\inf_{\theta\in\Theta} \int \left| \int_0^1 \partial_\theta p_{\alpha\theta+(1-\alpha)\theta_0}(x,y) d\alpha \right| dy \right)^2 \mu_{\theta_0}(x)dx \\ &\geq \int c(x)^2 \mu_{\theta_0}(x)dx > 0.\end{aligned}$$

The integrability (the upper bound) is easily obtain as p_θ is a Gaussian density function. That is, set $\theta' = \operatorname{argmax}_{\theta\in\Theta} |\partial_\theta p_\theta(x,y)|$, then from (8) and using the inequalities $(a+b)^2 \leq 2(a^2+b^2)$ and $2|ab| \leq (a^2+b^2)$,

$$\begin{aligned}&\int \left(\inf_{\theta\in\Theta} \int \left| \int_0^1 \partial_\theta p_{\alpha\theta+(1-\alpha)\theta_0}(x,y) d\alpha \right| dy \right)^2 \mu_{\theta_0}(x)dx \\ &\leq 2 \int\int\int_0^1 \frac{p^2_{\alpha\theta+(1-\alpha)\theta_0}(x,y)}{m_0^2}\end{aligned}$$

[1]Note that the solution $X^x_{(m)}(\theta)$ is twice continuously differentiable in θ, since from the definition of the Euler–Maruyama approximation, the OU process is polynomial in θ and the kernel $K(x)$ is infinitely differentiable in x.

$$\times \left(M_1^2 + 16(t-s)^2(|y|^4 + |x|^4) + \frac{32\left(y^4 + x^4\right)}{m_0^2} M_1^2 \right) d\alpha dy \mu_{\theta_0}(x) dx < \infty.$$

Here $\theta' = \alpha\theta + (1-\alpha)\theta_0$ and $E_{\theta_0}[X_0^{2k}] = \frac{k!}{(4\theta)^k}$.

Now $\mu_{\theta_0}(x) > 0$ for all $x \in \mathbf{R}$. Therefore, it is enough to prove that, for all $x \in \mathbf{R}$,

$$\inf_{\theta \in \Theta} \int \left| \int_0^1 \partial_\theta p_{\alpha\theta + (1-\alpha)\theta_0}(x, y) d\alpha \right| dy > 0.$$

To this end, we argue by contradiction. We assume that

$$\inf_{\theta \in \Theta} \int \left| \int_0^1 \partial_\theta p_{\alpha\theta + (1-\alpha)\theta_0}(x, y) d\alpha \right| dy = 0.$$

This is equivalent that, for all $x \in \mathbf{R}$, there exists some $\theta^* = \theta^*(x)$ such that

$$\int \left| \int_0^1 \partial_\theta p_{\alpha\theta^* + (1-\alpha)\theta_0}(x, y) d\alpha \right| dy = 0.$$

Then, for all $x \in \mathbf{R}$, there exists some $\theta^* = \theta^*(x)$ such that for all $y \in \mathbf{R}$,

$$\left| \int_0^1 \partial_\theta p_{\alpha\theta^* + (1-\alpha)\theta_0}(x, y) d\alpha \right| = 0.$$

This means that for all $x \in \mathbf{R}$, there exists some $\theta^* = \theta^*(x)$ such that for all $y \in \mathbf{R}$, $p_{\theta^*}(x, y) = p_{\theta_0}(x, y)$. As both density functions are Gaussian then the point where the maximum is taken has to be the same. Therefore, the mean values are equal. Similarly, if we take y equal to the common mean we obtain that the variances have to be equal. Then analyzing the variance function, we deduce that it is decreasing in θ and thus $\theta^* = \theta_0$. □

By using the similar argument, we obtain the identifiability condition for $\bar{p}^N$.

Proposition 5 *The Euler-Maruyama approximation of the OU process satisfies Assumption* (A) (4) *for* $\bar{p}^N$.

Proof Set

$$B := \int \left\{ \inf_\theta \inf_N \int \frac{|\bar{p}_\theta^N(x, y) - \bar{p}_{\theta_0}^N(x, y)|}{|\theta - \theta_0|} dy \right\}^2 \mu_{\theta_0}(x) dx \in (0, +\infty).$$

As before, it is easy to prove that $B < +\infty$. To show that $B > 0$, we argue by contradiction. If $B = 0$ then, from the assumption that $\operatorname{supp} \mu_\theta(x) = \mathbf{R}$, we have, for all $x \in \mathbf{R}$,

$$\inf_{\theta}\inf_{N} \int \frac{|\bar{p}_{\theta}^{N}(x,y) - \bar{p}_{\theta_0}^{N}(x,y)|}{|\theta - \theta_0|} dy = 0.$$

Then for all $x \in \mathbf{R}$, there exists some sequence $\theta_n = \theta_n(x)$ such that

$$\lim_{n\to\infty}\inf_{N} \int \frac{|\bar{p}_{\theta_n}^{N}(x,y) - \bar{p}_{\theta_0}^{N}(x,y)|}{|\theta_n - \theta_0|} dy = 0.$$

Also, for all $x \in \mathbf{R}$, there exists some sequence $\theta_n = \theta_n(x)$ such that there exists some sequence $N_n = N_n(x, \theta_n)$ satisfying

$$\lim_{n\to\infty} \int \frac{|\bar{p}_{\theta_n}^{N_n}(x,y) - \bar{p}_{\theta_0}^{N_n}(x,y)|}{|\theta_n - \theta_0|} dy = 0.$$

Using the mean value theorem, we obtain the following convergence

$$\lim_{n\to\infty} \int \left| \int_0^1 \partial_\theta \bar{p}_{\alpha\theta_n + (1-\alpha)\theta_0}^{N_n}(x,y) d\alpha \right| dy = 0.$$

Hence we obtain the desired conclusion. □

Note that from Lemma 1, we have $E[X_0^{2k}] = \frac{(2k)!}{(4\theta)^k k!}$ and, from (8), we have

$$E\left[X_t^{2k}\right] \le C_k \left(\left(\frac{1 - e^{-2\theta t}}{\theta} \right)^k + \theta^{-k} \right). \tag{12}$$

Proposition 6 *For the OU process and its Euler–Maruyama approximation, the regularity condition* (1) *of Assumption* (A) (5) *holds.*

Proof Using (12), we obtain

$$\begin{aligned}
&\sup_{\theta\in\Theta} \int\int (\ln p_\theta(y,z))^{12} p_{\theta_0}(y,z)\mu_{\theta_0}(y) dy dz \\
&= \sup_{\theta\in\Theta} E\left[\left(-\frac{1}{2}\log\left(2\pi\sigma_\Delta^2(\theta)\right) - \frac{\left(X_\Delta(\theta_0) - X_0(\theta_0)e^{-\theta\Delta}\right)^2}{2\sigma_\Delta^2(\theta)} \right)^{12} \right] \\
&\le C \left\{ \sup_{\theta\in\Theta} \log^{12}\left(2\pi\sigma_\Delta^2(\theta)\right) + \sup_{\theta\in\Theta} \sigma_\Delta^{-24}(\theta) E\left[X_\Delta(\theta_0)^{24} + X_0(\theta_0)^{24} e^{-24\theta\Delta} \right] \right\} \\
&< \infty.
\end{aligned} \tag{13}$$

Now $\sigma_\Delta^2(\theta) = \frac{1}{2\theta}(1 - e^{-2\theta\Delta})$. Note that

$$\sigma_\Delta^2(\theta) \ge \frac{1}{2\beta}(1 - e^{-2\alpha\Delta}) > 0$$

and also $\sigma_\Delta^2(\cdot) \in C_b^\infty([\alpha,\beta])$. Furthermore let $m(\theta) = e^{-\theta\Delta}$. Note that $m(\cdot) \in C_b^\infty([\alpha,\beta])$. Hence, by using similar arguments as in the above calculations, we obtain (1), for $q_\theta = p_\theta$ and $i = 1, 2$. We can also obtain the integrability conditions for $q_\theta = p_\theta$.

Therefore, as a random variable $\bar{X}_{(m)}^{x,(1)}(\theta,\cdot)$ has the density $\bar{p}_\theta^N(x,y)$ at y (see (11)) then

$$E\left[|\bar{X}_{(m)}^{x,(1)}(\theta,\cdot)|^{2k}\right] \le C_k\left(\left(\frac{(1-\frac{\theta\Delta}{m})^{2m}-1}{\theta(\frac{\theta}{m}-2)}+h^2\right)^k + x^{2k}\left(1-\frac{\theta\Delta}{m}\right)^{2km}\right).$$

From Lemma 10 in the Appendix, (12) and (13), we get

$$\begin{aligned}
&\sup_N \sup_{\theta\in\Theta} \int\int \left(\log \bar{p}_\theta^N(y,z)\right)^{12} p_{\theta_0}(y,z)\mu_{\theta_0}(y)\,dydz \\
&\le C \sup_N \sup_{\theta\in\Theta} \left\{ \left(\log\left(\frac{1}{\sqrt{2\pi\sigma^2(m,\theta,h)}}\right)\right)^{12} \right. \\
&\qquad \left. + C' \frac{E[X_\Delta(\theta_0)^{24}] + E[X_0(\theta_0)^{24}]\mu(m,\theta)^{24}}{2^{12}\sigma^{24}(m,\theta,h)} \right\} < \infty.
\end{aligned}$$

Moreover, for $i = 1, 2$, as in the above, we obtain (1) for $q_\theta = \bar{p}_\theta^N$. Hence we obtain our conclusions. □

Now, we check the second regularity condition of Assumption (A) (5).

Proposition 7 *For the OU process and its Euler–Maruyama approximation, the regularity condition* (2) *of Assumption* (A) (5) *holds.*

Proof For $q_\theta = p_\theta$, we have

$$\begin{aligned}
&\int\int \left(-\frac{1}{2}\log\left(2\pi\sigma_\Delta^2(\theta)\right) - \frac{\left(z - ye^{-\theta\Delta}\right)^2}{2\sigma_\Delta^2(\theta)}\right) \bar{p}_{\theta_0}^N(y,z)\mu_{\theta_0}(y)\,dydz \\
&= -\frac{1}{2}\log\left(2\pi\sigma_\Delta^2(\theta)\right) - \frac{\sigma^2(m,\theta_0,h) + (2\theta_0)^{-1}\left(\mu(m,\theta_0) - e^{-\theta\Delta}\right)^2}{2\sigma_\Delta^2(\theta)}.
\end{aligned}$$

Therefore, the result follows because $\sigma_\Delta^2(\theta)$ is twice continuously differentiable and the above quantities are uniformly bounded in m.

Next, we will check equation (2) for $q_\theta = \bar{p}_\theta^N$, Then, as before,

$$\int\int \left\{\log\left(\frac{1}{\sqrt{2\pi\sigma^2(m,\theta,h)}}\right) - \frac{(y - x\mu(m,\theta))^2}{2\sigma^2(m,\theta,h)}\right\} \bar{p}_{\theta_0}^N(y,z)\mu_{\theta_0}(y)\,dydz$$

$$= \log\left(\frac{1}{\sqrt{2\pi\sigma^2(m,\theta,h)}}\right) - \frac{1}{2}.$$

Therefore, the property follows as in the previous case. □

Next, we consider the third regularity condition of Assumption (A) (5).

Proposition 8 *For the OU process and its Euler–Maruyama approximation, the regularity condition* (3) *of Assumption* (A) (5) *holds.*

Proof For $i = 0, 1$, we have

$$\int\int\left(-\frac{1}{2}\frac{\partial^i}{\partial\theta^i}\log\left(2\pi\sigma_\Delta^2(\theta)\right) - \frac{\partial^i}{\partial\theta^i}\frac{\left(z - ye^{-\theta\Delta}\right)^2}{2\sigma_\Delta^2(\theta)}\right)\bar{p}_{\theta_0}^N(y,z)\mu_{\theta_0}(y)dydz$$

$$= -\frac{1}{2}\frac{\partial^i}{\partial\theta^i}\log\left(2\pi\sigma_\Delta^2(\theta)\right) - \left(\frac{\partial^i}{\partial\theta^i}\frac{1}{2\sigma_\Delta^2(\theta)}\right)E\left[\left(\bar{X}_{(m)}^{X_0,(1)}(\theta_0,\cdot) - X_0e^{-\theta\Delta}\right)^2\right]$$

$$- \frac{E\left[\frac{\partial^i}{\partial\theta^i}(\bar{X}_{(m)}^{X_0,(1)}(\theta_0,\cdot) - X_0e^{-\theta\Delta})^2\right]}{2\sigma_\Delta^2(\theta)}.$$

If we expand the last expectation in the above expression, it is clear that

$$\frac{\partial^i}{\partial\theta^i}E\left[\left(\bar{X}_{(m)}^{X_0,(1)}(\theta_0,\cdot) - X_0e^{-\theta\Delta}\right)^2\right] = E\left[\frac{\partial^i}{\partial\theta^i}\left(\bar{X}_{(m)}^{X_0,(1)}(\theta_0,\cdot) - X_0e^{-\theta\Delta}\right)^2\right].$$

Hence the last property of Assumption (A) (5) follows for $q_\theta = p_\theta$. A similar proof also applies to $q_\theta = \bar{p}_\theta^N$. □

3.3 *Assumption (A) (6)*

3.3.1 Parameter Tuning of Assumption (A) (6)-(a)

If we choose $0 < c_1 < \alpha$ then the moment hypothesis of (ii) in Sect. 2.3,

$$E[e^{c_1|Y_1|^2}] < \infty,$$

is satisfied since Y_1 has the Gaussian distribution. Furthermore, as we may take $a_N = \sqrt{c_2 \ln N}$ with $c_1 > \frac{2}{c_2}$, then condition (ii) in Sect. 2.3 holds.

From the explicit expression (10) of the OU process, we obtain the following derivatives of the Euler–Maruyama approximation of the OU process.

$$\partial_x X_{(m)}^x(\theta) = (1 - \theta\Delta t)^m,$$

$$\partial_\theta X^x_{(m)}(\theta) = -mx\Delta t(1-\theta\Delta t)^{m-1} - \Delta t\sum_{i=0}^{m-2}(m-1-i)\Delta_i W(1-\theta\Delta t)^{m-2-i},$$

$$\partial_\theta\partial_x X^x_{(m)}(\theta) = -m\Delta t(1-\theta\Delta t)^{m-1}, \tag{14}$$

$$\partial_\theta^2 X^x_{(m)}(\theta) = m(m-1)\Delta t^2(1-\theta\Delta t)^{m-2} + \Delta t^2\sum_{i=0}^{m-3}(m-1-i)(m-2-i)\Delta_i W(1-\theta\Delta t)^{m-3-i}.$$

Lemma 2 *For any $j \in \mathbf{N}$, we have*

$$\sup_{m\geq j\vee(\theta\Delta)}\sup_{(\mathbf{x},\theta)\in B^N} a_N^{-1}\left|x\frac{\partial^j}{\partial\theta^j}(1-\theta\Delta t)^m\right| < \infty.$$

Proof From the definition of B^N, it is clear that $\sup_{(\mathbf{x},\theta)\in B^N} a_N^{-1}|x| \leq 1$. Next, we have

$$\left|\frac{\partial^j}{\partial\theta^j}(1-\theta\Delta t)^m\right| = \left|(-1)^j 1\left(1-\frac{1}{m}\right)\cdots\left(1-\frac{j-1}{m}\right)\Delta^j(1-\theta\Delta t)^{m-j}\right|$$
$$\leq \Delta^j\left(1-\frac{\theta\Delta}{m}\right)^{m-j}.$$

Set $y = -\frac{m}{\theta\Delta}$. For any m and j such that $m \geq j\vee(\theta\Delta)$, we have

$$\left(1-\frac{\theta\Delta}{m}\right)^{m-j} = \left\{\left(1+\frac{1}{y}\right)^y\right\}^{-\frac{\theta\Delta(m-j)}{m}} \leq e^{-\frac{\theta\Delta(m-j)}{m}} \leq 1, \tag{15}$$

where we used Lemma 9. Hence the conclusion follows. □

For a differentiable function $h(\theta,t)$, we set $U(\theta) := \int_0^\Delta h(\theta,s)\,dW_s$. Then

$$U'(\theta) = \int_0^\Delta \frac{\partial}{\partial\theta}h(\theta,s)\,dW_s.$$

Lemma 3 *We assume that there exists some positive constant $C(\Delta)$, which depends on Δ, such that*

$$\sum_{j=0}^{1}\sup_{\substack{\theta\in[\alpha,\beta]\\ t\in[0,\Delta]}}\left|\frac{\partial^j}{\partial\theta^j}h(\theta,t)\right| \leq C(\Delta).$$

Then, for $p \in \mathbf{N}$, we have

$$E\Big[\sup_{\theta\in[\alpha,\beta]}|U(\theta)|^{2p}\Big] \leq C(\Delta)^{2p}\Delta^p(1+(\beta-\alpha)^{2p})\frac{(2p)!}{p!}.$$

Proof Note that

$$U(\theta) = U(\alpha) + \int_\alpha^\theta U'(\rho)\,d\rho, \quad \text{a.s.}$$

From Hölder's inequality and Fubini's theorem, we obtain

$$E\Big[\sup_{\theta\in[\alpha,\beta]} |U(\theta)|^{2p}\Big]$$
$$\le 2^p\left(E\Big[|U(\alpha)|^{2p}\Big] + (\beta-\alpha)^{2p-1}\int_\alpha^\beta E\Big[|U'(\rho)|^{2p}\Big]d\rho\right). \quad (16)$$

Note that $U(\theta)$ and $U'(\theta)$ have the Gaussian distribution with mean 0 and variance $\int_0^\Delta h(\theta,s)^2 ds$ and $\int_0^\Delta (\frac{\partial}{\partial\theta}h(\theta,s))^2 ds$, respectively. Then, from moment properties of the Gaussian distribution, we have that

$$E\big[|U(\alpha)|^{2p}\big] = \left(\int_0^\Delta h(\alpha,s)^2 ds\right)^p \frac{(2p)!}{2^p p!} \le \left(C(\Delta)^2\Delta\right)^p \frac{(2p)!}{2^p p!},$$
$$E\big[|U'(\rho)|^{2p}\big] = \left(\int_0^\Delta \left(\frac{\partial}{\partial\theta}h(\rho,s)\right)^2 ds\right)^p \frac{(2p)!}{2^p p!} \le \left(C(\Delta)^2\Delta\right)^p \frac{(2p)!}{2^p p!}.$$

Finally, we have

$$E\Big[\sup_{\theta\in[\alpha,\beta]} |U(\theta)|^{2p}\Big] \le \left(C(\Delta)^2\Delta\right)^p \frac{(2p)!}{p!}\left(1+(\beta-\alpha)^{2p}\right)$$

and thus we obtain the desired inequality. □

We note that

$$\sum_{i=0}^{m-1} \Delta_i W(1-\theta\Delta t)^{m-1-i} = \int_0^\Delta h_m(\theta,s)dW_s,$$

where $h_m(\theta,t) = (1-\theta\Delta t)^{m-1-i}$ for $t\in[t_i,t_{i+1})$ and $i=0,1,\ldots,m-1$. Also, we have

$$\frac{\partial}{\partial\theta}h_m(\theta,t) = (m-1-i)(-\Delta t)(1-\theta\Delta t)^{m-2-i},$$

for $t\in[t_i,t_{i+1})$, $i=0,1,\ldots,m-2$ and $=0$ for $t\in[t_{m-1},t_m]$, $i=m-1$. Note that from (15), we have, for $m\ge\theta\Delta$,

$$|h_m(\theta,t)|\le 1 \quad \text{and} \quad \left|\frac{\partial}{\partial\theta}h_m(\theta,t)\right| \le \Delta.$$

Next, we observe that

$$\sum_{i=0}^{m-2} \Delta t(m-1-i)\Delta_i W(1-\theta\Delta t)^{m-2-i} = \int_0^{\Delta} h_m^{(1)}(\theta,s)dW_s,$$

where

$$h_m^{(1)}(\theta,t) = \begin{cases} \Delta t(m-1-i)(1-\theta\Delta t)^{m-2-i}, \\ \quad \text{for } t\in[t_i,t_{i+1}),\ i=0,1,\dots,m-2, \\ =0, \quad \text{for } t\in[t_{m-1},t_m]. \end{cases}$$

Moreover, we have

$$\frac{\partial}{\partial\theta}h_m^{(1)}(\theta,t) = \begin{cases} -\Delta t^2(m-1-i)(m-2-i)(1-\theta\Delta t)^{m-3-i}, \\ \quad \text{for } t\in[t_i,t_{i+1}),\ i=0,1,\dots,m-3, \\ =0, \quad \text{for } t\in[t_{m-2},t_m]. \end{cases}$$

Then, as before, from (15), we obtain, for $m\geq\theta\Delta$,

$$|h_m^{(1)}(\theta,t)|\leq\Delta \quad \text{and} \quad \left|\frac{\partial}{\partial\theta}h_m^{(1)}(\theta,t)\right|\leq\Delta^2.$$

As above, we consider

$$\sum_{i=0}^{m-3} \Delta t^2(m-1-i)(m-2-i)\Delta_i W(1-\theta\Delta t)^{m-3-i} = \int_0^{\Delta} h_m^{(2)}(\theta,s)dW_s,$$

where

$$h_m^{(2)}(\theta,t) = \begin{cases} \Delta t^2(m-1-i)(m-2-i)(1-\theta\Delta t)^{m-3-i}, \\ \quad \text{for } t\in[t_i,t_{i+1}),\ i=0,1,\dots,m-3, \\ =0, \quad \text{for } t\in[t_{m-2},t_m]. \end{cases}$$

We now have

$$\frac{\partial}{\partial\theta}h_m^{(2)}(\theta,t) = \begin{cases} -\Delta t^3(m-1-i)(m-2-i)(m-3-i)(1-\theta\Delta t)^{m-4-i}, \\ \quad \text{for } t\in[t_i,t_{i+1}),\ i=0,1,\dots,m-4, \\ =0, \quad \text{for } t\in[t_{m-3},t_m]. \end{cases}$$

Consequently, from (15), we get, for $m\geq\theta\Delta$,

$$|h_m^{(2)}(\theta,t)|\leq\Delta^2 \quad \text{and} \quad \left|\frac{\partial}{\partial\theta}h_m^{(2)}(\theta,t)\right|\leq\Delta^3.$$

Lemma 4 *For* $H_m(\theta,t) = h_m(\theta,t),\ h_m^{(1)}(\theta,t),\ h_m^{(2)}(\theta,t)$, *we have, for* $p \in \mathbf{N}$,

$$\sup_{m\in\mathbf{N}} E\left[\sup_{\theta\in[\alpha,\beta]}\left|\int_0^{\Delta} H_m(\theta,s)\,dW_s\right|^{2p}\right] < +\infty.$$

Proof From the calculations preceding the lemma, we see that H_m satisfies the assumption of Lemma 3 as we take $C(\Delta) = 1 \vee \Delta^3$.

An application of Lemma 3 yields

$$E\left[\sup_{\theta\in[\alpha,\beta]}\left|\int_0^{\Delta} H_m(\theta,s)\,dW_s\right|^{2p}\right] \le C(\Delta)^{2p}\Delta^p(1+(\beta-\alpha)^{2p})\frac{(2p)!}{p!},$$

where the right-hand side does not depend on m. To complete the proof, it suffices to take sup with respect to $m \in \mathbf{N}$ for the left-hand side. □

From the above lemmas and explicit formulas (14), we obtain the following two results.

Lemma 5 *For all* $p \ge 1$ *and* $k \in \mathbf{N}$, *we have*

$$\sup_{N\in\mathbf{N}} E\left[\left(a_N^{-1}\sup_{(\mathbf{x},\theta)\in B^N}\left|V_{(m)}^{x}(\theta;\cdot)\right|\right)^{p}\right] < \infty,$$

for $V_{(m)}^x(\theta;\omega) = X_{(m)}^x(\theta;\omega), \partial_x X_{(m)}^x(\theta;\omega), \partial_\theta X_{(m)}^x(\theta;\omega), \partial_\theta\partial_x X_{(m)}^x(\theta;\omega)$ *and* $\partial_\theta^2 X_{(m)}^{x,(k)}(\theta;\omega)$.

Proposition 9 (Moment conditions of (iii), (iv), (vi), (viii) and (ix) in Sect. 2.3) *For all* $p \ge 1$, *we have* $\sup_{N\in\mathbf{N}} E[|T_N(\cdot)|^p] < +\infty$, *for* $T_N(\omega) = Z_{3,N}(\omega),\ Z_{4,N}(\omega),\ \dot{Z}_{4,N}(\omega),\ \dot{Z}_{6,N}(\omega)$.

From the above result, we obtain the required integrability conditions for $Z_{3,N}(\omega)$, $Z_{4,N}(\omega)$, $\dot{Z}_{4,N}(\omega)$ and $\dot{Z}_{6,N}(\omega)$. Therefore, we can take $r_3, \dot{r}_6, \dot{q}_6$ large enough, so as conditions (5) and (6) are met.

Proposition 10 (Parameter conditions (5) and (6)) *If* $\alpha_1 > 8\alpha_2 + 1 + \frac{2\varphi_2 c_2}{\Delta}$ *then there exist some* $r_3, \dot{r}_6, \dot{q}_6, \gamma_3, \dot{\gamma}_6$ *such that conditions* (5) *and* (6) *are satisfied.*

3.3.2 Parameter Tuning of Assumption (A) (6)-(b)

In this section, we consider the parameter tuning (b) of Assumption (A) (6). Set

$$q(y,z;\mu,\sigma^2) = \frac{1}{\sqrt{2\pi\sigma^2}}e^{-\frac{(z-y\mu)^2}{2\sigma^2}}.$$

Then we can represent the densities $p_{\theta_0}(y,z)$ and $\bar{p}^N_{\theta_0}(y,z)$ as

$$q(y,z,\mu_\Delta(\theta_0),\sigma^2_\Delta(\theta_0)) \quad \text{and} \quad q(y,z,\mu(m,\theta_0),\sigma^2(m,\theta_0,h)),$$

respectively, where $\mu_\Delta(\theta_0)=e^{-\theta_0\Delta}$. By applying the mean value theorem and Lemma 8, we obtain

$$\begin{aligned}
&|p_{\theta_0}(y,z)-\bar{p}^N_{\theta_0}(y,z)| \\
&\quad \le |\mu_\Delta(\theta_0)-\mu(m,\theta_0)| \int_0^1 \left|\partial_\mu q\left(y,z,\gamma\mu_\Delta(\theta_0)+(1-\gamma)\mu(m,\theta_0),\sigma^2_\Delta(\theta_0)\right)\right| d\gamma \\
&\qquad + \left|\sigma^2_\Delta(\theta_0)-\sigma^2(m,\theta_0,h)\right| \\
&\qquad \times \int_0^1 \left|\partial_{\sigma^2} q\left(y,z,\mu(m,\theta_0),\gamma\sigma^2_\Delta(\theta_0)+(1-\gamma)\sigma^2(m,\theta_0,h)\right)\right| d\gamma \\
&\quad \le C(\alpha,\beta,\Delta)\frac{1}{m}\int_0^1 \left|\partial_\mu q\left(y,z,\gamma\mu_\Delta(\theta_0)+(1-\gamma)\mu(m,\theta_0),\sigma^2_\Delta(\theta_0)\right)\right| d\gamma \\
&\qquad + C(\alpha,\beta,\Delta)\left(\frac{1}{m}+h^2\right) \\
&\qquad \times \int_0^1 \left|\partial_{\sigma^2} q\left(y,z,\mu(m,\theta_0),\gamma\sigma^2_\Delta(\theta_0)+(1-\gamma)\sigma^2(m,\theta_0,h)\right)\right| d\gamma. \qquad (17)
\end{aligned}$$

Next, we consider the derivatives of q with respect to μ,σ^2. Assume that

$$0<\mu_{min}\le\mu\le\mu_{max} \quad \text{and} \quad 0<\sigma^2_{min}\le\sigma^2\le\sigma^2_{max}.$$

From Lemma 6, we have, for $c>1$,

$$\left|\partial_\mu q(y,z;\mu,\sigma^2)\right| \le \frac{|yz|+y^2\mu_{max}}{\sigma^2_{min}}\frac{1}{\sqrt{2\pi\sigma^2_{min}}}\exp\left(-\frac{\frac{c-1}{c}z^2-(c-1)(\mu_{max}y)^2}{2\sigma^2_{max}}\right)$$

and

$$\begin{aligned}
\left|\partial_{\sigma^2} q(y,z;\mu,\sigma^2)\right| &\le \left\{\frac{1}{2\sigma^2_{min}}\frac{1}{\sqrt{2\pi\sigma^2_{min}}}+\frac{4(z^2+y^2\mu^2_{max})}{2(\sigma^2_{min})^2}\frac{1}{\sqrt{2\pi\sigma^2_{min}}}\right\} \\
&\quad \times \exp\left(-\frac{\frac{c-1}{c}z^2-(c-1)(\mu_{max}y)^2}{2\sigma^2_{max}}\right).
\end{aligned}$$

Next, for $0<\gamma<1$, we have

$$\gamma\mu_\Delta(\theta_0)+(1-\gamma)\mu(m,\theta_0)\le\gamma e^{-\theta_0\Delta}+(1-\gamma)e^{-\theta_0\Delta}\le e^{-\theta_0\Delta}$$

and

$$\frac{1}{2\beta}(1-e^{-2\alpha\Delta}) \leq \gamma\sigma_\Delta^2(\theta_0) + (1-\gamma)\sigma^2(m,\theta_0,h)$$
$$\leq \frac{1}{2\alpha}(1-e^{-2\beta\Delta}) + C(k,\Delta,\alpha) + 1,$$

where $C(k, \Delta, \alpha)$ is the constant defined in formula (18) in the Appendix. Therefore, we may take

$$\mu_{max} = e^{-\theta_0\Delta}, \quad \sigma_{min}^2 = \frac{1}{2\beta}\left(1-e^{-2\alpha\Delta}\right),$$
$$\sigma_{max}^2 = \frac{1}{2\alpha}\left(1-e^{-2\beta\Delta}\right) + C(k,\Delta,\alpha) + 1.$$

Then we have

$$(17) \leq C(\alpha,\beta,\Delta)\frac{1}{\sqrt{2\pi\sigma_{min}^2}}\left\{\frac{|yz| + y^2\mu_{max}}{\sigma_{min}^2} + \frac{1}{2\sigma_{min}^2} + \frac{4(z^2 + y^2\mu_{max}^2)}{2(\sigma_{min}^2)^2}\right\}$$
$$\times \exp\left(-\frac{\frac{c-1}{c}z^2 - (c-1)(\mu_{max}y)^2}{2\sigma_{max}^2}\right)\left(\frac{1}{m} + h^2\right).$$

Then we need the following parameter tuning condition: $(\frac{1}{m} + h^2)\sqrt{N} \leq C$, where C is a constant. Note that $h = C_2 N^{-\alpha_2}$ therefore we require that $\alpha_2 \geq \frac{1}{2}$. Furthermore, $m \geq \sqrt{N}$. Finally, we check the following integrability condition

$$\sup_N \sup_{\theta\in[\alpha,\beta]} \int\int \left|\frac{\partial}{\partial\theta} \ln \bar{p}_\theta^N(y,z)\right| \left\{\frac{|yz| + y^2\mu_{max}}{\sigma_{min}^2} + \frac{1}{2\sigma_{min}^2} + \frac{4(z^2 + y^2\mu_{max}^2)}{2(\sigma_{min}^2)^2}\right\}$$
$$\times \exp\left(-\frac{\frac{c-1}{c}z^2 - (c-1)(\mu_{max}y)^2}{2\sigma_{max}^2}\right)\mu_{\theta_0}(y)dydz < \infty.$$

Note that $\mu_\theta(y)$ is the density of $N(0, \frac{1}{2\theta})$ law and that we have an explicit expression for $\frac{\partial}{\partial\theta} \ln \bar{p}_\theta^N(y,z)$, which is a second degree polynomial in y, z. As the parameters, $\sigma^2(m,\theta,h)$ and $\mu(m,\theta)$ satisfy Lemma 10, the above integrability condition is satisfied. From the above calculations, we obtain the following result.

Proposition 11 *In the OU process and its Euler–Maruyama approximation case, for* $\alpha_2 \geq \frac{1}{2}$ *and* $m(N) \geq \sqrt{N}$, *Assumption* (A) (6)-(b) *holds.*

3.3.3 Parameter Tuning of Assumption (A) (6)-(c)

We now consider the parameter tuning (c) of Assumption (A) (6). Note that in order to verify this condition, we can concretely calculate

$$\left| \frac{\partial}{\partial \theta} \ln \bar{p}_\theta^N(y,z) - \frac{\partial}{\partial \theta} \ln p_\theta(y,z) \right|.$$

It suffices to analyze separately each term and use Lemma 8 together with Lemma 10. Then we obtain some polynomial function $g^N(y,z) = g(y,z)$ with respect to y, z, so that Assumption (A) (6)-(c) is satisfied. In particular, if Y_0 and Y_1 have the Gaussian distribution, it is clear that the integrability condition $E[|g(Y_0, Y_1)|^4] < +\infty$ is satisfied. Then we have

Proposition 12 *In the OU process and its Euler–Maruyama approximation case, Assumption* (A) (6)-(c) *holds.*

Recall that $\Theta = [\alpha, \beta]$ $(0 < \alpha < \beta < 2)$, $n = C_1 N^{\alpha_1}$, $h = C_2 N^{-\alpha_2}$ and

$$\inf_{(\mathbf{x},\theta)\in B^N} \bar{p}_\theta^N(x,y) \geq \left(c\sqrt{2c + \Delta N^{\frac{4cc_2}{\Delta}}} \right)^{-1},$$

where $B^N = \{(x,y,\theta);\ |(x,y)| \leq \sqrt{c_2 \ln N}\}$. We take a prior density function so that $\pi(\theta) > 0$ on Θ and a kernel function K as the Gaussian kernel. Finally, we obtain the following theorem for the OU process and its Euler-Maruyama approximation.

Theorem 3 *Assume* $\alpha_1 > 8\alpha_2 + 1 + \frac{4\varphi_2 c_2}{\Delta}$, $\alpha_2 \geq \frac{1}{2}$ *and* $m \geq \sqrt{N}$. *Then there exist some positive finite random variables* Ξ_1 *and* Ξ_2 *such that for* $f \in C^1(\Theta)$, *we have*

$$|E_N[f] - f(\theta_0)| \leq \frac{\Xi_1}{\sqrt{N}} \quad a.s. \quad \text{and} \quad \left|\hat{E}_{N,m}^n[f] - f(\theta_0)\right| \leq \frac{\Xi_1}{\sqrt{N}} \quad a.s.,$$

and thus

$$\left|E_N[f] - E_{N,m}^n[f]\right| \leq \frac{\Xi_1 + \Xi_2}{\sqrt{N}} \ a.s.$$

Appendix

Here we give some lemmas, which are used in the parameter tuning sections.

Lemma 6 *For* $c > 1$, *we have*

(i). $(x+y)^2 \leq \frac{c}{c-1}x^2 + cy^2$,
(ii). $\frac{c-1}{c}x^2 + (c-1)y^2 \leq (x-y)^2$.

The proofs are based on Young's lemma, which follows from simple calculations.

Lemma 7 *For* $m \geq 2\beta\Delta$, *we have* $|(1-\frac{\theta\Delta}{m})^m - e^{-\theta\Delta}| \leq e^{-\alpha\Delta}(\beta\Delta)^2\frac{1}{m}$.

From this lemma, we obtain

Lemma 8 *For* $k=0,1$ *and* $m \geq 2\beta\Delta$, *we have the following estimations;*

(i). $|\frac{\partial^k}{\partial\theta^k}(\sigma^2_\Delta(\theta) - \sigma^2(m,\theta,h))| \leq C(\alpha,\beta,\Delta)\{\frac{1}{m} + h^2\mathbf{1}(k=0)\}$,

(ii). $|\frac{\partial^k}{\partial\theta^k}(e^{-\theta\Delta} - \mu(m,\theta))| \leq C(\alpha,\beta,\Delta)\frac{1}{m}$,

where $C(\alpha,\beta,\Delta)$ *is some positive constant.*

Lemma 9 *For* $m > \beta\Delta$, *we have* $\left(1-\frac{\theta\Delta}{m}\right)^m \leq e^{-\theta\Delta}$.

Proof Set $f(x) = (1+\frac{1}{x})^x$. Then $f(x)$ is an increasing function for $-\infty < x < -1$ and $\lim_{x\to-\infty} f(x) = e$. The conclusion now follows. □

Lemma 10 *For* $k \in \mathbf{N}\cup\{0\}$,

(i).

$$\sup_{m\geq\max(\frac{k}{2},\beta\Delta)}\sup_{\theta\in\Theta}\left|\frac{\partial^k}{\partial\theta^k}\mu(m,\theta)\right| = \sup_{m\geq\max(\frac{k}{2},\beta\Delta)}\sup_{\theta\in\Theta}\left|\frac{\partial^k}{\partial\theta^k}(1-\frac{\theta\Delta}{m})^m\right|$$
$$\leq (2\Delta)^k 3^{2\beta\Delta} < +\infty,$$

(ii).

$$\sup_{0\leq h\leq 1}\sup_{m\geq\max(\frac{k}{2},\beta\Delta)}\sup_{\theta\in\Theta}\left|\frac{\partial^k}{\partial\theta^k}\sigma^2(m,\theta,h)\right|$$
$$= \sup_{0\leq h\leq 1}\sup_{m\geq\max(\frac{k}{2},\beta\Delta)}\sup_{\theta\in\Theta}\left|\frac{\partial^k}{\partial\theta^k}\frac{(1-\frac{\theta\Delta}{m})^{2m}-1}{\theta(\frac{\theta}{m}-2)} + h^2\mathbf{1}_{\{k=0\}}\right|$$
$$\leq C(k,\Delta,\alpha) + 1 < +\infty,$$

(iii).

$$\inf_{0\leq h\leq 1}\inf_{m\geq\max(\frac{k}{2},\beta\Delta)}\inf_{\theta\in\Theta}\left|\sigma^2(m,\theta,h)\right|$$
$$= \inf_{0\leq h\leq 1}\inf_{m\geq\max(\frac{k}{2},\beta\Delta)}\inf_{\theta\in\Theta}\left|\frac{(1-\frac{\theta\Delta}{m})^{2m}-1}{\theta(\frac{\theta}{m}-2)} + h^2\right| \geq \frac{2(1-e^{-2\alpha\Delta})}{3\beta} > 0,$$

where the positive constant $C(k,\Delta,\alpha)$ *is defined in the proof.*

Proof Now $\mu(m,\theta)=(1-\frac{\theta\Delta}{m})^m$ and set $D_\theta^k=\frac{\partial^k}{\partial\theta^k}$. Note that from Lemma 9, we have $0\le\mu(m,\theta)\le e^{-\theta\Delta}\le\sup_\theta e^{-\theta\Delta}=e^{-\alpha\Delta}$. Note that

$$D_\theta^k\mu(m,\theta)=(2m)(2m-1)\cdots(2m-(k-1))\left(1-\frac{\theta\Delta}{m}\right)^{2m-k}\left(-\frac{\Delta}{m}\right)^k.$$

Then

$$\begin{aligned}&D_\theta^{k+1}\mu(m,\theta)\\&\quad=(2m)(2m-1)\cdots(2m-(k-1))(2m-k)\left(1-\frac{\theta\Delta}{m}\right)^{2m-(k+1)}\left(-\frac{\Delta}{m}\right)^{k+1}.\end{aligned}$$

Moreover, for $2m\ge k$, we have

$$\sup_m\sup_\theta|D_\theta^k\mu(m,\theta)|\le\sup_m\left\{\frac{(2m\Delta)^k}{m^k}\left(1+\frac{\beta\Delta}{m}\right)^{2m-k}\right\}\le(2\Delta)^k3^{2\beta\Delta}.$$

Hence we obtain (i).

Recall that $\sigma^2(m,\theta)=\frac{\mu(m,\theta)^2-1}{\theta(\frac{\theta}{m}-2)}$. From the Leibnitz formula, we have

$$\begin{aligned}|D_\theta^k\sigma^2(m,\theta)|\le\sum_{i=0}^k C_{k,i}\sup_m\sup_\theta\left|D_\theta^i(\mu(m,\theta)^2)\right|\sup_m\sup_\theta\left|D_\theta^{k-i}\frac{1}{\theta(\frac{\theta}{m}-2)}\right|\\+\sup_m\sup_\theta\left|D_\theta^k\frac{1}{\theta(\frac{\theta}{m}-2)}\right|.\end{aligned}$$

From the above, the Leibnitz formula and the binomial theorem, we obtain, for $i=0,1,\dots,k$,

$$\begin{aligned}\sup_m\sup_\theta\left|D_\theta^i(\mu(m,\theta)^2)\right|&\le\sup_m\sup_\theta\left|\Delta^i\left(1-\frac{\theta\Delta}{m}\right)^{2m-i}\sum_{j=0}^i\binom{i}{j}\right|\\&\le\Delta^ie^{-\alpha\Delta}\sum_{j=0}^i\binom{i}{j}<\infty.\end{aligned}$$

Moreover, for all $i=0,1,\dots,k$, we have, from the binomial theorem,

$$\sup_m\sup_\theta\left|D_\theta^i\frac{1}{\theta(\frac{\theta}{m}-2)}\right|\le\sum_{j=0}^iC_{i,j}\frac{j!}{\alpha^{j+1}}\frac{(i-j)!}{2^{i-j}}.$$

Then we have

$$\sup_m \sup_{\theta\in[\alpha,\beta]} \left|D_\theta^k \sigma^2(m,\theta)\right|$$
$$\le \sum_{i=0}^{k} C_{k,i} \left\{ \Delta^i e^{-\alpha\Delta} \sum_{j=0}^{i} \binom{i}{j} \right\} \left\{ \sum_{j=0}^{k-i} C_{k-i,j} \frac{j!}{\alpha^{j+1}} \frac{(k-i-j)!}{2^{k-i-j}} \right\}$$
$$+ \sum_{j=0}^{k} C_{k,j} \frac{j!}{\alpha^{j+1}} \frac{(k-j)!}{2^{k-j}} =: C(k,\Delta,\alpha) < \infty, \tag{18}$$

so that (ii) holds.

Finally, for $m \ge \beta\Delta$, we have

$$\sigma^2(m,\theta) \ge \frac{1-e^{-2\theta\Delta}}{\theta\left(2-\frac{\theta}{2\theta}\right)} = \frac{2}{3\theta}\left(1-e^{-2\theta\Delta}\right) \ge \frac{2}{3\beta}\left(1-e^{-2\alpha\Delta}\right) > 0$$

and thus (iii) is valid. Here for $m \ge \beta\Delta$, $0 \le \left(1-\frac{\theta\Delta}{m}\right)^m \le e^{-\theta\Delta} \le e^{-\alpha\Delta}$, and for $m \ge \beta\Delta$,

$$\frac{2(1-e^{-2\alpha\Delta})}{3\beta} \le \frac{(1-\frac{\theta\Delta}{m})^{2m}-1}{\theta(\frac{\theta}{m}-2)} \le \frac{1}{\alpha(2-\beta)}.$$

Thus the proof is complete. □

References

1. Ait-Sahalia, Y., Mykland, P.A.: Estimators of diffusions with randomly spaced discrete observations: a general theory. Ann. Stat. **32**(5), 2186–2222 (2004)
2. Bain, A., Crisan, D.: Fundamentals of Stochastic Filtering. Springer, New York (2009)
3. Cano, J.A., Kessler, M., Salmeron, D.: Approximation of the posterior density for diffusion processes. Stat. Probab. Lett. **76**(1), 39–44 (2006)
4. Del Moral, P., Jacod, J., Protter, P.: The Monte Carlo method for filtering with discrete-time observations. Probab. Theory Relat. Fields **120**, 346–368 (2001)
5. Doukhan, P.: Mixing; Properties and Examples. Lecture Notes in Statistics, vol. 85. Springer, Berlin (1994)
6. Jacod, J.: Parametric inference for discretely observed non-ergodic diffusions. Bernoulli **12**(3), 383–401 (2006)
7. Kelly, L., Platen, E., Sorensen, M.: Estimation for discretely observed diffusions using transform functions. Stochastic methods and their applications. J. Appl. Probab. **41A**, 99–118 (2004)
8. Kessler, M.: Estimation of an ergodic diffusion from discrete observations. Scand. J. Stat. **24**(2), 211–229 (1997)
9. Kohatsu-Higa, A., Vayatis, N., Yasuda, K.: Tuning of a Bayesian estimator under discrete time observations and unknown transition density (2013, submitted)
10. Roberts, G.O., Stramer, O.: On inference for partially observed nonlinear diffusion models using the Metropolis-Hastings algorithm. Biometrika **88**, 603–621 (2001)
11. Yoshida, N.: Estimation for diffusion processes from discrete observation. J. Multivar. Anal. **41**(2), 220–242 (1992)

Multiasset Derivatives and Joint Distributions of Asset Prices

Ilya Molchanov and Michael Schmutz

Abstract Several of multiasset derivatives like basket options or options on the weighted maximum of assets exhibit the property that their prices determine uniquely the underlying asset distribution. Related to that the question how to retrieve this distributions from the corresponding derivatives quotes will be discussed. On the contrary, the prices of exchange options do not uniquely determine the underlying distributions of asset prices and the extent of this non-uniqueness can be characterised. The discussion is related to a geometric interpretation of multiasset derivatives as support functions of convex sets. Following this, various symmetry properties for basket, maximum and exchange options are discussed alongside with their geometric interpretations and some decomposition results for more general payoff functions.

Keywords Multiasset derivative · Exchange option · Lévy process · Probabilistic symmetries · Zonoid

Mathematics Subject Classification (2010) 60E07 · 60G51 · 91G20

1 Introduction

A portfolio of d assets over a finite time horizon is mathematically described by a d-dimensional stochastic price process $(S_{1t}, \dots, S_{dt})_{t\in[0,T]}$ defined on and adapted to a filtered probability space $(\Omega, \mathfrak{F}, (\mathfrak{F}_t)_{t\in[0,T]}, \mathbf{P})$, where $T > 0$ is a finite maturity time and $(\mathfrak{F}_t)_{t\in[0,T]}$ is assumed to satisfy the usual conditions. A financial derivative

I. Molchanov (✉) · M. Schmutz
Department of Mathematical Statistics and Actuarial Science, University of Bern, Sidlerstrasse 5, 3012 Bern, Switzerland
e-mail: ilya.molchanov@stat.unibe.ch

M. Schmutz
e-mail: michael.schmutz@stat.unibe.ch

Y. Kabanov et al. (eds.), *Inspired by Finance*, DOI 10.1007/978-3-319-02069-3_20,

can often be defined by a certain $\mathfrak{F}_T$-measurable non-negative random variable C. The classical pricing approach in frictionless markets[1] for the financial derivative described by C is to postulate a certain semimartingale model for the underlying asset price process satisfying a certain no-arbitrage type condition [18, 19, 31], so that the pricing problem essentially boils down to choosing an appropriate equivalent martingale measure and then computing the discounted conditional expectation of the payoff C, see e.g. [20]. A very recent concise summary of results of the theory of arbitrage can be found in [2]. However, choosing the measure in incomplete markets is absolutely crucial. In view of that many different strategies for choosing an appropriate martingale measure have been developed during the last fifteen years, see e.g. [11–13, 21, 24, 25, 27, 28, 36, 51] and the literature cited therein.

In markets where options are more or less liquidly traded, their prices can be used as a source of information for deriving a pricing measure for other, illiquid derivatives. This approach especially reflects practitioners point of view of choosing the martingale measure by calibration, but becomes also more and more popular at least for single-asset derivatives, see e.g. [14–16, 33, 45].

For a wide variety of financial derivatives the random vector containing the terminal asset prices $(S_{T1}, \dots, S_{Td})$ is of particular importance. Hence, interesting and quite fundamental questions being related to model calibration are, how "efficient" some financial instruments reflect the distribution of $(S_{T1}, \dots, S_{Td})$ and under what circumstances one can extract further information about the price process. In the sequel we will briefly discuss and partially complement some answers to these questions. As far as the second question is concerned the research is still at the beginning, even in the one-dimensional setting, however, it seems that the notion of symmetry plays an important role, see e.g. [10, 43]. Unfortunately, the existing classical financial literature often does not carefully distinguish between symmetry and duality. In view of that and somehow in a similar spirit as the very recent work of Profeta, Roynette and Yor [43], we will give a geometric interpretation of financial symmetries showing that they can be interpreted as geometric symmetries while the duality translates to the geometric notion of reflection. We hope that this will give a deeper insight to financial symmetries paving the way for future work.

Furthermore, in view of the fact that multivariate financial derivatives become more and more popular in theory and applications, we will focus on the multivariate case whenever possible. In relation to that note e.g. that recently there has been a liquid market in structured products, particularly in Europe. At the moment the majority of the trades still occur over the counter, but more and more trades are also organised at exchanges, especially at the quite new European exchange for structured products, Scoach. Structured products quite often involve equity indices and sometimes several purpose-built shares. In a recent paper Carr and Laurence [9] further stress that "all major banks stand ready to provide over-the-counter quotes on customised baskets", so that it seems to be interesting to analyse the "information" about the underlying asset prices being reflected in these quotes.

[1] For markets with transaction costs we refer to [32] and the literature cited therein.

2 Basket Options and Options on the Maximum of Several Assets

The random vector of the terminal prices for d assets can be written as

$$(S_{T1}, \ldots, S_{Td}) = (F_1\eta_1, \ldots, F_d\eta_d),$$

where $\eta = (\eta_1, \ldots, \eta_d)$ represents integrable stochastic price change-factors and $(F_1, \ldots, F_d)$ is the vector of forward prices that can be computed deterministically by the time zero (spot) prices and the carrying costs (e.g. financing or storage costs, presumably deterministic) of the assets. Under a risk-neutral measure the components of η have unit expectations.

A multiasset financial product of the European type has the payoff determined by the terminal prices of the assets. The most important payoff function

$$f_b(u_0, u_1, \ldots, u_d) = \left(\sum_{i=1}^{d} u_i\eta_i + u_0\right)_+ = \left(\sum_{i=1}^{d} \tilde{u}_i F_i \eta_i + u_0\right)_+ \tag{1}$$

defines a *basket option*, where $(x)_+ = \max(0, x)$ and $\tilde{u}_1, \ldots, \tilde{u}_d$ stand for the weights of the different assets in the basket. We stress the dependence on the weights $u_1, \ldots, u_d \in \mathbb{R}$ with included forward prices, since they are important for the analysis of the random price change vector η. The absolute value of u_0 is called a *strike*. Positive values of u_0 indicate *put* options, negative values of u_0 correspond to *call* options, while $u_0 = 0$ yields options to *exchange* some risky assets. Sometimes basket options with possibly positive and negative weights attached to the risky assets are called generalised basket options.

Further popular multivariate derivatives include call and put options on the (weighted) maximum or minimum of several assets. For instance, calls on the weighted *maximum* are defined by the payoff

$$\left(\bigvee_{l=1}^{d} u_l\eta_l - k\right)_+ = f_m(k, u_1, \ldots, u_d) - k, \quad k, u_1, \ldots, u_d \geq 0,$$

where $\vee$ denotes the maximum operation and

$$f_m(u_0, u_1, \ldots, u_d) = u_0 \vee \bigvee_{l=1}^{d} u_l\eta_l \tag{2}$$

is the derivative on the maximum of d (weighted) risky assets together with a riskless bond.

The *support function* of a nonempty compact convex set K in $\mathbb{R}^d$ is defined as

$$h_K(u) = \max\{(u_1x_1 + \cdots + u_dx_d) : x \in K\}.$$

It is well known that support functions are characterised by their sublinearity property. In many cases payoffs from multiasset derivatives are sublinear and so become

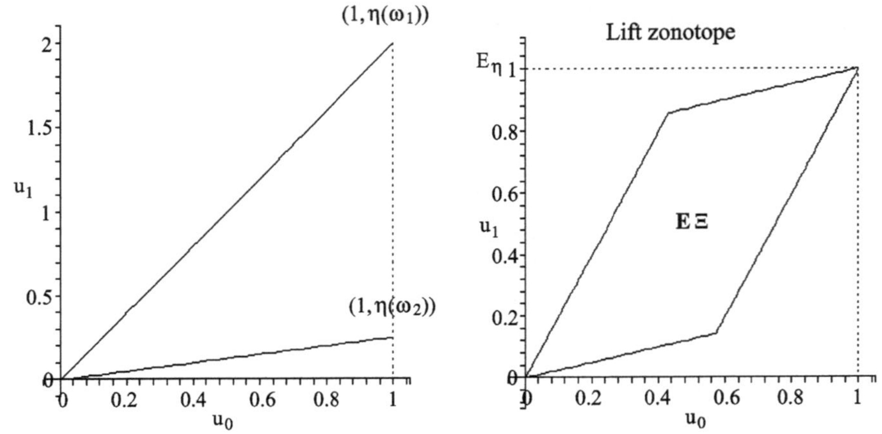

Fig. 1 The segment $\Xi = [(0, 0), (1, \eta)]$ for η taking two values with positive probabilities and the lift zonoid $Z_\eta = \mathbf{E}\Xi$

support functions of *random convex sets* determined by η. For instance, considered a function of $(u_0, u_1, \ldots, u_d)$, the basket payoff (1) is the support function of a *segment* in $\mathbb{R}^{d+1}$ with end-points at the origin and $(1, \eta)$. The payoff function f_m from (2) with non-negative arguments is the support function of a *crosspolytope* in $\mathbb{R}^{d+1}$, which is the convex hull of the origin, one basis vector scaled by the strike and all other basis vectors scaled by $\eta_1, \ldots, \eta_d$ respectively. In these both cases the payoffs are support functions of a random closed set Ξ in $(d + 1)$-dimensional space, see [37].

The price of an option is determined by taking the expectation of the discounted payoff with respect to a risk-neutral measure. Translated into the language of support functions (at least for the two basic families of payoffs), it corresponds to the expected support function of the random closed set Ξ up to discounting. It is known that the expected support function $\mathbf{E}h_\Xi(u)$ is a support function itself, namely that of the *expectation* $\mathbf{E}\Xi$ of Ξ, see [37, Sect. 2.1]. If Ξ has a discrete distribution, then $\mathbf{E}\Xi$ equals the Minkowski sum of possible realisations of Ξ weighted by their probabilities. Figure 1 shows the expectation of a random segment with two possible values. In general, Minkowski (elementwise) sum of segments are called zonotopes, while *zonoids* are limits of zonotopes in the Hausdorff metric. In other words, zonoids are expectations of random sets being segments. Zonoids are one of basic objects in convex geometry, see [49] and references therein.

Example 1 (Lift zonoid of η) Let Ξ be the segment that joins the origin and the point $(1, \eta)$ in $\mathbb{R}^{d+1}$. The expectation of Ξ is called the *lift zonoid* of η, is denoted by Z_η, and so satisfies $h_{Z_\eta}(u) = \mathbf{E}h_\Xi(u) = \mathbf{E}f_b(u)$ for all $u \in \mathbb{R}^{d+1}$. In the single asset case Z_η is a planar set. It is well known [49] that all centrally symmetric planar compact convex sets are zonoids, while this is not the case in dimension 3 and more. This fact already suggests an important dimensional effect that appears when dealing with more than one asset.

For two integrable random vectors η, η', define the *lift zonoid order* (see [42, Chap. 8]) by

$$\eta \preceq_{lz} \eta' \quad \text{if} \quad Z_\eta \subset Z_{\eta'} .$$

In the univariate case, this order coincides with the convex order, i.e. $\eta \preceq_{lz} \eta'$ if and only if $\mathbf{E} f(\eta) \le \mathbf{E} f(\eta')$ for all convex functions f with existing expectations. In this case prices of all European derivatives with convex payoffs written on $F\eta'$ are higher than those written on $F\eta$, see also [26, Cor. 2.62]. In the multivariate case the lift zonoid order is equivalent to the convex-linear order, i.e. $\eta \preceq_{lz} \eta'$ if and only if $\mathbf{E}\psi(l(\eta)) \le \mathbf{E}\psi(l(\eta'))$ for all convex ψ and real-valued linear l such that the expectations exist.

Example 2 (Zonoid of η) The expected payoffs from exchange options define a convex set Z^o_η called the zonoid of η, i.e.

$$h_{Z^o_\eta}(u) = \mathbf{E}(u_1\eta_1 + \cdots + u_d\eta_d)_+ , \quad u \in \mathbb{R}^d .$$

This object is of little interest if $d = 1$, since for a positive random variable η we obtain $Z^o_\eta = [0, \mathbf{E}\eta]$.

Example 3 (Lift max-zonoids and max-zonoids of η) The expectation of the random crosspolytope in $\mathbb{R}^{d+1}$ that has f_m as its support function is a convex set M_η called the *lift max-zonoid* of η. Max-zonotopes are sums of crosspolytopes (instead of segments used to construct a zonotope) in $\mathbb{R}^d$, max-zonoids are limits of max-zonotopes, while the lifting corresponds to extending η by an extra coordinate being one. The *max-zonoid* M^o_η of η is a convex set in $\mathbb{R}^d$ that appears as the expectation of the crosspolytope being the convex hull of the origin and the basis vectors scaled by $\eta_1, \ldots, \eta_d$. Note that max-zonoids have been introduced in [38] in order to characterise the dependency structure of multivariate extreme value distributions. Since financial quantities are non-negative, it is often useful to restrict the support function to $\mathbb{R}^{d+1}_+$ or to $\mathbb{R}^d_+$ as appropriate.

The max-zonoid M_η defines a norm in $\mathbb{R}^{d+1}$ by setting

$$\|x\|_\eta = h_{M_\eta}\big((|x_0|, \ldots, |x_d|)\big) , \quad x = (x_0, \ldots, x_d) \in \mathbb{R}^{d+1} .$$

The unit ball $\hat{M}_\eta$ in this norm for a log-normal random variable η is shown in Fig. 2. In the single asset log-normal risk-neutral (Black-Scholes) case the corresponding norm of (k, F) with $k, F > 0$ is given by

$$\|(k, F)\|_{\mathrm{BS}} = F\Phi\Big(\lambda + \frac{1}{2\lambda}\log\frac{F}{k}\Big) + k\Phi\Big(\lambda + \frac{1}{2\lambda}\log\frac{k}{F}\Big) , \tag{3}$$

where $\lambda = \frac{1}{2}\sigma\sqrt{T}$. Notably, expression (3) appears in the literature on extreme values [30] in relation to the limit distribution of coordinatewise maxima for triangular arrays of bivariate Gaussian vectors with correlation $\varrho(n)$ that approaches one with rate $(1 - \varrho(n))\log n \to \lambda^2 \in [0, \infty]$ as $n \to \infty$.

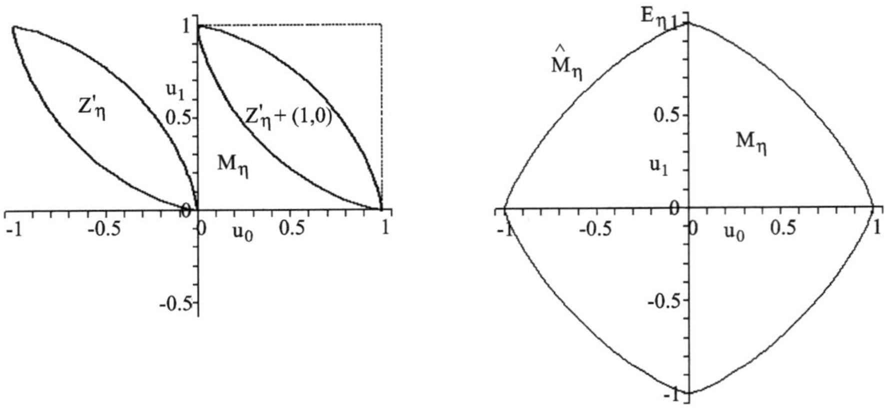

Fig. 2 Relation between M_η and Z_η in the single asset case and the unit ball $\hat{M}_\eta$ for log-normal η with mean one and volatility $\sigma = 0.5$ calculated for $T = 1$

In the single asset case the lift max-zonoid M_η can be directly obtained as the convex hull of the origin and the lift zonoid Z_η reflected with respect to the line $\{(u_0, u_1) : u_0 = 0\}$ and translated by $(1, 0)$, see Fig. 2 and [39, Lemma 3.2].

Example 4 (ℓ_p-zonoids) Another family of zonoid-type bodies for integrable η can be defined by taking expectation of rescaled ℓ_p-balls with $p \in (1, \infty]$. The corresponding expected payoff (and so the support function of the corresponding convex set) is given by

$$\mathbf{E}\big(u_0^p + u_1^p \eta_1^p + \cdots + u_d^p \eta_d^p\big)^{1/p} = \mathbf{E}\big\|(u_0, u_1\eta_1, \ldots, u_d\eta_d)\big\|_p$$

for $u_0, u_1, \ldots, u_d \geq 0$, where $\|\cdot\|_p$ is the ℓ_p-norm. This function can be extended to the whole $\mathbb{R}^{d+1}$ by taking its value at $(|u_0|, |u_1|, \ldots, |u_d|)$ and so yields the support function of a convex set. The obtained convex body is called the ℓ_p-lift zonoid of η. If $p = \infty$, one recovers the expectation of f_m from (2) and the corresponding lift max-zonoid. The non-lifted ℓ_p-zonoid appears if the strike and the corresponding zero coordinate are neglected.

3 Characterisation of the Distribution of the Underlying Asset Prices

It is known [42] that the lift zonoid of a random vector η determines uniquely the distribution of η. This is easily seen by noticing that $h_{Z_\eta}(u_0, u_1, \ldots, u_d)$ as a function of u_0 is the stop-loss transform of the scalar product $(u_1\eta_1 + \cdots + u_d\eta_d)$ that determines its distribution and so the distribution of η itself, since $u_1, \ldots, u_d$ are arbitrary.

Thus prices of all basket options written on the assets described by η determine uniquely the (joint) distribution of η. This uniqueness result does not rely on the existence of a probability density and even holds for η with possibly negative values.[2] Versions of this statement are particularly well known in the univariate case, see e.g. the classical articles [6, 44], but also the multivariate statement was (more or less explicitly) noted in various generalities and formulations with various proofs and explanations related to different fields of mathematics, see e.g. [4, 9, 17, 29, 34, 35, 39, 40, 42, 50].

It is worth noticing that the prices of *all* basket options are not required for the unique characterisation of the distribution. The put-call parity yields that it suffices to work with only put or call options. Furthermore, it suffices to consider options with any given and fixed non-vanishing strike. However, if the strike vanishes, then the characterisation is no longer unique. In other words, the (non-lifted) zonoid Z^o_η does not uniquely determine the distribution of η. It is shown in [41] that two positive integrable random vectors $\eta = e^\xi$ and $\eta^* = e^{\xi^*}$ (coordinatewisely) share the same zonoid, i.e. $\mathbf{E}(\langle u, \eta\rangle)_+ = \mathbf{E}(\langle u, \eta^*\rangle)_+$ for all $u \in \mathbb{R}^d$ if and only if $\mathbf{E} f(\eta) = \mathbf{E} f(\eta^*)$ for each positive-1-homogeneous function f. In particular, two *positive* one-dimensional random variables share the same zonoid if and only if they have the same expectation. An equivalent characterisation in terms of the characteristic functions of ξ and ξ^* is presented in [41].

It is proved in [40, Th. 2.1(ii)] and [39, Th. 3.1] that the lift max-zonoid M_η of an integrable random vector η determines uniquely the distribution of η, i.e. the expected payoffs of options on the maxima of weighted assets and the riskless bond (and also puts or calls on the weighted maxima) uniquely determine the joint distribution of the risky assets and so prices of all other payoffs.

The max-zonoid M^o_η does not uniquely characterise the distribution of η. The following result shows that the extent of this non-uniqueness is exactly the same for exchange and max-options. Define the family of functions $\tilde{\varkappa}_j : (0,\infty)^d \mapsto (0,\infty)^{d-1}$ acting as

$$\tilde{\varkappa}_j(x) = \left(\frac{x_1}{x_j}, \dots, \frac{x_{j-1}}{x_j}, \frac{x_{j+1}}{x_j}, \dots, \frac{x_d}{x_j}\right), \quad j = 1, \dots, d.$$

Consider an integrable $(0,\infty)^d$-valued random vector η and define the probability measure $\mathbf{Q}^j$ for $j = 1, \dots, d$ by

$$\frac{d\mathbf{Q}^j}{d\mathbf{Q}} = \frac{\eta_j}{\mathbf{E}\eta_j}, \quad j = 1, \dots, d, \tag{4}$$

where $\mathbf{E}\eta_j = 1$, $j = 1, \dots, d$, in the risk-neutral setting.

Proposition 1 *Let $\eta = e^\xi$ and $\eta^* = e^{\xi^*}$ be integrable random vectors. Then $M^o_\eta = M^o_{\eta^*}$ if and only if $Z^o_\eta = Z^o_{\eta^*}$.*

---

[2] In relation to this it is stressed e.g. in [20] that a general analysis of financial markets should also consider situations where prices, at least for some instruments, can be negative.

Proof Assume that $M^o_\eta = M^o_{\eta^*}$, i.e.

$$h_{M^o_\eta}(u) = \mathbf{E}\Big(0 \vee \bigvee_{l=1}^{d} u_l \eta_l\Big) = \mathbf{E}\Big(0 \vee \bigvee_{l=1}^{d} u_l \eta^*_l\Big) = h_{M^o_{\eta^*}}(u) \tag{5}$$

for all $u \in \mathbb{R}^d_+$. By choosing $u = e_i$ this implies $\mathbf{E}\eta_i = \mathbf{E}\eta^*_i$ for all i. Change measure $\mathbf{Q}$ to $\mathbf{Q}^1$ and $\mathbf{Q}^{1*}$ using respectively η_1 and η^*_1 as the density normalised by the (equal) expectations in order to see that (5) yields

$$c\mathbf{E}_{\mathbf{Q}^1}\Big(0 \vee u_1 \vee \bigvee_{l=2}^{d} u_l \frac{\eta_l}{\eta_1}\Big) = c\mathbf{E}_{\mathbf{Q}^{1*}}\Big(0 \vee u_1 \vee \bigvee_{l=2}^{d} u_l \frac{\eta^*_l}{\eta^*_1}\Big)$$

for all $u \in \mathbb{R}^d_+$ so that the distribution of $\tilde{\varkappa}_1(\eta)$ under $\mathbf{Q}^1$ coincides with the distribution of $\tilde{\varkappa}_1(\eta^*)$ under $\mathbf{Q}^{1*}$ as having the same lift max-zonoid. Then the lift zonoids are also equal, so that for all $u \in \mathbb{R}^d$ we have

$$c\mathbf{E}_{\mathbf{Q}^1}\Big(\sum_{k=2}^{d} u_k \frac{\eta_k}{\eta_1} + u_1\Big)_+ = c\mathbf{E}_{\mathbf{Q}^{1*}}\Big(\sum_{k=2}^{d} u_k \frac{\eta^*_k}{\eta^*_1} + u_1\Big)_+ ,$$

being equivalent to

$$\mathbf{E}\big(\langle u, \eta\rangle\big)_+ = \mathbf{E}_{\mathbf{Q}}\big(\langle u, \eta^*\rangle\big)_+$$

for all $u \in \mathbb{R}^d$, i.e. $Z^o_\eta = Z^o_{\eta^*}$ as having the same support functions. The converse statement can be proved by a similar argument. □

The ℓ_p-lift zonoid with $p > 1$ introduced in Example 4 uniquely characterises the distribution of integrable η with positive components. Indeed, then the function

$$g(t) = \mathbf{E}(t + \bar{\eta})^{1/p}, \quad t > 0,$$

is known for $\bar{\eta} = (u_1^p \eta_1^p + \cdots + u_d^p \eta_d^p)$. The dominated convergence theorem yields that this function is differentiable, so that

$$g'(t) = \mathbf{E}(t + \bar{\eta})^{1/p-1}.$$

Define $\rho = 1 - 1/p$ and consider the known function

$$\mathbf{E}(t + \bar{\eta})^{-\rho} = \frac{1}{\Gamma(\rho)}\mathbf{E}\int_0^\infty x^{\rho-1} \mathrm{e}^{-x(t+\bar{\eta})} \mathrm{d}x = \frac{1}{\Gamma(\rho)}\int_0^\infty x^{\rho-1} \mathrm{e}^{-xt} \mathbf{E}\mathrm{e}^{-x\bar{\eta}} \mathrm{d}x .$$

This function is the Laplace transform of $x^{\rho-1}\mathbf{E}\mathrm{e}^{-x\bar{\eta}}$ and so determines it uniquely. In turn, from it one can uniquely retrieve the distribution of $\bar{\eta}$. A variant of the Cramér-Wold device for non-negative random vectors implies that the distribution of $(\eta_1^p, \ldots, \eta_d^p)$ is known, so the distribution of $(\eta_1, \ldots, \eta_d)$.

Arguing as in Proposition 1, it is possible to show that two non-lifted ℓ_p-zonoids are equal if and only if the non-lifted zonoids are equal.

4 Recovery of Asset Distributions from Option Prices

In view of the uniqueness results one should be able to recover the distribution of the asset prices from the prices of options. In particular, Henkin and Shananin [29] implicitly discuss this recovery problem for basket puts. They implicitly relate put prices to non-complete integral transforms, so that the absolutely continuous case is directly related to the (non-complete) Radon transform whose inversion is considered under certain regularity assumptions, for details we refer to [29, Sect. 4] and [9] who address a similar problem. Other results from [29] can be used to get some insight into multivariate static hedging possibilities in certain particular situations, in relation to this see also [4].

The following result shows how to derive the assets distribution from the prices of max-options. For simplicity of notation we formulate the results for the call setting. A closely related observation can be found in [52].

Proposition 2 *Assume a risk-neutral setting with assets prices η having a continuous distribution function. Then the distribution of η is given by*

$$
\begin{aligned}
&\mathbf{Q}(\eta_i \le v_i,\ i = 1, \dots, d)\\
&\quad = 1 + \frac{\partial}{\partial k}\mathbf{E}\Bigl(\bigvee_{i=1}^{d} u_i\eta_i - k\Bigr)_{+}\Big|_{u_i = k/v_i, i=1,\dots,d}\\
&\quad = \frac{\partial}{\partial k}\mathbf{E}\Bigl(\bigvee_{i=1}^{d} u_i\eta_i \vee k\Bigr)\Big|_{u_i = k/v_i, i=1,\dots,d}, \quad \textit{for } v_1, \dots, v_d > 0.
\end{aligned}
$$

Proof Define $\bar\eta = \bigvee_{l=1}^{d} u_l\eta_l$ for fixed $u_1, \dots, u_d > 0$. Note that the continuity of the joint distribution of η implies that the distribution of $\bar\eta$ is continuous. Then

$$
\mathbf{E}(\bar\eta - k)_+ = \mathbf{E}\bar\eta - k + \int_0^k \mathbf{Q}(\bar\eta \le s)\mathrm{d}s\,.
$$

Differentiation yields that

$$
\mathbf{Q}(\bar\eta \le k) = 1 + \frac{\partial}{\partial k}\mathbf{E}(\bar\eta - k)_+\,.
$$

The statement follows by substituting $u_i = k/v_i$ for $i = 1, \dots, d$. □

While the non-discounted prices of a wide variety of calls or puts on the weighted maximum or weighted minimum of two assets easily yield the joint risk-neutral distribution function, one has to keep in mind that these expressions involve derivatives of market data that calls for the use of regularisation methods.

5 Symmetry Properties and Basket Options

In view of the fact that under certain symmetry or "quasi-symmetry" properties, some path-dependent options can be semi-statically hedged by European options, see e.g. [1, 5, 7, 8, 10, 23, 39, 41, 47] and the literature cited therein, one can conjecture that under these assumptions it will also be possible to extract some information about the price development until maturity from certain European option prices.

Indeed for example Carr and Lee [10, Sect. 5] discuss the possibility to use a version of the one-dimensional European put-call symmetry in order to extract the distribution of a certain (path-dependent) first passage time (presuming that the corresponding barrier is hit) from European option prices (which are determined by the marginal distribution). In the recent book by Profeta, Roynette and Yor [43] (classical) European put-call symmetry is the starting point to relate vanilla option prices to certain last passage times. In view of that we will give a geometric analysis of this and closely related financial symmetries, yielding a clear distinction from duality.

Since lift zonoids of random price vectors uniquely determine the joint distribution of the prices, geometric symmetries of lift zonoids can be translated into symmetries or parities for European option prices. For instance, each lift zonoid of η is centrally symmetric (with respect to the point $(\frac{1}{2}, \dots, \frac{1}{2})$ in a risk-neutral setting), since it appears as the sum of segments that are centrally symmetric themselves. The central symmetry means that the values of the support function of $Z_\eta - (\frac{1}{2}, \dots, \frac{1}{2})$ in opposite directions coincide, whence

$$
\begin{aligned}
\mathbf{E} f_b(-k, u) = h_{Z_\eta}\big((-k, u)\big) &= h_{Z_\eta - (\frac{1}{2}, \dots, \frac{1}{2})}(-k, u) - \frac{1}{2}k + \frac{1}{2}\mathbf{E}\langle \eta, u\rangle \\
&= h_{Z_\eta - (\frac{1}{2}, \dots, \frac{1}{2})}\big((k, -u)\big) + \frac{1}{2}k - \frac{1}{2}\mathbf{E}\langle \eta, u\rangle - k + \mathbf{E}\langle \eta, u\rangle \\
&= h_{Z_\eta}\big((k, -u)\big) - k + \mathbf{E}\langle \eta, u\rangle = \mathbf{E} f_b(k, -u) + \mathbf{E}\langle \eta, u\rangle - k\,,
\end{aligned}
$$

i.e. we arrive at the classical European call-put parity.

While the price of American options as function of the strike and the forward can be also interpreted as the support function of a convex set, such a set is usually not centrally symmetric, correspondingly the put-call parity usually does not hold for American options, cf. Fig. 3.

Plane symmetries of lift zonoids with respect to its last d coordinates are equivalent to the (possibly partial) exchangeability of the random price vector. Probabilistically, these symmetries correspond to the invariance of the expectation $\mathbf{E}(u_0 + u_1\eta_1 + \cdots + u_d\eta_d)_+$ with respect to swaps of any u_i and u_j for $i, j = 1, \dots, d$ and any fixed non-zero u_0.

A number of further symmetries appear if the strike u_0 is included in the swaps. For instance, one can observe that in the single-asset case, such a symmetry amounts to $\mathbf{E}(u_0 + u_1\eta_1)_+ = \mathbf{E}(u_0\eta_1 + u_1)_+$. This is exactly the (univariate) *put-call symmetry* property, sometimes also called classic put-call symmetry, see [10], meaning that a call with strike k and forward F equals in value to the put with strike F and

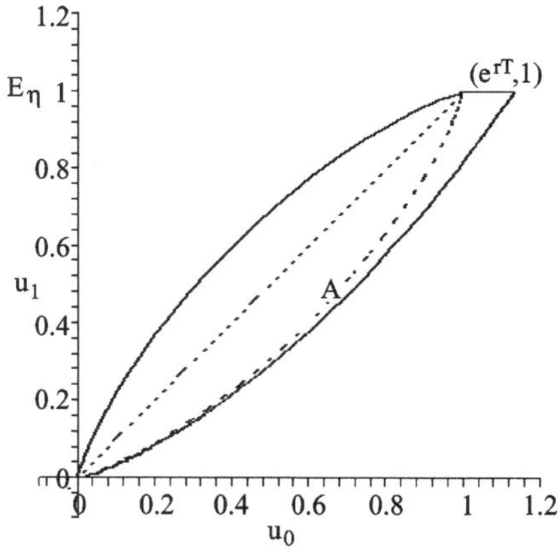

Fig. 3 An approximation of the payoff set A related to American options for the Black-Scholes economy with volatility $\sigma = 0.5$, interest rate $r = 0.12$, dividend yield $q = 0$ and maturity $T = 1$

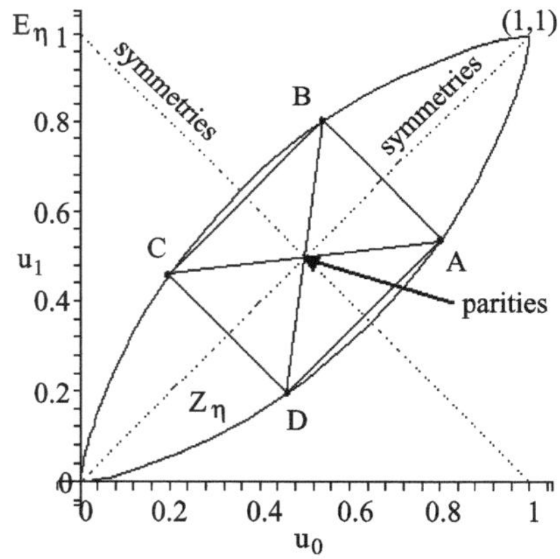

Fig. 4 Symmetries of Z_η for self-dual η and their financial interpretations

forward k. It is easy to check that this symmetry holds in the classical Black-Scholes case as e.g. observed in [3, 5].

The put-call symmetry is used to create so-called semi-static hedges for barrier options, since, roughly speaking, it is possible to switch between call and puts at the time when a certain barrier is crossed and so the option is either knocked-in or knocked-out, see [8, 10]. In view of the fact that under certain regularity assumptions the boundary of the lift zonoid can be parametrised with the help of the non-discounted prices of binary- and normalised gap options (gap options in the sense of [8]), the put-call symmetry gives rise to several equivalent symmetries formulated for binary and gap options, see Fig. 4. Since the lift zonoid uniquely determines the distribution of η, the classical put-call symmetry also implies a symmetry property for arbitrary payoff functions $f : \mathbb{R}_+ \mapsto \mathbb{R}_+$ (or for integrable payoff functions $f : \mathbb{R}_+ \mapsto \mathbb{R}$) given by $\mathbf{E} f(\eta) = \mathbf{E}[\eta f(1/\eta)]$, for details concerning this implication we refer to [40]. This implication also yields that classic put-call symmetry is in fact equivalent to several other (at the first glance more restrictive) symmetry properties, e.g. given in [10, Th. 2.5].

The multiasset generalisation of the put-call symmetry property can be formulated for each particular asset (or numeraire), see [40, Th. 2.4] for several equivalent formulations of this property. Namely, η is self-dual with respect to the ith numeraire if the distribution of η under $\mathbf{Q}$ is identical to the distribution of

$$\tilde{\eta}^i = \varkappa_i(\eta) = \left(\frac{\eta_1}{\eta_i}, \dots, \frac{\eta_{i-1}}{\eta_i}, \frac{1}{\eta_i}, \frac{\eta_{i+1}}{\eta_i}, \dots, \frac{\eta_d}{\eta_i}\right)$$

under $\mathbf{Q}^i$. As a direct consequence of Lemma 1 this is the case if and only if Z_η is symmetric with respect to the hyperplane $\{(u_0, u) \in \mathbb{R}^{d+1} : u_0 = u_i\}$, being again equivalent to the invariance of the expected basket payoff with respect to the swap of the strike u_0 and the weight u_i of the ith asset. It is shown in [40] that the symmetries of lift zonoids are equivalent to symmetries of lift-max-zonoids.

While plane symmetries of lift zonoids are equivalent to financial symmetries, the *reflection* with respect to a plane corresponds to a *dual market* transition. The permutation of the zero-coordinate with the ith coordinate of a vector $(u_0, u) \in \mathbb{R}^{d+1}$ is denoted by

$$\pi_{0i}(u_0, u) = (u_i, u_1, \dots, u_{i-1}, u_0, u_{i+1}, \dots, u_d).$$

If $B \subset \mathbb{R}^{d+1}$, then the mapping $\pi_{0i}(B)$ is the reflection of B at the hyperplane $\{(u_0, u) \in \mathbb{R}^{d+1} : u_i = u_0\}$. The *dual* lift zonoid $Z_{\tilde{\eta}^i}$ with respect to the ith numeraire or coordinate η_i is defined for the random vector $\tilde{\eta}^i = (\tilde{\eta}^i_1, \dots, \tilde{\eta}^i_d) = \varkappa_i(\eta)$ and the probability measure $\mathbf{Q}^i$, see (4). The following result relates reflections in higher dimensional spaces to the multivariate duality principle at maturity. The duality principle in a general semi-martingale setting is studied in [22].

Lemma 1 *Let $\eta = (\eta_1, \dots, \eta_d)$ be an integrable $(0,\infty)^d$-valued random vector with $\mathbf{E}\eta_i = 1$ for a fixed $i \in \{1, \dots, d\}$. Then $Z_{\tilde{\eta}^i} = \pi_{0i}(Z_\eta)$ and $M_{\tilde{\eta}^i} = \pi_{0i}(M_\eta)$.*

Proof For $(u_0, u) \in \mathbb{R}^{d+1}$ we have

$$\begin{aligned}
h_{Z_{\tilde{\eta}^i}}(u_0, u) &= \mathbf{E}_{\mathbf{Q}^i}\left(\sum_{l=1}^{d} u_l \tilde{\eta}^i_l + u_0\right)_+ = \mathbf{E}_{\mathbf{Q}^i}\left(\sum_{l=1,\, l\neq i}^{d} u_l \frac{\eta_l}{\eta_i} + \frac{u_i}{\eta_i} + u_0\right)_+ \\
&= \mathbf{E}\left[\left(\sum_{l=1,\, l\neq i}^{d} u_l \frac{\eta_l}{\eta_i} + \frac{u_i}{\eta_i} + u_0\right)_+ \eta_i\right] \\
&= \mathbf{E}\left(\sum_{l=1,\, l\neq i}^{d} u_l \eta_l + u_i + u_0 \eta_i\right)_+ \\
&= h_{Z_\eta}\big(\pi_{0i}(u_0, u)\big) = h_{\pi_{0i}(Z_\eta)}(u_0, u).
\end{aligned}$$

□

6 Symmetries of Exchange and Max-Options

Plane symmetries of (non-lifted) zonoids correspond to the swap of the ith and jth assets under the condition that the strike u_0 is set to zero, i.e. in the setting of exchange options. For two assets (and integrable η) it means

$$\mathbf{E}(u_1\eta_1 + u_2\eta_2)_+ = \mathbf{E}(u_2\eta_1 + u_1\eta_2)_+ \quad \text{for every} \quad (u_1, u_2) \in \mathbb{R}^2 . \tag{6}$$

An integrable random vector η is said to be *ij-swap-invariant* if the expected payoffs from the exchange options with weights u and $\pi_{ij}(u)$ are identical, where π_{ij} swaps the ith and jth component of a vector. This swap-invariance property is clearly weaker than the ij-exchangeability of η (i.e. identity of distributions of η and $\pi_{ij}(\eta)$), which can be characterised as the invariance of $\mathbf{E} f_b(u_0, u)$ with respect to the swap of u_i and u_j for all $u \in \mathbb{R}^d$ and all $u_0 \in \mathbb{R}$.

Lemma 2 *An integrable $(0,\infty)^d$-valued random vector η is ij-exchangeable for $i, j \in \{1, \dots, d\}$, $i \neq j$, if and only if the lift zonoid Z_η of η satisfies $\pi_{ij}(Z_\eta) = Z_\eta$, or, equivalently, the lift max-zonoid M_η of η satisfies $\pi_{ij}(M_\eta) = M_\eta$.*

Proof The definition of the lift zonoid and the π_{ij}-invariance of the expected payoff f_b yield that for all $(u_0, u) \in \mathbb{R}^{d+1}$

$$h_{Z_\eta}(u_0, u) = \mathbf{E} f_b(u_0, u) = \mathbf{E} f_b\big(\pi_{ij}(u_0, u)\big) = h_{Z_\eta}\big(\pi_{ij}(u_0, u)\big) = h_{\pi_{ij}(Z_\eta)}(u_0, u) ,$$

whence Z_η and $\pi_{ij}(Z_\eta)$ coincide as having identical support functions. The result for lift max-zonoids has a similar proof. □

Note that in the risk-neutral setting all bivariate log-normal random variables are swap-invariant.

We now give the geometric interpretation of the fact that the ij-swap invariance is related to the self-duality in a lower-dimensional space. Denote

$$f_{\mathrm{b}}^{\mathrm{o}}(u) = \left(\sum_{l=1}^{d} u_l \eta_l\right)_+ = \big(\langle u, \eta\rangle\big)_+ , \quad u \in \mathbb{R}^d . \tag{7}$$

Let us consider an integrable $(0,\infty)^d$-valued random vector η and recall that $h_{Z_\eta^{\mathrm{o}}}(u) = \mathbf{E} f_{\mathrm{b}}^{\mathrm{o}}(u)$ for $u \in \mathbb{R}^d$. The lift zonoid of $\tilde{\varkappa}_j(\eta)$ under the probability measure $\mathbf{Q}^j$ defined by (4) is denoted by $Z^{\mathbf{Q}^j}_{\tilde{\varkappa}_j(\eta)}$. Note that we do not a priori assume here that η appears from a normalised (multivariate) martingale.

Proposition 3 *Let $\eta = (\eta_1, \dots, \eta_d)$ be an integrable $(0,\infty)^d$-valued random vector with $\mathbf{E}\eta_j = 1$ for a fixed $j \in \{1, \dots, d\}$. Then $Z_\eta^{\mathrm{o}} = Z^{\mathbf{Q}^j}_{\tilde{\varkappa}_j(\eta)}$ and $M_\eta^{\mathrm{o}} = M^{\mathbf{Q}^j}_{\tilde{\varkappa}_j(\eta)}$, where the jth coordinate is the additional "lifting" coordinate of $Z^{\mathbf{Q}^j}_{\tilde{\varkappa}_j(\eta)}$.*

Proof For every $u \in \mathbb{R}^d$

$$h_{Z_\eta^{\circ}}(u) = \mathbf{E}\big(\langle u, \eta\rangle\big)_+ = \mathbf{E}\left(\sum_{l=1}^{d} u_l \eta_l\right)_+ = \mathbf{E}_{\mathbf{Q}^j}\left(\sum_{l=1, l\neq j}^{d} u_l \frac{\eta_l}{\eta_j} + u_j\right)_+$$

$$= \mathbf{E}_{\mathbf{Q}^j}\big(\big\langle u, \tilde{\varkappa}_j(\eta)\big\rangle + u_j\big)_+ = h_{Z^{\mathbf{Q}^j}_{\tilde{\varkappa}_j(\eta)}}(u),$$

i.e. the convex bodies Z_η° and $Z^{\mathbf{Q}^j}_{\tilde{\varkappa}_j(\eta)}$ coincide as having equal support functions. The proof for max-zonoids is similar. □

Therefore, we have that symmetry of the zonoid of η with respect to the hyperplane $\{u \in \mathbb{R}^d : u_i = u_j\}$ for $i \neq j$ is equivalent to the ij-swap-invariance of η under $\mathbf{Q}$ and to the self-duality with respect to the ith numeraire of $\tilde{\varkappa}_j(\eta)$ under $\mathbf{Q}^j$.

The probably most interesting financial interpretation of Proposition 3 is that the zonoid (i.e. the exchange option prices) of two different currencies can be extracted from the price quotes of vanilla options in a foreign derivative market. The zonoid of η equals the projection of the lift zonoid on its last d coordinates, see [42, Sect. 2.2]. Furthermore, by [42, Cor. 2.25] we have that the lift zonoid of a marginal measure is the corresponding projection of the lift zonoid (onto the planes, spanned by basis vectors $\{e_0, e_1\}$, $\{e_0, e_2\}$ respectively). Hence, in markets where vanilla options are traded liquidly in domestic and foreign markets we have a natural source for information about the joint distribution. Note that e.g. in a risk-neutral setting log-normal models are characterised by these three projections. In view of the fact that there seems to be an increasing interest in the financial community in using partial information about the dependency structure for getting improved model-free bounds for two-asset options, see [52], this observation could pave the way for an alternative insight into this problematic.

Based on univariate approaches presented e.g. in [7, 10] quasi-self-dual random vectors, being closely related to self-dual random vectors, have been introduced and their distributions were characterised in [40]. This concept turned out to be helpful to extend the application range of the self-duality property and to incorporate carrying costs in applications in the area of semi-static hedging strategies. In order to handle potentially unequal carrying costs in typical applications in the area of swap-invariance based semi-static hedging strategies a further weakening of the swap-invariance property by means of the power transformation is analysed in [41]. The corresponding random vectors are called *quasi*-swap-invariant.

7 Joint Symmetries

Consider now the case where random vectors possess the highest degree of invariance, namely, when lifted (or non-lifted) zonoids are invariant with respect to swap of any two coordinates. The results for (lifted) max-zonoids are identical in view of Proposition 1 and [40, Th. 2.4].

Following [40] an integrable random vector η with positive components is called *jointly self-dual* if it is self-dual with respect to all numeraires. Then the expected payoff f_b is invariant with respect to any permutation of their arguments. In particular, then η is *exchangeable*, i.e. its distribution is invariant under any permutation of its components. All components of η are then self-dual random variables, see [40, Cor. 3.1]. The joint swap-invariance is yet weaker and means that the expected payoffs from (7) are invariant with respect to swaps of any components of u. Geometrically, it is possible to summarise these symmetry properties in the order of weakening as follows.

(1) η is jointly self-dual if and only if Z_η is symmetric with respect to each hyperplane $\{(u_0, u_1, \dots, u_d) \in \mathbb{R}^{d+1} : u_i = u_j\}$ for all $i, j = 0, \dots, d, i \neq j$.
(2) η is exchangeable if and only if Z_η is symmetric with respect to each hyperplane $\{(u_0, u_1, \dots, u_d) \in \mathbb{R}^{d+1} : u_i = u_j\}$ for all $i, j = 1, \dots, d, i \neq j$.
(3) η is jointly swap-invariant if and only if Z_η^o (i.e. the projection of Z_η) is symmetric with respect to each hyperplane $\{(u_1, \dots, u_d) \in \mathbb{R}^d : u_i = u_j\}$ for all $i, j = 1, \dots, d, i \neq j$.

In view of the remarkably rich literature about exchangeability property, its relation to the joint self-duality and swap-invariance seems to be of a certain theoretical interest. However, so far the joint self-duality or joint swap-invariance do not appear to be particularly important for applications, where considerably less restrictive conditions often suffice, see [40, 41, 48].

Example 5 (Log-normal distribution, Black–Scholes setting) We now illustrate the difference between joint self-duality, exchangeability and joint swap-invariance for the multivariate log-normal $\eta = e^\xi$ in a risk-neutral setting. Assume that $\log \eta$ is normal with mean μ and the covariance matrix $A = (a_{ij})_{ij=1}^d$. In order to ensure that all components of η are related to a martingale measure, assume that $\mu = -\frac{1}{2}(a_{11}, \dots, a_{dd})$.

Then η is exchangeable if and only if

$$\mu = -\frac{\sigma^2}{2}(1, \dots, 1) \quad \text{and} \quad A = \sigma^2 \begin{pmatrix} 1 & \alpha & \cdots & \alpha \\ \alpha & 1 & \cdots & \alpha \\ \vdots & \vdots & \vdots & \vdots \\ \alpha & \alpha & \cdots & 1 \end{pmatrix}, \tag{8}$$

where $(1-d)^{-1} \le \alpha \le 1$ in order to ensure that A is non-negative definite. Furthermore, η is jointly self-dual if (8) holds with $\alpha = \frac{1}{2}$, see [40, Ex. 4.5]. Finally, η is jointly swap-invariant if

$$a_{li} - a_{lj} = \frac{1}{2}(a_{ii} - a_{jj}) \tag{9}$$

for all $l \neq i, j$ and $l, i, j = 1, \dots, d$.

In a risk-neutral *bivariate* setting all log-normal distributions are swap-invariant and may well be non-exchangeable. Also in higher dimensional cases it is possible to construct jointly swap-invariant and non-exchangeable random vectors, as the following example shows.

Example 6 Consider i.i.d. standard normal random variables $Z_0, Z_1, \dots, Z_d$. For constants $c_0, \dots, c_d$ define

$$\zeta_i = Z_i + \sum_{k=0}^{d} c_k Z_k\,, \quad i = 1, \dots, d\,.$$

Hence,

$$\mathrm{Var}(\zeta_i) = \sum_{k=0}^{d} c_k^2 + 2c_i + 1\,,$$

$$\mathrm{Cov}(\zeta_i, \zeta_j) = \sum_{k=0}^{d} c_k^2 + c_i + c_j$$

for $i, j = 1, \dots, d$. Thus, $\eta = \mathrm{e}^{\xi}$ with $\xi_i = \zeta_i + \gamma_i$, $i = 1, \dots, d$, and suitably chosen constants $\gamma_1, \dots, \gamma_d$ is risk-neutral, satisfies (9), i.e. is jointly swap-invariant, but not exchangeable unless $c_1 = \cdots = c_d$.

Example 7 (ℓ_p-ball) The ℓ_p ball in $\mathbb{R}^{d+1}$ is symmetric with respect to all hyperplanes $u_i = u_j$, $i, j = 0, 1, \dots, d$, $i \neq j$. The value of the corresponding option would be equal to the discounted ℓ_p-norm of the weight vector. It is shown in [38] that M_η being intersection of the ℓ_p-ball with $\mathbb{R}_+^{d+1}$ is a max-zonoid.[3] Proposition 2 yields that

$$\mathbf{Q}(\eta_1 \le v_1, \dots, \eta_d \le v_d) = \left(1 + \sum_{i=1}^{d} v_i^{-p}\right)^{\frac{1}{p}-1}, \quad v = (v_1, \dots, v_d) \in (0, \infty)^d\,,$$

is the cumulative distribution function of a jointly self-dual η.

8 Combinations, Lift Zonoids and General Univariate European Derivatives

A combination of basket payoffs can be defined by introducing a (possibly signed) measure μ on $\mathbb{R}^{d+1}$, so that $\mu(du_0, du)$ is the weight attached to the payoff $f_b(u_0, u)$ with $(u_0, u) \in \mathbb{R}^{d+1}$. Without loss of generality and in view of the homogeneity of the payoff function assume that μ is supported by the unit sphere $\mathbb{S}^d$

[3] Note that here we deal with max-zonoids and not the ℓ_p-zonoids from Example 4.

in $\mathbb{R}^{d+1}$. The finiteness of μ guarantees that the combination has a finite payoff for all η, while the positivity of η makes it possible to relax the finiteness condition on μ. The payoff from the so-defined combination is given by

$$g(\eta) = \int_{\mathbb{S}^d} (u_0 + u_1\eta_1 + \cdots + u_d\eta_d)_+ \mu(\mathrm{d}u_0, \mathrm{d}u)\,. \tag{10}$$

If μ is non-negative, then $g(x) = h_L((1,x))$ is convex in x and is the support function of a convex body L being a zonoid in $\mathbb{R}^{d+1}$.

The expected payoff then becomes

$$\mathbf{E}g(\eta) = \int_{\mathbb{S}^d} \mathbf{E} f_b(u_0,u)\mu(\mathrm{d}u_0, \mathrm{d}u) = \int_{\mathbb{S}^d} h_{Z_\eta}\big((u_0,u)\big)\mu(\mathrm{d}u_0, \mathrm{d}u)\,.$$

Integrals of the support function have a particularly nice geometric interpretation if the integration measure is the surface measure of a certain convex body K, i.e. $\mu(\Gamma)$ equals the area of the set on the boundary of K where normals belong to Borel set $\Gamma \subset \mathbb{S}^d$. Then

$$\mathbf{E}g(\eta) = V(Z_\eta, K, \ldots, K)$$

is the mixed volume [49, Sect. 5.1]. Minkowski's inequality [49, Th. 6.2.1] yields that

$$\big(\mathbf{E}g(\eta)\big)^d = V(Z_\eta, K, \ldots, K)^d \geq V_d(K)^{d-1} V_d(Z_\eta)\,,$$

where V_d is the d-dimensional volume. The equality holds if and only if K and Z_η are homothetic, which is the case exactly if μ is proportional to the distribution of $(1,\eta)$ projected onto the unit sphere.

For the rest of this section we discuss the particularly important single asset case and explain the roles of the involved convex bodies, in particular the area measure induced by the lift zonoid. Since the lift zonoid of η determines the distribution of η, the arbitrage prices of all derivatives based on the terminal asset price $S_T = F\eta$ become functionals of Z_η. In the single asset case it is possible to represent these derivatives as integrals with respect to the area measure of Z_η in a model independent manner. Recall that the *area measure* $\mathsf{S}_1(L,\Gamma)$ generated by the planar convex body L is a Borel measure on the unit circle $\mathbb{S}^1$ in the plane, such that for a Borel $\Gamma \subset \mathbb{S}^1$, the measure equals the one-dimensional Hausdorff measure of all boundary points of L with normal vectors belonging to Γ. For more details see [49, Chaps. 4 and 5].

Theorem 1 *If $f : [0,\infty) \mapsto \mathbb{R}$ is a payoff function satisfying $\mathbf{E}|f(S_T)| < \infty$, then*

$$\mathbf{E}f(S_T) = \frac{1}{2}\int_{\mathbb{S}^1} \tilde{f}(u)\mathsf{S}_1(Z_\eta, du)\,, \tag{11}$$

where $\tilde{f} : \mathbb{S}^1 \mapsto \mathbb{R}$ is any function on the unit circle such that

$$\tilde{f}(u) = |u_1| f(-Fu_0/u_1)\,, \quad u = (u_0,u_1) \in \mathbb{S}^1\,,\ u_0u_1 < 0\,,$$

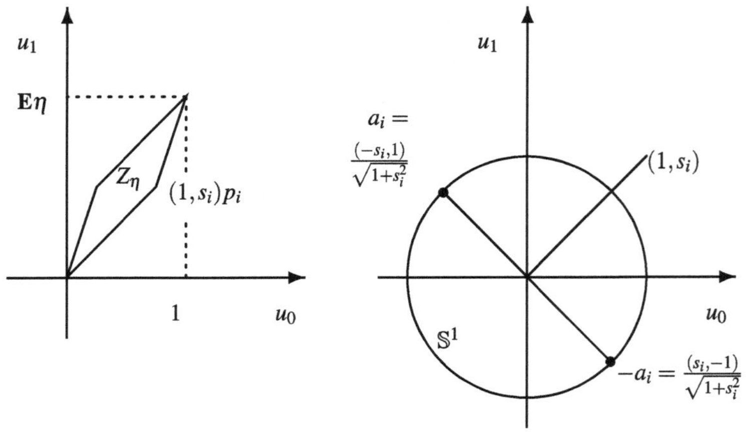

Fig. 5 Sketch of the proof of Theorem 1

and F is the theoretical forward price.

Proof The construction of Z_η yields that the boundary length of all points with normals from $B' = \{u = (u_0, u_1) \in \mathbb{S}^1 : u_0 u_1 \geq 0\}$ vanishes, i.e. $\mathsf{S}_1(Z_\eta, B') = 0$. This shows that the way $\tilde{f}$ is defined on B' does not matter.

Assume that η has a discrete distribution with a finite set of atoms $s_1, \ldots, s_k \geq 0$ and the corresponding probabilities $p_1, \ldots, p_k$. Let denote $L_i = [(0,0),(1,s_i)]$ for $i = 1, \ldots, k$. The area measure $\mathsf{S}_1(L_i, du)$ has atoms at $\{\pm a_i\}$ of mass $m_i = \sqrt{1+s_i^2}$ each, where $a_i = m_i^{-1}(-s_i, 1)$, see Fig. 5. Since the area measure of order one is Minkowski linear, $\mathsf{S}_1(Z_\eta, u)$ is the atomic measure with atoms $\{\pm a_i\}$ of weights $p_i m_i$, $i = 1, \ldots, k$. Since $\tilde{f}$ is even on $\mathbb{S}^1 \setminus B'$,

$$\frac{1}{2}\int_{\mathbb{S}^1} \tilde{f}(u)\mathsf{S}_1(Z_\eta, \mathrm{d}u) = \frac{1}{2}\int_{\mathbb{S}^1\setminus B'} \tilde{f}(u)\mathsf{S}_1(Z_\eta, \mathrm{d}u)$$
$$= \sum_{i=1}^{k} \tilde{f}(a_i)p_i m_i = \sum_{i=1}^{k} f(Fs_i)p_i = \mathbf{E}f(S_T).$$

Finally, (11) is obtained by approximating of η with discrete random variables. In other words, (11) holds for zonotopes Z_η, and thereupon for zonoids, since they are limits of zonotopes in the Hausdorff metric. □

In the absolutely continuous case with continuously differentiable and non-vanishing probability density, the above proof can also be carried over using the principal radii of curvature of the boundary. These principal radii of curvature can also be used to describe hedge parameters, see [46, Sect. 6.3].

If the integrand $\tilde{f}$ in (11) is the support function of a convex body $\tilde{L}$, then the integral can be interpreted as the *mixed area* of $\tilde{L}$ and Z_η, i.e.

$$\frac{1}{2}\int_{\mathbb{S}^1} h_{\tilde{L}}(u)\,\mathsf{S}_1(Z_\eta, du) = V(\tilde{L}, Z_\eta),$$

where the right-hand side is a functional of two convex sets that satisfy

$$V_2(L+M) = V_2(L) + V_2(M) + 2V(L, M)$$

with $V_2(\cdot)$ being the area, see [49, Chap. 5] for an introduction to a rich theory that concerns such functionals. The set $\tilde{L}$, called payoff set, determine the geometry of the payoff.

Example 8 Consider the straddle with the payoff function $f(S_T) = |F\eta - k|$. Then $\tilde{f}(u) = |Fu_0 + ku_1|$ is the support function of the line segment with end-points $\pm(F, k)$, i.e. $\tilde{L} = [-(F, k), (F, k)]$, so that $\mathbf{E}|S_T - k| = V(\tilde{L}, Z_\eta)$.

Example 9 If $\tilde{f}(u) = \sqrt{u_0^2 + u_1^2}$, then the related payoff function

$$f(S_T) = \sqrt{1 + (S_T/F)^2}$$

corresponds to the payoff set $\tilde{L} = B(0, 1)$.

Acknowledgements The authors are grateful to Thorsten Rheinländer for inspiring discussions. This work was supported by the Swiss National Science Foundation Grant Nr. 200021-126503.

References

1. Bardos, C., Douady, R., Fursikov, A.: Static hedging of barrier options with a smile: an inverse problem. ESAIM Control Optim. Calc. Var. **8**, 127–142 (2002)
2. Barndorff-Nielsen, O.E., Shiryaev, A.N.: Change of Time and Change of Measure. World Scientific, Singapore (2010)
3. Bates, D.S.: The skewness premium: option pricing under asymmetric processes. Adv. Futures Options Res. **9**, 51–82 (1997)
4. Baxter, M.: Hedging in financial markets. ASTIN Bull. **28**, 5–16 (1998)
5. Bowie, J., Carr, P.: Static simplicity. Risk **7**, 45–49 (1994)
6. Breeden, D.T., Litzenberger, R.H.: Prices of state-contingent claims implicit in options prices. J. Bus. **51**, 621–651 (1978)
7. Carr, P., Chou, A.: Hedging complex barrier options. Working paper, NYU's, Courant Institute and Enuvis Inc. (2002)
8. Carr, P., Ellis, K., Gupta, V.: Static hedging of exotic options. J. Finance **53**, 1165–1190 (1998)
9. Carr, P., Laurence, P.: Multi-asset stochastic local variance contracts. Math. Finance **21**, 21–52 (2011)
10. Carr, P., Lee, R.: Put-call symmetry: extensions and applications. Math. Finance **19**, 523–560 (2009)

11. Chan, T.: Pricing contingent claims on stocks driven by Lévy processes. Ann. Appl. Probab. **9**, 504–528 (1999)
12. Choulli, T., Hurd, T.R.: The role of Hellinger processes in mathematical finance. Entropy **3**, 150–161 (2001)
13. Cont, R., Tankov, P.: Financial Modelling with Jump Processes. Chapman & Hall/CRC, London (2004)
14. Cont, R., Tankov, P.: Non-parametric calibration of jump-diffusion option pricing models. J. Comput. Finance **7**, 1–49 (2004)
15. Cont, R., Tankov, P.: Retrieving Lévy processes from option prices: regularization of an ill-posed inverse problem. SIAM J. Control Optim. **45**, 1–25 (2007)
16. Crépey, S.: Calibration of the local volatility in a trinomial tree using Tikhonov regularization. Inverse Probl. **19**, 91–127 (2003)
17. d'Aspremont, A., El Ghaoui, L.: Static arbitrage bounds on basket option prices. Math. Program. **106**, 467–489 (2006)
18. Delbaen, F., Schachermayer, W.: A general version of the fundamental theorem of asset pricing. Math. Ann. **300**, 463–520 (1994)
19. Delbaen, F., Schachermayer, W.: The fundamental theorem of asset pricing for unbounded stochastic processes. Math. Ann. **312**, 215–250 (1998)
20. Delbaen, F., Schachermayer, W.: The Mathematics of Arbitrage. Springer, Berlin (2005)
21. Delbaen, F., Schachermayer, W.: The variance-optimal martingale measure for continuous processes. Bernoulli **2**, 81–105 (1996)
22. Eberlein, E., Papapantoleon, A., Shiryaev, A.N.: On the duality principle in option pricing: semimartingale setting. Finance Stoch. **12**, 265–292 (2008)
23. El Karoui, N., Jeanblanc, M.: Options exotiques. Finance **20**, 49–67 (1999)
24. El Karoui, N., Rouge, R.: Pricing via utility maximization and entropy. Math. Finance **10**, 259–276 (2000)
25. Esche, F., Schweizer, M.: Minimal entropy preserves the Lévy property: how and why. Stoch. Process. Appl. **115**, 299–337 (2005)
26. Föllmer, H., Schied, A.: Stochastic Finance. An Introduction in Discrete Time, 2nd edn. De Gruyter, Berlin (2004)
27. Frittelli, M.: The minimal entropy martingale measure and the valuation problem in incomplete markets. Math. Finance **10**, 39–52 (2000)
28. Goll, T., Rüschendorf, L.: Minimax and minimal distance martingale measures and their relationship to portfolio optimization. Finance Stoch. **5**, 557–581 (2001)
29. Henkin, G.M., Shananin, A.A.: Bernstein theorems and Radon transform. Application to the theory of production functions. In: Gel'fand, I.M., Gindikin, S.G. (eds.) Mathematical Problems of Tomography, pp. 189–223. Amer. Math. Soc., Providence (1990)
30. Hüsler, J., Reiss, R.D.: Maxima of normal random vectors: between independence and complete dependence. Stat. Probab. Lett. **7**, 283–286 (1989)
31. Kabanov, Y.M.: On the FTAP of Kreps-Delbaen-Schachermayer. In: Kabanov, Y.M., Rozovskii, B.L., Shiryaev, A.N. (eds.) Statistics and Control of Random Processes. The Liptser Festschrift. Proceedings of Steklov Mathematical Institute Seminar, pp. 191–203. World Scientific, Singapore (1997)
32. Kabanov, Y.M., Safarian, M.: Markets with Transaction Costs. Mathematical Theory. Springer, Berlin (2009)
33. Kindermann, S., Mayer, P.A.: On the calibration of local jump-diffusion market models. Finance Stoch. **15**, 685–724 (2011)
34. Koshevoy, G.A., Mosler, K.: Lift zonoids, random convex hulls and the variability of random vectors. Bernoulli **4**, 377–399 (1998)
35. Lipton, A.: Mathematical Methods for Foreign Exchange: A Financial Engineer's Approach. World Scientific, Singapore (2001)
36. Miyahara, Y., Fujiwara, T.: The minimal entropy martingale measures for geometric Lévy processes. Finance Stoch. **5**, 509–531 (2003)
37. Molchanov, I.: Theory of Random Sets. Springer, London (2005)

38. Molchanov, I.: Convex geometry of max-stable distributions. Extremes **11**, 235–259 (2008)
39. Molchanov, I., Schmutz, M.: Geometric extension of put-call symmetry in the multiasset setting. Tech. rep., University of Bern, Bern (2008). arXiv:0806.4506 [math.PR]
40. Molchanov, I., Schmutz, M.: Multivariate extensions of put-call symmetry. SIAM J. Financ. Math. **1**, 396–426 (2010)
41. Molchanov, I., Schmutz, M.: Exchangeability type properties of asset prices. Adv. Appl. Probab. **43**, 666–687 (2011)
42. Mosler, K.: Multivariate Dispersion, Central Regions and Depth. The Lift Zonoid Approach. Lect. Notes Statist., vol. 165. Springer, Berlin (2002)
43. Profeta, C., Roynette, B., Yor, M.: Option Prices as Probabilities. A New Look at Generalized Black-Scholes Formulae. Springer, Heidelberg (2010)
44. Ross, S.A.: Options and efficiency. Q. J. Econ. **90**, 75–89 (1976)
45. Samperi, D.: Calibrating a diffusion pricing model with uncertain volatility: regularization and stability. Math. Finance **12**, 71–87 (2002)
46. Schmutz, M.: Zonoid options. Master's thesis, Institute of Mathematical Statistics and Actuarial Science, University of Bern, Bern (2007)
47. Schmutz, M.: Semi-static hedging for certain Margrabe type options with barriers. Tech. rep., University of Bern, Bern (2008). arXiv:0810.5146 [math.PR]
48. Schmutz, M.: Semi-static hedging for certain Margrabe type options with barriers. Quant. Finance **11**, 979–986 (2011)
49. Schneider, R.: Convex Bodies. The Brunn–Minkowski Theory. Cambridge University Press, Cambridge (1993)
50. Shananin, A.A.: To the theory of production functions. In: Models and Algorithms of the Programmed Planning Method, pp. 24–50. Comp. Center AN SSSR, Moscow (1979). In Russian
51. Shiryaev, A.N.: Essentials of Stochastic Finance: Facts, Models, Theory. World Scientific Publishing, Singapore (1999)
52. Tankov, P.: Improved Fréchet bounds and model-free pricing of multi-asset options. J. Appl. Probab. **48**, 389–403 (2011)

Pricing of Volume-Weighted Average Options: Analytical Approximations and Numerical Results

Alexander A. Novikov, Timothy G. Ling, and Nino Kordzakhia

Abstract The volume weighted average price (VWAP) over rolling number of days in the averaging period is used as a benchmark price by market participants and can be regarded as an estimate for the price that a passive trader will pay to purchase securities in a market. The VWAP is commonly used in brokerage houses as a quantitative trading tool and also appears in Australian taxation law to specify the price of share-buybacks of publically-listed companies. Most of the existing literature on VWAP focuses on strategies and algorithms to acquire market securities at a price as close as possible to VWAP. In our setup the volume process is modeled via a shifted squared Ornstein-Uhlenbeck process and a geometric Brownian motion is used to model the asset price. We derive the analytical formulae for moments of VWAP and then use the moment matching approach to approximate a distribution of VWAP. Numerical results for moments of VWAP and call-option prices have been verified by Monte Carlo simulations.

Keywords Asian option · Moment matching · Volume process · Geometric Lévy model

Mathematics Subject Classification (2010) 91G20

1 Introduction

A volume weighted average price (VWAP) occurs frequently in finance. It is used as a benchmark price by market participants and can be regarded as an estimate for the

A.A. Novikov (✉) · T.G. Ling
University of Technology, Sydney, Australia
e-mail: Alex.Novikov@uts.edu.au

T.G. Ling
e-mail: Timothy.G.Ling@student.uts.edu.au

N. Kordzakhia
Macquarie University, Sydney, Australia
e-mail: nino.kordzakhia@mq.edu.au

Y. Kabanov et al. (eds.), *Inspired by Finance*, DOI 10.1007/978-3-319-02069-3_21,

price that a passive trader will pay to purchase a security in a market. The VWAP is commonly used in brokerage houses to assess the performance of a trader and has applications in algorithmic trading (see [3], vol. 4). The VWAP also appears in Australian taxation law as part of determining the price of share buy-backs in publicly listed companies [18].

Suppose that in a given time interval (day, week, etc.) there are N transactions involving shares of a particular company. Let S_i and U_i denote the price and trading volume pertinent to transaction $i \in \{1 \dots N\}$. There are a number of ways to define the VWAP (see e.g. [12]), the standard definition is

$$\mathrm{VWAP} = \frac{\sum_{i=1}^{N} S_i U_i}{\sum_{i=1}^{N} U_i}.$$

Most of the existing literature on VWAP focuses on strategies and algorithms to execute orders as close as possible to the VWAP price (see for e.g. [2, 8, 9] and [12]).

Calculating the VWAP moments is not a simple task because it involves computing the moments of a ratio of two random variables, say $\frac{Y}{Z}$. To the best of our knowledge, there exists only one paper which discusses VWAP options, see Stace [17]. A moment matching approach was used in [17] to find a lognormal approximation for the call option via the approximation of VWAP first and second moments using the following approximations for computing the moments:

$$\mathbb{E}\frac{Y}{Z} \approx \frac{\mathbb{E}Y}{\mathbb{E}Z} - \frac{\mathrm{Cov}(Y, Z)}{(\mathbb{E}Z)^2} + \frac{\mathbb{E}Y}{(\mathbb{E}Z)^3}\mathrm{Var}(Z) \ ,$$

$$\mathrm{Var}\frac{Y}{Z} \approx \left(\frac{\mathbb{E}Y}{\mathbb{E}Z}\right)^2 \left(\frac{\mathrm{Var}\, Y}{(\mathbb{E}Y)^2} + \frac{\mathrm{Var}\, Z}{(\mathbb{E}Z)^2} - 2\frac{\mathrm{Cov}(Y, Z)}{\mathbb{E}Y\mathbb{E}Z}\right) .$$

This approximation is based on a truncated Taylor series expansion, see [14]. In [17] the author used a continuous time setting for VWAP with a geometric Brownian motion for S_t and a CIR model for U_t. It was shown in [17] that approximations for the first and second moments of VWAP can be found by solving a large system (nineteen!) of ordinary differential equations.

Our contribution presented here consists in the derivation of *exact* analytical formulas for the first and second moments of a continuous-time VWAP process under the assumption that the volume process is modeled by a shifted squared Ornstein-Uhlenbeck process (which is close by nature to a CIR process) and the asset price is a geometric Brownian motion. As in [17] we assume in our paper that S_t and U_t are independent but this assumption can be removed by slightly more lengthy calculations. It is important to note that our setting can be easily extended to the case of a geometric Levy model for the asset price.

Section 2 describes the VWAP model and contains a summary of the moment-matching approach.

In Sect. 3 we find analytical formulae for the first and second moments of the VWAP via the calculation of the Laplace transform of the integral of the squared

Ornstein-Uhlenbeck process. Calculations of this type (which are based only on using the Girsanov transformation and do not involve solving any PDEs or ODEs) have been done in the context of the calibration of an Ornstein-Uhlenbeck process in [16]; see also the exposition of these results in [10].

In Sect. 4 the derived above formulae for moments of the VWAP are used for computing the drift and volatility parameters of a matching lognormal process. Further we compute the prices of VWAP call-options using the Black-Scholes formula. Note that the moment matching method is frequently used for approximating Asian options, see e.g. [5].

We also provide a comparison with Monte Carlo simulations and results showing improvements over our lognormal approximation when the Generalized Inverse Gaussian (GIG) distribution (which required matching of three moments) is used instead of the lognormal distribution. Properties of the GIG distribution are discussed in [7]. Our choice of approximating with the GIG is motivated by results in the papers by Dufresne [4] and also Milevsky and Posner [13], in the latter the authors demonstrated via numerical examples that the Reciprocal Gamma Distribution, which is a particular case of the GIG, is well suited for approximating Asian options on stocks with large volatilities.

We deliberately omit here any discussions of arbitrage pricing, hedging and calibration of the model using real data. These topics deserve special considerations which we plan to provide elsewhere.

2 The VWAP Model and the Moment Matching Approach

We assume the usual framework of a probability space $(\Omega, \mathcal{F}, \mathbb{P})$ equipped with filtration $\mathcal{F}_t$, $t \geq 0$. Let S_t denote the asset price at time t with known mean and covariance functions and let U_t be the volume (quantity) of assets that are traded at time t. Stace in [17] used mean reverting processes (CIR and Brennan-Schwartz processes) for modeling the trade volume. Here we adopt the following mean-reverting process to model the trade volume,

$$U_t = X_t^2 + \delta, \qquad dX_t = \lambda(a - X_t)dt + v\,dW_t, \quad X_0 = a,$$

where $\delta \geq 0$ and $\lambda > 0$. The Ornstein-Uhlenbeck process X_t has representation

$$X_t = a + v\xi_t \tag{1}$$

where ξ_t is a standard Ornstein-Uhlenbeck process satisfying the SDE

$$d\xi_t = -\lambda \xi_t dt + dW_t, \quad \xi_0 = 0. \tag{2}$$

In the symmetric case, when $\delta = 0$ and $a = 0$, the process $U(t)$ is a particular case of the Cox-Ingersoll-Ross (CIR) process, [15].

Further we assume that S_t and U_t are independent for any $t \geq 0$. The continuous time analog of the VWAP is given by

$$A_T = \frac{\int_0^T S_t U_t dt}{V_T}$$

where $V_T = \int_0^T U_t dt$.

The moment matching approach is a method whereby a number of moments of the process A_t at the time T is set equal to the corresponding moments of a chosen approximating process. The resulting set of equations then allows us to derive the parameters of the approximating process.

As an example, to match A_t to a lognormal process $\tilde{S}_t$ with drift $\tilde{\mu}$ and volatility $\tilde{\sigma}$ we require only the first two moments of A_T. We recall that the mean and variance of $\tilde{S}_t$ are given by

$$\mathbb{E}\tilde{S}_t = \tilde{S}_0 e^{\tilde{\mu}t}, \qquad \mathrm{Var}(\tilde{S}_t) = \tilde{S}_0^{\,2} e^{2\tilde{\mu}t}\big(e^{\tilde{\sigma}^2 t} - 1\big).$$

Making the substitutions $\mathbb{E}\tilde{S_T} = \mathbb{E}A_T$ and $\mathbb{E}\tilde{S_T}^{\,2} = \mathbb{E}A_T^2$ allows us to obtain the parameter values $\tilde{\mu}$ and $\tilde{\sigma}$. In the next section we describe our approach for obtaining the VWAP moments.

3 Computing the VWAP Moments

3.1 The VWAP First Moment

To find the mean of the VWAP process we first note that due to the assumption of independence of processes S_t and U_t

$$\mathbb{E}A_T = \mathbb{E}\frac{\int_0^T S_t U_t dt}{\int_0^T U_t dt} = \int_0^T \mathbb{E}S_t \mathbb{E}\frac{U_t}{\int_0^T U_t dt} dt. \tag{3}$$

Given the joint Laplace transform

$$\Phi(z, r, q) = \mathbb{E}\exp\{-zU_t - rU_s - qV_T\} \tag{4}$$

and assuming that

$$\mathbb{E}\frac{U_t}{V_T} < \infty \tag{5}$$

we can compute $\mathbb{E}A_T$ as follows. First we note that

$$\frac{\partial}{\partial z}\Phi(z, 0, q)|_{z=0} = \mathbb{E}\frac{\partial}{\partial z}e^{-zU_t - qV_T}|_{z=0} = -\mathbb{E}U_t e^{-qV_T}. \tag{6}$$

Integrating both sides of (6) with respect to q over $[0, \infty)$, we have

$$-\int_0^\infty \frac{\partial}{\partial z}\Phi(z,0,q)|_{z=0}dq = \int_0^\infty \mathbb{E}U_t e^{-qV_T}dq = \mathbb{E}\frac{U_t}{V_T}. \tag{7}$$

It follows from (3) that

$$\mathbb{E}A_T = -\int_0^T \mathbb{E}S_t\left(\int_0^\infty \frac{\partial}{\partial z}\Phi(z,0,q)|_{z=0}dq\right)dt.$$

Next we find $\Phi(z,0,q)$. We note that the derivation of $\Phi(z,r,q)$ which is required for computing $\mathbb{E}A_T^2$ is very similar in essence albeit involving lengthier calculations.

We have

$$\begin{aligned}\Phi(z,0,q) &= \mathbb{E}\exp\left\{-zU_t - q\int_0^T U_t dt\right\}\\ &= \exp\{-z\delta - q\delta T\}\Psi(z,q)\end{aligned}$$

where

$$\Psi(z,q) = \mathbb{E}\exp\left\{-zX_t^2 - q\int_0^T X_t^2 dt\right\}.$$

The next step consists in elimination of the term $\int_0^T X_t^2 dt$ using the change of measure. For convenience we define the stochastic exponent

$$\mathcal{E}_T(\lambda) = \exp\left\{-\lambda\int_0^T W_t dW_t - \lambda^2/2\int_0^T W_t^2 dt\right\}$$

where W_t is a standard Brownian motion. Using the Girsanov theorem (see details in [10]), we obtain

$$\begin{aligned}\Psi(z,q) &= \mathbb{E}\mathcal{E}_T(\lambda)\exp\left\{-z(a+vW_t)^2 - q\int_0^T (a+vW_t)^2 dt\right\}\\ &= \mathbb{E}\exp\left\{-z(a+vW_t)^2 - q\int_0^T \left(a^2+2vaW_t\right)dt\right.\\ &\quad \left. -\lambda\int_0^T W_t dW_t - \left(\lambda^2/2+qv^2\right)\int_0^T W_t^2 dt\right\}.\end{aligned}$$

Set

$$\kappa = \sqrt{\lambda^2+2qv^2}.$$

Since $\int_0^T W_t dW_t = (W_T^2 - T)/2$ we have

$$\Psi(z,q) = \mathbb{E}\mathcal{E}_T(\kappa)\exp\left\{-z(a+vW_t)^2 - q\int_0^T \left(a^2+2vaW_t\right)dt\right.$$

$$- \frac{(\lambda-\kappa)(W_T^2 - T)}{2}\bigg\}$$

and using again the Girsanov theorem,

$$= \mathbb{E}\exp\bigg\{-z(a+vY_t)^2 - q\int_0^T \big(a^2 + 2vaY_t\big)dt - \frac{(\lambda-\kappa)(Y_T^2 - T)}{2}\bigg\}$$

where Y_t is a standard Ornstein–Uhlenbeck process with parameter κ admitting the representation $Y_t = e^{-\kappa t}\int_0^t e^{\kappa s}dW_s$. After some simplification we obtain

$$\Psi(z,q) = \exp\bigg\{-za^2 - qa^2T + \frac{(\lambda-\kappa)T}{2}\bigg\}\gamma(z,q) \tag{8}$$

where

$$\gamma(z,q) = \mathbb{E}\exp\bigg\{-2zvaY_t - 2qva\int_0^T Y_s ds - zv^2Y_t^2 + \frac{(\kappa-\lambda)Y_T^2}{2}\bigg\}.$$

To compute $\gamma(z,q)$, we condition over the filtration $\mathcal{F}_t$,

$$\begin{aligned}
&\gamma(z,q)\\
&= \mathbb{E}\bigg[\mathbb{E}\bigg(\exp\bigg\{-2zvaY_t - 2qva\int_0^T Y_s ds - zv^2Y_t^2 + \frac{(\kappa-\lambda)Y_T^2}{2}\bigg\}|\mathcal{F}_t\bigg)\bigg]\\
&= \mathbb{E}\bigg[e^{\xi}\mathbb{E}\bigg(\exp\bigg\{-2qva\int_t^T Y_s ds + \frac{(\kappa-\lambda)Y_T^2}{2}\bigg\}|\mathcal{F}_t\bigg)\bigg]
\end{aligned} \tag{9}$$

where $\xi = -2zvaY_t - 2qva\int_0^t Y_s ds - zv^2Y_t^2$ and is $\mathcal{F}_t$-measurable. Using the fact that Y_t is a Markov process, the conditional expectation in (9) can be expressed as

$$\mathbb{E}\bigg(\exp\bigg\{-2qva\int_t^T Y_s ds + \frac{(\kappa-\lambda)Y_T^2}{2}\bigg\}|Y_t\bigg). \tag{10}$$

Set $X_1 = a\int_t^T Y_s ds$, $X_2 = Y_T$, $X_3 = Y_t$ and $\sigma_{ij} = \mathrm{Cov}(X_i, X_j)$ for $i,j \in \{1,2,3\}$ (see the appendix for the calculation of the covariances). Because Y_t is an Ornstein–Uhlenbeck process, X_1, X_2, and X_3 are Gaussian random variables and so together they form a multivariate normal distribution. Then the distribution of $\left[\begin{smallmatrix} X_1 \\ X_2 \end{smallmatrix}\right]$ given $X_3 = z$ is a multivariate normal distribution with the mean vector and covariance matrix given by

$$\mu = \begin{bmatrix} \mu_1 + \frac{\sigma_{13}}{\sigma_{33}}(z-\mu_3) \\ \mu_2 + \frac{\sigma_{23}}{\sigma_{33}}(z-\mu_3) \end{bmatrix}, \qquad \Sigma = \begin{bmatrix} \sigma_{11} - \frac{\sigma_{13}^2}{\sigma_{33}} & \sigma_{12} - \frac{\sigma_{13}\sigma_{23}}{\sigma_{33}} \\ \sigma_{12} - \frac{\sigma_{13}\sigma_{23}}{\sigma_{33}} & \sigma_{22} - \frac{\sigma_{23}^2}{\sigma_{33}} \end{bmatrix}$$

respectively. So to compute the conditional expectation of (10), we can find

$$\mathbb{E}\left(\exp\left\{-2qvX_1 + \frac{(\kappa-\lambda)X_2^2}{2}\right\} \mid X_3 = z\right)$$
$$= \int_{-\infty}^{\infty}\int_{-\infty}^{\infty} \exp\left\{-2qvx + \frac{(\kappa-\lambda)y^2}{2}\right\} f_{X_1,X_2|X_3}(x,y \mid z)\,dxdy \qquad (11)$$

where $f_{X_1,X_2|X_3}(x,y \mid z)$ is the density function of $\left[\begin{smallmatrix} X_1 \\ X_2 \end{smallmatrix}\right]$ given $X_3 = z$.

Computation of the double integral of (11) essentially requires us to solve

$$\int_{-\infty}^{\infty}\int_{-\infty}^{\infty} \exp\{-Ax^2 - By^2 + Cx + Dy + Fxy + G\}\,dxdy \qquad (12)$$

where A, B, C, D, F and G are constants. Under the condition $F^2 < 4AB$ the solution to (12) is

$$2\pi \exp\left\{\frac{BC^2 + D(AD + CF)}{4AB - F^2} + G\right\}\left(4AB - F^2\right)^{-1/2}.$$

Using this result, in addition to performing a number of symbol manipulations in Mathematica, we can rewrite (10) as

$$\mathbb{E}\left(\exp\left\{-2qva\int_t^T Y_s ds + \frac{(\kappa-\lambda)Y_T^2}{2}\right\} \mid \mathcal{F}_t\right) = \exp\{HY_t^2 + JY_t + L\}$$

where the constants H, J and L are known. Mathematica expressions for these constants are too long to reproduce here; we may supply the corresponding Mathematica code on request. This in turn allows us to express $\gamma(z,q)$ of Eq. (9) as another double integral of the same form as Eq. (12). This leads to a closed-form expression for the joint Laplace transform $\Phi(z,0,q)$ where its partial derivative with respect to z may be computed analytically.

Remark The Laplace transform of V_T given by $\Phi(0,0,q)$ was originally derived in [16]. In particular, the following expression was obtained in [16] (see also Sect. 17.3 in [10]) for the case $a = 0$ and $q \geq 0$:

$$g(q) = \mathbb{E}\exp\left\{-qv^2\int_0^T \xi_s^2 ds\right\} = \left[\frac{2\kappa e^{\lambda T}}{(\kappa-\lambda)e^{-\kappa T} + (\kappa+\lambda)e^{\kappa T}}\right]^{1/2},$$

where the process ξ_s is defined in (2) and $\kappa = \sqrt{\lambda^2 + 2qv^2}$. In view of Andersen's Lemma ([1], see also Sect. 2.10 in [6]) and taking into account equation (1) we have for any $X_0 = a$ and $x > 0$

$$\mathbb{P}\left\{\int_0^T X_s^2 ds < x\right\} \leq \mathbb{P}\left\{v^2\int_0^T \xi_s^2 ds < x\right\}.$$

This implies the following estimate for any $p > 0$

$$\mathbb{E}V_T^{-p} = \frac{1}{\Gamma(p)} \int_0^\infty q^{p-1}\Phi(0,0,q)dq \le \frac{1}{\Gamma(p)} \int_0^\infty q^{p-1} g(q)dq.$$

Since $g(q) = O(e^{-\kappa T})$ as $q \to \infty$ this estimate implies

$$\mathbb{E}V_T^{-p} < \infty.$$

When $\delta > 0$ this result is, of course, trivial. Since U_t is a shifted squared Gaussian process we have also $\mathbb{E}U_t^p < \infty$ for any $p > 0$. Using the Hölder inequality we obtain that for any $p > 0$

$$\mathbb{E}\frac{U_t^p}{V_T^p} < \infty$$

and so condition (5) holds.

3.2 Computing the Second Moment

The VWAP second moment is given by

$$\mathbb{E}A_T^2 = \mathbb{E}\frac{(\int_0^T S_t U_t dt)^2}{(\int_0^T U_t dt)^2} = \int_0^T \int_0^T \mathbb{E}S_t S_u \, \mathbb{E}\frac{U_t U_u}{V_T^2} \, dt du.$$

Given the Laplace transform $\Phi(z, r, q)$ in (4) we can compute $\mathbb{E}A_T^2$ as follows:

$$\begin{aligned} \frac{\partial}{\partial z}\Phi(z,r,q)\Big|_{z=0} &= -\mathbb{E}U_t e^{-zU_t - rU_s - qV_T}\big|_{z=0} \\ \Longrightarrow \quad \frac{\partial}{\partial z}\frac{\partial}{\partial r}\Phi(z,r,q)\Big|_{z=r=0} &= \mathbb{E}U_t U_s e^{-zU_t - rU_s - qV_T}\big|_{z=r=0}. \end{aligned} \tag{13}$$

Now multiply both sides of (13) by q and integrating with respect to q over $[0, \infty)$:

$$\begin{aligned} \int_0^\infty q\left(\frac{\partial}{\partial z}\frac{\partial}{\partial r}\Phi(z,r,q)\Big|_{z=r=0}\right)dq &= \int_0^\infty q\mathbb{E}U_t U_s e^{-qV_T}\, dq \\ &= \mathbb{E}U_t U_s \int_0^\infty q e^{-V_T q}\, dq \\ &= \mathbb{E}\frac{U_t U_s}{V_T^2}. \end{aligned}$$

So we have

$$\mathbb{E}\frac{U_t U_s}{V_T^2} = \int_0^\infty q \frac{\partial}{\partial z}\frac{\partial}{\partial r}\Phi(z,r,q)\Big|_{z=r=0} dq \tag{14}$$

and so the second moment is given by

$$\mathbb{E}A_T^2 = \int_0^T \int_0^T \int_0^\infty q\mathbb{E}S_t S_s \left(\frac{\partial}{\partial z}\frac{\partial}{\partial r}\Phi(z,r,q)\Big|_{z=r=0} \right) dq dt ds. \quad (15)$$

Further calculations of $\Phi(z,r,q)$ are similar to the case $\Phi(z,0,q)$ and thus are omitted here. We must note that all our analytical results have been implemented in the Mathematica software package and fully verified using Monte Carlo simulations (see Sect. 4).

3.3 Generalized Inverse Gaussian Distribution

The GIG distribution has the density function

$$p(x) = \frac{(a/b)^{p/2}}{2K_p(\sqrt{ab})} x^{p-1} \exp\left\{ -\frac{(ax + b/x)}{2} \right\}, \quad x > 0,$$

where $a > 0$, $b > 0$, p is a real number and K_p is a modified Bessel function of the second kind. Its ith moment is given by

$$m_i = \left(\frac{b}{a}\right)^{i/2} \frac{K_{p+i}(\sqrt{ab})}{K_p(\sqrt{ab})}.$$

Given the first three VWAP moments $\mathbb{E}A_T$, $\mathbb{E}A_T^2$ and $\mathbb{E}A_T^3$, the matching of moments gives a system of three nonlinear equations

$$m_i = \mathbb{E}A_T^i, \quad i = 1, 2, 3,$$

with three unknowns parameters a, b and p to be found.

4 Numerical Results

We have implemented the method based on lognormal and GIG approximations using geometric Brownian motion to model the asset price. All calculations of the first and second moments were performed symbolically leading to exact expressions for Eqs. (6) and (13). The subsequent multiple integrals were computed numerically by standard methods (we used the *NIntegrate* function in Mathematica) and are very fast. Monte Carlo simulation was used to estimate the third VWAP moment for pricing with the GIG. We note that all our Monte Carlo simulations were performed using $n = 1{,}000{,}000$ trajectories and 500 discretization points over $[0, T]$.

The parameters for asset price and volume were chosen to be as similar as possible to those in Stace's paper [17] and it is worth mentioning that the computed

Table 1 Numerical values of $\mathbb{E}A_T$ and Monte Carlo simulation of $\mathbb{E}A_T$ for different stock price volatility σ

σ	$\mathbb{E}A_T$[a]	$\hat{\mathbb{E}}A_T$	MC std. error	Rel. error (%)
0.1	115.68	115.67	0.0068	0.0095
0.2	115.68	115.69	0.0136	0.0104
0.3	115.68	115.67	0.0205	0.0061
0.4	115.68	115.70	0.0276	0.0190
0.5	115.68	115.62	0.0349	0.0501

[a]Note that σ does not enter into the computation of $\mathbb{E}A_T$, which leads to unchanging values for this column

Table 2 Numerical values of $\mathbb{E}A_T^2$ and Monte Carlo simulation of $\mathbb{E}A_T^2$ for different stock price volatility σ

σ	$\mathbb{E}A_T^2$	$\hat{\mathbb{E}}A_T^2$	MC std. error	Rel. error (%)
0.1	13427.92	13429.3	1.5802	0.0103
0.2	13566.90	13567.5	3.2362	0.0044
0.3	13803.32	13802.3	5.0657	0.0074
0.4	14144.68	14147.9	7.1639	0.0228
0.5	14602.11	14598.6	9.7013	0.0240

parameter values are quite similar to Stace's results. Our parameter choices give rise to

$$dS_t = (0.1)S_t dt + \sigma S_t dW_t, \quad S_0 = 110,$$

for the asset price and

$$U_t = X_t^2,$$
$$dX_t = 2(22 - X_t)dt + 5dW_t, \quad X_0 = 22,$$

for the volume dynamics. We set $T = 1$. Tables 1 and 2 display a range of computed moments and the corresponding simulated values. Figure 1 displays call option prices for different strike values (K) and stock price volatilities (σ) for both lognormal and GIG approximations. Figure 2 shows computed prices arising from the different methods with σ ranging over $[0.2, 0.5]$. Relative error plots comparing approximated prices with simulated counterparts are presented in Fig. 3.

For $K = 100$ it can be seen that the relative error of the lognormal approximation stays within 1 % in the volatility range $[0.2, 0.37]$ with a tendency to increase. The relative error for the GIG approximation stays beneath 0.8 % over the entire range of volatilities. The lognormal relative error behaves similarly for $K = 110$, however we note that the GIG approximation struggles for small σ. In fact, for $\sigma < 0.13$ the GIG approximation is not suitable due to the instability of the numerical calculations. These findings for VWAP options conform with a classical market observation for Asian options, namely, for a small volatility of underlying process and near at-the money options the log-normal approximation outperforms others, [11].

Further we use Monte Carlo simulation to estimate the third moment of VWAP. However, an analytical formula can be obtained in a similar way for calculating the third moment of VWAP.

Fig. 1 Call option prices for different strike values K and stock price volatility σ

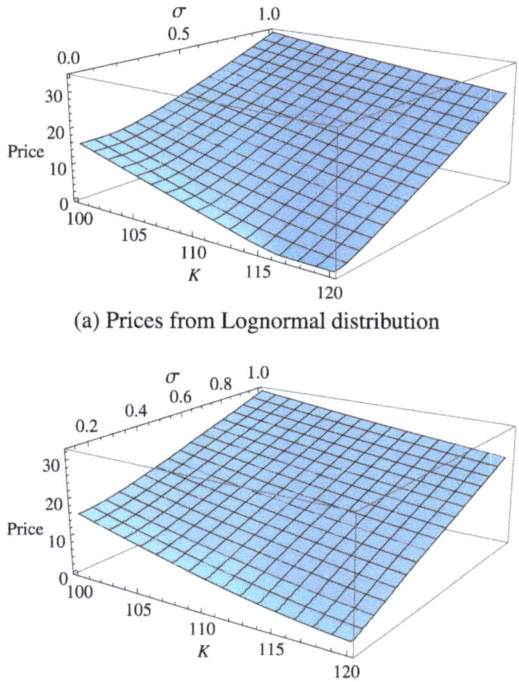

(a) Prices from Lognormal distribution

(b) Prices from GIG distribution

Fig. 2 Call option prices for $K = 100$ and different stock price volatility σ

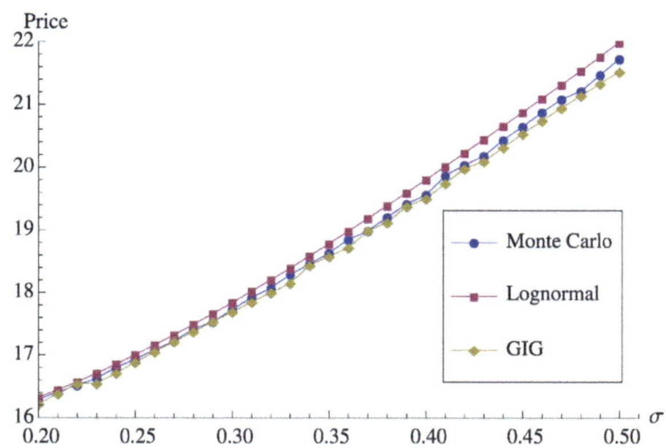

Fig. 3 Relative error of option prices as a function of σ. The solid line represents Lognormal error and the dashed line is the GIG error

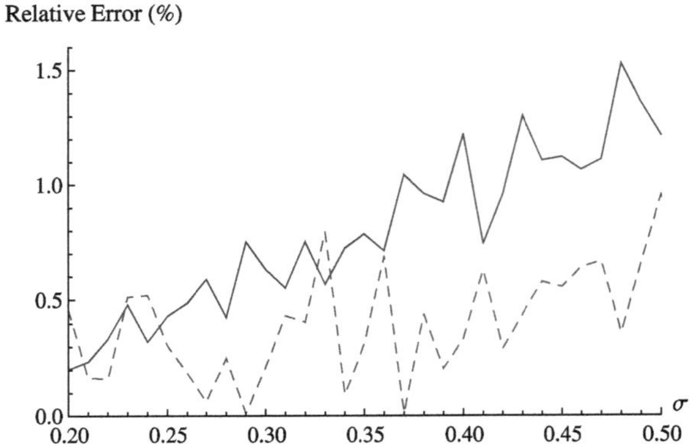

(a) $K = 100$

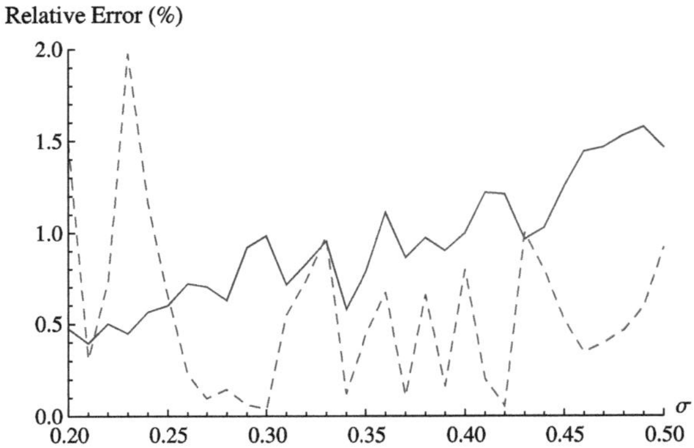

(b) $K = 110$

Acknowledgements The authors are thankful to Dr Alex Radchik and Dr Volf Frishling for useful discussions. The first author acknowledges the support of Australian Research Council's Discovery Projects funding scheme (project number 0880693).

Appendix

Here we provide details on the calculation of the different covariance functions required in this paper.

To compute the double integral of Eq. (11) we need

$$\sigma_{11} = \text{Cov}\left(a\int_t^T Y_s ds, a\int_t^T Y_s ds\right) = a^2\mathbb{E}\int_t^T\int_t^T Y_s Y_u du ds$$
$$= a^2\int_t^T\left(\int_t^s \mathbb{E}Y_s Y_u du + \int_s^T \mathbb{E}Y_s Y_u du\right)ds$$

$$= a^2 \int_t^T \left(\int_t^s \frac{1}{2\kappa} e^{-\kappa(u+s)} \left(e^{2\kappa u} - 1\right) du + \int_s^T \frac{1}{2\kappa} e^{-\kappa(u+s)} \left(e^{2\kappa s} - 1\right) du \right) ds$$

where the last expression can be easily computed in a symbolic package such as Mathematica.

Next

$$\sigma_{22} = \mathrm{Cov}(Y_T, Y_T) = \frac{1}{2\kappa}\left(1 - e^{-2\kappa T}\right),$$

$$\sigma_{33} = \mathrm{Cov}(Y_t, Y_t) = \frac{1}{2\kappa}\left(1 - e^{-2\kappa t}\right),$$

$$\sigma_{12} = \mathrm{Cov}\left(a \int_t^T Y_s ds, Y_T \right) = \int_t^T \frac{a}{2\kappa} e^{-\kappa(s+T)} \left(e^{2\kappa s} - 1\right) ds,$$

$$\sigma_{13} = \mathrm{Cov}\left(a \int_t^T Y_s ds, Y_t \right) = \int_t^T \frac{a}{2\kappa} e^{-\kappa(s+t)} \left(e^{2\kappa t} - 1\right) ds,$$

$$\sigma_{23} = \mathrm{Cov}(Y_T, Y_t) = \frac{1}{2\kappa} e^{-\kappa(t+T)} \left(e^{2\kappa t} - 1\right).$$

References

1. Andersen, T.: The integral of a symmetric unimodal function over a symmetric convex set and some probability inequalities. Proc. Am. Math. Soc. **6**(2), 170–176 (1955)
2. Białkowski, J., Darolles, S., Fol, G.L.: Improving vwap strategies: a dynamic volume approach. J. Bank. Finance **32**(9), 1709–1722 (2008)
3. Cont, R.: Encyclopedia of Quantitative Finance. Wiley, Chichester (2010)
4. Dufresne, D.: The distribution of a perpetuity, with applications to risk theory and pension funding. Scand. Actuar. J., 39–79 (1990)
5. Fusai, G., Roncoroni, A.: Implementing Models in Quantitative Finance: Methods and Cases. Springer, Berlin (2008)
6. Ibragimov, I., Khasminskii, R.: Statistical Estimation: Asymptotic Theory. Springer, Berlin (1981)
7. Jørgensen, B.: Statistical Properties of the Generalized Inverse Gaussian Distribution vol. 9. Springer, New York (1982)
8. Kakade, S.M., Kearns, M., Mansour, Y., Ortiz, L.E.: Competitive algorithms for vwap and limit order trading. In: EC '04: Proceedings of the 5th ACM Conference on Electronic Commerce. ACM, New York (2004)
9. Konishi, H.: Optimal slice of a vwap trade. J. Financ. Mark. **5**(2), 197–221 (2002)
10. Liptser, R.S., Shiryaev, A.N.: Statistics of Random Processes. II, vol. 6. Springer, Berlin (2001)
11. Lo, K.H., Wang, K., Hsu, M.F.: Pricing European Asian options with skewness and kurtosis in the underlying distribution. J. Futures Mark. **28**(6), 598–616 (2008)
12. Madhavan, A.N.: VWAP Strategies, pp. 32–39. Transaction Performance: The Changing Face of Trading. Institutional Investor, Inc. (2002)
13. Milevsky, M.A., Posner, S.E.: Asian options, the sum of lognormals, and the reciprocal gamma distribution. J. Financ. Quant. Anal. **33**(3), 409–422 (1998)
14. Mood, A.M., Graybill, F.A., Boes, D.C.: Introduction to the Theory of Statistics, 3rd edn. McGraw-Hill, New York (1974)
15. Musiela, M., Rutkowski, M.: Martingale Methods in Financial Modelling. Springer, New York (2005)

16. Novikov, A.A.: Estimation of the parameters of diffusion processes. Studia Sci. Math. Hung. **7**, 201–209 (1972)
17. Stace, A.W.: A moment matching approach to the valuation of a volume weighted average price option. Int. J. Theor. Appl. Finance **10**(1), 95–110 (2007)
18. Woellner, R., Barkoczy, S., Murphy, S., Evans, C.: Australian Taxation Law, 19th edn. CCH, Australia, North Ryde (2009)

A Class of Homothetic Forward Investment Performance Processes with Non-zero Volatility

Sergey Nadtochiy and Thaleia Zariphopoulou

Abstract We study forward investment performance processes with non-zero forward volatility. We focus on the class of homothetic preferences in a single stochastic factor model. The forward performance process is represented in a closed-form via a deterministic function of the wealth and the stochastic factor. This function is, in turn, given as a distortion transformation of the solution to a linear ill-posed problem. We analyze the solutions of this problem in detail. We, also, provide two examples for specific dynamics of the stochastic factor, specifically, log-mean reverting and Heston-type dynamics.

Keywords Forward investment performance · Hamilton–Jacobi–Bellman equation · Distortion transformation · Widder theorem · Heston model

Mathematics Subject Classification (2010) 91G20

1 Introduction

This paper is a contribution to the recently developed approach of forward investment performance measurement (see [8] and [9]). This approach allows for dynamic update of the investor's performance criterion and offers an alternative to the classical maximal expected utility objective which is defined only at a single instant. The underlying object is a stochastic process, the so called forward investment performance process, which is defined for all times. Its key properties are the supermartingality at admissible self-financing policies and martingality at an optimum.

S. Nadtochiy (✉)
Department of Mathematics, University of Michigan, Ann Arbor MI 48109, USA
e-mail: sergeyn@umich.edu

T. Zariphopoulou
Oxford-Man Institute, University of Oxford, Oxford, UK
e-mail: thaleia.zariphopoulou@oxford-man.ox.ac.uk

T. Zariphopoulou
Departments of Mathematics and IROM, McCombs School of Business, The University of Texas at Austin, Austin, TX, USA

Y. Kabanov et al. (eds.), *Inspired by Finance*, DOI 10.1007/978-3-319-02069-3_22,

Constructing such a process is a formidable task, for the underlying stochastic optimization problem is formulated "forward in time" and might be ill-posed.

In [10], a stochastic partial differential equation was introduced which the forward performance process is expected to satisfy. In many aspects, this SPDE is the stochastic analogue of the deterministic Hamilton–Jacobi–Bellman equation for the classical (backward) case. There are several elements which make the study of the SPDE and the derivation of analogous verification results hard. Indeed, one has to specify the appropriate class of initial conditions and, also, address the ill-posedness and the possible degeneracy of the equation.

Besides these issues, one also has to specify the correct family of forward performance volatility processes. These processes are chosen by the investor and constitute one of the novel elements of the forward investment theory. They are exogenous inputs for the volatility term of the SPDE. Note that their classical analogue is uniquely determined due to the static nature of the utility criterion (see Remark 3 herein).

To date, existence and uniqueness of solutions to the forward SPDE have not been established and the related verification results are still lacking. General results have been produced only for the case of zero volatility (see [9]). Under this rather strong assumption, the performance process is monotone in time (decreasing) and can be represented as a compilation of a deterministic function and the market input (see (16)). This form, however, is not any more valid when the investor allows for volatility in his criterion.

Herein, we do not study general questions but only analyze a family of forward processes and construct specific examples. Moreover, we concentrate on the class of homothetic criteria. We are motivated to look at this family because it offers the closest analogue of the classical value function under power utilities.

The market consists of one riskless asset and a stock whose dynamics are affected by a stochastic factor, denoted by Y_t. The latter is imperfectly correlated with the stock which makes the market incomplete. Such a model arises frequently when one assumes predictability of returns and/or stochastic volatility.

The homotheticity assumption suggests a separable form for the candidate processes with one of the components depending exclusively on the stochastic factor. In turn, the assumptions on the model dynamics suggest that the latter component is a process, denoted by $V(Y_t, t)$, that can be represented as a function of the stochastic factor and time. Constructing the function $V(y, t)$ is the main goal of this paper together with, as mentioned earlier, the specification of the correct initial condition and the appropriate class of volatility processes.

A distortion transformation on $V(y, t)$ yields a linear equation with a potential term. The forward in time nature of the underlying stochastic optimization problem makes this linear equation ill-posed. Specifying its nonnegative solutions is, to our knowledge, an open problem. Indeed, the only known case for which necessary and sufficient conditions for nonnegative solutions of such problems have been established is when the potential term is absent. This is the celebrated Widder's theorem. Herein, we study the more general case and provide results in this direction. A special case of these results yields one part (sufficiency) of Widder's theorem.

Once the form of the function $V(y,t)$ is specified, we are able to construct an admissible volatility process. This process is, also, taken to be homothetic in the wealth argument. A solution to the forward performance SPDE is then readily obtained.

Finally, we provide two concrete examples. In the first example, the stochastic volatility is taken to be a mean reverting process satisfying linear SDE, while in the second it satisfies Heston-type dynamics. In both cases, we calculate explicitly the appropriate initial condition and the volatility process as well as the associated forward performance process. We, also, study the robustness of the latter when its volatility vanishes and we compare it with its zero-volatility counterpart.

The paper is organized as follows. In Sect. 2, we describe the model and recall the investment performance criterion. In Sect. 3, we focus on homothetic performance processes and provide some preliminary informal results for the form of candidate processes. In Sect. 4, we study the underlying linear equation. We conclude in Sect. 5 where we present the two examples.

2 The Stochastic Factor Model and Investment Performance Measurement

The market consists of a risky and a riskless asset. The risky asset is a stock whose price S_t, $t \geq 0$, is modeled as a diffusion process solving

$$dS_t = \mu(Y_t)S_t dt + \sigma(Y_t)S_t dW_t^1, \tag{1}$$

with $S_0 > 0$. The stochastic factor Y_t, $t \geq 0$, satisfies

$$dY_t = b(Y_t)dt + d(Y_t)\big(\rho dW_t^1 + \sqrt{1-\rho^2}dW_t^2\big), \tag{2}$$

with $Y_0 = y$, $y \in \mathbb{R}$. The process $W_t = (W_t^1, W_t^2)$, $t \geq 0$, is a standard 2-dimensional Brownian motion, defined on a filtered probability space $(\Omega, \mathscr{F}, \mathbb{P})$. The underlying filtration is $\mathscr{F}_t = \sigma(W_s : 0 \leq s \leq t)$. It is assumed that the correlation coefficient $\rho \in (-1, 1)$.

The coefficients μ, σ, b and d satisfy the appropriate continuity and Lipschitz conditions such that the above system of equations has a unique strong solution. It is, also, assumed that $\sigma(y) > 0$, $y \in \mathbb{R}$.

The riskless asset, the savings account, offers constant interest rate $r > 0$.

We introduce the process, frequently called the market price of risk,

$$\lambda(Y_t) = \frac{\mu(Y_t) - r}{\sigma(Y_t)}. \tag{3}$$

Starting with an initial endowment x, the investor invests at future times in the riskless and risky assets. The present value of the amounts allocated in the two accounts are denoted, respectively, by π_t^0 and π_t. The present value of her investment is, then, given by $X_t^\pi = \pi_t^0 + \pi_t$, $t > 0$. We will refer to X_t^π as the discounted wealth. Using (1) we easily deduce that it satisfies

$$dX_t^\pi = \sigma(Y_t)\pi_t\big(\lambda(Y_t)dt + dW_t^1\big). \tag{4}$$

The investment strategies will play the role of control processes and are taken to satisfy the standard assumption of being self-financing. Such a portfolio, π_t, is deemed admissible if, for $t > 0$, $\pi_t \in \mathscr{F}_t$, $E_{\mathbb{P}}(\int_0^t \sigma^2(Y_s)\pi_s^2 ds) < +\infty$ and the associated discounted wealth satisfies the state constraint $X_t^\pi \geq 0$, $t \geq 0$. We will denote the set of admissible strategies by $\mathscr{A}$.

Stochastic factors have been used in portfolio choice to model asset predictability and stochastic volatility. A detailed survey of asset allocation models with a single stochastic factor can be found in [16] and we refer the reader therein for a complete bibliography.

2.1 Forward Investment Performance Process

The performance of implemented investment strategies is typically measured in terms of optimizing an expected criterion of the generated wealth. In the academic literature, this criterion is predominantly given by the investor's utility function (see, for example, the seminal papers [6] and [7]). One, then, chooses an investment horizon, say T, and a utility function at this time, $U_T(x)$, and maximizes, over all admissible self-financing strategies, the expected utility of terminal wealth. Such problems have been widely studied under rather general assumptions on market coefficients and constitute one of the cornerstones in modern mathematical portfolio management theory.

There is, however, a limitation in this setting. Indeed, the performance criterion is not dynamic in the sense that, from one hand, it cannot be revised at any previous investment time, $t < T$, and, from the other, it cannot be extended at any time $t > T$. One could say that the terminal utility criterion corresponds to a static objective. This does not mean that the associated value function is time independent, an obviously wrong conclusion. Rather, we state that it is the criterion per se that is static, for it is (pre)specified for only one time instant, T.

Recently, one of the authors and M. Musiela introduced an alternative approach which bypasses these shortcomings. The associated criterion is developed in terms of a family of stochastic processes defined on $[0, +\infty)$ and indexed by the wealth argument. It is called the forward investment performance process. Its key properties are its martingality at an optimum and its supermartingality away from it. These are in accordance with the analogous properties of the value function process which stem out from the Dynamic Programming Principle. However, in contrast to the existing framework, the risk preferences are specified for today and not at a (possibly remote) future time. As we will see in the upcoming analysis, one of the fundamental questions in this approach is the correct specification of the initial conditions in order for the relevant stochastic optimization problem to be well posed (see, for example, Propositions 6 and 8 herein and [9]).

For completeness, we provide the definition of the forward investment process below but we refer the reader to [8] and [9] for details. We recall that $\mathscr{F}_t$, $t \geq 0$, is the filtration generated by $W_t = (W_t^1, W_t^2)$, $t \geq 0$, and $\mathscr{A}$ the set of admissible policies.

Definition 1 An $\mathscr{F}_t$-progressively measurable process $U(x,t)$ is a forward investment performance if for $t \geq 0$ and $x \geq 0$:

(i) the mapping $x \to U(x,t)$ is concave and increasing,
(ii) for each portfolio process $\pi \in \mathscr{A}$, $E_{\mathbb{P}}(U(X_t^\pi, t))^+ < \infty$, and

$$E_{\mathbb{P}}\big(U\big(X_s^\pi, s\big)|\mathscr{F}_t\big) \leq U\big(X_t^\pi, t\big), \quad s \geq t, \tag{5}$$

(iii) there exists a portfolio process $\pi^* \in \mathscr{A}$, for which

$$E_{\mathbb{P}}\big(U\big(X_s^{\pi^*}, s\big)|\mathscr{F}_t\big) = U\big(X_t^{\pi^*}, t\big), \quad s \geq t. \tag{6}$$

While the above definition might appear like a pedantic rephrase of the Dynamic Programming Principle it is actually not. Indeed, it gives rise to a forward in time stochastic optimization problem which belongs to the family of the so called "ill-posed" problems. Such problems are notoriously difficult with regards to their well-posedeness, stability and finiteness of solutions. Herein, we do not address this question but, rather, construct specific examples. Specifying forward processes that satisfy the above definition is an open problem and is currently under investigation by the authors and others (see, for example, [1, 3, 10], and [17]).

2.2 The Forward Performance SPDE

Recently, it was shown in [10] and [16] that a sufficient condition for a (sufficiently smooth) process $U(x,t)$ to be a forward performance is that it satisfies a stochastic partial differential equation (see (7) below). For the single stochastic factor model we examine herein, Proposition 2 in [16] takes the following form.

Proposition 1 (i) *Let $U(x,t)$ be an $\mathscr{F}_t$-progressively measurable process such that the mapping $x \to U(x,t)$ is strictly increasing and concave. Let, also, $U(x,t)$ be a solution to the stochastic partial differential equation*

$$dU(x,t) = \frac{1}{2}\frac{(\lambda(Y_t)U_x(x,t) + a_x^1(x,t))^2}{U_{xx}(x,t)}dt + \big(a(x,t)\cdot dW_t\big), \tag{7}$$

where $a(x,t) = (a^1(x,t), a^2(x,t))$ is an $\mathscr{F}_t$-progressively measurable process. Then $U(x,t)$ is a forward investment performance process.

(ii) *Consider the process* π_t^*, $t \geq 0$, *given by*

$$\pi_t^* = -\frac{\lambda(Y_t)U_x(X_t^*, t) + a_x^1(X_t^*, t)}{\sigma(Y_t)U_{xx}(X_t^*, t)}, \tag{8}$$

where X_t^*, $t \geq 0$, *solves*

$$dX_t^* = \sigma(Y_t)\pi_t^*\big(\lambda(Y_t)dt + dW_t^1\big), \tag{9}$$

with $X_0^* = x$, $x \geq 0$. *If* $\pi_t^* \in \mathscr{A}$ *and* (9) *has a strong solution, then* π_t^* *and* X_t^* *are optimal.*

As it is shown in [10], the same stochastic partial differential equation emerges in the classical formulation of the optimal portfolio choice problem. Indeed, fix an investment horizon, say T, and recall the traditional value function process, denoted by $V(x, t; T)$ and defined as

$$V(x, t; T) = \sup_{\mathscr{A}_T} E_{\mathbb{P}}\big(U(X_T)|\mathscr{F}_t, X_t = x\big),$$

with $\mathscr{A}_T$ being the direct analogue of $\mathscr{A}$ in $[0, T]$. Let us now assume that there exists a smooth enough function, say $v(x, y, t)$ such that the representation

$$V(x, t; T) = v(x, Y_t, t) \tag{10}$$

holds. We note that the existence and regularity of such a function has not been established to date, expect for special utilities.

The associated Hamilton-Jacobi-Bellman (HJB) equation is then given (informally) by

$$v_t + \max_{\pi}\left(\frac{1}{2}\sigma^2(x)\pi^2 v_{xx} + \pi\big(\mu(y)v_x + \rho\alpha(y)\sigma(y)v_{xy}\big)\right) + \frac{1}{2}d^2(y)v_{yy} + b(y)v_y, \tag{11}$$

with $v(x, y, T) = U(x)$.

Using the representation (10) and expanding the process $v(x, Y_t, t)$ yield,

$$dv(x, Y_t, t) = \left(v_t(x, Y_t, t) + \frac{1}{2}d^2(Y_t)v_{yy}(x, Y_t, t) + b(Y_t)v_y(x, Y_t, t)\right)dt + \rho d(Y_t)v_y(x, Y_t, t)dW_t^1 + \sqrt{1-\rho^2}d(Y_t)v_y(x, Y_t, t)dW_t^2.$$

Using that $v(x, y, t)$ satisfies (11) and rearranging terms, we deduce that

$$dv(x, Y_t, t) = \frac{1}{2}\frac{(\lambda(Y_t)v_x(x, Y_t, t) + \rho d(Y_t)v_{xy}(x, Y_t, t))^2}{v_{xx}(x, Y_t, t)}dt$$
$$+ \rho d(Y_t)v_y(x, Y_t, t)dW_t^1 + \sqrt{1-\rho^2}d(Y_t)v_y(x, Y_t, t)dW_t^2.$$

From (10) we, then, deduce that the value function process, which now plays the role of the (backward) investment performance, satisfies the same SPDE as in (7). Specifically, for $0 \le t < T$, the process $V(x, t; T)$ satisfies the equation

$$dV(x, t; T) = \frac{1}{2}\frac{(\lambda(Y_t)V_x(x, t; T) + a_x^1(x, t; T))^2}{V_{xx}(x, t; T)}dt + \big(a(x, t; T) \cdot dW_t\big)$$

with terminal condition $V(x, T; T) = U(x)$ and the components of volatility process given by

$$a^1(x, t; T) = \rho d(Y_t)v_y(x, Y_t, t) \quad \text{and} \quad a^2(x, t; T) = \sqrt{1-\rho^2}d(Y_t)v_y(x, Y_t, t). \tag{12}$$

Its is worth noticing that the terminal data suggest that $\lim_{t\to T} a^i(x, t; T) = 0$.

Remark 1 It is important to notice three fundamental differences between the classical (backward) and the forward cases. Firstly, in the backward optimal investment model, we are given a terminal condition while in the forward an initial one. Secondly, in the former case, the performance process satisfies $V(x, T) \in \mathscr{F}_0$ while in the latter, $U(x, t) \in \mathscr{F}_t$. Finally, in the backward case, there is *no* flexibility in choosing the volatility coefficients, for they are *uniquely* obtained from the Itô decomposition of the value function process while in the forward case, the volatility process is up to the investor to choose. How the investor should make this choice is one of the main challenges in the new approach.

2.3 The Zero Volatility Case

An important class of forward performance processes are the ones that correspond to the choice of zero volatility, $a(x, t) \equiv 0$, $t \ge 0$. We easily see, using the concavity of the forward process and (7), that these processes are decreasing in time. Despite the strong assumption on the volatility, these processes yield a rich family of performance criteria which compile in an intuitively pleasing way the dynamic risk profile of the investor and the information coming from the evolution of the investment opportunity set, as (16) below indicates. They also provide an important benchmark when volatility is not zero, as it is discussed in Propositions 7 and 9 herein. They are extensively studied in [8] and [9], and we refer the reader therein for the proofs of the results that follow. Herein, we only state the main result and discuss some insights about the admissibility of the candidate initial conditions. Because all involved functions are smooth, we will not refer to their specific regularity (see [9]).

Theorem 1 *Let $u_0 : \mathbb{R}^+ \to \mathbb{R}$ be strictly increasing and concave and such that the function $h_0 : \mathbb{R} \to \mathbb{R}^+$ defined by*

$$u_0'\big(h_0(x)\big) = e^{-x} \tag{13}$$

can be represented as the Laplace transform of a finite positive Borel measure, denoted by ν, namely,

$$h_0(x) = \int_0^\infty e^{xy}\nu(dy), \tag{14}$$

such that $h_0(x) < \infty$, for all $x \in \mathbb{R}$. Let, also, $u : \mathbb{R}^+ \times (0, \infty) \to \mathbb{R}$ be a strictly concave and increasing in the spatial argument function satisfying

$$u_t = \frac{1}{2}\frac{u_x^2}{u_{xx}}, \tag{15}$$

and $u(x, 0) = u_0(x)$. Then, with $\lambda(Y_t)$, $t \geq 0$, as in (3), *the process*

$$U(x,t) = u\left(x, \int_0^t \lambda^2(Y_s)ds\right) \tag{16}$$

is a forward investment performance.

Relations (13) and (14) demonstrate the admissibility condition for a candidate initial condition, $u_0(x)$. Specifically, the inverse of its first derivative must be represented via a Laplace transform as in (14).

In [9] (see, also [1]) the following is shown. Let $h : \mathbb{R}\times[0,\infty) \to \mathbb{R}^+$ be given by the "dynamic" analogue of (17), namely,

$$h(x,t) = \int_0^\infty e^{xy-\frac{1}{2}y^2 t}\nu(dy). \tag{17}$$

Then, the solution $u(x,t)$ of (15) satisfies

$$u_x\big(h(x,t),t\big) = e^{-x+\frac{t}{2}}, \tag{18}$$

while $h(x,t)$ solves the backward heat equation

$$h_t + \frac{1}{2}h_{xx} = 0.$$

The reader is invited to compare (13) and its "dynamic" analogue (18) as well as the role of the measure ν as the essential defining element in generating solutions for positive times. Generalizations of some of these results is one of the main contributions herein (see Sect. 4).

3 Homothetic Forward Investment Performance Processes

We concentrate on forward investment performance processes which are homothetic in the spatial variable. We are motivated to do so for two reasons. Firstly, these processes are the natural analogues of the popular power utilities. Secondly, as the analysis will indicate, the homogeneity assumption allows for significant tractability and closed form solutions.

To this end, we are looking for initial conditions and volatility processes which produce well defined solutions, $U(x,t)$, to (7) that have the property

$$U(kx,t) = k^{\gamma} U(x,t), \tag{19}$$

for all $t \geq 0$ and $k \in \mathbb{R}^{+}$, with $0 < \gamma < 1$. We easily deduce that the forward processes must be of the multiplicative form

$$U(x,t) = \frac{x^{\gamma}}{\gamma} K_t, \tag{20}$$

where the multiplicative process K_t, $t \geq 0$, is to be determined[1] but does not depend on the spatial variable x. In the sequel, we will further restrict the class of solutions by looking at factors that depend functionally on time and the current state of the stochastic factor (see (24)).

Note that (20) tells us that the only admissible initial conditions are of the form

$$u_0(x) = \frac{x^{\gamma}}{\gamma} K_0. \tag{21}$$

3.1 The Zero-Volatility Homothetic Case

We recall the homothetic time-monotone performance process. We will revert to this case later in the analysis when we investigate their robustness of the forward process for vanishing volatilities (see Propositions 7 and 9).

Proposition 2 *Assume that* $a(x,t) \equiv 0$, $t \geq 0$, *in* (7) *and let the initial condition be as in* (21). *Then, the forward performance process is given by*

$$U(x,t) = \frac{x^{\gamma}}{\gamma} K_0 \exp\left(\int_0^t \frac{1}{2} \frac{\gamma}{\gamma - 1} \lambda^2(Y_s) ds \right), \tag{22}$$

for $x \geq 0$ *and* Y_t, $t \geq 0$, *solving* (2).

---

[1]For convenience, we introduce the factor $1/\gamma$. Moreover, we do not consider the case $\gamma < 0$, which can be analyzed with similar, albeit more tedious computationally arguments.

Proof The claim follows from (16) and the fact that the function

$$u(x,t) = \frac{x^\gamma}{\gamma} e^{\frac{1}{2}\frac{\gamma}{\gamma-1}t} K_0, \quad x \ge 0,$$

solves the nonlinear equation (15) with initial condition $u(x,0) = \frac{x^\gamma}{\gamma} K_0$. □

3.2 Non-zero Volatility Homothetic Case

We now focus our attention to the case of non-zero volatility coefficients, which is the main topic herein. As mentioned earlier, the underlying problem is to specify an initial condition, $u_0(x)$, a (non-zero) volatility process, $a(x,t)$, and a process $U(x,t)$, such that the latter solves (7) with $U(x,0) = u_0(x)$. Moreover, we will be looking at processes with the Markovian structure

$$U(x,t) = \frac{x^\gamma}{\gamma} K(Y_t, t), \tag{23}$$

which corresponds to the factor in (20) to be of the functional form

$$K_t = K(Y_t, t), \tag{24}$$

for an appropriately chosen function $K : \mathbb{R}\times[0,\infty) \to \mathbb{R}^+$. Such processes constitute the simplest extension of their zero volatility counterparts.

We start with an informal analysis. To this end, let us make the distortion transformation[2]

$$K(y,t) = \big(v(y,t)\big)^\delta \tag{25}$$

with the power δ given by

$$\delta = \frac{1-\gamma}{1-\gamma+\rho^2\gamma}. \tag{26}$$

Combining (23) and (25) , and plugging in (7) yields that the process in (23), indeed, satisfies (7), provided that, from one hand, the function $v : \mathbb{R}\times[0,\infty) \to \mathbb{R}^+$ solves the linear problem

$$v_t + \frac{1}{2}d^2(y)v_{yy} + \left(b(y) + \rho\frac{\gamma}{1-\gamma}\lambda(y)d(y)\right)v_y + \frac{1}{2\delta}\frac{\gamma}{1-\gamma}\lambda^2(y)v = 0, \tag{27}$$

with initial condition

$$v(y,0) = \big(K(y,0)\big)^{1/\delta}, \tag{28}$$

[2] Solutions of similar structure were produced for the traditional value function in [15].

and, from the other, the volatility process is set to be $a(x,t) = (a^1(x,t), a^2(x,t))$ with

$$a^1(x,t) = \rho\delta \frac{x^\gamma}{\gamma} d(Y_t) v_y(Y_t,t) \big(v(Y_t,t)\big)^{\delta-1} \tag{29}$$

and

$$a^2(x,t) = \sqrt{1-\rho^2}\,\delta \frac{x^\gamma}{\gamma} d(Y_t) v_y(Y_t,t) \big(v_y(Y_t,t)\big)^{\delta-1}. \tag{30}$$

The calculations are routine but tedious and are, thus, omitted.

What the above shows is that, in order to construct a solution to (7), it suffices to construct a well defined solution to the initial problem (27) and for the appropriate initial condition (28). This is the subject of investigation in the next section.

4 Non-negative Solutions to an Ill-Posed Heat Equation with a Potential

We consider the backward linear Cauchy problem

$$H_t + \frac{1}{2} a_1^2(x) H_{xx} + a_2(x) H_x + a_3(x) H = 0, \tag{31}$$

for $(t,x) \in (0,+\infty) \times \mathbb{R}$, and initial condition $H(x,0) = H_0(x)$.

The coefficients, a_1, a_2 and a_3 satisfy the following conditions: $a_1(x) > 0$ and is twice continuously differentiable, $a_2(x)$ is continuously differentiable, and $a_3(x)$ is continuous.

We are interested in characterizing the set of non-negative solutions, $H(x,t)$, to the above equation as well as the set of initial conditions, $H_0(x)$, for which (31) has a well-defined solution.

The first step in the analysis of solutions of (31) is to put the equation in the so-called *canonical form.* To this end, consider the change of variables (see, for example, Sect. 4.3 of [14]) $Z : \mathbb{R} \to \mathbb{R}$, given by

$$Z(x) = \sqrt{2} \int_\zeta^x \frac{dz}{a_1(z)}, \tag{32}$$

for some fixed $\zeta \in \mathbb{R}$. In turn, introduce the function $F : \mathbb{R} \times [0,\infty) \to \mathbb{R}^+$ defined as

$$F(z,t) = H\big(X(z), t\big) e^{\frac{1}{2}\int_\zeta^z b(z')dz'} \tag{33}$$

where

$$b(z) = \sqrt{2} \frac{a_2(X(z))}{a_1(X(z))} - \frac{1}{\sqrt{2}} a_1'\big(X(z)\big), \tag{34}$$

with $X : \mathbb{R} \to \mathbb{R}$ given by $X(z) = Z^{-1}(z)$.

In the new variables, Eq. (31) takes the canonical form

$$F_t + F_{zz} + q(z)F = 0,$$

where $q : \mathbb{R} \to \mathbb{R}$ is a continuous function given by

$$q(z) = -\frac{1}{4}b^2(z) - \frac{1}{2}b'(z) + a_3\big(X(z)\big), \tag{35}$$

with $b(z)$ as in (34).

The aim is, then, to specify the class of admissible initial conditions, $F_0 : \mathbb{R} \to \mathbb{R}^+$, and the associated nonnegative solutions $F : \mathbb{R} \times [0, +\infty) \to \mathbb{R}^+$, for the initial value problem

$$(IV) \begin{cases} F_t + F_{zz} + q(z)F = 0, \\ F(z, 0) = F_0(z). \end{cases} \tag{36}$$

A common approach in analyzing the set of solutions to linear time-homogeneous parabolic pdes is to consider the associated *Sturm-Liouville* problem.

In the context of the problems of financial mathematics, the use of Sturm-Liouville theory is, for example, demonstrated in [2] and [11].

Denoting by $f(z, .)$ the Laplace transform of $F(z, .)$, we obtain

$$f_{zz}(z, \lambda) + \big(\lambda + q(z)\big) f(z, \lambda) = f_0(z). \tag{37}$$

We remind the reader that the calculations that follow are, for the moment, formal.

The homogeneous version of (37) is (with a slight abuse of notation),

$$f_{zz}(z, \lambda) + \big(\lambda + q(z)\big) f(z, \lambda) = 0. \tag{38}$$

The following result shows how to generate solutions to (36) using (38). This result is, in many aspects, similar to Widder's theorem (see [18]) which holds for the case $q(z) \equiv 0$ and provides necessary and sufficient conditions for constructing positive solutions to (36). We recall this theorem and provide some comments in the sequel (see Sect. 4.1).

We note that, to our knowledge, an extension to Widder's theorem for non-zero potentials, as the case we study herein, is still lacking. The result below offers only a sufficient condition for constructing positive solutions to (36) but not a necessary one. A further study in this direction can be found in [12].

Proposition 3 *Let us assume that* $\{\psi(., p, \lambda)\}_{(p,\lambda) \in \mathscr{P} \times \Lambda}$ *is a family of solutions to the homogeneous equation* (38), *parameterized by* (p, λ), *where* $\Lambda \subset \mathbb{R}$ *is a Borel set and* $\mathscr{P}$ *is an abstract measurable space. Let us, also, assume that, for each* $z \in \mathbb{R}$, *the function* $\psi(z, ., .)$ *is a nonnegative measurable function on* $\mathscr{P} \times \Lambda$ *and that* ξ *is a measure on* $\mathscr{P} \times \Lambda$, *such that*

$$\sup_{(z,t) \in \mathscr{K}} \left(\int_{\mathscr{P} \times \Lambda} (1 + \lambda^2) e^{t\lambda} \psi(z, p, \lambda) \xi(dp, d\lambda) \right) < \infty, \tag{39}$$

for any compact set $\mathscr{K} \subset \mathbb{R} \times [0, \infty)$.

Let $F_0 : \mathbb{R} \rightarrow \mathbb{R}^+$ be defined by

$$F_0(z) = \int_{\mathscr{P} \times \Lambda} \psi(z, p, \lambda) \xi(dp, d\lambda). \tag{40}$$

Then, Eq. (36), *equipped with the above initial condition has a nonnegative solution, $F(z, t)$, given by*

$$F(z, t) = \int_{\mathscr{P} \times \Lambda} \psi(z, p, \lambda) e^{t\lambda} \xi(dp, d\lambda). \tag{41}$$

Proof It can be verified by direct computation that the function $F(z, t)$ satisfies (36). Therefore, we only need to show that F and its derivatives exist and are continuous, and that we can interchange the differentiation and integration in (41). These statements will follow from repeated applications of Fubini's theorem.

To this end, we first observe that $F(z, t)$ is well defined, for the corresponding integral converges absolutely due to the integrability assumption (39).

Using (38), we have, for $z \in \mathbb{R}$, that

$$\begin{aligned}
&\int_0^z \int_0^{z'} \int_{\mathscr{P} \times \Lambda} \left| \psi_{xx}(x, p, \lambda) \right| e^{t\lambda} \xi(dp, d\lambda) dx dz' \\
&\quad = \int_0^z \int_0^{z'} \int_{\mathscr{P} \times \Lambda} \left| \lambda + q(x) \right| \left| \psi(x, p, \lambda) \right| e^{t\lambda} \xi(dp, d\lambda) dx dz' < \infty,
\end{aligned}$$

as it follows from (39) and the continuity of the potential coefficient $q(z)$. Thus, we can interchange the order of integration to obtain

$$\begin{aligned}
&\int_0^z \int_0^{z'} \int_{\mathscr{P} \times \Lambda} \psi_{xx}(x, p, \lambda) e^{t\lambda} \xi(dp, d\lambda) dx \\
&\quad = \int_{\mathscr{P} \times \Lambda} \big(\psi(z, p, \lambda) - \psi(0, p, \lambda) - z \psi_z(0, p, \lambda) \big) e^{t\lambda} \xi(dp, d\lambda).
\end{aligned}$$

Notice that the integral in the right hand side above is absolutely convergent, because such is the integral in the left hand side. In addition, because of (39), the integral $\int_{\mathscr{P} \times \Lambda} (\psi(z, p, \lambda) - \psi(0, p, \lambda)) e^{t\lambda} \xi(dp, d\lambda)$ also converges absolutely. Therefore, the function $e^{t\lambda} \psi_z(0, p, \lambda)$ is absolutely integrable with respect to $\xi(dp, d\lambda)$. We, easily, deduce that, for some constant c_1,

$$\int_0^z \int_0^{z'} \int_{\mathscr{P} \times \Lambda} \psi_{xx}(x, p, \lambda) e^{t\lambda} \xi(dp, d\lambda) dx dz' = F(z, t) - F(0, t) - c_1 z,$$

for all $(z, t) \in \mathbb{R} \times [0, \infty)$.

Let $\phi(z, t)$ be given by

$$\phi(z, t) = \int_0^z \int_0^{z'} \int_{\mathscr{P} \times \Lambda} \psi_{xx}(x, p, \lambda) e^{t\lambda} \xi(dp, d\lambda) dx dz'.$$

Then,

$$\phi(z,t) = F(z,t) - F(0,t) - c_1 z$$

and, by construction, it is continuously differentiable in z, with absolutely continuous derivative. Therefore, the same holds for $F(z,t)$, and, for almost all $z \in \mathbb{R}$, we have

$$F_z(z,t) = c_1 + \phi_z(z,t)$$

and

$$F_{zz}(z,t) = \phi_{zz}(z,t) = \int_{\mathcal{P}\times\Lambda} \psi_{zz}(z,p,\lambda) e^{t\lambda} \xi(dp,d\lambda).$$

Following similar arguments, we can show that, for any fixed $z \in \mathbb{R}$, the function $F(z,.)$ is absolutely continuous on $[0,\infty)$, and, in turn,

$$F_t(z,t) = \int_{\mathcal{P}\times\Lambda} \lambda\psi(z,p,\lambda) e^{t\lambda} \xi(dp,d\lambda),$$

for (almost all) $t \geq 0$.

It remains to show that the partial derivatives are continuous in $(z,t) \in \mathbb{R}\times[0,\infty)$. We start with $F_t(z,t)$. Let $(z,t), (z',t') \in \mathbb{R}\times[0,\infty)$. Then,

$$\begin{aligned}
&\left|\int_{\mathcal{P}\times\Lambda} \lambda\big(\psi(z,p,\lambda)e^{t\lambda} - \psi\big(z',p,\lambda\big)e^{t'\lambda}\big)\xi(dp,d\lambda)\right| \\
&\quad\leq \int_{\mathcal{P}\times\Lambda} |\lambda| e^{t\lambda}\psi(z,p,\lambda)\big|1 - e^{\lambda(t'-t)}\big|\xi(dp,d\lambda) \\
&\qquad + \int_{\mathcal{P}\times\Lambda} |\lambda| e^{t'\lambda}\big|\psi(z,p,\lambda) - \psi\big(z',p,\lambda\big)\big|\xi(dp,d\lambda).
\end{aligned} \tag{42}$$

We estimate the above integrals separately. We first observe that, for some constant c_2, the first integral satisfies

$$\begin{aligned}
&\int_{\mathcal{P}\times\Lambda} |\lambda| e^{t\lambda}\psi(z,p,\lambda)\big|1 - e^{\lambda(t'-t)}\big|\xi(dp,d\lambda) \\
&\quad\leq c_2\big|t - t'\big| \int_{\mathcal{P}\times\Lambda} \lambda^2 e^{t\lambda}\psi(z,p,\lambda)\xi(dp,d\lambda).
\end{aligned}$$

The expression in the right hand side above converges to zero as $t' \to t$, since the integral therein is finite, due to (39). For the second integral in (42) we have

$$\begin{aligned}
&|\lambda| e^{t'\lambda}\big|\psi(z,p,\lambda) - \psi\big(z',p,\lambda\big)\big| \\
&\quad= |\lambda| e^{t'\lambda}\left|\int_z^{z'}\int_0^{x'} \psi_{xx}(x,p,\lambda)\,dx\,dx' + \big(z' - z\big)\psi_z(0,p,\lambda)\right|.
\end{aligned}$$

We readily deduce that the left hand side above is absolutely integrable with respect to ξ, uniformly over t' changing on a compact set in $[0, \infty)$. Therefore, the right hand side has the same property. On the other hand, (39) yields that

$$\int_{\mathscr{P}\times\Lambda}\int_z^{z'}\int_0^{x'}|\lambda|e^{t'\lambda}\big|\psi_{xx}(x,p,\lambda)\big|dx\,dx'\xi(dp,d\lambda)$$
$$=\int_{\mathscr{P}\times\Lambda}\int_z^{z'}\int_0^{x'}|\lambda|e^{t'\lambda}\big|\lambda+q(x)\big|\psi(x,p,\lambda)dx\,dx'\xi(dp,d\lambda)$$

is bounded, uniformly on t' changing on a compact set. Therefore, the function $\lambda e^{t'\lambda}\psi_z(0,p,\lambda)$ is absolutely integrable with respect to $\xi(dp,d\lambda)$, uniformly over t' varying on a compact set. We, then, deduce that

$$\int_{\mathscr{P}\times\Lambda}|\lambda|e^{t'\lambda}\big|\psi(z,p,\lambda)-\psi\big(z',p,\lambda\big)\big|\xi(dp,d\lambda)$$
$$\le c_3\int_{\mathscr{P}\times\Lambda}\int_z^{z'}\int_0^{x'}\big(1+\lambda^2\big)e^{t'\lambda}\psi(x,p,\lambda)dx\,dx'\xi(dp,d\lambda)$$
$$+\big|z'-z\big|\int_{\mathscr{P}\times\Lambda}|\lambda|e^{t'\lambda}\big|\psi_z(0,p,\lambda)\big|\xi(dp,d\lambda)$$
$$\le c_3\big|z-z'\big|\big(|z|+\big|z'\big|\big)\sup_{x\in[z,z']}\int_{\mathscr{P}\times\Lambda}\big(1+\lambda^2\big)e^{t'\lambda}\psi(x,p,\lambda)\xi(dp,d\lambda)$$
$$+\big|z'-z\big|\int_{\mathscr{P}\times\Lambda}|\lambda|e^{t'\lambda}\big|\psi_z(0,p,\lambda)\big|\xi(dp,d\lambda).$$

The above integrals are bounded uniformly over t' changing on a compact set, and, therefore, the above right hand side converges to zero, as $(z',t')\to(z,t)$.

Working along similar arguments, we obtain the continuity in (z,t) of the function $\int_{\mathscr{P}\times\Lambda}\psi(z,p,\lambda)e^{t\lambda}\xi(dp,d\lambda)$. We easily conclude. □

The above result shows how one can construct solutions to Eq. (36) directly from the appropriate initial condition. It is not, however, always clear how to actually construct a nonnegative solution to (38). This is what we explore next.

For the rest of the analysis, we focus on the class of coefficients $q(z)$ which are bounded from above. We remind the reader that the term $q(z)$ represents the negative of a potential term, as the latter appears in the literature. A natural assumption for potentials is that they are bounded from below: notice, for example, that the assumption of nonnegativity of the "killing rate" in [4] is another way of saying that the corresponding potential is nonnegative.

Proposition 4 *Let us assume that there exists $\bar{\lambda}\in\mathbb{R}$, such that the potential term in* (36) *satisfies $q(z)\le\bar{\lambda}$, $z\in\mathbb{R}$, and denote $\mathscr{D}=(-\infty,-\bar{\lambda})$. Then, the following statements hold*:

(i) *Assume that there exists* $L_1 \in \mathbb{R}$, *such that*

$$\int_0^\infty |q(z) - L_1| dz < \infty.$$

Then, for any $\lambda \in \mathscr{D}$, *there exists a unique solution of* (38), *denoted by* $\psi^{(1)}(.,\lambda)$, *which is square integrable over* $(0,\infty)$ *and satisfies* $\psi^{(1)}(0,\lambda) = 1$. *Moreover, for each* $z \in \mathbb{R}$, *the function* $\psi^{(1)}(z,.)$ *is nonnegative and continuous on* $\mathscr{D}$.

Let, also, μ_1 *be a Borel measure on* $\mathscr{D}$, *satisfying*

$$\sup_{(t,z)\in\mathscr{K}} \left(\int_{\mathbb{R}} (1+\lambda^2) e^{t\lambda} \psi^{(1)}(z,\lambda) \mu_1(d\lambda) \right) < \infty,$$

for any compact set $\mathscr{K} \subset [0,\infty) \times \mathbb{R}$, *and define the function* $F_0^{(1)} : \mathbb{R} \to \mathbb{R}^+$ *by*

$$F_0^{(1)}(z) = \int_{\mathbb{R}} \psi^{(1)}(z,\lambda) \mu_1(d\lambda). \tag{43}$$

Then, Eq. (36) *has a nonnegative classical solution, say* $F^{(1)}(z,t)$, *given by*

$$F^{(1)}(z,t) = \int_{\mathbb{R}} \psi^{(1)}(z,\lambda) e^{t\lambda} \mu_1(d\lambda), \tag{44}$$

satisfying $F^{(1)}(z,0) = F_0^{(1)}(z)$.

(ii) *Assume that there exists* $L_2 \in \mathbb{R}$, *such that*

$$\int_{-\infty}^0 |q(z) - L_2| dy < \infty. \tag{45}$$

Then, for any $\lambda \in \mathscr{D}$, *there exists a unique solution of* (38), *denoted by* $\psi^{(2)}(.,\lambda)$, *which is square integrable over* $(-\infty,0)$ *and satisfies* $\psi^{(2)}(0,\lambda) = 1$. *Moreover, for each* $z \in \mathbb{R}$, *the function* $\psi^{(2)}(z,.)$ *is nonnegative and continuous on* $\mathscr{D}$.

Let, also, μ_2 *be a Borel measure on* $\mathscr{D}$, *satisfying*

$$\sup_{(t,z)\in\mathscr{K}} \left(\int_{\mathbb{R}} (1+\lambda^2) e^{t\lambda} \psi^{(2)}(z,\lambda) \mu_2(d\lambda) \right) < \infty,$$

for any compact set $\mathscr{K} \subset [0,\infty) \times \mathbb{R}$, *and define the function* $F_0^{(2)} : \mathbb{R} \to \mathbb{R}^+$ *given by*

$$F_0^{(2)}(z) = \int_{\mathbb{R}} \psi^{(2)}(z,\lambda) \mu_2(d\lambda). \tag{46}$$

Then, Eq. (36) *has a nonnegative classical solution, say* $F^{(2)}(z,t)$, *given by*

$$F^{(2)}(z,t) = \int_{\mathbb{R}} \psi^{(2)}(z,\lambda) e^{t\lambda} \mu_2(d\lambda), \tag{47}$$

satisfying $F^{(2)}(z,0) = F_0^{(2)}(z)$.

(iii) *Let the above assumptions hold in both* (i) *and* (ii). *Then, problem* (36), *equipped with the initial condition* $F_0(z) = F_0^{(1)}(z) + F_0^{(2)}(z)$, *with* $F_0^{(1)}(z)$ *and* $F_0^{(2)}(z)$ *given, respectively, by* (43) *and* (46), *has a nonnegative classical solution, say* $F(z,t)$, *given by*

$$F(z,t) = F^{(1)}(z,t) + F^{(2)}(z,t),$$

with $F^{(1)}(z,t)$ *and* $F^{(2)}(z,t)$ *as in* (44) *and* (47), *respectively.*

Proof We only establish part (i), for part (ii) follows along the same arguments using a change of variables "$z \mapsto -z$" and part (iii) follows trivially from parts (i) and (ii).

We start with some elementary transformations which will facilitate the upcoming analysis. To this end, fix $\delta > 0$, and consider all (possibly complex) numbers λ, satisfying $\mathfrak{Re}(\lambda) < -\delta - \bar{\lambda}$. Let $\varepsilon \in (0, \frac{1}{2}\delta(1 - e^{-2\sqrt{\delta}}))$ and $N \geq 0$ satisfying $\int_N^\infty |q(z) - L_1| dy < \varepsilon$, and introduce the change of variables

$$\tilde{\lambda} = \lambda + L_1 \quad \text{and} \quad \tilde{q}(z) = q(z+N) - L_1.$$

It, then, follows that a function $f(z,\lambda)$, is a solution to (38), if and only if the function $g(z,\tilde{\lambda})$, defined by

$$g(z,\tilde{\lambda}) = f(z+N, \tilde{\lambda} - L_1)$$

satisfies the homogeneous problem

$$g_{zz}(z,\tilde{\lambda}) + \big(\tilde{\lambda} + \tilde{q}(z)\big) g(z,\tilde{\lambda}) = 0. \tag{48}$$

Let H_δ be the set

$$H_\delta = \big\{ z \in \mathbb{C} \,|\, \mathfrak{Re}(z) < -\delta - \bar{\lambda} + L_1 \big\}.$$

It is clear that $L_1 \leq \bar{\lambda}$ and, therefore, $-\delta - \bar{\lambda} + L_1 < 0$.

We proceed as follows. We first establish that for any $\tilde{\lambda} \in H_\delta$, there exists a square integrable solution, say $\chi(z,\tilde{\lambda})$, to the above equation (48), for $z \in [0,\infty)$. Then, we show that this solution can be extended to the entire set $\mathbb{R}$ and that it is the unique (up to a multiplicative factor) such solution that is square integrable. We conclude showing that $\chi(z,\tilde{\lambda})$ does not change its sign.

To this end, let $\tilde{\lambda} \in H_\delta$ and consider the following integral equation for functions of $z \in [0,+\infty)$,

$$\begin{aligned} \chi(z,\tilde{\lambda}) = e^{iz\sqrt{\tilde{\lambda}}} &- \frac{1}{2i\sqrt{\tilde{\lambda}}} \int_0^z e^{i(z-x)\sqrt{\tilde{\lambda}}} \tilde{q}(x) \chi(x,\tilde{\lambda}) dx \\ &- \frac{1}{2i\sqrt{\tilde{\lambda}}} \int_z^\infty e^{i(x-z)\sqrt{\tilde{\lambda}}} \tilde{q}(x) \chi(x,\tilde{\lambda}) dx. \end{aligned} \tag{49}$$

Herein, we choose a version of the "square root" which generates a continuous mapping from $\mathbb{C} \setminus [0, \infty)$ to the upper half plane.

It is, then, easy to see that if the above equation has a solution $\chi(., \tilde{\lambda})$, then, it is twice continuously differentiable and solves (48).

On the other hand, it is shown in Sect. 6.2 (p. 119) of [14] that the iterative scheme

$$\chi_1(z, \tilde{\lambda}) = e^{iz\sqrt{\tilde{\lambda}}}$$

and

$$\chi_{n+1}(z, \tilde{\lambda}) = e^{iz\sqrt{\tilde{\lambda}}} - \frac{1}{2i\sqrt{\tilde{\lambda}}} \int_0^z e^{i(z-x)\sqrt{\tilde{\lambda}}} \tilde{q}(x) \chi_n(x, \tilde{\lambda}) dx$$
$$- \frac{1}{2i\sqrt{\tilde{\lambda}}} \int_z^\infty e^{i(x-z)\sqrt{\tilde{\lambda}}} \tilde{q}(x) \chi_n(x, \tilde{\lambda}) dx, \quad \text{for } n \geq 1,$$

converges to the solution of (49), $\chi(., \tilde{\lambda})$.

In particular, it is shown in formulas (6.2.5) and (6.2.6) therein that the approximating terms satisfy

$$\left|\chi_{n+1}(z, \tilde{\lambda}) - \chi_n(z, \tilde{\lambda})\right| \leq \left(\frac{\varepsilon}{2\delta}\right)^n \left|e^{iz\sqrt{\tilde{\lambda}}}\right|,$$

and, hence, the convergence is uniform in $\tilde{\lambda}$ changing on any compact set in H_δ. This, in turn, yields that the function $\chi(z, .)$ is holomorphic in H_δ, for any $z \in [0, \infty)$. Moreover, the following estimate holds

$$\left|\chi(z, \tilde{\lambda})\right| \leq \frac{|e^{iz\sqrt{\tilde{\lambda}}}|}{1 - \varepsilon/(2\delta)}.$$

We easily deduce that $\chi(., \tilde{\lambda})$ solves (48) and that it is square integrable on $[0, +\infty)$.

Next, we extend $\chi(., \tilde{\lambda})$ to the entire set $\mathbb{R}$. To this end, notice that any solution of (48) can be uniquely represented as a linear combination of two solutions, say, $\varphi(z, \tilde{\lambda})$ and $\theta(z, \tilde{\lambda})$, satisfying

$$\varphi(0, \tilde{\lambda}) = 0 \quad \text{and} \quad \varphi_z(0, \tilde{\lambda}) = -1,$$

and

$$\theta(0, \tilde{\lambda}) = 1 \quad \text{and} \quad \theta_z(0, \tilde{\lambda}) = 0.$$

Therefore, one obtains the representation

$$\chi(z, \tilde{\lambda}) = K_1(\tilde{\lambda}) \theta(z, \tilde{\lambda}) + K_2(\tilde{\lambda}) \varphi(z, \tilde{\lambda}),$$

for some functions K_1 and K_2. On the other hand, differentiating (49) and applying the dominated convergence theorem yield that $\chi_z(z,.)$ is continuous in H_δ, for any $z \in [0,\infty)$. Notice, also, that

$$K_1(\tilde{\lambda}) = \chi(0,\tilde{\lambda}) \quad \text{and} \quad K_2(\tilde{\lambda}) = -\chi_z(0,\tilde{\lambda}),$$

and, hence, the functions K_1 and K_2 are continuous in H_δ. It also follows—see for example Theorem 1.5 in Sect. 1.5 of [14]—that $\varphi(z,.)$ and $\theta(z,.)$ are entire functions (holomorphic in $\mathbb{C}$), for any $z \in \mathbb{R}$. Combining the above, we conclude that $\chi(z,.)$ is continuous in H_δ.

Next, we establish that $\chi(z,\tilde{\lambda})$ is the unique (up to a multiplicative factor) square integrable solution to (48). We argue by contradiction. To this end, assume that, for some $\tilde{\lambda} \in H_\delta$, there exists a solution to (48), which is square integrable over $(0,\infty)$ and linearly independent of $\chi(.,\tilde{\lambda})$. Then, this solution, together with $\chi(.,\tilde{\lambda})$, will span the space of all solutions to (48). Hence, every solution is square integrable over $(0,+\infty)$. However, from Eq. (5.3.1) in Sect. 5.3 of [14], we obtain the following representation of φ,

$$\varphi(z,\tilde{\lambda}) = \frac{e^{-iz\sqrt{\tilde{\lambda}}}}{2i\sqrt{\tilde{\lambda}}}\Bigg(-1 + e^{2iz\sqrt{\tilde{\lambda}}} - \int_0^z e^{i(z-x)\sqrt{\tilde{\lambda}}}e^{iz\sqrt{\tilde{\lambda}}}\varphi(x,\tilde{\lambda})\tilde{q}(x)dx \\ + \int_0^z e^{ix\sqrt{\tilde{\lambda}}}\varphi(x,\tilde{\lambda})\tilde{q}(x)dx\Bigg).$$

Using the above representation and Lemma 5.2 in Sect. 5.2 of [14], we obtain the estimate (given in the last equation on page 98 in Sect. 5.3 therein),

$$\big|\varphi(z,\tilde{\lambda})\big| \le \frac{1}{\delta}\exp\left(\frac{\varepsilon}{\delta}\right)\big|e^{-iz\sqrt{\tilde{\lambda}}}\big|.$$

Using the above estimate we obtain, for $z \ge 1$, that

$$\left|e^{2iz\sqrt{\tilde{\lambda}}} - \int_0^z e^{i(z-x)\sqrt{\tilde{\lambda}}}e^{iz\sqrt{\tilde{\lambda}}}\varphi(x,\tilde{\lambda})\tilde{q}(x)dx + \int_0^z e^{ix\sqrt{\tilde{\lambda}}}\varphi(x,\tilde{\lambda})\tilde{q}(x)dx\right| \\ \le e^{-2\sqrt{\delta}} + 2\frac{\varepsilon}{\delta}\exp\left(\frac{\varepsilon}{\delta}\right) < 1,$$

where the last inequality follows from the choice of ε as in the beginning of the proof.

Thus, from the above representation of φ, we conclude that, for all $z \ge 1$,

$$\big|\varphi(z,\tilde{\lambda})\big| \ge \frac{c_1}{2i\sqrt{\tilde{\lambda}}}\big|e^{-iz\sqrt{\tilde{\lambda}}}\big|.$$

However, sending $z \to \infty$, we have $\lim_{z\to\infty} |e^{-iz\sqrt{\tilde{\lambda}}}| = \infty$, which contradicts the square integrability of $\varphi(z,\tilde{\lambda})$ over $(0,\infty)$, and we easily conclude that $\chi(.,\tilde{\lambda})$ is the unique solution to (48) that is square integrable (up to a multiplicative constant).

Next, we show that $\chi(.,\tilde{\lambda})$ does not change the sign. Indeed, notice that because $\tilde{\lambda}+\tilde{q}(z)<0$, for all $\tilde{\lambda}\in H_\delta$ and $z\in\mathbb{R}$, any solution to (48) is concave on the intervals on which it is negative, and it is convex on the intervals on which it is positive.

Fix, now, some $\tilde{\lambda}\in H_\delta$ and assume that there is $z_0\in\mathbb{R}$, such that $\chi(z_0,\tilde{\lambda})=0$. We know that, if $\chi_z(z_0,\tilde{\lambda})=0$, then, due to the uniqueness of a solution to the Eq. (48) with a given pair of initial conditions, the function $\chi(.,\tilde{\lambda})$ has to be identically zero. This, however, is not possible since the function identically equal to zero does not satisfy (49). Therefore, without loss of generality, we assume that $\chi_z(z_0,\tilde{\lambda})<0$. Then, there exist $\varepsilon'>0$ and $z'>z_0$, such that $\chi(z',\tilde{\lambda})=-\varepsilon'<0$ and $\chi(.,\tilde{\lambda})<-\varepsilon'$, in some right neighborhood of z'. This, in turn, implies that $\chi(z,\tilde{\lambda})<-\varepsilon'$ for all $z>z'$, since, otherwise the concavity of the function $\chi(.,\tilde{\lambda})$ in the interval $[z',\inf(z>z'\mid\chi(z,\tilde{\lambda})=-\varepsilon')]$ will be violated. On the other hand, if $\chi(z,\tilde{\lambda})<-\varepsilon'$, for all $z>z'$, we then obtain a contradiction to the square integrability of $\chi(.,\tilde{\lambda})$, for $z\in(0,\infty)$. Similarly, we arrive to a contradiction if we assume that $\chi_z(z_0,\tilde{\lambda})>0$.

Combining the above we deduce that the function $\chi(.,\tilde{\lambda})$ does not change its sign on $\mathbb{R}$. Therefore, the function $\psi^{(1)}(z,\lambda)$, defined as

$$\psi^{(1)}(z,\lambda)=\frac{\chi(\lambda+L_1,z-N)}{\chi(\lambda+L_1,-N)},$$

is well defined for all $\lambda\in(-\infty,-\delta-\bar{\lambda})$ and $z\in\mathbb{R}$. Moreover, it is uniquely characterized as a solution to (48), which is square integrable over $(0,+\infty)$ and satisfies $\psi^{(1)}(0,\lambda)=1$. We have, also, shown that $\psi^{(1)}(z,\lambda)>0$ and, moreover, it is continuous as a function of λ, changing on $(-\infty,-\delta-\bar{\lambda})$, for any $z\in\mathbb{R}$. Notice that, since $\delta>0$ is arbitrary, the above properties extend to all $\lambda\in(-\infty,-\bar{\lambda})$.

Finally, we apply Proposition 3 to conclude that (44) is well defined and it is a solution to (36) with the initial condition (43). □

4.1 The Backward Heat Equation

When $a_1\equiv\frac{1}{2}$, $a_2\equiv 0$, and $a_3\equiv 0$, the Eq. (31) reduces to the well-known backward heat equation, presented earlier in Sect. 2.3 and rewritten below to ease the presentation (we also denote the solution by F to preserve the above notation). As mentioned earlier, its non-negative solutions are completely characterized by the celebrated Widder's theorem given, for completeness, below. Its proof can be found in Chap. XI in [18].

Theorem 2 (Widder's) *Consider the heat equation*

$$F_t+\frac{1}{2}F_{xx}=0 \tag{50}$$

for $(x,t) \in \mathbb{R}\times(0,\infty)$*. A function* $F:\mathbb{R}\times(0,\infty)\to\mathbb{R}^+$ *is a solution to the above if and only if it can be represented as*

$$F(x,t)=\int_{\mathbb{R}} e^{zx-\frac{1}{2}z^2t}\nu(dz) \tag{51}$$

where ν *is a positive finite Borel measure.*

As the above theorem shows, the *only* functions that can serve as initial conditions to (50) are given by a bilateral Laplace transform of the underlying measure ν, namely,

$$F(x,0)=\int_{\mathbb{R}} e^{xz}\nu(dz), \tag{52}$$

given that the above integral converges for any $x\in\mathbb{R}$. We next show how the results proved herein can be used to obtain one direction of the above theorem. Specifically, we show how formula (51) can be obtained[3] by using the construction approach provided in Proposition 4.

Proposition 5 *Let* $F:\mathbb{R}\times[0,\infty)\to\mathbb{R}^+$ *be given by*

$$F(x,t)=\int_{\mathbb{R}} e^{xy-\frac{1}{2}y^2t}\nu(dy),$$

where ν *is a positive Borel measure, such that the above integral is finite for* $t=0$ *and all* $x\in\mathbb{R}$*. Then,* F *is a nonnegative solution of* (50)*, satisfying initial condition* (52).

Proof Rewrite equation (50) for $G(x,t)=F(x,2t)$. Then, we obtain Eq. (36) with $q\equiv 0$. Applying Proposition 4 with $L_1=L_2=\bar{\lambda}=0$, we conclude that the corresponding solutions $\psi^{(1)}$ and $\psi^{(2)}$ are given, respectively, by

$$\psi^{(1)}(x,\lambda)=e^{-ix\sqrt{\lambda}} \quad\text{and}\quad \psi^{(2)}(x,\lambda)=e^{ix\sqrt{\lambda}}.$$

Then, Eq. (36) has a nonnegative solution, say $G(x,t)$, for any initial condition of the form

$$G_0(x)=\int_{-\infty}^{0} e^{-ix\sqrt{\lambda}}\mu_1(d\lambda)+\int_{-\infty}^{0} e^{ix\sqrt{\lambda}}\mu_2(d\lambda),$$

where μ_1 and μ_2 are Borel measures on $(-\infty,0)$, satisfying the integrability conditions in parts (i) and (ii) of Proposition 4, respectively.

[3]Of course, one can easily verify that (51) indeed solves (50). The aim is, however, to develop a general approach for equations of the general form (36).

Notice that we, then, have

$$
\begin{aligned}
G(x,t) &= \int_{-\infty}^{0} e^{-ix\sqrt{\lambda}+t\lambda}\mu_1(d\lambda) + \int_{-\infty}^{0} e^{ix\sqrt{\lambda}+t\lambda}\mu_2(d\lambda) \\
&= \int_{0}^{+\infty} e^{xs-ts^2}\tilde{\mu}_1(ds) + \int_{0}^{+\infty} e^{-xs-ts^2}\tilde{\mu}_2(ds) \\
&= \int_{\mathbb{R}} e^{xs-ts^2}\big(\tilde{\mu}_1(ds)\mathbf{1}_{\mathbb{R}_+}(s) + \tilde{\mu}_2\big(d(-s)\big)\mathbf{1}_{\mathbb{R}_-}(s)\big),
\end{aligned}
$$

where

$$
\tilde{\mu}_1 = \mu_1 \circ m^{-1} \quad \text{and} \quad \tilde{\mu}_2 = \mu_2 \circ m^{-1},
$$

with $m(s) = \sqrt{-s}$.

It is easy to see that μ_1 and μ_2 satisfy the corresponding integrability conditions if and only if the above integral is finite for $t = 0$ and all $x \in \mathbb{R}$.

Reverting to the original variables, we obtain $F(x,t) = G(x, t/2)$, and note that we have proved the statement of the proposition for all measures ν, which satisfy the appropriate integrability conditions and have no mass at zero. Finally, we notice that if ν is a Dirac delta-function at zero, then the resulting function F is identically equal to one, and, therefore, solves (50). Using the linearity of (50), we conclude the proof. □

5 Examples

In this section we present two examples of processes satisfying the forward SPDE (7). For this, we apply the methodology developed in the previous section and the form of the candidate solutions. We do not, however, derive or study the associated optimal policy and optimal wealth processes. Such questions will be presented in a future paper in which a more general class of solutions will be considered (see [13]).

5.1 Mean Reverting Stochastic Volatility

We assume that the coefficients in (1) and (2) take, respectively, the form

$$
\mu(y) = \mu \quad \text{and} \quad \sigma(y) = (\mu - r)e^{-y} \tag{53}
$$

and

$$
b(y) = c_1 e^y + c_2 \quad \text{and} \quad d(y) = d, \tag{54}
$$

for $y \in \mathbb{R}$, and c_1, c_2, d, μ and r constants with $d > 0$ and $c_1 < 0$. An extra assumption on the ratio $|c_1|/d$ will be imposed in the sequel. For the other constants, we assume, without loss of generality, that $\mu > r > 0$ and $c_2 \geq 0$.

Under (53) and (54), Eqs. (1) and (2) become

$$dS_t = S_t\mu dt + S_t(\mu - r)e^{-Y_t}dW_t^1 \tag{55}$$

and

$$dY_t = \big(c_1e^{Y_t} + c_2\big)dt + d\big(\rho dW_t^1 + \sqrt{1-\rho^2}dW_t^2\big), \tag{56}$$

with $S_0 > 0$ and $Y_0 \in \mathbb{R}$. The above choice of the stochastic factor corresponds to a stock volatility

$$N_t = (\mu - r)e^{-Y_t} \tag{57}$$

which satisfies

$$dN_t = \left(|c_1|(\mu - r) + \left(\frac{d^2}{2} - c_2\right)N_t\right)dt - dN_t dW_t, \tag{58}$$

and, hence, if c_2 is large enough, exhibits mean reverting behavior. One can easily show that the above equation, and, consequently, the system consisting of (55) and (56), has a unique strong solution.

Next, we use the change of variables introduced at the beginning of Sect. 4, in order to derive a canonical form of Eq. (27). Recall that in this case, we have

$$a_1(y) = \frac{1}{2}d^2, \qquad a_2(y) = e^y\left(c_1 + \rho d\frac{\gamma}{1-\gamma}\right) + c_2, \qquad a_3(y) = \frac{1}{2\delta}\frac{\gamma}{1-\gamma}e^{2y}.$$

To this end, rescaling time, from t to $d^2t/2$, and applying the change of variables described at the beginning of Sect. 4, we get that the function $g : \mathbb{R}\times[0,\infty) \to \mathbb{R}^+$ defined by

$$g(y,t) = v\left(y, \frac{2}{d^2}t\right)\exp\left(C_2 + \frac{c_2}{d^2}y - C_2e^y\right),$$

with v introduced in Sect. 3.2 and the constants C_1 and C_2 as

$$C_1 = \frac{1}{d^2}\left(\frac{c_1^2}{d^2} + \rho\frac{2c_1}{d}\frac{\gamma}{1-\gamma} - \frac{\gamma}{1-\gamma}\right) \quad \text{and} \quad C_2 = \frac{1}{d}\left(\frac{|c_1|}{d} - \rho\frac{\gamma}{1-\gamma}\right), \tag{59}$$

needs to satisfy the linear equation

$$g_t + g_{yy} + q(y)g = 0, \tag{60}$$

with initial condition

$$g(y,0) = \exp\left(C_2 + \frac{c_2}{d^2}y - C_2e^y\right)\big(K(y)\big)^{1/\delta}, \tag{61}$$

where the distortion power δ is as in (26) and the potential term is given by

$$q(y) = -C_1e^{2y} + C_2\left(1 + \frac{2c_2}{d^2}\right)e^y - \frac{c_2^2}{d^4}. \tag{62}$$

It is further assumed that $|c_1|/d$ is large enough, so that both constants $C_1, C_2 > 0$.

We recall that, according to Proposition 3, one needs to represent the above initial condition as an integral over λ's of the nonnegative solutions to the corresponding Sturm-Liouville equation

$$\psi_{yy}(y,\lambda) + \big(\lambda + q(y)\big)\psi(\lambda, y) = 0, \tag{63}$$

with $q(y)$ given in (62). We, also, remind the reader that, herein, we are not looking for the entire class of solutions, but we seek to construct merely one solution. To this end, we first observe that the function

$$\varphi(y) = \exp\left(C_2 + \frac{c_2}{d^2}y - \sqrt{C_1}e^y\right),$$

satisfies (63) with $\lambda = 0$. Applying Proposition 3 with $\mathscr{P}$ being a singleton and $\Lambda = \{0\}$, we easily obtain that the same function is a solution for $t > 0$, i.e. the function $g : \mathbb{R}\times[0,\infty) \to \mathbb{R}^+$ given by

$$g(y,t) = \exp\left(C_2 + \frac{c_2}{d^2}y - \sqrt{C_1}e^y\right)$$

solves (60).

Therefore, if we choose the factor $K(y)$ to be

$$K(y) = \exp\big(\delta(C_2 - \sqrt{C_1})e^y\big),$$

we deduce that $g(y,0) = \varphi(y)$. Hence,

$$v(y,t) = \exp\big((C_2 - \sqrt{C_1})e^y\big),$$

and we easily conclude.

We summarize the above findings below.

Proposition 6 *Assume that the stock and the stochastic factor solve* (55) *and* (56). *Also, assume that the aforementioned assumptions on the involved coefficients hold and that the distortion power δ is as in* (26).

Define the process $a(x,t)$ by

$$a(x,t) = \left(\frac{x^\gamma}{\gamma}\rho Z_t, \frac{x^\gamma}{\gamma}\sqrt{1-\rho^2 Z_t}\right) \tag{64}$$

where

$$Z_t = d\delta(C_2 - \sqrt{C_1})\exp\big(Y_t + \delta(C_2 - \sqrt{C_1})(e^{Y_t} - e^{Y_0})\big) \tag{65}$$

and the constants C_1 and C_2 are as in (59).

Moreover, let the initial condition $u_0 : \mathbb{R}^+ \to \mathbb{R}^+$ *be*

$$u_0(x) = \frac{x^\gamma}{\gamma}.$$

Then, the process $U(x,t)$ *given by*

$$U(x,t) = \frac{x^\gamma}{\gamma} \exp\big(\delta(C_2 - \sqrt{C_1})(e^{Y_t} - e^{Y_0})\big) \tag{66}$$

solves the forward performance SPDE (7) *with the above performance volatility process* $a(x,t)$ *and initial condition* $U(x,0) = u_0(x)$.

Next, we study the behavior of the forward investment performance process as the forward volatility vanishes. This occurs when the coefficient $d \to 0$.

Proposition 7 *Let* $U^{(d)}(x,t)$ *be the forward investment performance process given in* (66). *Then, for each* $t > 0$,

(i) *the performance volatility process* $a(x,t)$ *(cf.* (64)*) satisfies a.s. for all* $x \geq 0$,

$$\lim_{d \to 0} a(x,t) = 0, \tag{67}$$

and

(ii) *the forward investment performance process satisfies a.s. for all* $x \geq 0$,

$$\lim_{d \to 0} U^{(d)}(x,t) = \frac{x^\gamma}{\gamma} \exp\left(-\frac{\gamma}{2c_1(1-\gamma)}\left(e^{Y_t^{(0)}} - e^{Y_0}\right)\right), \tag{68}$$

where $Y_t^{(0)}$ *is the solution to the deterministic problem*

$$dY_t^{(0)} = \left(c_1 e^{Y_t^{(0)}} + c_2\right) dt$$

with $Y_0^{(0)} = Y_0$.

Proof We first observe, using (59), that

$$\delta(C_2 - \sqrt{C_1}) = \frac{\gamma}{1-\gamma}\left(-c_1 - \rho\frac{d\gamma}{1-\gamma} + \sqrt{c_1^2 + 2\rho\frac{c_1 d\gamma}{(1-\gamma)} - \frac{\gamma d^2}{1-\gamma}}\right)^{-1}$$

and, in turn,

$$\lim_{d \to 0} \delta(C_2 - \sqrt{C_1}) = -\frac{\gamma}{2c_1(1-\gamma)} > 0.$$

Next, we recall that the process $N_t^{(d)}$, $t \geq 0$, defined in (57) solves the affine SDE (58), with $N_0^{(d)} = (\mu - r)e^{-Y_0}$. On the other hand, the solution of this equation can be represented explicitly (see, for example, Sect. 5.6 in [5]). From this explicit representation, it is easy to deduce that almost surely, for all $t > 0$,

$$\lim_{d\to 0} N_t^{(d)} = N_t^{(0)} = (\mu - r)e^{-Y_t^{(0)}}.$$

We easily obtain that $\lim_{d\to 0} Z_t = 0$, and using (64) and passing to the limit we obtain (67). Assertion (68) follows easily. □

5.2 *Heston-Type Stochastic Volatility*

We choose the model coefficients

$$\mu(y) = \mu \quad \text{and} \quad \sigma(y) = (\mu - r)\sqrt{y}$$

and

$$b(y) = c_1 y + c_2 \quad \text{and} \quad d(y) = d\sqrt{y},$$

for $y \in \mathbb{R}^+$. It is assumed that c_1, c_2, d, μ and r are constants, such that $r \geq 0$ and $c_2, d > 0$. In addition, without loss of generality, we assume that $\mu > r$. In order to prevent the process Y_t, $t \geq 0$, from hitting zero, we also assume that $d^2 < 2c_2$. An additional assumption on c_2/d will be made in the sequel.

Under the above assumptions, the stock and the stochastic factor processes (cf. (1) and (2)) satisfy

$$dS_t = S_t \mu dt + S_t(\mu - r)\sqrt{Y_t} dW_t^1 \tag{69}$$

and

$$dY_t = (c_1 Y_y + c_2)dt + d\sqrt{Y_t}\left(\rho dW_t^1 + \sqrt{1-\rho^2 dW_t^2}\right), \tag{70}$$

with $S_0, Y_0 > 0$. It is well known that the above system has a unique strong solution.

According to the methodology developed in Sect. 4, we perform the following change of variables in order to bring Eq. (27) in its canonical form. Specifically, in the notation of Sect. 4, we obtain

$$Z(y) = \frac{2\sqrt{2}}{d}\sqrt{y} \quad \text{and} \quad X(z) = Z^{-1}(z) = \frac{d^2}{8}z^2, \tag{71}$$

and introduce the function $g : \mathbb{R}^+ \times (0, \infty) \to \mathbb{R}^+$ given by

$$g(t, y) = \frac{1}{\sqrt{y}} \exp\left(\frac{c_1}{d^2}\left(y^2 \frac{d^2}{8} - 1\right) + C_2 \log\left(y^2 \frac{d^2}{8}\right)\right) v\left(\frac{\sigma^2}{8} y^2, t\right), \tag{72}$$

where v is as in (27), and the constants C_1 and C_2 are given by

$$C_1 = \left(\frac{c_2^2}{2d^2} + \frac{c_2 \rho \gamma}{d(1-\gamma)} + \frac{3d^2}{32} - \frac{c_2}{2} - \frac{\gamma}{2(1-\gamma)}(1 + d\rho)\right)\frac{8}{d^2}$$

and

$$C_2 = \frac{c_2}{d^2} + \frac{\rho\gamma}{d(1-\gamma)}.$$

We, also, conclude that g has to solve

$$g_t + g_{yy} + q(y)g = 0 \tag{73}$$

with initial condition

$$g(y,0) = \frac{1}{\sqrt{y}} \exp\left(\frac{c_1}{d^2}\left(y^2\frac{d^2}{8} - 1\right) + C_2 \log\left(y^2\frac{d^2}{8}\right)\right)\left(K\left(\frac{d^2}{8}y^2\right)\right)^{1/\delta}, \tag{74}$$

where the coefficient $q(y)$ is given by

$$q(y) = -\frac{c_1^2}{16}y^2 - C_1\frac{1}{y^2} - c_1 C_2,$$

We assume that c_2/d is large enough, so that $C_1 > -1/4$.

Elementary calculations yield that the functions $\psi^{(i)} : \mathbb{R}^+ \to \mathbb{R}^+$, $i = 1, 2$, given by

$$\psi^{(1)}(y) = e^{y^2 c_1/8} y^{1/2+\sqrt{C_1+1/4}} \quad \text{and} \quad \psi^{(2)}(y) = e^{y^2 c_1/8} y^{1/2-\sqrt{C_1+1/4}}$$

satisfy the corresponding Sturm–Liouville equation,

$$\frac{\partial}{\partial y^2}\psi(\lambda, y) + \big(\lambda + q(y)\big)\psi(\lambda, y) = 0, \tag{75}$$

with respective values λ_1 and λ_2 given by

$$\lambda_1 = \frac{c_1 c_2}{d^2} + \frac{c_1\rho\gamma}{d(1-\gamma)} - \frac{c_1(1+\sqrt{C_1+1/4})}{2}$$

and

$$\lambda_2 = \frac{c_1 c_2}{d^2} + \frac{c_1\rho\gamma}{d(1-\gamma)} - \frac{c_1(1-\sqrt{C_1+1/4})}{2}.$$

Next, we choose the factor $K : (0,\infty) \to (0,\infty)$ as

$$\begin{aligned} K(y) = {} & \left(\frac{2\sqrt{2}}{d}\right)^{\delta} y^{\delta(\frac{1}{2}-C_2)} \exp\left(\frac{c_1\delta}{d^2}\right) \\ & \times \left(k_1\left(\frac{2\sqrt{2}}{d}\right)^{\sqrt{C_1+1/4}} y^{\sqrt{C_1+1/4}/2} \right. \\ & \left. + k_2\left(\frac{2\sqrt{2}}{d}\right)^{-\sqrt{C_1+1/4}} y^{-\sqrt{C_1+1/4}/2}\right)^{\delta}, \end{aligned} \tag{76}$$

for any constants $k_1, k_2 \in [0, \infty)$. Then, the solution to the linear equation (73) is given by

$$g(y,t) = \sqrt{y} e^{y^2 c_1/8} \left(k_1 y^{\sqrt{C_1+1/4}} e^{\lambda_1 t} + k_2 y^{-\sqrt{C_1+1/4}} e^{\lambda_2 t}\right).$$

Consequently, we deduce that v is given by

$$\begin{aligned} v(y,t) &= \frac{2\sqrt{2}}{d} \exp\left(\frac{c_1}{d^2}\right) y^{1/2-C_2} \\ &\quad \times \left(k_1 \left(\frac{2\sqrt{2}}{d}\right)^{\sqrt{C_1+1/4}} y^{\sqrt{C_1+1/4}/2} e^{\lambda_1 t}\right. \\ &\quad \left. + k_2 \left(\frac{2\sqrt{2}}{d}\right)^{-\sqrt{C_1+1/4}} y^{-\sqrt{C_1+1/4}/2} e^{\lambda_2 t}\right). \end{aligned} \tag{77}$$

Summarizing the above, we have the following result.

Proposition 8 *Assume that the stock and the stochastic factor solve* (69) *and* (70). *Also, assume that the aforementioned assumptions on the involved coefficients hold and that the distortion power δ is as in* (26).

Define the process $a(x,t)$ by

$$a(x,t) = \left(\frac{x^\gamma}{\gamma} \rho Z_t, \frac{x^\gamma}{\gamma} \sqrt{1-\rho^2} Z_t\right) \tag{78}$$

where

$$Z_t = d\delta \sqrt{Y_t} \frac{v_y(Y_t,t)}{v(Y_0,0)} \left(\frac{v(Y_t,t)}{v(Y_0,0)}\right)^{\delta-1} \tag{79}$$

with $v : (0,+\infty) \times [0,+\infty) \to \mathbb{R}^+$ given by (77) *above.*

Moreover, consider the initial condition $u_0 : \mathbb{R}^+ \to \mathbb{R}^+$ given by

$$u_0(x) = \frac{x^\gamma}{\gamma}.$$

Then, the process

$$U_t(x) = \frac{x^\gamma}{\gamma} \left(\frac{v(Y_t,t)}{v(Y_0,0)}\right)^\delta, \tag{80}$$

satisfies the SPDE (7) *with the above performance volatility process $a(x,t)$ and initial condition $U(x,0) = u_0(x)$.*

Next, we study the behavior of the forward investment performance process in (80) as its volatility process $a(x,t)$ vanishes. For this, we will send the parameter $d \to 0$. Notice, however, that in the present case, if none of k_1 or k_2 is equal to zero, the particular choice of their values will affect the forward performance process. Therefore, for the sake of simplicity we assume that $k_2 = 0$.

Proposition 9 *Let $U^{(d)}(x,t)$ be the forward investment performance process given in* (80), *with $k_2 = 0$. Then, for each $t > 0$,*

(i) *the performance volatility process $a(x,t)$ (cf.* (78)) *satisfies a.s for all $x \geq 0$,*

$$\lim_{d\to 0} a(x,t) = 0,$$

(ii) *the forward performance process satisfies a.s. for all $x \geq 0$*

$$\lim_{d\to 0} U^{(d)}(x,t) = \frac{x^\gamma}{\gamma}\left(\frac{Y_t^{(0)}}{Y_0}e^{-c_1 t}\right)^{\delta(1-\frac{1}{2c_2}(\frac{\rho\gamma}{1-\gamma})^2+\frac{1}{2c_2}\frac{\gamma}{1-\gamma})},$$

where $Y_t^{(0)}$ is the solution to the deterministic problem

$$dY_t^{(0)} = \left(c_1 Y_t^{(0)} + c_2\right)dt,$$

with $Y_0^{(0)} = Y_0$.

Proof First, we make use of the assumption $c_2 > 0$ to obtain that for small enough $d > 0$ the following calculations are valid:

$$\begin{aligned}
A(d) &= \frac{1}{2}\sqrt{C_1 + 1/4} - C_2 \\
&= C_2\left(\sqrt{1 - \frac{(\frac{\rho\gamma}{1-\gamma})^2 + \frac{d^2}{4} - c_2 - \frac{\gamma}{1-\gamma}\rho d}{d^2 C_2^2}} - 1\right) \\
&= -\frac{(\frac{\rho\gamma}{1-\gamma})^2 + \frac{d^2}{4} - c_2 - \frac{\gamma(1+\rho d)}{1-\gamma}}{(c_2 + \frac{\gamma}{1-\gamma}\rho d)\sqrt{1 - \frac{(\frac{\rho\gamma}{1-\gamma})^2 + \frac{d^2}{4} - c_2 - \frac{\gamma}{1-\gamma}\rho d}{(\frac{c_2}{d} + \frac{\rho\gamma}{1-\gamma})^2}}}.
\end{aligned}$$

We, then, easily deduce that

$$\lim_{d\to 0} A(d) = \frac{1}{2} - \frac{1}{2c_2}\left(\frac{\rho\gamma}{1-\gamma}\right)^2 + \frac{1}{2c_2}\frac{\gamma}{1-\gamma}.$$

Finally, we note that because $c_1 > 0$, we have

$$\lim_{d\to 0} \lambda_1(d) = -c_1 + \frac{c_1}{2c_2}\left(\frac{\rho\gamma}{1-\gamma}\right)^2 - \frac{c_1}{2c_2}\frac{\gamma}{1-\gamma},$$

and therefore,

$$\lim \frac{v^{(d)}(y,t)}{v^{(d)}(Y_0,0)} = \left(\frac{y}{Y_0}e^{-c_1 t}\right)^{1-\frac{1}{2c_2}(\frac{\rho\gamma}{1-\gamma})^2+\frac{1}{2c_2}\frac{\gamma}{1-\gamma}}.$$

Using standard results for the CIR process, we deduce that there exists a modification of the family of processes $\{(Y_t^{(d)})_{t\geq 0}\}$, solving (70) for each $d > 0$, such that a.s for any $t \geq 0$,

$$\lim_{d\to 0} Y_t^{(d)} = Y_t^{(0)} = \left(\frac{c_2}{c_1} + Y_0\right)e^{c_1 t} - \frac{c_2}{c_1}.$$

We easily conclude. □

References

1. Barrier, F., Rogers, L.C., Tehranchi, M.: 2009, A characterization of forward utility functions. Preprint. http://www.statslab.cam.ac.uk/~mike/papers/forward-utilities.pdf
2. Carr, P., Nadtochiy, S.: Static hedging under time-homogeneous diffusions. SIAM J. Financ. Math. **2**(1), 794–838 (2011)
3. El Karoui, N., M'Rad, M.: 2010, Stochastic utilities with a given optimal portfolio: approach by stochastic flows. Preprint. arXiv:1004.5192
4. Itô, K., McKean, H.P. Jr: Diffusion Processes and Their Sample Paths (Classics in Mathematics), 2nd edn. Springer, Berlin (1974)
5. Karatzas, I., Shreve, S.: Brownian Motion and Stochastic Calculus, 2nd edn. Springer, Berlin (1998)
6. Merton, R.: Lifetime portfolio selection under uncertainty: the continuous-time case. Rev. Econ. Stat. **51**, 247–257 (1969)
7. Merton, R.: Optimum consumption and portfolio rules in a continuous-time model. J. Econ. Theory **3**, 373–413 (1971)
8. Musiela, M., Zariphopoulou, T.: Portfolio choice under dynamic investment performance criteria. Quant. Finance **9**, 161–170 (2009)
9. Musiela, M., Zariphopoulou, T.: Portfolio choice under space-time monotone performance criteria. SIAM J. Financ. Math. **1**, 326–365 (2010)
10. Musiela, M., Zariphopoulou, T.: Stochastic partial differential equations in portfolio choice. In: Chiarella, C., Novikov, A. (eds.) Contemporary Quantitative Finance, pp. 195–216. Springer, Berlin (2010)
11. Linetski, V., Davydov, D.: Pricing options on scalar diffusions: an eigenfunction expansion approach. Oper. Res. **51**(2), 185–209 (2003)
12. Nadtochiy, S., Tehranchi, M.: Optimal investment for all time horizons and Martin boundary of space-time diffusions (2013). arXiv:1308.2254
13. Nadtochiy, S., Zariphopoulou, T.: The SPDE for the forward investment performance process (2010). Work in progress
14. Titchmarsh, E.C.: In: Eigenfunction Expansions Associated with Second-Order Differential Equations, Clarendon, Oxford (1946)
15. Zariphopoulou, T.: A solution approach to valuation of unhedgeable risks. Finance Stoch. **5**, 61–82 (2001)
16. Zariphopoulou, T.: Optimal asset allocation in a stochastic factor model—an overview and open problems. Adv. Financ. Model. Radon Ser. Comput. Appl. Math. **8**, 427–453 (2009)
17. Zitkovic, G.: A dual characterization of self-generation and exponential forward performances. Ann. Appl. Probab. **19**(6), 2176–2210 (2008)
18. Widder, D.V.: The Heat Equation. Academic Press, San Diego (1975)

Solution of Optimal Stopping Problem Based on a Modification of Payoff Function

Ernst Presman

Abstract An optimal stopping problem of a Markov process with infinite horizon is considered. For the case of discrete time and finite number m of states Sonin proposed an algorithm which allows to find the value function and the stopping set in no more than $2(m-1)$ steps. The algorithm is based on a *modification of a Markov chain on each step*, related to the elimination of the states which definitely belong to the continuation set. To solve the problem with arbitrary state space and to have possibility of a generalization to a continuous time, the procedure was modified in Presman (Stochastics 83(4–6):467–475, 2011). The modified procedure was based on a sequential *modification of the payoff function* for *the same chain* in such a way that the value function is the same for both problems and the modified payoff function is greater than the initial one on some set and is equal to it on the complement. In this paper, we give some examples showing that the procedure can be generalized to continuous time.

Keywords Markov chain · Markov process · One-dimensional diffusion · Optimal stopping · Elimination algorithm

Mathematics Subject Classification (2010) 91B28 · 60G40 · 34K10

1 Discrete Time Case

We consider a time homogeneous Markov chain $Z = (Z_n)_{n \geq 0}$ defined on a filtered probability space $(\Omega, \mathscr{F}, (\mathscr{F}_n)_{n \geq 0}, \mathbf{P}_z)$ and taking values in a measurable space $(X, \mathscr{B})$. It is assumed that the chain Z starts at z under $\mathbf{P}_z$ for $z \in X$. It is also assumed that the mapping $z \mapsto \mathbf{P}_z(F)$ is measurable for each $F \in \mathscr{F}$. Denote by $\mathscr{P}$ the transition operator of Z, so that $E_z f(Z_1) = \mathscr{P} f(z)$ for any f, such that the corresponding expectation exists.

E. Presman (✉)
Central Economics and Mathematics Institute (CEMI), Russian Academy of Sciences (RAS), 47 Nakhimovsky prospect, Moscow, 117428, Russia
e-mail: presman@cemi.rssi.ru

Y. Kabanov et al. (eds.), *Inspired by Finance*, DOI 10.1007/978-3-319-02069-3_23,

A number β, $0 < \beta \leq 1$, and measurable payoff function $g(z)$ and cost function $c(z)$ are given. Stopping times are considered with respect to a sequence of σ-algebras $\mathscr{F}_n$, $n \geq 0$. Here β is a discount coefficient, $g(z)$ is a reward for stopping at point z, and $c(z)$ is a fee for the observation (both functions can take positive and negative values). The problem of optimal stopping consists, first, in finding the value function

$$V(z) = \sup_{\tau} V^{\tau}(z), \quad \text{where } V^{\tau}(z) = E_z\left[g(Z_\tau)\beta^{\tau} - \sum_{k=0}^{\tau-1} c(Z_k)\beta^k\right],$$

and the supremum is taken over all stopping times, and, second, in finding an optimal stopping time, i.e. the stopping time where the supremum is achieved.

It is well known that the case $0 < \beta < 1$ can be reduced to the case $\beta = 1$ by introducing an absorbing state, which we shall denote by e. The probability of transition to e from any state of X is equal to $1 - \beta$ and the new transition probabilities between states from X are equal to the old ones multiplied by β (see, for example, [5]). Then

$$E_z\left[g(Z_\tau)\beta^{\tau} - \sum_{k=0}^{\tau-1} c(Z_k)\beta^k\right] = \tilde{E}_z\left[g(Z_\tau) - \sum_{k=0}^{\tau-1} c(Z_k)\right],$$

where $\tilde{E}_z$ corresponds to the new transition probabilities.

Thus, in what follows we assume that $\beta = 1$.

Let us define an operator T as follows:

$$Tf(z) = -c(z) + \mathscr{P}f(z).$$

The operator T is called the reward operator.

It is well known that under natural assumptions (see, for example, [4], p. 12, condition (2.1.1)) the following statement holds (see [4], Theorem 1.11, Corollary 1.12 and Sect. 11 of Chap. 1; or [9], Sect. 14):

Theorem 1 (a) *The value function $V(z)$ is the minimal solution of the Bellman (optimality) equation*

$$V(z) = \max\big[g(z), TV(z)\big]. \tag{1}$$

(b) *Let $\tau^* = \inf\{n \geq 0 : \ Z_n \in D^*\}$ where the set $D^* = \{z : V(z) = g(z)\}$. If $\mathbf{P}_z[\tau^* < \infty] = 1$ for all $z \in X$, then the stopping time τ^* is an optimal one and $\tau^* \leq \tau'$ $\mathbf{P}_z$-a.s. for any z and any optimal stopping time τ'.*

(c) *The sequence $\tilde{V}_0(z) = g(z)$, $\tilde{V}_{k+1}(z) = \max[g(z), T\tilde{V}_k(z)]$ is nondecreasing and converges to $V(z)$.*

The set D^* is called the stopping set and the set $C^* = X \setminus D^* = \{z : V(z) > g(z)\}$ is called the continuation set.

It is said often that statement c) offers a constructive method for finding the value function $V(z)$ (see, for example, [4], p. 19). Nevertheless, if $P_z[\tau^* > a] > 0$ for some $z \in X$ and any $a < \infty$ then $\tilde{V}_k(z) \le \tilde{V}_{k+1}(z) < V(z)$ for all k.

If Z_n takes only a finite number m of values then Eq. (1) can be solved by linear programming (see, for example, [3]). But under such an approach the probabilistic meaning is lost and it is not clear how to generalize such an approach even to the countable case. For the case of a finite number m of states Sonin (see [10–13]) proposed an algorithm, which allows to find the value function and the stopping set in no more than $2(m-1)$ steps. The idea underlying this algorithm is as follows. Those points where the expected reward for doing one more step (which equals to $Tg(z)$) is larger than the reward for immediate stopping (which equals to $g(z)$) belong definitely to the continuation set. Therefore, the set C of such points can be eliminated and we can consider a new chain, with the new reduced state space $X \setminus C$ and new transition probabilities. These probabilities coincide with the distribution of the initial chain at the time of the first return to the new state space. They can be simply recalculated from the old ones. In the case of a finite number of states, after a finite number of steps we obtain the new chain and the new state space for which the reward for stopping—which equals the payoff function—is greater than or equal to the expected reward for doing one more step for all points. In such situations, the stopping set coincides with the final state space and the value function coincides with the reward for instant stopping. After that the value functions corresponding to the previous chains can be restored sequentially. The possibilities of generalization to the countable case in some situations were discussed in [13].

The following procedure was proposed in [6] for having the possibility for generalizing the approach to an arbitrary state space and to continuous time. Instead of modifying the chain, one needs at each step to modify the payoff function, changing it on the set C to the expected reward at the time of the first exit from C. The modified payoff function is greater than or equal to the initial one and the value function is the same for both problems. Sequentially repeating this step, one obtains an increasing sequence of sets, and the corresponding sequence of the modified payoff functions which converges nondecreasingly to the value function of the initial problem. In the case of a finite number of states the sequence C_k remains the same as in the Sonin's algorithm.

For simplicity of exposition, it was assumed in [6] that the following condition holds:

A. Functions $g(z)$ and $c(z)$ are bounded and there exists an absorbing state $e \in X$ and numbers $n_0 > 0,\ b < 1$, such that $\mathbf{P}_z\{Z_{n_0} = e\} \ge b > 0$ for any $z \in X$, and $g(e) = c(e) = 0$.

Remark 1 Condition **A** implies that the value function $V(z)$ is finite, $e \in D^*$ and therefore $\mathbf{P}_z[\tau^* < \infty] = 1$ for all $z \in X$. Hence Theorem 1 is applicable. The possibility of relaxing the condition **A** is discussed at the end of this section.

We consider sets $C \subset X$ and $D \subset X$ with or without indexes assuming that $D = X \setminus C$, and $C = X \setminus D$. Let I_C be an operator for multiplication by an indicator function of the set C, $I = I_X$.

Let $\tau_D,\ 0 \le \tau_D \le \infty$, be a random time when Z first time visits D. If $z \in D$ then $\tau_D = 0$. Denote

$$g_C(z) = E_z\left(g(Z_{\tau_D}) - \sum_{k=0}^{\tau_D - 1} c(Z_k)\right) I_{\{\tau_D < \infty\}}. \tag{2}$$

Main Lemma (see [6]) (a) *If $z \in C$ then $Tg_C(z) = g_C(z)$.*

(b) *If $C \subseteq \{z : Tg(z) \ge g(z)\}$ and condition* **A** *is fulfilled then $g(z) \le g_C(z) < \infty$ for $z \in C$ and $g_C(z) > g(z)$ if $z \in C$ and $Tg(z) > g(z)$.*

A version of this lemma was proved in [7].

Consider for the chain Z an optimal stopping problem with payoff function $g_C(z)$ and cost function $c(z)$.

Lemma 1 (see [6]) *Suppose that $C \subseteq \{z : Tg(z) \ge g(z)\}$, $C \subseteq C^*$ and condition* **A** *is fulfilled. Then the optimal stopping problem of the chain Z with payoff function $g_C(z)$ and cost function $c(z)$ has the same value function as the initial problem.*

Let us define now a sequence of sets C_k and functions $g_k(z)$, $k \ge 0$, as follows: $C_0 = \emptyset,\ g_0(z) = g(z)$, and if $C_l,\ g_l(z)$ are defined for $0 \le l \le k,\ k \ge 0$, then

$$C_{k+1} = C_k \bigcup \{z : g_k(z) < Tg_k(z)\}, \qquad g_{k+1}(z) = g_{k,C_{k+1}}(z),$$

where $g_{k,C_{k+1}(z)}$ is constructed from $g_k(z)$ using formula (2) as the expected reward at the time of the first visit to D_{k+1}. Note that by the strong Markov property and monotonicity of the sequence C_k, $k \ge 0$, the function $g_{k+1}(z)$ can be constructed using $g(z)$ instead of $g_k(z)$, so that $g_{k+1}(z) = g_{C_{k+1}}(z)$. Note also that if there exists k_0, such that $\{z : g_{k_0}(z) < Tg_{k_0}(z)\} = \emptyset$, then $g_k(z) = g_{k_0}(z)$, $C_k = C_{k_0}$ for $k \ge k_0$.

Now we can formulate the main theorem from [6].

Theorem 2 *If condition* **A** *is fulfilled then the sequence C_k, $k \ge 0$, does not decrease and tends to the continuation set C^* in the problem of optimal stopping of the Markov chain Z with payoff function $g(z)$ and cost function $c(z)$, and the sequence $g_k(z)$, $k \ge 0$, does not decrease and tends to the corresponding value function $V(z)$.*

Remark 2 Let X consists of $m < \infty$ states. As a rule $P_z[\tau^* > a] > 0$ for any $a < \infty$ at least for some $z \in X$ and one needs an infinite number of steps to obtain the value function using the constructive method. The proposed procedure ensures that the value function will be found in no more than for $(m-1)$ steps.

Remark 3 The statement of Theorem 2 is valid in essentially more general situation than under condition **A**. It seems that if the value function is finite and the probability to reach the stopping set is one for each point of X, then the result is true. The author plans to investigate this question in a future work.

2 Some Examples for One-Dimensional Diffusion

In this section, we shall consider the case of continuous time. The general theory of the optimal stopping and methods of constructing the value function can be found, for example, in [4], [8], [2]. The goal of this section is to demonstrate by some examples how the proposed procedure of a modification of the reward function can be generalized to the case of one-dimensional diffusion ξ_t with functional $E_z[g(\xi_\tau)]$. The idea is the same as in the discrete time case.

For any open interval C let denote by $g_C(z)$ the expected reward at the time of the first visit in the complement of C. Then $g_C(z) = g(z)$ for $z \in D = X \setminus C$ and $Lg_C(z) = 0$ for $z \in C$, where L is a differential operator corresponding to the diffusion (see, for example, [4] Sects. 4.5, 7.1). The operator L plays in continuous time the role of the operator $T - I$ in discrete time.

Lemma 2 *Suppose we found C such that $g_C(z) > g(z)$ on C. Then $C \in C^*$, where C^* is the continuation set, and the problem of optimal stopping with the payoff function $g_C(z)$ has the same value function as the initial one.*

Proof The proof is the same as the proof of Lemma 1. Indeed, the value function corresponding to $g_C(z)$ is greater than or equal to the value function corresponding to $g(z)$ since $g_C(z) \geq g(z)$. On the other hand for each τ we can define

$$\tau' := \begin{cases} \tau & \text{if } \xi_\tau \in D\,, \\ \inf[s : s > \tau,\ \ \xi_s \in D] & \text{if } \xi_\tau \in C\,. \end{cases}$$

Then $E_z[g(\xi_{\tau'})] = E_z[g_C(\xi_\tau)]$ and hence the value functions coincide. □

For the new payoff function we can try similarly to find intervals which definitely belong to C^*. Repeating this procedure we obtain finally a set $\tilde{C}$ and the modified payoff function $g_{\tilde{C}}(z)$ such that there is no point in $\tilde{D} = X \setminus \tilde{C}$ such that in the neighborhood of this point we can increase the reward. In this situation $\tilde{C} = C^*$ and $g_{\tilde{C}}(z) = V(z)$. In our examples, intervals which definitely belong to C^* are:

(a) one-side neighborhoods of points of discontinuity of $g(z)$;
(b) intervals where $Lg(z) > 0$;
(c) neighborhoods of points, where $g'_-(a) < g'_+(a)$;
(d) neighborhoods of points of singularities of the diffusion.

Example 1 We consider a standard Wiener process w_t with initial point in $(-1, 1)$, stopped at the points -1 and 1, with the functional $E_z[g(w_\tau)]$. We suppose that the set of discontinuities of functions $g(z)$, $g'(z)$, $g''(z)$ is finite, the set of isolated zeros of the function $g''(z)$ is also finite, the functions have left and right limits at the points of discontinuity, and $g(z)$ at the points of discontinuity is greater than or equal to the left or the right limit.

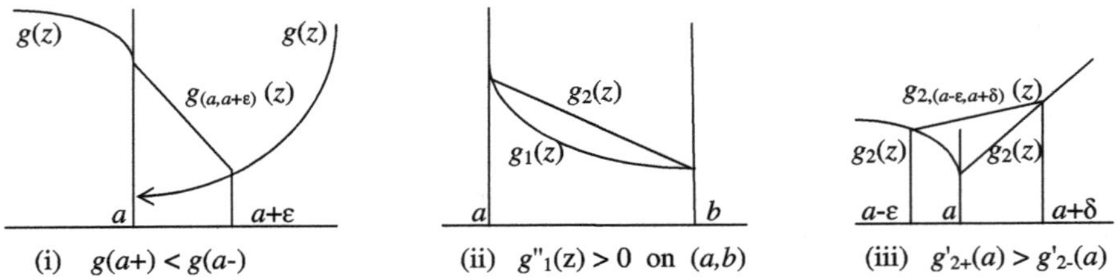

Fig. 1 Example 1

Recall (see, for example, [4] p. 145) that the differential operator corresponding to this process is $Lf(z) = (1/2)f''(z)$. For any interval (a, b) the expected reward at the time of the first exit from (a, b) is equal to

$$g_{(a,b)}(z) := g(a) + \frac{g(b) - g(a)}{b - a}(z - a) \quad \text{for } a \leq z \leq b\,.$$

The solution of the problem is well known (see, for example, [4] p. 146): the value function coincides with the minimal convex majorant of the payoff function. We propose the following procedure for constructing the value function.

(1) At the first stage, we change the payoff function in a neighborhood of each point of discontinuity $g(z)$. We change it in such a way that the new payoff function is continuous and the problem of optimal stopping with the new payoff function has the same value function as the initial problem.

Let $g(a) > \lim_{z\downarrow a} g(z)$ for some $a \in (-1, 1)$. Due to our assumptions about the function $g(z)$, this limit exists. We can choose $\varepsilon > 0$ such that there exist no points of change of sign of $g''(z)$, no points of discontinuity on the interval $(a, a + \varepsilon)$, and $g_{(a,a+\varepsilon)}(z) > g(z),\ z \in (a, a + \varepsilon)$ (see Fig. 1 (i)).

Therefore, the problem of optimal stopping with the payoff function $g_{(a,a+\varepsilon)}(z)$ has the same value function as the initial problem. The same situation holds for the points where $g(a) > \lim_{z\uparrow a} g(z)$. Now, we consider function $g_1(z)$, which is obtained from $g(z)$ using the earlier procedure for all points of discontinuity of $g(z)$, and we set $C_1 = \{z :\ g_1(z) > g(z)\}$. Note that the function $g_1(z)$ is continuous on $[-1; 1]$, functions $g_1'(z)$, $g_1''(z)$ have only a finite number of points of discontinuity, and the problem of optimal stopping with the payoff function $g_1(z)$ has the same value function as the initial problem.

(2) At the second stage, we change $g_1(z)$ on intervals, where $g_1''(z) > 0$. Let $C_2 = C_1 \bigcup \{z :\ g_1''(z) > 0\}$. Due to our assumptions about function $g(z)$, the set C_2 consists of the finite number of open intervals. Denote by A the set of such intervals. Let $g_2(z) = g_1(z)$ for $z \notin C_2$ and $g_2(z) = g_{1,(a,b)}(z)$ for $a \leq z \leq b$ and any $(a, b) \in A$, where, as earlier, $g_{1,(a,b)}(z)$ is the expected reward at the time of the first exit from (a, b) for the payoff function $g_1(z)$. Then $g_2(z) \geq g_1(z)$ (see Fig. 1(ii)), $C_2 \subseteq C^*$ and the problem with the functional $E_z[g_2(w_\tau)]$ has the same value function as the initial problem. Note that $g_2''(z) \leq 0$ for all points of continuity, the function $g_2(z)$ is continuous and the functions $g_2'(z)$, $g_2''(z)$ have only finite number of points of discontinuity.

Fig. 2 Smooth fitting interval near points u_1, u_4

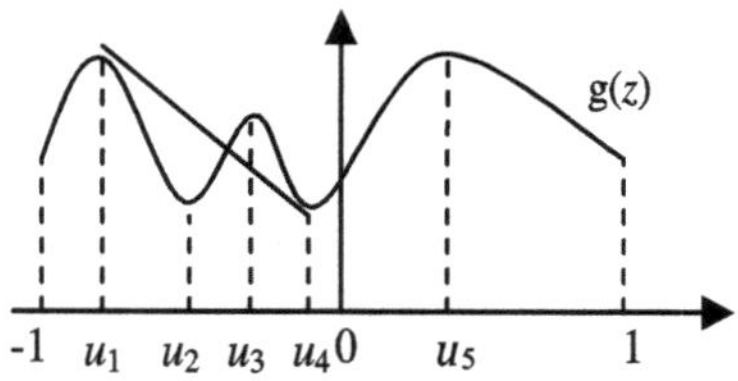

(3) Consider now the points of discontinuity of $g_2'(z)$. Denote by $g_{2+}'(z)$ the right and by $g_{2-}'(z)$ the left derivative of $g_2(z)$. The existence of these derivatives follows from our assumptions about function $g(z)$. If $g_{2-}'(a) < g_{2+}'(a)$, $-1 < a < 1$, then there exist $\varepsilon > 0$ and $\delta > 0$ such that $g_{2-}'(a-\varepsilon) < g_{2,(a-\varepsilon,a+\delta)}'(z) < g_{2+}'(a+\delta)$, $g_{2,(a-\varepsilon,a+\delta)}(z) > g_2(z)$ for $z \in (a-\varepsilon, a+\delta)$ (see Fig. 1(iii)). This follows from the fact that we can choose ε and δ in such a way that there are no points of change of sign of $g_2''(z)$ and no points of discontinuity on the interval $(a-\varepsilon, a+\delta)$.

The left (respectively right) derivative is not defined at the point -1 (respectively $+1$). For the convenience of exposition, we set $g_{2-}'(-1) = +\infty$, $g_{2+}'(1) = -\infty$. Since $g_2''(z) \leq 0$ for all points of continuity, by increasing ε and δ, we find that there exist minimal values of ε and δ – denote them by ε_1 and δ_1 – which satisfy inequalities: $g_{2-}'(a-\varepsilon_1) \geq g_{2,(a-\varepsilon_1,a+\delta_1)}'(z) \geq g_{2+}'(a+\delta_1)$ for $z \in (a-\varepsilon_1, a+\delta_1)$. This is an analog of the smooth fitting condition in the case of smooth $g(z)$.

Define $C_3 = C_2 \bigcup (a-\varepsilon_1, a+\delta_1)$ and consider $g_3(z) = g_{2,(a-\varepsilon_1,a+\delta_1)}(z)$ for $z \in (a-\varepsilon_1, a+\delta_1)$. It is obvious that $g_3(z) \geq g_2(z)$, $C_3 \subseteq C^*$, the problem with the functional $E_z[g_3(w_\tau)]$ has the same value function as the initial problem, and *the number of points of discontinuity of* $g_3'(z)$ *such that* $g_{3-}'(z) < g_{3+}'(z)$ *is less then the number of such points for* $g_2'(z)$. We can apply to $g_3(z)$ the same procedure as we applied to $g_2(z)$. Since the number of points where $g_{2-}'(z) < g_{2+}'(z)$ is finite, *after finite number of steps* we obtain a set $\tilde{C}$, and a function $\tilde{g}(z)$ such that:

(a) $\tilde{g}(z) \geq g(z)$, $\tilde{g}(z) = E_z[g(w_{\tilde{\tau}})]$, where $\tilde{\tau} = \inf\{t \geq 0 : w_t \notin \tilde{C}\}$,
(b) $\tilde{g}''(z) = 0$ for $z \in \tilde{C}$, $\tilde{g}_-'(z) \geq \tilde{g}_+'(z)$ for all $z \in (-1, 1)$, and $\tilde{g}''(z) \leq 0$ for all points of continuity.

It follows from a) that the value function is the same as in the problem of the optimal stopping with the payoff function $g(z)$ and with the payoff function $\tilde{g}(z)$. It follows from b) that $\tilde{g}(z)$ is convex and coincides with its minimal convex majorant. Consequently $\tilde{g}(z) = V(z)$ and $\tilde{C} = C^*$.

Remark 4 One can say that an interval (a, b) in the problem of Example 1 with a smooth $g(z)$ is a smooth fitting interval if the function $g_{(a,b)}(z)$ has the same derivative as $g(z)$ at points a and b. For example, one can construct six smooth fitting intervals for $g(z)$ in Fig. 2 with a, b near points u_1, u_3, or u_1, u_4, or u_1, u_5, or u_2, u_4, or u_2, u_5, or u_3, u_5.

Any smooth fitting interval gives a solution of the Stefan free-boundary problem. It can happen that such an interval has no relation to the set C^* and to check that the solution of the Stefan free-boundary problem coincides with the value function one usually applies a verification theorem. In the proposed procedure, we do not need to use a verification theorem. We believe that in an essentially more general situation for a regular diffusion, instead of a verification theorem, it suffices to prove that if the *payoff function* satisfies the conditions: $g'_-(z) \geq g'_+(z)$ for all z, and $Lg(z) \leq 0$ for all points of continuity, then $g(z) = V(z)$.

Remark 5 A method of a sequential construction of the value function for the case of discounting depending on the state of the process and piecewise constant non-decreasing payoff function taking finite number of values was considered in [1]. In that paper, the authors used an optimality equation and a variational inequality for the construction. In the present work, the proposed procedure does not employ the optimality equation.

Example 2 We consider a Wiener process $w_{1,t}$ on the interval $[-1;1]$ with the absorbtion at points -1 and 1, a partial reflection with a coefficient α, $0 < |\alpha| < 1$, at the point 0, and the functional $E_z[g(w_{1,\tau})]$ where $g(z)$ is the same function as in Example 1. The partial reflection means that $P_0[w_{1,t} > 0] = (1+\alpha)/2 + o(t)$ as $t \to 0$.

The differential operator corresponding to this process is $L_1 f(z) = (1/2) f''(z)$ for $z \neq 0$ with the condition $(1+\alpha) f'_+(0) - (1-\alpha) f'_-(0) = 0$.

We first use the same procedure as in Example 1 for interval $[-1,0]$ assuming that points $z=-1$ and $z=0$ are absorbing. Next, we use the same procedure for the interval $[0,1]$. As a result, we obtain the continuous function $g_1(z)$ and the set C_1 such that $C_1 \subseteq C^*$ and the problem of optimal stopping with functional $E_z[g_1(w_{1,\tau})]$ has the same value function as the initial problem. The set C_1 consists of the final number of open intervals, the functions $g'_1(z)$ and $g''_1(z)$ have only a finite number of points of discontinuity, $g''_1(z) = 0$ for $z \in C_1$, $g''_1(z) \leq 0$ for all points of continuity, $g'_{1-}(z) \geq g'_{1+}(z)$ for all $z \in (-1,0)$ and $z \in (0,1)$. So $g_1(z)$ is concave for $z \in (-1,0)$ and $z \in (0,1)$.

Let us consider the point $z = 0$. For any a, b, $-1 \leq a < b \leq 1$, denote by $g_{(a,b)}(z)$ the expected reward at the time of the first exit from (a,b) for the payoff function $g_1(z)$. If either $b \leq 0$ or $a \geq 0$ then, due to our construction, $g_{(a,b)}(z) \leq g_1(z)$ for all $z \in [-1,1]$. If $a < 0$, $b > 0$, then the function $g_{(a,b)}(z)$ satisfies the conditions: $L_1 g_{(a,b)}(z) = 0$ for $z \in (a,0) \bigcup (0,b)$, $g_{(a,b)}(a) = g_1(a)$, $g_{(a,b)}(b) = g_1(b)$, $(1+\alpha) g'_{(a,b)}(+0) - (1-\alpha) g'_{(a,b)}(-0) = 0$.

Therefore, if $a < 0$, $b > 0$, then

$$g_{(a,b)}(z) = \begin{cases} [g_{(a,b)}(0)(z-a) - g_1(a)z]/(-a) & \text{for } z \in (a,0), \\ [g_1(b)z + g_{(a,b)}(0)(b-z)]/b & \text{for } z \in (0,b), \\ \frac{b(1-\alpha)g_1(a) - a(1+\alpha)g_1(b)}{b(1-\alpha) - a(1+\alpha)} & \text{for } z = 0. \end{cases} \tag{3}$$

If $(1+\alpha) g'_{1+}(0) - (1-\alpha) g'_{1-}(0) \leq 0$ then from the above-mentioned concavity of $g_1(z)$ and (3) it follows that $g_{(a,b)}(z) \leq g_1(z)$ for all $-1 \leq a, b, z \leq 1$, and consequently $g_1(z) = V(z)$ and $C_1 = C^*$.

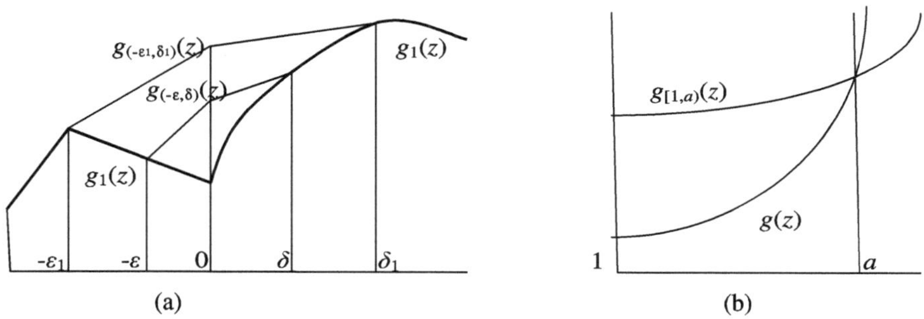

Fig. 3 Examples 2 and 3

If $(1+\alpha)g'_{1+}(0) - (1-\alpha)g'_{1-}(0) > 0$ then there exist $\varepsilon > 0$ and $\delta > 0$ such that $g'_{1-}(-\varepsilon) < g'_{+(-\varepsilon,\delta)}(-\varepsilon)$, $g'_{-(-\varepsilon,\delta)}(\delta) < g'_{1+}(\delta)$, $g_{(-\varepsilon,\delta)}(z) > g_1(z)$ for $z \in (-\varepsilon, \delta)$ (see Fig. 3(a)).

Since $g_1(z)$ is concave for $z \in (-1, 0)$ and $z \in (0, 1)$, increasing ε and δ leads to the existence of minimal values of ε and δ—denote them by ε_1 and δ_1—such that $g'_{1-}(\delta_1) \geq g'_{(-\varepsilon_1,\delta_1)}(-0)$, $g'_{(-\varepsilon_1,\delta_1)}(+0) \geq g'_{1+}(\delta_1)$. Note that for each fixed z the function $g_{(-\varepsilon,\delta)}(z)$, as a function on ε and δ increases on ε and δ for $\varepsilon < \varepsilon_1$, $\delta < \delta_1$. As a result we have $C^* = C_1 \bigcup (-\varepsilon_1, \delta_1)$, $V(z) = g_1(z)$ for $z \notin (-\varepsilon_1, \delta_1)$, $V(z) = g_{(-\varepsilon_1,\delta_1)}(z)$ for $z \in (-\varepsilon_1, \delta_1)$.

Example 3 Geometric Brownian motion x_t on $[1; \infty]$ with parameters $(-r, \sigma)$, a killing intensity λ, a reflection at the point 1 and with the functional $E_z[g(x_\tau)]$. We assume that the function $g(z)$ satisfies the same conditions of continuity and differentiability as in Example 1 and the set of the isolated zeros of the function $L_2 g(z) := \frac{\sigma^2 z^2}{2} g''(z) - rzg'(z) - \lambda g(z)$ is finite. Let $\kappa_+ > 1$ and $\kappa_- < 0$ be the solutions of the equation $\kappa^2 - (1 + \frac{2r}{\sigma^2})\kappa - \frac{2\lambda}{\sigma^2} = 0$. We assume also that:

(a) $\lim_{z\to\infty} |g(z)| z^{-\kappa_+} < \infty$,
(b) $L_2 g(z) < 0$ for $z \geq z_1 \geq 1$, and (c) $g'(1) > 0$.

It is well known (see, for example, [4] formula (26.1.18)) that the differential operator corresponding to this process is $L_2 f(z)$ with boundary condition $f'(1) = 0$. We start by investigating the behavior of $g(z)$ at point 1. Denote by $g_{[1,a)}(z)$ the expected reward at the time of the first exit from $[1, a)$. Then $g_{[1,a)}(z) = g(z)$ for $z \geq a$ and for $z \in (1, a)$ it satisfies the equation $L_2 g_{[1,a)}(z) = 0$, with boundary conditions $g_{[1,a)}(a) = g(a)$, $g'_{[1,a)}(1) = 0$. Therefore,

$$g_{[1,a)}(z) := \frac{g(a)(\kappa_+ z^{\kappa_-} - \kappa_- z^{\kappa_+})}{\kappa_+ a^{\kappa_-} - \kappa_- a^{\kappa_+}} \quad \text{for } z \in [1, a). \tag{4}$$

It follows from (c) and the conditions on function $g(z)$ that if $a - 1$ is small enough then $g'_{[1,a)}(a) < g'(a)$ and $g_{[1,a)}(z) > g(z)$ for $z \in [1, a)$ (see Fig. 3(b)).

Thus $[1, a) \subseteq C^*$ and the problem with the functional $E_z[g_{[1,a)}(w_\tau)]$ has the same value function as the initial problem. Now we shall use the same procedure as in Example 1, but we shall change $g_{[1,a)}(z)$ on each interval (b,c) from C^* to a function $f(z) = B_1 z^{\kappa_-} + B_2 z^{\kappa_+}$, where B_1 and B_2 are chosen from the condition $f(b) = g_{[1,a)}(b)$, $f(c) = g_{[1,a)}(c)$ in case $b > 1$ and $f'(b) = 0$, $f(c) = g_1(c)$ in case $b = 1$, which coincides with the expected reward at the time of the first exit from (b,c). After a finite number of steps, we obtain the stopping set and the value function. It is simple to check that from conditions (a) and (b) it follows that the value function is finite and the set C^* is bounded. Note that the case $g(z) = z$ corresponds to the Russian option (see [4], Sect. 26). Since in this case $L_2 g(z) = -(r+\lambda)z < 0$, the only point where we can locally increase the payoff function without changing the value function is the point $z = 1$, and one has only one step. The optimal value a^* in (4) can be found, as before, from the condition $a^* = \{\inf a : g'_{[1,a)}(a) \geq g'(a) \equiv 1\}$.

Example 4 We consider a standard Wiener process w_t with an initial point $z \in (-\infty, \infty)$ and a functional $E_z[e^{-\lambda\tau} g(w_\tau)]$. Such problem is equivalent to the problem with functional $E_z[g(\tilde{w}_\tau)]$, where $\tilde{w}_t$ is a standard Wiener process with a killing intensity λ. The differential operator corresponding to this process is

$$L_3 f(z) = (1/2) f''(z) - \lambda f(z).$$

For the sake of simplicity we suppose that $g(0) = 0$, $L_3 g(z) < 0$ for $z \neq 0$, and $g'_+(0) = b > 0 > g'_-(0) = a$. The payoff function $g(z) = az$ for $z \leq 0$ and $g(z) = bz$ for $z \geq 0$ satisfies these conditions.

Since $L_3 g(z) < 0$ for all $z \neq 0$, and $a = g'_-(0) < g'_+(0) = b$, the only point where we can locally increase payoff function without changing the value function is the point $z = 0$.

Let $\tau(c,d)$ be the time of the first visit to the complement of the interval (c,d), where $-\infty < c < d < \infty$. As earlier, we define the expected reward at the time $\tau(c,d)$ as

$$g_{(c,d)}(z) = E_z g(w_{\tau(c,d)}).$$

Then

$$L_3 g_{(c,d)}(z) \equiv 0 \quad \text{for } z \in (c,d), \quad g_{(c,d)}(c) = g(c), \quad g_{(c,d)}(d) = g(d). \tag{5}$$

Let $\mathbf{B}$ be the set of intervals such that $c < 0 < d$ and

$$a = g'_-(c) \leq g'_{+,(c,d)}(c), \qquad g'_{-,(c,d)}(d) \leq g'_+(d) = b. \tag{6}$$

We shall use the following properties of $\mathbf{B}$, which are valid in essentially more general situations. They follow from the fact that $L_3 g(z) \leq 0$ for all points of continuity of $g''(z)$. The proof of these properties is analogous to the proof of step 3 in Example 1.

(1) If $(c,d) \in \mathbf{B}$ then $g_{(c,d)}(z) > g(z)$, $(c,d) \subseteq C^*$ and the problem with the payoff function $g_{(c,d)}(z)$ has the same value function as the problem with the payoff function $g(z)$.

(2) If $c < 0 < d$ and $c, |d|$ are small enough then $(c, d) \in \mathbf{B}$ and both inequalities in (6) are strong.

(3) If $(c, d) \in \mathbf{B}$ and the first (or the second) inequality in (6) is strong, then there exists $c_1 < c$ (or $d_1 > d$) such that $(c_1, d) \in \mathbf{B}$ (or $(c, d_1) \in \mathbf{B}$) and $g_{(c_1,d)}(z) > g_{(c,d)}(z)$ for $z \in (c_1, d)$ (or $g_{(c,d_1)}(z) > g_{(c,d)}(z)$ for $z \in (c, d_1)$).

Let (c^*, d^*) be the minimal interval for which $c^* < 0 < d^*$ and $g'_-(c^*) \geq g'_{+,(c^*,d^*)}(c^*)$, $g'_{-,(c^*,d^*)}(d^*) \geq g'_+(d^*)$.

(4) If $|c^*|$, $d^* < \infty$ then $(c^*, d^*) \in \mathbf{B}$ and $g'_-(c^*) = g'_{+,(c^*,d^*)}(c^*)$, $g'_{-,(c^*,d)^*}(d^*) = g'_+(d^*)$.

In case $|c^*|$, $d^* < \infty$, the function $g_{(c^*,d^*)}(z)$ is smooth and $L_3 g(c^*, d^*)(z) \leq 0$ for all $z \neq c^*, d^*$. Using a standard method one can show that the value function in the problem of optimal stopping with payoff function $g_{(c^*,d^*)}(z)$ coincides with $g_{(c^*,d^*)}(z)$. It follows from here that in the initial problem the value function coincides with $g_{(c^*,d^*)}(z)$, $(c^*, d^*) = C^*$ and $\tau(c^*, d^*)$ is the optimal stopping time. So, we need just to construct the values c^*, d^*.

Let us consider the case $g(z) = az$ for $z \leq 0$ and $g(z) = bz$ for $z \geq 0$. Without loss of generality, we may and do suppose that $\lambda = 1/2$. Consider the function

$$\psi(z, c, d) = bd\frac{\sinh(z-c)}{\sinh(d-c)} + ac\frac{\sinh(d-z)}{\sinh(d-c)}.$$

This function satisfies Eq. (5) and, thus,

$$\psi(z, c, d) = g_{(c,d)}(z) \quad \text{for } z \in (c, d).$$

The values c^*, d^* are the roots of the system of equations

$$\psi'_z(c, c, d) = a, \qquad \psi'_z(d, c, d) = b,$$

which can be written in the form

$$bd - ac\cosh(d-c) = a\sinh(d-c)\,, \tag{7}$$

$$bd\cosh(d-c) - ac = b\sinh(d-c)\,. \tag{8}$$

If $a = -b$ then $c^* = -d^*$ and (7) follows from (8). From (8) and the equalities $a = -b$, $c^* = -d^*$ we get $bd^*(\cosh(2d^*) - 1) = b\sinh(2d^*)$. It is easy to show that this equation has a unique root d^*, which is the same for all values of b.

Let $a \neq -b$. The system (7)–(8) can be rewritten as

$$b^2 d + a^2 c = ab(d+c)\cosh(d-c)\,, \tag{9}$$

$$b^2 d^2 - a^2 c^2 = ab(d+c)\sinh(d-c)\,, \tag{10}$$

or, using the equality $\cosh^2(x) - \sinh^2(x) = 1$, as

$$\left(b^2 d + a^2 c\right)^2 - \left(b^2 d^2 - a^2 c^2\right)^2 = a^2 b^2 (d+c)^2\,, \tag{11}$$

$$b^2d^2 - a^2c^2 = ab(d + c)\sinh(d - c). \tag{12}$$

Equation (11) can be represented in the form

$$(b^2d^2 - a^2c^2)\big[(b^2d^2 - a^2c^2) - (b^2 - a^2)\big] = 0. \tag{13}$$

Since $a \neq -b$, the solution $bd = ac$ of (13) contradicts to (12). So the optimal values c^*, d^* are the roots of the system

$$b^2d^2 - a^2c^2 = b^2 - a^2, \tag{14}$$

$$b^2 - a^2 = ab(d + c)\sinh(d - c). \tag{15}$$

Solving (14) with respect to c and substituting the result into (15), we obtain the equation with respect to d^*, which has a unique positive solution.

Remark 6 We believe that the proposed procedure can be extended to a much more general situation, as well as to the multi-dimensional case.

Acknowledgements The author would like to thank V.I. Arkin, A.D. Slastnikov for useful discussions, I.M. Sonin, Yu.M. Kabanov and anonymous referees for very valuable remarks and suggestions, and one of the referees for drawing his attention to the paper [1].

This work was partly supported by RFBR grant 10-01-00767-a.

References

1. Bronstein, A.L., Hughston, L.P., Pistorius, M.R., Zervos, M.: Discretionary stopping of one-dimensional Ito diffusions with a staircase reward function. J. Appl. Probab. **43**, 984–996 (2006)
2. Dayanik, S., Karatzas, I.: On the optimal stopping problem for one-dimensional diffusions. Stoch. Process. Appl. **107**, 173–212 (2003)
3. Feldman, R., Valdez-Flores, C.: Applied Probability and Stochastic Processes. PWS, Boston (1995)
4. Peskir, P., Shiryaev, A.N.: Optimal Stopping and Free-Boundary Problems. Birkhauser, Basel (2006)
5. Presman, E.L.: On Sonin's algorithm for solution of the optimal stopping problem. In: Proceedings of the Fourth International Conference on Control Problems (January 26–30, 2009), pp. 300–309. Institute of Control Sciences (2009)
6. Presman, E.L.: A new approach to the solution of optimal stopping problem in a discrete time. Stochastics **83**(4–6), 467–475 (2011)
7. Presman, E.L., Sonin, I.M.: On optimal stopping of random sequences modulated by Markov chain. Theory Probab. Appl. **54**(3), 534–542 (2009)
8. Salminen, P.: Optimal stopping of one-dimensional diffusions. Math. Nachr. **124**, 85–101 (1985)
9. Shiryayev, A.N.: Statistical Sequential Analysis: Optimal Stopping Rules. Nauka, Moscow (1969) (in Russian). English translation of the second edition: Shiryayev, A.N.: Optimal Stopping Rules, Springer, Berlin, 1978
10. Sonin, I.M.: Two simple theorems in the problems of optimal stopping. In: Proc. 8th INFORMS Applied Probability Conference, Atlanta, Georgia, p. 27 (1995)

11. Sonin, I.M.: The elimination algorithm for the problem of optimal stopping. Math. Methods Oper. Res. **49**, 111–123 (1999)
12. Sonin, I.M.: The state reduction and related algorithms and their applications to the study of Markov chains, graph theory and the optimal stopping problem. Adv. Math. **145**, 159–188 (1999)
13. Sonin, I.M.: Optimal stopping of Markov chains and recursive solution of Poisson and Bellman equations. In: Kabanov, Yu., Liptser, R., Stoyanov, J. (eds.) From Stochastic Calculus to Mathematical Finance. The Shiryaev Festschrift, pp. 609–621. Springer, Berlin (2006)

A Stieltjes Approach to Static Hedges

Michael Schmutz and Thomas Zürcher

Abstract Static hedging of complicated payoff structures by standard instruments becomes increasingly popular in finance. The classical approach is developed for quite regular functions, while for less regular cases, generalized functions and approximation arguments are used. In this note, we discuss the regularity conditions in the classical decomposition formula due to P. Carr and D. Madan (in Jarrow ed, Volatility, pp. 417–427, Risk Publ., London, 1998) if the integrals in this formula are interpreted as Lebesgue integrals with respect to the Lebesgue measure. Furthermore, we show that if we replace these integrals by Lebesgue–Stieltjes integrals, the family of representable functions can be extended considerably with a direct approach.

Keywords Absolute continuity · Bounded variation · Static hedging · Stieltjes integral

Mathematics Subject Classification (2010) 91G20 · 26A42 · 26A45 · 26A46 · 26A48 · 26A51

1 Introduction

It is well known that sufficiently regular payoff functions depending on the terminal asset price can be statically hedged by taking buy and hold positions in bonds, forwards, and lots of vanilla options. Due to various reasons, in particular static

---

A large part of the research was carried out while the second author was a postdoctoral researcher at the Mathematical Institute, University of Bern, Sidlerstrasse 5, 3012 Bern, Switzerland.

M. Schmutz (✉)
Mathematical Statistics and Actuarial Science, University of Bern, Sidlerstrasse 5, 3012 Bern, Switzerland
e-mail: michael.schmutz@stat.unibe.ch

T. Zürcher
Department of Mathematics and Statistics, University of Jyväskylä, P.O. Box 35 (MaD), 40014 Jyväskylä, Finland
e-mail: thomas.t.zurcher@jyu.fi

Y. Kabanov et al. (eds.), *Inspired by Finance*, DOI 10.1007/978-3-319-02069-3_24,

hedging, related to semi-static hedging or valuation, the decomposition of complicated payoff functions has become increasingly popular in finance during the recent years, see e.g. [1–3, 6–9, 12, 15, 17].

The aim of this note is a deeper mathematical analysis of Carr-Madan's well-known formula obtained in a different context in [9]. We discuss some regularity aspects, and in particular, we show with an easy argument that if the integral expressions in the formula are interpreted as Lebesgue integrals with respect to the Lebesgue measure with locally integrable weights, then the structure of the hedge for a continuous payoff function f already implies a certain differentiability property. However, if we change the integral from the Lebesgue to the Lebesgue–Stieltjes integral, i.e. the difference of two Lebesgue integrals with respect to certain Lebesgue–Stieltjes measures, the analogous representation holds for a considerably richer family of payoff functions, quite similar to the one considered by Carr and Lee [8] or to the one considered by Baldeaux and Rutkowski [3, 4], where, however, a slightly different approach is used, and slightly different representations are stated.

2 Static Hedging with the Lebesgue Measure

In what follows, we use the abbreviation $\int_a^b f(x)\,\mathrm{d}x = \int_{[a,b]} f(x)\mathcal{L}(\mathrm{d}x)$, for non-negative a and b, $a \leq b$, for integrals with respect to the Lebesgue measure $\mathcal{L}$. Furthermore, we will denote the right (left) derivative (presumed to exist) of a function $f\colon I \to \mathbb{R}$ by f_r' (f_l'), where I stands for any interval. For $S \subset \mathbb{R}$, we say that $f: S \to \mathbb{R}$ is *differentiable* on S if f can be extended to a differentiable function on an open set $U \supset S$. Note that if f is defined on $\mathbb{R}_+ = [0, \infty)$, differentiable on $(0, \infty)$, and the right derivative of f in 0 is finite, then f is differentiable on $\mathbb{R}_+$. Throughout measure theoretical considerations, we are following the convention $0 \cdot \infty = 0$.

Originally, and in a different context, the formulated assumptions on the payoff functions were twice differentiability, related to a derivation of the formula based on generalized functions. Bakshi and Madan [2] and Carr and Madan [10] require the payoff functions to be two times *continuously* differentiable. Here, we assume that $f : \mathbb{R}_+ \to \mathbb{R}$ is continuously differentiable with f' being locally absolutely continuous,[1] i.e. absolutely continuous on the compact intervals $[a, b]$ for all $a < b$ (notation $f \in \mathrm{AC}_{\mathrm{loc}}$). Similarly to e.g. [17, 20], we can respectively use for the first three equalities [16, Th. 7.1.34, Th. 7.1.47] and the formula $xf'(x) = \int_a^x xf''(t)\,\mathrm{d}t + xf'(a)$ (resulting in the needed generality e.g. from [16, Th. 7.1.15]) in order to see that for any $c \in \mathbb{R}_+$ and *whenever* $c \leq x$

$$f(x) = f(c) + \int_c^x f'(k)\,\mathrm{d}k = f(c) + xf'(x) - cf'(c) - \int_c^x kf''(k)\,\mathrm{d}k$$

[1] Recall that following e.g. [16, Def. 7.1.4], a finite function f defined on a closed interval $[a, b]$ is absolutely continuous on $[a, b]$ (notation $f \in \mathrm{AC}[a, b]$) if, for every $\varepsilon > 0$, there exists a $\delta > 0$ such that $\sum_{k=1}^n |f(b_k) - f(a_k)| < \varepsilon$ for any $a \leq a_1 < b_1 \leq a_2 < b_2 \leq \cdots \leq a_n < b_n \leq b$ for which $\sum_{k=1}^n (b_k - a_k) < \delta$.

$$= f(c) + \int_c^x x f''(k)\,\mathrm{d}k + x f'(c) - c f'(c) - \int_c^x k f''(k)\,\mathrm{d}k$$

$$= f(c) + f'(c)(x-c) + \int_c^x f''(k)\big((x-k)_+ - (k-x)_+\big)\,\mathrm{d}k$$

$$= f(c) + f'(c)(x-c) + \int_c^x f''(k)(x-k)_+\,\mathrm{d}k$$

$$= f(c) + f'(c)(x-c) + \int_c^\infty f''(k)(x-k)_+\,\mathrm{d}k\,. \tag{1}$$

By applying the same theorems and analogous arguments for the case $x < c$, we arrive at

$$f(x) = f(c) + f'(c)(x-c) + \int_0^c f''(k)(k-x)_+\,\mathrm{d}k\,. \tag{2}$$

Since the term $\int_0^c f''(k)(k-x)_+\,\mathrm{d}k$ vanishes for $c \le x$, we can add this integral to (1), while if $x < c$, the integral $\int_c^\infty f''(k)(x-k)_+\,\mathrm{d}k$ vanishes, so that we can add this integral to (2). Hence, in both cases, we arrive for any $c \in \mathbb{R}_+$ at the well-known decomposition formula

$$f(x) = f(c) + f'(c)(x-c) + \int_c^\infty f''(k)(x-k)_+\,\mathrm{d}k + \int_0^c f''(k)(k-x)_+\mathrm{d}k \tag{3}$$

for $x \in \mathbb{R}_+$, see e.g. the literature cited in the introduction. Its different original proof is presented in [9]. Note that for $x \in \mathbb{R}_+$ the integrands in (3) are only non-vanishing on bounded sets.

An economical interpretation of (3) is that if c is the current forward price, f can be statically hedged with bonds, forwards, and lots of vanilla options (with vanilla options being out of and at the money in a certain sense). For practical implementation of static hedges the problem of the existence of only finitely many liquid strikes needs also to be addressed, see [1, 20] and the literature cited therein. Besides of choosing c to be the forward price, it is also quite popular to set $c = 0$, see e.g. [12] in order to get a decomposition related to valuation problems, or e.g. [2, 6, 10] for special cases of hedges with particularly simple structure. This choice is clearly possible under the assumption that $f : \mathbb{R}_+ \to \mathbb{R}$ is continuously differentiable with f' being locally absolutely continuous. The obvious problems that occur if $f'(c) = \infty$ along with the more subtle problems appearing in Example 1 below show how important proper conditions on the functions are, in particular near the boundary, and in particular, if we want to include the popular special case $c = 0$.

Let us give an example that demonstrates that for the representation formula (3), for $c = 0$, we cannot omit the continuity of the second derivative without assuming that the first derivative is locally absolutely continuous. On the other hand, note that our assumptions do not guarantee that the function f is two times differentiable everywhere (but due to [16, Th. 7.1.15, Th. 7.1.47], this is not needed in view of the absolute continuity of the first derivative). Recall that in this section, the integrals are interpreted as Lebesgue integrals with respect to $\mathcal{L}$. The interpretation is *different* in several other papers concerning the decomposition formula.

Example 1 This example is based on the typical example, which shows that the Lebesgue integral does not recover a function from its derivative (without additional assumptions), see e.g. [14, p. 107]. Here, let us define $g: \mathbb{R}_+ \to \mathbb{R}$ by

$$g(t) = \begin{cases} 0, & t = 0, \\ t^2 \sin(\frac{1}{t^2}), & 0 < t. \end{cases}$$

We leave it to the reader to verify that g is differentiable with

$$g'(t) = \begin{cases} 2t \sin(\frac{1}{t^2}) - \frac{2}{t} \cos(\frac{1}{t^2}), & t > 0, \\ 0, & t = 0. \end{cases}$$

Note that the derivative is unbounded on $[0, 1]$. We define $f: \mathbb{R}_+ \to \mathbb{R}$ as follows

$$f(x) = \int_0^x g(t)\,\mathrm{d}t \quad \text{for } x \geq 0\,.$$

Since g is differentiable, it is continuous on $[0, b]$ for every $b > 0$, so that f is differentiable on $(0, b)$ and $f'(x) = g(x)$ in $(0, b)$. One easily verifies that f is differentiable on $\mathbb{R}_+$ with $f'(x) = g(x)$. Hence, f is indeed twice differentiable with discontinuous second derivative. We claim now that the representation (3) does not hold for $c = 0$ and $x > 0$ since $f''(k)(x-k)_+$ (as a function in k) is not integrable in the Lebesgue sense. Assume by way of contradiction that the integral is finite. Then

$$\int_0^\infty f''(k)(x-k)_+\,\mathrm{d}k = \int_0^x f''(k)(x-k)\,\mathrm{d}k.$$

We note that for $0 < k < x$

$$\begin{aligned} f''(k)(x-k) &= \left(2k \sin\left(\frac{1}{k^2}\right) - \frac{2}{k} \cos\left(\frac{1}{k^2}\right)\right)(x-k) \\ &= 2kx \sin\left(\frac{1}{k^2}\right) - 2k^2 \sin\left(\frac{1}{k^2}\right) + 2\cos\left(\frac{1}{k^2}\right) - \frac{2x}{k} \cos\left(\frac{1}{k^2}\right). \end{aligned}$$

Since the first three summands are measurable and bounded on $[0, x]$, and we are integrating over a compact set, it suffices to show that the last summand is not integrable. We leave it to the reader to verify that both integrals, over the positive and the negative part of the last summand, are infinite. Hence, the representation formula cannot hold in the classical Lebesgue sense. Often, only payoff functions with values in $\mathbb{R}_+$ are considered. We leave it to the reader to modify the above example in order to obtain an example given by an $\mathbb{R}_+$-valued function without representation (3) for $c = 0$.

Furthermore, note that for locally integrable weights, already the structure of the hedge immediately implies some *necessary* regularity for a continuous function f, if the integral expressions have the interpretation of *Lebesgue integrals* with respect to $\mathcal{L}$. Specifically, assume that for arbitrary c, there exist locally integrable g_1 and g_2 along with h such that

$$f(x) = f(c) + h(c)(x-c) + \int_c^\infty g_1(k)(x-k)_+ \,\mathrm{d}k + \int_0^c g_2(k)(k-x)_+ \,\mathrm{d}k\,. \quad (4)$$

For $x > c$, the second integral vanishes. We conclude further

$$\left|\frac{1}{x-c}\int_c^\infty g_1(k)(x-k)_+ \,\mathrm{d}k\right| = \left|\frac{1}{x-c}\int_c^x g_1(k)(x-k)\,\mathrm{d}k\right|$$
$$\leq \int_c^x \left|g_1(k)\right| \frac{|x-k|}{|x-c|}\,\mathrm{d}k \leq \int_c^x \left|g_1(k)\right| \mathrm{d}k\,.$$

As x tends to c, this term vanishes. Since $h(c)(x-c)/(x-c)$ tends to $h(c)$ as x approaches c, we conclude that

$$f_r'(c) = \lim_{x\to c+} \frac{f(x)-f(c)}{x-c} = h(c)\,.$$

This means that the right derivative of f exists at c and equals $h(c)$. If $c = 0$, the differentiability at c follows. Otherwise, the analogous argument for sequences approaching c from the left shows that $f_l'(c)$ exists and is $h(c)$ as well. Hence, f is differentiable at c with $f'(c) = h(c)$ in the classical sense. Thus, if the hedge of a continuous payoff function f is of the form (4) holding for any non-negative c and any $x \in \mathbb{R}_+$, we immediately get the differentiability of f.

Note that many classical option strategies have payoff functions that are not differentiable at some points. A classical approach to handle the resulting problems relies on generalized functions, see e.g. [3, 4, 8].

3 Static Hedging with Lebesgue–Stieltjes Integrals

In this section, we study a variant of (3) with respect to the Lebesgue–Stieltjes integral. If at a point a, a function f has a unique limit on the right, this limit will be denoted by $f(a+)$, similarly, $f(a-)$ will stand for the unique limit on the left. We prove the following result, which again includes the particularly popular special case of $c = 0$.

Theorem 1 *Assume that $f\colon \mathbb{R}_+ \to \mathbb{R}$ is the difference of two convex functions whose right derivatives in 0 are finite. Suppose that $c \in \mathbb{R}_+$. Then*

$$f(x) = f(c) + f_r'(c)(x-c) + \int_{(c,\infty)} (x-k)_+ \,\mathrm{d}f_r'(k) + \int_{[0,c]} (k-x)_+ \,\mathrm{d}f_r'(k)$$

for all $x \in \mathbb{R}_+$.

The integral expressions are of Lebesgue–Stieltjes type, which will briefly be explained below. For the price of loosing the guarantee of the existence of the popular decomposition based on calls without puts, the assumptions on the behavior of the right derivatives are sometimes relaxed in representations based on generalized functions, see e.g. [3, 4, 8], yielding other merits. Later, we will modify the involved Lebesgue–Stieltjes measures so that the above restrictions on the boundary behavior can also be relaxed to a certain extent based on the direct Stieltjes approach. Related to similar approaches based on generalized functions, we point out the importance of addressing boundary anomalies of convex functions on $\mathbb{R}_+$. Before we prove Theorem 1, we first have to collect some known results. In order for our arguments to work, we need that $f\colon \mathbb{R}_+ \to \mathbb{R}$ is locally absolutely continuous, and that f' has a representative being locally of bounded variation[2]. In view of that, we start with the following well-known result.

Theorem 2 (See e.g. [16], Th. 7.1.18) *Let $f : [a,b] \to \mathbb{R}$. Then f is absolutely continuous on $[a,b]$ if and only if there exists h, integrable on $[a,b]$, such that*

$$f(x) = f(a) + \int_a^x h(t)\,\mathrm{d}t\,. \tag{5}$$

It follows that $f' = h$ a.e.

The following result is an immediate consequence of this theorem and of one of the statements of Theorem A in [18, p. 23].

Theorem 3 *Suppose $f\colon [a,b] \to \mathbb{R}$ is given. If $f = g - h$ is the difference of two convex functions g and h such that $g'_r(a)$, $g'_l(b)$, $h'_r(a)$, and $h'_l(b)$ are all finite, then f is absolutely continuous and its derivative has a representative that is of bounded variation.*

From Theorem B in [18, p. 5], we obtain an existence result for the one sided derivatives.

Theorem 4 *If $f\colon I \to \mathbb{R}$ is defined on an interval and convex, then $f'_l(x)$ and $f'_r(x)$ exist for each x in the interior I° of I and are increasing on I°.*

The following results are parts of Theorem 6.1.3 and Theorem 6.1.7 in [16], respectively.

---

[2]For the following, see Definition 6.1.2 in [16]. Let $f\colon [a,b] \to \mathbb{R}$. Assume that $P = \{x_0, x_1, \ldots, x_{n(P)}\}$ is a partition of the interval $[a,b]$. If

$$T_f[a,b] = \sup_P \sum_{k=1}^{n(P)} \big| f(x_k) - f(x_{k-1}) \big| < \infty,$$

where the supremum is taken over all partitions P of $[a,b]$, then f is said to be of *bounded variation* on $[a,b]$, for short $f \in \mathrm{BV}[a,b]$. If $f\colon \mathbb{R} \to \mathbb{R}$ or $f\colon \mathbb{R}_+ \to \mathbb{R}$ is such that the restriction of f to $[a,b]$ is in $\mathrm{BV}[a,b]$ for all $a < b$, then f is said to be *locally of bounded variation* ($f \in \mathrm{BV}_{\mathrm{loc}}$).

Theorem 5 *If* $f\colon [a,b] \to \mathbb{R}$ *is monotonic then* $f \in \mathrm{BV}([a,b])$.

Theorem 6 *If* $f \in \mathrm{BV}([a,b])$ *then* f *is bounded on* $[a,b]$.

As already mentioned, in this note we will use the Lebesgue–Stieltjes integral. Here, we have to stress that there are other, non-equivalent, approaches to the Stieltjes integral, where some of them have serious defects. For a brief comparing summary we refer to [5, App. H].

As on p. 220 in [22], given a nondecreasing function $v\colon \mathbb{R} \to \mathbb{R}$, we define a function of intervals as $\nu(\emptyset) = 0$ and

$$\nu\big([a,b]\big) = v(b) - v(a).$$

For $v\colon \mathbb{R}_+ \to \mathbb{R}$ nondecreasing with $v(0)$ being finite, we extend this function to $\mathbb{R}$ by setting $v(x) = v(0)$ for all $x < 0$. Let $\mathcal{I}$ denote the class consisting of $\emptyset$ and all closed intervals $I = [a,b]$ (where $a < b$), so $\nu\colon \mathcal{I} \to \overline{\mathbb{R}}_+$ is now defined. For $E \subset \mathbb{R}$, we set

$$\mu^*(E) = \inf \sum_n \nu(I_n), \tag{6}$$

where the infimum is taken over all sequences $\{I_n\}$ from $\mathcal{I}$ such that $E \subset \cup I_n^\circ$ is contained in the union of the interiors I_n° of the intervals I_n, for more details we refer to [22].

From Theorem 4-10 II in [22], we obtain the following result.

Theorem 7 *Above defined* μ^* *is an outer measure, and all Borel sets are* μ^**-measurable. If* $I = [a,b]$ *is a closed interval, then* $\mu^*(I) = v(b+) - v(a-)$, *and especially*

$$\mu^*\big(\{a\}\big) = \mu^*\big([a,a]\big) = v(a+) - v(a-).$$

Denote by $\mathcal{F}$ the collection of μ^*-measurable sets. Then, $\mathcal{F}$ is a σ-algebra, and μ^* is countably additive on $\mathcal{F}$, see e.g. [13, Th. 5.2.5]. For the restriction of μ^* to $\mathcal{F}$, we will simply write μ, i.e. μ is a measure on $\mathcal{F}$ containing the Borel σ-algebra.

Definition 1 (Lebesgue–Stieltjes integral) Let $v\colon \mathbb{R} \to \mathbb{R}$ be monotonically increasing. We denote by μ the measure corresponding to v as described in the above derivation. If $f\colon \mathbb{R} \to \mathbb{R}$ is such that

$$\int_{\mathbb{R}} f \,\mathrm{d}\mu$$

exists (Lebesgue integral with respect to the measure μ), then we denote it by

$$\int_{\mathbb{R}} f \,\mathrm{d}v$$

and call it the *Lebesgue–Stieltjes integral* of f with respect to v.

If v is of (locally) bounded variation, i.e. $T_v[a,b] < \infty$ for all $a < b$, then we define the total variation function of v as the increasing function given by

$$Tv(x) = \begin{cases} v(0) + \sup\{\sum|v(x_k) - v(x_{k-1})|,\ 0 \le x_0 < \cdots < x_n \le x,\ x \ge 0\}, \\ v(0) - \sup\{\sum|v(x_k) - v(x_{k-1})|,\ x \le x_0 < \cdots < x_n \le 0,\ x \le 0\}. \end{cases}$$

By Theorem 6.4.5 in [16], and the proof of Theorem 6.1.15 in [16]

$$v_1(x) = \frac{1}{2}\big(Tv(x) + v(x)\big), \qquad v_2(x) = \frac{1}{2}\big(Tv(x) - v(x)\big) \tag{7}$$

are monotone increasing and $v = v_1 - v_2$. Now we set

$$\int_{\mathbb{R}} f\,\mathrm{d}v = \int_{\mathbb{R}} f\,\mathrm{d}v_1 - \int_{\mathbb{R}} f\,\mathrm{d}v_2, \tag{8}$$

provided that the right hand side makes sense.

If $A \subset \mathbb{R}$ is measurable with respect to the measures induced by v_1 and v_2, we set

$$\int_A f\,\mathrm{d}v = \int_{\mathbb{R}} f\chi_A\,\mathrm{d}v.$$

The key result we will need from the Lebesgue–Stieltjes integral theory is the integration by parts formula. The following result is an immediate consequence of Theorem III.14.1 in [19].

Theorem 8 *If v and w are two functions of bounded variation, we have for every interval $I = [a,b]$*

$$\int_{[a,b]} v(t)\,\mathrm{d}w(t) + \int_{[a,b]} w(t)\,\mathrm{d}v(t) = v(b+)w(b+) - w(a-)v(a-), \tag{9}$$

provided that at each point of I either one at least of the functions v and w is continuous.

If $f\colon \mathbb{R} \to \mathbb{R}$ and $v\colon \mathbb{R} \to \mathbb{R}$ satisfy $f, v \in \mathrm{BV}_{\mathrm{loc}}$, then f is measurable with respect to the measures μ_1 and μ_2 induced by v_1 and v_2, respectively. Furthermore, by Theorem 6, we have that f is bounded on $[a,b]$, so that by noticing that $[a,b]$ has finite μ_i-measure, $i = 1,2$, we obtain that $\int_{[a,b]} f\,\mathrm{d}v$ is finite for (finite) $a \le b$.

With the help of above definitions and results, we can now prove Theorem 1.

Proof of Theorem 1 By assumption, we can write $f = g_1 - g_2$ for $g_i : \mathbb{R}_+ \to \mathbb{R}$ being convex functions on $\mathbb{R}_+$, $i = 1,2$. The adapted statement for the case $I = \mathbb{R}_+$ of Theorem 4 is that $(g_i)'_r(0)$ exist at least in the infinite sense and $(g_i)'_r$ are increasing on $\mathbb{R}_+$, see [18, Chap. I, Sect. 11]. Hence, with the assumed finiteness of

$(g_i)'_r(0)$, we can extend $(g_i)'_r$ to $\mathbb{R}$ by $(g_i)'_r(0)$ for all $t < 0$ and with (the adapted) Theorem 4, these functions are increasing, so that Theorem 5 and Theorem 6 yield the finiteness of $(g_i)'_r(a)$ and $(g_i)'_l(b)$ for all $[a,b]$. Hence, Theorem 3 yields that $f \in \mathrm{AC}_{\mathrm{loc}}$. Furthermore, since the extended $(g_i)'_r$ are increasing, we obtain from Theorem 5 that they are locally of bounded variation, so that $f'_r \in \mathrm{BV}_{\mathrm{loc}}$ since $\mathrm{BV}_{\mathrm{loc}}$ is a linear space, see e.g. [16, p. 142]. As a consequence of Theorem 2, f' exists a.e. on $\mathbb{R}_+$ where clearly $f' = f'_r$ holds, i.e. f'_r is a representative of f' on $\mathbb{R}_+$ being in $\mathrm{BV}_{\mathrm{loc}}$.

Let us assume first that $c \le x$. We start by preparing some equalities that we will need. Applying Theorem 8 for the functions $v(x) = f'_r(x)$ and $w(x) = x$, we have

$$\int_{[c,x]} f'_r(k)\,\mathrm{d}k + \int_{[c,x]} k\,\mathrm{d}f'_r(k) = x f'_r(x+) - c f'_r(c-). \tag{10}$$

Noting that the measure corresponding to a constant function is trivial and using integration by parts for $v(x) = 1$ and $w(x) = f'_r(x)$, we obtain

$$f'_r(x+) - f'_r(c-) = \int_{[c,x]} \mathrm{d}f'_r(k) + \int_{[c,x]} f'_r(k)\,\mathrm{d}1 = \int_{[c,x]} \mathrm{d}f'_r(k). \tag{11}$$

Let us now prove the representation formula. We start with Theorem 2 and use the fact that if v is the identity, then the corresponding measure is the Lebesgue measure. Hence,

$$f(x) = f(c) + \int_{[c,x]} f'_r(k)\,\mathrm{d}k\,.$$

Using first (10) and then (11), we obtain

$$\begin{aligned}\int_{[c,x]} f'_r(k)\,\mathrm{d}k &= x f'_r(x+) - c f'_r(c-) - \int_{[c,x]} k\,\mathrm{d}f'_r(k)\\ &= x\left(f'_r(c-) + \int_{[c,x]} \mathrm{d}f'_r(k)\right) - c f'_r(c-) - \int_{[c,x]} k\,\mathrm{d}f'_r(k)\\ &= f'_r(c-)(x-c) + \int_{[c,x]} (x-k)\,\mathrm{d}f'_r(k). \end{aligned} \tag{12}$$

Let us write

$$\int_{[c,x]} (x-k)\,\mathrm{d}f'_r(k) = \int_{[c,x]} (x-k)_+\,\mathrm{d}f'_r(k) - \int_{[c,x]} (k-x)_+\,\mathrm{d}f'_r(k). \tag{13}$$

Note that by the assumption $c \le x$, all the following integrals vanish:

$$\int_{(x,\infty)} (x-k)_+\,\mathrm{d}f'_r(k), \qquad -\int_{[c,x]} (k-x)_+\,\mathrm{d}f'_r(k), \quad \text{and} \quad \int_{[0,c]} (k-x)_+\,\mathrm{d}f'_r(k).$$

It follows that

$$f(x) = f(c) + f_r'(c-)(x-c) + \int_{[c,\infty)} (x-k)_+ \, \mathrm{d}f_r'(k) + \int_{[0,c]} (k-x)_+ \, \mathrm{d}f_r'(k).$$

By noticing that

$$\int_{\{c\}} (x-k)_+ \, \mathrm{d}f_r'(k) = (x-c)\big(f_r'(c+) - f_r'(c-)\big),$$

the claim follows in the case $c \leq x$.

The proof for $x < c$ is very similar as before, so we skip some details. We have

$$f(x) = f(c) - \int_{[x,c]} f_r'(k) \, \mathrm{d}k.$$

As before

$$\begin{aligned}\int_{[x,c]} f_r'(k) \, \mathrm{d}k &= c f_r'(c+) - x f_r'(x-) - \int_{[x,c]} k \, \mathrm{d}f_r'(k) \\ &= c f_r'(c+) - x\left(f_r'(c+) - \int_{[x,c]} \mathrm{d}f_r'(k)\right) - \int_{[x,c]} k \, \mathrm{d}f_r'(k).\end{aligned}$$

Hence,

$$f(x) = f(c) + f_r'(c+)(x-c) + \int_{[x,c]} (k-x) \, \mathrm{d}f_r'(k).$$

Further,

$$\int_{[x,c]} (k-x) \, \mathrm{d}f_r'(k) = \int_{[x,c]} (k-x)_+ \, \mathrm{d}f_r'(k) - \int_{[x,c]} (x-k)_+ \, \mathrm{d}f_r'(k).$$

Since $x < c$, the integrals

$$\int_{[0,x)} (k-x)_+ \, \mathrm{d}f_r'(k), \qquad -\int_{[x,c]} (x-k)_+ \, \mathrm{d}f_r'(k), \quad \text{and} \quad \int_{(c,\infty)} (x-k)_+ \, \mathrm{d}f_r'(k)$$

all vanish. We can now use that the right derivatives of the g_is are finite in 0 in order to obtain a *suitable* convex extension (not ∞) to $\mathbb{R}$ so that the right continuity for f_r' (including $x = 0$) follows e.g. by [21, Th. 1.5.2]. □

An obvious question now is, how restrictive our assumptions on the boundary behavior are, which are, as already mentioned, often relaxed in approaches based on generalized functions. And it turns out that the assumptions are not completely harmless. E.g. the square root function defined by $f(x) = \sqrt{x}$ (satisfying that $-f$ is convex) clearly does not satisfy the conditions of Theorem 1, and it is also clear that this function cannot be represented for every $x \in \mathbb{R}_+$ without puts, i.e. by choosing

$c=0$. Related to that, note that this function is two times continuously differentiable on $(0,\infty)$ but not on $\mathbb{R}_+$ so that we could not use this function in place of the Counterexample 1. This fact shows that besides of clearly defining the meaning of the integral, it is also important to clearly identify the meaning of (continuous) differentiability when using static-hedging formulas. More generally, since $(g_i)'_+(0)$, $i=1,2$, are finite, we have that the functions g_i are Lipschitz on any $[0,b]$, $b>0$, see [18, Sect. 11]. However, for the example of the square root, it is not hard to see, that the representation is possible if c is restricted on $(0,\infty)$ being the interior of $\mathbb{R}_+$, since the integrand of the puts "tempers" the behavior of the second derivative near 0. This observation can be extended quite considerably.

We start by analyzing the boundary behavior at 0 of the right derivative of convex and continuous functions on $\mathbb{R}_+$.

Lemma 1 *Let $g\colon \mathbb{R}_+\to\mathbb{R}$ be convex and continuous. Then*

$$\lim_{x\to 0+} g'_r(x)x = 0.$$

Proof Let $x_1<x_2<x_3$ be positive real numbers. Then

$$\frac{g(x_2)-g(x_1)}{x_2-x_1}\le\frac{g(x_3)-g(x_2)}{x_3-x_2}. \tag{14}$$

Setting $x_1=x$ and $x_3=2x$ and multiplying by x, we obtain

$$\frac{g(x_2)-g(x)}{x_2-x}x\le\frac{g(2x)-g(x_2)}{2x-x_2}x\,.$$

We let x_2 tend (from the right) to x and obtain

$$g'_r(x)x\le g(2x)-g(x).$$

Now, we set $x_1=x/2$, $x_2=x$ in (14) and multiply again by x to obtain

$$\frac{g(x)-g(x/2)}{x/2}x\le\frac{g(x_3)-g(x)}{x_3-x}x.$$

We let x_3 tend (from the right) to x and obtain

$$2\big(g(x)-g(x/2)\big)\le g'_r(x)x.$$

Now $2(g(x)-g(x/2))\le g'_r(x)x\le g(2x)-g(x)$ and the continuity of g at 0 gives the claim. □

Assume again that $g\colon\mathbb{R}_+\to\mathbb{R}$ is convex and continuous. We define a sequence $(g_n)_n$ of functions defined on $\mathbb{R}$ by

$$g_n(x)=\begin{cases} g'_r(x), & x\ge\frac{1}{n},\\ g'_r(1/n), & \text{otherwise}.\end{cases} \tag{15}$$

This gives us a sequence $(\mu_n^*)_n$ of corresponding Borel outer measures.

Theorem 9 *For each $A \subset \mathbb{R}$, $\mu^*(A) = \lim_{n\to\infty} \mu_n^*(A)$ exists (we include the case $\mu^*(A) = \infty$ here) and defines a Borel outer measure μ^*. If A satisfies* $\inf A > 0$, *then there exists $N \in \mathbb{N}$ such that $\mu^*(A) = \mu_n^*(A)$ for each $n \geq N$.*

Proof Let us fix $A \subset \mathbb{R}$. We first want to show that the sequence $(\mu_n^*(A))_n$ is monotone increasing, verifying the existence of $\mu^*(A)$. Given a closed interval $I = [a, b]$, it easily follows that $g_{n+1}(a) \leq g_n(a)$. Note that if $g_{n+1}(b) < g_n(b)$, then it follows that $b \leq 1/n$ and hence $g_n(b) - g_n(a) = 0$. In conclusion, we have in any case,

$$g_n(b) - g_n(a) \leq g_{n+1}(b) - g_{n+1}(a).$$

It follows that $\mu_n^*(A)$ is monotone increasing and hence the limit exists.

It is clear that $\mu^*(A) \geq 0$ and $\mu^*(\emptyset) = 0$. Let us now assume that $A \subset \cup_i A_i$. Then

$$\mu^*(A) = \lim_{n\to\infty} \mu_n^*(A) \leq \lim_{n\to\infty} \sum_k \mu_n^*(A_k) \leq \lim_{n\to\infty} \sum_k \lim_{n\to\infty} \mu_n^*(A_k) = \sum_k \mu^*(A_k)$$

verifying that μ^* is an outer measure. That each Borel sets is measurable follows easily.

Let us now assume that $\inf A > 0$. Determine $N \in \mathbb{N}$ such that

$$\frac{1}{N-1} \leq \inf A.$$

We note that in the computation of $\mu_N^*(A)$, we can additionally require that the left endpoints of the intervals in the covering are larger than $1/N$. It follows that the value $\mu^*(A) = \mu_n^*(A)$ for all $n \geq N$. □

A measure μ is again obtained by restricting μ^* to the collection $\mathcal{F}$ of μ^* measurable sets. Note that e.g. all continuous functions are measurable with respect to μ.

Let $f\colon \mathbb{R}_+ \to \mathbb{R}$ be the difference of two continuous convex functions, $f = g - h$, where $g, h\colon \mathbb{R}_+ \to \mathbb{R}$. Denote by $\mu_{g_r'}$ ($\mu_{h_r'}$) the measures obtained from g (h) by the construction of outer measures given in Theorem 9. Furthermore, assume that $F\colon \mathbb{R}_+ \to \mathbb{R}$ is measurable with respect to the σ-algebras of the $\mu_{g_r'}$- and $\mu_{h_r'}$-measurable sets. We introduce the following notation

$$\int_{\mathbb{R}_+} F \, \mathrm{d}\tilde{f}_r' = \int_{\mathbb{R}_+} F \, \mathrm{d}\mu_{g_r'} - \int_{\mathbb{R}_+} F \, \mathrm{d}\mu_{h_r'},$$

provided that the right hand side makes sense. Note that in the context of Theorem 1, this definition yields an equivalent representation since the Stieltjes integral on bounded sets does not depend on the way we split f_r' into a difference of monotonically increasing functions.

Proposition 1 *Assume that* $f\colon \mathbb{R}_+ \to \mathbb{R}$ *is the difference of two continuous convex functions* $g, h\colon \mathbb{R}_+ \to \mathbb{R}$. *Suppose that* $c \in (0,\infty)$. *Then*

$$f(x) = f(c) + f_r'(c)(x-c) + \int_{(c,\infty)} (x-k)_+ \,\mathrm{d}\tilde{f}_r'(k) + \int_{[0,c]} (k-x)_+ \,\mathrm{d}\tilde{f}_r'(k) \quad (16)$$

for all $x \in \mathbb{R}_+$.

Proof Since the difference of two representations of the from (16), is again of the form (16), it suffices to prove that the representation holds for g. By Theorem 4, $g_r'(1/n)$ exists and is finite for every interval $I_n = [1/n, \infty)$. Furthermore, for every $x > 0$, there is an $N \in \mathbb{N}$ such that for all $n \geq N$ we have $1/(n-1) < x$. Since $\mu_{g_r'}$ coincides with μ_{g_n} on I_{n-1} for $n \geq N$, we obtain by a slight modification of Theorem 1 and by $0 < 1/(n-1) < x$ that

$$\begin{aligned} g(x) &= g(c) + g_r'(c)(x-c) + \int_{(c,\infty)} (x-k)_+ \,\mathrm{d}\mu_{g_n}(k) + \int_{[\frac{1}{(n-1)},c]} (k-x)_+ \,\mathrm{d}\mu_{g_n}(k) \\ &= g(c) + g_r'(c)(x-c) + \int_{(c,\infty)} (x-k)_+ \,\mathrm{d}\mu_{g_r'}(k) + \int_{[0,c]} (k-x)_+ \,\mathrm{d}\mu_{g_r'}(k), \end{aligned}$$

where we have w.l.o.g. assumed that also $1/(n-1) < c$ holds and where μ_{g_n} stands for the measure obtained from g by (15). Hence, it remains to consider the limiting case $x \downarrow 0$. For $c > 0$ we can assume w.l.o.g. that $0 < x \leq c$. Along with $x > 0$, we obtain

$$\begin{aligned} g(x) &= g(c) + g_r'(c)(x-c) + \int_{[0,c]} (k-x)_+ \,\mathrm{d}\mu_{g_r'}(k) \\ &= g(c) + g_r'(c)(x-c) + \int_{(x,c]} (k-x) \,\mathrm{d}\mu_{g_r'}(k), \end{aligned}$$

or equivalently

$$g(x) - g(c) - g_r'(c)(x-c) = \int_{(x,c]} (k-x) \,\mathrm{d}\mu_{g_r'}(k).$$

By letting $x \downarrow 0$, we see that the l.h.s. of this equation is clearly finite, and so is the r.h.s., i.e.

$$\lim_{x\to 0+} \int_{(x,c]} (k-x) \,\mathrm{d}\mu_{g_r'}(k) = \lim_{x\to 0+} \left(\int_{(x,c]} k \,\mathrm{d}\mu_{g_r'}(k) - x g_r'(c+) + g_r'(x+)x \right)$$

exists and is finite. Applying Lemma 1 (and the right continuity of g_r' at $x > 0$, see [18], p. 7), we obtain that

$$\lim_{x\to 0+} \int_{(x,c]} k \,\mathrm{d}\,\mu_{g_r'}(k) = \lim_{x\to 0+} \int_{[0,c]} k \chi_{(x,c]} \,\mathrm{d}\,\mu_{g_r'}(k)$$

exists and is finite. As a consequence of monotone convergence and by using $\int_{\{0\}} k \, d\mu_{g'_r}(k) = 0$, we obtain that

$$\lim_{x \to 0+} \int_{[0,c]} k\chi_{(x,c]} \, d\mu_{g'_r}(k) = \int_{[0,c]} k \, d\mu_{g'_r}(k) ,$$

exists and is finite. □

Example 2 Assume that $0 < k_0$ and $f \colon \mathbb{R}_+ \to \mathbb{R}$ is defined as $f(x) = (x - k_0)_+$, which is certainly a convex function. Then $f(0) = f'_r(0) = 0$ and

$$f'_r(x) = \begin{cases} 0, & 0 \leq x < k_0, \\ 1, & k_0 \leq x. \end{cases}$$

Let μ be the measure induced by (the extended) $v = f'_r$. By Theorem 7, we obtain that $\mu(\{k_0\}) = f'_r(k_0+) - f'_r(k_0-) = 1$. Furthermore, from [22, Sect. 4.10], cf. also [11, p. 50], it follows that $\mu([0, k_0)) = f'_r(k_0-) - f'_r(0-)$, where $f'_r(0-) = f'_r(0)$ (as defined above), so that $\mu([0, k_0)) = 0 - 0 = 0$, and it also follows that

$$\mu\big((k_0, \infty)\big) = f'_r(\infty) - f'_r(k_0+) = 1 - 1 = 0 .$$

Finally we note that $\mu(\{0\}) = 0 = \mu((0, k_0))$. Hence, with $f(0) = f'_r(0+) = 0$ and by letting $c = 0$, we obtain from Theorem 1 that

$$f(x) = \int_{(0,\infty)} (x - k)_+ \, d f'_r(k) + \int_{\{0\}} (k - x)_+ \, d f'_r(k) = (x - k_0)_+ ,$$

as expected.

Example 3 Let us assume that $0 < k_0$ and $f \colon \mathbb{R}_+ \to \mathbb{R}$ is defined as $f(x) = (k_0 - x^2)_+$. We can rewrite this as

$$f(x) = \begin{cases} k_0 - x^2, & x \leq \sqrt{k_0}, \\ 0, & \text{otherwise.} \end{cases}$$

The function f is the difference of the two convex functions $f_i \colon \mathbb{R}_+ \to \mathbb{R}$ defined as

$$f_1(x) = \begin{cases} k_0, & x < \sqrt{k_0}, \\ x^2, & x \geq \sqrt{k_0}, \end{cases}$$

and $f_2(x) = x^2$, so that $(f_i)'_r(0) = 0$, $i = 1, 2$, and $f'_r(0) = 0$. We obtain

$$f'_r(x) = \begin{cases} -2x, & 0 \leq x < \sqrt{k_0}, \\ 0 & x \geq \sqrt{k_0}. \end{cases}$$

We note that $f_r'(x) = v_1(x) - v_2(x)$, where

$$v_1(x) = \begin{cases} 0, & 0 \le x < \sqrt{k_0}, \\ 2\sqrt{k_0}, & x \ge \sqrt{k_0}, \end{cases}$$

$$v_2(x) = \begin{cases} 2x, & 0 \le x < \sqrt{k_0}, \\ 2\sqrt{k_0}, & x \ge \sqrt{k_0}, \end{cases}$$

are monotonically increasing functions. Denote by μ_i, $i = 1, 2$, the measures induced by (the extended) v_i, $i = 1, 2$. Hence,

$$\begin{aligned} \mu_1\big([0, \sqrt{k_0})\big) &= v_1(\sqrt{k_0}-) - v_1(0-) = 0, \\ \mu_1\big(\{k_0\}\big) &= v_1(\sqrt{k_0}+) - v_1(\sqrt{k_0}-) = 2\sqrt{k_0}, \\ \mu_1\big((\sqrt{k_0}, \infty)\big) &= v_1(\infty-) - v_1(\sqrt{k_0}+) = 0, \end{aligned}$$

and clearly $\mu_1(\{0\}) = 0 = \mu_1((0, \sqrt{k_0}))$. Furthermore, we have for all x, with $0 \le a \le x \le b < \sqrt{k_0}$ that

$$\begin{aligned} \mu_2\big([a, b]\big) &= v_2(b+) - v_2(a-) = 2(b - a) = 2\mathcal{L}\big([a, b]\big), \\ \mu_2\big([a, \sqrt{k_0}]\big) &= v_2(\sqrt{k_0}+) - v_2(a-) = 2(\sqrt{k_0} - a) = 2\mathcal{L}\big([a, \sqrt{k_0}]\big), \end{aligned}$$

along with $\mu_2((\sqrt{k_0}, \infty)) = v_2(\infty-) - v_2(\sqrt{k_0}+) = 0$.

In view of that and by noticing that $f(0) = k_0$, $f_r'(0+) = 0$, the formula in Theorem 1 for $c = 0$ reads

$$\begin{aligned} f(x) &= k_0 + \int_{(0,\infty)} (x - k)_+ \, \mathrm{d} f_r'(k) \\ &= k_0 + \int_{[0,\infty)} (x - k)_+ \mu_1(\mathrm{d}k) - \int_{[0,\infty)} (x - k)_+ \mu_2(\mathrm{d}k) \,. \end{aligned}$$

If $x \le \sqrt{k_0}$, then the first integral in above sum vanishes (note that for $x = \sqrt{k_0}$, the integrand for $k = \sqrt{k_0}$ vanishes in the first integral), the second one gives $-x^2$. If $x > \sqrt{k_0}$, then the second integral evaluates to $-2(\sqrt{k_0}x - \frac{1}{2}k_0)$ and the first to $2\sqrt{k_0}(x - \sqrt{k_0})$. Hence, we get f back.

In the above examples, it was easy to verify the assumptions of Theorem 1. For more complicated cases, Theorem D on p. 26, and even more explicitly, Theorem B on p. 24 in [18] can be helpful.

Acknowledgements The authors are grateful to Katrin Fässler, Ilya Molchanov, Jean-Francois Renaud, and Thorsten Rheinländer for helpful hints and discussions. This work was supported by the Swiss National Science Foundation Grant Nr. 200021-126503 and PBBEP3_130157.

References

1. Albrecher, H., Mayer, P.: Semi-static hedging strategies for exotic options. In: Kiesel, R., Scherer, M., Zagst, R. (eds.) Alternative Investments and Strategies, pp. 345–373. World Scientific, Singapore (2010)
2. Bakshi, G., Madan, D.: Spanning and derivative-security valuation. J. Financ. Econ. **55**, 205–238 (2000)
3. Baldeaux, J., Rutkowski, M.: Static replication of univariate and bivariate claims with applications to realized variance swaps. Working paper, University of New South Wales (2007)
4. Baldeaux, J., Rutkowski, M.: Static replication of forward-start claims and realized variance swaps. Appl. Math. Finance **17**, 99–131 (2010)
5. Bartle, R.G.: A Modern Theory of Integration. AMS, Rhode Island (2001)
6. Carr, P., Chou, A.: Breaking barriers. Risk **10**, 139–145 (1997)
7. Carr, P., Chou, A.: Hedging complex barrier options. Working paper, NYU's, Courant Institute and Enuvis Inc (2002)
8. Carr, P., Lee, R.: Put-call symmetry: extensions and applications. Math. Finance **19**, 523–560 (2009)
9. Carr, P., Madan, D.B.: Towards a theory of volatility trading. In: Jarrow, R. (ed.) Volatility, pp. 417–427. Risk Publications, London (1998)
10. Carr, P., Madan, D.B.: Optimal positioning in derivative securities. Quant. Finance **1**, 19–37 (2001)
11. Carter, M., van Brunt, B.: The Lebesgue-Stieltjes Integral. A Practical Introduction. Springer, New York (2000)
12. Cont, R., Tankov, P.: Financial Modelling with Jump Processes. Chapman & Hall/CRC, London (2004)
13. Gerald, E.: Measure, Topology, and Fractal Geometry, 2nd edn. Springer, New York (2008)
14. Gordon, R.A.: The Integrals of Lebesgue, Denjoy, Perron, and Henstock. AMS, Rhode Island (1994)
15. Henry-Labordère, P.: Analysis, Geometry, and Modeling in Finance. Advanced Methods in Option Pricing. Chapman & Hall, Boca Raton (2009)
16. Kannan, R., Krueger, C.K.: Advanced Analysis on the Real Line. Springer, New York (1996)
17. Lipton, A.: Mathematical Methods for Foreign Exchange: A Financial Engineer's Approach. World Scientific, Singapore (2001)
18. Roberts, A.W., Varberg, D.E.: Convex Functions. Academic Press, New York (1973)
19. Saks, S.: Theory of the Integral, 2nd edn. Hafner, New York (1937)
20. Schmutz, M., Zürcher, T.: Static replications with traffic light options. Accepted for publication in J. Futures Mark.; Early View: http://onlinelibrary.wiley.com/doi/10.1002/fut.21621/full
21. Schneider, R.: Convex Bodies. The Brunn–Minkowski Theory. Cambridge University Press, Cambridge (1993)
22. Taylor, A.E.: General Theory of Functions and Integration. Blaisdell, Waltham (1965)

Optimal Stopping of Seasonal Observations and Projection of a Markov Chain

Isaac M. Sonin

Abstract We consider the recently solved problem of Optimal Stopping of Seasonal Observations and its more general version. Informally, there is a finite number of dice, each for a state of "underlying" finite MC. If this MC is in a state k, then k-th die is tossed. A Decision Maker (DM) observes both MC and the value of a die, and at each moment of discrete time can either continue observations or to stop and obtain a discounted reward. The goal of a DM is to maximize the total expected discounted reward. This problem belongs to an important class of stochastic optimization problems—the problem of optimal stopping of Markov chains (MCs). The solution was obtained via an algorithm which is based on the general, so called, State Elimination algorithm developed by the author earlier. An important role in the solution is played by the relationship between the fundamental matrix of a transient MC in the "large" state space and the fundamental matrix for the modified underlying transient MC. In this paper such relationship is presented in a transparent way using the general concept of a projection of a Markov model. The general relationship between two fundamental matrices is obtained and used to clarify the solution of the optimal stopping problem.

Keywords Markov chain · Optimal stopping · Elimination algorithm · Seasonal observations

Mathematics Subject Classification (2010) 60G42 · 60J10 · 82B35

1 Introduction

The problem described below was formulated in [7] and dubbed as Optimal Stopping of *Seasonal Observations*. The solution was published recently in [5]. The goal of this note is to introduce the notion of a *projection* of a Markov chain (MC),

I.M. Sonin (✉)
Department of Mathematics and Statistics, University of North Carolina at Charlotte, Charlotte, NC 28223, USA
e-mail: imsonin@uncc.edu

Y. Kabanov et al. (eds.), *Inspired by Finance*, DOI 10.1007/978-3-319-02069-3_25,

which is of interest in its own right, and using this concept to obtain one of the key equalities in [5] in a more general form.

Seasonal observations. Suppose that (U_n), $n \geq 0$ is a MC with values in a finite set $B = \{1, 2, \ldots, m\}$ and known transition matrix $U = \{u(s,k), s, k \in B\}$. Suppose that there are m different "dice", each die for a state in B, and the probability that k-th die takes value $j \in Z = \{1, 2, \ldots\}$ is $f(j|k), k \in B, j \in Z$. If at the moment n the MC (U_n) takes value k, then the k-th die is tossed and a Decision Maker (DM) observes both U and the value j obtained. At each moment $n = 0, 1, 2, \ldots$ a DM can either continue observations or to stop and obtain a discounted reward $\beta^n g(k, j)$, where β is a discount factor, $0 < \beta \leq 1$, and $g(k, j)$ is the *terminal reward function.* The goal of a DM is to maximize the total expected discounted reward. This problem can be generalized if one introduces a *one step cost function* $c(k)$, but for simplicity we assume that $c(k) = 0$ for all k. Formally, we assume that a DM observes MC (Z_n) with values in $X = B \times Z$ and with transition probabilities $p(x, y) \equiv p(s, i; k, j) = u(s, k) f(j|k), s, k \in B, i, j \in Z$. Thus, these probabilities depend only on the first "horizontal" coordinate of a state $x = (s, i)$. We can represent this relationship symbolically by the "factorization equality"

$$P = U \times F, \tag{1}$$

where U is $m \times m$ stochastic matrix and $F = \{f(\cdot|k), k \in B\}$ is a vector of distributions on Z.

2 Optimal Stopping of MC

The problem described above belongs to an important class of stochastic optimization problems—the *problem of optimal stopping (OS) of MC*, where a DM observing a MC, has two possible actions at each moment of discrete time: to continue observations or to stop, and then to obtain a terminal reward. Formally, such a problem is specified by a tuple $M = (X, P, c, g, \beta)$, where X is a state space, $P = \{p(x, y)\}$ is a transition matrix, $c(x)$ is a *one step cost function*, $g(x)$ is a *terminal reward function*, and β is a discount factor, $0 < \beta \leq 1$. We call such a model *OS model* and a tuple $M = (X, P)$, we call a *Markov model*. The *value function* $v(x)$ for OS model is defined as $v(x) = \sup_{\tau \geq 0} E_x[\sum_{i=0}^{\tau-1} \beta^i c(Z_i) + \beta^\tau g(Z_\tau)]$, where the sup is taken over all stopping times $\tau \leq \infty$. To simplify our presentation we will assume that $c(x) = 0$ and $v(x) < \infty$ for all x.

It is well-known that in stochastic optimization problems the discounted case can be treated as undiscounted if an absorbing point e is introduced and the transition probabilities are modified as follows:

$$p^\beta(x, y) = \beta p(x, y), \quad x, y \in X, \qquad p^\beta(x, e) = 1 - \beta, \qquad p^\beta(e, e) = 1.$$

In other words, with probability β the Markov chain "survives" and with complimentary probability it transits to an absorbing state e. More than that, for our method it is convenient and important to consider a more general situation when the constant β can be replaced by the probability of "survival", that is by the function $\beta(x) = P_x(Z_1 \neq e)$, $0 \leq \beta(x) \leq 1$. Further we will assume that this transformation is made and we skip the superscript β, using again notation P_x and E_x.

Let $Pf(x)$ be the *averaging operator*, $Pf(x) = \sum_y p(x, y) f(y)$. It is well-known that the value function v is a minimal solution of a corresponding *Bellman (optimality) equation* $v = \max(g, c + Pv)$. Let $A \subset B \times Z$, that is $A = \{A(k)\}$, $A(k) \subset Z$, $k \in B$ and let us denote by $F(A(k)|k) = \sum_{j \in A(k)} f(j|k)$ and by $F_d(A)$ the $m \times m$ diagonal matrix $F_d(A) = (\delta_{sk} F(A(k)|k))$, $s, k \in B$. The complement of a set D_* is denoted by S_*. The following theorem was proved in [5].

Theorem 1 *There is a vector* $d^* = (d_1^*, \ldots, d_m^*)$ *such that*

(a) *an optimal stopping time* τ^* *is the moment of first visit of the Markov chain* Z *to the set* $\{e\} \cup S^*$, *where*

$$S^* = \{z = (k, j) : k \in B, \ j \in S^*(k)\}, \qquad S^*(k) = \{j : g(k, j) \geq d_k^*\};$$

(b) *the value function satisfies the equation*

$$v(x) = g(x), \quad x \in S^*, \qquad v(x) = d_k^* > g(k, j), \quad x = (k, j) \in D^* = X \setminus S^*, \tag{2}$$

and d^* *satisfies the equation*

$$d_s^* = \sum_{k \in B} l^*(s, k) \sum_{j \in D^*(k)} g(k, j) f(j|k), \tag{3}$$

where the matrix $L^* = \{l^*(s, k), \ s, k \in B\}$ *is defined by the equality*

$$L^* = \left[I - U F_d(D^*)\right]^{-1} U. \tag{4}$$

The proof of Theorem 1 is obtained via an algorithm which allows one to find the vector d^*, and, therefore, to construct the value function and the optimal stopping set in a finite number of steps. This algorithm is based on the general, so called, State Elimination (SE) algorithm developed by the author earlier and described in [8] (see also [9]). This algorithm has some features in common with the so called State Reduction (SR) approach used in computational MCs and which is exemplified by works of Grassmann, Taksar, Heyman [1] and Sheskin [6], who independently developed GTH/S algorithm to calculate the invariant distribution for an ergodic MC. The explanation of this approach is given in [9]. We first briefly describe this approach and afterwards we explain the SE algorithm. Our notations in these sections are slightly different than those used in the original author's papers.

3 Recursive Calculation of Characteristics of MC and the State Reduction (SR) Approach

Let us assume that a *Markov model* $M = (X, P)$ is given and let $D \subset X$, $S = X \setminus D$. Then the matrix $P = \{p(x, y)\}$ can be decomposed as the first matrix below

$$P = \begin{bmatrix} Q & T \\ R & P_0 \end{bmatrix}, \qquad P'_S = \begin{bmatrix} 0 & NT \\ 0 & P_S \end{bmatrix} \tag{5}$$

where the substochastic matrix Q describes the transitions inside of D, P_0 describes the transitions inside of S and so on. Let us introduce the sequence of Markov times $\tau_0, \tau_1, \dots, \tau_n, \dots$, the moments of zero, first, and so on, return of (Z_n) to the set S, i.e., $\tau_0 = 0$, $\tau_{n+1} = \min\{k > \tau_n : Z_k \in S\}$. Let us consider the sequence of random variables $Y_n = Z_{\tau_n}$, $n = 0, 1, 2, \dots$, $Z_0 \in S$. The strong Markov property and standard probabilistic reasoning imply the following basic lemma of the SR approach which probably should be credited to Kolmogorov and Doeblin.

Lemma 1 (a) *The sequence* (Y_n) *is a Markov chain in the model* $M_S = (S, P_S)$, *where* $S = X \setminus D$ *and*

(b) *the transition matrix* $P_S = \{p_S(x, y), x, y \in S\}$ *is given by the formula*

$$P_S = P_0 + RV = P_0 + RN_DT, \tag{6}$$

where $V = N_DT$ *is the matrix of distribution of the MC at the moment of first return to* S, *and* $N_D = N$ *is the fundamental matrix for the substochastic matrix* $Q = \{p(x, y), x, y \in D\}$.

We remind that $N = \sum_{n=0}^{\infty} Q^n = (I - Q)^{-1}$, where I is the $|D| \times |D|$ identity matrix. This representation is proved, for example, in the classical text of Kemeny and Snell, [3]. This matrix N satisfies also the equality

$$N = I + QN = I + NQ. \tag{7}$$

An important case is when the set D consists of one nonabsorbing point z. In this case formula (6) takes the form

$$p_S(x, \cdot) = p(x, \cdot) + p(x, z)n(z)p(z, \cdot), \tag{8}$$

where $n(z) = 1/(1 - p(z, z))$. According to this formula, each row-vector of the new stochastic matrix P_S is a linear combination of two rows of P (with the z-column deleted). This transformation corresponds formally to one step of the Gaussian elimination method. This matrix P_S describes the behavior of MC with values in a set S, or we can extend this matrix to the full size $X \times X$ matrix P'_S, see the second matrix in (5), assuming that MC (Y_n) can have an initial point in set D also. But in both cases, to obtain the matrix P_S, we need to study the behavior of the related transient MC with values in D.

The matrix N, a fundamental matrix for this transient MC with transition matrix Q, has the following well known probabilistic interpretation,

$$N = \{n(x,y), x, y \in D\}, n(x,y) = E_x \sum_{n=0}^{\tau_S} I_y(Z_n),$$

where τ_S is the moment of the first visit to S, i.e. $\tau_S = \min(n \geq 0 : x_n \in S)$ (moment of first exit from D), i.e. the expected number of visits to y starting from x till τ_S. In this case, i.e. when the transition matrix P is changed in such a way that S become an absorbing set, we shall say that MC (Z_n) is *stopped* at $S = X \setminus D$, and we shall denote this new MC as (Z_n^D).

The recursive calculation of the second fundamental matrix, for the ergodic MC was described in [10].

If an initial Markov model $M_1 = (X_1, P_1)$, is finite, $|X_1| = k$, and only one point is eliminated each time, then a sequence of stochastic matrices $(P_n), n = 2, \ldots, k$, can be calculated recursively on the basis of formula (8). Generally, a set of points D can be eliminated using formula (6). In both cases such sequence of stochastic matrices provides an opportunity to calculate many characteristics of the initial Markov model M_1 recursively starting from some reduced model M_s, $1 < s \leq k$.

4 State Elimination (SE) Algorithm

In this section we describe briefly the SE algorithm (for the case of $c(x) = 0$). Let an OS model $M = (X, P, g)$, be given, and suppose that an optimal stopping set $S^* = \{x : v(x) = g(x)\}$ does exists. Let a subset $D \subset \{x : g(x) < Pg(x)\}$. Since $g(x) \leq v(x)$, and $Pg(x) \leq Pv(x)$ the optimality equation implies that $D \cap S^* = \emptyset$. It was proved in [8] that the optimal stopping set in the reduced OS model $M_S = (X_S = X \setminus D, P_S, g)$ will be the same as in the initial OS model and the value functions will be the same for all points in X_S. After that we can repeat the process by eliminating points in a set $D' \subset \{x : g(x) - P_S g(x) < 0\}$ and so on. If at some stage after k steps, with $D_1 = D, D_2 = D' \cup D_1$ and so on, we obtain that $g(x) - P_{S_k} g(x) \geq 0$ for all remaining points, then $S^* = S_k = X \setminus D_k$. For the finite space X this algorithm solves the OS problem in no more than $|X|$ steps, and allows us also to find the distribution of the MC at the moment of stopping in an optimal stopping set S^*. Recently E. Presman modified this idea and applied to the case of OS in continuous time, see [4].

5 Projection of MC and Seasonal Observations

We note that the matrix $[I - UF_d(D^*)]^{-1}$ from formula (4) is the *fundamental* matrix for the transient MC obtained from the underlying MC (U_n) by modifying its transition matrix U. An important role in the proof of Theorem 1 is played by the relationship between the fundamental matrix of a transient MC in the state space X and the fundamental matrix for the modified transient MC in the state space B. This relationship can be presented in a transparent way using the concept of *projection of a Markov model*, and, correspondingly of projection of a MC.

Let $M_i = (X_i, P_i)$ be two Markov models, $i = 1, 2$ and let $h : X_1 \longrightarrow X_2$ be a mapping. If (Z_n) is a MC in model M_1 then generally random sequence $(U_n), U_n = h(Z_n)$ is not a MC in the model M_2. In [2] Howard introduced a notion of a "mergeable" Markov chain when the random sequence (U_n) is a MC. In terms of two models, the model M_1 is mergeable if the transitional probabilities in these models for any $x, x' \in h^{-1}(s) \subset X_1$ and any $s, k \in X_2$ satisfy the following equality: $\sum_{y \in h^{-1}(k)} p_1(x, y) = \sum_{y \in h^{-1}(k)} p_1(x', y)$. If these two Markov models have terminal reward functions $g_1(x), x \in X_1$, $g_2(k), k \in X_2$ and terminal reward function g_1 is also "mergeable", i.e. if $g_1(x) = g_2(h(x))$ for all $x \in h^{-1}(k), k \in X_2$ then of course the solution of the OS in M_1 can be reduced to the solution in M_2, but this is a trivial situation. To be able to consider the OS problem for the seasonal observations we need a stronger assumption.

We say that the model M_2 is a *projection* of a model M_1 (under h) if the transitional probabilities in these models satisfy the following property for all $x, y \in X_1$,

$$p_1(x, y) = p_2\big(h(x), h(y)\big) f_1\big(y|h(y)\big), \tag{9}$$

where $f_1(y|t)$ is a probability distribution on a set $h^{-1}(t) = \{y \in X_1 : h(y) = t\}$, defined for each $t \in X_2$. In other words, the state space X_1 is partitioned into classes $T_t = h^{-1}(t)$, $t \in X_2$ and transitions from the state x in the class T_s to the state y in T_k depend only on s, k and y but not on x. The reader may think about the model M_1 as a "large", basic model and about the model M_2 as a "small", more manageable model. It is easy to check that if (Z_n) is a MC in the model M_1 then the random sequence (U_n), $U_n = h(Z_n)$ is a MC in model M_2.

To simplify our presentation we will assume that the sets X_1 and X_2 are discrete and that the Markov model M_2 has an absorbing state e. Let $|X_2| = m + 1$, where $m \leq \infty$ is the number of proper states, i.e. $x \neq e$. Let $D \subset X_1$, $S = X_1 \setminus D$. We consider MC (Z_n^D) stopped at $S = X \setminus D$. According to the SE algorithm if a set D should be eliminated then in order to find the matrix P_S by formula (6), we have to find the fundamental matrix $N_{1,D} = \{n_{1,D}(x, y), x, y \in D\}$.

To accomplish this goal we will introduce MC (U_n^D) in the model M_2, the "projection" of MC (Z_n^D), defined by the equality $U_n^D = h_D(Z_n^D)$, where $h_D(x) = h(x)$ if $x \in D$ and $h_D(X) = e$ if $x \in X_1 \setminus D$. In Theorem 2 we will relate the fundamental matrix $N_{2,D}$ for this MC with the matrix $N_{1,D}$.

If P is an $m \times m$ stochastic matrix, $D \subset X_1$ and $F_d(D)$ is the $m \times m$ diagonal matrix with elements $F(D(k)) = \sum_{j \in D(k)} f_1(j|k)$, $D(k) = D \cap h^{-1}(k)$, then we denote substochastic matrix $P_D = P F_d(D)$ and we denote the fundamental matrix for P_D as $(I - P_D)^{-1} = N_D$.

We denote $M_{2,D} = (X_2, P_{2,D}^*)$ a Markov model obtained from model M_2 as follows. The state space is the same, X_2 and the transition matrix $P_{2,D}^* = P_2 F^*(D)$, where $F^*(D)$ is the $(m+1) \times (m+1)$ matrix, which has in the upper left corner the $m \times m$ diagonal matrix $F_d(D)$ described above, and the last column of matrix $F^*(D)$ contains entries $f^*(s,e) = 1 - \sum_k p_2(s,k)F(D(k)), s \neq e, f^*(e,e) = 1$. In other words, in this model the transitional probabilities are: $p_{2,D}(s,k) = p_2(s,k)F(D(k))$, for $k \neq e$, and $p_{2,D}(s,e) = p_2(s,e) + \sum_k p_2(s,k)F(S(k))$. We denote the $m \times m$ upper left corner of matrix $P^*_{2,D}$ by $P_{2,D}$. According to the definition of $P^*_{2,D}$, we have $P_{2,D} = P_2 F_d(D) \equiv P_D$. This is a substochastic matrix for the transient MC in model $M_{2,D}$ with absorption in e.

Let us consider $N_{2,D} = \{n_{2,D}(s,k), s,k \in X_2, s,k \neq e\}$, the fundamental matrix for $P_{2,D}$. The following theorem holds.

Theorem 2 *If* (Z_n) *is a Markov chain in model* M_1 *and* $D \subset X_1$ *then*

(a) *the random sequence* (U_n), $U_n = h(Z_n)$ *is a MC in model* M_2 *with the transition matrix* P_2; *the random sequence* (U_n^D), $U_n^D = h_D(Z_n^D)$ *is a MC in model* $M_{2,D}$ *with the transition matrix (for the proper states)* $P^*_{2,D}$ *described above*;

(b) *the fundamental matrices in the original and the projected models,* $N_{1,D}$ *and* $N_{2,D}$ *are related by the equalities valid for all* $x, y \in D \subset X_1, s, k \in X_2, s, k \neq e$,

$$n_{1,D}(x,y) = n_{2,D}(s,k) f_1(y|k)/F\big(D(k)\big), s = h(x), k = h(y); \tag{10}$$

(c) *stochastic matrix* $P_{1,S}$ *has factorization*

$$P_{1,S} = U_{2,S} \times F_S, \tag{11}$$

where $F_S = \{f_{1,S}(y|k) = f_1(y|k)/F(S(k))\}, k \in X_2$ *and*

$$U_{2,S} = P_{2,S} + P_{2,D}\big(I - P_{2,D}\big)^{-1} P_{2,S} = N_{2,D} P_{2,S}. \tag{12}$$

Proof We omit the proof of point (a) which can be obtained using standard probability reasoning. To prove (b) note that by the definition of a fundamental matrix for a MC (Z_n^D) stopped at $S = X_1 \setminus D$, we have

$$n_{1,D}(x,y) = E_{1,x} \sum_{n=0}^{\infty} I_y\big(Z_n^D\big) = \sum_{n=0}^{\infty} P_{1,x}\big(Z_n^D = y\big).$$

According to (9) we have

$$\begin{aligned} P_{1,x}\big(Z_n^D = y\big) &= P_{2,s}\big(U_n^D = k\big) P_{1,x}\big(Z_n^D = y | h\big(Z_n^D\big) = k\big) \\ &= P_{2,s}\big(U_n^D = k\big) f_1(y|k)/F\big(D(k)\big). \end{aligned}$$

Using the equality $n_{2,D}(s,k) = \sum_{n=0}^{\infty} P_{2,s}(U_n^D = k)$, we obtain (10).

Point (c). Using the formula (6), factorization (1), (9), and introducing the notations $x = (s,x')$, $y = (k,y')$, $z = (l,z')$ and $v = (t,v')$, we have

$$p_{1,S}(x,y) = p_1(x,y) + \sum_z p_1(x,z) \sum_v n_{1,D}(z,v) p_1(v,y)$$

$$= p_2(s,k) f_1\big(y'|k\big) + \sum_l p_2(s,l) \sum_{z'\in D(l)} f_1\big(z'|l\big) \sum_v n_{1,D}(z,v) p_2(t,k) f_1\big(y'|k\big).$$

Using point (b), i.e. replacing $n_{1,D}(z,v)$ by $n_{2,D}(l,t) f_1(y'|t)/F_1(D(t))$, and using the equality $\sum_{z'\in D(t)} f_1(z'|t) = F(D(t)), t \in X_2$, we have

$$\sum_{v=(t,v')} n_{1,D}(z,v) p_2(t,k) = \sum_t n_{2,D}(l,t) \sum_{v'\in D(t)} \frac{f_1(v'|t)}{F(D(t))} p_2(t,k)$$
$$= \sum_t n_{2,D}(l,t) p_2(t,k). \tag{13}$$

From the equalities $\sum_{z'\in D(l)} f_1(z'|l) = F(D(l)), l \in X_2$, $P_{2,D} = P_2 F_{1,d}(D)$, and (13), we obtain finally

$$p_{1,S}(x,y) = \Big[p_2(s,k) F\big(S(k)\big) + \sum_l p_{2,D}(s,l) \sum_t n_{2,D}(l,t) p_2(t,k) F\big(S(k)\big) \Big] \times \frac{f_1(y'|k)}{F(S(k))}.$$

The expression in square brackets in matrix notation is $P_{2,S} + P_{2,D}(I - P_{2,D})^{-1} P_{2,S}$, which equals the last term in (12) by the first equality in (7). The expression outside of square brackets corresponds to the term F_S. Theorem 2 is proved. □

6 Open Problem

Let $M_i = (X_i, P_i)$ be two Markov models, $i = 1,2$ and let $h : X_1 \longrightarrow X_2$ be a mapping. An open problem is to find all relationships between the transitional probabilities in these two models such that the solution of the OS problem for the "large" model M_1 can be simplified using the projection model M_2. For example, a potential candidate is the case when the transition probabilities for all $x, y \in X_1$ satisfy

$$p_1(x,y) = p_2(s,k) \sum_{i=1}^{N} \alpha_i(s,k) f_1(y|k,i), \tag{14}$$

where $s = h(x), k = h(y), \alpha_i(s,k) \geq 0, \sum_{i=1}^N \alpha_i(s,k) = 1, s,k \in X_2$. In other words, instead of one die for each state of $k \in X_2$, there are sets of N dice, and transitions are defined using randomization over these sets.

Acknowledgements The author would like to thank Joe Quinn, Ernst Presman, and an anonymous referee for valuable comments.

References

1. Grassmann, W.K., Taksar, M., Heyman, D.: Regenerative analysis and steady state distributions for Markov chains. Oper. Res. **33**(5), 1107–1116 (1985)
2. Howard, R.: Dynamic Probabilistic Systems. Markov Models. Wiley, New York (1971)
3. Kemeny, J., Snell, L.: Finite Markov Chains. Springer, Berlin (1960, 1983)
4. Presman, E.: The solution of optimal stopping problem based on a modification of a payoff function (2010). This volume
5. Presman, E., Sonin, I.: On optimal stopping of random sequences modulated by Markov chain. Theory Probab. Appl. **54**(3), 534–542 (2010)
6. Sheskin, T.: A Markov chain partitioning algorithm for computing steady state probabilities. Oper. Res. **33**(1), 228–235 (1985)
7. Sonin, I.: The optimal stopping of "seasonal observations". In: Proc. 11th INFORMS Appl. Prob. Conf., p. 18. New York (2001)
8. Sonin, I.: The elimination algorithm for the problem of optimal stopping. Math. Methods Oper. Res. **4**(1), 111–123 (1999)
9. Sonin, I.: The state reduction and related algorithms and their applications to the study of Markov chains, graph theory and the optimal stopping problem. Adv. Math. **145**(2), 159–188 (1999)
10. Sonin, I., Thornton, J.: Recursive algorithm for the fundamental group inverse matrix of a Markov chain from an explicit formula. SIAM J. Matrix Anal. Appl. **23**(1), 209–224 (2001)

MIX
Papier aus verantwortungsvollen Quellen
Paper from responsible sources
FSC® C105338

If you have any concerns about our products,
you can contact us on
ProductSafety@springernature.com

In case Publisher is established outside the EU,
the EU authorized representative is:
Springer Nature Customer Service Center GmbH
Europaplatz 3, 69115 Heidelberg, Germany

Printed by Libri Plureos GmbH
in Hamburg, Germany